ADVANCED CALCULUS

INTERNATIONAL SERIES IN PURE AND APPLIED MATHEMATICS

E. H. Spanier, G. Springer, and P. J. Davis.
Consulting Editors

AHLFORS: Complex Analysis
BUCK: Advanced Calculus
BUSACKER AND SAATY: Finite Graphs and Networks
CHENEY: Introduction to Approximation Theory
CHESTER: Techniques in Partial Differential Equations
CODDINGTON AND LEVINSON: Theory of Ordinary Differential Equations
CONTE AND DE BOOR: Elementary Numerical Analysis: An Algorithmic Approach
DENNEMEYER: Introduction to Partial Differential Equations and Boundary Value Problems
DETTMAN: Mathematical Methods in Physics and Engineering
GOLOMB AND SHANKS: Elements of Ordinary Differential Equations
HAMMING: Numerical Methods for Scientists and Engineers
HILDEBRAND: Introduction to Numerical Analysis
HOUSEHOLDER: The Numerical Treatment of a Single Nonlinear Equation
KALMAN, FALB, AND ARBIB: Topics in Mathematical Systems Theory
LASS: Vector and Tensor Analysis
MCCARTY: Topology: An Introduction with Applications to Topological Groups
MONK: Introduction to Set Theory
MOORE: Elements of Linear Algebra and Matrix Theory
MOURSUND AND DURIS: Elementary Theory and Application of Numerical Analysis
PEARL: Matrix Theory and Finite Mathematics
PIPES AND HARVILL: Applied Mathematics for Engineers and Physicists
RALSTON AND RABINOWITZ: A First Course in Numerical Analysis
RITGER AND ROSE: Differential Equations with Applications
RITT: Fourier Series
RUDIN: Principles of Mathematical Analysis
SHAPIRO: Introduction to Abstract Algebra
SIMMONS: Differential Equations with Applications and Historical Notes
SIMMONS: Introduction to Topology and Modern Analysis
SNEDDON: Elements of Partial Differential Equations
STRUBLE: Nonlinear Differential Equations

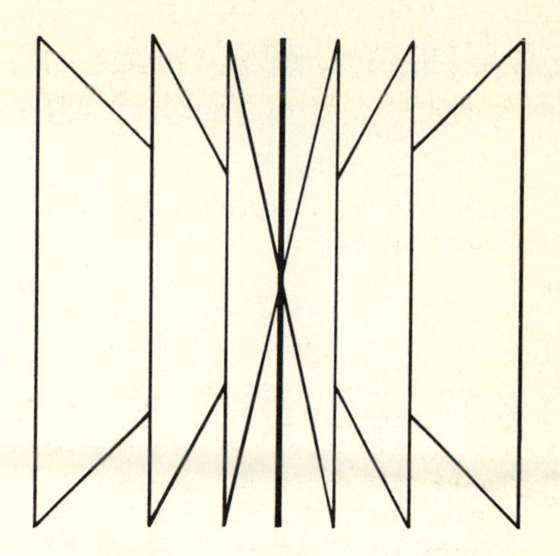

ADVANCED CALCULUS

THIRD EDITION

R. Creighton Buck
Professor of Mathematics
University of Wisconsin

with the collaboration of
Ellen F. Buck

McGraw-Hill Book Company

New York • St. Louis
San Francisco • Auckland
Bogotá • Düsseldorf
Johannesburg • London
Madrid • Mexico
Montreal • New Delhi
Panama • Paris
São Paulo • Singapore
Sydney • Tokyo
Toronto

ADVANCED CALCULUS

1234567890 DODO 783210987

This book was set in Times Roman. The editors were A. Anthony Arthur and Shelly Levine Langman; the production supervisor was Dennis J. Conroy. New drawings were done by J & R Services, Inc.
R. R. Donnelley & Sons Company was printer and binder.

Library of Congress Cataloging in Publication Data

Buck, Robert Creighton, date
Advanced calculus.

(International series in pure and applied mathematics)
Bibliography: p.
Includes index.
1. Calculus. 2. Mathematical analysis.
I. Buck, Ellen F., joint author. II. Title.
QA303.B917 1978 515 77-2859
ISBN 0-07-008728-8

Contents

Preface

The present edition covers essentially the same topics as the second edition, but with certain omissions, a few additions, and a considerable amount of rearrangement. We hope that these changes will make the material more accessible to students and more flexible for instructors' use.

For the first half of the book, we assume that students have learned to use elementary calculus, but are not yet experienced in the techniques of proof and rigorous reasoning. For the second half, we assume that they may be taking concurrently (or have already completed) an elementary computational course in matrices and systems of linear equations. (Appendix 3 contains an abbreviated summary of this material.)

The general objectives of the text are unchanged:

1. To introduce students to the language, fundamental concepts, and standard theorems of analysis so that they may be prepared to read appropriate mathematical literature on their own.
2. To develop analytical and numerical techniques for attacking problems that arise in applications of mathematics.
3. To revisit certain portions of elementary calculus, this time with attention to the underlying logical relationship of fundamental notions of analysis such as continuity and convergence.
4. To give a systematic, modern approach to the differential and integral calculus of functions and transformations in several variables, including an introduction to the useful theory of differential forms.
5. To display the structure of analysis as a subject in its own right, and not solely as a tool, for the benefit of those students whose interests lean

toward research in mathematics and its applications, without sacrificing intelligibility to abstraction.

The earlier editions have been used very successfully with a wide spectrum of students, most often at the junior or senior level with students majoring in science, engineering, or mathematics, but also with graduate students in mathematics or other fields whose undergraduate preparation was not adequate for more advanced courses. We hope that the changes we have made in this edition will serve the needs of these students even better.

In revising the book, we have changed the order of topics and subdivided some of the chapters in order to make the first half of the book more unified and complete in itself; in particular, Chapter 3 incorporates the material on differentiation of functions of several variables, which formerly appeared much later. We have also changed the emphasis in Chapter 1 so that topological concepts and properties appear more as useful tools in the study of analysis, than as topics for separate study in themselves. (In teaching this material, we had found it all too easy to spend time in this direction!)

The long chapter on series, in the earlier edition, is now two chapters, and a brief treatment of Fourier series has been added. Throughout the text we have also added a good many problems, and dropped others, in order to improve the mix of easy and challenging exercises.

Finally, we have added a short chapter on numerical methods (which can be taken up any time after Chapter 3), and included additional illustrations of numerical techniques in other chapters, to give students a chance to see the connection between computational algorithms and theoretical analysis.

In addition, we have included a few new pictures which we hope will help students acquire good geometric intuition. (We take this opportunity to thank those artists at McGraw-Hill who have worked with us to achieve such high-quality illustrations.)

Textbooks and research papers tend to be very different in style and structure. Hans Freudenthal, a gifted teacher, researcher, and expositor, has described the latter: "This is the way we write our mathematical papers. We conceal the train of thought which led us to the result." Instead, in this book, we have tried to provide both motivation and insight, so that the reader will emerge having acquired both skill and understanding.

We are very grateful to the large number of persons who have in the past sent us comments and suggestions. A partial list is inadequate, but we would like to mention particularly Professors H. F. Lowig, P. E. Miles, B. J. Pettis, and R. S. Spira.

R. Creighton Buck
Ellen F. Buck

To the Student

We hope that you approach this book with some degree of curiosity and commitment; the title "Advanced Calculus" might equally well have been "Basic Analysis" or even "Introduction to Mathematics for Applications." Your background ought to include the following:

1. A knowledge of the usual elementary (nonhonors) calculus, including some work with analytical geometry of space, double integrals, and partial derivatives.
2. An interest in learning more about this, either because of its applications or for its own sake.

In the second half of the book, we assume a slight familiarity with matrices; for those for whom this topic is new, there is a condensed summary in Appendix 3. It is not necessary for you to have had a course in differential equations or vector analysis.

No text intended for a one-year course can cover all the useful topics in analysis. Our selection criteria led us to concentrate on certain ideas and techniques that seemed most needed for further work in mathematics and its applications. Sketches of further developments and other digressions appear in the appendixes, some of which may be helpful reading from time to time. This applies in particular to Appendix 1, which explains some of the mysteries involved in logic and theorem proving.

All of Chapter 10 deals with numerical analysis; in addition, there are many exercises scattered throughout the book which ask for numerical answers or computational thinking. In some of these, you may find it convenient to use a small pocket calculator to carry out the calculations.

The text is intended to be read and not used solely as a source for problem assignments. We have provided answers for about half the exercises, and in some cases, cryptic hints which may make sense only if you have already tried the problem. The hardest exercises are so indicated by the usual★.

AD ASTRA PER ASPERA!

R. Creighton Buck
Ellen F. Buck

ADVANCED
CALCULUS

CHAPTER
ONE
SETS AND FUNCTIONS

1.1 INTRODUCTION

The most important difference between elementary calculus and advanced calculus is that, in the latter, we begin to explore some of the more complicated types of problems and techniques that can arise when we deal with functions of more than one variable. In particular, one has to become familiar with certain elementary geometrical ideas, not only in the plane and in 3-space, but also in spaces of higher dimension.

We all possess rather clear intuitive ideas as to the meaning of certain geometric terms such as line, plane, circle, angle, distance, and so on, acquired from our earlier work in mathematics. To a less uniform degree, we have also acquired the ability to visualize objects in space and to answer geometric questions about them without actually having to see or touch the objects themselves.

In complicated cases, we have learned to work from inaccurate two-dimensional diagrams of the true situation; the perception of perspective in a drawing is another illustration of the same adaptive mechanism. (Example: Can the two curves in space which are shown in Fig. 1-1 be separated without cutting either?)

Almost from the start, in the study of functions of several variables, we must deal with spaces of higher dimension than three; in such cases, it is a great convenience to be able to use geometric terminology and ideas. In

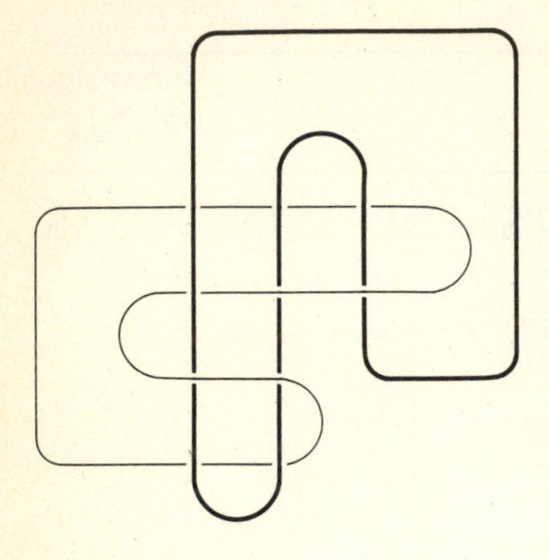

Figure 1-1 Linked or unlinked?

higher dimensions, where our intuition is less trustworthy, we must make more use of analytical and algebraic tools than we do of pictures.

The basic tools of classical analysis—limits, integration, differentiation—ultimately depend on the special properties of the real field **R**, and these in turn on the discoveries in logic and set theory made during the last century. It is not our intention to give this sequential axiomatic development, but instead to get on with the more concrete aspects of the calculus of functions of several variables; reasoning and rigor are still important, for here one meets new ideas and complications, and one needs to know more urgently than before the limits of validity of definitions and techniques. Readers who are interested in the axiomatic treatment of **R** and the various deductive chains that tie the whole theory together will find this discussed in Appendix 2.

We assume that most of elementary calculus is familiar background; in particular, unless indicated otherwise, you are free to use the customary tools of elementary one-variable calculus, such as differentiation, curve tracing, integration, and so on, in doing exercises.

1.2 R AND $\mathbf{R}^n$

We start by describing some of the notation and basic concepts which we will use.

The set of all real numbers is denoted by **R**; geometrically we identify this with an ordinary line and regard each real number x as specifying the point on this line with coordinate x. We also refer to such an axis as 1-space. The ordinary cartesian plane is denoted by $\mathbf{R}^2$, and space by $\mathbf{R}^3$; $\mathbf{R}^2$ is the set of all ordered pairs (a, b), and $\mathbf{R}^3$ the set of ordered triples (a, b, c), where the coordinates may be any real numbers in **R**. We also call $\mathbf{R}^2$ and $\mathbf{R}^3$ 2-space and 3-space, respectively.

Points, whether they be in the plane or in space, will most often be denoted by a single letter: the point p, the point Q, the point x. If we are

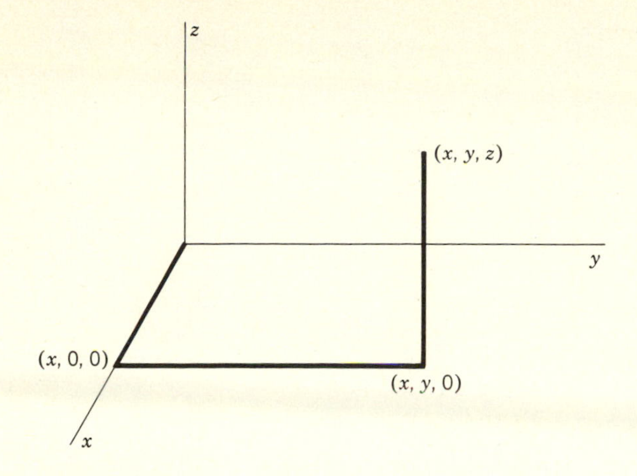

Figure 1-2

specifically dealing with a point in the **plane,** we might want to represent it in coordinate form, although we do not have to; furthermore, we shall not be bound to the use of any particular prescribed letters for the coordinates but may choose them to suit our convenience,

$$p = (x, y) \qquad Q = (a, b) \qquad P = (t, x)$$

The same will be true in space; here, we might write $p = (x, y, z)$, but we might equally well write $p = (u, v, w)$ or $p = (t, x, y)$, depending upon our purpose.

For this reason, we shall not speak of the X or the Y coordinates of a point, but rather speak of the "first coordinate" or the "second coordinate," and so on. For example, the letter "a" denotes the second coordinate of the point (b, a), and t is the first coordinate of the point (t, u, v).

When we wish to interpret $\mathbf{R}^3$ graphically, we use the familiar coordinate axis diagram shown in Fig. 1-2.

The road to generalization is obvious: **n space,** which we denote by $\mathbf{R}^n$, is the set of all n-tuples of real numbers. A typical point in $\mathbf{R}^n$ can be denoted by

$$p = (a_1, a_2, \ldots, a_n)$$

although, again, other labels for the coordinates may often be used. Thus, 4 space is simply the set of all (x, y, z, w), where x, y, z, w can be any real numbers. By analogy, the **origin** in $\mathbf{R}^n$ is the special point

$$\mathbf{0} = (0, 0, 0, \ldots, 0)$$

We do not know any convenient way to picture $\mathbf{R}^n$ for $n > 3$ that gives the same intuitive understanding of the geometry of n space as that conveyed by Fig. 1-2 when $n = 3$.

The mathematical usefulness of spaces of dimension higher than 3 will become much clearer later. For the present, we give several examples to show how they arise in applications. (Additional discussion of the relationship between mathematics and its applications may be found in Appendix 4.)

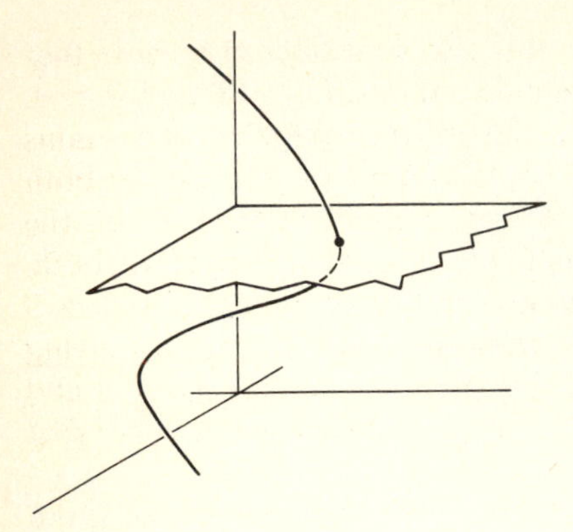

Figure 1-3

Suppose that we have five oil tanks, with capacities C_1, C_2, C_3, C_4, C_5. Suppose that, at the present time, tank k contains x_k gal of oil. Certain inequalities must hold,

$$0 \le x_k \le C_k \qquad k = 1, 2, 3, 4, 5 \tag{1-1}$$

We can represent the complete state of the system by a single point in 5-space,

$$P_0 = (x_1, x_2, x_3, x_4, x_5)$$

The set of all possible states of the collection of tanks is the set of all such points P, whose coordinates are limited only by the inequalities (1-1). (Can you visualize this set as a box in $\mathbf{R}^5$?) If we begin to use oil from each tank, not necessarily at the same rate, the point representing the current state will move from P_0 along some curve toward the origin (0, 0, 0, 0, 0), which represents the empty state. Various questions about the way the oil is being used can be phrased as geometric questions about the shape of this curve.

Perhaps more familiar is the use of 4-space to image a physical point moving in 3-space by plotting the points (x, y, z, t), where (x, y, z) is the location of the moving point at time t. If we compress the three dimensions of space into two, then we can picture the four-dimensional world of space-time in $\mathbf{R}^3$. The "history," or world line, of the moving particle then becomes a curve in 3-space. If the "time" axis is vertical, the space location of the particle at a particular moment of time t_0 is found by intersecting the world line of the particle by the horizontal plane $t = t_0$ (see Fig. 1-3). If we accept the view that a particle cannot be in two different locations at the same time, then such a horizontal plane cannot cut a world line twice. In particular, the world line of a single particle cannot have the shape of a circle.

Further geometric insight into the geometry of $\mathbf{R}^n$ for $n > 3$ can be obtained from any of the following references in the Reading List at the end of the book: [26, 49, 50, 52]. (We especially recommend the two stories by Robert Heinlein.)

If S is a set of points (or of other objects), the notation $p \in S$ will mean that p is a **member of** S. We use $A \subset B$ to indicate that the set A is a **subset** of the set B, meaning that every member of A is also a member of B; note that $A \subset B$ does not mean that A and B must be different sets. If $A \subset B$ and $B \subset A$, then $A = B$, for they have exactly the same members. If $A \neq B$, then A contains a member that does not belong to B, or B contains a member not in A; both may happen. The **union** of two sets A and B is $A \cup B$, and consists of all the points that belong either to A or to B, including those that belong to both. Note that $A \subset A \cup B$ and $B \subset A \cup B$. The **intersection** of A and B is $A \cap B$ and consists of just the points that belong to both A and B. The set having no members is denoted by $\varnothing$ and is called the **empty set,** or **null** set. If A and B have no members in common, then $A \cap B = \varnothing$; in this case, we also say that A and B are **disjoint.**

All the geometric properties of sets in $\mathbf{R}^n$ depend on basic properties of the set $\mathbf{R}$ of all real numbers. The algebraic properties of $\mathbf{R}$ can be summarized by saying that $\mathbf{R}$ is a **field.** This means that the operations called addition and multiplication are defined for any pair of real numbers, and that a familiar list of algebraic rules applies, including the following:

(i) Addition and multiplication are commutative and associative:

$$a + b = b + a \quad \text{and} \quad ab = ba \quad \text{for all } a, b \in \mathbf{R}$$

$$(a + b) + c = a + (b + c) \quad \text{and} \quad (ab)c = a(bc) \quad \text{for all } a, b, c \in \mathbf{R}$$

(ii) $a(b + c) = ab + ac$ for all $a, b, c \in \mathbf{R}$

(iii) Additive and multiplicative inverses exist; in general, the equation

$$x + a = b$$

has a unique solution x for any $a, b \in \mathbf{R}$, and the equation

$$ax = b$$

has a unique solution x for any $a, b \in \mathbf{R}$, provided that $a \neq 0$.

Since we are assuming familiarity with the ordinary rules of algebra, and experience in calculating with numbers, we do not go into this aspect of $\mathbf{R}$ further. The complete list of field axioms will be found in Appendix 2.

The real field also has an **order** relation, $<$, which obeys such laws as the following:

(iv) For any x and y in $\mathbf{R}$, exactly one of the following holds:

$$x = y, x < y, y < x.$$

(v) If $x < y$ and $y < z$, then $x < z$.

(vi) If $x < y$, then $x + z < y + z$ for any $z \in R$.

(vii) If $x < y$ and $z > 0$, then $xz < yz$.

This ordering of $\mathbf{R}$ is used in many ways. For example, a subset $S \subset \mathbf{R}$ is said to be **bounded** if there are numbers b and c such that $b \leq x \leq c$ for all

$x \in S$. Any suitable number c is called an **upper bound** for S, and any suitable b is called a **lower bound** for S.

R has two special subsets, the **integers**

$$Z = \{0, \pm 1, \pm 2, \pm 3, \ldots\}$$

and the **rational** numbers

$$Q = \left\{\text{all numbers } \frac{m}{n} \text{ where } n \neq 0 \text{ and } m, n \in Z\right\}$$

As a subset of **R**, the set of integers is an infinite set that is unbounded and uniformly dispersed. Any real number lies between two consecutive integers; given $x \in \mathbf{R}$, there is a unique integer, usually denoted by $[x]$ and called the **greatest integer** in x, such that $[x] = n$ where

$$n \leq x < n + 1$$

The set Z also has the important property that any bounded set of integers is finite, and thus has a smallest member and a biggest member.

The set Q of rational numbers behaves quite differently. As a subset of **R**, Q is dense, meaning that given any two real numbers b and c with $b < c$, there are infinitely many rational numbers r with $b < r < c$. Moreover, a bounded set of rational numbers need not have either a biggest or a smallest member. If S is the set of all rational numbers r with $0 < r < 1$, then for any $r \in S$ there is an integer n such that $1/n < r < (n - 1)/n$; since $1/n$ and $(n - 1)/n$ also belong to S, no member of S can be either the largest member of S or the smallest.

In Sec. 1.7, we will discuss this again in connection with one further very important property of **R** itself called the **least-upper-bound property,** which lies at the heart of many of the basic theorems in analysis. For completeness, we state the LUB property now, but defer further explanations until later. Put very briefly, the real numbers have the property that while a bounded set S does not itself have to have either a largest member or a smallest member, the set of upper bounds for S must have a smallest member, and the set of lower bounds for S must have a largest member. As we shall see, this fact provides a useful replacement for the maximum or minimum member of a bounded set of numbers when these do not exist.

We have used the words "finite" and "infinite" above without explanation. The notion of cardinal number, applied to arbitrary sets, is an important part of modern set theory which we do not choose to discuss here; some elementary aspects are touched on in Appendix 1. For our purposes, an intuitive understanding is sufficient. A set is either finite or infinite; if it is infinite, it is either **countable,** meaning that its members can be paired one-to-one with the set of positive integers and thus labeled as

$$S = \{p_1, p_2, p_3 \ldots\}$$

or it is **noncountable,** in which case such a labeling is not possible. The set of all real numbers is noncountable, but the set Q is still countable.

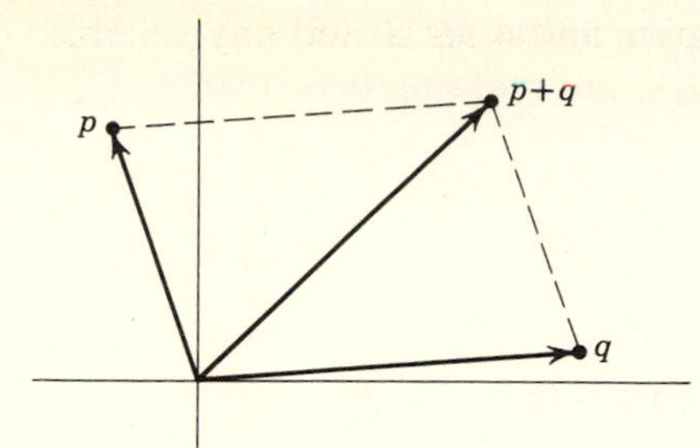

Figure 1-4 Addition of points.

We return now to n space, $\mathbf{R}^n$. The algebraic operations available in $\mathbf{R}$ immediately give a way to introduce certain algebraic operations for points. We add points by adding corresponding coordinates. If $p = (x_1, x_2, \ldots, x_n)$ and $q = (y_1, y_2, \ldots, y_n)$, then their **sum** is the point

$$p + q = (x_1 + y_1, x_2 + y_2, \ldots, x_n + y_n) \tag{1-2}$$

We also define the product of a point and a real number. Given the point p above, and any number $\beta \in R$, we set

$$\beta p = (\beta x_1, \beta x_2, \ldots, \beta x_n) \tag{1-3}$$

To illustrate these definitions, let $p = (2, 1, -3)$ and $q = (3, 0, 4)$. Then, $p + q = (5, 1, 1)$, $2p = (4, 2, -6)$, $-q = (-1)q = (-3, 0, -4)$, and $p - q = p + (-q) = (-1, 1, -7)$.

These algebraic operations have simple geometric interpretations in the plane and in space; here, we speak of addition of "vectors" and multiplication of vectors by "scalars." The first step is to represent a point P in the plane by the directed line segment (arrow) which starts at the origin and ends at P. Then, the addition of points defined above corresponds exactly to the parallelogram rule for adding position vectors (see Fig. 1-4). Likewise, multiplication of points by real numbers corresponds to expansion or contraction of the associated vector by the appropriate factor, with negative numbers effecting reversal (see Fig. 1-5). The picture in 3-space is similar.

It is now easily verified directly from the algebraic properties of $\mathbf{R}$ itself that these algebraic operations on points on $\mathbf{R}^n$ satisfy a list of simple rules.

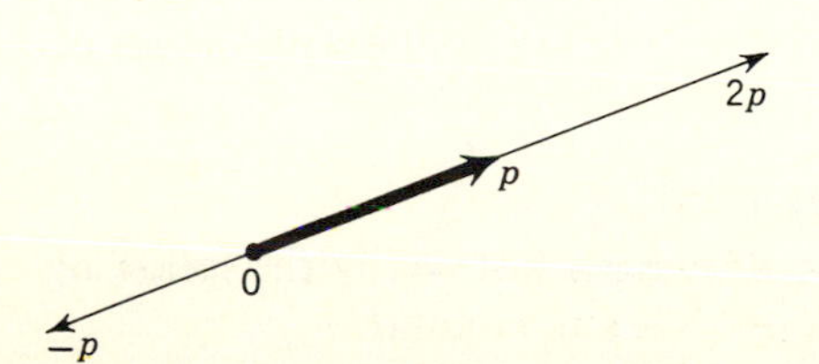

Figure 1-5

(V1) Addition of points is commutative and associative.

(V2) If p and q are points and α and β are real numbers, then

$$\alpha(p + q) = \alpha p + \alpha q$$

$$(\alpha + \beta)p = \alpha p + \beta p$$

$$\alpha(\beta p) = (\alpha\beta)p \tag{1-4}$$

(V3) The special point $\mathbf{0}$ has the property that

$$p + \mathbf{0} = p$$

(V4) The real numbers 0 and 1 have the property that

$$(0)p = \mathbf{0} \qquad \text{and} \qquad (1)p = p$$

While the statements in this list apply to $\mathbf{R}^n$, they also describe a standard mathematical structure called a **vector space** over $\mathbf{R}$. A set V is called a vector space with real scalars if there is an operation of addition defined for pairs of elements of V, and a product defined for any $p \in V$ and any real number $\lambda \in \mathbf{R}$, such that $p + q$ and λp belong to V for every choice of p, q in V and $\lambda \in R$, so that V1 and V2 hold, and there is a special element $\mathbf{0}$ in V such that V3 and V4 also hold. In $\mathbf{R}^n$ the role of $\mathbf{0}$ is filled by the origin, $(0, 0, \ldots, 0)$, so that, by analogy, $\mathbf{0}$ is often called the origin of V.

The preceding discussion, starting with formulas (1-4), can now be summarized by saying that $\mathbf{R}^n$, with the operations defined in (1-2) and (1-3), is a vector space over $\mathbf{R}$.

Please note that we do not choose to introduce an order relation among points in $\mathbf{R}^n$ for $n \geq 2$ comparable to that which holds among the real numbers themselves. The reason is merely that there is no *simple* order relation that turns out to be sufficiently useful. Furthermore, there is no general way to define multiplication of two points in $\mathbf{R}^n$ that will yield a point for the product, and that will obey the usual laws of algebra except when $n = 2$, and in a partial way for $n = 3$, 4, and 8. The reasons behind this form an interesting chapter in modern algebra, and are treated in detail in the article by Curtis in reference [16] in the Reading List.

However, there is a different sort of multiplication operation for points in $\mathbf{R}^n$ that is extremely useful and that applies for any n. This is called the scalar or inner product; the former name is used because the product of two points in $\mathbf{R}^n$ is a real number or scalar, not another point. Given two points $p = (x_1, x_2, \ldots, x_n)$ and $q = (y_1, y_2, \ldots, y_n)$, their **scalar product** (sometimes also called their dot product) is the number

$$p \cdot q = x_1 y_1 + x_2 y_2 + \cdots + x_n y_n \tag{1-5}$$

Thus, if $p = (2, -1, 4)$ and $q = (-3, 5, 4)$, $p \cdot q = 5$.

Using the familiar algebraic laws for $\mathbf{R}$, it is easy to verify the following formulas:

$$(\beta p)\cdot q = \beta(p\cdot q) = \beta(q\cdot p) \tag{1-6}$$

$$p\cdot(q_1+q_2) = p\cdot q_1 + p\cdot q_2 \tag{1-7}$$

where p, q, q_1, and q_2 are points in $\mathbf{R}^n$ and β is a real number. In the next section, we will discuss the geometric meaning of the scalar product and obtain further properties.

The vector space operations in $\mathbf{R}^n$—and even in the plane or in space—make it possible to restate many familiar geometric ideas in algebraic terms. Recall, for example, the midpoint formula from plane analytical geometry. If $P = (x_1, y_1)$ and $Q = (x_2, y_2)$, then the midpoint of the line segment PQ is the point

$$R = \left(\frac{x_1+x_2}{2}, \frac{y_1+y_2}{2}\right) \tag{1-8}$$

In algebraic form, we can now write $R = \frac{1}{2}P + \frac{1}{2}Q = \frac{1}{2}(P+Q)$. If we make use of a familiar result about parallelograms, we can find a very simple proof of the following well-known property of quadrilaterals: *Let A, B, C, and D be any four points in the plane and form a quadrilateral by joining them in that order; then, the midpoints of the four sides are the vertices of a parallelogram.* (Two cases are shown in Fig. 1-6.)

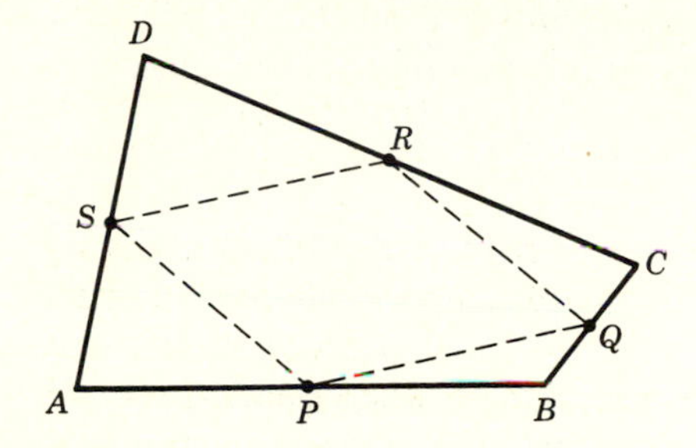

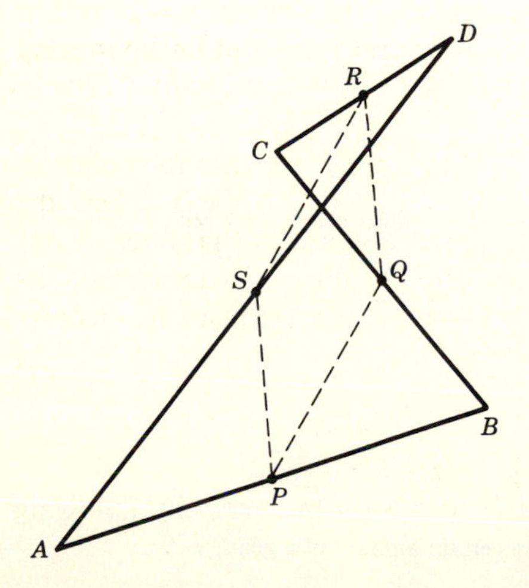

Figure 1-6

As above, the four midpoints are

$$P = \frac{A+B}{2}, \qquad Q = \frac{B+C}{2}, \qquad R = \frac{C+D}{2}, \qquad S = \frac{D+A}{2}$$

To show that $PQRS$ is a parallelogram, we use the following: *A quadrilateral is a parallelogram if and only if its diagonals bisect each other.* It is clear that this is the same as asking that the midpoints of the diagonals coincide. But the midpoint of the diagonal PR is

$$\frac{P+R}{2} = \frac{1}{2}\left(\frac{A+B}{2} + \frac{C+D}{2}\right) = \frac{A+B+C+D}{4}$$

and the midpoint of diagonal QS is

$$\frac{Q+S}{2} = \frac{1}{2}\left(\frac{B+C}{2} + \frac{D+A}{2}\right) = \frac{B+C+D+A}{4}$$

and since these are equal, we have proved the theorem.

Note that in Exercise 15 below, an algebraic proof is given for the italicized statement about parallelograms that we used above, based on the definition of a parallelogram as a four-sided polygon whose opposite sides are parallel. (One person's definition is another person's theorem!) Does this mean that the complete argument can be given algebraically without bringing in geometry? Does this mean that the theorem about the midpoints of the sides of a quadrilateral holds for one sitting in 4-space or n-space?

EXERCISES†

1 How would you describe the world lines of two particles that collide and destroy each other?

2 Construct a world-line diagram for the motion of two elastic balls of different mass that move along a line toward each other, collide, and rebound.

★3 Draw a sketch to illustrate the following events: A photon vanishes, giving rise to two particles, one an electron and one a positron. The electron moves off in one direction, the positron in another. The positron strikes another electron, and the two annihilate each other, giving rise to a photon which travels off. Could this be the history of only one particle?

4 If A and B are sets and $A \subset B$, what are $A \cup B$ and $A \cap B$?

5 For any sets A and B, let $A - B$ be the set of those things which belong to A but do not belong to B. What is $A - (A - B)$? Is it true that $C \cap (A - B) = (C \cap A) - (C \cap B)$?

6 (*a*) Which is larger, $[\sqrt{243}/3]$ or $[12/\sqrt{5}]$?

(*b*) Find a rational number between $\sqrt{37}$ and $\sqrt{39}$.

7 Solve for P in each of the following equations:

(*a*) $(2, 1, -3) + P = (0, 2, 4)$

(*b*) $(1, -1, 4) + 2P = 3P + (2, 0, 5)$

† A star ★ indicates an exercise which requires a certain amount of ingenuity.

8 Solve for the points P and Q if

$$2P + 3Q = (0, 1, 2)$$
$$P + 2Q = (1, -1, 3)$$

9 Solve for P and Q if

$$3P + Q = (1, 0, 1, -4)$$
$$P - Q = (2, 1, 2, 3)$$

10 Let $A = (1, 1, 3)$ and $B = (2, -1, 1)$. Can you find a point p such that $p \cdot A = 0$ and $p \cdot B = 0$?

11 Draw a diagram to illustrate that the associative law of addition, $p + (q + r) = (p + q) + r$, holds for the operation of addition of vectors.

12 The "center of gravity" of the triangle with vertices at A, B, and C is the point $\frac{1}{3}(A + B + C)$. Show that the center of gravity of a triangle is always the same as that of the triangle formed by the midpoints of its sides.

13 What is the center of gravity of the triangle whose vertices are $(1, 2, -4, 1)$, $(2, 0, 5, 2)$, $(0, 4, 2, -3)$?

14 Show that any three noncollinear points can be the midpoints of the sides of a unique triangle.

15. Using the definition given at the end of this section, give an algebraic proof that a four-sided polygon is a parallelogram if and only if the diagonals bisect each other.

16. Show that the four points A, B, C, D are the vertices of a parallelogram if and only if $A + C = B + D$, or $A + B = C + D$, or $A + D = B + C$.

17. In the rules (i), (ii), (iii) which were given for real numbers, there was no mention of subtraction. Formulate a set of rules concerning subtraction, and then check these with the development given in Appendix 2.

18. Show that the rules in (1-4) hold, by direct use of the definition of addition of points and the previously given rules about real numbers.

19 Using the order properties (iv) to (vii) for the real field, derive the following additional properties:

(*a*) For any $a, b \in \mathbf{R}$, if $a < b$, then $-b < -a$.

(*b*) For any $x \in \mathbf{R}$, $x^2 \geq 0$, with equality only if $x = 0$.

(*c*) For any real x and y, if $x^2 + y^2 = 0$, then $x = y = 0$. Does this extend to more terms?

20 (*a*) If $0 < m < n$, show that $m^2 < n^2$ and $1/n < 1/m$.

(*b*) If $m < n < 0$, show that $m^2 > n^2$ and $1/m > 1/n$.

21 Suppose that a, b, A, B are all >0. Is it always true that

$$\frac{a + b}{A + B} \leq \frac{a}{A} + \frac{b}{B}$$

22 Using the coordinate representation for points and the dot product, prove the identities (1-6) and (1-7).

23 Show that the set Z of all integers is countable.

1.3 DISTANCE

In the plane, the distance from the point (x, y) to the origin is $\sqrt{x^2 + y^2}$. In $\mathbf{R}^n$ we adopt the obvious analog. If $p = (x_1, x_2, \ldots, x_n)$, we define what we shall call its **norm** to be the number

$$|p| = \sqrt{p \cdot p} = \sqrt{x_1^2 + x_2^2 + x_3^2 + \cdots + x_n^2} \tag{1-9}$$

and we interpret this as the distance in $\mathbf{R}^n$ from p to the origin. If p and q are two points in n space, we take $|p - q|$ to be the **distance** between them. This agrees with the familiar distance formula both in the plane and in space. If $p = (x, y, z)$ and $p_0 = (x_0, y_0, z_0)$, then $p - p_0 = (x - x_0, y - y_0, z - z_0)$ and

$$|p - p_0| = \sqrt{(x - x_0)^2 + (y - y_0)^2 + (z - z_0)^2}$$

When we specialize this general treatment to $\mathbf{R}^1$, it takes on a slightly different appearance but is still familiar. If we write

$$p = (x) = x$$

then the formula for norm becomes

$$|p| = |x| = \sqrt{x^2}$$

We recall that, when c is a positive number, $\sqrt{c}$ is always the positive number whose square is c; thus, $\sqrt{x^2}$ is not always x. Indeed, if x is negative, $\sqrt{x^2} = -x$. Thus, the formula for norm reduces in 1-space to the ordinary **absolute value**

$$|x| = \begin{cases} x & \text{if } x \geq 0 \\ -x & \text{if } x < 0 \end{cases}$$

Again, $|x - y|$ is the absolute value of the difference of x and y, and this is in fact the distance on the line between the points with coordinate x and with coordinate y. We see that we can use the notation $|p|$ and $|p - q|$ in all the spaces $\mathbf{R}^n$, $n = 1, 2, \ldots$.

We note that $|p| \geq 0$ and that $|p| = 0$ only if $p = \mathbf{0}$. Furthermore, the norm function obeys certain simple identitites.

(1-10) For any point p, $\quad |p| = |-p|$

For any real number $\lambda \geq 0$, $\quad |\lambda p| = \lambda |p|$

(1-11) More generally, for any real number λ and any p,

$$|\lambda p| = |\lambda| \, |p|$$

From these others can be derived; for example, it is evident that $|p - q| = |q - p|$.

Finally, the norm function obeys the following triangle inequality: *For any points p and q in $\mathbf{R}^n$,*

$$|p + q| \leq |p| + |q| \tag{1-12}$$

In the plane and in space, this relation has a simple geometric interpretation in terms of the sides of a triangle, as shown in Fig. 1-7. However, we can no longer draw pictures in $\mathbf{R}^n$ when $n > 3$, and it is necessary to have an analytical proof of (1-12) that does not depend in any way upon geometric arguments. If we restate (1-12) using (1-9), we must show that

$$\sqrt{(x_1 + y_1)^2 + (x_2 + y_2)^2 + \cdots + (x_n + y_n)^2} \leq \sqrt{x_1^2 + \cdots + x_n^2} + \sqrt{y_1^2 + y_2^2 + \cdots + y_n^2}$$

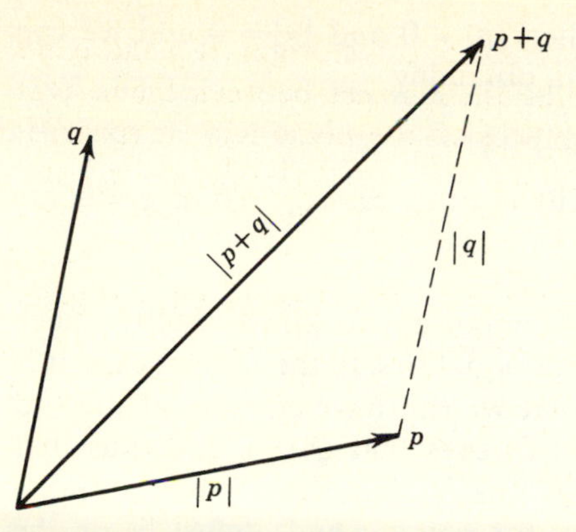

Figure 1-7 The triangle law.

for any choice of the $2n$ real numbers $x_1, x_2, \ldots, x_n, y_1, y_2, \ldots, y_n$. If we square both sides of this conjectured inequality and do some cancelling, we are led to conjecture another inequality, namely

$$x_1 y_1 + x_2 y_2 + \cdots + x_n y_n \le \sqrt{x_1^2 + x_2^2 + \cdots + x_n^2}\,\sqrt{y_1^2 + y_2^2 + \cdots + y_n^2}$$

This new inequality is valid, as we will shortly prove, and it is both important and useful in its own right; it is variously credited to Schwartz, Cauchy, and Bunyakovski. We restate this result in terms of norms and scalar products.

Theorem 1 (The Schwarz Inequality) *Let p and q be any points in n space,*

(1-13) *then* $$p \cdot q \le |p||q|$$

Take real numbers α and β, as yet unspecified, and form the point $Q = \alpha p - \beta q$. Clearly, a norm cannot be strictly negative, so $|Q| \ge 0$. Writing $|Q|^2$ as $Q \cdot Q$, we have

$$\begin{aligned} 0 \le Q \cdot Q &= Q \cdot (\alpha p - \beta q) \\ &= \alpha Q \cdot p - \beta Q \cdot q \\ &= \alpha(\alpha p - \beta q) \cdot p - \beta(\alpha p - \beta q) \\ &= \alpha^2 p \cdot p - \alpha\beta q \cdot p - \beta\alpha p \cdot q + \beta^2 q \cdot q \cdot q \end{aligned}$$

which, by (1-6) and (1-9), we can rewrite as

$$0 \le \alpha^2 |p|^2 + \beta^2 |q|^2 - 2\alpha\beta p \cdot q$$

or
$$2\alpha\beta p \cdot q \le \alpha^2 |p|^2 + \beta^2 |q|^2$$

This must hold for any choices of the numbers α and β. Suppose that we pick $\alpha = |q|$ and $\beta = |p|$, so that this becomes

$$2|p||q|(p \cdot q) \le |q|^2 |p|^2 + |p|^2 |q|^2 = 2|p|^2 |q|^2$$

If neither p nor q is the origin $\mathbf{0}$, then $|p| \neq 0$ and $|q| \neq 0$ and we can divide by the positive number $2|p||q|$, obtaining

$$p \cdot q \leq |p||q|$$

and we have proved the inequality. The omitted cases, $p = \mathbf{0}$ or $q = \mathbf{0}$, are trivial. ▮†

In inequalities such as this, it is often important to know exactly under what circumstances (if any) they can become equalities. In the present case, it is easy to trace back the argument to see that we can have equality at the end only if we had $|Q| = 0$ at the start. But this says that $Q = \mathbf{0}$, and thus that $\alpha p = \beta q$. The geometric interpretation of this is that the vectors p and q differ only in length, and not in direction; the points p and q must lie on the same line through the origin.

We also note that if we replace p by $-p$, (1-13) yields the inequality $-p \cdot q \leq |p||q|$, so that a more general statement of the Schwartz inequality is

$$-|p||q| \leq p \cdot q \leq |p||q| \tag{1-14}$$

We can now return to the triangle inequality, and show that distance in n space satisfies this intuitive requirement.

Corollary *For any points p and q in n space,*

$$|p + q| \leq |p| + |q| \tag{1-15}$$

and equality holds only if p and q lie on the same half line from the origin. ▮

We have

$$\begin{aligned} |p + q|^2 &= (p + q) \cdot (p + q) \\ &= p \cdot p + p \cdot q + q \cdot p + q \cdot q \\ &= |p|^2 + |q|^2 + 2p \cdot q \\ &\leq |p|^2 + |q|^2 + 2|p||q| = \{|p| + |q|\}^2 \end{aligned}$$

and the result follows.

The Schwarz inequality (1-14) is geometrically evident in the case of the plane or 3-space because of a simple interpretation of the scalar product of vectors. Suppose that p and q are points in space, and represent them by position vectors as shown in Fig. 1-8. Let θ be the angle between them. Then, the trigonometric law of cosines shows easily that

$$p \cdot q = |p||q| \cos \theta \tag{1-16}$$

† Following the lead of others, we use the sign ▮ to signify the end of a proof.

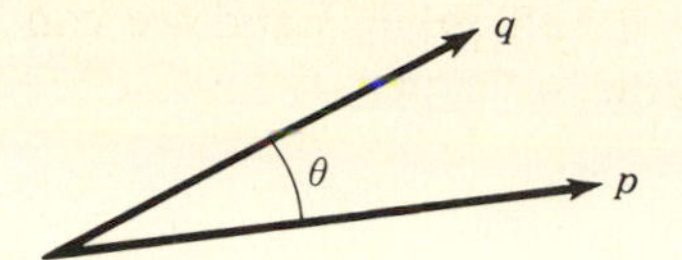

Figure 1-8

Thus, when n is 2 or 3, the Schwarz inequality merely asserts the obvious fact that $|\cos \theta|$ cannot be larger than 1.

When $n \geq 4$, Fig. 1-8 is no longer convincing, and the notion of the angle between two lines in n space may raise doubts; we therefore turn things around and use (1-14), which we know to hold for any choice of n, to define the notion of angle. Given two points p and q in n space, neither of which is the origin, we define the **angle** between them subtended at the origin to be the unique θ between 0 and π such that

(1-17) $$\cos \theta = \frac{p \cdot q}{|p||q|}$$

We need the Schwarz inequality to be sure that the number on the right side remains between -1 and 1, so that a value of θ can always be found.

When $p \cdot q = 0$, then the angle θ must be $\pi/2$. In this case we say that the points p and q are **orthogonal** and write $p \perp q$. Geometrically, this means that the position vectors from the origin to p and q form a right angle. For example, $(1, 2, -3, 1)$ and $(5, 1, 2, -1)$ are orthogonal in 4-space.

In the plane, a line has the equation

$$Ax + By = C$$

and in space, the equation of a plane is

$$Ax + By + Cz = D$$

In n space, it is natural to introduce the term **hyperplane** for the set of all points $p = (x_1, x_2, \ldots, x_n)$ that satisfy an equation of the form

(1-18) $$b_1x_1 + b_2x_2 + \cdots + b_nx_n = C$$

where the numbers b_i and C are specified and at least one of the b_i is different from 0. This equation can be written in a more condensed way as $v \cdot p = C$, setting $v = (b_1, b_2, \ldots, b_n) \neq \mathbf{0}$. Suppose that C is 0. Then, the equation of the hyperplane becomes $v \cdot p = 0$, so that the hyperplane consists of all the points p that are orthogonal to the given point v. In 3-space, the picture would be that shown in Fig. 1-9, and v would be the vector normal to the plane. By analogy, we use the same description in n space, and call v the **normal vector** to the hyperplane whose equation is $v \cdot p = 0$.

Planes that do not pass through the origin can be treated in a similar way. In 3-space, the general plane through $p_0 = (x_0, y_0, z_0)$ has the equation

$$A(x - x_0) + B(y - y_0) + C(z - z_0) = 0$$

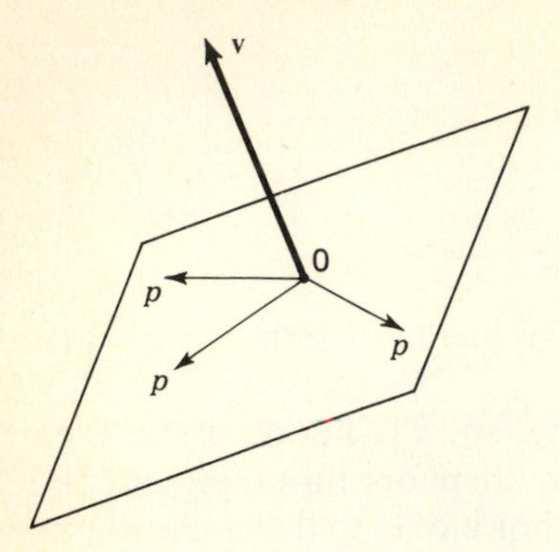

Figure 1-9

which can be restated in the form $v \cdot (p - p_0) = 0$ with $v = (A, B, C)$, and pictured as in Fig. 1-10. We therefore adopt the same formalism, and say that in n space the hyperplane through a point p_0 normal to the vector v consists of all points p such that $v \cdot (p - p_0) = 0$. Note that this equation agrees with (1-18) if C is chosen as the constant $v \cdot p_0$.

Just as a plane in 3-space is a two-dimensional space, a hyperplane in n space is an $(n - 1)$-dimensional object, isomorphic to $\mathbf{R}^{n-1}$ itself; "isomorphic," meaning "same form," is a technical mathematical term, used here informally. As an illustration, in 4-space the hyperplanes are each isomorphic to ordinary 3-space; thus, in the conventional space-time picture of the universe, the time axis is regarded as a line orthogonal to a hyperplane that represents ordinary three-dimensional space.

At the other end of the dimensional scale, we can also discuss lines in n-space, which will be one-dimensional geometric objects. In 3-space, we describe a **line** by parametric equations such as

$$
\begin{aligned}
x &= x_0 + at \\
y &= y_0 + bt \qquad -\infty < t < \infty \\
z &= z_0 + ct
\end{aligned}
\tag{1-19}
$$

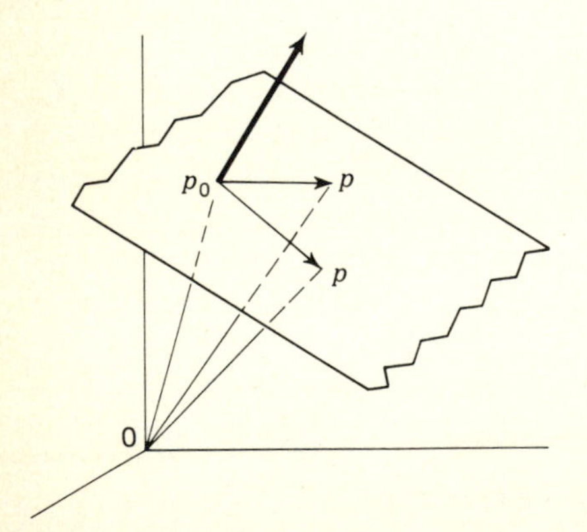

Figure 1-10

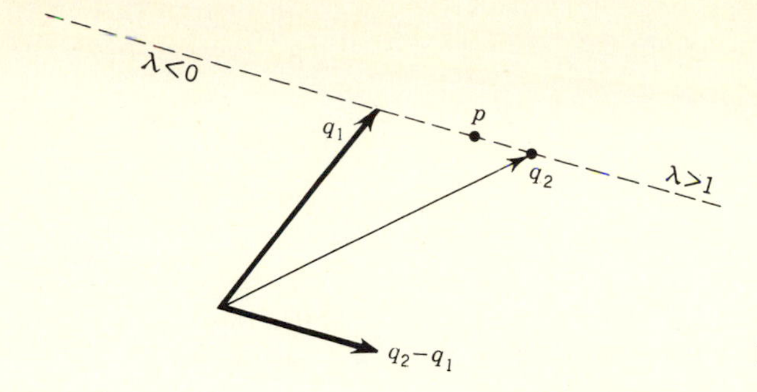

Figure 1-11

This is the line through $p_0 = (x_0, y_0, z_0)$ in the direction of the vector $v = (a, b, c) \neq 0$. Setting $p = (x, y, z)$, we can condense these equations to the single equation

$$p = p_0 + vt \qquad -\infty < t < \infty \tag{1-20}$$

We adopt the same equation in n space with $v = (b_1, b_2, \ldots, b_n)$.

We can also approach the topic of lines in n space more geometrically, avoiding coordinates entirely. A line L will be determined completely by any distinct pair of points on it. If q_1 and q_2 lie on L, they divide L into three portions, the interval between them and the unbounded segments to either side. We characterize each algebraically. As shown in Fig. 1-11, any point on L is obtained by adding to q_1 a positive or negative scalar multiple of the vector $v = q_2 - q_1$. Thus, the general point on the line L will have the form

$$\begin{aligned} p &= q_1 + \lambda v \\ &= (1 - \lambda)q_1 + \lambda q_2 \end{aligned} \tag{1-21}$$

for some choice of $\lambda \in \mathbf{R}$. Note that if $\lambda = \frac{1}{2}$, the point p is the midpoint of the segment between q_1 and q_2. More generally, if $0 < \lambda < 1$, p is the unique point on this segment whose distance from q_1 is exactly λ times the distance from q_1 to q_2. We verify this, showing that the sum of the distances from p to q_1 and q_2 is $|q_1 - q_2|$. We have, using (1-21) and the fact that $\lambda > 0$ and $1 - \lambda > 0$,

$$\begin{aligned} |p - q_1| &= |(1 - \lambda)q_1 + \lambda q_2 - q_1| \\ &= |\lambda(q_2 - q_1)| = \lambda |q_1 - q_2| \end{aligned}$$

$$\begin{aligned} |p - q_2| &= |(1 - \lambda)q_1 + \lambda q_2 - q_2| \\ &= |(1 - \lambda)(q_1 - q_2)| \\ &= (1 - \lambda)|q_1 - q_2| \end{aligned}$$

and adding, we have

$$|p - q_1| + |p - q_2| = |q_1 - q_2|$$

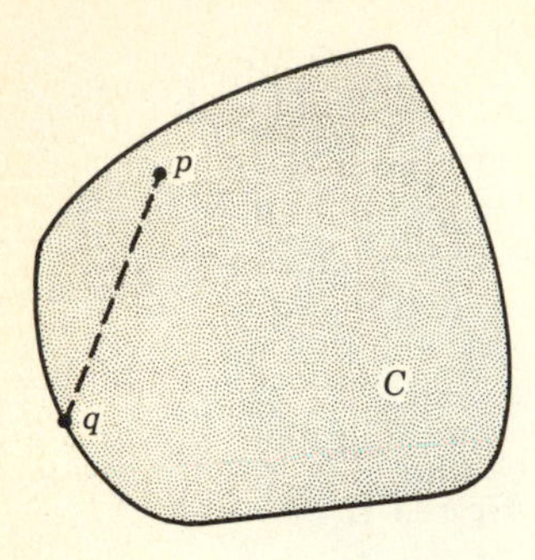

Figure 1-12 Convexity.

In a similar manner, it is possible to show that if $\lambda > 1$, the point p lies on the portion of the line L that is beyond q_2, and if $\lambda < 0$, on the portion beyond q_1. Finally, we note that formula (1-21) is in fact the same as the parametric equation (1-20), substituting q_1 for p_0, $q_2 - q_1$ for v, and λ for t.

Another important geometric concept which is conveniently described in terms of the ideas of the present section and which is suitable for n space is that of convexity. In the plane, a region C is said to be **convex** if it always contains the line segment joining any two points in the region (see Fig. 1-12). This definition is used in space as well and carries over at once to $\mathbf{R}^n$.

Definition 1 *A set C in n space is convex if it has the property that, whenever two points p and q are in C, then so are all points of the form*

$$\lambda p + (1 - \lambda)q \qquad 0 < \lambda < 1 \tag{1-22}$$

An important example of a convex set in n space is the solid spherical **ball**, which we define as follows:

$$\begin{aligned} B(p_0, r) &= \{\text{all } p \text{ with } |p - p_0| < r\} \\ &= \text{the open ball, center } p_0\text{, radius } r \end{aligned} \tag{1-23}$$

In 3-space, this is the interior of an ordinary sphere; in the plane, it is a round disc without the edge; in 1-space, $B(x_0, r)$ is the real interval consisting of the numbers x that obey $x_0 - r < x < x_0 + r$.

Let us show that the ball $B(\mathbf{0}, r)$ is convex. Suppose that p and q lie in B, so that $|p| < r$ and $|q| < r$. Choose any λ, $0 < \lambda < 1$; we must show that the point $\lambda p + (1 - \lambda)q$ lies in B. We calculate its distance from $\mathbf{0}$. Using the triangle inequality, we have

$$\begin{aligned} |\lambda p + (1 - \lambda)q - \mathbf{0}| &\le |\lambda p| + |(1 - \lambda)q| \\ &\le \lambda|p| + (1 - \lambda)|q| \\ &< \lambda r + (1 - \lambda)r = r \end{aligned}$$

EXERCISES

1 For $n = 1, 2$, and 3 in turn, plot the set of points p in $\mathbf{R}^n$ where
(*a*) $|p| < 1$ (*b*) $|p| \geq 1$ (*c*) $|p| = 1$.

2 Let $A = (4, 2)$. Graph the set of points p in the plane for which
(*a*) $|p| < |p - A|$ (*b*) $|p| + |p - A| = 6$ (*c*) $|p| + |p - A| \leq 4$.

3 Sketch the set of points (x, y) where
(*a*) $|x + 2y| \leq x - y$
(*b*) $(x^2 - y)(x - y^2) < 0$

4 Show that $|p_1 + p_2 + p_3 + \cdots + p_n| \leq |p_1| + |p_2| + \cdots + |p_n|$.

5 Prove that $|p - q| \geq |p| - |q|$.

6 If $p = (u, v, w)$, show that

(*a*) $|p| \leq |u| + |v| + |w|$
(*b*) $|u| \leq |p|, |v| \leq |p|, |w| \leq |p|$

7 Use the law of cosines in the plane and the properties of the norm and scalar product to verify that $p \cdot q = |p||q| \cos \theta$.

8 Show that the three points $A = (2, -1, 3, 1)$, $B = (4, 2, 1, 4)$, and $C = (1, 3, 6, 1)$ form a triangle with two equal angles. Find its area.

9 Find the equation of the hyperplane in 4-space which goes through the point $p_0 = (0, 1, -2, 3)$ perpendicular to the vector $\mathbf{a} = (4, 3, 1, -2)$.

10 If the angle between two hyperplanes is defined as the angle between their normals, are the hyperplanes $3x + 2y + 4z - 2w = 5$ and $2x - 4y + z + w = 6$ orthogonal?

11 Write the parametric equations of the line through $(2, 3, -1, 1)$ which is perpendicular to the hyperplane $3x + 2y - 4z + w = 0$.

12 Where does the line through $q_1 = (1, 0, 1, 0)$ and $q_2 = (0, 1, 0, 1)$ intersect the two hyperplanes of Exercise 10?

13 Given a triangle with vertices at A, B, C, show that the point $R = \frac{1}{3}(A + B + C)$ lies on each of the medians (the line from a vertex to the midpoint of the opposite side).

14 Formulate and prove an analogous property for the tetrahedron with vertices at A, B, C, D.

***15** In the triangle ABC, join A to a point $\frac{1}{3}$ of the way from B toward C, join B to a point $\frac{1}{3}$ of the way from C toward A, and join C to a point $\frac{1}{3}$ of the way from A toward B. Express the vertices of the smaller triangle thus formed in terms of A, B, and C.

16 Let l be the line determined by the two points p and q. Let $P = \lambda p + (1 - \lambda)q$. Show that, when $\lambda > 1$, $|P - p| + |p - q| = |P - q|$, and interpret this geometrically.

17 Show that the intersection of two convex sets is convex but that the union of convex sets does not have to be convex.

1.4 FUNCTIONS

The notion of **function** is essentially the same as that of mapping. A numerical-valued function f assigns to each point p in its domain a single real number $f(p)$ called the value of f at p. The rule of correspondence may be given by a formula such as

$$f(p) = x^2 - 3xy \qquad \text{for any } p = (x, y)$$

or by several formulas, as in

$$f(x, y) = \begin{cases} x & \text{when } x > y \\ x^2 + y & \text{when } x \le y \end{cases}$$

or by a geometrical description,

$f(p)$ is the distance from p to the point (4, 7)

or even by an assumed physical relationship,

$f(p)$ is the temperature at the point p

In all these cases, it is important to bear in mind that the function f itself is the rule or mapping, while $f(p)$ is the value which f assigns to p. It is also useful to think of functions in terms of an idealized computer; each function is then described by a specific software program or algorithm, and $f(p)$ is the output corresponding to a data input p.

The **domain** of a function f is the set of objects to which it may be applied and for which $f(p)$ is defined. It is always possible to increase the domain of a function by giving definitions for $f(p)$ for points p that were not in the original domain. This can be done quite arbitrarily in many ways; however, when the extension of the function is required to retain some of the properties that the function had on its original domain (e.g., smoothness, continuity, differentiability), then it becomes much more difficult and is indeed an interesting mathematical problem.

Real-valued functions are often classified according to the nature of their domains. If $f(p)$ is defined for points p in a region D in the plane, then we may write p as (x, y) and $f(p)$ as $f(x, y)$ and may refer to f as a function of **two real variables.** Similarly, when D is a set in 3-space, we may write $f(x, y, z)$ for $f(p)$ and say that f is a function of three real variables. In general, f is said to be a function of ***n*** **real variables** if the domain of f is a set in $\mathbf{R}^n$. In all these cases, it is still very helpful at times to think of f as a function of a single point p and to write $f(p)$ for its value. To see how this may happen, suppose that $f(p)$ is the temperature at the point p on a thin, curved wire S. The function f has S for its domain and might thus be said to be a function of three real variables. However, it is clearly much more convenient to think of f as a function of a variable point p which is confined to the wire. Again, if we are interested in the distribution of electric charge on a curved metal shell, we would be led to work with a numerical-valued function whose domain of definition is the set of points on the shell.

It is a serious mistake to think that functions must always have numerical values. Many of the most useful functions have values that are points or vectors. The motion of a particle may be described by setting up a correspondence between moments of time and points in space, by means of a function f, with $P = f(t)$. Here, the domain of f will be an interval on the time line, and the values of f will be points in 3-space (see Fig. 1-13). If the moving particle is subject to a magnetic field and we want to indicate the direction and

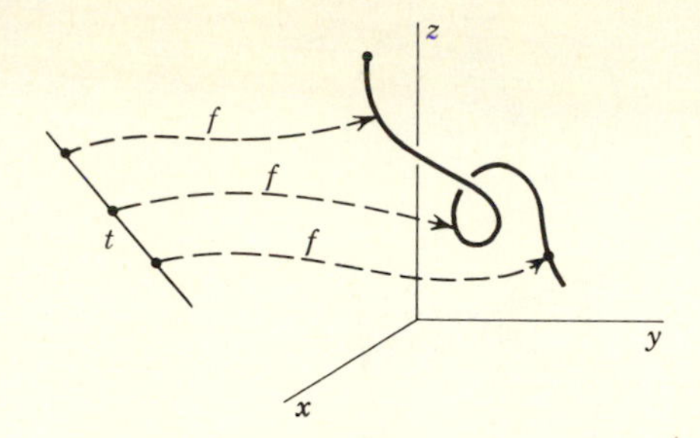

Figure 1-13 $f: \mathbf{R}^1 \to \mathbf{R}^3$

magnitude of the force field at each position of the particle, we may be led to work with a function F which is defined for points on the track of the particle and whose values are vectors (see Fig. 1-14).

More generally, whenever we are dealing with a **mapping** from objects in one set A to objects in a second set B, we are dealing with a function. We can represent this type of situation by a diagram, as in Fig. 1-15, and sometimes indicate it as

$$f: A \to B \tag{1-24}$$

The set A is still the domain of f, and the set B will contain the set of values (or range) of f. This concept of function is so broad that it permits mathematicians to label things as functions, and to work with them, which do not at all resemble the simple class of numerical-valued functions.

However, our interest in this book will be with less abstract functions. Stated succinctly, the main theme of this book is the study of functions whose domains are in $\mathbf{R}^n$ and whose values are in $\mathbf{R}^m$. To most of these functions, we can attach geometric interpretations that help us to work with them better.

We have already illustrated this for functions from $\mathbf{R}$ into $\mathbf{R}^n$, as in Fig. 1-13. Any such function is called a **curve,** since it corresponds to the familiar concept of the parametric description of a curve in n space. For example, a curve in the plane would be described by the equations

$$\begin{cases} x = \phi(t) \\ y = \psi(t) \end{cases} \qquad a \le t \le b$$

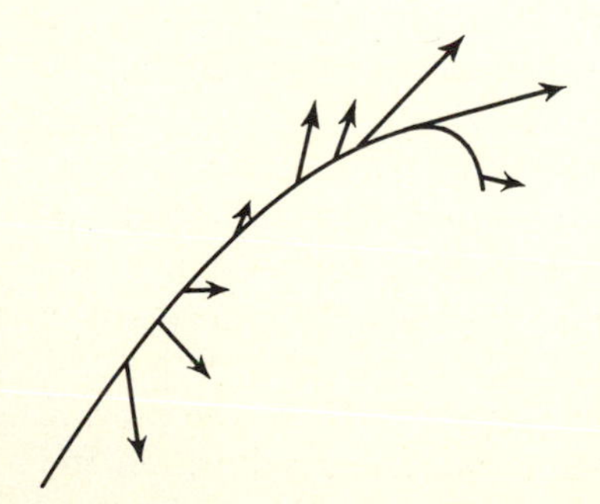

Figure 1-14

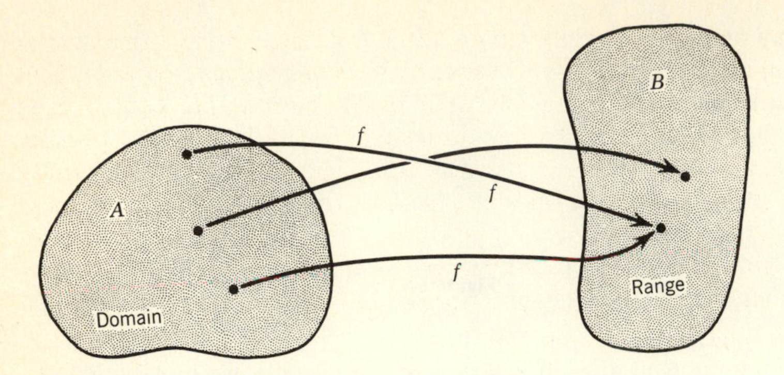

Figure 1-15 $f: A \to B$

But these in turn define a function f from the interval $a \leq t \leq b$ in $\mathbf{R}^1$ into $\mathbf{R}^2$, with

$$f(t) = (x, y) = (\phi(t), \psi(t))$$

In a similar way, the set of equations

$$\begin{aligned} x &= \phi(u, v) \\ y &= \psi(u, v) \qquad (u, v) \in D \\ z &= \theta(u, v) \end{aligned}$$

is a mapping from the domain D of the plane into 3-space. It therefore represents what we call a **surface** in $\mathbf{R}^3$ and is in fact a single function F from D into $\mathbf{R}^3$, given by

$$F(u, v) = (x, y, z) = (\phi(u, v), \psi(u, v), \theta(u, v))$$

For example, consider the set of equations

$$\text{(1-25)} \qquad \begin{cases} x = u + v \\ y = u - v + 1 \\ z = u^2 \end{cases} \qquad \begin{matrix} 0 \leq u \leq 1 \\ 0 \leq v \leq 1 \end{matrix}$$

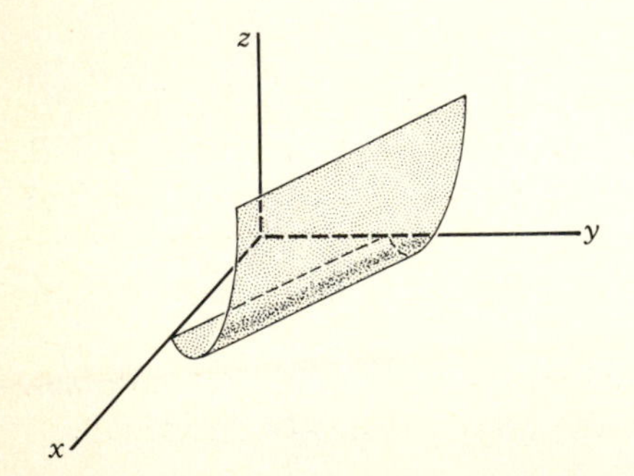

Figure 1-16

This describes a mapping F from a square in the UV plane into 3-space. As the point $p = (u, v)$ moves throughout the square, the image point $F(p) = (x, y, z)$ moves in space, tracing out the shape shown in Fig. 1-16. Thus, $F(1, 0) = (1, 2, 1)$ and $F(0, 1) = (1, 0, 0)$. Other similar pictures will be found in Chap. 8, especially Fig. 8-17, 8-18, and 8-20, which show more complicated examples. The study of curves and surfaces is one of the more difficult areas of analysis, and some aspects of this are treated there as an application of the tools to be developed.

We will also study functions that map a portion of $\mathbf{R}^n$ into $\mathbf{R}^n$. An illustration is the function F from 3-space into 3-space described by the formula $F(x, y, z) = (u, v, w)$ where

$$\begin{cases} u = x - y \\ v = y^2 + 2z \\ w = yz + 3x^2 \end{cases}$$

For example, we have $F(1, 2, 1) = (-1, 6, 5)$ and $F(1, -1, 3) = (2, 7, 0)$.

All these cases can be subsumed under one general formula. A mapping F from $\mathbf{R}^n$ into $\mathbf{R}^m$ has the form $y = F(x)$, where we write $x = (x_1, x_2, \ldots, x_n)$ and $y = (y_1, y_2, \ldots, y_m)$ and where

$$\begin{cases} y_1 = f(x_1, x_2, \ldots, x_n) \\ y_2 = g(x_1, x_2, \ldots, x_n) \\ \cdots\cdots\cdots\cdots\cdots\cdots \\ y_m = k(x_1, x_2, \ldots, x_n) \end{cases} \tag{1-26}$$

Here, $f, g, h, \ldots, k$ are m specific real-valued functions of the n real variables $x_1, x_2, \ldots, x_n$. Such functions as F are often called **transformations** to emphasize their nature; the study of their properties is one of the central topics of later chapters in this book.

Side by side with the view of a function as a mapping $A \to B$, there is also the equally important and useful idea of its **graph.** If f is a function of one variable, with domain $D \subset \mathbf{R}^1$, then the graph of f is the set of all points (x, y) in the plane, with $x \in D$ and $y = f(x)$. The graph of a function of two variables is the set of points (x, y, z), with (x, y) in the domain of the function f and with $z = f(x, y)$. Generalizing this, if f is a function on A into B,

$$f: A \to B$$

then the graph of f is the set of all ordered pairs (a, b), with $a \in A$ and $b = f(a)$. It is customary to use the term **cartesian product** of A and B, written as $A \times B$, to denote the class of all possible ordered (a, b), with $a \in A$ and $b \in B$. The graph of f is therefore a special subset of $A \times B$. By analogy, (a, b) is often called a point in $A \times B$, and the graph of f can be visualized as something like a curve in the space $A \times B$ (see Fig. 1-17).

Let us apply this to a numerical-valued function f of three real variables. Suppose that the domain of f is a set $D \subset \mathbf{R}^3$. Since f is numerical-valued, its

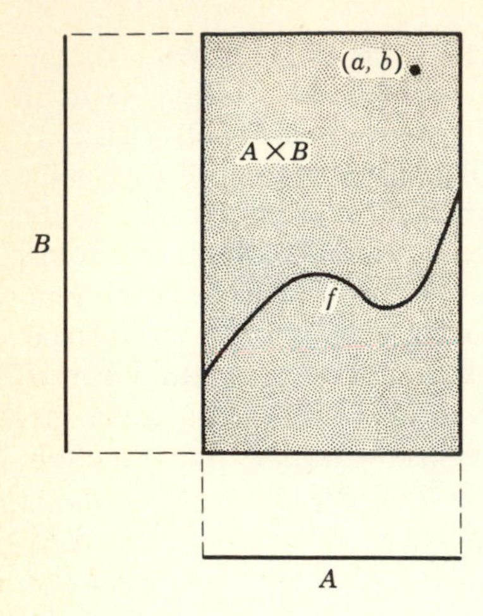

Figure 1-17

range lies in $\mathbf{R}^1$, and f is a mapping from part of $\mathbf{R}^3$ into $\mathbf{R}^1$. According to the general discussion above, the graph of f consists of the ordered pairs (a, b), with $a \in D$ and $b = f(a)$. Setting $a = (x, y, z)$, we have $b = f(a) = f(x, y, z)$. The graph of f is therefore a subset of $\mathbf{R}^3 \times \mathbf{R}^1$. But this is essentially 4-space, for $(a, b) = (x, y, z, b)$. The graph of f is then the set in 4-space consisting of all points (x, y, z, b) with $b = f(x, y, z)$ and $(x, y, z) \in D$. More generally, if F is a function on $\mathbf{R}^n$ to $\mathbf{R}^m$, its graph will be a subset of $\mathbf{R}^n \times \mathbf{R}^m$, which is essentially the same as $\mathbf{R}^{n+m}$. Thus, the graph of a mapping of the plane into the plane will be a set in 4-space.

As can be seen from these examples, the actual geometric construction of the graph of a function becomes impossible as soon as the number of variables involved, either in the domain or in the range, becomes large enough. However, the corresponding mental picture is still a very useful intuitive guide (Fig. 1-17).

A graph is not the only way to get a useful global picture of the nature of a function. For example, solely for illustration, we might think of a function of three variables as telling us the temperature at each point throughout a region of space. We can then think of each point p in the domain of f as having a number attached to it and, in terms of this, get some feeling for the way the values of the function change as we move p around in this region. For example, let

$$F(x, y, z) = (x - 1)^2 + (y + 1)^2 + z^2$$

Thinking of $F(p)$ as the temperature at p, we observe that $F(p)$ is the square of the distance from p to $(1, -1, 0)$ and notice that $F(p)$ is smallest at this point and increases as we move away from it in any direction. The equithermal surfaces, where F is constant, are spheres, with $(1, -1, 0)$ as center. With this as a guide, we can use geometric language to speak of the graph of F in 4-space; it is a bowl-shaped object, parabolic in cross section.

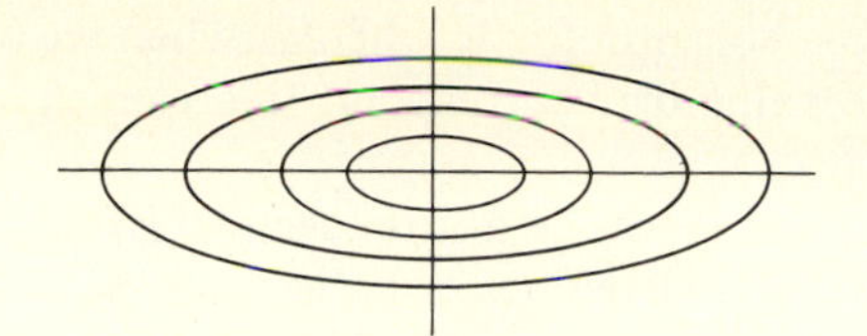

Figure 1-18

The study of the set of points where a given function takes on specific values is often a very useful way to examine the function's behavior. For a function of two variables f, the set of points where $f(x, y) = C$ is called the **level line** of value (or height) C. Thus, the level lines of the function $f(x, y) = x^2 + 9y^2$ are the family of ellipses shown in Fig. 1-18. In Chap. 3, we display the level lines of several more complicated functions in Figs. 3-13 and 3-14. The level lines of such functions are also a help in visualizing the graphs of the functions. This, of course, is nothing more than the standard process by which a topographic map is converted into a three-dimensional scale model.

For a function of three variables $F(x, y, z)$, the set of points where $F(p) = C$ will usually be a surface in 3-space. Thus, one speaks of the **level surfaces** of F. These too help in understanding the behavior of F, but this time they are of no assistance in constructing the graph of F.

Something similar can be done with functions of four variables, for one of the variables can be regarded as time; if the value of the function is regarded as specifying temperature, then we are dealing with a temperature distribution throughout space which also varies with time, and the concept of "level sets" is replaced by equithermal surfaces in space which change shape with time.

Many special properties of a function are reflected in simple geometrical properties of its graph. A numerical function f defined on the line is said to be **strictly increasing** if $f(x_1) < f(x_2)$ whenever $x_1 < x_2$; this means that the graph of f is a curve that rises as we move along it from left to right. Again, a function of n variables is said to be **midpoint convex** if it obeys the special relation

$$f(\tfrac{1}{2}p + \tfrac{1}{2}q) \le \tfrac{1}{2}f(p) + \tfrac{1}{2}f(q) \tag{1-27}$$

for all points p and q in its domain. This is easily seen to be satisfied if the set of points above the graph of f is a convex set (see Exercise 10).

Elementary texts in calculus sometimes leave the impression that any equation in x and y "defines y as a function of x." This must be both qualified and explained. What is usually meant is that, given an equation

$$E(x, y) = 0 \tag{1-28}$$

one is often able to "solve for y," at least in theory. Without some restrictions, this is certainly false; the equation

$$y^2 + (x - y)(x + y) - 1 = 0$$

certainly cannot be "solved for y." Moreover, solution of an equation such as

(1-28) does not usually yield a single unique solution for y, whereas when we write $y = f(x)$, we require that exactly one value of y correspond to a chosen value of x.

The meaning of the original statement can be made a little clearer. If the function E is suitably restricted, then Eq. (1-28) defines a collection (with possibly just one member) of functions f such that, if f is one of them, then

$$E(x, f(x)) = 0$$

for all x in the domain of f.

The geometric point of view is very helpful here. Let S be the graph of Eq. (1-28), that is, all points (x, y) for which $E(x, y) = 0$. The set S may be single-valued, meaning that it has the property that no two distinct points in S have the same first coordinate. If so, then S itself is the graph of a function, and we have "solved" (1-28). In general, S will not be single-valued; some vertical lines will cut it twice. However, the set S will contain as subsets many single-valued sets, and each of these will be the graph of a function which "solves" Eq. (1-28). We return to this topic when we take up the implicit function theorems and try to find the needed conditions for good solutions to exist.

So far, the only numerical-valued functions were real-valued. However, there is nothing to prevent one from dealing with functions whose values are complex numbers, such as

$$f(t) = (2 + 3i)t^2 - 4i\,t + (1 - i)$$

All can be reduced to the study of real-valued functions, however, for we can write any complex-valued function F in the form

$$F(p) = G(p) + i\,H(p)$$

where G and H are real-valued functions. In our example,

$$f(t) = (2t^2 + 1) + i(3t^2 - 4t - 1)$$

Recalling that the complex number $a + bi$ can be plotted as the point (a, b) in the plane, we see that a complex-valued function F can also be regarded as a function whose values lie in $\mathbf{R}^2$.

It also turns out to be useful to consider functions whose values are more abstract objects. For example, in the study of systems of differential equations, many techniques are simplified if one works with functions whose values are matrices. An example of such a function is

$$F(t) = \begin{bmatrix} 3t^2 + 1 & 5t - 2 & t^3 + 7t \\ 2t + 1 & t^2 + 1 & t + 5 \end{bmatrix}$$

We will meet matrix-valued functions later on in this book, where they emerge as the natural form for the derivative of a transformation from n space into m space.

Finally, in Chap. 8 we will meet functions whose domain itself is a set of functions. You have already met such functions, for differentiation itself is an example; d/dx is a function that can be applied to functions such as $x^3 - 3x^2$, giving for the value another function $3x^2 - 6x$.

There is one special class of functions that is so important that it deserves separate mention. In your earlier work with mathematics, you have already met the term "infinite sequence" and the notation $\{a_n\}$. By this, one is to understand that to each positive integer n has been assigned a specific number a_n in some determinable manner. If we examine this critically, it is evident that we are again dealing with a function whose domain is the set $Z^+ = \{1, 2, 3, 4, \ldots\}$ of all positive integers and whose values are numbers. Explicitly, the **sequence** symbolized by $\{a_n\}$ is the function mapping $n \to a_n$.

Taking the next step, a function on Z^+ to the plane will be a sequence of points in the plane, and a sequence in $\mathbf{R}^n$ is defined to be any function from Z^+ into n space. We can denote it either by $\{p_n\}$ or by: $n \to p_n$. The values of this function are the points p_n, which it is usual to call the **terms** of the sequence. They do not have to be all different. The sequence given by $a_n = (-1)^n$ has only two distinct terms, 1 and -1. We shall use the word "trace" to describe the exact range of the function, the set of distinct points that appear as terms in the sequence.

Other types of sequences will also be important in all the work to follow. For example, we will often deal with *sequences of sets*, such as $\{D_n\}$ where

$$D_n = \{\text{all points } p \text{ with } |p| \le n\}$$

Each of the terms in this sequence is a ball, and as n increases they expand (see Exercise 15).

Again, we will devote much space to the study of **sequences of functions;** a sample of this is the sequence $\{f_n\}$ where

$$f_n(x) = nx^2e^{-nx}$$

While each of these examples could be formalized as a function on Z^+ taking its values in either the class of all sets or the class of all functions, we prefer to treat this more informally.

EXERCISES

1 Give the domain of definition of each function f defined below, and describe or sketch its graph:

(*a*) $f(x) = 1/(1 + x^2)$ (*b*) $f(x, y) = 4 - x^2 - y^2$

(*c*) $f(x) = x/(x - 1)$ (*d*) $f(x, y) = 1/(x^2 - y^2)$

(*e*) $f(x, y) = \begin{cases} 1 & \text{for } x < y \\ 0 & \text{for } x = y \\ \frac{1}{2} & \text{for } x > y \end{cases}$

2 Let $f(x) = x^2 + x$, $g(x, y) = xy$, and $h(x) = x + 1$. What are:
(*a*) $f(g(1, 2))$ (*b*) $h(f(3))$
(*c*) $g(f(1), h(2))$ (*d*) $g(f(x), h(y))$
(*e*) $g(h(x), f(x))$ (*f*) $f(g(x, h(y)))$
(*g*) $f(f(x))$

3 (*a*) If $F(x) = x^2 + x$ and $G(s) = s + s^2$, are F and G different functions?
(*b*) If $F(x, y) = x^2 + y$ and $G(x, y) = x + y^2$, are F and G different functions?

4 (*a*) What is the natural domain of the function $g(x) = \sqrt{2 - x}$?
(*b*) What is the natural domain of the function $f(x) = \sqrt{x - 3} + \sqrt{2 - x}$?

***5** Sketch F for which $F(1/(x - 1)) = x/(x + 1)$, $\quad x \neq 1$.

6 Sketch the level curves of the function described by $f(x, y) = x^2 - y^2$.

7 Sketch the level curves for f when
(*a*) $f(x, y) = y^2 - x$ (*b*) $f(p) = |p| - 1$
(*c*) $f(p) = \begin{cases} 1 & \text{when } |p| < 1 \\ x - y & \text{when } |p| \geq 1 \end{cases}$

8 Sketch the level surfaces for the function $f(x, y, z) = x^2 + y^2 - z^2$.

9 Let $F(x, y, z, t) = (x - t)^2 + y^2 + z^2$. By interpreting this as the temperature at the point (x, y, z) at time t, see if you can get a feeling for the behavior of the function.

10 If f is a function of two variables, show that if the set of points above the graph of f is a convex set, then f satisfies (1-27).

11 What can you say about the problem of solving for y in the equation $x^3 - y^3 + x - y = 0$? How many real functions on $-\infty < x < \infty$ does this equation define?

***12** Given $E(x, y) = x^2 - y^2$, how many different functions f are there that are "defined by the equation $E(x, y) = 0$ so that $y = f(x)$"?

13. Find the first six terms of the sequence defined by $a_n = (-2)^{n+1} + (-3)^n$

14 Given $x_n = 3n + (-1)^n(n - 5) + 7$,
(*a*) Calculate $x_1, x_2, \ldots, x_{10}$.
(*b*) Find all the numbers that ever appear twice in the entire sequence.
*(*c*) Do any terms appear three times?

15 What is $\bigcup_1^\infty D_n$ where $D_n = \{\text{all points } p \text{ with } |p| \leq n\}$?

***16** Show that the collection of all functions defined on a set D, with values in $\mathbf{R}^3$, is a vector space.

1.5 TOPOLOGICAL TERMINOLOGY

The role of elementary topology is to give precision and structure to a variety of intuitive concepts such as "nearness" and "neighborhood," "limits" and "continuity." When these concepts are used in the plane or in n space, they rest ultimately upon certain basic properties of the real number system $\mathbf{R}$, and upon properties of the distance function $|p - q|$, which gives the distance between two points p and q. We start with a sequence of definitions.

The word "near" in everyday use is ambiguous. The rational number 355/113 is very near to π, since they differ by less than 3×10^{-7}, and Alpha Centauri is one of our nearest stellar neighbors, at a distance of more than 3×10^{13} kilometers. We shall say that the set of all points p that are **near** a point p_0 is those that obey $|p - p_0| < \delta$, for some choice of $\delta > 0$. This

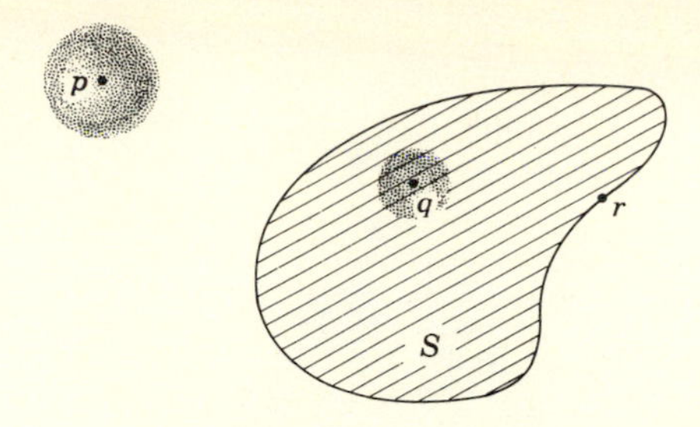

Figure 1-19

set is just the open ball $B(p_0, \delta)$ of radius δ centered at p_0; usually, δ will be thought of as a very small number.

Each point p in n space has one of three relationships to a given set S: It is either interior to S, exterior to S, or on the boundary of S. We proceed to make these words precise.

A point q is said to be **interior** to a set S if all the points sufficiently near to q also belong to S. Thus, for some choice of δ the ball $B(p_0, \delta)$ is a subset of S. This is illustrated in Fig. 1-19.

The **interior of a set** S is the subset of S consisting of all points q that are interior to S.

A set S is called **open** if every point in S is interior; in this case, S coincides with its own interior.

We have already used the word "open" as applied to the open ball. Let us show that it was used correctly by proving that every point in the set $S = B(\mathbf{0}, r)$ is interior. If $p_0 \in S$, then $|p_0 - \mathbf{0}| < r$. Choose $\delta > 0$, so that $|p_0| + \delta < r$. We claim that the ball $B(p_0, \delta/2)$ is a subset of S (see Fig. 1-20). For, if $p \in B(p_0, \delta/2)$, then $|p - p_0| < \delta/2$ and

$$\begin{aligned} |p - \mathbf{0}| &= |(p - p_0) + p_0| \leq |p - p_0| + |p_0| \\ &< \delta/2 + |p_0| < \delta/2 + r - \delta = r - \delta/2 < r \end{aligned}$$

and $p \in S$.

A similar calculation can be used to show that the set of points p such that $|p - p_0| > r$ is open, for any p_0 and any $r \geq 0$. This is the region outside a sphere of radius r.

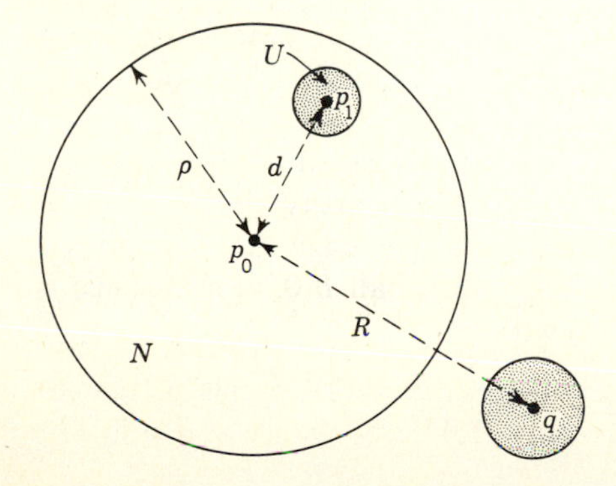

Figure 1-20

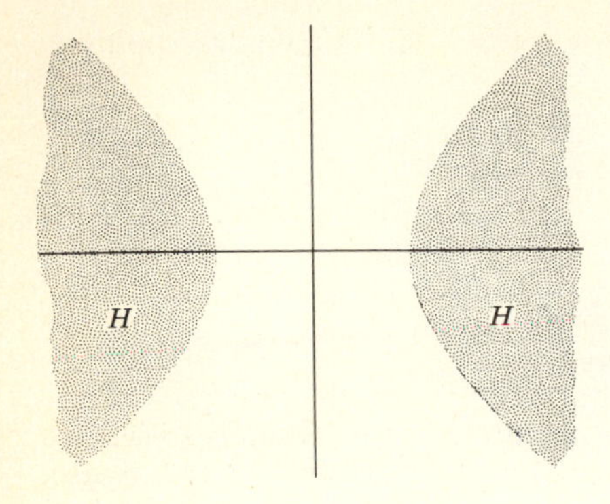

Figure 1-21 $x^2 - y^2 > 1$

Other examples of open sets are the set of points (x, y) in the plane with $x^2 - y^2 - 1 > 0$ (see Fig. 1-21) and the half plane

$$\{\text{all } (x, y) \text{ with } x > 0\}$$

In the next chapter, we discover why open sets arise as the set of points p where a continuous function is strictly positive.

On the line, which is $\mathbf{R}^1$, we see that according to the calculations above, the interval of x such that $b < x < c$ is an open set.

A point p that is not in a set S is said to be **exterior** to S if all the points sufficiently near to p are also outside the set S. Thus, if p is exterior to S there must be some open ball centered at p that is disjoint from S. In Fig. 1-19 the point p is exterior to S.

A set S is said to be **closed** in $\mathbf{R}^n$ if the **complement** of S—that is, the set of points in n space that are not in S—is an open set. Accordingly, S is a closed set if every point outside S is in fact exterior to S. The "closed" ball with center p_0 and radius r is the set of all p with $|p - p_0| \leq r$. It is a closed set because its complement, the set of p where $|p - p_0| > r$, is an open set.

On the line, an interval such as the x with $b \leq x \leq c$ is a closed set (and called a closed interval). In the plane, the set of points (x, y) with $y - x^2 = 0$ is a closed set, as is the set where $xy \geq 0$.

Finally, a **boundary point** for a set S is a point that is neither interior to S nor exterior to S. Accordingly, a point p_0 that is a boundary point for S must be near to S and near to the complement of S at the same time. Every open ball centered at p_0 must contain at least one point of S and at least one point not in S. In Fig. 1-19, r is a boundary point for S.

The collection of all boundary points for a set S is called the **boundary** (or frontier) of the set, and will be denoted by bdy(S); in many simple cases, it will agree with the intuitive notion of boundary as the edge that separates a set from its complement. The boundary of the open ball $B(\mathbf{0}, r)$ in 3-space is the sphere consisting of all points (x, y, z) with

$$x^2 + y^2 + z^2 = r^2$$

The **closure** of a set S is formed by adjoining to S all its boundary points. Thus

$$\bar{S} = \text{closure of } S = S \cup \text{bdy}(S)$$

The closure of the open ball $B(\mathbf{0}, r)$ is the closed ball of the same radius.

The reader is warned not to confuse the terms "boundary" and "bounded." A set S in n space is called **bounded** if there is a number M such that $|p| < M$ for all $p \in S$; any larger value of M will also serve, and the set S is thus a subset of the ball $B(\mathbf{0}, M)$. An unbounded set is one that is too big to fit inside any ball $B(\mathbf{0}, r)$, no matter how large r is. For example, the entire first quadrant in the plane, consisting of the (x, y) with $x > 0$ and $y > 0$, is unbounded. Note that an unbounded set can have a boundary, as in this example, where the boundary of the quadrant consists of the positive X axis and the positive Y axis and $\mathbf{0}$.

While this sequence of definitions has been based on the use of balls to describe "nearness," it is often convenient to use other sets as well. Any set $\mathcal{N}$ is called a **neighborhood** of p_0 if p_0 is interior to $\mathcal{N}$. Any open ball about p_0 is a neighborhood of p_0, and any neighborhood of p_0 contains an open ball about p_0. However, neighborhoods do not have to have any particular shape, and we can use square neighborhoods if this is more convenient. In all the definitions above, the word "ball" can be replaced by neighborhood; for example, p is a boundary point for S if every neighborhood of p contains both a point in S and a point not in S.

Another illustration may be helpful in understanding the concepts of boundary and boundary point. Let us examine the set A, pictured in Fig. 1-22, which is described formally by

$$A = \{\text{all points } p \in \mathbf{R}^2 \text{ with } 0 < |p| \leq 1\} \cup \{\text{the point } (0, 2)\}$$

If the definitions are applied with care, we find that the boundary of A is in three separated pieces and consists of the circumference, where $|p| = 1$, and the two points $(0, 0)$ and $(0, 2)$. The interior of A is the set of points p with $0 < |p| < 1$; the closure of A is the set consisting of the point $(0, 2)$, together with the unit disk, $|p| \leq 1$.

This last example also illustrates another simple topological idea. We say that a point p is a **cluster point** for a set S if every neighborhood about p contains infinitely many points of the set S. The terms "limit point" and "accumulation point" are also used. In contrast, a point p in S is said to be

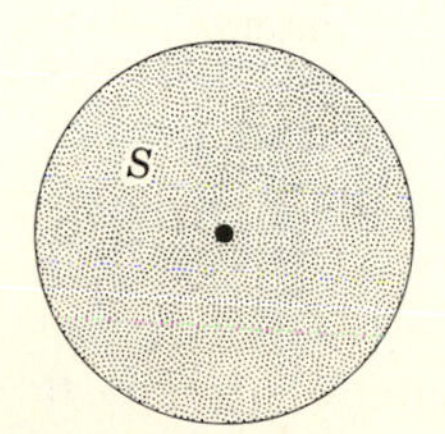

Figure 1-22

isolated if there is a neighborhood about p which contains no other point of S. Note that any interior point of a set is also a cluster point, since a ball always contains infinitely many points. Every point in the closure of a set is either an isolated member of the set or a cluster point for the set. In the example in Fig. 1-22, (0, 2) is an isolated point, but (0, 0) is a cluster point.

The subject of this book is analysis and not elementary topology; the topological terminology and concepts appear in this chapter because they will be useful tools in the study of functions on $\mathbf{R}^n$ that will follow in subsequent chapters. Our purpose is therefore not to study all the interconnections and implications of the various definitions, but to explain them and state certain useful relations. The following list summarizes a number of these basic properties, to be used whenever they are found helpful.

(1-29)

(i) If A and B are open sets, so are $A \cup B$ and $A \cap B$.
(ii) The union of any collection of open sets is open, but the intersection of an infinite number of open sets need not be open.
(iii) If A and B are closed sets, so are $A \cup B$ and $A \cap B$.
(iv) The intersection of any collection of closed sets is a closed set, but the union of an infinite number of closed sets need not be closed.
(v) A set is open if and only if its complement is closed.
(vi) The interior of a set S is the largest open set that is contained in S.
(vii) The closure of a set S is the smallest closed set that contains S.
(viii) The boundary of a set S is always a closed set and is the intersection of the closure of S and the closure of the complement of S.
(ix) A set S is closed if and only if every cluster point for S belongs to S.
(x) The interior of a set S is obtained by deleting every point in S that is on the boundary of S.

Each of these can be verified by a proof based on the definitions given above. We present this only for the first two assertions in the list, to show the nature of the proofs.

PROOF OF (i) Suppose that A and B are open sets. To show that $A \cup B$ is open, suppose that p_0 belongs to $A \cup B$. Then p_0 is in A or it is in B. In either case, p_0 is the center of an open ball that is a subset of A or a subset of B, since A and B are themselves open and p_0 must be interior to one of them. This open ball is then a subset of $A \cup B$, and p_0 is therefore interior to $A \cup B$. Every point of $A \cup B$ is therefore interior to $A \cup B$, and $A \cup B$ is open. The proof that $A \cap B$ is open is slightly different. Let p_0 be any point in $A \cap B$; we must show that p_0 is interior to $A \cap B$.

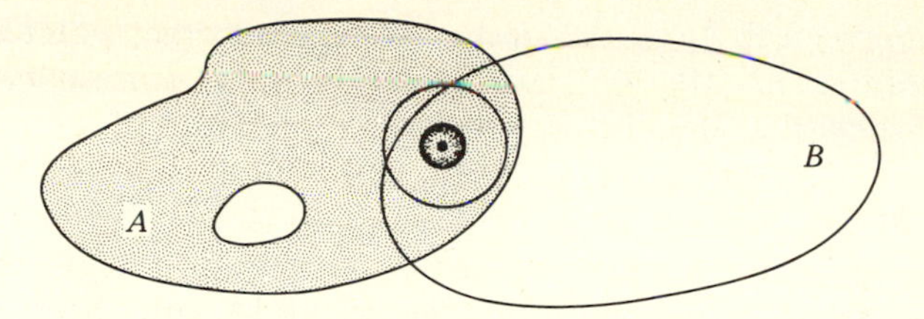

Figure 1-23

Since $p_0 \in A$ and A is open, p_0 is interior to A. There is then a δ_1 such that the open ball $B(p_0, \delta_1)$ is a subset of A. In the same way, we conclude that there must be a δ_2 such that the open ball $B(p_0, \delta_2) \subset B$. Let δ be the smaller of the two numbers δ_1, δ_2. Then, the ball $B(p_0, \delta)$ is a subset of both A and B, and thus a subset of $A \cap B$. Accordingly, p_0 is interior to $A \cap B$, and since p_0 was an arbitrarily selected point of $A \cap B$, $A \cap B$ is open.

Such detailed verbal arguments are tedious when carried to an extreme. The essence of the second can be conveyed by the diagram given in Fig. 1-23, showing the step involving the two open balls. Note, for example, that this same diagram supplies the argument supporting the following statement: *If $\mathcal{N}_1$ and $\mathcal{N}_2$ are neighborhoods of p, so is $\mathcal{N}_1 \cap \mathcal{N}_2$.* Rather than construct detailed verbal arguments for the remainder of the assertions in the list above, we suggest that the reader construct examples and diagrams which will illuminate and support their truth.

We choose to discuss (ii) further, however, because it brings in something new, the behavior of an infinite collection of sets. First observe that assertion (i) can be extended easily to any finite collection of open sets; if $A_1, A_2, \ldots, A_n$ are open sets, so are the sets

$$\bigcup_1^n A_k = A_1 \cup A_2 \cup A_3 \cup \cdots \cup A_n$$

$$\bigcap_1^n A_k = A_1 \cap A_2 \cap A_3 \cap \cdots \cap A_n$$

Suppose now that we have an infinite (countable) collection of open sets, $A_1, A_2, A_3, \ldots$. We wish to prove that their union, $A = \bigcup_1^\infty A_k$, is an open set. The argument differs little from that used for a finite collection. If $p \in A$, then there is at least one of the sets, say A_k, to which p belongs. Since that set is open, p is surrounded by a ball which lies entirely inside A_k; however, this ball will also be a subset of A, since $A_k \subset A$, and p is therefore interior to A, which must then be open. (Note that this argument does not really depend on having a *countable* collection of open sets; thus, we are justified in saying that the union of any collection of open sets is open.)

The second half of assertion (ii) states that we cannot prove the corresponding assertion for the intersection of open sets. To show this, all that is needed is a counterexample. Let A_k be the open disc in the plane, centered at the origin and of radius $1 + 1/k$. We have $A_1 \supset A_2 \supset A_3 \supset \cdots$, and a moment's thought will show that the intersection

$$B = \bigcap_1^\infty A_k = A_1 \cap A_2 \cap A_3 \cap \cdots$$

is the closed disc of radius 1, definitely not an open set. (It is instructive to see where the proof for a finite collection of open sets breaks down.)

The purpose of such a list of topological facts and definitions is to give a language in which to describe new aspects of sets and functions with mathematical precision. As an illustration of this, consider the intuitive notion of a set that is "connected." Asked to explain the meaning of this word, one might point to the set H shown in Fig. 1-21 as one that is certainly not connected, and then to the set B in Fig. 1-23 as a sample of one that is connected. With enough samples, perhaps the precise meaning of the term will be clear, but a careful mathematical definition would be better.

In fact, there are two slightly different aspects of the intuitive idea. The set H in Fig. 1-21 would certainly be said to be disconnected, since it consists of two separated pieces. This aspect is captured very adequately in the following.

Definition 2 *Two disjoint sets A and B, neither empty, are said to be* **mutually separated** *if neither contains a boundary point of the other. A set is* **disconnected** *if it is the union of separated subsets, and is called* **connected** *if it is not disconnected.*

The scope of this definition can be realized by contrasting the set S consisting of all (x, y) with $x^2 - y^2 < 0$, and the set T that is obtained from S by adjoining all the points (x, y) with $y = |x|$. The former is an open set, shown in Fig. 1-24, and it is made up of two separated pieces, the top half where $y > 0$ and the bottom half where $y < 0$. We would certainly conclude that S is dis-

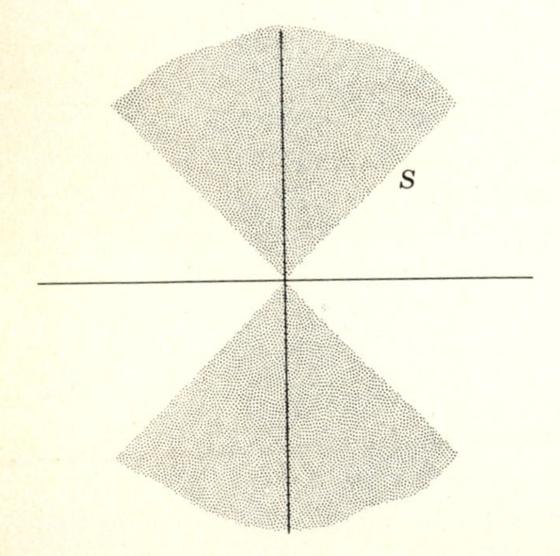

Figure 1-24

connected. However, we would not wish to call the set T disconnected. T is the union of two disjoint subsets, the top half A consisting of all (x, y) with $y \geq |x|$, and the bottom half B comprising those (x, y) with $y < |x|$. However, these subsets are not mutually separated, since the origin $(0, 0)$ belongs to A but is a boundary point of B. Note how slight is the difference in appearance of the set S and the set T, which differs only in containing the boundary of A.

The second aspect of the intuitive meaning of connected is brought out by saying that the set B in Fig. 1-23 is connected "because it is possible to travel between any two points of the set without ever leaving the set," a statement that is clearly false for the set H in Fig. 1-21. It is technically more difficult to make this concept precise, since we must explain "travel between." We choose to interpret this to mean that there is a path or curve which lies in the given set and joins the two chosen points. However, the notion of "curve" itself has pitfalls, to be resolved in later chapters, and so we adopt for the present a temporary version that has no ambiguity.

Definition 3 *A set S is said to be* **polygon connected** *if, given any two points p and q in S, there exists a chain of line segments in S which abut and form a path starting at p and ending at q.*

This definition has the virtue that is relatively easy to check, given a set S and a specific pair of points p and q. On the other hand, Definition 2 is easy to use in confirming that a specific set is disconnected, but very difficult to use to check that a set is connected, since one must prove that no decomposition of the set into mutually separated pieces is possible. Indeed, it is not until Sec. 1.7 that we will be able to prove that the line itself is connected, while it is obvious that the line is *polygon* connected. In Appendix 2, we will give an example to show that the difficulty is real and not artificial by exhibiting a set that can be shown to be connected, but which is not path connected.

For the present, we show that such difficulties do not arise if we work with open sets, as in fact we shall for much of the work to follow.

Theorem 2 *Any open connected set is polygon connected.*

Let $\mathcal{O}$ be an open connected set, and take any two points q_1 and q_2 in $\mathcal{O}$. We will show that there is a polygon path in $\mathcal{O}$ that joins q_1 to q_2. Let A be the set of all points $p \in \mathcal{O}$ that can be joined to q_1 by a finite polygon path lying entirely in $\mathcal{O}$. Clearly, $q_1 \in A$; indeed, since $\mathcal{O}$ is open and all points p in a small ball centered at q_1 are in $\mathcal{O}$ and can be joined to q_1 along a radius, this ball is a subset of A. We prove that A is open. Let p_0 be any point in A. Since $\mathcal{O}$ is open, there is an open ball $\mathcal{N}$ about p_0 that lies entirely in $\mathcal{O}$ (see Fig. 1-25). Clearly, any point in $\mathcal{N}$ can be reached from p_0 by a segment in $\mathcal{N}$. Since p_0 is in A, there is a polygon path in $\mathcal{O}$ from q_1 to p_0. Putting these together, we are able to connect q_1 in $\mathcal{O}$ to any point in $\mathcal{N}$. Hence, all points in $\mathcal{N}$ lie in A, proving that p_0 is interior to A and A is

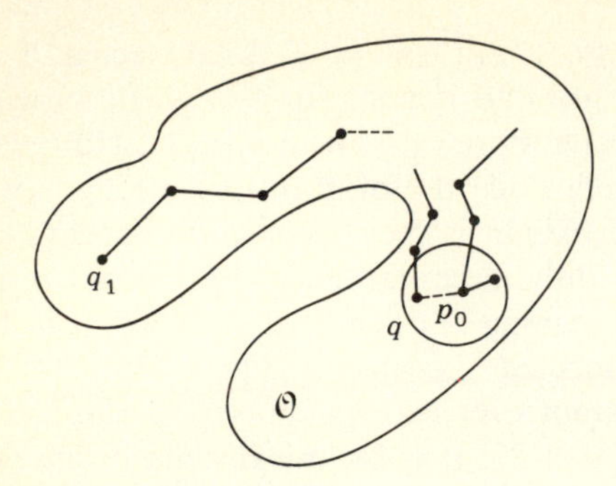

Figure 1-25

open. Now, consider instead the set B consisting of all points $p \in \mathcal{O} - A$, that is, the points of $\mathcal{O}$ that *cannot* be reached from q_1 by a polygon path in $\mathcal{O}$. Suppose now that $p_0 \in B$. As before, $\mathcal{O}$ being open, there is an open ball $\mathcal{N}$ about p_0 lying in $\mathcal{O}$. If there were a point q in $\mathcal{N}$ that could be reached from q_1, then q could then be joined to p_0 in $\mathcal{N}$ along the radius of the ball, producing a polygon path from q_1 to p_0. Since this is impossible, no point q in $\mathcal{N}$ lies in A, and $\mathcal{N}$ must be a subset of B. Thus, p_0 is interior to B, and B too is open. We thus have $\mathcal{O} = A \cup B$ where A and B are disjoint and both open. This contradicts the hypothesis that $\mathcal{O}$ is connected, unless one of the sets A or B is empty. Since A contains q_1, as well as other points near q_1, B is empty. Thus, there is no point in $\mathcal{O}$ that cannot be reached from q_1 by a polygon path in $\mathcal{O}$, and $\mathcal{O}$ is polygon connected. ▌

In Chap. 2, we will prove the converse of this, showing that every path-connected set is also connected. This will depend on the fact that the line and any closed interval $[b, c]$ are each connected.

EXERCISES

1 By quoting appropriate definitions and statements, verify the following assertions. Sketches may be helpful.

(i) The set $W = \{\text{all } p = (x, y) \text{ with } 1 \le |p| \le 2\}$ is closed, bounded, connected, and bdy(W) is the disconnected set, consisting of two circles of radius 1 and 2.

(ii) The set $R = \{\text{all } (x, y) \text{ with } x \ge 0\}$ is closed, unbounded, connected, and its boundary is the vertical axis.

(iii) The set $T = \{\text{all } (x, y) \text{ with } |x| = |y|\}$ is closed, connected, unbounded, has empty interior, and bdy(T) = T.

(iv) The set $Q = \{\text{all } (x, y) \text{ with } x \text{ and } y \text{ integers}\}$ is closed, unbounded, infinite, countable, disconnected, and its boundary is itself.

2 Tell which of the properties described in this section apply to the set of points (x, y) such that:

(*a*) $x^2 + y^2 = 1$ (*b*) $x^2 + y \ge 0$
(*c*) $x = y$ (*d*) $x > 1$
(*e*) $xy > 0$ (*f*) $y = |x - 1| + 2 - x$

3 The same as Exercise 2 for the set of points (x, y, z) such that:

(*a*) $x^2 + y^2 + z^2 > 4$ (*b*) $x^2 + y^2 \leq 4$

(*c*) $xy > z$ (*d*) $(x - y)^2 = z^2$

4 The same as Exercise 2 for the set of points x on the line such that:

(*a*) $x(x - 1)^2 > 0$ (*b*) $x(x - 1)(x + 1)^2 \leq 0$

5 Let $S = \{\text{all } (x, y) \text{ with } x \text{ and } y \text{ rational numbers}\}$.

(*a*) What is the interior of S?

(*b*) What is the boundary of S?

6 What are the cluster points for the set

$$S = \left\{\text{all } \left(\frac{1}{n}, \frac{1}{m}\right) \text{ with } n = 1, 2, \ldots, m = 1, 2, \ldots\right\}$$

7 Produce an unbounded infinite set with no cluster points.

8 By constructing an example, show that the union of an infinite collection of closed sets does not have to be closed.

9 Let C be a closed set and V an open set. Then $C - V$ is closed and $V - C$ is open. Verify this statement, using statements in (1-29).

10 Construct pictures to show that each of the following is false.

(i) If $A \subset B$, then $\text{bdy}(A) \subset \text{bdy}(B)$.

(ii) $\text{bdy}(S)$ is the same as the boundary of the closure of S.

(iii) $\text{bdy}(S)$ is the same as the boundary of the interior of S.

(iv) The interior of S is the same as the interior of the closure of S.

11 Let S be any bounded set in n space. Is the closure of S a bounded set?

12 What is the relationship of $\text{bdy}(A \cap B)$ to $\text{bdy}(A)$ and $\text{bdy}(B)$?

13 (*a*) Why should the empty set $\varnothing$ be called both open and closed?

(*b*) Why, in the plane, is the set $\{\text{all } (x, y), x^2 + y^2 \geq 0\}$ both open and closed?

14 Let $U_n = \{\text{all } p = (x, y) \text{ with } |p - (0, n)| < n\}$. Show that the union of all the open sets U_n, for $n = 1, 2, 3, \ldots$, is the open upper half plane.

15 Let A and B be connected sets in the plane which are not disjoint. Is $A \cap B$ necessarily connected? Is $A \cup B$ necessarily connected?

16 Show that the complement of the set in Exercise 5 is a polygon connected set. Is it an open set?

17 (*a*) Is the interior of a connected set necessarily connected?

*(*b*) Is the closure of a connected set necessarily connected?

18 (*a*) Show that any two disjoint nonempty open sets are mutually separated.

(*b*) Show that any two disjoint nonempty closed sets are mutually separated.

***19** When a set is not connected, the separate pieces that it comprises are called **components.** How many components are there to the set $S = \{\text{all } (x, y) \text{ with } x^2 - 3xy + y^3 < 0\}$?

1.6 SEQUENCES

Before introducing additional terms that are descriptive of the properties of certain types of sets of points, we show that those properties already stated are useful in discussing some standard parts of analysis. We take up the topic of sequences and convergence. Recall that a sequence $\{p_n\}$ of points in any space is a special type of function f defined on the set $Z^+ = \{1, 2, 3, \ldots\}$ whose value at n is $f(n) = p_n$. The trace of a sequence $\{p_n\}$ in a space is the set A of all

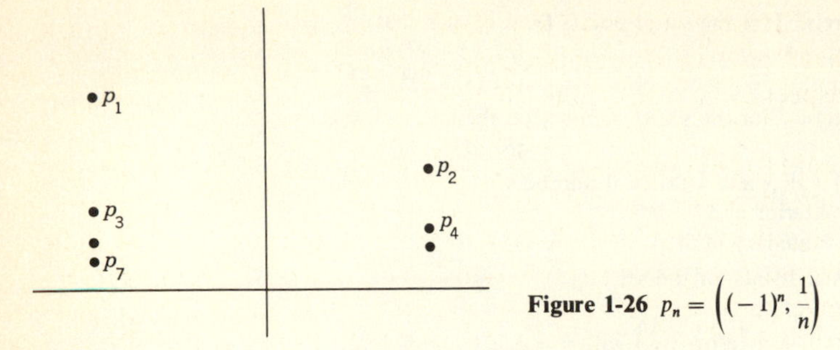

Figure 1-26 $p_n = \left((-1)^n, \frac{1}{n}\right)$

the points that appear as terms in the sequence;

Thus $$A = \{\text{all } p_n \text{ for } n = 1, 2, 3, \ldots\}$$

For example, the sequence in the plane given by $p_n = ((-1)^n, 1/n)$ has for its trace the infinite set shown in Fig. 1-26, while the trace of the sequence on the line, $q_n = 3 + (-1)^n 2$ is the set $\{1, 5\}$. A sequence is said to be **bounded** if its trace is a bounded set. Thus, for $\{p_n\}$ to be bounded, there must exist a number M such that $|p_n| < M$ for all $n = 1, 2, 3, \ldots$. The sequence whose terms are displayed in Fig. 1-26 is a bounded sequence.

The concept of a convergent sequence is basic to analysis.

Definition 4 *A sequence $\{p_n\}$ converges to the point p if and only if every neighborhood about p contains all the terms p_n for sufficiently large indices n; to any neighborhood U about p, there corresponds an index N such that $p_n \in U$ whenever $n \geq N$.*

Under these circumstances, it is customary to call p the **limit** of $\{p_n\}$, and to write $\lim_{n\to\infty} p_n = p$ or $p_n \to p$, which is often read: "p_n **approaches** p as n increases." If the terms of the sequence were displayed as lights which flashed in succession, convergence would be observed by noticing that the points p_n tend to concentrate more and more closely around the point p, for later values of n, and that if we construct a spherical neighborhood U of radius $\rho > 0$ about p, then eventually (when $n > N$) all flashes appear inside U and none appear outside. Note that the approach to p does not have to be from any particular direction, nor does the distance from p_n to p have to decrease uniformly at each change in n. For example, the sequence in $\mathbf{R}^1$ described by

$$p_k = \frac{2 + (-1)^k}{k}$$

has as its terms in order the numbers $1, \frac{3}{2}, \frac{1}{3}, \frac{3}{4}, \frac{1}{5}, \frac{3}{6}, \frac{1}{7}, \frac{3}{8}, \ldots$ and converges to 0. (Given $\rho > 0$, choose N so that $N > 3/\rho$; then, $|p_k - 0| < \rho$ whenever $n > N$.)

A word of caution about notation: The symbol "∞" used in connection with limits is neither a number nor a point on the line, but merely part of the expression "$n \to \infty$." It need not have any meaning out of context. A test of this is the fact that "∞" does not appear in the statement of Definition 2 above. (Another aspect of the use of the symbol ∞ will be found at the end of Appendix 2.)

A sequence is said to be **convergent** if there is a point to which it converges. (As will be seen later, there are many times when it is possible to be sure that a sequence is convergent even if one does not know the point to which it converges.) A convergent sequence cannot converge to two distinct limits. For if $\{p_n\}$ were to converge to both p and q with $p \neq q$, then we could choose spherical neighborhoods about p and q that are disjoint; but, all but a finite number of the p_n would have to be inside each of these neighborhoods, which is impossible.

A sequence that is not convergent is called **divergent.** This can happen in a number of different ways. The sequences given by

$$a_n = n^2$$
$$b_n = (-1)^n n$$
$$c_n = 1 + (-1)^n$$

are each divergent, as is the sequence pictured in Fig. 1-26.

It is evident from the definitions that the convergence or divergence of a sequence is not affected if one changes any finite number of its terms. In this sense, we are dealing with something that is a property of the "tails" of a sequence, the images of the sets

$$I_n = \{n, n+1, n+2, \ldots\}$$

In order to show that a specific sequence is convergent, the only method we have available to us (at this stage) is to guess the limit and then check this by a certain amount of calculation. As an illustration, consider the sequence

$$n \to p_n = \left(\frac{1}{n}, \frac{n}{n+1}\right)$$

By one method or another (for example, by plotting some points of its trace), we are led to guess that this sequence converges to the point $p = (0, 1)$. We want to show that $\lim p_n = (0, 1)$. Taking for U a neighborhood about p of radius $\varepsilon > 0$, we want to have $|p_n - p| < \varepsilon$ for all sufficiently large n. Substituting the coordinates of p_n and p, we wish to have

$$\left|\left(\frac{1}{n}, \frac{n}{n+1}\right) - (0, 1)\right| < \varepsilon$$

or
$$\frac{1}{n^2} + \left(\frac{n}{n+1} - 1\right)^2 = \frac{1}{n^2} + \frac{1}{(n+1)^2} < \varepsilon^2$$

for all n larger than some number N. This probably seems self-evident;

however, to give a formal argument (and to get some idea of how large N must be), we proceed thus: Since $1/n^2 + 1/(n+1)^2 < 2/n^2$, we can obtain our goal by choosing N so that $2/N^2 < \varepsilon^2$. Hence, we can choose N to be $(\sqrt{2})/\varepsilon$. Any larger value of N, such as $2/\varepsilon$ or $20/\varepsilon^2$, would do as well. ▮

We now turn to the proof of several theorems about limits.

Theorem 3 *Any convergent sequence is bounded.*

Let $\lim_{n\to\infty} p_n = p$. If we take U to be a neighborhood of radius 1, centered on p, then there will be only a finite number of terms of the sequence outside U. U is bounded, and any finite set is bounded; so the trace of the sequence is bounded. ▮

Theorem 4 *If* $\lim_{n\to\infty} p_n = p$ *and* $\lim_{n\to\infty} q_n = q$, *then*

$$\lim_{n\to\infty} (p_n + q_n) = p + q$$

Given $\varepsilon > 0$, we may assume that there are numbers N' and N'' such that $|p_n - p| < \varepsilon$ whenever $n > N'$ and $|q_n - q| < \varepsilon$ whenever $n > N''$. Take N to be any number larger than both N' and N''. If we have $n > N$, then both the above inequalities hold, so that (by using the triangle inequality)

$$\begin{aligned} |p_n + q_n - (p+q)| &= |p_n - p + q_n - q| \\ &\le |p_n - p| + |q_n - q| \\ &< \varepsilon + \varepsilon = 2\varepsilon \end{aligned}$$

This says that, except for a finite number of terms at the start, all terms of the sequence $\{p_n + q_n\}$ lie in a neighborhood of radius 2ε, centered at the point $p + q$. Since ε can be any positive number, we have shown that this is true of every neighborhood of $p + q$, and thus $\lim_{n\to\infty} (p_n + q_n) = p + q$. ▮

The topological properties of a set in n space can often be explored most easily by means of sequences. For example, the following result affords an alternative way to define closure.

Theorem 5 *The closure of a set S in $\mathbf{R}^n$ is the set of all limits of converging sequences of points from S.*

Let $\{p_n\}$ be a sequence of points in S, with $\lim p_n = p$. We shall first show that p belongs to the closure of S. This is certainly true if p is either in S or in the boundary of S. Could p be exterior to S? If so, then there would be a small neighborhood U about p such that U contains no points of S. In particular, U contains none of the points p_n, contradicting the fact that $\{p_n\}$ converges to p. Hence, p lies in the closure of S.

To complete the proof, we must show that every point in the closure of S can be the limit of a converging sequence from S. Let q be such a point.

If $q \in S$, then, clearly, we can take $p_n = q$ for all n and the sequence $\{p_n\}$ converges to q. Suppose therefore that q is in the closure of S, and not in S. The point q must then be a cluster point for S; every neighborhood of q contains points of S. Take any convenient decreasing sequence of neighborhoods of q, such as the sets

$$U_n = \{\text{all } p \text{ with } |p - q| < 1/n\}$$

and choose any point p_n in $U_n \cap S$. Then, it is evident that $\{p_n\}$ is a sequence in S which converges to q. ▮

Corollary 1 *Every point in the boundary of a set S is simultaneously the limit of a converging sequence of points in S and a converging sequence of points in the complement of S.*

For, by item (viii) in the list of basic topological facts (1-29), bdy(S) is the intersection of the closure of S and the closure of the complement of S.

Corollary 2 *A set S is closed if and only if it contains the limit of every converging sequence $\{p_n\}$ whose terms lie in S.*

For sequences, the analog of the term "cluster point" is **limit point.** A point p is a limit point for the sequence $\{p_n\}$ if every neighborhood of p contains infinitely many of the terms p_n; since many of the terms could be identical, we give a formal definition.

Definition 5 *p is a limit point for $\{p_n\}$ if given any neighborhood $\mathcal{N}$ about p, there is an infinite set of integers A such that $p_n \in \mathcal{N}$ for all $n \in A$.*

Note that the set A depends on $\mathcal{N}$, and may change if $\mathcal{N}$ is decreased in size.

The difference between "limit of a sequence" and "limit point of a sequence" is subtle; for the former to hold, *all* the terms p_n must lie in $\mathcal{N}$ for n beyond some index n_0, while for the second to hold, p_n must lie in $\mathcal{N}$ only for some infinite sequence of values of n, which may in fact be widely separated with none consecutive. As a result, a sequence can be divergent and still have limit points. The sequence $p_n = ((-1)^n, 1/n)$ has $(1, 0)$ and $(-1, 0)$ as limit points, as seen from Fig. 1-26. Again, the sequence defined by

$$a_n = n + 1 + (-1)^n\left(n + \frac{1}{n}\right)$$

has exactly one limit point, and yet is divergent. For, we see that

$$a_n = \begin{cases} 2n + 1 + \dfrac{1}{n} & \text{when } n \text{ is even} \\[2ex] 1 - \dfrac{1}{n} & \text{when } n \text{ is odd} \end{cases}$$

It is evident that 1 is a limit point for $\{a_n\}$, and since the sequence is unbounded, it is divergent. Moreover, there are no other limit points, for there is no number other than 1 about which terms of the sequence cluster.

This concept, limit point, becomes easier to work with if we also introduce the related idea **subsequence.** Given a sequence $\{p_n\}$, and any increasing sequence of integers

$$n_1 < n_2 < n_3 < \cdots$$

we can define a new sequence $\{q_k\}$ by

$$q_k = p_{n_k}$$

This is called a subsequence of $\{p_n\}$; its terms are terms of p_n and are p_{n_1}, p_{n_2}, p_{n_3}, ... For example, the (divergent) sequence $1, \frac{1}{2}, 1, \frac{1}{3}, 1, \frac{1}{4}, \ldots$ has, among its subsequences, the sequence $1, \frac{1}{2}, \frac{1}{3}, \frac{1}{4}, \ldots$, as well as the sequence $1, \frac{1}{4}, \frac{1}{9}, 1, \frac{1}{16}, \frac{1}{25}, 1, \ldots$.† Then, it is easy to give an argument similar to that in the proof of Theorem 5 to show the following.

Theorem 6 *The limit points of a sequence $\{p_n\}$ are exactly the limits of the converging subsequences of $\{p_n\}$.*

While we have discussed convergence for sequences of points in n space, it is clear that the problem of checking the statement $\lim_{n \to \infty} p_n = p$ can always be reduced to a problem dealing only with *real* sequences. One way to do this is to observe that $\{p_n\}$ converges to p if and only if

$$\lim_{n \to \infty} |p_n - p| = 0$$

A better method is based on the following simplifying observation.

Theorem 7 *A sequence in n space is convergent if and only if each of the real sequences obtained from its coordinates is convergent. Thus, in 3-space, $p_n = (a_n, b_n, c_n)$ defines a convergent sequence if and only if $\{a_n\}$, $\{b_n\}$, and $\{c_n\}$ are each convergent real sequences.*

PROOF IN $\mathbf{R}^3$ Let $p = (a, b, c)$. Then, using Exercise 6 in Sec. 1.3,

$$|p_n - p| \le |a_n - a| + |b_n - b| + |c_n - c|$$

$$|a_n - a| \le |p_n - p|$$

$$|b_n - b| \le |p_n - p|$$

and

$$|c_n - c| \le |p_n - p|$$

† Strictly speaking, a description of a sequence is not complete until a formula can be given for the general term or until a recursive rule is given which specifies the terms in some unique way. In practice, however, it may be sufficient for communications purposes to list enough of the terms to make the law of formation evident. This brings in matters of experience and judgment which could lead to conflicting choices; the terms 2, 3, 5, 8 might with equal justice be followed by 12, 17, 23, 30 or by 13, 21, 34, 55.

Hence, $\lim_{n\to\infty} p_n = p$ if and only if $\lim_{n\to\infty} a_n = a$, $\lim_{n\to\infty} b_n = b$, and $\lim_{n\to\infty} c_n = c$. ▮ (The proof in n space follows the same pattern.)

The advantage of thus being able to reduce questions about the convergence of a sequence of points to questions dealing only with sequences of real numbers is that one is back on more familiar ground. For example, one may use the following standard facts, whose proofs we include only for completeness.

Theorem 8 *If* $\lim_{n\to\infty} a_n = A$ *and* $\lim_{n\to\infty} b_n = B$, *then* $\lim_{n\to\infty}(a_n + b_n)$ *exists and is* $A + B$.

PROOF This is merely Theorem 4 for $\mathbf{R}^1$.

Theorem 9 *If* $\lim_{n\to\infty} a_n = 0$ *and* $\{b_n\}$ *is bounded, then*

$$\lim_{n\to\infty} a_n b_n = 0$$

PROOF Suppose that $|b_n| < M$ for all n. Given any neighborhood $\mathcal{N}$ about 0, choose ε small enough so that if $|c| < M\varepsilon$, then c lies in $\mathcal{N}$. Since $\lim_{n\to\infty} a_n = 0$, there is an n_0 such that $|a_n| < \varepsilon$ for all $n > n_0$. Then, for these values of n, $|a_n b_n| < |a_n|M < M\varepsilon$, and $a_n b_n$ lies in $\mathcal{N}$. (Note that it is only necessary to know that $|b_n| < M$ holds for sufficiently large n.) ▮

Corollary *If* $\lim_{n\to\infty} a_n = A$, *then for any real number* c,

$$\lim_{n\to\infty} ca_n = cA$$

PROOF Take $b_n = c$ for all n, and observe that $\lim_{n\to\infty} (a_n - A) = 0$. ▮

Theorem 10 *If* $\lim a_n = A$ *and* $\lim b_n = B$, *then*

$$\lim_{n\to\infty} a_n b_n = AB$$

PROOF Observe that

$$a_n b_n = (a_n - A)b_n + Ab_n$$

and use the fact that $\lim_{n\to\infty} (a_n - A) = 0$, $\{b_n\}$ is bounded, and $\lim_{n\to\infty} Ab_n = AB$. ▮

Theorem 11 *Let* $\lim_{n\to\infty} a_n = A \neq 0$. *Then, there is an* n_0 *such that* $1/|a_n| < 2/|A|$ *for all* $n > n_0$; *thus, the sequence* $\{1/a_n\}$ *is defined and bounded for all sufficiently large* n.

PROOF Using $\varepsilon = |A|/2$, choose n_0 so that $|A - a_n| < |A|/2$ for all $n > n_0$. Then, for $n > n_0$ we have

$$|A| = |A - a_n + a_n| \le |A - a_n| + |a_n| < \frac{|A|}{2} + |a_n|$$

which implies that $|a_n| > |A| - |A|/2$ and $1/|a_n| < 2/|A|$. ▮

Corollary *If* $\lim_{n\to\infty} a_n = A \neq 0$, *then* $\lim_{n\to\infty} 1/a_n = 1/A$.

PROOF $1/a_n - 1/A = (A - a_n)(1/Aa_n)$ and $\lim_{n\to\infty} (A - a_n) = 0$ and $1/(Aa_n)$ is bounded by $2/|A|^2$ for large n, so that $\lim_{n\to\infty} (1/a_n - 1/A) = 0$. ▮

Theorem 12 *If* $\lim_{n\to\infty} a_n = A$ *and* $\lim_{n\to\infty} b_n = B$, *and* $B \neq 0$, *then*

$$\lim_{n\to\infty} a_n/b_n = A/B$$

As will be shown in the next section, every bounded sequence of real numbers has at least one limit point, using the term as it was given in Definition 5 above. If such a sequence is not convergent, it will have at least two limit points. The largest limit point of $\{a_n\}$ is called the **limit superior** or **upper limit** of the sequence, and is denoted by $\limsup_{n\to\infty} a_n$. Similarly, the smallest limit point of $\{a_n\}$ is called the **limit inferior** or **lower limit,** and is denoted by $\liminf_{n\to\infty} a_n$. Of course, if the sequence $\{a_n\}$ is convergent, then the upper and lower limits coincide with $\lim_{n\to\infty} a_n$; otherwise, we always have $\liminf a_n <$ $\limsup a_n$. We can use these at times to work with a bounded sequence that is not known to be convergent; however, *lim sup* and *lim inf* do not obey the same rules as the ordinary limit operation, and are thus somewhat harder to work with (see Exercises 27 and 28).

These results, many of which may have been familiar, enable us to discuss the convergence of more complicated sequences in terms of the behavior of known simpler sequences. For example, a starting point is the fact that $\lim_{n\to\infty} 1/n = 0$. (We remark that the truth of this goes back to the basic fact that the rational numbers are dense in the real numbers; given any $\varepsilon > 0$, there is an integer N such that $0 < 1/N < \varepsilon$, so that $|1/n - 0| < \varepsilon$ for all $n > N$.) We can use this to show that

$$\lim_{n\to\infty} \frac{5n^2 - 3n + 7}{6n^2 + n + 2} = \frac{5}{6}$$

by writing

$$\frac{5n^2 - 3n + 7}{6n^2 + n + 2} = \frac{5 - 3/n + 7/n^2}{6 + 1/n + 2/n^2} \to \frac{5 - 0 + 0}{6 + 0 + 0} = \frac{5}{6}$$

Again, given $a_n = \sqrt{n+1} - \sqrt{n}$, we write

$$\sqrt{n+1} - \sqrt{n} = \frac{(\sqrt{n+1} - \sqrt{n})(\sqrt{n+1} + \sqrt{n})}{\sqrt{n+1} + \sqrt{n}}$$

$$= \frac{1}{\sqrt{n+1} + \sqrt{n}} < \frac{1}{\sqrt{n}}$$

and since $\lim_{n\to\infty} 1/\sqrt{n} = 0$ (see Exercise 11), we conclude (see Exercise 9) that $\lim_{n\to\infty} \sqrt{n+1} - \sqrt{n} = 0$.

These examples illustrate the general approach to be used. One uses algebraic manipulation, inequalities, the standard theorems on limits (Theorems 8, 9, 10, and 12), and other simple relations such as those in the exercises to convert the original limit problem into another one that has been already understood and solved. The stage at which a mathematical problem is considered to have an obvious solution depends on the experience and insight of the solver. The test is always how easy it is to supply a formal proof.

For example, it is undoubtedly seems obvious that $\lim_{n\to\infty} 2^{-n} = 0$. A formal argument can be given by observing that the inequality $2^{-n} \le 1/n$ holds for all $n \ge 1$.

However, is it equally obvious that

$$\lim_{n\to\infty} \frac{n^3}{2^n} = 0 \quad ?$$

If it is claimed that $n^3/2^n \le 1/n$, which is equivalent to the assertion that $n^4 \le 2^n$, then a little testing shows this is not immediately clear; thus, for $n = 10$ we have $10^4 = 10{,}000$, which is larger than $2^{10} = 1024$. However, the following is indeed true; the proof is a standard use of **mathematical induction.** (A fuller discussion of this technique of proof is found in Appendix 1.)

Lemma 1 *$n^4 \le 2^n$ for all $n \ge 16$.*

We observe first that $16^4 = (2^4)^4 = 2^{16}$, so the stated inequality holds when $n = 16$. We show that if it holds for $n = N$, then it must hold for $n = N + 1$. We then conclude that the inequality must hold for $n = 17, 18, 19$, etc.

Suppose $N^4 \le 2^N$ and $N \ge 16$. By the binomial theorem

$$(N+1)^4 = N^4 + 4N^3 + 6N^2 + 4N + 1$$
$$< N^4 + 4N^3 + 6N^3 + 4N^3 + N^3 = N^4 + 15N^3$$

But, $N \ge 16$ and the original inequality has been assumed for $n = N$, so that

$$(N+1)^4 < N^4 + N(N^3) = 2N^4$$
$$\le 2(2^N) = 2^{N+1}$$

But, this is exactly the form the stated inequality takes when $n = N + 1$. Hence, $n^4 \leq 2^n$ for all $n \geq 16$. ▌

We can now use this to dispose of the previous limit problem. Since

$$\frac{n^3}{2^n} = \left(\frac{n^4}{2^n}\right)\frac{1}{n} \leq \frac{1}{n}$$

we may conclude that $n^3/2^n \to 0$.

Is it now evident that $n^{10}/2^n \to 0$ as n increases? To give a similar proof of this would require that we first show that $n^{11} \leq 2^n$ for all sufficiently large n. To prove this by induction, we need a starting point, and it happens that the first n for which this holds is $n = 67$. Even less "obvious" is the statement

$$\text{For any choice of the number } r,\ \lim_{n\to\infty} \frac{n^r}{2^n} = 0$$

since there does not seem to be a way to start an inductive argument if you do not have a specific value for r.

There is a way to avoid all these difficulties. If we want to prove that $n^{11} \leq 2^n$ for large n, then it is equivalent to prove that $11 \log n \leq n \log 2$, or that $(\log n)/n \leq (\log 2)/11$. [As we have stated earlier, we feel free to use all the standard computational techniques from one-variable calculus.] This leads us to the following result.

Lemma 2 $\lim_{n\to\infty} (\log n/n) = 0$.

Given n, choose k so that $(k-1)^2 \leq n < k^2$. It is easily checked that the inequality $k^2 < 2^{k-1}$ holds for all $k \geq 7$. Thus, for any $n \geq (6)^2 = 36$, we have

$$\begin{aligned}\log n \leq \log(k^2) < \log(2^{k-1}) &= (k-1)\log 2 \\ &\leq \sqrt{n}\log 2\end{aligned}$$

and we have shown that

$$\frac{\log n}{n} \leq \frac{\sqrt{n}\log 2}{n} = \frac{\log 2}{\sqrt{n}} \to 0. \quad ▌$$

Theorem 13 *For any $b > 1$ and any number r,*

$$\lim_{n\to\infty} \frac{n^r}{b^n} = 0$$

Since $(\log n)/n \to 0$, choose N so that

$$0 < \frac{\log n}{n} < \frac{\log b}{r+1} \qquad \text{for all } n > N$$

Then, $(1 + r)\log n < n \log b$, or $n^{r+1} < b^n$, and $n^r/b^n < 1/n \to 0$, which proves the assertion. ▮

All these illustrations have dealt with *positive* sequences that approach 0. The next illustration shows a different technique.

Theorem 14 $\lim_{n\to\infty} n^{1/n} = 1$.

Set $a_n = n^{1/n}$ and note that $a_n \geq 1$. If we set $a_n = 1 + c$, so that $c \geq 0$, then

$$(1 + c)^n = (n^{1/n})^n = n$$

By the binomial theorem,

$$(1 + c)^n = 1 + nc + \frac{n(n-1)}{2}c^2 + \cdots$$

and in particular, since $c \geq 0$, throwing away all but two terms gives

$$(1 + c)^n \geq 1 + \frac{n(n-1)}{2}c^2$$

or

$$n \geq 1 + \frac{n(n-1)}{2}c^2$$

and this yields $c^2 \leq 2/n$, and $0 \leq c \leq \sqrt{2}/\sqrt{n}$. But, $a_n = 1 + c$, so that $0 \leq a_n - 1 \leq \sqrt{2}/\sqrt{n}$, and we have proved $\lim_{n\to\infty} a_n = 1$. ▮

Another very important method for proving that a sequence of real numbers converges is based on the order properties of **R**. A sequence $\{a_n\}$ is said to be **increasing** if

$$a_1 \leq a_2 \leq a_3 \leq \cdots$$

and **decreasing** if

$$a_1 \geq a_2 \geq a_3 \geq \cdots$$

It is said to be **monotonic** if it is one or the other, and **strictly** monotonic if consecutive terms are never equal. The following statement describes a fundamental property of the real number system:

Every bounded monotonic sequence is convergent

Because this is so basic to the rest of this chapter and to later chapters, and because its nature is ambivalent, regarded both as an axiom and a theorem, we postpone a discussion of it until the next section. However, we give three applications of it here to show its usefulness.

Consider a sequence $\{x_n\}$ defined by the following:

$$x_1 = 1$$

(1-30)
$$x_{n+1} = x_n + \frac{1}{x_n} \qquad \text{for } n \geq 1$$

Note first that this differs from previous examples of sequences in that no explicit formula is given for the nth term of the sequence. Of course, it is easy to use this to calculate the terms of the sequence: $x_1 = 1$, $x_2 = 2$, $x_3 = 2 + \frac{1}{2} = 2.5$, $x_4 = 2.9$, $x_5 = 3.24482 \cdots$, $x_6 = 3.55301 \cdots$, $x_7 = 3.83446 \cdots$, and so on. Does this sequence converge?

Often, it is possible to guess the limit of a converging sequence from the numerical values. This is the case, for example, with the sequence $n^{1/n}$ we discussed in Theorem 14, where the first seven terms were

$$1,\ 1.41,\ 1.44,\ 1.41,\ 1.38,\ 1.35,\ 1.32$$

By the time we reach large values of n, we obtain numbers strongly suggesting that the limit is 1. Thus, when $n = 100$ we have 1.047 and when $n = 200$ we have 1.027. This process is not very helpful with the sequence $\{x_n\}$. Calculation yields $x_{50} = 10.0839$ and $x_{100} = 14.143$, which suggests that the sequence is divergent, but the data are not conclusive since we do not know how the sequence will behave for much larger values of n.

The monotone sequence property settles this at once. By an easy induction argument, $x_n \geq 1$ for all n, and hence $\{x_n\}$ is monotonic increasing. Suppose it were a bounded sequence. It would then *have* to be convergent, having a limit, $\lim_{x \to \infty} x_n = L$, for some real number $L > 1$. It would then necessarily be true that $\lim_{x \to \infty} (x_n + 1/x_n) = L + 1/L$, which, by (1-30) is the same as the assertion

$$\lim_{x \to \infty} x_{n+1} = L + \frac{1}{L}$$

But, the sequence with terms $x_2, x_3, \ldots$ must have the same limit as the sequence $\{x_n\}$, namely L. Thus, the number L must satisfy the equation

$$L = L + \frac{1}{L}$$

which is impossible. Thus, we can conclude that the sequence $\{x_n\}$ is unbounded. This means that for sufficiently large indices n, x_n becomes as large as you like. (In Exercise 15, a method is suggested for determining more precisely how fast this sequence grows.)

Our second example is the sequence $a_n = (1 + 1/n)^n$. We have $a_1 = 2$, $a_2 = 2.25$, $a_3 = 2.37037 \cdots$, $a_4 = 2.44140 \cdots$, so that the sequence appears to be increasing. This can be verified by a tedious algebraic computation, and it can also be shown that $\{a_n\}$ is bounded, so that it would then follow that $\lim_{n \to \infty} a_n$ exists. The limit is the important number $e = 2.718281 \cdots$, which is the base of the natural logarithm. However, we will use a different method to show that $\{a_n\}$ converges.

Consider instead the sequence $b_n = (1 + 1/n)^{n+1}$. Computation suggests that $\{b_n\}$ is a decreasing sequence. If this is true, then, since $b_n > 0$ for all n, $\{b_n\}$ must converge. To show that $b_n > b_{n+1}$, we examine the quotient b_n/b_{n+1}, which is

$$\frac{\left(1 + \frac{1}{n}\right)^{n+1}}{\left(1 + \frac{1}{n+1}\right)^{n+2}} = \left(\frac{n+1}{n}\right)^{n+1}\left(\frac{n+1}{n+2}\right)^{n+2}$$

$$= \left(\frac{n^2 + 2n + 1}{n^2 + 2n}\right)^{n+1}\left(\frac{n+1}{n+2}\right)$$

$$= \left(1 + \frac{1}{n^2 + 2n}\right)^{n+1}\left(\frac{n+1}{n+2}\right)$$

We need a simple but useful inequality in order to simplify this.

Lemma 3 *For any integer $m > 0$ and any $x > 0$,*

$$(1 + x)^m > 1 + mx$$

This comes at once from the observation that

$$(1 + x)^m = 1 + mx + \frac{m(m-1)}{2}x^2 + \cdots > 1 + mx$$

Using this, we can continue the estimate of the size of b_n/b_{n+1}, obtaining first

$$\left(1 + \frac{1}{n^2 + 2n}\right)^{n+1} > 1 + (n+1)\left(\frac{1}{n^2 + 2n}\right) = \frac{n^2 + 3n + 1}{n^2 + 2n}$$

and then

$$\frac{b_n}{b_{n+1}} > \left(\frac{n^2 + 3n + 1}{n^2 + 2n}\right)\left(\frac{n+1}{n+2}\right) = \frac{n^3 + 4n^2 + 4n + 1}{n^3 + 4n^2 + 4n}$$

$$> 1$$

This shows that $b_n > b_{n+1}$, and that $\{b_n\}$ is a decreasing bounded sequence, and therefore a convergent sequence, by the monotonic sequence property. Hence, $\lim_{n\to\infty} b_n = L$ exists; note that we know only that $L \geq 0$. [The monotonic sequence property guarantees that a bounded sequence converges, but tells little about the limit.] This is enough to tell us that the original sequence $\{a_n\}$ converges, for $a_n = (1 + 1/n)^{-1}b_n$, and that $\lim_{n\to\infty} (1 + 1/n) = 1$. Hence, $\{a_n\}$ converges to the same limit as does $\{b_n\}$. ∎

As a final example, we shall investigate the convergence of a familiar algorithm for computing square roots. Suppose that we want to find $\sqrt{A}$ and that we have already made an estimate x. Then, the algorithm supplies a new and better estimate

$$x' = \frac{1}{2}\left(x + \frac{A}{x}\right)$$

As an illustration, let us find approximations for $\sqrt{2}$. Taking 1 as the first estimate, we find for the second estimate

$$x' = \tfrac{1}{2}(1 + \tfrac{2}{1}) = 1.50$$

Using this as the old estimate, and applying the algorithm, we get a new estimate,

$$\begin{aligned} x' &= \frac{1}{2}\left(1.50 + \frac{2}{1.50}\right) = \frac{1}{2}(1.50 + 1.333) \\ &= 1.416\,666 \end{aligned}$$

One more repetition of this loop gives

$$\begin{aligned} x' &= \frac{1}{2}\left(1.416\,666 + \frac{2}{1.416\,666}\right) \\ &= \tfrac{1}{2}(1.416\,666 + 1.411\,765) \\ &= 1.414\,215\,5 \end{aligned}$$

It is clear that this algorithm is actually a process for generating sequences. If we label an initial estimate for $\sqrt{A}$ as x_1, then the next term is

$$x_2 = \frac{1}{2}\left(x_1 + \frac{A}{x_1}\right)$$

and in general

$$x_n = \frac{1}{2}\left(x_{n-1} + \frac{A}{x_{n-1}}\right) \tag{1-31}$$

We would like to show that $\lim_{n\to\infty} x_n = \sqrt{A}$; we would also like to learn something about the rate of convergence, so that we can estimate the accuracy obtained in using x_n as an approximation to $\sqrt{A}$.

Theorem 15 *If $A > 0$ and $x_1 > 0$, the sequence defined by* (1-31) *converges rapidly to $\sqrt{A}$.*

We shall prove that, after the first term, the sequence $\{x_n\}$ is monotonic decreasing. We have $x_2 = \frac{1}{2}(x_1 + A/x_1)$. Since A and x_1 are positive, so is

x_2. Moreover,

$$(x_2)^2 - A = \frac{1}{4}\left(x_1^2 + 2A + \frac{A^2}{x_1^2}\right) - A$$
$$= \frac{1}{4}\left(x_1^2 - 2A + \frac{A^2}{x_1^2}\right)$$
$$= \frac{1}{4}\left(x_1 - \frac{A}{x_1}\right)^2 > 0$$

It therefore follows that x_2 is larger than $\sqrt{A}$, no matter how x_1 was chosen. This is true for the remaining terms as well. By the same calculation,

$$(x_n)^2 - A = \frac{1}{4}\left(x_{n-1} + \frac{A}{x_{n-1}}\right) - A$$
$$= \frac{1}{4}\left(x_{n-1} - \frac{A}{x_{n-1}}\right)^2 > 0$$

and $x_n > \sqrt{A}$.

To prove that the sequence is decreasing, we examine

$$x_n - x_{n+1} = x_n - \frac{1}{2}\left(x_n + \frac{A}{x_n}\right)$$
$$= \frac{1}{2}\left(x_n - \frac{A}{x_n}\right) = \frac{x_n^2 - A}{2x_n} > 0$$

Hence, $x_2 > x_3 > x_4 > \cdots \geq \sqrt{A}$. By the monotonic-sequence theorem, $\{x_n\}$ converges. Let its limit be $L \geq \sqrt{A}$. Now, using Theorems 8 and 12, we have

$$\lim_{n\to\infty} \frac{1}{2}\left(x_n + \frac{A}{x_n}\right) = \frac{1}{2}\left(L + \frac{A}{L}\right)$$

Since this is the same as computing $\lim_{n\to\infty} x_{n+1}$, the value of the limit must again be L, and we have $L = \frac{1}{2}(L + A/L)$. From this, we have $L^2 = A$, and since $L \geq \sqrt{A} > 0$, $L = \sqrt{A}$. ▮

The rate of convergence of the sequence $\{x_n\}$ to A is quite rapid. We can see this by comparing $|x_n - \sqrt{A}|$ and $|x_{n+1} - \sqrt{A}|$. We have

$$x_{n+1} - \sqrt{A} = \frac{1}{2}\left(x_n + \frac{A}{x_n}\right) - \sqrt{A}$$
$$= \frac{x_n^2 + A - 2\sqrt{A}\,x_n}{2x_n}$$
$$= \frac{(x_n - \sqrt{A})^2}{2x_n} < \frac{(x_n - \sqrt{A})^2}{2\sqrt{A}}$$

This shows that the error decreases at a rate much faster than geometric, because of the squaring; ignoring the factor $2\sqrt{A}$, one would expect to see the error go down according to the pattern .1, .01, .0001, .000 000 01, .000 000 000 000 000 1.

To illustrate this, we have computed the sequence of approximations for $\sqrt{25}$, starting with $x_1 = 6$, and have obtained

$$x_2 = 5\frac{1}{12} = 5.0833$$

$$x_3 = 5\frac{1}{1464} = 5.000\,683\,1$$

$$x_4 = 5\frac{1}{214\,358\,88} = 5.000\,000\,047$$

The monotonic-sequence property is not the only method that can be used to show that a sequence is convergent without having to guess its limit in advance. We conclude this section by making a few remarks about what is usually called the Cauchy convergence criterion—although, as will be explained at the end of Sec. 1.7, this name is somewhat inappropriate. We first define the term "Cauchy sequence."

Definition 6 *A sequence $\{p_n\}$ of points is said to be a* **Cauchy sequence** *if, corresponding to any $\varepsilon > 0$, a number N (depending upon ε) can be found such that $|p_n - p_m| < \varepsilon$ whenever both n and m are larger than N. This condition is also written as*

$$\lim_{m,\,n \to \infty} |p_n - p_m| = 0$$

At first glance, the connection between this definition and convergence is not clear. However, it is an easy exercise to show that any convergent sequence is also a Cauchy sequence (Exercise 32). What *is* useful is the converse of this, namely, the **Cauchy convergence criterion:** *In n space, any Cauchy sequence is convergent.*

The proof that this holds requires more mathematical tools than we have at present, and so it is postponed to Sec. 1.7. It will follow from the monotonic-sequence property, which will also be justified later. Instead, we now show several simple applications of it.

Suppose that a sequence is defined by the formula

$$x_n = \int_1^n \frac{\cos t}{t^2}\,dt$$

It is easy to see that $\{x_n\}$ is a bounded sequence, for

$$|x_n| \le \int_1^n \frac{|\cos t|}{t^2}\,dt \le \int_1^n \frac{1}{t^2}\,dt = -\frac{1}{t}\Big|_1^n$$
$$\le (-1/n) - (-1) = 1 - 1/n \le 1$$

for all $n = 1, 2, 3, \ldots$. From this fact alone we learn nothing, since the sequence $\{x_n\}$ does not happen to be monotonic. However, if we calculate the difference of x_n and x_m, with $n > m$,

$$x_n - x_m = \int_1^n \frac{\cos t}{t^2}\,dt - \int_1^m \frac{\cos t}{t^2}\,dt$$

$$= \int_m^n \frac{\cos t}{t^2}\,dt$$

Hence,

$$|x_n - x_m| \le \int_m^n \frac{|\cos t|}{t^2}\,dt \le \int_m^n t^{-2}\,dt$$

$$\le -\frac{1}{t}\Big|_m^n = \frac{1}{m} - \frac{1}{n}$$

Since $\lim_{m,\,n\to\infty} |x_n - x_m| = 0$, the sequence $\{x_n\}$ converges. (Note that these calculations give us no information about the numerical value of $\lim_{n\to\infty} x_n$. This example also illustrates some of the techniques we assume to be familiar from elementary calculus, e.g., that $|\cos t| \le 1$ for all t, and that the indefinite integral of $1/t^2$ is $-1/t$.)

One of the most common mistakes made in the use of the Cauchy criterion is to overlook the fact that the indices m and n in $|p_n - p_m| \to 0$ must be treated as unrelated and independent. For example, it is not sufficient to have the consecutive terms of a sequence become closer and closer as you move out. Thus, with $x_n = \sqrt{n}$, $\lim_{n\to\infty} |x_{n+1} - x_n| = 0$, and yet $\{x_n\}$ is unbounded and therefore it is divergent. Nor is it enough to have $\lim_{n\to\infty} |x_{\phi(n)} - x_n| = 0$, even if the values of $\phi(n)$ are considerably larger than n. (In Exercise 22 you will meet an example where $\lim_{n\to\infty} |x_{n^2} - x_n| = 0$, and yet $\{x_n\}$ is a divergent sequence that is monotonic.) In understanding the definition of "Cauchy sequence" (Definition 6), the key is that after N has been chosen, the only restrictions on n and m are $n > N$, $m > N$; thus, each may be as large as you like, independent of the choice of the other.

In the light of these remarks, the following result may seem contradictory!

Theorem 16 *A sequence $\{p_n\}$ is a Cauchy sequence, and therefore convergent, if the sequence $|p_{n+1} - p_n|$ of distances between consecutive terms approaches 0 fast enough. Specifically, if $|p_{n+1} - p_n| < Ac^n$ where $0 \le c < 1$, for all n, then $\lim_{n\to\infty} p_n$ exists.*

We shall prove that $\{p_n\}$ is a Cauchy sequence. For this, we need to estimate $|p_n - p_m|$ for large values of n and m; we may as well assume that m is the larger and write $m = n + k$. In order to make the argument clearer, we do the calculation first for the special case of $k = 3$, and then in general. We have

$$p_n = p_{n+3} = p_n - p_{n+1} + p_{n+1} - p_{n+2} + p_{n+2} - p_{n+3}$$

so that

$$\begin{aligned}|p_n - p_{n+3}| &\le |p_n - p_{n+1}| + |p_{n+1} - p_{n+2}| + |p_{n+2} - p_{n+3}| \\ &\le c^n + c^{n+1} + c^{n+2} \\ &\le c^n(1 + c + c^2) = c^n \frac{1 - c^3}{1 - c} \\ &\le \frac{1}{1 - c} c^n\end{aligned}$$

In general, we have

$$\begin{aligned}|p_n - p_{n+k}| &\le |p_n - p_{n+1}| + \cdots + |p_{n+k-1} - p_{n+k}| \\ &\le c^n + c^{n+1} + \cdots + c^{n+k-1} \\ &\le c^n(1 + c + c^2 + \cdots + c^{k-1}) \\ &\le c^n \frac{1 - c^k}{1 - c} < \frac{1}{1 - c} c^n\end{aligned}$$

Hence, for any n and m, with $m > n$, we have

$$|p_n - p_m| < \frac{1}{1 - c} c^n$$

Now, since $c < 1$, $\lim_{n \to \infty} c^n = 0$ (see Theorem 13). In particular, given any $\varepsilon > 0$, we can choose N large enough so that $c^N/(1 - c) < \varepsilon$. Then, if $n > N$ and $m > N$, we have $|p_n - p_m| < \varepsilon$. In other words, $\lim_{n, m \to \infty} |p_n - p_m| = 0$, and the sequence $\{p_n\}$ converges. ▌

EXERCISES

1 Show that the sequence defined by $p_n = (n, 1/n)$ does not converge.

2 Show that the sequence described by $p_n = \left(\frac{n+1}{n}, \frac{(-1)^n}{n}\right)$ converges.

3 Let $|p_{n+1} - q| \le c|p_n - q|$ for all n, where $c < 1$. Show that $\lim_{n \to \infty} p_n = q$.

4 Let $\{p_n\}$ and $\{q_n\}$ be sequences in 3-space with $p_n \to p$ and $q_n \to q$. Prove that $\lim_{n \to \infty} p_n \cdot q_n = p \cdot q$.

***5** Starting at the origin in the plane, draw a polygonal line as follows: Go 1 unit east, 2 units north, 3 units west, 4 units south, 5 units east, 6 units north, and so on. Find a formula for the nth vertex of this polygon.

6 Among the following sequences, some are subsequences of others; determine all those which are so related.

(*a*) $1, -1, 1, -1, \ldots$ (*b*) $1, 1, -1, 1, 1, -1, \ldots$
(*c*) $1, \frac{1}{2}, \frac{1}{3}, \frac{1}{4}, \frac{1}{5}, \ldots$ (*d*) $1, \frac{1}{4}, \frac{1}{9}, \frac{1}{16}, \frac{1}{25}, \ldots$
(*e*) $1, 0, \frac{1}{2}, 0, \frac{1}{3}, 0, \frac{1}{4}, 0, \ldots$

7 Exhibit a sequence having exactly three limit points. Can a sequence have an infinite number of limit points? No limit points? Could a divergent sequence have exactly one limit point?

8 Discuss the behavior of the sequence $\{a_n\}$, where

$$a_n = n + 1 + 1/n + (-1)^n n$$

9 If $a_n \le x_n \le b_n$ and $\lim_{n\to\infty} a_n = \lim_{n\to\infty} b_n = L$, show that $\lim_{n\to\infty} x_n = L$. (This is sometimes called the "sandwich" property.)

10 Find a formula for the sequence that begins with 1/2, 1/5, 1/10, 1/17, 1/26, ..., and show that it converges to 0.

11 Prove that $\lim_{n\to\infty} 1/\sqrt{n} = 0$.

12 What are the correct hypotheses for the truth of the following assertion? If $\lim_{n\to\infty} a_n = A$, then $\lim_{n\to\infty} \sqrt{a_n} = \sqrt{A}$.

13 Prove that $\lim_{n\to\infty} c^{1/n} = 1$ for any $c > 1$ by setting $a_n = c^{1/n} - 1$, and then deriving the estimate $0 \le a_n \le (c - 1)/n$.

14. Investigate the convergence of the sequence

$$a_n = \sqrt{n^2 + n} - n$$

15 Show that the sequence $\{x_n\}$ defined by the recursive formula:

$$x_1 = 1, \qquad x_{n+1} = x_n + 1/x_n \qquad \text{for } n > 1$$

obeys the inequality $x_n > \sqrt{n}$ for all $n \ge 2$.

16 Define a sequence $\{x_n\}$ by

$$x_1 = 1, \qquad x_{n+1} = x_n + \sqrt{x_n} \qquad \text{for } n > 1$$

(*a*) Prove that $\{x_n\}$ is unbounded.
(*b*) Prove that if, for some N, $x_N \le N^2/4$, then $x_{N+1} \le (N + 1)^2/4$.
(*c*) Is there a value of n for which $x_n \le n^2/4$?
(You will have to do some calculation. Use a pocket calculator; that's what they are for!)
(*d*) Show that $x_n \ge n^2/9$ for all $n \ge 1$.
*(*e*) Show that $\lim_{n\to\infty} x_n/n^2 = \frac{1}{4}$.

17 Let $a_n = \dfrac{1 \cdot 3 \cdot 5 \cdots (2n - 1)}{2 \cdot 4 \cdot 6 \cdots 2n}$

(*a*) Prove that $\{a_n\}$ is convergent.
*(*b*) Can you determine $\lim_{n\to\infty} a_n$?

18 Let $x_1 = 1$, $x_2 = 3$, and define all later terms recursively by $x_n = (x_{n-1} + x_{n-2})/2$. Thus, $x_3 = 2$, $x_4 = 5/2$. Is the sequence $\{x_n\}$ monotonic? Does it converge?

19 Let $a_1 = 1$, $a_2 = 2$, and $a_{n+2} = (4a_{n+1} - a_n)/3$. Show that $\{a_n\}$ converges.

20 Define the sequence $\{x_n\}$ by $x_1 = a$, $x_2 = b$, and $x_{n+2} = (1 + x_{n+1})/x_n$. Investigate the convergence of $\{x_n\}$. [*Hint:* It may help to try some numerical values.]

21 Let $a_1 = 0$, $a_2 = 1$, and $\qquad a_{n+2} = \dfrac{na_{n+1} + a_n}{n + 1}$

(*a*) Calculate the value of a_6 and a_7.
(*b*) Prove that $\{a_n\}$ converges.
*(*c*) Show that $\lim_{n\to\infty} a_n = 1 - e^{-1}$.

22 Show that $x_n = \sqrt{\log \log n}$, for $n \ge 3$ defines a divergent increasing sequence such that

$$\lim_{n\to\infty} (x_{n^2} - x_n) = 0$$

23 Let $\{x_n\}$ be a bounded real sequence and set $\beta = \limsup_{n\to\infty} x_n$. Show that for any $\varepsilon > 0$, $x_n \le \beta + \varepsilon$ holds for all but a finite number of n, and $x_n \ge \beta - \varepsilon$ holds for infinitely many n. What are the analogous statements about $\liminf_{n\to\infty} x_n$?

24 If $b \le x_n \le c$ for all but a finite number of n, show that $b \le \liminf_{n\to\infty} x_n$ and $\limsup_{n\to\infty} x_n \le c$.

25 If $\{x_n\}$ is a bounded sequence with $\liminf_{n\to\infty} x_n = \limsup_{n\to\infty} x_n$, show that $\{x_n\}$ converges.

26 Find $\limsup_{n\to\infty} a_n$ and $\liminf_{n\to\infty} a_n$ when:

(*a*) $a_n = (-1)^n$ (*b*) $a_n = (-1)^n\left(2 + \dfrac{3}{n}\right)$

(*c*) $a_n = \dfrac{n + (-1)^n(2n+1)}{n}$ (*d*) $a_n = \sin\left(n\dfrac{\pi}{3}\right)$

27 If $\limsup_{n\to\infty} a_n = A$ and $\limsup_{n\to\infty} b_n = B$, must it be true that

$$\limsup_{n\to\infty} (a_n + b_n) = A + B \quad ?$$

***28** Show that, for any bounded sequences a_n and b_n,

$$\liminf_{n\to\infty} a_n + \liminf_{n\to\infty} b_n \le \liminf_{n\to\infty} (a_n + b_n)$$

and that

$$\limsup_{n\to\infty} (a_n + b_n) \le \limsup_{n\to\infty} a_n + \limsup_{n\to\infty} b_n$$

***29** Let $\{a_n\}$ be any sequence of numbers converging to 0, and let σ_n be the sequence of **arithmetic means** (averages),

$$\sigma_n = \frac{a_1 + a_2 + a_3 + \cdots + a_n}{n}$$

Prove that $\lim_{n\to\infty} \sigma_n = 0$.

30 If we start with $x_1 = 2$, how far must we go with the square root algorithm to get $\sqrt{2}$ accurate to 10^{-50}? 10^{-100}?

31 Define a sequence of points thus: Starting at the origin, move 1 unit east, then $\frac{1}{2}$ unit north, then $\frac{1}{4}$ unit east, then $\frac{1}{8}$ unit north, then $\frac{1}{16}$ unit east, and so on. (*a*) Does the sequence of vertices converge? (*b*) Can you find the "end" of this polygon?

32 Show that a convergent sequence $\{p_n\}$ must be a Cauchy sequence.

***33** Let $A = (0, 1)$ and $B = (1, 0)$. Let P_1 be any point in the plane, and construct a sequence $\{P_n\}$, with P_1 as its first term, as follows: Let Q_1 = midpoint of AP_1 and P_2 = midpoint of BQ_1; then, let Q_2 = midpoint of AP_2 and P_3 = midpoint of BQ_2, and so on. Prove that $\{P_n\}$ converges.

34 Show that every Cauchy sequence $\{p_n\}$ is bounded.

35 Let $p_n = (x_n, y_n, z_n)$. Show that if $\{p_n\}$ is Cauchy, so are $\{x_n\}$, $\{y_n\}$, and $\{z_n\}$.

36 (*a*) Explain why $\lim_{m,n\to\infty} \dfrac{3n + 5m}{4n^2 + 7m^2} = 0$.

(*b*) Is it true that
$$\lim_{m,n\to\infty} \frac{2n + m}{3n + 5m^2} = \frac{2}{3}$$

37 Let $a_k > 0$ for all k and suppose that

$$\lim_{m,n\to\infty} a_m/a_n = 1$$

Prove that $\{a_k\}$ converges.

38 Show that Theorems 10 and 12 still hold if the sequences $\{a_n\}$ and $\{b_n\}$ are sequences of complex numbers.

1.7 CONSEQUENCES OF THE MONOTONIC-SEQUENCE PROPERTY

Historically, plane geometry probably arose from the efforts of people engaged in an empirical study of the properties of lines, circles, and triangles who first observed certain repeated patterns and then used these to obtain others; much later, someone (perhaps Euclid?) codified the material, decided which should be axioms and which theorems, and presented it as an elegant deductive structure, with no misleading instances of circular reasoning.

In this section, we discuss briefly one similar aspect of analysis, and then proceed to a series of very important applications of it in mathematics.

In Sec. 1.2, we mentioned the algebraic axioms for the real numbers ("**R** is a field") and the existence of the order relation $<$ and some of its properties. There are other aspects of **R** we did not discuss; for example, there are properties that have the effect of asserting that certain equations have solutions—for example, $x^2 = 2$. Our belief in this clearly has much to do with our understanding of the nature of the real numbers themselves. If we think of them as unending decimals, then one root of $x^2 - 2 = 0$ may be thought of as 1.41421356 ..., which in turn may be regarded as the limit of a monotonic sequence 1, 1.4, 1.41, 1.414, 1.4142, ... Another approach is suggested by the example at the end of the preceding section. There, an algorithm was obtained for generating a monotonic decreasing sequence whose limit would have to be $\sqrt{2}$.

In both of these approaches, the existence of the real number $\sqrt{2}$ depends on our belief in the truth of the monotonic-sequence property; it is therefore natural to list this as one of the basic axioms for the real numbers: *Any bounded monotonic sequence of real numbers is convergent.*

However, what is *axiom* and what is *theorem* is often an arbitrary choice, since one is often able to derive each logically from the other. Sometimes one can go a step further and make all the axioms of a mathematical system into theorems by basing the system entirely upon some other system. Thus, analytical geometry makes it possible to base all of plane geometry upon the properties of the real numbers. This is also possible with **R** itself, throwing it back onto modern set theory. With this, we end this digression; refer to Appendix 2 for further details and references.

We adopt the position that the monotonic-sequence property stated above is one of the basic axioms for **R**, and adjoin it to earlier lists. (In the last section, we saw how useful it is in working with sequences of real numbers.)

Earlier, in Sec. 1.2, we mentioned another basic property of **R**, also connected with the order relation $<$, there called the *least-upper-bound property*. Recall that any bounded set of integers must have a largest and a smallest member, but that this need not be true of a bounded set of real numbers; thus, the set of all x with $2 < x < 3$ has neither a largest member nor a smallest member, while the bounded set

$$A = \left\{ \text{all } \frac{n+1}{n+3} \text{ for } n = 0, 1, 2, \ldots \right\}$$

has $\frac{1}{3}$ as its smallest member, but does not have a maximum member. The least-upper-bound property (LUB property), applied to any bounded set S, guarantees the existence of two real numbers which, for many purposes, serve to replace the maximum or minimum of S, if either is missing. The formal definition is as follows: Let S be a bounded subset of $\mathbf{R}$, and let

(1-32) $$\mathscr{U} = \textit{the set of upper bounds for } S = \{\text{all } c \text{ such that } x \leq c \text{ for every } x \in S\}$$

(1-33) $$\mathscr{L} = \textit{the set of lower bounds for } S = \{\text{all } b \text{ such that } b \leq x \text{ for every } x \in S\}$$

Then, the **LUB property** asserts that the set $\mathscr{U}$ always has a minimum

(1-34) $$\sup(S) = \text{the smallest member of } \mathscr{U}$$

and the set $\mathscr{L}$ always has a maximum

(1-35) $$\inf(S) = \text{the largest member of } \mathscr{L}$$

The notation "sup" is meant to suggest "supremum," and "inf," "infimum"; $\sup(S)$ is also called the **least upper bound** of S, written $\text{lub}(S)$, and $\inf(S)$ is called the **greatest lower bound** of S, written $\text{glb}(S)$; $\sup(S)$ and $\inf(S)$ always belong to the closure of S.

Since some sets of real numbers do *not* have maxima or minima, why do the sets $\mathscr{U}$ and $\mathscr{L}$? This follows from the monotonic-sequence property; the proof is not easy.

Theorem 17 *If S is any bounded nonempty set of real numbers,* $\sup(S)$ *and* $\inf(S)$ *exist.*

To show how such a result is obtained, we assume that all the numbers in S are positive, and prove only that $\inf(S)$ exists. We must therefore show that the set $\mathscr{L}$ defined in (1-33) has a maximum member, L; we obtain L as the limit of a monotonic increasing sequence whose terms are chosen from $\mathscr{L}$. We start by choosing any number s in the set S. Then, since $\mathscr{L}$ is the set of lower bounds for the set S, $u \leq s$ for every $u \in \mathscr{L}$; since we have assumed that S contains no negative numbers, $0 \in \mathscr{L}$. We construct our monotonic sequence recursively, choosing the terms as positive rationals, each less than s.

Choose a_0 as the largest integer in $\mathscr{L}$. Clearly, $a_0 \geq 0$. Then, choose a_1 as the largest integer n such that $n/2 \in \mathscr{L}$. Noting that $a_0 = (2a_0)/2$, we see that $2a_0$ is one of these possible values of n, so that $2a_0 \leq a_1$. In general, we choose a_k as the largest integer n such that $n/2^k \in \mathscr{L}$. We therefore have $a_k/2^k$ in $\mathscr{L}$, for $k = 0, 1, 2, \ldots$. We move on to a_{k+1}, which is to be the largest integer n such that $n/2^{k+1}$ is in $\mathscr{L}$, and note that $(2a_k)/2^{k+1} = a_k/2^k$; thus we see that $2a_k$ is a possible value of n and $2a_k \le a_{k+1}$. Accordingly, we have constructed a sequence $\{a_n/2^n\}$ whose terms are rational numbers and are all in the set $\mathscr{L}$, and such that

$$a_0 \le \frac{a_1}{2} \le \frac{a_2}{4} \le \frac{a_3}{8} \le \cdots \le \frac{a_n}{2^n} \le \cdots \le s$$

As a bounded monotonic sequence, this converges, and we set $L = \lim_{n\to\infty} a_n/2^n$. If x is any member of S, then since $a_n/2^n$ belongs to $\mathscr{L}$, $a_n/2^n \le x$ for every n and $L \le x$; but since this holds for every $x \in S$, L is a lower bound for S and $L \in \mathscr{L}$. The only remaining task is to show that L is the largest member of $\mathscr{L}$. Suppose that there were a number $b \in \mathscr{L}$ with $L < b$. Since, for any k, $a_k/2^k \le L < b$, we have $a_k \le 2^k L < 2^k b$. If we take k sufficiently large, there will be an integer N between $2^k L$ and $2^k b$, with $a_k < N$. Accordingly, the integer a_k could not have been the largest integer n such that $n/2^k \in \mathscr{L}$, since N would have the same property. Thus, there cannot exist a number $b \in \mathscr{L}$ with $b > L$, and L is the largest member of $\mathscr{L}$. ▌

The following result may give an easier and more plausible reason for accepting the truth of the LUB property.

Theorem 18 *If the line is connected, then the* LUB *property holds.*

(Note that the word "connected" is used in the special technical sense given in Definition 2 in Sec. 1.5.)

Suppose that S is a bounded nonempty set that does not have a greatest lower bound. If $\mathscr{L}$ is the set of all lower bounds for S, then this hypothesis means that $\mathscr{L}$ does not have a maximum member. Take any $x_0 \in \mathscr{L}$. Clearly, all $x_1 < x_0$ lie in $\mathscr{L}$. There must also exist $x_2 \in \mathscr{L}$ with $x_0 < x_2$, because x_0 is not the largest member of $\mathscr{L}$. Thus, the interval $[x_1, x_2]$ is a neighborhood of x_0 lying inside $\mathscr{L}$. This proves $\mathscr{L}$ to be open. Let $\mathscr{V}$ be all the numbers not in $\mathscr{L}$. If $y_0 \in \mathscr{V}$, then y_0 is not a lower bound for S, and there must exist $s \in S$ with $s < y_0$. Then, $s < y$ for all y near y_0, and $\mathscr{V}$ is open. Finally, $\mathscr{V}$ is not empty, since $S \ne \varnothing$, and if $s_0 \in S$, any $y > s_0$ belongs to $\mathscr{V}$. Thus, the whole real line is $\mathscr{L} \cup \mathscr{V}$ where these sets are disjoint and both open, showing that **R** is not connected. ▌

One final remark about the axiomatics of the real numbers: Any of the three properties—LUB, monotonic-sequence, or the connectedness of **R**—can be derived from the others, so that each could equally well be taken as a basic axiom for **R**. There are also other alternative routes through the country called Foundations, some of which are indicated in the roadmap in Appendix 2.

We now explore the usefulness of some of these various concepts, and start with some geometric applications. Intuitively, the diameter of a set S ought to be the maximum distance between any two points in S. Unfortunately, in many cases such a maximum does not exist; consider, for example, an open disc in the plane. The solution is to replace "maximum" by "supremum," using the LUB property.

Definition 7 *If S is any bounded set in n space, its* **diameter** *is given by*

$$\text{diam}\,(S) = \sup\,\{\textit{all numbers } |p - q| \textit{ for } p, q \in S\}$$

Again, how should one define the distance between two sets, or the distance from a point to a set? Intuitively, the distance from a point p to a set S ought to be the distance from p to the nearest point of S, but if the set is open, no nearest point will exist. The solution is to replace "minimum" by "inf."

Definition 8 *If S is a nonempty set in n space, and p is any point in n space, the distance from p to S is*

$$\text{dist}\,(p, S) = \inf\,\{\textit{all } |p - q| \textit{ for } q \in S\}$$

Definition 9 *If A and B are sets in n space, neither empty, then the distance between them is*

$$\text{dist}\,(A, B) = \inf\,\{\textit{all } |p - q| \textit{ for } p \in A, q \in B\}$$

While dist (A, B) may not be achieved as the minimum $|p - q|$ for an actual pair of points in the sets A and B, the fact that inf (S) is always a number in the closure of S implies that there must exist two sequences of points, $p_n \in A$ and $q_n \in B$, such that $\lim_{n\to\infty} |p_n - q_n| = \text{dist}\,(A, B)$. However, as seen in Exercise 6, neither $\{p_n\}$ nor $\{q_n\}$ itself need be a convergent sequence.

To illustrate the way one works with inf and sup, we prove a form of the triangle inequality for distances between sets.

Theorem 19 *Given any three sets A, B, and C in n space, with B bounded, the following relation holds:*

$$\text{dist}\,(A, C) \le \text{dist}\,(A, B) + \text{dist}\,(B, C) + \text{diam}\,(B) \tag{1-36}$$

Before reading the proof, we suggest that you make your own picture to illustrate the meaning of this relation.

Let $\alpha = \text{dist}\,(A, B)$ and $\beta = \text{dist}\,(B, C)$. Given any $\varepsilon > 0$, we may choose points $p \in A$, $q \in C$, and b_1, $b_2 \in B$ so that $|p - b_1| < \alpha + \varepsilon$, and

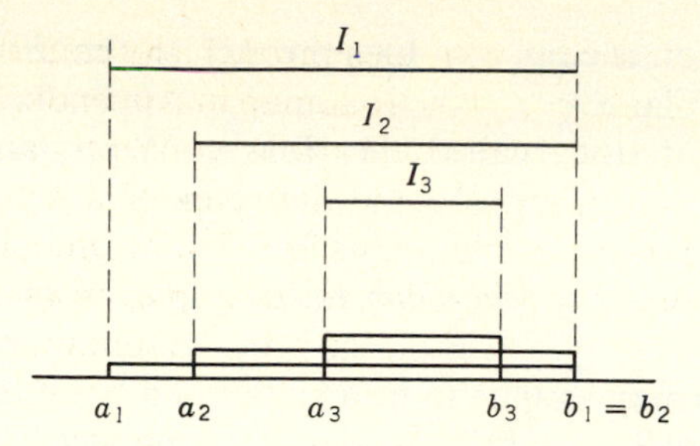

Figure 1-27 Nested intervals.

$|b_2 - q| < \beta + \varepsilon$. Using the ordinary triangle law for distance, we have

$$(1\text{-}37) \qquad |p - q| \le |p - b_1| + |b_1 - b_2| + |b_2 - q| \\ \le \alpha + \beta + 2\varepsilon + |b_1 - b_2|$$

From the definition of dist (A, C) and diam (B), we have dist $(A, C) \le |p - q|$ and $|b_1 - b_2| \le$ diam (B). Thus, (1-37) yields

$$\text{dist}\,(A, C) \le \alpha + \beta + 2\varepsilon + \text{diam}\,(B)$$

Since this now holds for arbitrarily small ε, let $\varepsilon \downarrow 0$ and have (1-36). ∎

We next turn to a very useful consequence of the monotonic-sequence property.

Theorem 20 (nested intervals) *Let $\{I_n\}$ be a sequence of (nonempty) bounded closed intervals on the line which are monotonic decreasing ("nested") in the sense that $I_1 \supset I_2 \supset I_3 \supset \cdots$. Then, $\bigcap_1^\infty I_n \ne \varnothing$, so that there must exist at least one point p that lies in all the I_n. (If the length of I_n approaches 0 as n increases, the intersection of the I_n is a set with exactly one point.)*

Let $I_n = [a_n, b_n]$. Then, the fact that these intervals are nested implies that, as shown in Fig. 1-27,

$$a_k \le a_{k+1} \le b_{k+1} \le b_k \qquad \text{for each } k$$

Thus, $\{a_n\}$ is an increasing sequence bounded above by b_1, and $\{b_n\}$ is a decreasing sequence bounded below by a_1. Accordingly, $\lim_{n\to\infty} a_n = a$ exists and $\lim_{n\to\infty} b_n = b$ exists, and

$$a_n \le a \le b \le b_n \qquad \text{for all } n$$

This means that the interval $[a, b]$ is a subset of each of the intervals I_n, and is exactly their intersection. Note that if $a = b$, this intersection is a single point, and this occurs when $\lim_{n\to\infty} (b_n - a_n) = 0$. ∎

As shown in Exercise 8, it is essential in this result that the intervals I_n be both closed and bounded, for otherwise it is possible to have an infinite nested sequence of intervals whose intersection is empty. In Exercise 9, Theorem 20 is extended to nested rectangles in the plane.

The following important result is often called the **Bolzano-Weierstrass** theorem for the line.

Theorem 21 *Every bounded infinite set of real numbers has a cluster point.*

Let S be a given set, which we may assume to be a subset of a bounded interval $I = [a, b]$. The midpoint of I, $(a + b)/2$, divides the interval I into two equal pieces. Since S is an infinite subset of I, one of these two pieces must contain infinitely many points of S. Choose one that does and label this piece I_1, noting that it is a closed bounded interval half the length of I, with $I_1 \subset I$. Repeat this process with I_1 to generate a subinterval $I_2 \subset I_1$ such that $I_2 \cap S$ is infinite. Continue this process, and obtain a nested sequence of closed bounded intervals $\{I_n\}$ such that for each value of n, $I_n \cap S$ is infinite. Applying Theorem 20, we see that $\bigcap_1^\infty I_n$ contains a single point p. We now prove that p is a cluster point for the original set S. Recall that this means that every neighborhood of p contains infinitely many points of S. Let $\mathcal{N}$ be any neighborhood about p. Since the intervals I_n close down on p, so that their respective endpoints form sequences converging to p, and since p is interior to $\mathcal{N}$, it follows that $I_n \subset \mathcal{N}$ for all sufficiently large n. But, every interval I_n contains infinitely many points of S, and so must the larger set $\mathcal{N}$. ∎

A very similar argument also proves the following.

Theorem 22 *Every bounded sequence of real numbers has a limit point, and therefore a converging subsequence.*

(We will shortly generalize Theorems 20, 21, and 22 to n space.)

We are also able now to justify the Cauchy convergence criterion, discussed at the end of Sec. 1.6.

Theorem 23 *Any Cauchy sequence of real numbers is convergent.*

Any Cauchy sequence is bounded (see Exer. 34, Sec. 1.6). If $\{x_n\}$ is a Cauchy sequence, and is divergent, then $\alpha = \liminf x_n$ and $\beta = \limsup x_n$ exist, and $\alpha < \beta$. Choose $\varepsilon < \beta - \alpha$, and then choose N so that $|x_n - x_m| < \varepsilon$ for all $n, m > N$. We can rewrite this inequality as

$$-\varepsilon < x_n - x_m < \varepsilon \qquad \text{all } n, m > N$$

and then rewrite the right half of this as

$$x_n < \varepsilon + x_m \qquad \text{all } n, m > N$$

Hold m fixed, and apply Exercise 24, Sec. 1.6 to obtain

$$\limsup x_n = \beta \le \varepsilon + x_m$$

Write this as $\beta - \varepsilon \le x_m$, holding for all $m > N$, and again apply the same exercise to obtain

$$\beta - \varepsilon \le \liminf x_n = \alpha$$

However, this implies $\beta - \alpha < \varepsilon$, which is false. Thus, $\alpha = \beta$ and $\{x_n\}$ is convergent. ▮

Corollary *In n space, any Cauchy sequence is convergent.*

In 3-space, suppose we have the sequence $\{p_n\}$ with $p_n = (x_n, y_n, z_n)$, and suppose it is Cauchy. Then, one easily sees (Exercise 35, Sec. 1.6) that $\{x_n\}$, $\{y_n\}$, and $\{z_n\}$ are Cauchy, and therefore convergent. ▮

We remark that the property of being a Cauchy sequence is one that can be described in any space in which there is a notion of distance. (Such a space is usually called a **metric** space; we will encounter a metric space in Chap. 6 that is different from the spaces $\mathbf{R}^n$.) However, in such spaces it is not always true that Cauchy sequences automatically converge, although they are always bounded. A space in which all Cauchy sequences converge is called **complete;** thus, this corollary is a proof that n space is complete.

EXERCISES

1 Show that the LUB property implies the monotonic-sequence property by proving the following: If $a_1 \le a_2 \le \cdots$ is a bounded increasing real sequence, then $\lim_{n\to\infty} a_n = L$ where $L = \sup\{a_1, a_2, a_3, \ldots\}$.

2 Fill in the missing details in the following proof that the LUB property implies that $\mathbf{R}$ is connected.

(*a*) Let $\mathbf{R} = A \cup B$ where A and B are mutually separated. Then, A and B are both open.

(*b*) With $a_0 \in A$ and $b_0 \in B$, assume $a_0 < b_0$; let I be the interval $a_0 \le x \le b_0$, and set $c = \sup(A \cap I)$. Then, c belongs to neither A nor B.

(*c*) Hence, a contradiction has been found, and $\mathbf{R}$ is connected.

(*d*) Modify the above to prove that an interval $[b, c]$ is connected.

3 If S is any bounded set of real numbers, show that the numbers $\inf(S)$ and $\sup(S)$ belong to the closure of S.

4 If $A \subset B$ and B is a bounded set in n space, show that $\operatorname{diam}(A) \le \operatorname{diam}(B)$. Can equality occur without having $A = B$?

5 Suppose that A, B, and C are three sets and that $A \subset B$. Show that $\operatorname{dist}(B, C) \le \operatorname{dist}(A, C)$.

6 Let A and B be closed sets in the plane defined by:

$$A = \{\text{all } (x, y) \text{ with } y \ge 2\}$$

$$B = \{\text{all } (x, y) \text{ with } x \ge 0 \text{ and } y \le x/(x+1)\}$$

(*a*) Find $d = \text{dist}(A, B)$.

(*b*) Show there does not exist $p \in A$, $q \in B$ with $|p - q| = d$.

(*c*) Find sequences $\{p_n\}$ in A and $\{q_n\}$ in B with $\lim_{n\to\infty} |p_n - q_n| = d$. Does either sequence converge?

7 Can one have two closed sets A and B which are disjoint (and not empty) and such that $\text{dist}(A, B) = 0$?

8 Show that the intersection $\bigcap_1^\infty I_n$ of the nested sequence of intervals $\{I_n\}$ is empty in the following cases:

(*a*) I_n is the open interval $0 < x < 1/n$.

(*b*) I_n is the (unbounded) closed interval $n \le x < \infty$.

9 Let $\{R_n\}$ be a sequence of closed bounded rectangles in the plane, with $R_1 \supset R_2 \supset R_3 \supset \cdots$; describe R_n by

$$R_n = \{\text{all } (x, y) \text{ with } a_n \le x \le b_n,\ c_n \le y \le d_n\}$$

Prove that $\bigcap_1^\infty R_n \ne \varnothing$.

10 Prove that a bounded sequence of real numbers that has exactly one limit point must be convergent. Is this still true if the sequence is unbounded?

***11** Show that every noncountable set of points in the plane must have a cluster point. Must it have more than one?

1.8 COMPACT SETS

We next move to a different aspect of the behavior of sets on the line and in space which unifies the treatment of many topics in analysis. In the proof of the Bolzano-Weierstrass theorem for a bounded interval on the line (Theorem 21), we used a simple partitioning device which divided the interval up into smaller intervals of arbitrarily short length. This technique, which allows one to regard a set as contained in the union of a finite collection of simple sets of arbitrarily small diameter, simplifies many arguments and suggests certain useful definitions.

Definition 10 *A collection $\mathcal{S}$ of open sets $\mathcal{O}_\alpha$ is said to be an open covering of the set S if $S \subset \bigcup \mathcal{O}_\alpha$. The covering is said to be a definite covering if $\mathcal{S}$ consists of only a finite number of open sets.*

Any set S in n space, for example, can be covered by an infinite collection of open balls. For example, choose any $\delta > 0$ and use the collection

$$\mathcal{S} = \{\text{all } B(p, \delta) \text{ for } p \in S\}$$

Since each point p in S is covered by its own ball $B(p, \delta)$, S is a subset of the union of all the balls $B(p, \delta)$. (Note that this will be a noncountable open covering of S in general.) A covering is obviously easier to work with if it is a *finite* covering, and it is therefore important to know if a given infinite open covering can be reduced to a finite covering by discarding most of the open sets. This property has turned out to be such an important one for analysis that it has received a special name.

Definition 11 *A set S is called* **compact** *if every open covering of S can be reduced to a finite covering. This means that there must exist a finite subcollection* $\mathcal{S}_0$ *of the original open sets which is still a covering of S.*

On the line, it is easily seen (as an application of the LUB property) that any closed bounded interval $[a, b]$ is compact. This statement is also called the **Heine-Borel** theorem for the line.

Theorem 24 *The interval* $[a, b]$ *is compact.*

Suppose that $\mathcal{S}$ is some given open covering of $[a, b]$. Since $[a, x]$ is a subset of $[a, b]$, $\mathcal{S}$ is also an open covering of each of the intervals $[a, x]$ for any x, $a \le x \le b$. Let A be the set of all x such that the covering $\mathcal{S}$ can be reduced to a finite covering of $[a, x]$. The theorem is proved if we can show that $b \in A$. Clearly, $a \in A$, since the left end point of the interval $[a, b]$ must lie in some open set $\mathcal{O}_1 \in \mathcal{S}$, and this single open set would be a finite reduction of $\mathcal{S}$ covering the degenerate interval $[a, a]$. Since the set A is bounded, $\sup(A) = \alpha$ exists. Since α is a point in the interval $[a, b]$, α lies in some open set $\mathcal{O}$ in the covering $\mathcal{S}$. Since α is interior to $\mathcal{O}$, all points near α (e.g., in an open interval about α) lie in $\mathcal{O}$. Since α is the smallest upper bound for A, there is $x \in A$ such that $x < \alpha$, and such that all points t with $x \le t \le \alpha$ lie in $\mathcal{O}$. Since $x \in A$, $\mathcal{S}$ has a finite reduction $\mathcal{S}_0$ that covers $[a, x]$. The single open set $\mathcal{O}$ covers $[x, \alpha]$. Thus, adjoining $\mathcal{O}$ to the collection $\mathcal{S}_0$ produces another finite subcollection of $\mathcal{S}$ that now covers $[a, \alpha]$, and $\alpha \in A$. If $\alpha < b$, then there are points y to the right of α that also lie in $\mathcal{O}$ and are such that the interval $[\alpha, y] \subset \mathcal{O}$. By the same argument, $\mathcal{S}_0$ in fact covers $[a, y]$, so that $y \in A$. However, this is impossible, since α was an upper bound for A, while $\alpha < y$. Hence, $\alpha = b$ and $b \in A$. ▮

Having shown that there are some simple sets that are compact, we state a number of theorems that show why the notion is important, and which generalize many of the previous results from the line to the plane, space, and n space. The first is the general **Heine-Borel** theorem.

Theorem 25 *The compact sets in n space are exactly those that are closed and bounded.*

The next two together are often called the **Bolzano-Weierstrass** theorem.

Theorem 26 *Any bounded infinite set in n space has a cluster point.*

Theorem 27 *Any bounded sequence in n space has a limit point, and thus a converging subsequence.*

The next generalizes the nested interval theorem.

Theorem 28 *If $C_1 \supset C_2 \supset C_3 \supset \cdots$ is a nested sequence of nonempty compact sets, then their intersection $\bigcap_1^\infty C_n$ is nonempty.*

Finally, the following result often makes it possible to work in n space with a finite subset of any bounded set, instead of the whole set, in much the same way as one does with a finite set of partition points on an interval.

Theorem 29 *If S is any bounded set in n space, and $\delta > 0$ is given, then it is possible to choose a finite set of points p_i in S such that every point $p \in S$ is within a distance δ of at least one of the points $p_1, p_2, \ldots, p_m$.*

Of these theorems, only the first has a proof that is not immediate. For this reason, we prove others first.

PROOF OF THEOREM 26 We prove that if S is an infinite subset of a compact set C, then S has a cluster point. If this were not the case, then every point $p \in C$ would have an open neighborhood $\mathcal{N}_p$ about p which contained only a finite number of points of S. The collection of all these sets $\mathcal{N}_p$ is an open covering of C and must have a finite reduction $\mathcal{S}_0$ which still covers C. Since $S \subset C$, $\mathcal{S}_0$ covers S. Since $\mathcal{N}_p \cap S$ is always finite, the set S itself is finite, contradicting the assumption that S was infinite. ▮

To prove Theorem 27, repeat the same argument, choosing $\mathcal{N}_p$ as a neighborhood of p such that $p_n \in \mathcal{N}_p$ for at most a finite number of n.

PROOF OF THEOREM 28 By Theorem 25 (or by Exercise 2) each set C_k is closed. Choose an open set U containing C_1, and thus all the C_n; for example, choose U as n space itself. Then, form the open sets $V_n = U - C_n$ by removing from U all the points of C_n. Since the C_n decrease, $V_1 \subset V_2 \subset V_3 \subset \cdots$. If we suppose that $\bigcap_1^\infty C_n$ is empty, then it follows that $\bigcup_1^\infty V_n = U$, and since $C_1 \subset U$, the sets V_i form an open covering of C_1. This covering has a finite reduction that still covers C_1. Hence, for some index N we have

$$C_1 \subset V_1 \cup V_2 \cup V_3 \cup \cdots \cup V_N = V_N = U - C_N$$

But, $C_N \subset C_1$, so that $C_N = \varnothing$, contradicting the hypothesis that none of the set C_n was empty. ▮

PROOF OF THEOREM 25 We have left the proofs that a compact set is necessarily both closed and bounded as Exercises 1 and 2. To prove the converse, we will also use Exercise 3, which asserts that a closed subset of a compact set is compact. The proof of Theorem 25 for subsets of the plane will then be completed if we prove that a closed square in the plane is

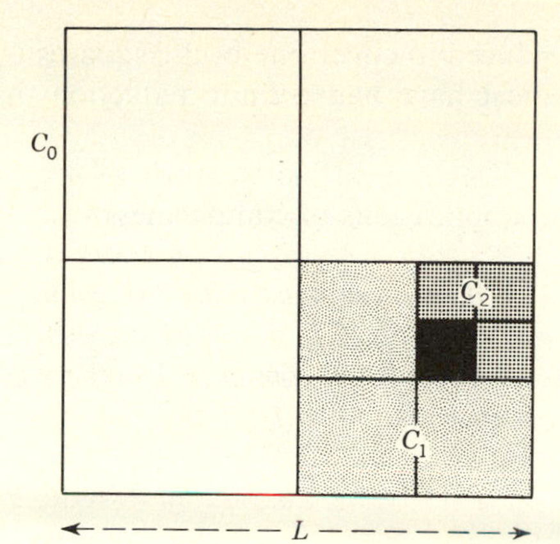

Figure 1-28

compact. The proof for n space follows the same pattern. We again use the device of partition used in the proof of Theorem 21 for the line.

Let C_0 be a closed square in the plane, and suppose that $\mathcal{S}$ is any open covering of C_0. We are to show that $\mathcal{S}$ has a finite reduction that still covers C_0. Suppose that it does not. Subdivide C_0 into four closed squares of half the side, as shown in Fig. 1-28. If $\mathcal{S}$ had a reduction for each of these four smaller squares, we could have obtained a finite reduction for C_0 by using those open sets to cover each of the smaller squares. Hence, one of these smaller squares—call it C_1—is such that $\mathcal{S}$ has no finite reduction covering it. Repeat this process with the square C_1, obtaining a still smaller square for which a finite reduction is impossible, and so on. The result is a nested sequence of closed squares $C_0 \supset C_1 \supset C_2 \supset C_3 \cdots$ such that none of them can be covered by any finite subcollection of the given open sets in $\mathcal{S}$. We now use the nested rectangle theorem for the plane, stated as Exercise 9 of Sec. 1.7 and proved in the same way as Theorem 20. We conclude that there is a point $\bar{p}$ common to all the C_n, and that these close down on $\bar{p}$ so that $\bigcap_1^\infty C_n$ consists just of the point $\bar{p}$. But, $\bar{p} \in C_0$ and must itself be covered by some single open set $\mathcal{O} \in \mathcal{S}$. However, for sufficiently large n, $C_n \subset \mathcal{O}$, as in

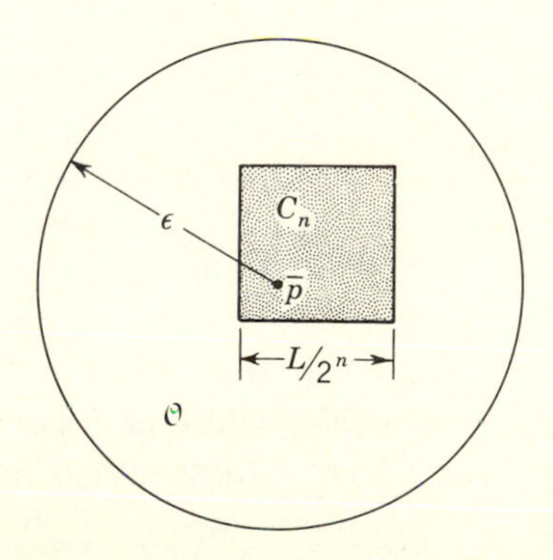

Figure 1-29

Fig. 1-29, and $\mathcal{S}$ has been finitely reduced to cover one of the squares C_n. This contradiction shows that $\mathcal{S}$ must have had a finite reduction that covered C_0 itself. ▮

We end this section with two useful applications of compactness.

Theorem 30 (*a*) *If C is a compact set, then there must exist two points p, $q \in C$ such that $|p - q| = \text{diam}(C)$.*
(*b*) *If A and B are closed sets, and A is compact, then there is a point $p \in A$ and $q \in B$ such that*

$$|p - q| = \text{dist}(A, B)$$

PROOF OF (*a*) Choose p_n and q_n in C with $\lim_{n\to\infty} |p_n - q_n| = \text{diam}(C)$. Since S is compact, $\{p_n\}$ has a subsequence $\{p_{n_k}\}$ that converges to some point $p \in C$. Since

$$|p_{n_k} - q_{n_k}| \le |p_{n_k} - p| + |p - q_{n_k}|$$

and since the left side of this converges to diam (C) while the right side is less than the sum of $|p_{n_k} - p|$, which approaches 0, and diam (C), we have

$$\lim_{k\to\infty} |p - q_{n_k}| = \text{diam}(C)$$

In a similar manner, the sequence $\{q_{n_k}\}$ has a limit point q in S, and by an analogous argument, $|p - q| = \text{diam}(C)$.

PROOF OF (*b*) Choose $p_n \in A$ and $q_n \in B$ such that

$$\lim_{n\to\infty} |p_n - q_n| = \text{dist}(A, B)$$

Since A is compact, $\{p_n\}$ has a limit point $p \in A$, and as in the proof of part (*a*), there is a sequence $\{n_k\}$ of indices such that $\{p_{n_k}\}$ converges to p and

$$\lim_{k\to\infty} |p - q_{n_k}| = \text{dist}(A, B)$$

Since

$$\begin{aligned} |q_{n_k}| &= |q_{n_k} - p + p| \\ &\le |q_{n_k} - p| + |p| \end{aligned}$$

the sequence $\{q_{n_k}\}$ is bounded. It must then have a limit point q which lies in B, since B is closed. Repeating the sort of argument used in part (a), we conclude that $|p - q| = \text{dist}(A, B)$. ▮

Corollary 1 *If A and B are disjoint closed sets, one of which at least is bounded, then they are a positive distance apart; there is a number $\delta > 0$ such that $|p - q| \ge \delta$ for all $p \in A$, $q \in B$.*

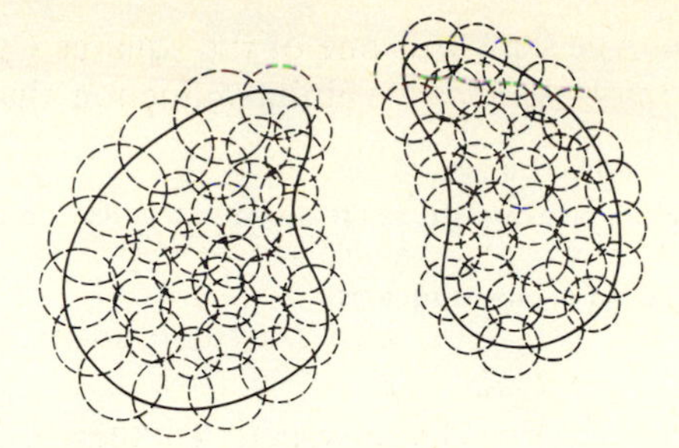

Figure 1-30

Corollary 2 *Let A and B be disjoint compact sets. Then, two disjoint open sets U and V can be chosen such that $A \subset U$ and $B \subset V$, and such that U and V are each the union of a finite number of open balls.*

The result is illustrated in Fig. 1-30. The practical importance of this result is that U and V are so simple in construction that their boundaries are each made up of a finite number of arcs of circles or portions of spheres. To prove Corollary 2, let $\delta = \text{dist}(A, B) > 0$. For each $p \in A$ and each $q \in B$, consider the open balls with center p or q and radius $\delta/3$. Note that these will be disjoint, since their centers are at least δ apart. As p varies over A, we obtain an open covering of A which can be reduced to a finite subcovering whose union is an open set U. Similarly, as q ranges over B, we obtain a covering of B which then yields a finite subcovering whose union is the set V. Note that $\text{dist}(U, V) \geq \delta/3$.

EXERCISES

1 Show that any compact set in n space must be bounded.

2 Show that any compact set in n space must be closed.

3 Show that every closed subset of a compact set is compact.

4 Fill in the details of the proof of Corollary 1 of Theorem 30.

5 Prove that a set S in n space is compact if and only if every sequence in S has a limit point that belongs to S.

6 Let A and B be compact sets on the line. Use Exercise 5 to show that their cartesian product $A \times B$ is a compact set in the plane.

7 (*a*) Prove Theorem 29, assuming that the set S is both closed and bounded.

(*b*) Prove Theorem 29, assuming only that S is bounded. [The difficulty lies in showing that the points p_i can be chosen in S itself.]

8 Must every bounded nonempty set in n space have a nonempty boundary?

9 For any set S in the plane, let

$$X(S) = \{\text{all } x \text{ for which } (x, y) \in S \text{ for some } y\}$$

This is called the projection of S into the X axis.

(*a*) If S is bounded, is $X(S)$ necessarily bounded?

(*b*) If S is closed, is $X(S)$ necessarily closed?

(*c*) If S is compact, show that $X(S)$ is necessarily compact.

10 (*a*) Show that the open unit disc in the plane can be expressed as the union of a collection of closed squares.

*(*b*) Can this be done with a countable collection of closed squares?

CHAPTER

TWO

CONTINUITY

2.1 PREVIEW

The purpose of this chapter is to clarify the intuitive notion of continuity, and to arrive at many of its most useful consequences. We start by discussing and comparing continuity of a function and uniform continuity of a function. The former is what is termed a local property, something that can be decided by examining a function on a neighborhood of each point of a set; the latter is a global property, and is a statement about the way a function behaves on an entire set. We then take up many of the properties of continuous functions which follow immediately from the definition of continuity, particularly those that describe the behavior of a continuous function on a set that is compact or connected. One familiar instance of the latter is the intermediate-value theorem, which is used informally by people more than any other theorem in analysis. ("If the temperature was 40° and now it's 100°, it must have been 70° sometime!")

With the concept of continuity as a tool, the discussions of limits and discontinuities in Secs. 2.5 and 2.6 become easier, especially when dealing with functions of several variables, where the behavior of a function near a boundary point of its domain can be so complicated. Section 2.7 is very short and deals with inverses for functions of one variable. The continuity of such inverses turns out to be an immediate consequence of the useful compact graph theorem of Sec. 2.4.

2.2 BASIC DEFINITIONS

Let f be a numerical-valued function, defined on a region D in the plane. Suppose that we interpret $f(p)$ to be the temperature at the point p. Then, the intuitive notion of continuity can be described by saying that the temperature on a small neighborhood of any point p_0 in D will vary only slightly from that at p_0; moreover, we feel that these variations can be made as small as we like by decreasing the size of the neighborhood. This behavior can be shown on a graph of f. In Fig. 2-1, we have shown the range of variation in the values of a function of one variable when x is confined to a neighborhood of a point x_0; note also that the size of neighborhood needed to attain the same limitation of variation may be smaller at another point.

Formally, we are led to the following.

Definition 1 *A numerical-valued function f, defined on a set D, is said to be* **continuous** *at a point $p_0 \in D$ if, given any number $\varepsilon > 0$, there is a neighborhood U about p_0 such that $|f(p) - f(p_0)| < \varepsilon$ for every point $p \in U \cap D$. The function f is said to be continuous on D if it is continuous at each point of D.*

The work of checking from this definition that a specific function is continuous can be easy or difficult, depending upon how the function has been described and how simple it is. As a start, let us show that the function F given by $F(x, y) = x^2 + 3y$ is continuous on the unit square S consisting

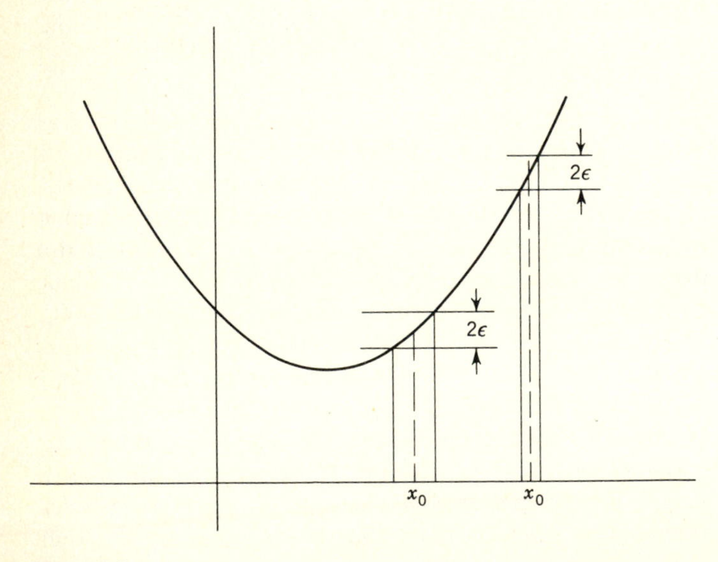

Figure 2-1

of those points (x, y) with $0 \le x \le 1$, $0 \le y \le 1$. Let $p_0 = (x_0, y_0)$ be any point in S. Then

$$\begin{aligned} F(p) - F(p_0) &= F(x, y) - F(x_0, y_0) \\ &= x^2 + 3y - x_0^2 - 3y_0 \\ &= x^2 - x_0^2 + 3y - 3y_0 \\ &= (x - x_0)(x + x_0) + 3(y - y_0) \end{aligned}$$

No matter where the points p and p_0 are located in S, it is always true that $0 \le x \le 1$ and $0 \le x_0 \le 1$, so that $x + x_0 \le 2$. Consequently,

$$|F(p) - F(p_0)| \le 2|x - x_0| + 3|y - y_0| \tag{2-1}$$

We have here an estimate for the variation of f near p_0. If we confine p to a small neighborhood of p_0, then $|x - x_0|$ and $|y - y_0|$ must both be small and so will $|F(p) - F(p_0)|$. We can make this argument more concrete. Suppose that we wish to find a neighborhood U about p_0 in which the variation of F is less than $\varepsilon = .03$. Take U to be a square neighborhood centered on p_0, with sides .01. Then, if $p \in U$ and lies in D, we must have $|x - x_0| < .005$ and $|y - y_0| < .005$, so that (2-1) gives

$$|F(p) - F(p_0)| < 2(.005) + 3(.005) = .025 < .03$$

In general, if we take U to be a square of side $(.4)\varepsilon$, then $|x - x_0| < (.2)\varepsilon$ and $|y - y_0| < (.2)\varepsilon$, so that

$$|F(p) - F(p_0)| < 2(.2)\varepsilon + 3(.2)\varepsilon = \varepsilon$$

This example exhibits another fundamental property of continuous functions. Let $\{p_n\}$ be a sequence of points in S, converging to a point $p_0 \in S$. Then, $\{F(p_n)\}$ will converge to $F(p_0)$. For, writing $p_n = (x_n, y_n)$, we have, from (2-1),

$$|F(p_n) - F(p_0)| < 2|x_n - x_0| + 3|y_n - y_0|$$

and since $\lim x_n = x_0$, $\lim y_n = y_0$, we conclude that the right side approaches 0 and $\lim F(p_n) = F(p_0)$. Another way to describe this property is to say that a continuous function **preserves convergence.**

Theorem 1 *Let f be continuous at $p_0 \in D$ and let $\{p_n\}$ be a sequence of points in the domain D of f, with* $\lim_{n\to\infty} p_n = p_0$. *Then,* $\lim_{n\to\infty} f(p_n) = f(p_0)$.

Given any $\varepsilon > 0$, choose a neighborhood U about p_0 such that $|f(p) - f(p_0)| < \varepsilon$ whenever $p \in U \cap D$. Since $p_n \in D$ and $\{p_n\}$ converges to p_0, we must have $p_n \in U$ for all sufficiently large n. Hence, there is an N such that $|f(p_n) - f(p_0)| < \varepsilon$ for all $n > N$. But this is precisely what we mean by saying that the sequence of numbers $\{f(p_n)\}$ converges to $f(p_0)$. ∎

It is also easy to see that "convergence preserving" is a characteristic property of continuous functions.

Theorem 2 *If a function f defined on D has the property that, whenever $p_n \in D$ and $\lim p_n = p_0 \in D$, then it follows that $\lim_{n\to\infty} f(p_n) = f(p_0)$, then f is continuous at p_0.*

It is easiest to prove this by an indirect argument. Assume that f is convergence preserving but that f is not continuous at p_0. If the definition of continuity (Definition 1) is read carefully, one sees that, in order for f *not* to be continuous at p_0, there must exist a particular value of $\varepsilon > 0$ such that no neighborhood U can be found to satisfy the required condition. (There is a routine for carrying out the logical maneuver of constructing the denial of a mathematical statement in a semimechanical fashion; those who have difficulty reasoning verbally will find it explained in Appendix 1.) Thus if f is not continuous at p_0, and we think of trying a specific neighborhood U, then it must fail because there is a point p of D in U with $|f(p) - f(p_0)| \geq \varepsilon$. If we try U in turn to be a spherical neighborhood of radius $1, \frac{1}{2}, \frac{1}{3}, \frac{1}{4}, \ldots$ and let U_n therefore be

$$U_n = \left\{\text{all } p \text{ with } |p - p_0| < \frac{1}{n}\right\}$$

then there must be a point $p_n \in U_n \cap D$ with $|f(p_n) - f(p)| > \varepsilon$. Since $p_n \in U_n$, $|p_n - p_0| < 1/n$ and $\lim p_n = p_0$. We have therefore produced a sequence $\{p_n\}$ in D that converges to p_0, but such that $f(p_n)$ does not converge to $f(p_0)$. This contradicts the assumed convergence-preserving nature of f and forces us to conclude that f was continuous at p_0. ∎

Theorem 2 is more useful as a way to show that a specific function is *not* continuous than it is as a way to show that a function *is* continuous. To use it for the latter purpose, one would have to prove something about $\{f(p_n)\}$ for *every* sequence $\{p_n\}$ converging to p_0, and there are infinitely many such sequences. However, if there is *one* sequence $\{p_n\}$ in D which converges to p_0, but for which $\{f(p_n)\}$ is divergent, then we know at once that f is not continuous at p_0. For example, let f be defined on the plane by

$$f(x, y) = \begin{cases} \dfrac{xy^2}{x^2 + y^4} & (x, y) \neq (0, 0) \\ 0 & x = y = 0 \end{cases} \tag{2-2}$$

We want to see if f is continuous at $(0, 0)$. Among the sequences that approach the origin, look at those of the form $p_n = (1/n, c/n)$. As c takes on different values, the sequence p_n will approach $(0, 0)$ along different lines, taking on all possible directions of approach. Since

$$f(p_n) = \frac{c/n^3}{\dfrac{1}{n^2} + \dfrac{c^4}{n^4}} = \frac{c}{1 + \dfrac{c^4}{n^2}} \frac{1}{n}$$

we find that $\lim_{n \to \infty} f(p_n) = 0$. It might seem from this that f is continuous at $(0, 0)$. However, if we try a new sequence, $q_n = (1/n^2, 1/n)$, we find that

$$f(q_n) = \frac{1/n^4}{\dfrac{1}{n^4} + \dfrac{1}{n^4}} = \frac{1}{2}$$

so that, while $\lim q_n = (0, 0)$, $\lim f(q_n) \neq f(0, 0)$. Thus, we have shown by the single sequence $\{q_n\}$ that f is not continuous at the origin.

In later sections, we shall need to work with continuous functions whose values are not numbers but points. Although any detailed study of these will be postponed until Chap. 7, it is convenient to give the corresponding definition here.

Definition 1′ *Let f be a function defined in a set D in n space, and taking its values in m space. Then, f is said to be continuous at a point $p_0 \in D$ if, given any neighborhood V about $f(p_0) = q_0$, there is a neighborhood U about p_0 such that $f(p) \in V$ whenever $p \in U \cap D$. The function (or mapping) f is said to be continuous on D if f is continuous at each point of D.*

This can be visualized as in Fig. 2-2; it should be clear that this again coincides with our intuitive ideas of continuity. The images (values) $f(p)$ can be restricted to a small neighborhood of $f(p_0)$ by confining p to an appropriately small neighborhood of p_0. It is also easy to show that the analogs of Theorems 1 and 2 hold; *f is continuous if and only if f is convergence preserving.*

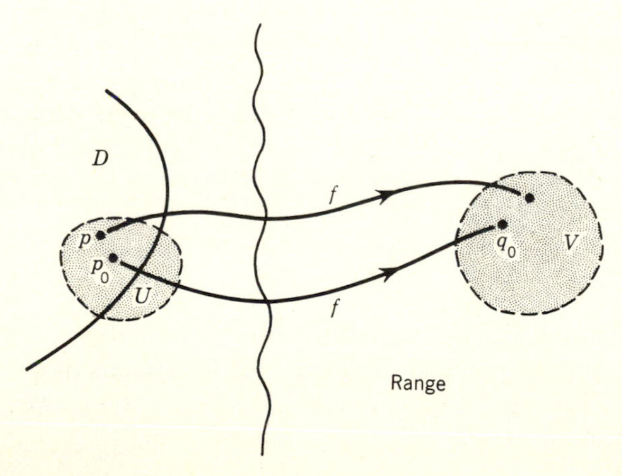

Figure 2-2 Continuity of a mapping f.

When we adopt the point of view that a function is a correspondence or mapping, we think of $f(p)$ as the image of the point p. In a similar fashion, we can speak of the **image of a set** under f. If S is any set in the domain D of f, then $f(S)$ is the set of points $f(p)$, for $p \in S$. Using this language, we can describe the relationship shown in Fig. 2-2 and Definition 1′ more simply: *Given any neighborhood V, there is a neighborhood U such that $f(U \cap D) \subset V$.*

We shall also speak of the **inverse image** or **pre-image** of a set S under a mapping f. Given any set S, its inverse image is the set

$$f^{-1}(S) = \{\text{all } p \in D \text{ with } f(p) \in S\} \tag{2-3}$$

It should be noted that the set S might be chosen in such a way that no point p in D satisfies $f(p) \in S$. In this case, $f^{-1}(S)$ will still be defined, but the result is $\varnothing$, the empty set. Relation (2-3) merely requires that the points in $f^{-1}(S)$ consist of all those in the domain of f whose images are in S, and if there are none, $f^{-1}(S)$ is empty. Some mathematical problems lend themselves to formulation in terms of these ideas. For example, instead of talking about the set of solutions x for the equation $f(x) = 0$, we could equally well ask for the inverse image of the set $\{0\}$. A striking example of this is the following reformulation of the idea of continuity.

Theorem 3 *A function f is continuous on an open set D if and only if the inverse image of every open set, under f, is open.*

Let S be any open set in the range space of f, and let R be its inverse image $f^{-1}(S)$. We must show that, if f is continuous, R is open. Take any point $p_0 \in R$; we wish to show that p_0 is interior to R. Since p_0 is in the inverse image, under f, of S, we know that $f(p_0)$ belongs to S. Also, since S is open, there is a neighborhood V about $f(p_0)$ which lies entirely inside S. Now, since f is continuous at p_0, we can find a neighborhood U about p_0 so that $f(p) \in V$ for every $p \in U \cap D$ (see Fig. 2-2). Since D itself is open, the set $U \cap D$ is a neighborhood of p_0 lying entirely in D. Moreover, the fact that $f(U \cap D) \subset V$, together with $V \subset S$, shows that $U \cap D \subset R$. Hence, we have found a neighborhood of p_0 composed only of points in R. p_0 is therefore interior to R, and R is open.

Conversely, if we suppose that f^{-1} has the property of carrying open sets back into open sets, it follows that f is continuous. Given any $p_0 \in D$ and any neighborhood V about $f(p_0)$, we look at the inverse image W of V. Since V is open, W is open. Since $f(p_0) \in V$,

$$p_0 \in f^{-1}(V) = W$$

Hence, p_0 is an interior point of W, and there must be a neighborhood U about p_0 which lies in W. But, all points of W are carried into points of V, under f, so that $f(W) \subset V$. Taking W for the required neighborhood U, we have shown that f is continuous at p_0. ▮

In this theorem, the apparent restriction that the domain of f be an open set can be overcome in several ways. One method is to introduce the notion of **relative topology.** If D is a set in n space, then one says that a subset $A \subset D$ is **open, relative to** D, when A is the result of intersecting D with an open set in n space. In the same way, a set B is called **closed, relative to** D, if $B = D \cap C$, where C is closed in n space. This is merely a device for restricting one's attention precisely to the points in a specific set D, but still using all the terminology and topology of n space. If we use this interpretation, then Theorem 3 can be given a broader meaning as follows: *A function f is continuous on a set D if and only if the inverse image of every open set under f is open relative to D.*

There is also a second way to overcome the limitation in the stated form of Theorem 3. We shall show later on that any continuous function f that is defined on a closed set D and takes values in $\mathbf{R}$, or indeed in $\mathbf{R}^m$, can be assumed in fact to have been defined and continuous on any convenient larger open set $\mathcal{O} \supset D$. This important result, known as the Tietze extension theorem, is proved in Chap. 6, and simplifies many arguments.

If we choose a set D, an arbitrary set in n space, then we can study the collection of all real-valued functions that are continuous on D. This is a very important mathematical object, the focus of much research work in the past several decades. The first questions that one asks about it deal with the sort of structure the whole collection possesses: what are the basic laws and operations that allow you to combine continuous functions and get continuous functions? The first result shows that this class of functions is a vector space, and in fact something more. (See also Appendix 6.)

Theorem 4 *The class of real-valued functions continuous on D forms an algebra. Explicitly, if f and g are continuous on D, so are their sum* $f + g$, *their product* fg, *and any scalar multiples* αf *or* $\alpha f + \beta g$, *where* α *and* β *are numbers. The quotient of two functions* f/g *is continuous at all those points* p *in* D *where* $g(p) \neq 0$.

We can derive this from the corresponding facts about convergent sequences that were proved in Sec. 1.6 if we also make use of Theorems 1 and 2 in the present section. Thus to prove that fg is continuous on D, we need show only that fg is convergence preserving. Given any $p_0 \in D$ and any sequence $\{p_n\}$ in D that converges to p_0, we know from the assumed continuity of f and g that $\lim_{n\to\infty} f(p_n) = f(p_0)$ and $\lim_{n\to\infty} g(p_n) = g(p_0)$. By the standard result on real sequences, we conclude that $\{f(p_n)g(p_n)\}$ must be a convergent sequence, and

$$\lim_{n\to\infty} f(p_n)g(p_n) = \lim_{n\to\infty} f(p_n) \lim_{n\to\infty} g(p_n)$$
$$= f(p_0)g(p_0)$$

The product function fg is therefore convergence preserving. ∎

The proof of all the remaining statements follows the same pattern.

Several consequences of this result are worth pointing out. First, we can conclude that all polynomial functions are everywhere continuous. For polynomials in one variable, this follows by noting that constant functions [e.g., the function c with $c(x) = 3$ for all x] and the identity function J [$J(x) = x$ for all x] are clearly continuous. Then, by addition and multiplication and by multiplication by scalars (i.e., coefficients), we construct the most general polynomial function. For polynomials in two variables, we would start from the basic functions $I(x, y) = x$ and $J(x, y) = y$, prove that these are continuous, and then build up the general polynomial from these. Second, we can conclude that all rational functions, for example,

$$R(x, y) = \frac{P(x, y)}{Q(x, y)}$$

are continuous everywhere except at the points where the denominator is 0. (In Sec. 2.6, we shall discuss the behavior of such functions near points where the denominator is 0; in some cases, we can restore continuity, and in others we cannot.)

There is a further way to combine continuous functions to yield continuous functions. It may be stated roughly as follows: *Continuous functions of continuous functions are continuous.* With more attention to the fine points of detail, we have the following:

Theorem 5 *Let the function g be continuous on a set D and f continuous on a set S. Suppose that $p_0 \in D$ and $g(p_0) = q_0 \in S$. Then, the composite function F, given by*

$$F(p) = f(g(p))$$

is continuous at p_0.

This can be proved easily by the scheme of using sequences and the sequence-preserving property. For variety, we choose to prove this by using the neighborhood definition of continuity. The argument can be followed in the diagram of Fig. 2-3. Let $f(q_0) = c = F(p_0)$. Since f is continuous at q_0, we know that, for any $\varepsilon > 0$, there is a neighborhood V about q_0 such that $|f(q) - f(q_0)| < \varepsilon$ whenever $q \in V$ and q is in the domain S of f. Since $g(p_0) = q_0$ and g is continuous at p_0, we also know that there is a neighborhood U about p_0 such that $g(p) \in V$ for all points $p \in U \cap D$. Putting these remarks together, we see that, if we set $q = g(p)$, then

$$|f(g(p)) - f(g(p_0))| < \varepsilon$$

whenever $p \in U \cap D$. Since this can be rewritten as $|F(p) - F(p_0)| < \varepsilon$, we have shown that F is continuous at p_0. ∎

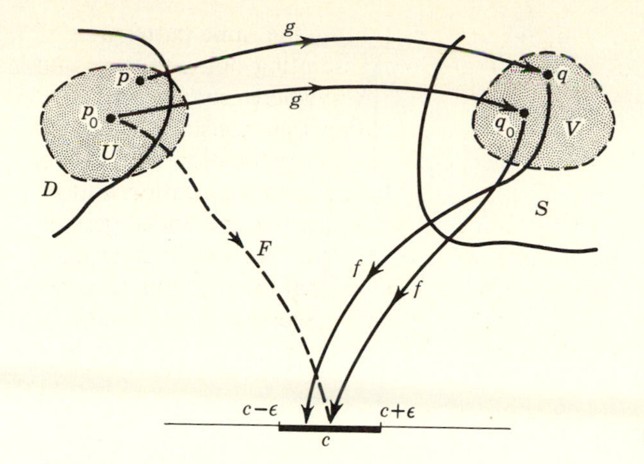

Figure 2-3

The continuity of composite functions is used constantly. It can even be used to derive Theorem 4 in a different way. Suppose that we were to verify directly from the definition that the functions A and M are continuous, where $A(x, y) = x + y$ and $M(x, y) = xy$. Then, the composite function theorem tells us at once that $f + g$ and fg are continuous on a set D if f and g are each separately continuous on D; for

$$\begin{aligned} f(p) + g(p) &= A(f(p), g(p)) \\ &= A(F(p)) \end{aligned}$$

and

$$\begin{aligned} f(p)g(p) &= M(f(p), g(p)) \\ &= M(F(p)) \end{aligned}$$

where F is the function from D to $\mathbf{R}^2$ given by $F(p) = (f(p), g(p))$. In the same fashion, the fact that $Q(x, y) = x/y$ is continuous on the set where $y \neq 0$ implies at once that f/g is continuous where $g(p) \neq 0$.

As a more practical example, what can we say about the continuity of a function such as $f(x, y) = x^2 y \csc(x + y)$? If we assume that the sine function is known to be continuous everywhere, then the composite function theorem and Theorem 4 show that f is continuous at all points (x, y) except for those where $\sin(x + y) = 0$, namely, the points on the family of lines with equation

$$x + y = n\pi \qquad n = 0, \pm 1, \pm 2, \ldots$$

How do we know that a function such as the sine function is continuous? The only honest answer is to say that it depends upon how this function was defined. A traditional definition of the sine function is to say: *For any* x, $\sin(x)$ *is the second coordinate of the point P on the unit circle whose distance from* $(1, 0)$, *measured along the circumference in the positive direction, is* x (Fig. 2-4). From the figure, it probably seems self-evident that $\sin(x)$ is near $\sin(x_0)$ whenever x is sufficiently near x_0. It has even been traditional to bolster this

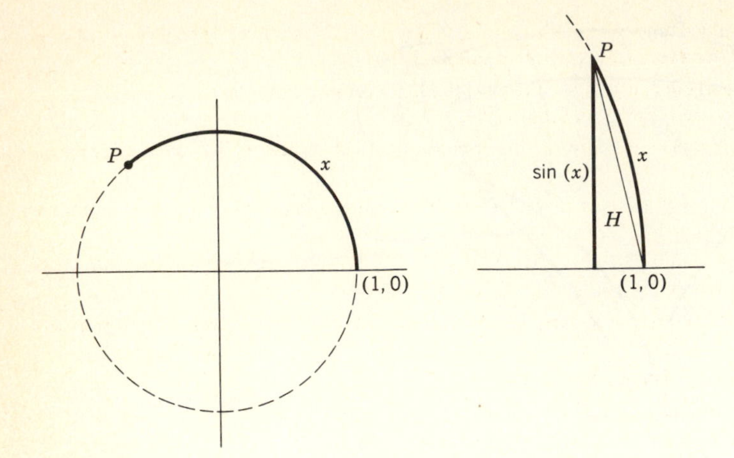

Figure 2-4

geometric argument by calculations involving arc lengths or areas of sectors to prove that lim sin $(x_n) = 0$ when x_n is a sequence approaching 0. However, this "proof by picture" needs much work before it is on a firm foundation. In particular, this definition of the sine function itself is a little shaky, since it rests upon an already assumed theory of arc length, particularly for the circle whose equation is $x = \cos\theta$, $y = \sin\theta$!

Such overt circularity of reasoning is clearly unsound. There are, of course, ways to present a treatment that avoids these defects. The trigonometric functions can be defined by means of infinite series, or by means of certain indefinite integrals, or even as solutions of certain differential equations. Since we are not attempting to construct a total connected logical development of all analysis, we do not adopt one of these; a suggestion of the first two will be found in several of the exercises in Secs. 5.4 and 6.5, and there is a complete treatment in the article by Eberlein [7] mentioned in the Reading List at the end of the book.

Henceforth, you may assume the continuity of any of the standard elementary functions of one-variable calculus, on its natural domain.

EXERCISES

1 Prove directly from Definition 1 that the function $M(x, y) = xy$ is continuous on any disk: $|(x, y)| \leq r$.

2 Prove that $Q(x, y) = x/y$ is continuous everywhere except on the line $y = 0$.

3 Check that the function defined in (2-2) is such that it is convergence preserving for all sequences of the form $p_n = (a/n^2, b/n^2)$.

4 Let f be defined by $f(x, y) = x^2y^2/(x^2 + y^2)$, with $f(0, 0) = 0$. By checking various sequences, test this for continuity at (0, 0). Can you tell whether or not it is continuous there?

5 Show that a real-valued function f is continuous in D if the set $S = \{\text{all } p \in D \text{ with } b < f(p) < c\}$ is open, relative to D, for every choice of the numbers b and c.

6 Using Theorem 4, where can you be sure that the function given by $F(x, y) = (x + y)/(x^2 - xy - 2y^2)$ is continuous?

7 Use the example $f(x, y) = x^2$ to show that a continuous function does not have to map an open set onto an open set.

***8** Use the example $f(x) = x^2/(1 + x^2)$ to show that a continuous function does not always have to map a closed set onto a closed set.

9 Using the assumed continuity properties of the sine function, what can you say about the set on which the function $g(x) = \csc(\sin(1/x))$ is continuous?

10 Show that f is continuous if and only if the inverse images of closed sets are closed sets relative to D.

11 Prove Theorem 5 using Theorems 1 and 2.

12 How are $f^{-1}(A \cap B)$ and $f^{-1}(A \cup B)$ related to $f^{-1}(A)$ and $f^{-1}(B)$?

13 Let $f: X \to Y$ be a function, and S and T arbitrary sets. Show that:

(*a*) $ff^{-1}(S) \subset S$ (*b*) $T \subset f^{-1}f(T)$

14 Let $F(x, y)$ be continuous on the square $|x| \le 1$, $|y| \le 1$. Where is the function $f(x) = F(x, c)$ continuous?

15 Formulate the definition of continuity for a complex-valued function f. Show that f is continuous if and only if its real and imaginary parts are continuous functions.

16 (*a*) Setting $z = x + iy = (x, y)$, consider the function f defined from complex numbers to complex numbers ($\mathbf{R}^2$ to $\mathbf{R}^2$) by $f(z) = z^2 + (1 - i)z + 2$, and show that it is continuous everywhere.

(*b*) What can you say about the continuity of the function f where:

(i) $f(z) = \dfrac{1}{z}$ (ii) $f(z) = \dfrac{z}{z^2 + 1}$

17 Show that a mapping $y = F(x)$ on a set D in $\mathbf{R}^n$ to $\mathbf{R}^m$ as given in (1-26) is continuous if and only if the component functions $f, g, h, \ldots, k$ are continuous on D.

2.3 UNIFORM CONTINUITY

We now make a closer study of the concept of continuity of a real-valued function, particularly with respect to the way the behavior of a function near one point may differ from that near another. As we shall see, the contrast here is between properties that are local and properties that are global.

Suppose we return to the special polynomial function

$$F(x, y) = x^2 + 3y$$

which we studied at the start of the preceding section. There, we showed that F is continuous on the unit square with vertices at $(0, 0)$, $(0, 1)$, $(1, 0)$, $(1, 1)$. From the more general theorems we proved in the last section, we know in fact that F is continuous in the whole plane. To obtain this directly, we would again write

$$F(x, y) - F(x_0, y_0) = (x - x_0)(x + x_0) + 3(y - y_0)$$

If we are given a positive number ε, we must find a neighborhood U about $p_0 = (x_0, y_0)$ so that $|F(p) - F(p_0)| < \varepsilon$ for all points $p \in U$. If we take U to be a square box of side 2δ, centered on p_0, then we have $|x - x_0| < \delta$ and $|y - y_0| < \delta$ for any $p \in U$. The number δ is as yet undetermined, but we can agree to take it smaller than 1. Then, when $p \in U$, we have

$$\begin{aligned} |x + x_0| &= |x - x_0 + 2x_0| \le |x - x_0| + 2|x_0| \\ &\le 1 + 2|x_0| \end{aligned}$$

Hence, for any point $p \in U$, we have

$$\begin{aligned} |F(p) - F(p_0)| &\le |x - x_0||x + x_0| + 3|y - y_0| \\ &\le \delta(1 + 2|x_0|) + 3\delta \\ &\le (4 + 2|x_0|)\delta \end{aligned}$$

In order to prove that F is continuous at p_0, we wish to make the *left* side less than ε for every $p \in U$. We can achieve this if we can choose δ so that the *right* side of this inequality is less than ε. Thus, we are led to choose δ as a number obeying

$$0 < \delta < \frac{\varepsilon}{4 + 2|x_0|} \tag{2-4}$$

With this selected, we can be sure that $|F(p) - F(p_0)| < \varepsilon$ for all $p \in U$, and F is continuous at each point p_0.

Now, in examining what we have done, we notice that the size of the neighborhood U that we selected depended not only upon ε (as was to be expected) but also upon x_0, and thus upon the point p_0. Indeed, as the point p_0 is selected farther and farther to the right, $|x_0|$ will increase and the value of δ will decrease. If we were trying to show that F is continuous on a specific bounded set D, then, since the point p_0 must lie in D, there would be an upper bound on the values of $|x_0|$. Accordingly, we would then be able to select a value for δ which is smaller than all the values given in formula (2-4), and we would be able to choose a single-sized neighborhood that works for *all* positions of p_0 in D at once. This in fact was the case when we first examined this function in the previous section, for there we found that $\delta = (.2)\varepsilon$ worked for all points p_0 in the square. This special phenomenon of being able to pick δ independent of p_0 is one to which a special name is given.

Definition 2 *We say that a function f is* **uniformly continuous** *on a set E if and only if, corresponding to each $\varepsilon > 0$, a number $\delta > 0$ can be found such that $|f(p) - f(q)| < \varepsilon$ whenever p and q are in E, and $|p - q| < \delta$.*

There is a vivid way to describe this property of a function that may help to emphasize its characteristic features. Suppose that we invent a measuring device with two movable prongs and a meter which will indicate the temperature difference of the prongs. Suppose that we have a sheet of metal whose temperature varies from point to point, and that the temperature at p is given by $f(p)$. If we place one prong at a point p on the sheet and the other at q, the meter will read the value $|f(p) - f(q)|$. The meaning of the uniform continuity of f is simply that, given a value for $\varepsilon > 0$, we can set the prongs at a fixed separation of at most δ and be sure that, wherever they are placed in the region E, the meter will read no higher than ε.

The term "uniformly continuous" is always used in conjunction with a set, which must also be specified. In the example of the function F above, we showed that it was uniformly continuous on the unit square S and indeed that it was uniformly continuous on *any* bounded set. It can be shown that it is not uniformly continuous on the whole plane.

For another example, let $f(x) = 1/x$. We shall show that f is continuous on the open interval $0 < x < 1$ but is not uniformly continuous there. We first write

$$|f(x) - f(x_0)| = \left|\frac{1}{x} - \frac{1}{x_0}\right| = \frac{|x_0 - x|}{xx_0}$$

To prove continuity at x_0, which may be any point with $0 < x_0 < 1$, we wish to make $|f(x) - f(x_0)|$ small by controlling $|x - x_0|$. If we decide to consider only numbers δ obeying $\delta < x_0/2$, then any point x such that $|x - x_0| < \delta$ must also satisfy $x > x_0/2$, and $xx_0 > x_0^2/2$. Thus, for such x, $|f(x) - f(x_0)| < \delta/xx_0 < 2\delta/x_0^2$. Given $\varepsilon > 0$, we can ensure that $|f(x) - f(x_0)| < \varepsilon$ by taking δ so that $\delta \leq (x_0^2/2)\varepsilon$. Thus, f is continuous at each point x_0 with $0 < x_0 < 1$. If f were uniformly continuous there, then a number $\delta > 0$ could be so chosen that $|f(x) - f(x')| < 1$ for every pair of points x and x' between 0 and 1 with $|x - x'| \leq \delta$. To show that this is not the case, we consider the special pairs, $x = 1/n$ and $x' = \delta + 1/n$. For these, we have $|x - x'| = \delta$ and

$$|f(x) - f(x')| = \left|n - \frac{1}{\delta + 1/n}\right| = \frac{n\delta}{\delta + 1/n}$$

No matter how small δ is, n can be chosen so that this difference is larger than 1; for example, any n bigger than both $1/\delta$ and 3 will suffice.

A plausible (but incorrect) graphical argument can also be given. Examination of the graph of f shows that its slope becomes arbitrarily steep as we move toward the vertical axis, and this would seem to imply that f cannot be uniformly continuous. This reasoning has an analytical counterpart; the choice $\delta = (x_0^2/2)\varepsilon$ which we have made above becomes arbitrarily small as x_0 moves toward 0, so that "no positive number can be found which works for all $x_0 > 0$." To see the flaw in these arguments, consider F, where $F(x) = \sqrt{x}$. Since the graph of F also becomes arbitrarily steep as we move toward the vertical axis, the first argument would seem to show that this function too is not uniformly continuous on the interval $0 < x < 1$. We seem to reach the same conclusion if we repeat the second line of reasoning. To see this, we first prove continuity. Given points x and x_0 with $0 < x < 1$, $0 < x_0 < 1$, we have

$$|F(x) - F(x_0)| = |\sqrt{x} - \sqrt{x_0}| = \left|\frac{(\sqrt{x} - \sqrt{x_0})(\sqrt{x} + \sqrt{x_0})}{\sqrt{x} + \sqrt{x_0}}\right|$$

$$= \frac{|x - x_0|}{\sqrt{x} + \sqrt{x_0}} \le \frac{|x - x_0|}{\sqrt{x_0}}$$

Given ε, we select δ so that $\delta \le \sqrt{x_0}\,\varepsilon$; then, if $|x - x_0| < \delta$,

$$|F(x) - F(x_0)| < \frac{\delta}{\sqrt{x_0}} < \varepsilon$$

This shows that F is continuous on the interval $0 < x < 1$. The choice of δ as $\sqrt{x_0}\,\varepsilon$ suggests that F is not uniformly continuous there, since this number becomes arbitrarily small as x_0 approaches 0. However, F in fact *is* uniformly continuous. To prove this, we must make a more careful choice of δ. Assume that we have chosen δ, and again estimate the difference $|F(x) - F(x_0)|$. Two cases arise, depending on the size of x_0. If $0 < x_0 < \delta$, and $|x - x_0| < \delta$, then $0 < x < 2\delta$ and we have

$$|F(x) - F(x_0)| = |\sqrt{x} - \sqrt{x_0}| \le \sqrt{x} + \sqrt{x_0}$$
$$\le \sqrt{2\delta} + \sqrt{\delta} < 3\sqrt{\delta}$$

If $x_0 \ge \delta$, then, using the original estimate of the difference,

$$|F(x) - F(x_0)| \le \frac{|x - x_0|}{\sqrt{x_0}} < \frac{\delta}{\sqrt{\delta}} = \sqrt{\delta} < 3\sqrt{\delta}$$

No matter how δ is chosen, we have shown that, if $|x - x_0| < \delta$, then $|F(x) - F(x_0)| < 3\sqrt{\delta}$. This at once proves that F is uniformly continuous, since given ε we may choose $\delta = \varepsilon^2/9$ and have $|F(x) - F(x_0)| < \varepsilon$ whenever x and x_0 obey $x > 0$, $x_0 > 0$, and $|x - x_0| < \delta$. In fact, as is easily seen, these calculations show that $F(x) = \sqrt{x}$ is uniformly continuous on the unbounded interval consisting of all x, $x > 0$.

When is a continuous function uniformly continuous? The following basic theorem supplies a partial answer.

Theorem 6 *If E is a compact set and f is continuous on E, then f is necessarily uniformly continuous on E. In particular, any continuous function defined on a closed and bounded set E in n space, and taking values in m space, is uniformly continuous on E.*

There are a number of proofs for this result. We give one here and indicate another in the exercises. We shall again argue indirectly, proving that, if f were not uniformly continuous on E, then f could not be continuous on E. Our first problem is to understand the meaning of "f is not uniformly continuous on E." This may be clearer if we put the statements side by side.

f is uniformly continuous on $E \equiv$
For any $\varepsilon > 0$, there is a $\delta > 0$ such that whenever p and q lie in E and $|p - q| < \delta$, then $|f(p) - f(q)| < \varepsilon$.

f is not uniformly continuous on $E \equiv$
There is some $\varepsilon > 0$ such that, for any $\delta > 0$, there exist points p and q in E with $|p - q| < \delta$ but such that $|f(p) - f(q)| \geq \varepsilon$.

(Again, a brief explanation of the mechanical procedure for finding the negative of complicated statements is given in Appendix 1.)

If, therefore, it were true that f is not uniformly continuous on E, then we could select the special value of ε mentioned above, and then take δ in turn to have the values $1, \frac{1}{2}, \frac{1}{3}, \frac{1}{4}, \ldots$; for each choice of δ, we can produce a pair of points in E, say, p_n and q_n for the choice of δ as $1/n$, such that $|p_n - q_n| < \delta = 1/n$, and with $|f(p_n) - f(q_n)| \geq \varepsilon$. We next show that this cannot be done if f is continuous on E. Because the set E is compact, we apply the Bolzano-Weierstrass theorem (Theorem 26, Sec. 1.7) to conclude that the sequence $\{p_n\}$ must have a limit point $p_0 \in E$ and a subsequence $\{p_{n_k}\}$ converging to p_0. Since the terms of the sequence $\{q_n\}$ are progressively closer and closer to the terms of the sequence $\{p_n\}$, we would expect $\{q_{n_k}\}$ also to converge to p_0. Indeed, we have

$$\begin{aligned}
|q_{n_k} - p_0| &= |q_{n_k} - p_{n_k} + p_{n_k} - p_0| \\
&< |q_{n_k} - p_{n_k}| + |p_{n_k} - p_0| \\
&< \frac{1}{n_k} + |p_{n_k} - p_0| \to 0
\end{aligned}$$

But, if f is continuous on E, it is convergence preserving and we must have

$$\lim_{k \to \infty} (f(p_{n_k}) - f(q_{n_k})) = f(p_0) - f(p_0) = 0$$

which contradicts the fact that $|f(p_n) - f(q_n)| \geq \varepsilon > 0$ for all n. ∎

This result also has a direct proof which uses the covering property of a compact set to obtain a value of δ that depends on ε, but not upon the location of the points p and q in E (see Exercises 11 and 12).

A standard method for presenting the numerical values of a function of one variable is to tabulate them at a constant interval distance Δx. Thus, a short table of the sine function might list the values of $\sin \theta$ for every choice of θ in degrees between $\theta = 0°$ and $\theta = 90°$, and $\sin \theta$ would be estimated for other choices of θ between these extremes by linear interpolation. It is clear that this process can be described by saying that one starts with a finite number of points on the graph of a given function, and then approximates

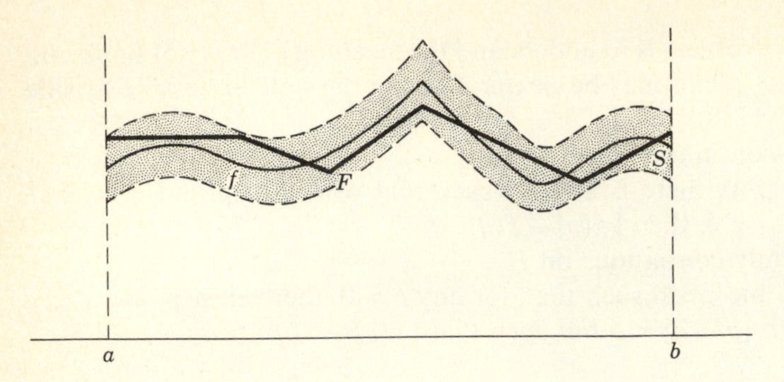

Figure 2-5 $\|f - F\|_{[a,b]} \leq \varepsilon$

the function between these by line segments. A practical problem arises immediately: How close should the tabulated values of x be in order that the interpolated values of function not have more than an assigned error?

In such a problem, we are dealing with two functions, one of which is regarded as an approximation to the other; in the example quoted, a piecewise linear function is used as an approximation to the sine function on the interval $[0°, 90°]$. We need a precise notion of degree of approximation.

Definition 3 *We say that a function F is a uniform ε-approximation to a function f, on a set E, if*

$$|f(p) - F(p)| \leq \varepsilon \qquad \textit{all } p \in E$$

If f and F are functions of one variable and E is an interval $[a, b]$, this has a simple geometric interpretation. Let S be the band of vertical width 2ε which is obtained by moving a vertical line segment of length 2ε along the graph of f, its midpoint being kept on the curve (Fig. 2-5). Then, F is an ε-approximation to f if the graph of F lies completely in the set S. By analogy with the situation in euclidean spaces, one says that F lies in an ε neighborhood of f, and one writes

$$\|f - F\|_E \leq \varepsilon$$

Here $\| \quad \|_E$ is a special notation which is explained by

$$\|g\|_E = \sup\{|g(p)| \text{ with } p \in E\} \tag{2-5}$$

the least upper bound (or maximum if attained) of the values of $|g|$ on E. Hence, $\|F - f\|_E$ is the maximum separation between the graphs of f and F, over the set E. $\|g\|$ is read "the norm of g."

One of the key results in this subject is the Weierstrass approximation theorem, which states that any continuous function of one variable defined on a closed bounded interval $[a, b]$ can be approximated uniformly there, within ε, by a polynomial. We shall not prove this or its generalizations to functions of several variables; instead we shall prove a simpler result dealing with approximation by *piecewise linear functions.* The graph of such a function is a polygonal line, and the result we prove is directly connected with the problem of construction of numerical tables.

Theorem 7 *If f is a continuous function on the closed and bounded interval $[a, b]$, then, for any $\varepsilon > 0$, there is a piecewise linear function F which approximates f uniformly within ε on the interval.*

As the name implies, a piecewise linear function is one whose graph is a polygon, consisting of a finite number of straight-line segments. As ε decreases, the number of segments we have to use to get the function F may have to increase, in order to match the behavior of f. The method of proof is extremely natural. We select points $P_0, P_1, \ldots, P_N$ on the graph of the given function f and then construct F by joining these with line segments. The only trick is to be sure that the points P_k are close enough.

We shall decide upon the size of the number N later. Divide the interval $[a, b]$ into N equal subintervals at points

$$a = x_0 < x_1 < x_2 < \cdots < x_N = b$$

Let P_k be the point (x_k, y_k), where $y_k = f(x_k)$. Notice that P_k is on the graph of f. Now, define the function F on each of the subintervals in turn,

$$F(x) = \frac{(x_{k+1} - x)y_k + (x - x_k)y_{k+1}}{x_{k+1} - x_k} \tag{2-6}$$

for all x with $x_k \le x \le x_{k+1}$. Notice that $F(x)$ is of the form $A + Bx$ on this interval, so that its graph is a straight line. Also, putting $x = x_k$ in (2-6) gives $F(x_k) = y_k = f(x_k)$, while $F(x_{k+1}) = y_{k+1} = f(x_{k+1})$. Thus, this portion of the graph of F goes from P_k to P_{k+1}.

We now wish to estimate the difference between $F(x)$ and $f(x)$ at an arbitrary point x between x_k and x_{k+1}. Clearly, $F(x)$ lies somewhere between y_k and y_{k+1}, since its graph is a straight line. We do not know much about the shape of f between x_k and x_{k+1}, but we know with certainty that $|F(x) - f(x)|$ cannot be larger than the bigger of the numbers $|f(x) - y_k|$ and $|f(x) - y_{k+1}|$. Since y_k and y_{k+1} are the values of f at points in the interval $[x_k, x_{k+1}]$, neither of these numbers in turn can be larger than the biggest separation of values $|f(x) - f(x')|$, for arbitrary placement of x' between x_k and x_{k+1}.

We are now ready to choose the integer N. Given ε, we make use of the uniform continuity of f to choose $\delta > 0$ so that $|f(x) - f(x')| < \varepsilon$ whenever x and x' are two points anywhere on the interval $[a, b]$ with $|x - x'| < \delta$. Choose N large enough so that $(b - a)/N < \delta$. This has the effect of making each of the subintervals shorter than δ. The last remark in the paragraph just above now applies, since the points x and x' there must obey $|x - x'| < \delta$. Hence, we conclude that $|F(x) - f(x)| < \varepsilon$. With this done in each subinterval $[x_k, x_{k+1}]$, it is clear that we have proved this uniformly throughout the entire interval $[a, b]$, and $\|F - f\| < \varepsilon$. ∎

It is interesting to note that the process given in (2-6) for obtaining the values of the approximating function F at points x intermediate between x_k and x_{k+1} is nothing more nor less then ordinary linear interpolation, as it is standardly done in mathematical tables.

For functions of two or more variables, a similar process can be used, obtaining approximate values for a function f by interpolating from the values at a discrete set of points. In two variables, for example, one can use linear interpolation in triangles to replace the formula given in (2-6) (see Exercise 10).

In order to apply these methods to construct uniform approximations to a given function f, one must know a value of δ that is appropriate for a given ε. There is one case in which this step is simple.

Definition 4 *A function f is said to obey a* **Lipschitz** *condition on the set D if there is a constant M such that*

$$|f(p) - f(q)| \leq M|p - q|$$

for every choice of p and q in D.

When this happens, it is clear that f is uniformly continuous on D and that we may choose $\delta = \varepsilon/M$. For, if $|p - q| < \delta$, then $|f(p) - f(q)| \leq M\delta \leq \varepsilon$. In Sec. 3.2 we will see that any function of one variable that has a continuous derivative on an interval $[a, b]$ obeys a Lipschitz condition on that interval. An analogous result will also be proved later for functions of several variables.

EXERCISES

1 Show that $F(x, y) = x^2 + 3y$ is not uniformly continuous on the whole plane.

2 Prove that the function $f(x) = 1/(1 + x^2)$ is uniformly continuous on the whole line.

3 Let f and g each be uniformly continuous on a set E. Show that $f + g$ is uniformly continuous on E.

4 Let A and B be disjoint sets, and let f be continuous on A and continuous on B. When is it continuous on $A \cup B$?

5 Let A and B be disjoint closed sets and suppose that f is uniformly continuous on each.

(*a*) Show that f is necessarily uniformly continuous on $A \cup B$ if A is compact.

(*b*) Show that f need not be uniformly continuous on $A \cup B$ if neither A nor B is compact.

6 If f is uniformly continuous on D, show that it has the property that if p_n, $q_n \in D$ and $|p_n - q_n| \to 0$, then $|f(p_n) - f(q_n)| \to 0$.

7 Let D be a bounded set and let f be uniformly continuous on $D \subset \mathbf{R}^n$. Prove that f is bounded on D.

8 Let f be a function defined on a set E which is such that it can be uniformly approximated within ε on E by functions F that are uniformly continuous on E, for every $\varepsilon > 0$. Show that f must itself be uniformly continuous on E.

9 Using the special notation explained in (2-5), prove that, if f and g are defined on a set E, then

$$\|f + g\|_E \le \|f\|_E + \|g\|_E$$

10 Let p_1, p_2, p_3 be the vertices of a triangle D in the plane.

(*a*) Show that any point p in D can be expressed as

$$p = \alpha_1 p_1 + \alpha_2 p_2 + \alpha_3 p_3 \qquad \text{where } \alpha_i \ge 0 \text{ and } \alpha_1 + \alpha_2 + \alpha_3 = 1$$

(*b*) If f is a function defined on D with Lipschitz constant M, use part (*a*) to define a function F on D by

$$F(p) = \alpha_1 f(p_1) + \alpha_2 f(p_2) + \alpha_3 f(p_3)$$

and show that $\|F - f\|_D \le M \operatorname{diam}(D)$.

Exercises 11 *and* 12 *constitute a different proof of Theorem* 6.

11 Let f be continuous on the interval $[a, b]$. Given $\varepsilon > 0$ and a point t in the interval, choose $\rho = \rho(t)$ so that, if $|x - t| < \rho$, then $|f(x) - f(t)| < \varepsilon$. Let U_t be the symmetric interval centered on t of radius $\frac{1}{2}\rho(t)$. Show that there are points $t_1, t_2, \cdots, t_m$ such that the sets U_{t_j} together cover the interval $[a, b]$.

12 (*Continuation of Exercise* 11.) Let ρ_0 be the smallest of the numbers $\rho(t_j)$, and let x' and x'' be any two points of the interval $[a, b]$ with $|x' - x''| < \frac{1}{2}\rho_0$. Show that there must be one of the points t_j such that $|x' - t_j| < \rho(t_j)$ and $|x'' - t_j| < \rho(t_j)$.

Conclude that $|f(x') - f(x'')| < 2\varepsilon$ and hence that f is uniformly continuous on $[a, b]$.

2.4 IMPLICATIONS OF CONTINUITY

Many properties of a continuous function—e.g., that it is bounded and attains a maximum—seem intuitively obvious and to require no formal proof, if one thinks only of those smooth functions whose graphs are easy to sketch with a pencil. However, there are continuous functions whose graphs are so complicated that they have tangent lines nowhere; an example will be discussed in Sec. 6.2. Moreover, if one considers any function of three variables, its graph becomes a set in 4-space, and there intuitive arguments become much less convincing. It is therefore important to see that these useful properties which hold for every continuous function do in fact follow logically from the definition of continuity and do not depend upon intuitive geometric reasoning.

It is helpful to introduce the words "local" and "global" to contrast two types of situations that frequently arise. If we are considering a given set D, then we say that any specific property holds **locally** at $p_0 \in D$ if it is true at p_0 and at all points p sufficiently near p_0; thus, there will be an open ball B about p_0 and the property will hold for all $p \in B \cap D$. On the other hand, a property that holds at all points in D is said to hold **globally** in D.

Here is a useful example of this viewpoint.

Theorem 8 *Let f be a real-valued function defined and continuous on a set D in n space. If $p_0 \in D$ and $f(p_0) > 0$, then f is locally positive at p_0. Indeed, there must be a neighborhood U about p_0 and an $\varepsilon > 0$ such that $f(p) > \varepsilon$ for all $p \in U \cap D$.*

Set $\varepsilon = (\frac{1}{2})f(p_0)$. Since f is continuous at p_0, choose U so that $|f(p) - f(p_0)| < \varepsilon$ for all $p \in U \cap D$. Accordingly, $f(p) - f(p_0) > -\varepsilon$, and hence

$$f(p) > f(p_0) - \varepsilon = 2\varepsilon - \varepsilon = \varepsilon \quad \blacksquare$$

This result is a special case of the following more general theorem; to see the connection, take $c = 0$, and observe that p_0 will lie in an open set and must be interior to it.

Theorem 9 *Let f be continuous on a set D, and let c be any number; then, the set R of points $p \in D$ for which $f(p) > c$ is an open set, relative to D, and the set G of points p where $f(p) = c$ is a closed set, relative to D.*

This is an immediate consequence of Theorem 3. Let V be the open unbounded interval consisting of those numbers y with $y > c$. Since f is continuous, the inverse image $f^{-1}(V)$ of V must be open, relative to D. Clearly, $R = f^{-1}(V)$. The second part is treated in the same way by observing that G is the inverse image of the closed set $\{c\}$. $\blacksquare$

This result also explains the fact, noted earlier, that formulas such as

$$S = \{\text{all } (x, y) \text{ with } x^2 - 3xy^3 + y^5 > 1\}$$

always describe open sets, and formulas such as

$$T = \{\text{all } (x, y) \text{ with } x^5y^2 + 7x^2y^4 = 7\}$$

always describe closed sets. (See also Exercises 1 and 2.)

A function f is said to be **bounded** on a set S if the image $f(S)$ is a bounded set. There must then be a number M such that $|f(p)| \le M$ for all $p \in S$. A function can be continuous on a set without being bounded there. Examples are $f(x) = 1/x$ on the open interval $0 < x < 1$ and $g(x) = x^2$ on the closed interval $0 \le x < \infty$.

Theorem 10 *Let f be a real-valued function defined and continuous on a compact set D. Then, f is bounded on D.*

Let us first prove that any continuous function is locally bounded on its domain. Given $p_0 \in D$, take $\varepsilon = 1$ and use the assumed continuity of f to choose a neighborhood U about p_0 so that $|f(p) - f(p_0)| < 1$ for all $p \in U \cap D$. Clearly, we then have $|f(p)| < 1 + |f(p_0)|$ for all such p, and f is locally bounded at p_0. We now use the compactness of D to obtain a single bound for $|f(p)|$ that works for all p in D. The fact that f is locally bounded everywhere in D shows that given any $p \in D$, there is an open set $\mathcal{O}_p$ about p and a number M_p with $|f(q)| < M_p$ for all points $q \in \mathcal{O}_p$. The sets $\mathcal{O}_p$ form an open covering of D. Since D is compact, this covering can be reduced to a finite covering by sets $\mathcal{O}_{p_k}$ for $k = 1, 2, \ldots, m$. In each of these sets, f is bounded, and every point q in D lies in one of these sets. Thus, if M is the largest of the numbers M_{p_k}, we have $|f(q)| < M$ for all $q \in D$. ▮

The conclusion of this theorem can also be stated in the useful form: *If f is continuous on the compact set D, then $f(D)$ is a bounded set.* In this form, the proof given above also holds if f is a function taking values in m space. (See also the statement of Theorem 13 below.)

A real-valued function can be bounded and continuous on a set without attaining a maximum value on the set. This is true, for example, of the function $f(x) = x^2$ on the open interval $0 < x < 1$ and on the unbounded closed interval $0 \le x < \infty$ for the function $f(x) = x/(1 + x)$. Again, the situation is different for compact sets.

Theorem 11 *If S is compact, and f is continuous on S, then f takes a maximum and a minimum value somewhere on S.*

By the previous theorem, f is bounded on S. There then exist numbers B and b, with $b \le f(p) \le B$, for all $p \in S$. Let M be the smallest such upper bound for the values of f on S and m the largest lower bound. We must have $m \le f(p) \le M$ for all $p \in S$ as before, but now M cannot be decreased, nor m increased. If there is a point p_0 in S with $f(p_0) = M$, then M is the maximum value for f on S, and it is attained in S. If there is no point p_0 with $f(p_0) = M$, then $f(p) < M$ for all $p \in S$. In this case, set $g(p) = 1/(M - f(p))$. Since the denominator is never 0, g is continuous on S and must therefore be bounded there. If we suppose that $g(p) \le A$ for all $p \in S$, then we would have

$$\frac{1}{M - f(p)} \le A \qquad \text{all } p \in S$$

and therefore

$$f(p) \le M - \frac{1}{A}$$

This, however, contradicts the fact that M was the smallest upper bound for the values of f on S, and we conclude that this case does not occur and there is a point $p_0 \in S$ with $f(p_0) = M$. In a similar fashion, one may show that there is a point in S where f has the value m, and a minimum is attained. ▮

There is an alternative approach to Theorems 10 and 11 which makes their meaning a little clearer, and also permits us to generalize them immediately for vector-valued functions and transformations. While such functions will be studied more fully in later chapters, it seems appropriate to mention these generalizations here.

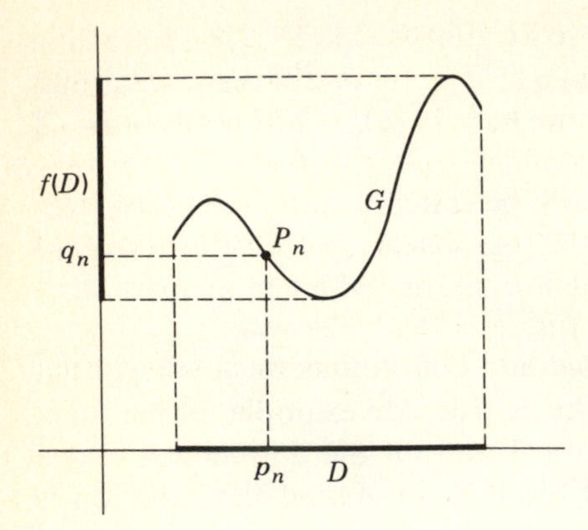

Figure 2-6

The key idea is to look at functions from a more geometric viewpoint, and first obtain the following useful result.

Theorem 12 *Let f be a function defined on a compact set D in n space and taking values in m space, and let G be its graph. Then, f is continuous on D if and only if G is compact.*

While the proof is valid for a function f from n space into m space, we suggest that it is easier to follow in terms of the usual picture for a real-valued function of one variable defined on an interval (see Fig. 2-6). In general, the graph of f is the set

$$G = \{\text{all } P = (p, q) \text{ where } q = f(p) \text{ and } p \in D\}$$

First, suppose that f is continuous and prove that G is compact. Since D is compact, it is bounded, and there is a number B such that $|p| < B$ for all $p \in D$. Likewise, by Theorem 10, $f(D)$ is bounded, and for some number M we have $|f(p)| < M$ for all $p \in D$. Accordingly, if P is any point in G.

$$\begin{aligned} |P| = |(p, q)| &= |(p, 0) + (0, q)| \le |(p, 0)| + |(0, q)| \\ &\le |p| + |q| = |p| + |f(p)| \le B + M \end{aligned}$$

Thus, the graph G is a bounded set.

To show that G is also closed, let $P_n = (p_n, q_n)$ be any sequence of points in G, and suppose that it converges to a point (p, q). It is then true that $\lim_{n\to\infty} p_n = p$ and $\lim_{n\to\infty} q_n = q$. To prove G closed, we need only show that the point (p, q) is in G. But, since $\{p_n\}$ lies in D, p lies in the closure of D; but D, being compact, is closed, so $p \in D$. Since (p_n, q_n) is in the graph of f, $q_n = f(p_n)$, and since f is continuous on D, $\lim_{n\to\infty} f(p_n) = f(p)$, which may be restated as $\lim_{n\to\infty} q_n = f(p)$. However, $\lim_{n\to\infty} q_n = q$, so that $q = f(p)$, showing that $(p, q) \in G$.

To prove the other half of Theorem 12, we assume that G is compact and show that f must be continuous. Suppose that this were false. Then, there would be some point p_0 in D, and a sequence $\{p_n\}$ in D with $p_n \to p_0$, and an $\varepsilon > 0$ such that $|f(p_n) - f(p_0)| > \varepsilon$ for all n. Put $q_n = f(p_n)$ and consider the points $p_n = (p_n, q_n)$ in G, for $n = 1, 2, \ldots$. Note that $p_n \to p_0 \in D$, but that $|q_n - f(p_0)| > \varepsilon$. Because G is compact, the sequence $\{P_n\}$ must have a subsequence $\{P_{n_k}\}$ that converges to some point (p, q) in G. Accordingly, $\lim_{n\to\infty} p_{n_k} = p$ and $\lim_{n\to\infty} q_{n_k} = q$. However, since $\{p_{n_k}\}$ is a subsequence of $\{p_n\}$, which itself converges to p_0, we have $p = p_0$. Because the point (p, q) is in G, which is the graph of f, $q = f(p) = f(p_0)$. However, $\lim_{n\to\infty} q_{n_k} = f(p_0)$ and $|q_n - f(p_0)| > \varepsilon$ are not both possible; we conclude that f must have been continuous on D. ▮

In connection with this result, Exercise 4 shows that Theorem 12 does not hold if "G is compact" is replaced by "G is closed."

As our first application of this theorem, we obtain the following.

Theorem 13 *If f is continuous on the compact set D, then $f(D)$ is also a compact set.*

Because f is continuous and D is compact, the graph of f is a compact set G. The set $f(D)$ is the projection of the set G (see Fig. 2-6). By Exercise 9, Sec. 1.8, the projection of a compact set is compact. Hence $f(D)$ is compact. ▮

This result contains both Theorem 10 and Theorem 11. Since $f(D)$ is compact, $f(D)$ is closed and bounded. The latter shows that there must exist a number M with $|f(p)| < M$ for all $p \in D$. The latter, together with the fact that the sup and inf of any set of real numbers belong to the closure of that set, shows that $f(p)$ must achieve the values

$$\text{Maximum of } f(p) = \sup\,(f(D))$$

$$\text{Minimum of } f(p) = \inf\,(f(D))$$

Another application of the compact graph theorem will be given in Sec. 2.6, in connection with the problem of extending the definition of a function defined on an open set $\mathcal{O}$ to the boundary of $\mathcal{O}$ in such a way that it remains continuous there.

We now turn to what is usually called the **intermediate value theorem.** We state this for real-valued functions.

Theorem 14 *Let S be a connected set, and let f be continuous on S. Let the numbers a and b be any two values of f on S, and suppose that $a < c < b$. Then, there is a point $p \in S$ with $f(p) = c$.*

The one-variable form of this, with S an interval, is especially plausible. Here, it states that if the graph of f lies below the line $y = c$ at one point in the interval, and above the line at another, then it must intersect the line at some intervening point. The proof of the general theorem is based upon Theorem 9 and the definition of a connected set. Suppose that $f(p)$ is never c, for any point $p \in S$. Then, we always have either $f(p) > c$ or $f(p) < c$. Let U be the set of $p \in S$ with $f(p) > c$, and V the set where $f(p) < c$. These sets together must cover S. Since f is continuous, both U and V are open, relative to S. Moreover, since there is a point $p_1 \in S$ with $f(p_1) = a$, the set V contains p_1 and is therefore not empty; likewise, U is not empty. The existence of two such sets contradicts the assumption that S was connected; so we know that, somewhere in S, f must take the value c. ▌

This proof was again indirect; the argument proves that there must be a solution to the equation $f(p) = c$, but it gives no help in finding a solution. One reason to study such proofs in detail is to see if ways can be found to make them constructive, so that one has an algorithm for locating a point p in S with $f(p) = c$. Another approach to this result is found in Exercise 15, which will be recognized as a familiar process for locating the roots of an polynomial equation.

This theorem too has a more general form which applies to continuous functions on n space with values in m space.

Theorem 15 *Let f be continuous on D, and let D be connected. Then, the image set $f(D)$ is also connected.*

Suppose that $f(D) = A \cup B$, where A and B are mutually separated. Since no point of A is arbitrarily near a point of B, A is open relative to $f(D)$. So is B. Since f is continuous, $f^{-1}(A)$ and $f^{-1}(B)$ must be open, relative to D. Hence, $D = f^{-1}(A) \cup f^{-1}(B)$, where these sets are disjoint and mutually separated. However, since D was assumed connected, one of these sets must be empty. If $f^{-1}(A)$ is empty, A is empty. A similar observation applies to $f^{-1}(B)$, and it follows that $f(D)$ is not disconnected, and must be connected. ▌

This general result also contains the usual intermediate value theorem (Theorem 14). We see this as follows. Suppose that f is a real-valued function defined on a set D in n space, and that f takes the values a and b somewhere in D. Let c be a number with $a < c < b$. If D is connected and f continuous, $f(D)$ is a connected subset of $\mathbf{R}$, and must therefore be an interval; but since $f(D)$ contains the points a and b, it must also contain c. Thus, f takes the value c somewhere in D.

We can also use Theorem 15 to clarify our understanding of connectedness. In Sec. 1.5, we discussed two intuitive notions of this concept before defining "connected." We can now make the notion of "pathwise connected" precise, and also prove the converse of Theorem 2, Sec. 1.5.

Definition 5 *A set S is pathwise connected if every pair of points p, q in S can be joined by a continuous path γ lying entirely in S. Specifically, this requires that there be a continuous function $\gamma(t)$, defined for $0 \le t \le 1$ such that $\gamma(0) = p$ and $\gamma(1) = q$ and such that $\gamma(t) \in S$ for all t.*

We see that "polygon connected" is merely a special case of this, arising when the function $\gamma(t)$ is especially simple.

Theorem 16 *Any pathwise connected set S is connected.*

Suppose that $S = A \cup B$, where A and B are mutually separated and nonempty. Choose $p \in A$ and $q \in B$, and then join p and q in S by a continuous path γ. The trace of γ is the set $\gamma(I)$, where I is the interval [0, 1]. Since I is connected, so is $\gamma(I)$. Set $A_0 = A \cap \gamma(I)$ and $B_0 = B \cap \gamma(I)$. Since these are subsets of A and B, respectively, A_0 and B_0 are also mutually separated. Moreover, p and q lie in $\gamma(I)$, so that A_0 contains p and B_0 contains q and neither set is empty. But $\gamma(I) = A_0 \cup B_0$, contradicting the fact that $\gamma(I)$ is connected. ▮

The intermediate value theorem lies at the heart of many "intuitively obvious" mathematical facts dealing with the behavior of quantities that "change continuously." We have given a number of these as exercises; in several which involve physical or geometrical quantities, you should assume that the appropriate function is continuous.

We give two illustrations now that are less recreational.

Theorem 17 *No continuous function can map the open unit square S onto the interval [0, 1] in a one-to-one fashion, although this is possible for discontinuous functions.*

Suppose that ϕ is a continuous real-valued function defined on S. We show that there must be two points $q_1 \ne q_2$ in S with $\phi(q_1) = \phi(q_2)$. We may assume that we have p_1 and p_2 in S with $\phi(p_1) < \phi(p_2)$, for otherwise we are done. Choose c with $\phi(p_1) < c < \phi(p_2)$ and take two different paths α and β in S which go from p_1 to p_2. Then, on each there must be a point q where $\phi(q) = c$. Calling these points q_1 and q_2 we clearly have $q_1 \ne q_2$ and $\phi(q_1) = \phi(q_2)$. ▮

There is a standard example of a discontinuous function ϕ that is a 1-to-1 map of S into the unit interval. Given any point $(x, y) \in S$, write each coordinate in decimal form:

$$x = .x_1 x_2 x_3 \cdots$$

$$y = .y_1 y_2 y_3 \cdots$$

and then define ϕ by

$$\phi(x, y) = .x_1 y_1 x_2 y_2 x_3 y_3 x_4 \cdots$$

If one is careful to avoid decimal representations that terminate in an endless sequence of 9s, ϕ can be seen to be a real-valued function, defined for $0 \le x < 1$, $0 \le y < 1$, which takes distinct values at every distinct pair of points (x, y). It is also easily seen to be discontinuous.

Our second illustration is a familiar observation from elementary calculus, used there frequently but without proof.

Theorem 18 *Let f be a real-valued continuous function defined on the interval $I = [a, b]$, and suppose that f is 1-to-1 on I. Then, f is strictly monotonic on I.*

The hypothesis means that if x_1 and x_2 are points in I with $x_1 \ne x_2$, then $f(x_1) \ne f(x_2)$. Clearly, we have either $f(x_1) < f(x_2)$ or the reverse inequality; we show that this inequality always holds in the same direction, determined by the values of f at the endpoints of $[a, b]$. We may suppose that $f(a) < f(b)$. Take any x with $a < x < b$. If $f(x) > f(b)$, there must be t with $a < t < x$ such that $f(t) = f(b)$. If $f(x) < f(a)$, there must be s with $x < s < b$ and $f(s) = f(a)$. We conclude that $f(a) < f(x) < f(b)$ for all x between a and b. Now consider x_1 and x_2 with $a < x_1 < x_2 < b$, and suppose that $f(x_1) > f(x_2)$. Then, there must exist s with $x_2 < s < b$ and $f(s) = f(x_1)$. We conclude that $f(x_1) < f(x_2)$ for every such pair of points, and f is strictly increasing on I. (If we had started with $f(a) > f(b)$, f would have been strictly decreasing.) ▌

EXERCISES

1 Show that if f is continuous on D, then the set of points p where $f(p) \le C$ is closed relative to D.

2 Let $S = \left\{\text{all } (x, y) \text{ with } 3 - \dfrac{x}{y} - \dfrac{y}{x} \le 1\right\}$.

(*a*) Is S a closed set in the plane?
(*b*) Does this conflict with Exercise 1?

3 Let $F(x, y)$ be a polynomial in x and y, and let

$$A = \{\text{all } (x, y),\ F(x, y) \ge 0\}$$

$$B = \{\text{all } (x, y),\ F(x, y) = 0\}$$

(*a*) Show that bdy $(A) \subset B$.
(*b*) Is it always true that B is exactly the boundary of A?

***4** Show by an example that the graph of a function defined on the interval $0 \le x \le 1$ can be a closed set, without the function f being continuous.

5 Let f and g be continuous on the interval $[0, 1]$, and suppose that $f(x) = g(x)$ for every rational number $x = a/b$ in this interval. Prove that $f = g$.

6 Use the intermediate value theorem to prove that any polynomial of odd degree with real coefficients has at least one real root.

7 Let f and g be continuous on $[0, 1]$ and suppose that $f(0) < g(0)$ and $f(1) > g(1)$. Prove that there is a point x, $0 < x < 1$, with $f(x) = g(x)$. Can you illustrate this by a picture?

8 Show that any function that is locally constant on an open connected set D is in fact constant on D.

9 Prove that a set S is disconnected if and only if there is a real-valued function f that is continuous on S but takes only the values 2 and 3 on S.

10 Give a mathematical argument to show that a heated wire in the shape of a circle must always have two diametrically opposite points with the same temperature.

11 Give a mathematical argument to show that any compact convex set can be divided into four subsets of the same area by two perpendicular cuts.

12 Show that any real-valued function defined on the set consisting of the nonnegative X, Y, and Z axes must take the same value at least twice.

13 Five line segments meet at a point. Show that any continuous real-valued function defined on this set must take the same value three times.

14 (*a*) Show that any heated tetrahedron must have three points located on its edges or vertices that have the same temperature.

(*b*) Can you prove there must actually be four such points with the same temperature?

15 Suppose that f is continuous on $[a, b]$ and that $f(a)f(b) < 0$. Prove Theorem 14 by filling in the details of the following argument.

(*a*) Apply the process of repeated bisection to construct two sequences $\{a_n\}$ and $\{b_n\}$ such that $a \le a_n < b_n \le b$, with $b_n - a_n \to 0$, and $f(a_n)f(b_n) \le 0$.

(*b*) Show that $\lim_{n\to\infty} a_n = c$ where $a < c < b$ and $f(c) = 0$.

16 Let $F(x, y) = (x - y)^2$. Then show that $\max_{0\le x\le 1} \min_{0\le y\le 1} F(x, y) = 0$ and $\min_{0\le y\le 1} \max_{0\le x\le 1} F(x, y) = \frac{1}{4}$.

17 Let $F(x, y)$ be continuous on the square

$$S = \{\text{all } (x, y),\ |x| \le 1,\ |y| \le 1\}$$

Let $\max_{|x|\le 1} \min_{|y|\le 1} F(x, y) = A$ and $\min_{|y|\le 1} \max_{|x|\le 1} F(x, y) = B$. Prove that $A \le B$ is always true.

18 Suppose that f is a complex-valued continuous function defined for $0 \le t \le 1$. Suppose that $f(0) = -1$ and $f(1) = 1$. Does there have to be a value of t with $f(t) = 0$? Explain.

2.5 LIMITS OF FUNCTIONS

The concept of continuity was introduced in this chapter by comparing the value of a function f at a point x_0 with the values which f takes on a small interval about x_0. In this section, we examine the notion of convergence for functions which is implicit in this and which is analogous to the notion of convergence for sequences. The concept of the limiting value for a function at a point x_0 is particularly important when x_0 is a point at which f has not been defined, or where f is not continuous.

We shall discuss functions of one variable first and treat the general case later. We assume for the present that functions f are defined on some neighborhood of a point $x_0 = b$; whether or not f is defined at b itself turns out to be irrelevant for the discussion of limits.

Definition 6 *We write* $\lim_{x \to b} f(x) = L$ *if, corresponding to any* $\varepsilon > 0$, *there is a number* δ *such that*

$$|f(x) - L| < \varepsilon \qquad \text{whenever } 0 < |x - b| < \delta \tag{2-7}$$

Note that the last inequality rules out the choice $x = b$. It is customary to use the term "**deleted neighborhood of** p" to describe the set obtained by taking a neighborhood of p and removing p from it. Using this terminology, this definition could be rephrased as: $\lim_{x \to b} f(x) = L$ if, given any $\varepsilon > 0$, there is a deleted neighborhood γ of b such that (2-7) holds for all x in γ.

It should also be pointed out that there is a strong analogy between expressions involving the notation "$x \to b$" and the notation "$n \to \infty$," visible in both definitions and theorems; thus, compare "$\lim_{n \to \infty} a_n$" with "$\lim_{x \to b} f(x)$," and refer back to the appropriate definitions, noting also that just as we have $x \neq b$, we have $n \neq \infty$. Further instances of this analogy will appear later.

As an illustration of the calculation of a limit, let f be described by

$$f(x) = \begin{cases} \dfrac{x^3 - 1}{x - 1} & x \neq 1 \\ 2 & x = 1 \end{cases}$$

and consider $\lim_{x \to 1} f(x)$. Computation gives

$$f(1.1) = 3.31 \qquad f(1.01) = 3.0301$$

and we are led to guess that $\lim_{x \to 1} f(x) = 3$. We can check this by the definition. We must estimate the difference $|f(x) - 3|$ for x near 1. When $x \neq 1$, we have

$$\begin{aligned} f(x) - 3 &= \frac{x^3 - 1}{x - 1} - 3 = \frac{x^3 - 3x + 2}{x - 1} \\ &= x^2 + x - 2 = (x - 1)(x + 2) \end{aligned}$$

When $x = 1, f(x) - 3 = f(1) - 3 = 2 - 3 = -1$. However, our aim is to make $f(x) - 3$ small whenever $0 < |x - 1| < \delta$, and this explicitly rules out $x = 1$; on this deleted neighborhood of 1, we have

$$|f(x) - 3| = |x - 1||x + 2|$$

If we again agree to use numbers δ smaller than 1, then the points x will be confined to the interval $[0, 2]$. For such x, $|x + 2| \leq 4$, and we have

$$|f(x) - 3| \leq |x - 1|4 < 4\delta$$

Given $\varepsilon > 0$, choose δ as $\varepsilon/4$; then, $|f(x) - 3| < \varepsilon$ whenever $0 < |x - 1| < \delta$, and $\lim_{x \to 1} f(x) = 3$.

Of course, if the function f is known to be continuous at b, no such work is necessary, for it is evident from the definition of continuity that we then have $\lim_{x \to b} f(x) = f(b)$. Thus, for continuous functions, calculation of limits is easy; one need merely substitute.

We could have used this fact to simplify the calculations above which were used to find the limit of this specific function. We make the simple observation that $x^3 - 1$ can be factored as $(x - 1)(x^2 + x + 1)$, so that we can write

$$f(x) = x^2 + x + 1 \qquad x \neq 1$$

In other words, there is a function P given by $P(x) = x^2 + x + 1$ (for all x) such that $f(x) = P(x)$ for all $x \neq 1$. Since f and P agree on a deleted neighborhood of $x = 1$, and since these are the only values of f which are used in evaluating $\lim_{x \to 1} f(x)$, we must have $\lim_{x \to 1} f(x) = \lim_{x \to 1} P(x)$. Clearly, P is continuous everywhere, being a polynomial, so that $\lim_{x \to 1} P(x) = P(1) = 3$. This trick of replacing a function f by another function P which agrees with it near a point b, but which is known to be continuous at b, makes the calculation of certain limits easy.

The theorems which were obtained in Sec. 1.6 for convergence of sequences (Theorems 8 to 12) each have their analog for limits of functions.

Theorem 19 *Assuming that f and g are each defined on a deleted neighborhood of $x = b$, and that $\lim_{x \to b} f(x) = A$ and $\lim_{x \to b} g(x) = B$, then it is true that*

$$\lim_{x \to b} (f(x) + g(x)) = A + B$$

$$\lim_{x \to b} f(x)g(x) = AB$$

and if $B \neq 0$

$$\lim_{x \to b} \frac{f(x)}{g(x)} = \frac{A}{B}$$

We postpone proofs of these until we take up the comparable results for functions of several variables.

As an illustration, to find

$$\lim_{x \to 2} \frac{2x^3 + 5x^2 - 8x - 20}{x^3 - 8} = L$$

we may write this function as

$$\frac{(x^2 - 4)(2x + 5)}{(x - 2)(x^2 + 2x + 4)} = \frac{(x - 2)(x + 2)(2x + 5)}{(x - 2)(x^2 + 2x + 4)}$$

and note that if the factor $(x - 2)$ is removed, the result is continuous at $x = 2$, arriving at

$$L = \lim_{x \to 2} \frac{(x + 2)(2x + 5)}{x^2 + 2x + 4} = \frac{(4)(9)}{12} = 3$$

In Sec. 3.2, we discuss L'Hospital's rule, often used in evaluating such limits; until then, we depend on other approaches.

The limit notation is used to describe certain types of behavior for functions. Another possible behavior is indicated by the expression $\lim_{x \to b} f(x) = \infty$. A word of caution: the symbol "∞" as used here is neither a number nor a point on the line. It is merely part of the expression and has no meaning out of context. A test of this is the fact that "∞" does not occur in the formal explanation of the whole expression which follows.

Definition 7 *We write* $\lim_{x \to b} f(x) = \infty$ *if and only if, corresponding to any positive number B, there is a number* $\delta > 0$ *such that* $f(x) > B$ *whenever x satisfies* $0 < |x - b| < \delta$.

As illustrations, we would write $\lim_{x \to 1} (x - 1)^{-2} = \infty$ and

$$\lim_{x \to 0} \frac{1}{\sin (x^2)} = \infty$$

We would *not* write $\lim_{x \to 0} 1/x = \infty$, since the function involved behaves differently when x is negative and near 0 from what it does when x is positive. However, we could write $\lim_{x \to 0} 1/|x| = \infty$.

Another useful modification of the limit notation is the following, which uses the symbol "∞" in a different way.

Definition 8 *We write*

$$\lim_{x \uparrow \infty} f(x) = L$$

whenever f is defined on some unbounded interval such as $0 < x < \infty$ *and, corresponding to any* $\varepsilon > 0$, *there is a number* x_0 *such that* $|f(x) - L| < \varepsilon$ *whenever* $x > x_0$.

For example, $\lim_{x \uparrow \infty} 1/x^2 = 0$, $\lim_{x \uparrow \infty} x/(x - 10) = 1$, while $\lim_{x \uparrow \infty} \sin x$ and $\lim_{x \uparrow \infty} 1/(x \sin x)$ do not exist.

These can be combined and modified in other useful ways. Without giving formal definitions for them, we illustrate several possibilities. We would write

$$\lim_{x \downarrow -\infty} e^x = 0 \qquad \lim_{x \uparrow \infty} e^x = \infty$$

$$\lim_{x \to 0} \log |x| = -\infty \qquad \lim_{x \uparrow \infty} (x \sin x + 2x) = \infty$$

but would not write

$$\lim_{x \uparrow \infty} x \sin x = \infty$$

$$\lim_{x \uparrow \infty} (x \sin x + x) = \infty$$

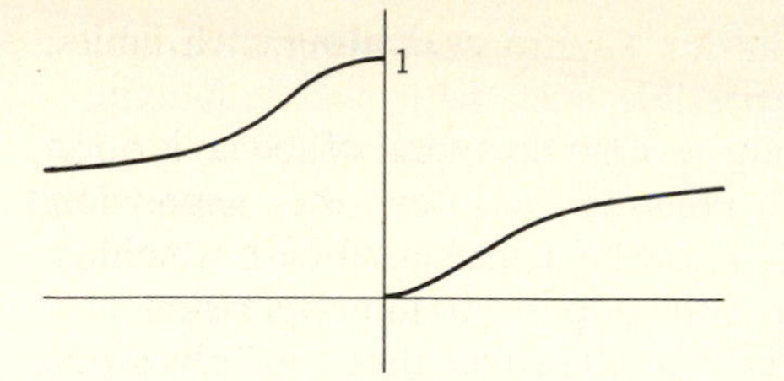

Figure 2-7 $f(x) = \dfrac{1}{1 + e^{1/x}}$

One additional refinement of considerable usefulness is the notion of a one-sided limit.

Definition 9 *We write*

$$\lim_{x \uparrow b} f(x) = L$$

if and only if, corresponding to a given $\varepsilon > 0$, *there is a number* $\delta > 0$ *such that* $|f(x) - L| < \varepsilon$ *whenever* $x < b$ *and* $|x - b| < \delta$, *that is, whenever* $b - \delta < x < b$.

This is often called the **left-hand limit** of $f(x)$ at b, or the limit of $f(x)$ as x approaches b from below. A **right-hand limit** is defined in a similar fashion. Both may exist when the usual two-sided limit does not, and the two-sided limit exists when and only when both left- and right-hand limits exist and are the same. The function described by

$$f(x) = \frac{1}{1 + e^{1/x}}$$

has left-hand limit 1 and right-hand limit 0 at the origin (see Fig. 2-7). Again, the one-sided limit notation can be combined with the other conventions already introduced. For example, the type of behavior exhibited by $g(x) = 1/x$ at the origin can now be described by writing

$$\lim_{x \downarrow 0} g(x) = \infty \qquad \lim_{x \uparrow 0} g(x) = -\infty$$

There is also an analog for the theorem about the convergence of bounded monotonic sequences. Recall that a function f is said to be bounded on a set S if there is a number B with $|f(p)| < B$ for all $p \in S$. For functions of one variable, we also speak of monotonic functions; f is **increasing** on a set E if $f(x_1) \le f(x_2)$ whenever x_1 and x_2 belong to E and $x_1 < x_2$ and **decreasing** if, under the same circumstances, we have $f(x_1) \ge f(x_2)$. We say that f is **monotonic** on E if f is either increasing on E or decreasing on E.

Theorem 20 *If* f *is bounded and monotonic on the open interval* $a < x < b$, *then* $\lim_{x \uparrow b} f(x)$ *and* $\lim_{x \downarrow a} f(x)$ *exist.*

We now turn to the study of limits for functions of several variables, where much more diverse behavior is possible even with simple functions. This is to be expected, since there are only two modes of approach to a point b on the line but many approaches to a point in the plane or in space. The general case is covered by the following: Let S be a set on which f is defined, and let p_0 be a cluster point for S. For simplicity, we assume that p_0 is not itself in S. Thus, p_0 is a boundary point of S. If S contains a deleted neighborhood of p_0, the reference to S in the statement below is usually omitted.

Definition 10 *We say that $f(p)$ converges to L as p approaches p_0 in S, written*

$$\lim_{p \to p_0} f(p) = L \qquad [p \in S]$$

if and only if, corresponding to any $\varepsilon > 0$ a number $\delta > 0$ can be found such that $|f(p) - L| < \varepsilon$ whenever $0 < |p - p_0| < \delta$ and $p \in S$.

An important special case of this arises when S is a line segment or an arc (curve) having p_0 as an endpoint. In these cases, the limit of $f(p)$ as p approaches p_0 from S reduces essentially to the computation of the limit of a function of one variable. For, let the arc be given by parametric equations: $x = \phi(t)$, $y = \psi(t)$, with $0 \le t \le 1$, and such that $\lim_{t \downarrow 0} \phi(t) = x_0$, $\lim_{t \downarrow 0} \psi(t) = y_0$, the coordinates of p_0. Then, setting $g(t) = f(\phi(t), \psi(t))$, we see that $\lim_{p \to p_0} f(p)$, $[p \in S]$, is exactly $\lim_{t \downarrow 0} g(t)$. As an illustration, the limit of $f(x, y)$ as (x, y) approaches the origin along the horizontal axis from the right becomes $\lim_{x \downarrow 0} f(x, 0)$, while the limit along the vertical axis from below is $\lim_{y \uparrow 0} f(0, y) = \lim_{t \downarrow 0} f(0, -t)$. If (x, y) approaches the origin along the ray of slope 1, we obtain $\lim_{t \downarrow 0} f(t, t)$.

The following simple result is often quite useful in discussing the existence of limits. It can be considered as an extension of Theorem 1, which asserted that continuous functions were convergence preserving.

Theorem 21 *If f is defined at all points of a neighborhood of p_0, except possibly at p_0 itself, and $\lim_{p \to p_0} f(p) = L$, then the limit of $f(p)$ exists for p approaching p_0 in any set S, and the limit is always L.*

The usual algebraic theorems for limits hold, either for approach in a deleted neighborhood of p_0 or in a general set S; for simplicity, we state only the first.

Theorem 22 *If $\lim_{p \to p_0} f(p) = A$ and $\lim_{p \to p_0} g(p) = B$, then*

$$\lim_{p \to p_0} (f(p) + g(p)) = A + B$$

$$\lim_{p \to p_0} f(p)g(p) = AB$$

and, if $B \neq 0$, $$\lim_{p \to p_0} \frac{f(p)}{g(p)} = \frac{A}{B}$$

The proofs of these can be patterned on those of the comparable theorems for limits of sequences. Alternatively, for example, the second can be obtained by observing that

$$f(p)g(p) - AB = (f(p) - A)(g(p) - B) + A(g(p) - B) + B(f(p) - A)$$

If $\mathscr{V}$ is a deleted neighborhood of p_0 on which $|f(p) - A| < \varepsilon$ and $|g(p) - B| < \varepsilon$, then on $\mathscr{V}$,

$$|f(p)g(p) - AB| < \varepsilon^2 + |A|\varepsilon + |B|\varepsilon$$

Since A and B are constants, it is clear that the number $\varepsilon^2 + |A|\varepsilon + |B|\varepsilon$ can be made arbitrarily small by choosing ε sufficiently small. Thus, $f(p)g(p)$ is seen to be arbitrarily near AB for p in $\mathscr{V}$, if $\mathscr{V}$ is chosen small enough.

A similar approach can be used for the limit of a quotient, starting from the identity

$$\frac{f(p)}{g(p)} - \frac{A}{B} = \frac{B(f(p) - A) - A(g(p) - B)}{Bg(p)}$$

and using the fact that the deleted neighborhood $\mathscr{V}$ about p_0 can be chosen so that, in addition, $|g(p)| > |B|/2$ for $p \in \mathscr{V}$; accordingly, one has

$$\left|\frac{f(p)}{g(p)} - \frac{A}{B}\right| \leq \frac{|B|\varepsilon + |A|\varepsilon}{|B|^2/2} \quad \blacksquare$$

It will be noticed that we do not state any result dealing with the limit of the composition $f(g(x))$ of two functions, parallel to the theorem on the continuity of $f(g(x))$. The absence of this is explained by Exercise 1.

The Cauchy convergence criterion for sequences does have an analog for functions.

Theorem 23 *Suppose that for any* $\varepsilon > 0$, *a deleted neighborhood* $\mathscr{V}$ *about* p_0 *can be chosen so that* $|f(p) - f(q)| < \varepsilon$ *for every choice of* p *and* q *in* $\mathscr{V}$. *Then* $\lim_{p \to p_0} f(p)$ *exists.*

We give a proof of this which throws the argument back onto the sequence case. Because of the hypothesis on f, we can choose a sequence δ_n, decreasing to 0, such that if p and q are any points with $0 < |p - p_0| < \delta_n$, $0 < |q - p_0| < \delta_n$, then $|f(p) - f(q)| < 1/n$. Take any sequence $\{p_n\}$ such that $0 < |p_n - p_0| < \delta_n$; clearly, $\{p_n\}$ converges to p_0. Set $a_n = f(p_n)$. Then, with p_n and p_m playing the role of p and q, we see that $|f(p_n) - f(p_m)| < 1/N$ for all n, $m > N$, and $\{a_n\}$ is a Cauchy sequence of real numbers. It must converge; let $L = \lim_{n \to \infty} a_n$. We will show that $\lim_{p \to p_0} f(p) = L$.

Given $\varepsilon > 0$, choose a deleted neighborhood $\mathscr{V}$ of p_0 as in the statement of the theorem. Choose k sufficiently large so that $p_k \in \mathscr{V}$ and $|a_k - L| < \varepsilon$. Let p be any point in $\mathscr{V}$. Then, $|f(p_k) - f(p)| < \varepsilon$, and, since $a_k = f(p_k)$, we have

$$\begin{aligned} |f(p) - L| &= |f(p) - f(p_k) + a_k - L| \\ &\le |f(p) - f(p_k)| + |a_k - L| \\ &< \varepsilon + \varepsilon = 2\varepsilon \end{aligned}$$

Thus, $f(p)$ is near L for all p in $\mathscr{V}$, and $\lim_{p \to p_0} f(p) = L$. ▮

The additional complexity introduced in going from functions of one variable to those of several variables can be seen from the following simple examples.

Consider first the function f defined everywhere in the plane, except at $(0, 0)$, by $f(x, y) = xy/\sqrt{x^2 + y^2}$, and let us study its behavior near the origin. When p lies on either axis, then $xy = 0$ and $f(p) = 0$. Thus, $f(p)$ approaches 0 as p approaches the origin along the axes. On the 45° line, where $y = x$, we have $f(x, x) = x^2/\sqrt{2x^2} = |x|/\sqrt{2}$, so that again $f(p)$ approaches 0. Is it true that $\lim_{p \to \mathbf{0}} f(p) = 0$? As a first step, we observe that $|x| \le |p|$ and $|y| \le |p|$, so that $|xy| \le |p|^2$ and

$$|f(p)| = \frac{|xy|}{\sqrt{x^2 + y^2}} \le \frac{|p|^2}{|p|} = |p|$$

Thus, given any $\varepsilon > 0$, we have $|f(p) - 0| < \varepsilon$ for all points p with $0 < |p - \mathbf{0}| < \delta$, for the choice $\delta = \varepsilon$. We have thus shown that $\lim_{p \to \mathbf{0}} f(p) = 0$.

Consider next the function g defined by

$$g(x, y) = \frac{xy}{x^2 + y^2} \qquad (x, y) \ne (0, 0)$$

Again, $g(p) = 0$ when p lies on either axis, so that $g(p)$ converges to 0 as p approaches the origin along either axis. This time, $\lim_{p \to \mathbf{0}} g(p)$ fails to exist, however, for the limit as p approaches $\mathbf{0}$ along the 45° line is not 0. To see this, we set $y = x$, have $g(p) = g(x, x) = x^2/2x^2 = \frac{1}{2}$, and find that the limit of $g(p)$ on this line is $\frac{1}{2}$.

The final illustration will show that the behavior of a function can be considerably more complicated. Put

$$F(x, y) = \frac{xy^2}{x^2 + y^4} \qquad (x, y) \ne (0, 0)$$

On the axes, $F(p) = 0$. On the line $y = x$, we have

$$F(p) = F(x, x) = \frac{x^3}{x^2 + x^4} = \frac{x}{1 + x^2}$$

and $\lim_{x \to 0} F(x, x) = 0$. In fact, we can show that $F(p)$ converges to 0 as p approaches the origin along *every* straight line. When $y = mx$,

$$F(p) = F(x, mx) = \frac{m^2 x}{1 + m^4 x^2}$$

and $\lim_{x \to 0} F(x, mx) = 0$. In spite of this, it is not true that $\lim_{p \to \mathbf{0}} F(p) = 0$. To show this, we produce a curve terminating at the origin, along which $F(p)$ does not converge to 0; this curve is the parabola $y^2 = x$, and $F(p) = F(y^2, y) = y^4/2y^4 = \frac{1}{2}$.

When f is a function of two real variables, the notation $\lim_{\substack{x \to x_0 \\ y \to y_0}} f(x, y)$ is often used in place of $\lim_{p \to p_0} f(p)$. This should not be confused with notion of an **iterated** limit, such as $\lim_{x \to x_0} \lim_{y \to y_0} f(x, y)$, in which we treat f as a function of x and y separately, rather than as a function of the point (x, y). For example,

$$\lim_{x \to 0} \left(\lim_{y \to 0} \frac{x^2}{x^2 + y^2} \right) = 1 \qquad \text{and} \qquad \lim_{y \to 0} \left(\lim_{x \to 0} \frac{x^2}{x^2 + y^2} \right) = 0$$

while $\lim_{\substack{x \to 0 \\ y \to 0}} x^2/x^2 + y^2$ fails to exist.

We may also discuss the behavior of a function "at infinity," that is, when $|p|$ is large.

Definition 11 *We write*

$$\lim_{|p| \to \infty} f(p) = L$$

if and only if, corresponding to each $\varepsilon > 0$, *a number* N *can be found such that* $|f(p) - L| < \varepsilon$ *whenever* $|p| \geq N$.

For example, if $f(x, y) = 1/(x^2 + y^2 + 1)$, then we may write

$$\lim_{|p| \to \infty} f(p) = 0$$

Again, if $f(x, y, z) = T + (x^2 + y^2 + z^2)^{-1/2}$ is the temperature at (x, y, z), we would say that the temperature "at infinity" is T, meaning that

$$\lim_{|p| \to \infty} f(p) = T.$$

EXERCISES

Note: Exercises are to be done without using L'Hospital's rule.

1 Let

$$f(x) = \begin{cases} 1 & \text{if } x \neq 0 \\ 2 & \text{if } x = 0 \end{cases}$$

$$g(x) = \begin{cases} 2 & \text{if } x \text{ is different from 1 or 2} \\ 3 & \text{if } x = 1 \\ 4 & \text{if } x = 2 \end{cases}$$

(*a*) Verify that $\lim_{x\to 0} f(x) = 1$, $\lim_{x\to 1} g(x) = 2$, $\lim_{x\to 0} g(f(x)) = 3$, and $g(f(0)) = 4$.
(*b*) Do the statements in (*a*) still hold if

$$f(x) = \begin{cases} x+1 & x \neq 0 \\ 2 & x = 0 \end{cases}$$

2 Find $\lim_{x\to 2} \dfrac{\sqrt[3]{x} - \sqrt[3]{2}}{\sqrt{x} - \sqrt{2}}$.

3 Find $\lim_{x\to 1} \dfrac{\dfrac{2}{x-1} + 3}{4 + \dfrac{5}{x^2 - \dfrac{3}{x+2}}}$

4 Find $\lim_{x\to 2} f(x)$ where $f(x) = \dfrac{x^3 + x^2 - 7x + 2}{2x^3 - 5x^2 + 6x - 8}$.

5 Prove that for any constant m, $\lim_{x\uparrow\infty} \dfrac{x^m}{e^x} = 0$.

6 Discuss the existence of:

(*a*) $\lim_{x\to 1} \dfrac{x^5 - 1}{x^4 - 1}$ (*b*) $\lim_{x\uparrow\infty} \dfrac{x}{1 + 3x}$

(*c*) $\lim_{x\uparrow 1} (1 - x)^{-1/2}$

7 Formulate precise definitions for:

(*a*) $\lim_{x\downarrow -\infty} f(x) = L$ (*b*) $\lim_{x\uparrow\infty} f(x) = \infty$

(*c*) $\lim_{x\downarrow b} f(x) = L$

8 Discuss the existence of:

(*a*) $\lim_{x\downarrow -\infty} \dfrac{x - 1}{\sqrt{1 + x^2}}$ (*b*) $\lim_{x\uparrow\infty} \dfrac{x - 1}{\sqrt{1 + x^2}}$

9 What is the form that Theorem 23 should take if f is a function of one variable and p_0 is replaced by "∞"?

10 If $c \neq 0$, show that $\lim_{x\to c} f(1/x) = \lim_{t\to 1/c} f(t)$ if either exists.

11 Is it always true that:

(*a*) $\lim_{x\uparrow\infty} f(x) = \lim_{t\downarrow 0} f\left(\dfrac{1}{t}\right)$ (*b*) $\lim_{x\downarrow 0} f(x) = \lim_{n\to\infty} f\left(\dfrac{1}{n}\right)$

12 Show that if ϕ is defined on a neighborhood of t_0 and continuous at t_0 with $\phi(t_0) = x_0$, then $\lim_{x\to x_0} f(x) = \lim_{t\to t_0} f(\phi(t))$.

13 Prove Theorem 20.

14 Let f be bounded on the interval $c < x < \infty$, and let $\lim_{x\uparrow\infty} g(x) = 0$. Prove that $\lim_{x\uparrow\infty} f(x)g(x) = 0$. Does this follow directly from Theorem 19?

***15** Can you formulate a definition for $\limsup_{x\to b} f(x)$? What properties would you expect this to have? For example, does Theorem 19 hold?

16 Discuss the existence of the following limits:

(*a*) $\lim_{p\to 0} \dfrac{x + y}{x^2 + y^2}$ (*b*) $\lim_{|p|\to\infty} \dfrac{x + y}{x^2 + y^2}$

(*c*) $\lim_{p\to 0} \dfrac{xy - z^2}{x^2 + y^2 + z^2}$ (*d*) $\lim_{|p|\to\infty} \dfrac{xy - z^2}{x^2 + y^2 + z^2}$

17 Following the pattern of Definition 11, formulate a definition for "$f(p)$ converges to L as p becomes infinite in the set S." Using this, discuss the behavior of $f(x, y) = \exp(x - y)$ when $|p|$ is large. (You may assume knowledge of the exponential function and its properties.)

18 If f is uniformly continuous on a set D in $\mathbf{R}^n$ and p_0 is a cluster point for D, show that $\lim_{p \to p_0} f(p)$ exists in D.

2.6 DISCONTINUITIES

A function that is not continuous at a point p_0 is said to be discontinuous there. The term **discontinuity** is used in two ways. The first refers to a point at which the function is defined but is not continuous. For example, consider the function f described by

$$f(p) = f(x, y) = \begin{cases} x^2 + y^2 & \text{when } |p| \le 1 \\ 0 & \text{when } |p| > 1 \end{cases}$$

This function is defined in the whole plane and is continuous there except at the points p with $|p| = 1$; each point of this circle is thus a discontinuity for f. If we consider f only on the set $E = \{\text{all } p \text{ with } |p| \le 1\}$, then f is continuous on E; the points of the circumference would not be considered discontinuities this time. On the other hand, if we were to restrict f to the set consisting of points p with $1 \le |p| \le 2$, it would not be continuous everywhere in this ring, for the points p with $|p| = 1$ would again be discontinuities for f.

In its second usage, the term discontinuity is also applied to points where a function is not defined. For example, the function described by $f(x) = 1/x$ might be said to be discontinuous (or to have a discontinuity) at $x = 0$.

Discontinuities can be further classified as **removable** and **essential.** If $f(p_0)$ is defined, and $L = \lim_{p \to p_0} f(p)$ exists but $L \ne f(p_0)$, then p_0 is a discontinuity for f; however, it may be "removed" by altering the definition for f at p_0. If we construct a new function F by setting $F(p) = f(p)$ for all p in the domain of f, except p_0, and setting $F(p_0) = L$, then F is now continuous at p_0. Again, if a function f is not defined at p_0, but $L = \lim_{p \to p_0} f(p)$ exists, then we may define $f(p_0)$ to be L and thus extend the domain of f to include p_0 so that f is continuous at p_0. In both these cases, we would say that p_0 was a removable discontinuity for f. When $\lim_{p \to p_0} f(p)$ does not exist, p_0 is said to be an essential discontinuity for f, since by no assignment of a value for $f(p_0)$ can we make f continuous there. For example, let

$$f(x) = x^x \qquad \text{when } x > 0$$

$$g(x) = \begin{cases} x & \text{when } x > 0 \\ 2 & \text{when } x = 0 \end{cases}$$

$$h(x) = \sin\left(\frac{1}{x}\right) \qquad \text{when } x > 0$$

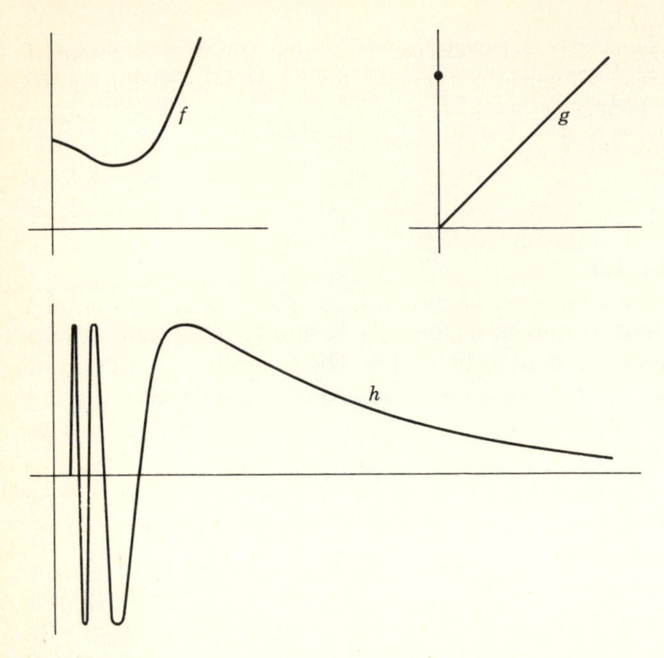

Figure 2-8

All are continuous on the open interval $0 < x$, and can be said to have discontinuities at the origin. However, this is a removable discontinuity for f and g, and an essential discontinuity for h (see Fig. 2-8). To explain this, we observe that f is not defined for $x = 0$, but that it may be shown that $\lim_{x \downarrow 0} f(x) = 1$; if we set $f(0) = 1$, the extended function is now continuous on the closed interval $0 \leq x$. The function g is defined at the origin, but $g(0) = 2 \neq 0 = \lim_{x \downarrow 0} g(x)$. If we alter g there so that $g(0) = 0$, then g too is continuous for $0 \leq x$. (Since this redefinition of g has produced a new function, a new letter such as "G" should be used to denote it; however, when the context is sufficiently clear, such precision is not usual.) The third function, h, is not defined at the origin, nor does $\lim_{x \downarrow 0} h(x)$ exist, and no choice for $h(0)$ will make h continuous there.

The function $f(x, y) = xy/(x^2 + y^2)$, which we have examined before, is continuous at all points of the plane except the origin, where it is not defined. Since $\lim_{p \to \mathbf{0}} f(p)$ fails to exist, we would say that the origin is an essential discontinuity for f. In contrast, the function

$$g(x, y) = \frac{x^2 y^2}{x^2 + y^2}$$

also undefined at $(0, 0)$, has this point as a removable discontinuity since $\lim_{p \to \mathbf{0}} g(p)$ exists. To check this, observe that $|x^2 y^2| \leq |p|^2 |p|^2 = |p|^4$ and $x^2 + y^2 = |p|^2$ so that $|g(p)| \leq |p|^4/|p|^2 = |p|^2$ and $\lim_{p \to \mathbf{0}} g(p) = 0$.

It is not always easy to tell when a discontinuity is removable or essential. It is an important research problem in mathematics to discover conditions under which it is possible to extend the definition of a function from a set E to a larger set so as to retain certain desirable properties such as continuity. For example, one question that could be asked is: When can a function f, defined and continuous on a set S, be extended to its closure as a continuous function? The following result gives one answer to this question.

Theorem 24 *Let S be a bounded set and E its closure. Then, a function f, continuous on S, can be extended continuously to E if and only if f is uniformly continuous on S.*

The hypothesis that f is *uniformly* continuous on S is a necessary one, for E, the closure of S, is itself bounded and thus compact, and if f can be defined on the boundary of S in such a way that it becomes continuous on all of E, then the resulting function must be uniformly continuous on E, and must therefore have been uniformly continuous on the subset S.

What remains to be proved, then, is that if f is uniformly continuous on S, such an extension of f is possible. We give one proof which depends on the compact graph theorem (Theorem 12 in Sec. 2.4), and indicate the skeleton of another which uses the Cauchy convergence criterion. The first has the advantage of elegance, but should be read with pencil and paper and suitable sketches. Let G be the graph of the given function f. Because f is uniformly continuous on S, which is bounded, f is bounded (Exercise 7, Sec. 2.3). Combining this with the fact that S is bounded, we see that G is a bounded set. The closure of G, $\overline{G}$, will then be a closed bounded set in an appropriately chosen euclidean space, and will therefore be compact. If we know that $\overline{G}$ is the graph of a function F, then F will be an extension of f that is defined on the set E, and because of the compact graph theorem, F is continuous on E.

The missing step is the proof that $\overline{G}$ is the graph of a function. Could a vertical line cut $\overline{G}$ in two distinct points? The points of G have the form $(p, f(p))$, for $p \in S$, and the points of $\overline{G}$ are limit points of these. The set $\overline{G}$ will not be the graph of a function if we can find two points (q, b) and (q, c) in $\overline{G}$ with $b \neq c$. Choose two sequences (q_n, u_n) and (p_n, v_n) in G, with the first converging to (q, b) and the second to (q, c). From the fact that G is the graph of f, $u_n = f(q_n)$ and $v_n = f(p_n)$. In addition, $\{p_n\} \to q$, $\{q_n\} \to q$, and $u_n = f(q_n) \to b$, $v_n = f(p_n) \to c$. However, $|p_n - q_n| \to |q - q| = 0$, so that (by Exercise 6, Sec. 2.3), $|f(p_n) - f(q_n)| \to 0$, and we conclude that $|c - b| = 0$ and $b = c$, completing the proof of Theorem 24. ▮

A second proof, which may seem more direct, starts by asking what value the continuous extension of f ought to have at any particular boundary point q of the set S. To find out, take any sequence $\{p_n\}$ in S which converges to q, and examine the values $v_n = f(p_n)$. Use the uniform continuity of f in S

to show that $\{v_n\}$ is a Cauchy sequence. This must converge, so we are led to the definition: $f(q) = \lim_{n\to\infty} v_n$. Use Exercise 6 of Sec. 2.3 to show that the value for $f(g)$ so chosen doesn't depend on what sequence $\{p_n\}$ was originally chosen. In this fashion, the function f has been extended to the closure of the original set S by defining it at every boundary point of S. All that remains is to prove that the extended function is continuous on E, and this can be done by invoking the compact graph theorem. ▮

A typical illustration of this theorem is the situation in which one has a function F defined on a bounded open set D, and one would like to be able to speak of the values of F on bdy(D); for example, $F(p)$ might be the temperature at the point p (see Exercise 7). Theorem 24 states that if F is uniformly continuous on S, then there is one and only one way to do this to keep the function F continuous. All points of the boundary of D are removable discontinuities.

For a one-variable illustration, let D be the open interval $0 < x < 1$. Given a function $f(x)$ defined on D, we want to determine the "correct" values for $f(0)$ and $f(1)$. In this case, they exist if and only if f is uniformly continuous on D, and then $f(0) = \lim_{x\downarrow 0} f(x)$ and $f(1) = \lim_{x\uparrow 1} f(x)$, and these limits will in fact exist.

It is natural to ask whether it is possible to extend functions still further, beyond the closure of the original domain, and still keep the functions continuous; we return to this in Sec. 6.2 where we sketch the proof of the Tietze extension theorem, which gives an affirmative answer.

EXERCISES

1 Discuss the continuity of the function f described by:

(*a*) $f(x) = \begin{cases} x \sin(1/x) & x \neq 0 \\ 0 & x = 0 \end{cases}$

(*b*) $f(x, y) = \dfrac{xy}{|x| + |y|}$ for $(x, y) \neq (0, 0)$

(*c*) $f(x, y) = \dfrac{x^2y^3}{x^4 + y^6}$ for $(x, y) \neq (0, 0)$

(*d*) $f(x, y) = \begin{cases} \dfrac{x^2 - y^2}{x - y} & \text{for } x \neq y \\ x - y & \text{when } x = y \end{cases}$

2 Let $f(x) = \begin{cases} 1 \text{ if } x \text{ is a rational number} \\ 0 \text{ if } x \text{ is an irrational number.} \end{cases}$ Is f continuous anywhere?

3 Let $f(x) = \begin{cases} 0 \text{ if } x \text{ is irrational} \\ 1/q \text{ if } x \text{ is the rational number } p/q \text{ in lowest terms.} \end{cases}$

Is f continuous anywhere?

4 For each of the following functions, find all the discontinuities and indicate any that are removable.

$$(a)\ f(x) = \frac{x}{\sin(5\cos x)} \qquad (b)\ f(x) = x + \frac{\sin x}{2x - \dfrac{4}{x-1}} \qquad (c)\ f(x) = \frac{1}{1 + e^{\sec x}}$$

5 For each of the following functions, find all the discontinuities and indicate any that are removable.

$$(a)\ F(x, y) = \frac{x + 2y}{\sin(x + y) - \cos(x - y)} \qquad (b)\ F(x, y) = x \sin\left(\frac{y}{x}\right)$$

6 Investigate the behavior of $F(x, y)$ at $(0, 0)$ if

$$(a)\ F(x, y) = \frac{x^2 y}{2x^2 + y^2} \qquad (b)\ F(x, y) = \frac{x^2 y}{3x^4 + 2y^2}$$

7 Suppose that there is a continuous distribution of temperature on the open square $|x| < 1$, $|y| < 1$. Is it possible to extend the temperature continuously to the boundary so that the temperature is 0 on the north edge and 100° on the other three edges?

8 The function

$$\exp\left(\frac{x^2 + y^2 - xy}{x^2 + y^2}\right) = f(x, y)$$

is continuous on the open first quadrant.

(*a*) Is it bounded there?

(*b*) Can f be extended continuously to the closed first quadrant?

★9 Let f be an increasing function on the interval $[0, 1]$. Show that f cannot have more than a countable number of discontinuities on this interval. (*Hint:* First look at the one-sided limits at a point x_0.)

2.7 INVERSES FOR FUNCTIONS OF ONE VARIABLE

The topic of the inverse of a function is one for which the geometrical point of view is particularly well suited. Let f be a function of one variable. Considered as a transformation from $\mathbf{R}^1$ into $\mathbf{R}^1$, f sends the point a into the point $b = f(a)$. A function g is called an **inverse** for f if g reverses the effect of f, sending b back into a, so that $g(f(a)) = a$. In most cases, it is not possible to find such a function g which has this property for all points x in the domain of f. As an illustration, let $f(x) = x^2$, $-\infty < x < \infty$. If f had an inverse g such that $g(f(x)) = x$ for all real numbers x, then $g(x^2) = x$ and it would be necessary to have both $g(4) = g(2^2) = 2$ and $g(4) = g((-2)^2) = -2$. Such ambiguity is impossible for a function, since the point 4 must have a unique image. However, the function f has an inverse g on the interval $0 < x < \infty$ and a second inverse h on the interval $-\infty < x < 0$, namely, the functions defined on the interval $0 < x < \infty$ by $g(x) = \sqrt{x}$ and $h(x) = -\sqrt{x}$.

In order for f to have a partial inverse associated with a subset S of its domain, all that is needed is to have f 1-to-1 on S. There then exists a function

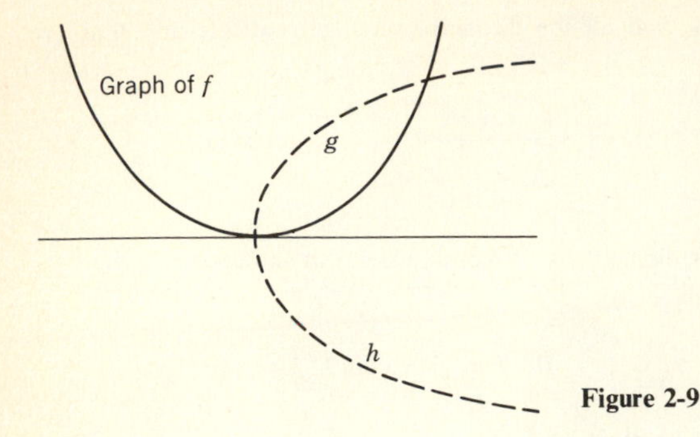

Figure 2-9

g defined on $f(S)$ such that $g(f(x)) = x$ for all $x \in S$, and $f(g(x)) = x$ for all $x \in f(S)$.

Let us examine the problem from the geometrical point of view. The graph of f is the set of points $(x, f(x))$, so that (a, b) is on the graph of f if and only if $b = f(a)$. If $g(b) = a$, then the point (b, a) must be on the graph of g. Let us introduce a special transformation R from $\mathbf{R}^2$ into itself which sends (x, y) into (y, x). It is easily seen that this can be regarded as a reflection of the plane about the line $y = x$, in $\mathbf{R}^2$. Let C be the image under R of the graph of f. If (a, b) is on the graph of f, $(b, a) \in C$. Thus, the graph of any function g which is an inverse for f must be part of the set C. Turning this around, any subset of C which is the graph of a function (i.e., any subset which is met no more than once by each vertical line) yields a particular inverse function for f.

Applying this to the example $f(x) = x^2$, we have given in Fig. 2-9 the graph of f, and its image C under the reflection R. The set C falls into two connected pieces, each of which is the graph of a function; the upper half is the function g, $g(x) = \sqrt{x}$, and the lower half the function h,

$$h(x) = -\sqrt{x}$$

Turning to a less trivial example, consider the function F given by $F(x) = \frac{3}{2}x - \frac{1}{2}x^3$. The graph of F and its reflection C are shown in Fig. 2-10. As indicated there, C can be split into three connected pieces, each of which is the graph of a function, and each of which therefore defines a function which is an inverse for F. The function g_1 is defined on the interval $-\infty < x < 1$, the function g_2 on $[-1, 1]$, and the function g_3 on the interval $-1 < x < \infty$. An accurate graph of F would enable us to read off the values of these functions, and thus tabulate them. (In this example, it is also possible to give analytical formulas for g_1, g_2, and g_3; see Exercise 3.)

Again, if $f(x) = \sin x$, then (see Fig. 2-11) the graph of the reflection C falls into an infinite number of connected pieces, each of which provides an inverse for f. Among these, one is usually singled out as shown and is called the principal inverse for the sine function, $g(x) = \arcsin(x)$.

This same geometric process can be used with any mapping f from n space into n space. If $S \subset \text{domain}(f)$ is a set on which f is 1-to-1, then the cor-

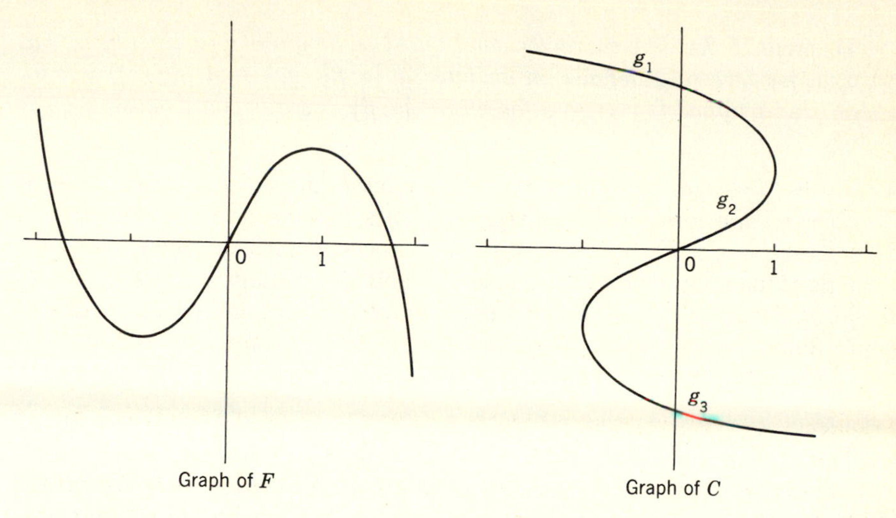

Figure 2-10

responding portion of the graph of f is a set of points (p, q), with $p \in S$, whose reflection C is the collection of points (q, p) that forms the graph of a partial inverse g for f. Since the graph of f is an n-dimensional set in $2n$ space, all this may be difficult to visualize when $n \geq 2$, but the process is still valid. What makes it particularly useful is the fact that if it is applied to a function f that is continuous, the process automatically yields partial inverses that are also continuous. For $n = 1$, this is established as follows.

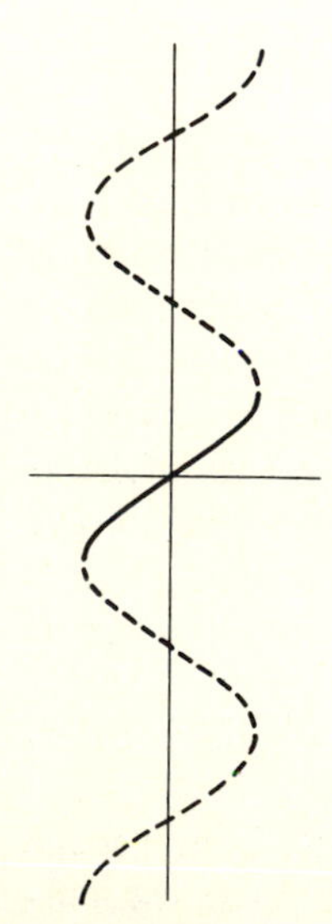

Figure 2-11

Theorem 25 *Let f be continuous and 1-to-1 on an interval $[a, b]$. Then, f has a unique inverse g defined on an interval $[\alpha, \beta]$ such that $g(f(x)) = x$ for all $x \in [a, b]$ and $f(g(x)) = x$ for all $x \in [\alpha, \beta]$, and g is continuous.*

By Theorem 18, f must be strictly monotonic on $[a, b]$. Since $[a, b]$ is compact and connected, its image $f([a, b])$ is the same and must be a closed interval, say $[\alpha, \beta]$. By the geometrical process described above, there is a unique function g which is inverse to f, and which maps $[\alpha, \beta]$ 1-to-1 onto $[a, b]$, reversing the mapping f. Let the graph of f be G, and the graph of g be C. Since f is continuous, G is compact. Since C is simply the reflection of G, C is also compact. Hence, g is continuous. ▮

The same proof can be used to show that any 1-to-1 mapping from a compact set S in n space into n space also has a continuous inverse. We will see later that this same geometric process can be used to discover other properties of the inverses of a given mapping f.

When $n = 1$, and we are studying a function f defined on an interval, it can be a simple matter to find subintervals on which f is 1-to-1, since f must be strictly monotonic there. Indeed, if f is differentiable there, then it will be recalled from elementary calculus that $f'(x)$ must be everywhere positive or everywhere negative on such an interval. This elementary result is a consequence of the mean value theorem (and will in fact be an exercise in the next chapter, Sec. 3.2); because this relationship between the monotonicity of $f(x)$ and the sign of $f'(x)$ is both so familiar and so useful, we remind you of it in connection with the exercises below.

EXERCISES

1 (*a*) Find an interval on which $f(x) = \dfrac{3x}{x + 4}$ is continuous and 1-to-1.

(*b*) Find a formula for the corresponding inverse.

2 (*a*) Find an interval on which $f(x) = x^3 - 3x^2 + 3x$ is continuous and 1-to-1.

(*b*) Find a formula for the corresponding inverse.

3 Show that a formula for the function g_1 of Fig. 2-10 is

$$g_1(x) = \begin{cases} [\sqrt{x^2 - 1} - x]^{1/3} - [\sqrt{x^2 - 1} + x]^{1/3}, & -\infty < x < -1 \\ 2 \cos\left(\frac{1}{3} \arccos(-x)\right), & -1 \le x \le 1 \end{cases}$$

4 Find inverses for the function f given by

$$f(x) = x^2 - 2x - 3$$

5 How many continuous inverses are there for the function described by

$$F(x) = x^3 + 3x$$

6 Is there any interval on which the function f described by

$$f(x) = 2x + |x| - |x + 1|$$

fails to have an inverse?

★**7** Show that a continuous function f cannot map the interval $[0, 1]$ onto itself exactly 2-to-1.

8 Investigate the existence of local and global inverses for the function $f(x) = Ax - \sin x$, for various values of A.

9 Let f be a continuous function defined on the interval $I = [a, b]$ which maps I onto I 1-to-1 and which is its own inverse.

(*a*) Show that, except for one possible function, f must be monotonic decreasing on I.

(*b*) What are the polynomial functions that are choices for f?

CHAPTER

THREE

DIFFERENTIATION

3.1 PREVIEW

In this chapter, we introduce and exploit the properties of the vector-valued derivative $\mathbf{D}f$ of a function of several variables. This is also called the gradient of f, and is related to the partial derivatives and directional derivatives of f. There is some review of the one-variable theory, including the mean value theorem and L'Hospital's rule. We also discuss Taylor's theorem for one and several variables, and several forms of the mean value theorem for functions of several variables.

We discuss chain rules for differentiation, and illustrate them in cases of complicated functional relationships and changes of variables; we think that the use of diagrams such as Fig. 3-7 makes this easier to follow.

We also discuss the applications of differentiation, treating extremal problems and the location and nature of critical points.

Any treatment of higher derivatives of functions of several variables is postponed to Chap. 7, since this requires the differentiation of vector-valued functions. However, note the brief remarks that follow the proof of Theorem 19.

3.2 MEAN VALUE THEOREMS AND L'HOSPITAL'S RULE

Certain simple properties of functions of one variable involve differentiation. Recall from elementary calculus that f is said to be **differentiable** (or to have a derivative) at x_0 if f is defined on a neighborhood of x_0 and if $f'(x_0)$ exists, defined by

$$(3\text{-}1) \qquad f'(x_0) = \lim_{x \to x_0} \frac{f(x) - f(x_0)}{x - x_0} = \lim_{h \to 0} \frac{f(x_0 + h) - f(x_0)}{h}$$

Another way to state this is to say that the function

$$g(x) = \frac{f(x) - f(x_0)}{x - x_0}$$

has x_0 as a removable discontinuity.

A frequent application of differential calculus is to so-called maximum-minimum problems. We say that f has a **local maximum** at x_0 if there is a neighborhood U about x_0 such that $f(x) \leq f(x_0)$ for all $x \in U$. The notion of a **local minimum** is defined similarly, and the term **extreme value** may be used to refer to either.

Theorem 1 *Let $f(x)$ be defined on a neighborhood of x_0 and have a local extreme value at x_0. If f is differentiable at x_0, then $f'(x_0) = 0$.*

We may assume that $f(x_0 + h) \leq f(x_0)$ for all h with $|h| < \delta$, and that $C = \lim_{h \to 0} (f(x_0 + h) - f(x_0))/h = f'(x_0)$ exists. This limit may also be computed by letting h approach 0 first from above, and then from below. Since the numerator of the fraction is never strictly positive, we find that C obeys the conditions $C \leq 0$ and $C \geq 0$, so that necessarily $C = 0$. ▮

It is important to keep in mind that $f'(x_0)$ need not be 0 if x_0 is an endpoint, rather than an interior point. Thus in using this theorem in the solution of a maximum-minimum problem, separate consideration must be given to the possibility of an *endpoint* extreme value.

One immediate consequence of Theorem 1 has acquired a special name.

Theorem 2 (Rolle's Theorem) *Let f be continuous on the interval $[a, b]$, and let $f'(x)$ exist for $a < x < b$. If $f(a) = f(b)$ then there is a point x_0 with $a < x_0 < b$ at which $f'(x_0) = 0$.*

If f is a constant function, any choice of the point x_0 will do. If f is not constant, then it must have either an interior minimum or an interior maximum at some point x_0 on the open interval $a < x < b$; and since f is differentiable there, $f'(x_0) = 0$. ▮

Corollary 1 *If f is differentiable on the interval $a < x < b$, then the zeros of f are separated by zeros of f'.*

Corollary 2 *Let f and g be continuous on $[a, b]$ and differentiable on $a < x < b$. Suppose that $f(a) = g(a)$ and $f(b) = g(b)$. Then, there is at least one point x_0 interior to $[a, b]$ such that $f'(x_0) = g'(x_0)$.*

The second of these, which follows from the theorem if we consider the function $f - g$, has a simple geometric meaning, as shown in Fig. 3-1; some

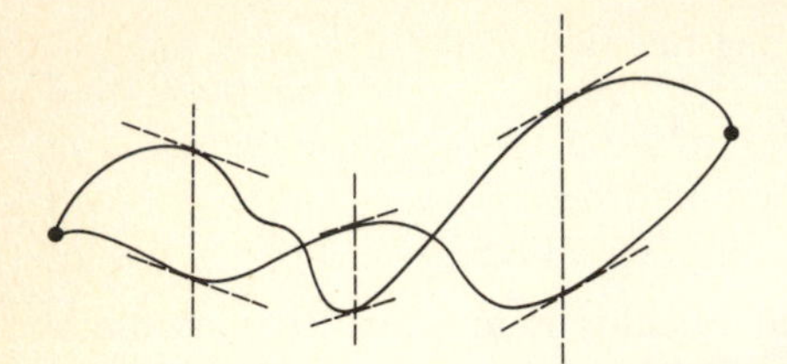

Figure 3-1

vertical line must cross the graphs of f and g at a point where their slopes are the same, between two points where the curves meet.

If the function g is specialized to be a straight line, then its slope will everywhere be

$$m = \frac{f(b) - f(a)}{b - a}$$

This case then gives the first of the following two very useful consequences of Rolle's theorem.

Theorem 3 (Mean Value Theorem) *Let f be continuous on $[a, b]$, and let $f'(x)$ exist for $a < x < b$. Then, at least one point x_0 exists interior to $[a, b]$ such that*

$$f(b) - f(a) = (b - a)f'(x_0) \tag{3-2}$$

Theorem 4 (General Mean Value Theorem) *Let f and g be continuous on $[a, b]$, and let $f'(x)$ and $g'(x)$ both exist for $a < x < b$. Then at least one point x_0 exists interior to $[a, b]$ such that*

$$[f(b) - f(a)]g'(x_0) = [g(b) - g(a)]f'(x_0) \tag{3-3}$$

The first of these has many familiar consequences. If $f'(x)$ is 0 on an interval, f is constant there (Exercise 1). If $f'(x)$ never changes sign on an interval, then f is monotonic there (Exercise 4).

If g is chosen to be the identity function, $g(x) = x$, then Theorem 4 becomes Theorem 3. To prove Theorem 4, we construct a special function F to which we apply Rolle's theorem. Let

$$F(x) = f(x) - Kg(x)$$

where K is a constant to be selected later. The function F is continuous on $[a, b]$ and differentiable in the interior. To apply Rolle's theorem, we want to have $F(a) = F(b)$. Forcing this, we must have

$$f(a) - Kg(a) = f(b) - Kg(b)$$

or $f(b) - f(a) = K(g(b) - g(a))$. Put aside for the moment the consideration of what we should do if $g(b) = g(a)$. Then, we can solve for K, and F satisfies the hypothesis for Rolle's theorem. There is then a point x_0 with

$F'(x_0) = 0$. Since $F'(x) = f'(x) - Kg'(x)$, this tells us that $f'(x_0) = Kg'(x_0)$, and substituting the value we found for K, we obtain (3-3). Suppose now that $g(b) - g(a) = 0$. We cannot now solve for K. However, we do not need to; for, examining (3-3), we see that it will hold if we can find an x_0 with $g'(x_0) = 0$, so that both sides of (3-3) will be 0. This we can do by applying Rolle's theorem to g. ▮

The geometric meaning of the general mean value theorem is very similar to that of the ordinary mean value theorem. If we assume that $g'(x)$ is never 0 on $[a, b]$, then from Rolle's theorem, we know that $g(b) \neq g(a)$, and we can rewrite (3-3) in the form

$$\frac{f(b) - f(a)}{g(b) - g(a)} = \frac{f'(x_0)}{g'(x_0)} \tag{3-4}$$

Let Γ be the curve in the plane whose parametric equation is $x = g(t)$, $y = f(t)$. As t moves along the interval $[a, b]$, the point (x, y) moves along Γ from the point $P = (g(a), f(a))$ to $Q = (g(b), f(b))$. The left side of (3-4) is the slope of the line joining P and Q. The right side of (3-4), which can also be written as $(dy/dt)/(dx/dt)$, is the slope of the curve. Thus, one meaning of Theorem 4 is that there must be a point on the curve Γ where its slope is the same as that of the line PQ (see Fig. 3-2).

We remark that (3-3) is a more general form of this result than is (3-4), since the latter requires the hypothesis that g' is never 0. For example, let $f(x) = x^2$ and $g(x) = x^3$, on the interval $[-1, 1]$. The left side of (3-4) is 0; however, since $f'(x)/g'(x) = (2x)/(3x^2) = 2/(3x)$, there is no choice of $x_0 \in [-1, 1]$ for which (3-4) holds. (This also provides a curve Γ for which the geometric interpretation of Theorem 4 fails; see Exercise 17.)

The analytical forms of these theorems can be modified. Since the point x_0 lies between a and b, we can write $x_0 = a + \theta h$, where $h = b - a$ and $0 < \theta < 1$. The conclusion of Theorem 3 would then read

$$f(a + h) = f(a) + hf'(a + \theta h)$$

The uses of the mean value theorems are of two sorts, one "practical" and the other "theoretical"; the distinctions here are quite subjective. As an

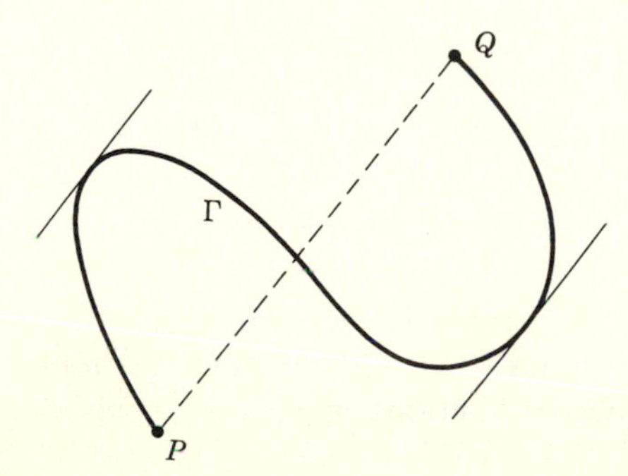

Figure 3-2 Mean value theorem.

example of the former, we shall use the mean value theorem to obtain a simple approximation to a special function.

Theorem 5 *When $u > 0$ and $v \geq 0$, $\sqrt{u^2 + v}$ may be replaced by $u + v/2u$, with an error of $v^2/4u^3$ at most.*

To illustrate this, $\sqrt{87} = \sqrt{81 + 6} \sim 9 + \frac{6}{18} = 9\frac{1}{3}$, with an error of at most $36/(4)(9)^3 \sim .012$. To prove this, put $f(x) = \sqrt{u^2 + x}$ so that $f(0) = u$, while $f(v)$ is the value we wish to estimate. By the mean value theorem, there is an x_0, $0 < x_0 < v$, with

$$\begin{aligned} f(v) &= f(0) + (v - 0)f'(x_0) \\ &= u + \frac{v}{2\sqrt{u^2 + x_0}} \end{aligned}$$

Since $x_0 > 0$, $\sqrt{u^2 + x_0} > u$, and we have shown that $f(v) < u + v/(2u)$. Thus, the approximation $u + v/(2u)$ is always larger than the true value of $\sqrt{u^2 + v}$. To estimate the error in the approximation, we observe that $x_0 < v$ so that $\sqrt{u^2 + x_0} < \sqrt{u^2 + v} < u + v/(2u)$. Hence,

$$\begin{aligned} f(v) &= u + \frac{v}{2\sqrt{u^2 + x_0}} \\ &> u + \frac{v}{2\left[u + \dfrac{v}{2u}\right]} = u + \frac{uv}{2u^2 + v} \end{aligned}$$

and thus

$$u + \frac{uv}{2u^2 + v} < \sqrt{u^2 + v} < u + \frac{v}{2u}$$

The error made in using the right-hand term as the approximate value is less than the difference

$$\left[u + \frac{v}{2u}\right] - \left[u + \frac{uv}{2u^2 + v}\right] = \frac{v^2}{(2u)(2u^2 + v)} < \frac{v^2}{4u^3} \quad \blacksquare$$

[The more exact methods made possible by Taylor's theorem, Sec. 3.5, show that the approximation $u + v/(2u)$ is accurate to within $v^2/(8u^3)$.]

An extremely useful result which arises from the general mean value theorem (Theorem 4) is known as L'Hospital's rule. It provides a simple procedure for the evaluation of limiting values of functions which are expressible as quotients.

Theorem 6 (L'Hospital's Rule) *Let f and g be differentiable on the interval $a \le x < b$, with $g'(x) \ne 0$ there. If*

(i)
$$\lim_{x \uparrow b} f(x) = 0 \qquad \lim_{x \uparrow b} g(x) = 0$$

or if

(ii)
$$\lim_{x \uparrow b} f(x) = \infty \qquad \lim_{x \uparrow b} g(x) = \infty$$

and if
$$\lim_{x \uparrow b} \frac{f'(x)}{g'(x)} = L$$

then
$$\lim_{x \uparrow b} \frac{f(x)}{g(x)} = L$$

The upper endpoint b may be finite or "∞", and L may be finite or "∞". Before proving the theorem, we give several illustrations of its use, and possible modifications.

To evaluate

$$\lim_{x \downarrow 0} \frac{1 - \cos(x^2)}{x^4}$$

we consider instead

$$\lim_{x \downarrow 0} \frac{2x \sin(x^2)}{4x^3} = \lim_{x \downarrow 0} \frac{\sin(x^2)}{2x^2} = \frac{1}{2}$$

By the theorem, this is also the value of the original limit.

Again, consider $\lim_{x \downarrow 0} x^x$. Since $\log(x^x) = x \log x$, we consider instead $\lim_{x \downarrow 0} x \log x = \lim_{x \downarrow 0} (\log x)/(1/x)$. We replace this by $\lim_{x \downarrow 0} (1/x)/(-1/x^2) = \lim_{x \downarrow 0} -x = 0$. Since the exponential function is continuous, we conclude that

$$\lim_{x \downarrow 0} x^x = \lim_{x \downarrow 0} \exp(x \log x) = \exp(0) = 1$$

We call your attention to two of the exercises which point out ways in which L'Hospital's rule can be *misapplied*. In Exercise 24, it is used to evaluate a limit (known not to exist) and finds the answer 0; in this case, one of the hypotheses of Theorem 6 has been overlooked. In Exercise 25 an incorrect result is again obtained, but this time the error lies in assuming that a converse of Theorem 6 is valid, namely, that if $\lim f'(x)/g'(x)$ does not exist, neither does $\lim f(x)/g(x)$.

PROOF We first prove Theorem 6 under hypothesis (i), the case usually described as "a fraction that is indeterminate of the form 0/0." We also suppose b finite, and can therefore set $f(b) = g(b) = 0$ and thus assume that f and g are each continuous on $[a, b]$. Using the form of the general mean value theorem given in (3-4), we take any x interior to $[a, b]$, and then can be sure that there is a point t with $x < t < b$ such that

$$\frac{f(b) - f(x)}{g(b) - g(x)} = \frac{f'(t)}{g'(t)}$$

Since $f(b) = g(b) = 0$, this implies

$$\frac{f(x)}{g(x)} - L = \frac{f'(t)}{g'(t)} - L$$

Since $\lim_{t \to b} f'(t)/g'(t) = L$, the right side (and thus the left) will approach 0 if we can force t toward b. Since t lies somewhere between x and b, we can achieve this by controlling x. Accordingly, given $\varepsilon > 0$ we can choose δ so that if $b - \delta < x < b$, then of necessity $|b - t| < \delta$ and $|f(x)/g(x) - L| < \varepsilon$, proving that $\lim_{x \to b} f(x)/g(x) = L$.

We give a different type of proof for Theorem 6 under the second hypothesis, (ii); here, one speaks of a "fraction that is indeterminate of the form ∞/∞." (As suggested by Exercise 26, there are essential geometric differences between the two cases.)

We start from the fact that $f(x) \to \infty$, $g(x) \to \infty$, and $f'(x)/g'(x) \to L$ as x approaches b. Since $g'(x)$ is never 0, we know that it must be positive and thus that g is strictly increasing; we may therefore assume $g(x) > 0$ for all x in $[a, b]$. Given $\varepsilon > 0$, choose x_0 so that if $x_0 < t < b$,

$$-\varepsilon < \frac{f'(t)}{g'(t)} - L < \varepsilon$$

Since $g'(t) > 0$, we may rewrite this as

$$(L - \varepsilon)g'(t) < f'(t) < (L + \varepsilon)g'(t) \tag{3-5}$$

The right half of this can be written

$$f'(t) - (L + \varepsilon)g'(t) < 0$$

which implies that the function $f(x) - (L + \varepsilon)g(x)$ is decreasing, since its derivative is negative on $[x_0, b]$. A decreasing function is necessarily bounded above, so that we have $f(x) - (L - \varepsilon)g(x) < B$. Divide by $g(x)$, which is positive, and arrive at

$$\frac{f(x)}{g(x)} < L + \varepsilon + \frac{B}{g(x)}$$

Since $g(x) \to \infty$ as $x \to b$, we see that for some x_1 near b, $f(x)/g(x) < L + 2\varepsilon$ for all x, $x_1 < x < b$.

If we return to (3-5) and carry out a similar calculation using the left half of the inequality, we obtain $f(x)/g(x) > L - 2\varepsilon$, and the combination proves that $\lim_{x \to b} f(x)/g(x) = L$. ▮

Monotonic functions are much easier to work with than other functions. For example, as will be shown later, the graph of a continuous monotonic function is always a curve of finite length, while this need not be true for a function which is merely continuous. A function that is piecewise monotonic is also well behaved; any polynomial function $P(x) = a_m x^m + a_{m-1}x^{m-1} + \cdots + a_0$ is piecewise monotonic (Exercise 6). For differentiable functions, this is easy to test, using the relationship between monotonicity and the sign of $f'(x)$. (However, in this connection, note Exercise 30.) A function f that is piecewise monotonic will possess partial inverses associated with portions of its domain on which it is 1-to-1. If f is also differentiable, and the derivative doesn't vanish, the inverse is also differentiable.

Theorem 7 *Let f be monotonic on $[a, b]$, and differentiable with $f'(x) \neq 0$ for $a < x < b$. Then g, the inverse of f, is defined on an interval $[\alpha, \beta]$, and g is differentiable in its interior, with $g'(\gamma) = 1/f'(g(\gamma))$ for all γ, $\alpha < \gamma < \beta$.*

We must show that $\lim_{x \to \gamma} (g(x) - g(\gamma))/(x - \gamma)$ exists. Set $g(x) = y$ and $g(\gamma) = c$. Then, $x = f(y)$, $\gamma = f(c)$, and since f and g are both continuous, the limit we must consider becomes

$$\lim_{y \to c} \frac{y - c}{f(y) - f(c)}$$

which we see at once to exist and be $1/f'(c) = 1/f'(g(\gamma))$. ∎

EXERCISES

1 If $f'(x) = 0$ for all x, $a < x < b$, show that f is constant there.

2 Use Rolle's theorem to prove that if $g(x)$ is a polynomial and if $g(a) = g'(a) = g''(a) = g^{(3)}(a) = 0$ and $g(b) = 0$, then there is a number c, $a < c < b$, with $g^{(4)}(c) = 0$.

3 If $f(x)$ is defined and $f'(x)$ exists for each x, $a < x < b$, prove that f is continuous there.

4 If $f(x)$ is defined and $f'(x)$ exists for x, $a < x < b$, show that

(*a*) If $f'(x) \geq 0$ for $a < x < b$, then f is monotonic there.

(*b*) If $f'(x) > 0$ for $a < x < b$, then f is strictly monotonic.

5 If $f(x)$ and $g(x)$ are functions both of which are differentiable at least three times, and each has at least 4 zeros, what can you say about the number of zeros of $F^{(2)}$ where $F(x) = f(x)g(x)$?

6 Show that any polynomial function $P(x)$ is piecewise monotonic.

7 Prove that if $f'(x) \to 0$ as $x \uparrow \infty$, $\lim (f(x + 1) - f(x)) = 0$.

8 Suppose that f is such that $|f(a) - f(b)| \leq M|a - b|^2$ for all $a, b \in \mathbf{R}$. Prove that f is a constant function.

9 Let f' exist and be bounded for $-\infty < x < \infty$. Prove that f is uniformly continuous on the line.

10 Let f'' exist and be negative on the interval $[0, 1]$. Show that if P and Q lie on the graph of f, then the line PQ is below the graph, between P and Q.

11 Left- and right-hand derivatives at a point x_0 for a function f are defined as

$$\lim_{x \uparrow x_0} \frac{f(x) - f(x_0)}{x - x_0} \quad \text{and} \quad \lim_{x \downarrow x_0} \frac{f(x) - f(x_0)}{x - x_0}$$

respectively. Show by examples that these may exist where the usual two-sided derivative does not. Can such a function be discontinuous at x_0?

12 Given $f(x) = 1/(1 - e^{1/x})$, which is defined for all $x \neq 0$,

(*a*) Discuss the existence of $f'(x)$ at $x = 0$ (see Fig. 2-7).

(*b*) Does this conflict with Exercise 3?

13 Let $f'(x)$ exist and be continuous for all x. Prove that f obeys a Lipschitz condition on every bounded interval.

14 If $f(x) > 0$, and $f''(x) \leq 0$ for $x > 0$,

(*a*) Show that $f'(x) \geq 0$ for $x > 0$.

(*b*) Does it follow that $f(x) \to \infty$ as $x \uparrow \infty$?

15 (*a*) If $f(x) \to \infty$ as $x \to a < \infty$, can f' be bounded?

(*b*) If $f(x) \uparrow \infty$ as $x \uparrow \infty$, does $f'(x) \to \infty$?

(*c*) If $f(x) \to \infty$ as $x \to a < \infty$, does $f'(x) \to \infty$?

16 If $P(x)$ is a polynomial such that $P(0) = 1$, $P(2) = 3$, and $|P'(x)| \leq 1$ for $0 \leq x \leq 2$, what can you say about $P(x)$? (Prove it!)

17 Plot the curve given by $y = f(t) = t^2$, $x = g(t) = t^3$, where $t \in [-1, 1]$. Show that this curve provides an example for which the geometric interpretation of Theorem 4 fails.

18 The general mean value theorem has a geometric formulation in terms of tangents to curves being parallel to chords. Is the corresponding result true for curves in space (Fig. 3-2)?

19 Let the sides of a right triangle be longer leg $= B$, shorter leg $= b$, hypotenuse $= H$. Let the smallest angle of the triangle be θ. Show that the old surveying estimate given by $\theta = 3b/(2H + B)$, is accurate to within .02 (radians). (*Hint:* Express b and B in terms of θ, and then estimate this expression.)

20 Show that for large x, $\arctan x \approx \pi/2 - 1/x$ and estimate the error.

21 Let $f'(x) \to A$ as $x \uparrow \infty$. Prove $f(x)/x \to A$.

22 (*a*) Use L'Hospital's rule for Exercises 2, 4, and 5 of Sec. 2.5.

(*b*) Should you use it for Exercise 3 of Sec. 2.5?

23 Evaluate

(*a*) $\displaystyle\lim_{x \to 0} \frac{1 - \cos(x^2)}{x^3 \sin x}$ (*b*) $\displaystyle\lim_{x \downarrow 0} x^x$

(*c*) $\displaystyle\lim_{x \to 0} \frac{\sin x + \cos x - e^x}{\log(1 + x^2)}$

24 It is clear that $\lim_{x \uparrow \infty} e^{-\sin(x)}$ does not exist. Write this as

$$\lim_{x \uparrow \infty} \frac{2x + \sin 2x}{(2x + \sin 2x)e^{\sin x}}$$

and apply L'Hospital's rule, reducing your result to

$$\lim_{x \uparrow \infty} \frac{4 \cos x}{(2x + 4 \cos x + \sin 2x)e^{\sin x}} = 0$$

How do you explain this apparent contradiction?

25 Let $f(x) = x^2 \sin(1/x)$ and $g(x) = \sin x$. Calculate $\lim_{x \to 0} (f(x)/g(x))$ by L'Hospital's rule. Does your result show that this limit fails to exist? (Check this by evaluating the limit in another way.)

26 Use $x = g(t)$, $y = f(t)$ as parametric equations of a curve Γ. Construct pictures to illustrate the geometric meaning of the two cases of L'Hospital's rule. (*Hint:* Recall that the slope of Γ at $t = t_0$ is $f'(t_0)/g'(t_0)$.)

27 Let $f(x) \to \infty$ and $g(x) \to \infty$ as $x \to \infty$.

(*a*) Show that if $f(x)/g(x) \to L$, then $\log f(x)/\log g(x) \to 1$ where $0 < L < \infty$.

(*b*) Does the converse hold? (*Hint:* L'Hospital's rule may not be helpful.)

28 Use the example $f(x) = x^2 \sin(1/x)$, suitably defined for $x = 0$, to prove that $f'(x)$ can exist everywhere but not be continuous.

29 Let f' exist for all x on $[a, b]$, and suppose that $f'(a) = -1$, $f'(b) = 1$. Prove that even if f' is not continuous, there must exist a number c, $a < c < b$, with $f'(c) = 0$.

30 It is plausible that if f is differentiable and $f'(x_0) \neq 0$, then f is monotonic on a neighborhood of x_0. Disprove this by a geometric analysis of the function $f(x) = 2x + x^2 \sin(1/x)$.

31 If f, f', f'' are continuous on $1 \leq x < \infty$, $f > 0$, and $f'' < 0$, show that $f' \geq 0$.

3.3 DERIVATIVES FOR FUNCTIONS ON R^n

In elementary calculus, partial derivatives of functions of several variables were introduced merely as the ordinary derivative of the functions of one variable obtained by treating all the remaining variables in turn as constants. Thus, if $w = f(x, y, z) = x^3y^2 + xz^4$, one wrote $\partial w/\partial x = 3x^2y^2 + z^4$, $\partial w/\partial y = 2x^3y$, $\partial w/\partial z = 4xz^3$. In this section, we will study the vector-valued derivative of a function of several variables. This is also called the gradient of f, the differential of f, and sometimes merely the (total) derivative of f. We will also obtain several forms of the mean value theorem for functions of several variables, and then explore some of the properties of the vector-valued derivative.

The process of differentiation is the means by which the intuitive notion of "rate of change" is made precise. For a function of one variable, there are only a limited variety of ways in which a change can be made: t may be either to the left of t_0 or to the right of t_0. Thus, we end up with only one main concept, *the* derivative $f'(t)$, and two minor variations, the left derivative and the right derivative (Exercise 11, Sec. 3.2).

When we turn to a function of several variables, the situation is considerably more complicated. We can move from a point p_0 in many different directions and, in each case, obtain a resultant change $F(p) - F(p_0)$, and it is intuitively clear that this will depend upon the direction as well as the distance $|p - p_0|$. A convenient approach to the notion of the differential of a function starts with that of a directional derivative.

In 1-space, we have only two directions, left and right (or ahead and behind). In 2-space, we are apt to think of angles as the standard way to describe directions. In 3-space, and in n space for any n, it is easier to say that a **direction** is merely any point β, with $|\beta| = 1$. Such a point lies on the boundary of the unit sphere; intuitively, we think of it as the unit vector starting at the origin and ending at the point β. For example, if we wish to move away from the point p_0 "in the direction β," we understand that we are to proceed from p_0 on the line segment toward the point $p_0 + \beta$. In general, the ray or half line starting at p_0 and pointing in the direction β consists of all the points $p_0 + t\beta$, for $0 \leq t$.

Suppose now that f is a real-valued function, defined and continuous on a neighborhood of p_0. Then, the rate of change of f at p_0 in the direction β, or the **directional derivative** of f at p_0 in the direction β, is defined to be

$$(3\text{-}6) \qquad (\mathbf{D}_\beta f)(p_0) = \lim_{t \to 0} \frac{f(p_0 + t\beta) - f(p_0)}{t}$$

As an illustration, let $f(x, y) = x^2 + 3xy$, $p_0 = (2, 0)$, and $\beta = (1/\sqrt{2}, -1/\sqrt{2})$. (Note that this specifies the direction $-45°$.)

Since $p_0 + t\beta = (2 + t/\sqrt{2}, -t/\sqrt{2})$, we have

$$\begin{aligned} f(p_0 + t\beta) &= (2 + t/\sqrt{2})^2 + 3(2 + t/\sqrt{2})(-t/\sqrt{2}) \\ &= 4 - \frac{2}{\sqrt{2}}t - t^2 \end{aligned}$$

Accordingly,

$$\begin{aligned} (\mathbf{D}_\beta f)(p_0) &= \lim_{t \to 0} \frac{\left(4 - \frac{2}{\sqrt{2}}t - t^2\right) - 4}{t} \\ &= -\frac{2}{\sqrt{2}} \end{aligned}$$

If we hold p_0 the same and vary β, the value of $(\mathbf{D}_\beta f)(p_0)$ need not remain the same. Intuitively, it is clear that reversal of the direction β ought to reverse the sign of the directional derivative. Indeed,

$$(\mathbf{D}_{-\beta} f)(p_0) = \lim_{t \to 0} \frac{f(p_0 - t\beta) - f(p_0)}{t}$$

If we put $\lambda = -t$, we have

$$\frac{f(p_0 - t\beta) - f(p_0)}{t} = -\frac{f(p_0 + \lambda\beta) - f(p_0)}{\lambda}$$

so that $(\mathbf{D}_{-\beta} f)(p_0) = -(\mathbf{D}_\beta f)(p_0)$, as conjectured.

The **partial derivatives** of a function f of n variables are the directional derivatives that are obtained by specializing β to be each of the **basic unit vectors** $(1, 0, 0, \ldots, 0)$, $(0, 1, 0, \ldots, 0)$, $\ldots$, $(0, 0, \ldots, 0, 1)$ in turn. There are a variety of notations in use; depending upon the circumstances, one may be more convenient than another, and the table below gives most of the more common ones. Since the case of three variables is typical, we treat this alone.

A preliminary word of caution and explanation is needed. It is customary to use certain notations in mathematics, even when this can lead to confusion or misunderstandings. In particular, this is true of partial derivatives; the special usage of "variables" in the table (for example, w_x and $\partial w/\partial x$) must be imitated with care.

$w = f(x, y, z)$

$\beta =$	(1, 0, 0)	(0, 1, 0)	(0, 0, 1)
$\mathbf{D}_\beta f =$	f_1	f_2	f_3
	$D_1 f$	$D_2 f$	$D_3 f$
	$\dfrac{\partial f}{\partial x}$	$\dfrac{\partial f}{\partial y}$	$\dfrac{\partial f}{\partial z}$
	f_x	f_y	f_z
	$\dfrac{\partial w}{\partial x}$	$\dfrac{\partial w}{\partial y}$	$\dfrac{\partial w}{\partial z}$
	w_x	w_y	w_z

Of these, the alternatives in the first two rows are less likely to cause confusion; the numerical subscripts refer to the coordinate variables in order (e.g., the *first* coordinate) and can be used regardless of what letter is used for the corresponding coordinate.

To obtain the formal definition of $f_1 = D_1 f$, we need only take $\beta = (1, 0, 0)$ in the definition of $\mathbf{D}_\beta f$. Thus, we have

$$f_1(x, y, z) = \lim_{t \to 0} \frac{f(x + t, y, z) - f(x, y, z)}{t}$$

From this, we see that f_1 may be obtained by treating $f(x, y, z)$ as a function of x alone, with y and z held constant, and then differentiating the resulting function of one real variable in the usual fashion. For this reason, we refer to f_1 as the partial of f with respect to the first variable.

To illustrate the use of these notations, take

$$f(x, y, z) = w = x^2 y + y^3 \sin(z^2)$$

Then,

$$f_1(x, y, z) = \frac{\partial w}{\partial x} = 2xy$$

$$f_2(x, y, z) = \frac{\partial w}{\partial y} = x^2 + 3y^2 \sin(z^2)$$

$$f_3(x, y, z) = \frac{\partial w}{\partial z} = 2y^3 z \cos(z^2)$$

Since a partial derivative of a function of several variables is again such a function, the operation may be repeated, and if the appropriate limits exist, we obtain higher partial derivatives.

$$f_{11}(x, y, z) = \frac{\partial^2 w}{\partial x^2} = \frac{\partial}{\partial x}\left(\frac{\partial w}{\partial x}\right) = 2y$$

$$f_{12}(x, y, z) = \frac{\partial^2 w}{\partial y\, \partial x} = \frac{\partial}{\partial y}\left(\frac{\partial w}{\partial x}\right) = 2x$$

$$f_{22}(x, y, z) = \frac{\partial^2 w}{\partial y^2} = \frac{\partial}{\partial y}\left(\frac{\partial w}{\partial y}\right) = 6y \sin (z^2)$$

$$f_{21}(x, y, z) = \frac{\partial^2 w}{\partial x\, \partial y} = \frac{\partial}{\partial x}\left(\frac{\partial w}{\partial y}\right) = 2x$$

More generally, when β is fixed, $\mathbf{D}_\beta f$ is again a function, and one can consider its directional derivatives, such as $\mathbf{D}_\alpha \mathbf{D}_\beta f$, where α is again a direction, possibly the same as β.

In order to implement the definition of $\mathbf{D}_\beta f$ at a point p_0, it is convenient to have f defined on a neighborhood of p_0. For this reason, we will differentiate functions mostly on open sets, and leave the consideration of differentiation at boundary points until later.

Definition 1 *Let f be defined and continuous on an open set $D \subset \mathbf{R}^n$. Then, f is said to be of class C^k in D if all the partial derivatives of f of order up to and including k exist and are continuous everywhere in D. The symbols C' and C'' are sometimes used instead of C^1 and C^2.*

It is important in many situations that the partial derivatives involved be continuous and not merely exist. In the next chapter, we show that if $f \in C''$ in the plane, the mixed derivatives f_{12} and f_{21} must be the same. More generally, if $f \in C^k$, then any two mixed kth-order partial derivatives involving the same variables are equal; in C^4, $f_{xyxx} = f_{xxxy}$. However, Exercise 11 gives an instance where $f_{12} \neq f_{21}$, even though the first partial derivatives f_1 and f_2 are continuous and all the second-order derivatives exist. There are also other differences between differentiation in one variable and in several variables. In the former case, a function must be continuous to have a derivative; in Exercise 4, an example is given of a discontinuous function for which f_1 and f_2 exist everywhere. However, if the partial derivatives are continuous, the function is also continuous. This follows from our next result, which is the first version of the mean value theorem for several variables.

Lemma 1 *Let $f \in C'$ in an open ball $B(p_0, r)$ about the point p_0 in n space. Let $p \in B$, and set*

$$p - p_0 = \Delta p = (\Delta x_1, \Delta x_2, \ldots, \Delta x_n)$$

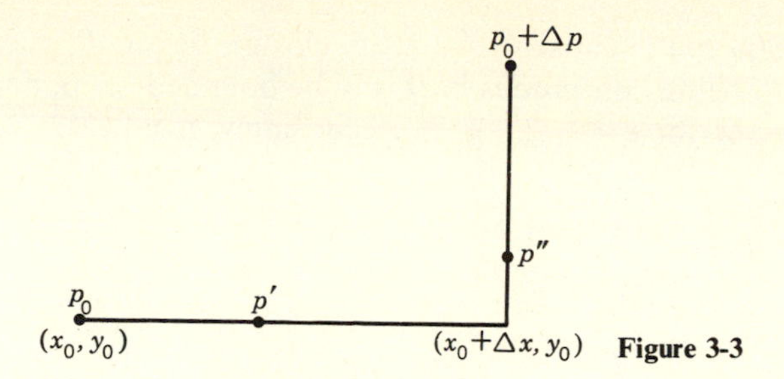

Figure 3-3

Then, there are points $p_1, p_2, \ldots, p_n$ in B such that

(3-7) $$f(p) - f(p_0) = f_1(p_1)\,\Delta x_1 + f_2(p_2)\,\Delta x_2 + \cdots + f_n(p_n)\,\Delta x_n$$

In this version, each of the partial derivatives f_i is computed at a different point p_i; later, we shall improve this by showing that the points p_i can all be replaced by a single properly chosen point p^* that lies on the line joining p_0 and p.

We prove the lemma only in the two-variable case, since this illustrates the general method. It is also more convenient to change the notation, putting $\Delta p = (\Delta x, \Delta y)$. Let q be the point $p_0 + (\Delta x, 0) = (x_0 + \Delta x, y_0)$, and write

$$f(p) - f(p_0) = [f(p) - f(q)] + [f(q) - f(p_0)]$$

Noting that in each bracketed term, only one variable has been altered, we apply the one-variable mean value theorem to each. Accordingly, there is an x' between x_0 and $x_0 + \Delta x$, and a y' between y_0 and $y_0 + \Delta y$, yielding points p' and p'' as indicated in Fig. 3-3, such that

$$\begin{aligned} f(q) - f(p_0) &= f(x_0 + \Delta x, y_0) - f(x_0, y_0) \\ &= \Delta x\, f_1(x', y_0) = f_1(p')\,\Delta x \end{aligned}$$

and

$$\begin{aligned} f(p) - f(q) &= f(x_0 + \Delta x, y_0 + \Delta y) - f(x_0 + \Delta x, y_0) \\ &= \Delta y\, f_2(x_0 + \Delta x, y') = f_2(p'')\,\Delta y \end{aligned}$$

Adding these, we have

$$f(p) - f(p_0) = f_1(p')\,\Delta x + f_2(p'')\,\Delta y$$

which is the two-variable form of (3-7). ▮

Corollary *If all the first partial derivatives of f exist and are continuous in an open set D, then f itself is continuous in D.*

The proof of the lemma did not use the continuity of f, since only the existence of each f_i was needed for the appropriate one-variable mean value theorem. If the ball $B(p_0, r)$ is chosen so that its closure lies in D, then each of the functions f_i, being continuous in D, will be bounded on B, and we will have $|f_i(p_i)| \le M$ for $i = 1, 2, \ldots, n$. Accordingly, from (3-7) we will have

$$\begin{aligned} |f(p) - f(p_0)| &\le M|\Delta x_1| + M|\Delta x_2| + \cdots + M|\Delta x_n| \\ &\le nM|p - p_0| \end{aligned}$$

and $\lim_{p \to p_0} f(p) = f(p_0)$. ▮

The role of a single comprehensive derivative of a function of several variables is filled by the vector-valued derivative of f, which we write as $\mathbf{D}f$. This is sometimes called the **total derivative** of f, to distinguish it from the numerical-valued partial derivatives of f. (Some mathematicians are beginning to write f' instead of $\mathbf{D}f$, but we prefer the latter to avoid possible confusion with the one-variable case.) When f is a function of three variables, $\mathbf{D}f$ is the same as the **gradient** of f, usually written ∇f.

Definition 2 *Let $f \in C'$ in an open set S in n space. Then, the derivative of f is the vector-valued function $\mathbf{D}f$ defined in S by*

$$\mathbf{D}f(p) = (f_1(p), f_2(p), f_3(p), \ldots, f_n(p)) \tag{3-8}$$

Note that $\mathbf{D}f$ is continuous in S, since each component function f_i is continuous in S. To illustrate the definition, if $f(x, y, z) = x^2y - y^3z^2$, then

$$\mathbf{D}f(x, y, z) = (2xy, x^2 - 3y^2z^2, -2y^3z)$$

For functions of one variable $\mathbf{D}f$ coincides with f'. Here, the derivative has a geometric meaning, explained by the familiar diagram showing a tangent to the graph of f and the statement that this line provides a good approximation to the function near the point of tangency. The corresponding analytic statement is the following: *If f is differentiable at x_0 and $\Delta x = x - x_0$, then the remainder function*

$$R = R(\Delta x) = f(x) - f(x_0) - f'(x_0)\,\Delta x \tag{3-9}$$

approaches 0 *faster than Δx, meaning explicitly that*

$$\lim_{x \to x_0} \frac{R(\Delta x)}{|\Delta x|} = 0$$

Equivalently, given $\varepsilon > 0$ there is a neighborhood $\mathcal{N}$ about x_0 such that $|R| < |\Delta x|\varepsilon$ for all x in $\mathcal{N}$.

Similar statements hold for functions of several variables, and the vector-valued derivative $\mathbf{D}f$, but the geometry is harder to visualize except in the special case of functions of two variables. In Fig. 3-4, we show the graph of such a function, and the tangent plane at the point $P = (0, 0, f(0, 0))$. Among

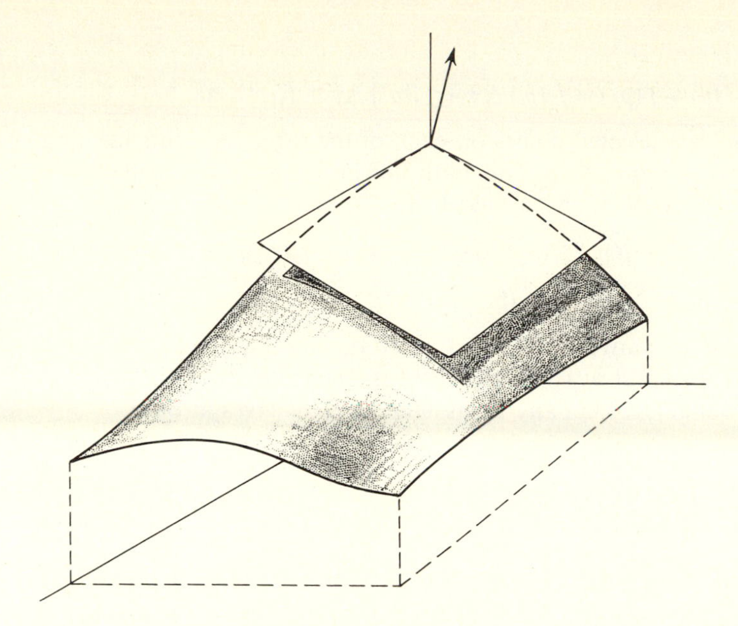

Figure 3-4 Tangent plane.

all the planes through P, the tangent plane fits f best on a neighborhood of $p_0 = (0, 0)$. The corresponding analytic statement for the general case is called the **local approximation theorem,** and is basic to the rest of the chapter. Note that as stated below, formula (3-10) closely resembles (3-9), except that the *scalar* product of the vectors $\mathbf{D}f(p_0)$ and Δp replaces the ordinary product of the numbers $f'(x_0)$ and Δx.

Theorem 8 *Let $f \in C'$ in an open set S. For any p_0, p in S, define the remainder function $R = R(p_0, p)$ by*

$$R = f(p) - f(p_0) - \mathbf{D}f(p_0) \cdot \Delta p \tag{3-10}$$

where $\Delta p = p - p_0$. Then, R approaches 0 faster than Δp, meaning that

$$\lim_{p \to p_0} \frac{R}{|\Delta p|} = 0 \tag{3-11}$$

Equivalently, for any $\varepsilon > 0$ there is a neighborhood $\mathcal{N}$ depending on both p_0 and ε such that $|R| < |\Delta p| \varepsilon$ for all $p \in \mathcal{N}$.

For a function of three variables, with $p = (x, y, z)$ and $p_0 = (x_0, y_0, z_0)$, formula (3-10) giving R translates into

$$R = f(p) - f(p_0) - f_1(p_0)\,\Delta x - f_2(p_0)\,\Delta y - f_3(p_0)\,\Delta z \tag{3-12}$$

We give the proof of Theorem 8 in this case; it is easily generalized to n variables. We first use the mean value theorem given in the previous lemma to write

$$f(p) - f(p_0) = f_1(p')\,\Delta x + f_2(p'')\,\Delta y + f_3(p''')\,\Delta z$$

where p', p'', p''' are selected points on each of the three segments forming a polygonal path from p_0 to p. This path lies in S if p lies in a sufficiently small neighborhood of p_0. Returning to (3-12), this yields

$$R = \{f_1(p') - f_1(p_0)\}\,\Delta x + \{f_2(p'') - f_2(p_0)\}\,\Delta y + \{f_3(p''') - f_3(p_0)\}\,\Delta z \tag{3-13}$$

Since the partial derivatives f_1, f_2, f_3 are continuous at p_0, we can choose a neighborhood $\mathcal{N}$ about p_0 so that each of the terms in (3-13) in brackets has absolute value less than any preassigned ε. This gives at once

$$\begin{aligned} |R| &\le \varepsilon|\Delta x| + \varepsilon|\Delta y| + \varepsilon|\Delta z| \\ &\le (3\varepsilon)|\Delta p| \end{aligned}$$

for any $p \in \mathcal{N}$.

Small changes in this proof yield a slightly better result. Suppose that E is a compact subset of S. Since the distance from E to the boundary of S is strictly positive, we can choose a larger compact set E_0, with $E \subset E_0 \subset S$. and $\delta > 0$, so that if $p_0 \in E$ and $|p - p_0| < \delta$, then $p \in E_0$. The partial derivatives f_i are continuous in E_0, and therefore uniformly continuous in E_0. One is then able to show that given any $\varepsilon > 0$, there is a radius ρ, depending only on ε, f, and the sets E and S, such that $|R| < |\Delta p|\varepsilon$ for any choice of p_0 in E, and any p with $|p - p_0| < \rho$. This fact can be described by saying that the limit in (3-11) is "uniformly for all $p_0 \in E$." ▌ (The general topic of uniform convergence will be taken up systematically in Chap. 6.)

There is also a converse to Theorem 8 which serves to characterize the vector derivative $\mathbf{D}f$.

Theorem 9 *Let f be continuous on a neighborhood of p_0, and suppose that there is a vector u such that*

$$\lim_{\Delta p \to \mathbf{0}} \frac{f(p_0 + \Delta p) - f(p_0) - u \cdot \Delta p}{|\Delta p|} = 0 \tag{3-14}$$

Then, the partial derivatives of f exist at p_0 and

$$u = \mathbf{D}f(p_0)$$

For, if $u = (u_1, u_2, \ldots, u_n)$, then by specializing Δp to approach $\mathbf{0}$ along each of the coordinate axes in turn, one finds that $u_i = f_i(p_0)$. ▌

Formula (3-14) can be taken as the basis for a general theory of differentiation. For example, f is said to be differentiable in an open set D if there is such a vector u corresponding to each point $p_0 \in D$, and u obeys (3-14). As

Theorem 9 shows, this will lead us back to the vector derivative $\mathbf{D}f$. This approach has advantages when we take up differentiation of general transformations from n-space into m-space, and we will adopt it in Chap. 7.

The approximation theorem has several immediate consequences that are very useful. The first relates the vector derivative $\mathbf{D}f$ to directional derivatives.

Theorem 10 *If $f \in C'$ in an open set S, then all its directional derivatives exist at any point $p \in S$, and $D_\beta f(p) = \beta \cdot \mathbf{D}f(p)$.*

Corollary *The derivative $\mathbf{D}f$ at a point p is a vector that points in the direction of the maximum rate of change of f at p, and whose length is the derivative of f in this direction.*

Apply the approximation theorem with $\Delta p = \lambda\beta$. Then,

$$f(p + \lambda\beta) - f(p) - \mathbf{D}f(p) \cdot \lambda\beta = R$$

so that

$$\frac{f(p + \lambda\beta) - f(p)}{\lambda} = \beta \cdot \mathbf{D}f(p) + \frac{R}{\lambda}$$

But, $\lambda = |\Delta p|$, since β is a unit vector, and $\lim_{\lambda \to 0} R/\lambda = 0$, which yields $\mathbf{D}_\beta f(p) = \beta \cdot \mathbf{D}f(p)$. ∎

The corollary arises from the simple observation that as the vector β varies, the maximum value of $\beta \cdot \mathbf{D}f(p)$ will occur when the angle between β and $\mathbf{D}f(p)$ is 0, and this maximum value will then be $|\mathbf{D}f(p)|$.

Many of the results in the preceding section have analogs for functions of several variables.

Theorem 11 *Let $f \in C'$ in an open set S, and suppose that f has a local maximum (minimum) at a point $p_0 \in S$. Then, $\mathbf{D}f(p_0) = \mathbf{0}$. Thus, all the partial derivatives of f vanish at p_0.*

Suppose p_0 is a local maximum for f. Thus, $f(p) \le f(p_0)$ for all p sufficiently near p_0. Accordingly, $\mathbf{D}_\beta f(p_0) \le 0$ for every choice of the direction β. Writing this as $\beta \cdot \mathbf{D}f(p_0) \le 0$ and then replacing β by $-\beta$, we have $(-\beta) \cdot \mathbf{D}f(p_0) \le 0$, so that $\beta \cdot \mathbf{D}f(p_0) = 0$. This cannot hold for all directions β unless $\mathbf{D}f(p_0) = \mathbf{0}$. ∎

Any point where $\mathbf{D}f = 0$ is said to be a **critical point** for f. Every local extremum of f occurs at a critical point; however, not every critical point gives an extremum, since some yield saddle points, as will be seen in Sec. 3.6.

Theorem 12 *If $f \in C'$ in an open set S which is also connected, and $\mathbf{D}f = \mathbf{0}$ everywhere in S, then f is constant in S.*

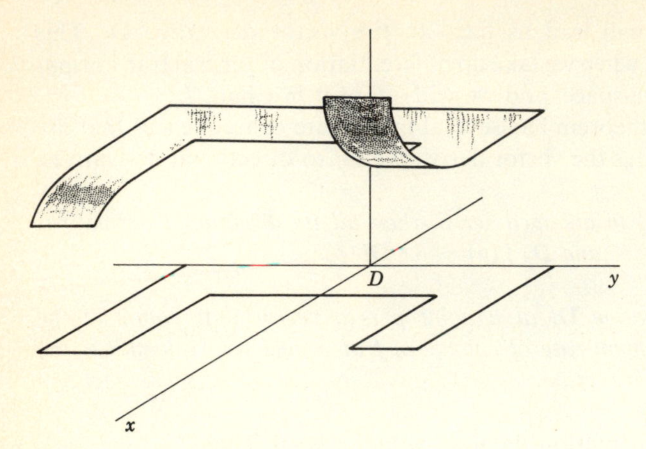

Figure 3-5

Using the mean value lemma, formula (3-7), and the fact that the partial derivatives $f_1, f_2, \ldots, f_n$ vanish everywhere in S, we must have $f(p) - f(p_0) = 0$ for all p sufficiently near p_0. Thus, f is locally constant in S. Since S is connected, it follows that f is globally constant in S (Exercise 8, Sec. 2.4). ▌

It is natural to ask what happens if just one of the partial derivatives is identically 0. Such is the case, for example, in the function f

$$f(x, y, z) = x^2 + y^2$$

where z is missing, and $\partial f/\partial z = 0$ at all points.

Theorem 13 *Let $w = f(x, y, z)$ where f is of class C' in a convex open set D, and let $\partial w/\partial z = 0$ throughout D. Then, z is missing, in the sense that*

$$f(x, y, z') = f(x, y, z'')$$

whenever (x, y, z') and (x, y, z'') are both in D.

The proof of this is again a simple consequence of the mean value lemma, by use of the fact that the entire segment between (x, y, z') and (x, y, z'') lies in D. The need for some restriction on D is seen by the two-variable example in Fig. 3-5. Here, D is not convex, and the graph is a function that is *locally* independent of y but not independent of y in all D.

EXERCISES

1 Find $f_1(x, y), f_2(x, y), f_{12}(x, y)$ if
  (a) $f(x, y) = x^2 \log (x^2 + y^2)$
  (b) $f(x, y) = x^y$.

2 With $f(x, y) = x^2y^3 - 2y$, find $f_1(x, y), f_2(x, y), f_2(2, 3)$, and $f_2(y, x)$.

3 Compute $\mathbf{D}f$ for each of the following functions at the given point:

(*a*) $f(x, y) = 3x^2y - xy^3 + 2$ at $(1, 2)$

(*b*) $f(u, v) = u \sin(uv)$ at $(\pi/4, 2)$

(*c*) $f(x, y, z) = x^2yz + 3xz^2$ at $(1, 2, -1)$.

4 (*a*) Let $f(x, y) = xy/(x^2 + y^2)$, with $f(0, 0) = 0$. Show that f_1 and f_2 exist everywhere, but that f is not of class C'.

(*b*) Does f have directional derivatives at the origin?

(*c*) Is f continuous at the origin?

5 Let a function f be defined in an open set D of the plane, and suppose that f_1 and f_2 are defined and bounded everywhere in D. Show that f is continuous in D.

6 Can you formulate and prove an analog for Rolle's theorem, for functions of two real variables?

7 Let f and g be of class C' in a compact set S, and let $f = g$ on bdy (S). Show that there must exist a point $p_0 \in S$ where $\mathbf{D}f(p_0) = \mathbf{D}g(p_0)$.

8 Find the derivative of $f(x, y, z) = xy^2 + yz$ at the point $(1, 1, 2)$ in the direction $(2/3, -1/3, 2/3)$.

9 Let $f(x, y) = xy$. Show that the direction of the gradient of f is always perpendicular to the level lines of f.

10 Show that each of the following obeys $\partial^2 u/\partial x^2 + \partial^2 u/\partial y^2 = 0$:

(*a*) $u = e^x \cos y$ (*b*) $u = \exp(x^2 - y^2) \sin(2xy)$

11 Let $f(x, y) = xy(x^2 - y^2)/(x^2 + y^2)$ with $f(0, 0) = 0$. Show that f is continuous everywhere, that f_1, f_2, f_{12}, and f_{21} exist everywhere, but $f_{12}(0, 0) \neq f_{21}(0, 0)$.

12 Find the directional derivative of $F(x, y, z) = xyz$ at $(1, 2, 3)$ in the direction from this point toward the point $(3, 1, 5)$.

13 If $F(x, y, z, w) = x^2y + xz - 2yw^2$, find the derivative of F at $(1, 1, -1, 1)$ in the direction $\beta = (4/7, -4/7, 1/7, -4/7)$.

14 Some economics students have been quoted as saying the following: $F(x_1, x_2, \ldots, x_n)$ is such that it does not change if you change only one variable, leaving the rest alone, but it does change if you make changes in two of them. What reaction would you give to such a statement?

3.4 DIFFERENTIATION OF COMPOSITE FUNCTIONS

A function may often be regarded as built up by composition from a number of other functions. If $f(x, y) = xy^2 + x^2$, $g(x, y) = y \sin x$, and $h(x) = e^x$, then a function F may be defined by

$$F(x, y) = f(g(x, y), h(x)) = ye^{2x} \sin x + y^2 \sin^2(x) \tag{3-15}$$

The introduction of additional variable symbols sometimes helps to clarify such relations. For example, an equivalent description of (3-15) is obtained by setting $w = F(x, y)$, and writing

$$\begin{aligned} w &= f(u, v) = uv^2 + u^2 \\ u &= g(x, y) = y \sin x \\ v &= h(x) = e^x \end{aligned} \tag{3-16}$$

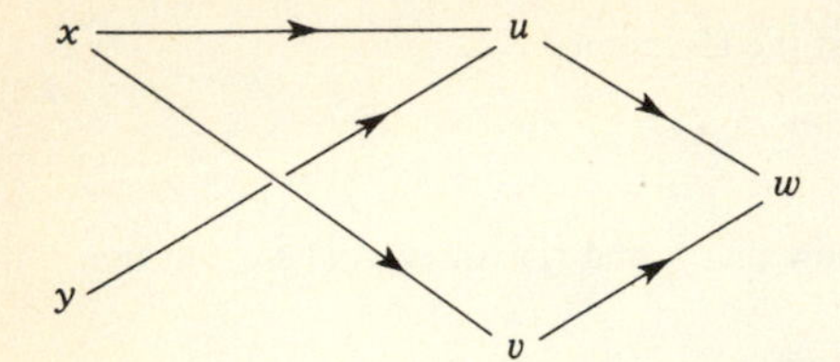

Figure 3-6

These equations express w in terms of x and y indirectly through the intermediate variables u and v. The interdependence involved in this particular example may also be indicated schematically, as in Fig. 3-6.

The main concern of this section is the theory and application of the rules for obtaining derivatives of such composite functions. The general case is postponed to Chap. 7. As an example of the so-called "chain rules of differentiation," we shall here prove a special case which illustrates the method of proof.

Theorem 14 *Let $F(t) = f(x, y)$, where $x = g(t)$ and $y = h(t)$. Here, g and h are assumed to be class C' on a neighborhood of t_0, and f of class C' on a neighborhood of $p_0 = (x_0, y_0)$ where $x_0 = g(t_0)$, $y_0 = h(t_0)$. Then, F is of class C' on a neighborhood of t_0, and*

$$F'(t) = f_1(p)g'(t) + f_2(p)h'(t) \tag{3-17}$$

where $p = (g(t), h(t))$.

The differentiation formula (3-17) is more lucid if we alter the notation and write $w = F(t)$, for (3-17) then becomes

$$\frac{dw}{dt} = \frac{\partial w}{\partial x}\frac{dx}{dt} + \frac{\partial w}{\partial y}\frac{dy}{dt}$$

To prove this, we must calculate $F(t + \Delta t) - F(t)$.
Set $\Delta x = g(t + \Delta t) - g(t)$, $\Delta y = h(t + \Delta t) - h(t)$. Then,

$$\begin{aligned} F(t + \Delta t) - F(t) &= f(x + \Delta x, y + \Delta y) - f(x, y) \\ &= f(p + \Delta p) - f(p) \\ &= \mathbf{D}f(p) \cdot \Delta p + R \\ &= f_1(p)\,\Delta x + f_2(p)\,\Delta y + R \end{aligned}$$

where $\lim_{\Delta p \to 0} |R|/|\Delta p| = 0$. Dividing by Δt, we have

$$\frac{F(t + \Delta t) - F(t)}{\Delta t} = f_1(p)\frac{\Delta x}{\Delta t} + f_2(p)\frac{\Delta y}{\Delta t} + \frac{R}{\Delta t} \tag{3-18}$$

As $\Delta t \to 0$, $\Delta x/\Delta t \to g'(t)$ and $\Delta y/\Delta t \to h'(t)$. Also, since $|\Delta p| \le |\Delta x| + |\Delta y|$,

$|\Delta p|/|\Delta t|$ is bounded and

$$\left|\frac{R}{\Delta t}\right| = \frac{|R|}{|\Delta p|}\frac{|\Delta p|}{|\Delta t|} \to 0$$

Thus, as $\Delta t \to 0$, (3-18) becomes (3-17). ▮

Using this, one can easily obtain more elaborate chain rules for calculating partial derivatives of composite functions. For example let $w = f(u, v)$ with $u = g(x, y)$, $v = h(x, y)$. This defines a function F by $w = F(x, y)$. To find the partial derivative. $F_1 = \partial w/\partial x$, we need the directional derivative of F along the X axis. This amounts to holding y constant and finding the derivative of the resulting function of x alone. Using the chain rule obtained in Theorem 14, we arrive at

$$F_1(x, y) = \frac{\partial w}{\partial x} = \frac{\partial w}{\partial u}\frac{\partial u}{\partial x} + \frac{\partial w}{\partial v}\frac{\partial v}{\partial x}$$

as expected; in a similar way, one finds

$$F_2(x, y) = \frac{\partial w}{\partial y} = \frac{\partial w}{\partial u}\frac{\partial u}{\partial y} + \frac{\partial w}{\partial v}\frac{\partial v}{\partial y}$$

Note that these could also have been written

$$F_1 = f_1 g_1 + f_2 h_1$$
$$F_2 = f_1 g_2 + f_2 h_2$$

To illustrate this, the example (3-16) yields

$$\frac{\partial w}{\partial x} = (v^2 + 2u)(y \cos x) + (2uv)(e^x)$$

and

$$\frac{\partial w}{\partial y} = (v^2 + 2u)(\sin x) + (2uv)(0)$$

The next illustration is somewhat more complicated; it also shows how the quotient notation for partial derivatives is sometimes ambiguous. Let w be related to x and y by the following equations:

$$w = f(x, u, v) \qquad u = (x, v, y) \qquad v = h(x, y) \tag{3-19}$$

The corresponding diagram is shown in Fig. 3-7. We see that the dependence of w upon x is complicated by the fact that x enters in directly, and also through u and v. Each path in the diagram joining x to w corresponds to a term in the formula for $\partial w/\partial x$, so that we obtain

$$\frac{\partial w}{\partial x} = \frac{\partial w}{\partial x} + \frac{\partial w}{\partial u}\frac{\partial u}{\partial x} + \frac{\partial w}{\partial v}\frac{\partial v}{\partial x} + \frac{\partial w}{\partial u}\frac{\partial u}{\partial v}\frac{\partial v}{\partial x} \tag{3-20}$$

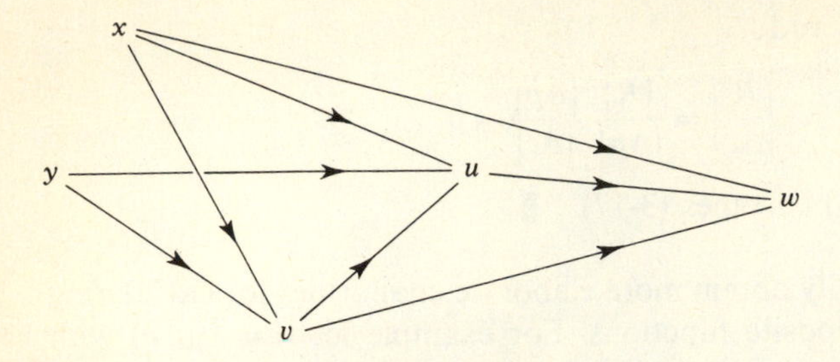

Figure 3-7

Since y enters in through v and u,

$$\frac{\partial w}{\partial y} = \frac{\partial w}{\partial u}\frac{\partial u}{\partial y} + \frac{\partial w}{\partial v}\frac{\partial v}{\partial y} + \frac{\partial w}{\partial u}\frac{\partial u}{\partial v}\frac{\partial v}{\partial y} \tag{3-21}$$

In both of these formulas, the partial derivatives must be understood in the correct context of Eqs. (3-19) above. In (3-20), for example, the first occurrence of "$\partial w/\partial x$" refers to the partial derivative of w, regarding it as "a function of the independent variables x and y." (This is often indicated by writing $\partial w/\partial x|_y$ to show that y is being held constant.) The second occurrence of "$\partial w/\partial x$", however, refers to the partial derivative of w regarded as a function of the independent variables x, u, v. The use of numerical subscripts helps to remove such ambiguity. We may write (3-20) and (3-21) in the alternative forms

$$\frac{\partial w}{\partial x} = f_1 + f_2 g_1 + f_3 h_1 + f_2 g_2 h_1$$

$$\frac{\partial w}{\partial y} = f_2 g_3 + f_3 h_2 + f_2 g_2 h_2$$

As another illustration of the use of chain rules, consider the following relationship:

$$w = F(x, y, t) \qquad x = \phi(t) \qquad y = \psi(t)$$

These express w in the form $w = f(t)$, as shown by the diagram in Fig. 3-8 and

$$\frac{dw}{dt} = \frac{\partial w}{\partial t} + \frac{\partial w}{\partial x}\frac{dx}{dt} + \frac{\partial w}{\partial y}\frac{dy}{dt} \tag{3-22}$$

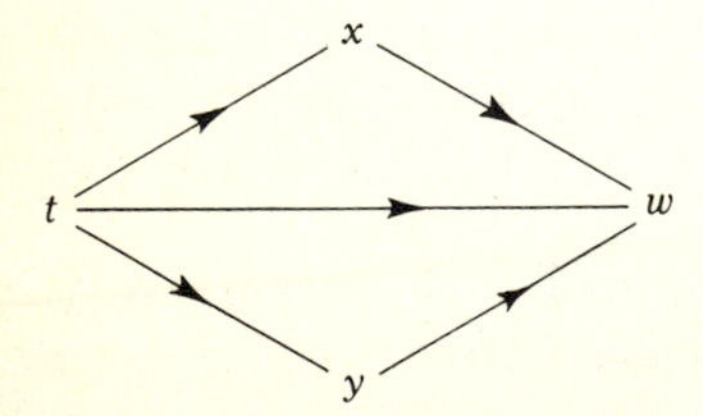

Figure 3-8

This procedure can also be used to compute higher derivatives too. For example, to find d^2w/dt^2, we rewrite (3-22) in the form

$$\frac{dw}{dt} = F_3(x, y, t) + F_1(x, y, t)\frac{dx}{dt} + F_2(x, y, t)\frac{dy}{dt}$$

and then differentiate a second time, getting

$$\frac{d^2w}{dt^2} = F_{33} + F_{31}\frac{dx}{dt} + F_{32}\frac{dy}{dt} + F_{11}\left(\frac{dx}{dt}\right)^2 + F_{12}\frac{dx}{dt}\frac{dy}{dt} + F_{13}\frac{dx}{dt}$$
$$+ F_{22}\left(\frac{dy}{dt}\right)^2 + F_{21}\frac{dx}{dt}\frac{dy}{dt} + F_{23}\frac{dy}{dt} + F_1\frac{d^2x}{dt^2} + F_2\frac{d^2y}{dt^2}$$

Assuming that F is of class C'', this may also be expressed in the form

$$\frac{d^2w}{dt^2} = \frac{\partial^2 w}{\partial t^2} + 2\left\{\frac{\partial^2 w}{\partial x\,\partial t}\frac{dx}{dt} + \frac{\partial^2 w}{\partial y\,\partial t}\frac{dy}{dt}\right\} + \frac{\partial^2 w}{\partial x^2}\left(\frac{dx}{dt}\right)^2$$
$$+ 2\frac{\partial^2 w}{\partial x\,\partial y}\frac{dx}{dt}\frac{dy}{dt} + \frac{\partial^2 w}{\partial y^2}\left(\frac{dy}{dt}\right)^2 + \frac{\partial w}{\partial x}\frac{d^2x}{dt^2} + \frac{\partial w}{\partial y}\frac{d^2y}{dt^2}$$

Another type of problem in which the chain rules prove useful is that of finding formulas for the derivatives of functions which are defined "implicitly." Consider the pair of equations

$$\text{(3-23)} \qquad \begin{cases} x^2 + ux + y^2 + v = 0 \\ x + yu + v^2 + x^2 v = 0 \end{cases}$$

If we give x and y numerical values, we obtain a pair of algebraic equations which have one or more solutions for u and v; for some choices of x and y, these solutions will be real, so that Eqs. (3-23) serve to define one or more functions f and g such that

$$\begin{cases} u = f(x, y) \\ v = g(x, y) \end{cases}$$

For example, if $x = 1$, $y = 1$, then (3-23) becomes

$$\begin{cases} u + v + 2 \qquad\quad = 0 \\ u + v^2 + v + 1 = 0 \end{cases}$$

which has the solutions $(u, v) = (-1, -1)$ and $(u, v) = (-3, 1)$. General theorems, which we shall discuss later in Sec. 7.6, show that there are functions f and g defined in a neighborhood $\mathcal{N}$ of $(1, 1)$ such that $f(1, 1) = -3$, $g(1, 1) = 1$, and such that (3-23) holds with these substitutions for all (x, y) in $\mathcal{N}$. Knowing the function f, it would then be possible to compute $\partial u/\partial x = f_1$ and find $f_1(1, 1) = \partial u/\partial x|_{(1,1)}$.

By the use of the chain rules, it is possible to compute such partial derivatives without carrying out the often difficult task of solving explicitly for the functions f and g. To achieve this for the specific example above, differentiate each of the equations in (3-23) with respect to x while holding y constant.

$$2x + u + x\frac{\partial u}{\partial x} + 0 + \frac{\partial v}{\partial x} = 0$$

$$1 + y\frac{\partial u}{\partial x} + 2v\frac{\partial v}{\partial x} + 2xv + x^2\frac{\partial v}{\partial x} = 0$$

Solving these for $\partial u/\partial x$, we find

$$\frac{\partial u}{\partial x} = \frac{1 - 2xv - 2uv - 2x^3 - ux^2}{2xv + x^3 - y}$$

If we set $x = 1$, $y = 1$, $u = -3$, $v = 1$, we find the desired value

$$\left.\frac{\partial u}{\partial x}\right|_{(1,\,1)} = f_1(1, 1) = 3$$

We can obtain a formula for the solution of a general class of such problems. Suppose that one is given two equations

$$\text{(3-24)} \qquad \begin{cases} F(x, y, u, v) = 0 \\ G(x, y, u, v) = 0 \end{cases}$$

which we may regard as solvable for u and v in terms of x and y. We wish to find $\partial u/\partial x$ and $\partial v/\partial x$. Holding y constant, differentiate (3-24) with respect to x.

$$F_1 + F_3\frac{\partial u}{\partial x} + F_4\frac{\partial v}{\partial x} = 0$$

$$G_1 + G_3\frac{\partial u}{\partial x} + G_4\frac{\partial v}{\partial x} = 0$$

Solving these, we obtain the desired formulas

$$\frac{\partial u}{\partial x} = -\frac{\begin{vmatrix} F_1 & F_4 \\ G_1 & G_4 \end{vmatrix}}{\begin{vmatrix} F_3 & F_4 \\ G_3 & G_4 \end{vmatrix}} = -\frac{F_1G_4 - F_4G_1}{F_3G_4 - F_4G_3}$$

$$\frac{\partial v}{\partial x} = -\frac{\begin{vmatrix} F_3 & F_1 \\ G_3 & G_1 \end{vmatrix}}{\begin{vmatrix} F_3 & F_4 \\ G_3 & G_4 \end{vmatrix}} = -\frac{F_3G_1 - F_1G_3}{F_3G_4 - F_4G_3}$$

Determinants of this special form are common enough to have acquired a special name and notation; they are called **Jacobians,** and the notation is illustrated by

$$\frac{\partial(A, B)}{\partial(s, t)} = \begin{vmatrix} \dfrac{\partial A}{\partial s} & \dfrac{\partial A}{\partial t} \\ \dfrac{\partial B}{\partial s} & \dfrac{\partial B}{\partial t} \end{vmatrix} \tag{3-25}$$

Using this, we can write the formulas above in simpler form as

$$\begin{aligned} \frac{\partial u}{\partial x} &= -\frac{\partial(F, G)}{\partial(x, v)} \Big/ \frac{\partial(F, G)}{\partial(u, v)} \\ \frac{\partial v}{\partial x} &= -\frac{\partial(F, G)}{\partial(u, x)} \Big/ \frac{\partial(F, G)}{\partial(u, v)} \end{aligned} \tag{3-26}$$

To apply these to the pair of equations given in (3-23), we take

$$F(x, y, u, v) = x^2 + ux + y^2 + v$$

and

$$G(x, y, u, v) = x + yu + v^2 + x^2 v$$

Then,

$$\frac{\partial F}{\partial x} = 2x + u \qquad \frac{\partial F}{\partial u} = x \qquad \frac{\partial F}{\partial v} = 1$$

$$\frac{\partial G}{\partial x} = 1 + 2xv \qquad \frac{\partial G}{\partial u} = y \qquad \frac{\partial G}{\partial v} = 2v + x^2$$

so that

$$\frac{\partial u}{\partial x} = -\begin{vmatrix} 2x + u & 1 \\ 1 + 2xv & 2v + x^2 \end{vmatrix} \div \begin{vmatrix} x & 1 \\ y & 2v + x^2 \end{vmatrix}$$

$$= -\frac{2xv + 2uv + 2x^3 + ux^2 - 1}{2xv + x^3 - y}$$

and

$$\frac{\partial v}{\partial x} = -\begin{vmatrix} x & 2x + u \\ y & 1 + 2xv \end{vmatrix} \div \begin{vmatrix} x & 1 \\ y & 2v + x^2 \end{vmatrix}$$

$$= -\frac{x + 2x^2 v - 2xy - uy}{2xv + x^3 - y}$$

Further formulas of this nature will be found in the exercises. It should be pointed out that the Jacobian which occurs in the denominator of both fractions in (3-26) is one whose nonvanishing will be sufficient to ensure that Eqs. (3-24) really do have a solution in the form

$$u = f(x, y), \; v = g(x, y)$$

(see Sec. 7.6).

A somewhat more difficult problem is the following, which is patterned after a type of situation which occurs in physical applications. A physical law or hypothesis is often formulated as a partial differential equation. If a change of variables is made, what is the corresponding form for the differential equation? We first examine a simple case. Suppose the differential equation is $\partial^2u/\partial x^2 - \partial^2u/\partial y^2 = 0$, and suppose we wish to make the substitution

$$\begin{cases} x = s + t \\ y = s - t \end{cases} \tag{3-27}$$

By the appropriate chain rule,

$$\frac{\partial u}{\partial x} = \frac{\partial u}{\partial s}\frac{\partial s}{\partial x} + \frac{\partial u}{\partial t}\frac{\partial t}{\partial x}$$

$$\frac{\partial u}{\partial y} = \frac{\partial u}{\partial s}\frac{\partial s}{\partial y} + \frac{\partial u}{\partial t}\frac{\partial t}{\partial y}$$

Solving (3-27), we have $s = (x + y)/2$ and $t = (x - y)/2$, so that

$$\frac{\partial u}{\partial x} = \frac{\partial u}{\partial s}\left(\frac{1}{2}\right) + \frac{\partial u}{\partial t}\left(\frac{1}{2}\right) = \frac{1}{2}\left(\frac{\partial}{\partial s} + \frac{\partial}{\partial t}\right)(u)$$

and

$$\frac{\partial u}{\partial y} = \frac{\partial u}{\partial s}\left(\frac{1}{2}\right) + \frac{\partial u}{\partial t}\left(-\frac{1}{2}\right) = \frac{1}{2}\left(\frac{\partial}{\partial s} - \frac{\partial}{\partial t}\right)(u)$$

Repeating this, and assuming that $u = F(x, y)$ with F of class C'',

$$\frac{\partial^2 u}{\partial x^2} = \frac{\partial}{\partial x}\left(\frac{\partial u}{\partial x}\right) = \frac{1}{2}\left(\frac{\partial}{\partial s} + \frac{\partial}{\partial t}\right)\left(\frac{\partial u}{\partial s}\frac{1}{2} + \frac{\partial u}{\partial t}\frac{1}{2}\right)$$

$$= \frac{1}{4}\left(\frac{\partial^2 u}{\partial s^2} + 2\frac{\partial^2 u}{\partial s\,\partial t} + \frac{\partial^2 u}{\partial t^2}\right)$$

and

$$\frac{\partial^2 u}{\partial y^2} = \frac{1}{2}\left(\frac{\partial}{\partial s} - \frac{\partial}{\partial t}\right)\left(\frac{\partial u}{\partial s}\frac{1}{2} - \frac{\partial u}{\partial t}\frac{1}{2}\right)$$

$$= \frac{1}{4}\left(\frac{\partial^2 u}{\partial s^2} - 2\frac{\partial^2 u}{\partial s\,\partial t} + \frac{\partial^2 u}{\partial t^2}\right)$$

Subtracting these, we find that

$$\frac{\partial^2 u}{\partial x^2} - \frac{\partial^2 u}{\partial y^2} = \frac{\partial^2 u}{\partial s\,\partial t}$$

so that the transformed differential equation is $\partial^2u/\partial s\,\partial t = 0$.

A more complicated problem of the same type is that of transforming the Laplace equation $\partial^2u/\partial x^2 + \partial^2u/\partial y^2 = 0$ into polar coordinates by the substitution

$$\begin{cases} x = r\cos\theta \\ y = r\sin\theta \end{cases}$$

Differentiate the first of these with respect to y, and the second with respect to x, regarding x and y as independent.

(3-28)
$$0 = \frac{\partial r}{\partial y} \cos\theta - \frac{\partial \theta}{\partial y} r \sin\theta$$
$$0 = \frac{\partial r}{\partial x} \sin\theta + \frac{\partial \theta}{\partial x} r \cos\theta$$

Also, $x^2 + y^2 = r^2$, so that $2r(\partial r/\partial x) = 2x$ and $2r(\partial r/\partial y) = 2y$, and therefore

(3-29)
$$\frac{\partial r}{\partial x} = \frac{x}{r} = \cos\theta \qquad \frac{\partial r}{\partial y} = \frac{y}{r} = \sin\theta$$

Putting these into (3-28), we obtain

(3-30)
$$\frac{\partial \theta}{\partial x} = -\frac{\sin\theta}{r} \qquad \frac{\partial \theta}{\partial y} = \frac{\cos\theta}{r}$$

Then
$$\frac{\partial u}{\partial x} = \frac{\partial u}{\partial r}\frac{\partial r}{\partial x} + \frac{\partial u}{\partial \theta}\frac{\partial \theta}{\partial x} = \cos\theta \frac{\partial u}{\partial r} - \frac{\sin\theta}{r}\frac{\partial u}{\partial \theta}$$
$$= \left(\cos\theta \frac{\partial}{\partial r} - \frac{\sin\theta}{r}\frac{\partial}{\partial \theta}\right) u$$

and
$$\frac{\partial u}{\partial y} = \frac{\partial u}{\partial r}\frac{\partial r}{\partial y} + \frac{\partial u}{\partial \theta}\frac{\partial \theta}{\partial y} = \sin\theta \frac{\partial u}{\partial r} + \frac{\cos\theta}{r}\frac{\partial u}{\partial \theta}$$
$$= \left(\sin\theta \frac{\partial}{\partial r} + \frac{\cos\theta}{r}\frac{\partial}{\partial \theta}\right) u$$

Iterating these,

$$\frac{\partial^2 u}{\partial x^2} = \left(\cos\theta \frac{\partial}{\partial r} - \frac{\sin\theta}{r}\frac{\partial}{\partial \theta}\right)\left(\cos\theta \frac{\partial u}{\partial r} - \frac{\sin\theta}{r}\frac{\partial u}{\partial \theta}\right)$$
$$= \cos^2\theta \frac{\partial^2 u}{\partial r^2} + \frac{\sin^2\theta}{r^2}\frac{\partial^2 u}{\partial \theta^2} - \frac{2\sin\theta\cos\theta}{r}\frac{\partial^2 u}{\partial r\,\partial \theta}$$
$$+ \frac{2\sin\theta\cos\theta}{r^2}\frac{\partial u}{\partial \theta} + \frac{\sin^2\theta}{r}\frac{\partial u}{\partial r}$$

$$\frac{\partial^2 u}{\partial y^2} = \left(\sin\theta \frac{\partial}{\partial r} + \frac{\cos\theta}{r}\frac{\partial}{\partial \theta}\right)\left(\sin\theta \frac{\partial u}{\partial r} + \frac{\cos\theta}{r}\frac{\partial u}{\partial \theta}\right)$$
$$= \sin^2\theta \frac{\partial^2 u}{\partial r^2} + \frac{\cos^2\theta}{r^2}\frac{\partial^2 u}{\partial \theta^2} + \frac{2\sin\theta\cos\theta}{r}\frac{\partial^2 u}{\partial r\,\partial \theta}$$
$$- \frac{2\sin\theta\cos\theta}{r^2}\frac{\partial u}{\partial \theta} + \frac{\cos^2\theta}{r}\frac{\partial u}{\partial r}$$

and adding, we find that

$$\frac{\partial^2 u}{\partial x^2} + \frac{\partial^2 u}{\partial y^2} = \frac{\partial^2 u}{\partial r^2} + \frac{1}{r^2}\frac{\partial^2 u}{\partial \theta^2} + \frac{1}{r}\frac{\partial u}{\partial r}$$

Thus, the equation $\partial^2 u/\partial x^2 + \partial^2 u/\partial y^2 = 0$ which governs the distribution of heat becomes

$$\frac{\partial^2 u}{\partial r^2} + \frac{1}{r^2}\frac{\partial^2 u}{\partial \theta^2} + \frac{1}{r}\frac{\partial u}{\partial r} = 0$$

in polar coordinates.

Finally, let us consider a special type of change-of-variable problem. Suppose that E, T, V, and p are four physical variables which are connected by two relations, of the form

$$\begin{cases} \phi(E, T, V, p) = 0 \\ \psi(E, T, V, p) = 0 \end{cases} \tag{3-31}$$

We suppose also that these may be solved for any pair of the variables in terms of the remaining two, i.e., any pair may be selected as independent. When V and T are independent, the physical theory supplies the following differential relation between the variables

$$\frac{\partial E}{\partial V} - T\frac{\partial p}{\partial T} + p = 0 \tag{3-32}$$

Suppose that we wish to shift our point of view, and regard p and T as the independent variables; what form does the physical relation (3-32) take now? To answer this, two techniques can be used. One is to assume that equations (3-31) have been solved for E and V in terms of the new independent variables p and T in the form

$$\begin{aligned} E &= f(p, T) \\ V &= g(p, T) \end{aligned} \tag{3-33}$$

Since the differential equation (3-32) involves $\partial E/\partial V$ and $\partial p/\partial T$, derivatives in which V and T were the independent variables, we obtain these from (3-33) by differentiating these two equations with respect to V and T.

$$\frac{\partial E}{\partial V} = f_1\frac{\partial p}{\partial V} \qquad \frac{\partial E}{\partial T} = f_1\frac{\partial p}{\partial T} + f_2$$

$$1 = g_1\frac{\partial p}{\partial V} \qquad 0 = g_1\frac{\partial p}{\partial T} + g_2$$

Solve these for the derivatives that appear in (3-32):

$$\frac{\partial E}{\partial V} = \left.\frac{\partial E}{\partial V}\right|_T = \frac{f_1}{g_1} \qquad \frac{\partial p}{\partial T} = \left.\frac{\partial p}{\partial T}\right|_V = -\frac{g_2}{g_1}$$

Substituting these into (3-32), we obtain

$$f_1 + Tg_2 + pg_1 = 0$$

and then, using (3-33) to convert this into more familiar notation, we see that the new form for the physical law (3-32) is

$$\frac{\partial E}{\partial p} + T\frac{\partial V}{\partial T} + p\frac{\partial V}{\partial p} = 0 \tag{3-34}$$

The second technique is a dual of this. In the context of (3-32), E and p were expressible in terms of V and T.

$$\begin{aligned} E &= F(V, T) \\ p &= G(V, T) \end{aligned} \tag{3-35}$$

Thus, the derivatives that appear in (3-32) were $\partial E/\partial V = F_1$ and $\partial p/\partial T = G_2$, and the relation (3-32) could have been written as $F_1 - TG_2 + p = 0$. Now treat p and T as independent, and differentiate (3-35) with respect to each, obtaining

$$\frac{\partial E}{\partial p} = F_1\frac{\partial V}{\partial p} \qquad \frac{\partial E}{\partial T} = F_1\frac{\partial V}{\partial T} + F_2$$

$$1 = G_1\frac{\partial V}{\partial p} \qquad 0 = G_1\frac{\partial V}{\partial T} + G_2$$

Solve these for F_1 and G_2 and substitute the result into the relation $F_1 - TG_2 + p = 0$, and one again arrives at (3-34).

EXERCISES

1 Construct schematic diagrams to show the following functional relationships, and find the indicated derivatives:

(*a*) $w = f(x, y, z)$, $x = \phi(t)$, $y = \psi(t)$, $z = \theta(t)$. Find dw/dt.

(*b*) $w = F(x, u, t)$, $u = f(x, t)$, $x = \phi(t)$. Find dw/dt.

(*c*) $w = F(x, u, v)$, $u = f(x, y)$, $v = g(x, z)$. Find $\partial w/\partial x$, $\partial w/\partial y$, and $\partial w/\partial z$.

2 If $w = f(x, y)$ and $y = F(x)$, find dw/dx and d^2w/dx^2.

3 When x, y, and z are related by the equation $x^2 + yz^2 + y^2x + 1 = 0$, find $\partial y/\partial x$ and $\partial y/\partial z$ when $x = -1$ and $z = 1$.

4 Let x, y, u, v be related by the equations $xy + x^2u = vy^2$, $3x - 4uy = x^2v$. Find $\partial u/\partial x$, $\partial u/\partial y$, $\partial v/\partial x$, $\partial v/\partial y$ first by implicit differentiation, and then by solving the equations explicitly for u and v.

5 Let $F(x, y, z) = 0$. Assuming that this can be solved for z in terms of (x, y), find $\partial z/\partial x$ and $\partial z/\partial y$.

6 Under the same assumptions as Exercise 5, find expressions for $\partial^2 z/\partial x^2$ and $\partial^2 z/\partial x\,\partial y$ in terms of F and its derivatives.

7 Let $F(x, y, z) = 0$. Prove that

$$\left.\frac{\partial z}{\partial y}\right|_x \left.\frac{\partial y}{\partial x}\right|_z \left.\frac{\partial x}{\partial z}\right|_y = -1$$

8 Let $F(x, y, t) = 0$ and $G(x, y, t) = 0$ be used to express x and y in terms of t. Find general formulas for dx/dt and dy/dt.

9 Let $z = f(xy)$. Show that this obeys the differential relation

$$x\left(\frac{\partial z}{\partial x}\right) - y\left(\frac{\partial z}{\partial y}\right) = 0$$

10 Let $w = F(xz, yz)$. Show that

$$x\frac{\partial w}{\partial x} + y\frac{\partial w}{\partial y} = z\frac{\partial w}{\partial z}$$

11 A function f is said to be homogeneous of degree k in a neighborhood $\mathcal{N}$ of the origin if $f(tx, ty) = t^k f(x, y)$ for all points $(x, y) \in \mathcal{N}$ and all t, $0 \le t \le 1$. Assuming appropriate continuity conditions, prove that f satisfies in $\mathcal{N}$ the differential equation

$$xf_1(x, y) + yf_2(x, y) = kf(x, y)$$

12 Setting $z = f(x, y)$, Exercise 11 shows that $x(\partial z/\partial x) + y(\partial z/\partial y) = 0$ whenever f is homogeneous of degree $k = 0$. Show that in polar coordinates this differential equation becomes simply $r(\partial z/\partial r) = 0$, and from this deduce that the general homogeneous function of degree 0 is of the form $f(x, y) = F(y/x)$.

13 If $z = F(ax + by)$, then $b(\partial z/\partial x) - a(\partial z/\partial y) = 0$.

14 If $u = F(x - ct) + G(x + ct)$, then

$$c^2\frac{\partial^2 u}{\partial x^2} = \frac{\partial^2 u}{\partial t^2}$$

15 If $z = \phi(x, y)$ is a solution of $F(x + y + z, Ax + By) = 0$, show that $A(\partial z/\partial y) - B(\partial z/\partial x)$ is constant.

16 Show that the substitution $x = e^s$, $y = e^t$ converts the equation

$$x^2\left(\frac{\partial^2 u}{\partial x^2}\right) + y^2\left(\frac{\partial^2 u}{\partial y^2}\right) + x\left(\frac{\partial u}{\partial x}\right) + y\left(\frac{\partial u}{\partial y}\right) = 0$$

into the equation $\partial^2 u/\partial s^2 + \partial^2 u/\partial t^2 = 0$.

17 Show that the substitution $u = x^2 - y^2$, $v = 2xy$ converts the equation $\partial^2 W/\partial x^2 + \partial^2 W/\partial y^2 = 0$ into $\partial^2 W/\partial u^2 + \partial^2 W/\partial v^2 = 0$.

18 Show that if p and E are regarded as independent, the differential equation (3-32) takes the form

$$\frac{\partial T}{\partial p} - T\frac{\partial V}{\partial E} + p\frac{\partial(V, T)}{\partial(E, p)} = 0$$

19 Let f be of class C'' in the plane, and let S be a closed and bounded set such that $f_1(p) = 0$ and $f_2(p) = 0$ for all $p \in S$. Show that there is a constant M such that $|f(p) - f(q)| \le M|p - q|^2$ for all points p and q lying in S.

***20** (*Continuation of Exercise* 19) Show that if S is the set of points on an arc given by the equations $x = \phi(t)$, $y = \psi(t)$, where ϕ and ψ are of class C', then the function f is constant-valued on S.

21 Let f be a function of class C' with $f(1, 1) = 1$, $f_1(1, 1) = a$, and $f_2(1, 1) = b$. Let $\phi(x) = f(x, f(x, x))$. Find $\phi(1)$ and $\phi'(1)$.

3.5 TAYLOR'S THEOREM

This important and useful result can be regarded either as a statement about the approximation of functions by polynomials, or as a generalization of the mean value theorem.

Suppose that a function f is given to us, and we wish to find a polynomial P which approximates f in a specified sense, for example, uniformly on some interval I. One familiar procedure is that of interpolation, or curve fitting. We choose points $x_1, x_2, \ldots, x_n$ on I, and determine P so that $P(x_i) = f(x_i)$, $i = 1, 2, \ldots, n$. If P has degree m, then there will be $m + 1$ coefficients to be determined, so that in general a polynomial of degree $n - 1$ must be used to fit f at n points. Once P has been computed, it remains a separate problem to study the accuracy of the approximation at points x of the interval I other than the points x_i. Another method for choosing a suitable polynomial P is to select one point x_0 in the interval (e.g., the midpoint) and then choose P to match f very closely at the point x_0. We first introduce a convenient notation.

Let $f \in C^n$ on an interval I about x_0. Among all the polynomials of degree n, there is exactly one which matches f at x_0 up through the nth derivative, so that

$$P^{(k)}(x_0) = f^{(k)}(x_0) \qquad k = 0, 1, \ldots, n \tag{3-36}$$

We shall call this the **Taylor polynomial** of degree n at x_0, and denote it by P_{x_0}. When $n = 1$, P_{x_0} is the first-degree polynomial which goes through $(x_0, f(x_0))$ and has there the same slope as does f; it is therefore just the line tangent to the graph of f, and

$$P(x) = f(x_0) + f'(x_0)(x - x_0)$$

When $n = 2$, P is a parabola which is tangent to f at $(x_0, f(x_0))$ and there has the same curvature as does f. Writing

$$P(x) = A + B(x - x_0) + C(x - x_0)^2$$

and imposing conditions (3-36), we see that

$$P(x) = f(x_0) + f'(x_0)(x - x_0) + f''(x_0)\frac{(x - x_0)^2}{2}$$

In general,

$$P_{x_0}(x) = f(x_0) + f'(x_0)(x - x_0) + \cdots + f^{(n)}(x_0)\frac{(x - x_0)^n}{n!}$$

We are concerned with the accuracy with which this polynomial approximates f at points of the interval I *away* from x_0; therefore we study the remainder $R_n(x) = f(x) - P_{x_0}(x)$. Taylor's theorem expresses this remainder in terms of the function f.

Theorem 15 (Taylor Remainder) *Let $f \in C^{n+1}$ on an interval I about $x = c$, and let $P_c(x)$ be the Taylor polynomial of degree n at c. Then, $f(x) = P_c(x) + R_n(x)$, for any $x \in I$, where the remainder R_n is given by*

$$R_n(x) = \frac{1}{n!}\int_c^x f^{(n+1)}(t)(t-c)^n\,dt \tag{3-37}$$

The key to the proof is to hold x fixed, and regard c as variable. Thus, consider the function

$$g(t) = P_t(x) = f(t) + f'(t)(x-t) + f''(t)(x-t)^2/2! + \cdots + f^{(n)}(t)(x-t)^n/n!$$

Note that $g(x) = f(x)$ and $g(c) = P_c(x)$. Hence,

$$R_n(x) = f(x) - P_c(x) = g(x) - g(c) = \int_c^x g'(t)\,dt$$

It is only necessary then to calculate g'; doing so, we find

$$\begin{aligned} g'(t) = f'(t) &+ \{f''(t)(x-t) - f'(t)\} \\ &+ \{f^{(3)}(t)(x-t)^2/2! - 2f''(t)(x-t)/2!\} \\ &\cdots\cdots\cdots\cdots\cdots\cdots\cdots\cdots \\ &+ \{f^{(n+1)}(t)(x-t)^n/n! - f^{(n)}(t)(x-t)^{n-1}/n!\} \end{aligned}$$

This sum "telescopes," resulting in the simple result

$$g'(t) = f^{(n+1)}(t)(x-t)^n/n!,$$

which immediately yields (3-37). ∎

Corollary 1 *If $|f^{(n+1)}(x)| \le M$ for all $x \in I$, then on I*

$$|R_n(x)| \le M\frac{(x-c)^{n+1}}{(n+1)!} \tag{3-38}$$

This useful estimate is obtained directly from (3-37), since

$$|R_n(x)| \le \frac{1}{n!}\int_c^x M(t-c)^n\,dt = \frac{1}{n!}\frac{M(x-c)^{n+1}}{n+1}$$

There are other alternative forms for the remainder $R_n(x)$. One of these, which depends upon applying a different estimate to the integral in (3-37), leads to the following statement of the relationship between $f(x)$ and the Taylor polynomials at $x = c$ which is perhaps more familiar than (3-37).

Corollary 2 *If $f \in C^{n+1}$ in a neighborhood I of c, then for any $x \in I$,*

$$\begin{aligned} f(x) = f(c) + f'(c)(x-c) &+ f''(c)\frac{(x-c)^2}{2!} \\ &+ \cdots + f^{(n)}(c)\frac{(x-c)^n}{n!} + R_n(x) \end{aligned} \tag{3-39}$$

where

$$R_n(x) = f^{(n+1)}(\tau)\frac{(x-c)^{n+1}}{(n+1)!} \tag{3-40}$$

and τ is an appropriately chosen point between x and c.

Note that if $n = 0$, this becomes the usual mean value theorem, discussed in Sec. 3.2. Thus, Taylor's theorem is a generalization of this. The equivalence of (3-37) and (3-40) follows from an exercise given in the next chapter (Exercise 5, Sec. 4.2), as well as from the simple argument in Exercise 8 of the present section.

For the Taylor polynomial to be a good approximation to f, uniformly on an interval I, the remainder R_n must be uniformly small there. The importance of the theorem lies in the fact that by having a formula for R_n, we are thereby able to estimate its size. For example, let us find a polynomial which approximates e^x on the interval $[-1, 1]$ accurately to within .005. Using (3-36) with $x_0 = 0$, we have

$$P(x) = f(0) + f'(0)x + f''(0)\frac{x^2}{2!} + \cdots + f^{(n)}(0)\frac{x^n}{n!}$$

which in our case is

$$P(x) = 1 + x + \frac{x^2}{2!} + \cdots + \frac{x^n}{n!}$$

The remainder, using the formula of Corollary 1, is

$$|e^x - P(x)| = |R_n(x)| \le M\frac{x^{n+1}}{(n+1)!}$$

where M is the maximum of e^x on $[-1, 1]$. Since x also lies in $[-1, 1]$, this yields $|R_n| \le e/(n+1)!$. In order to have accuracy to within .005, we choose $n = 5$, and we have found a polynomial

$$1 + x + \frac{x^2}{2} + \frac{x^3}{6} + \frac{x^4}{24} + \frac{x^5}{120}$$

with the desired property.

It is clear in this case that because of the rapid growth of the factorials $n!$, increasing the degree n of the Taylor polynomial will continue to improve the accuracy of the approximation on $[-1, 1]$, or in fact, on any bounded interval. This is a special property of the exponential function which is also shared on their domains by an important class of functions.

Definition 3 *A function f is said to be* ***analytic*** *at a point x_0 if there is an open interval I about x_0 on which f is of class C^∞ and such that $\lim_{n\to\infty} R_n(x) = 0$ for each $x \in I$.*

The functions e^x, $\sin x$, and $\cos x$ are analytic everywhere; $\sqrt{x}$ is analytic for each point $x_0 > 0$, and $x/(x^2 - 1)$ is analytic at each point of the intervals $x < -1$, $-1 < x < 1$, and $1 < x$. Since

$$R_n(x) = f(x) - P_{x_0}(x)$$

this property is equivalent to saying that the sequence of Taylor polynomials for f at x_0 converges to f in a neighborhood of x_0; as we shall see in Chap. 6, this enables us to say that analytic functions are those that are sums of their **Taylor series.** The theory of analytic functions is a separate and highly evolved branch of analysis; a sketch of this subject is given in Appendix 5, and references to fuller treatments will be found there.

There are functions of class C^∞ which are not analytic, and for which the Taylor polynomials are very poor approximations. An often used example is $f(x) = \exp(-1/x^2)$. When $x \neq 0$, f is continuous and infinitely differentiable. In fact, $f'(x) = 2x^{-3} \exp(-1/x^2)$, and induction shows that $f^{(n)}(x)$ has the form $x^{-3n}Q(x) \exp(-1/x^2)$, where Q is a polynomial of degree $2(n-1)$. Since $\lim_{x \to 0} f(x) = 0$, the discontinuity at the origin may be removed by setting $f(0) = 0$; the formula for $f^{(n)}$ shows that it is also true that $\lim_{x \to 0} f^{(n)}(x) = 0$ for $n = 1, 2, \ldots$. This allows us to show that f is also infinitely differentiable at $x = 0$, with $f^{(n)}(0) = 0$ for $n = 1, 2, \ldots$. For $n = 1$, $f'(0) = \lim_{h \to 0} f(h)/h$; applying L'Hospital's rule, we replace this by $\lim_{h \to 0} f'(h)$, which exists and has the value 0. Induction establishes the general result. The function f can be shown to be analytic for all $x \neq 0$. However, it is not analytic at $x = 0$, even though it is infinitely differentiable there. The Taylor polynomial of degree n at 0 is $0 + 0x + 0x^2/2 + \cdots + 0x^n/n!$ so that $R_n(x) = \exp(-1/x^2)$, which does not tend to 0 as n increases, on any neighborhood of 0. For this function, then the Taylor polynomials at the origin do not converge to the function.

This points up the fact that while the Taylor polynomials are easy to define, they are not always the best polynomials to use for approximating a given function. For example, on $[-1, 1]$, the Taylor polynomial $1 + x + x^2/2$ for e^x approximates it with an error of .22, while the special polynomial $.99 + 1.175\,x + .543\,x^2$ approximates e^x on the same interval with maximum error of only .04. The theory of approximation is currently a very active field of research, stimulated in part by the development of high-speed computers, and new methods for finding good polynomial and piecewise polynomial approximations are being discovered.

Even with an apparently well-behaved function, it is not always true that increasing the degree of the Taylor polynomial will extend the interval of approximation. Let $f(x) = 1/(1 + x^2)$, and construct the Taylor polynomial for f at $x_0 = 0$. It is easy to see that it is exactly

$$P(x) = 1 - x^2 + x^4 - x^6 + \cdots + (-1)^n x^{2n}$$

Moreover, by elementary algebra, we can compute the remainder exactly, obtaining

$$R_{2n}(x) = \frac{1}{1+x^2} - P(x) = (-1)^{n+1} \frac{x^{2n+2}}{1+x^2}$$

We see at once that $|R_{2n}(x)|$ is very small on the open interval $-1 < x < 1$ if n is sufficiently large. Thus, increasing the degree of P yields a much better approximation to f on this interval. However, it is also evident that $P(x)$ will never be a good approximation to f outside this interval, no matter how large n is chosen. Indeed, for a value such as $x = 2$, increasing n worsens the approximation! The explanation for this fact lies in the behavior of $f(x)$ for *complex* values of x, in particular the fact that it is not defined when $x = \sqrt{-1}$, for this fact influences the degree of approximation that the Taylor polynomial can achieve for real x. (This will be discussed in Chap. 6.)

We now turn to functions of several variables. The first objective is an improved form of the mean value theorem, promised in the last section.

Theorem 16 *Let $f \in C'$ in an open convex set S in n space. Then, for any points p_1 and p_2 in S, there is a point p^* lying on the segment joining them such that*

(3-41)
$$f(p_2) - f(p_1) = \mathbf{D}f(p^*) \cdot (p_2 - p_1)$$

The proof for a function of two variables exhibits the general method.

If we have $p_1 = (x, y)$ and $p_2 = (x + \Delta x, y + \Delta y)$, then this theorem asserts that there is a number λ, $0 < \lambda < 1$, such that

$$f(p_2) - f(p_1) = f_1(p^*)\,\Delta x + f_2(p^*)\,\Delta y$$

where $p^* = (x + \lambda\,\Delta x,\ y + \lambda\,\Delta y)$. To prove this, we construct a special function F of one variable

$$\begin{aligned} F(t) &= f(p_1 + t(p_2 - p_1)) \\ &= f(x + t\,\Delta x,\ y + t\,\Delta y) \end{aligned}$$

Applying the one-variable mean value theorem,

$$F(1) - F(0) = (1 - 0)F'(\lambda)$$

where λ is some point between 0 and 1. By the chain rules,

$$F'(t) = f_1\,\Delta x + f_2\,\Delta y$$

so that $F'(\lambda) = \mathbf{D}f(p^*) \cdot (\Delta x, \Delta y)$, where p^* is the point $(x + \lambda\,\Delta x, y + \lambda\,\Delta y)$, which lies on the line from p_1 to p_2. Since $F(1) - F(0) = f(p_2) - f(p_1)$, we have established (3-41). [Note that the convexity of S was used only to be sure that the point p^* belongs to S; all that is required for (3-41) is that the segment p_1p_2 lie in S.] ▮

If we apply Taylor's theorem to F, instead of merely the mean value theorem, we obtain the corresponding form of **Taylor's theorem** for functions of two variables.

Theorem 17 *Let f be of class C^{n+1} in a neighborhood of $p_0 = (x_0, y_0)$. Then, with $p = (x, y)$,*

$$f(p) = f(p_0) + \frac{(x - x_0)}{1} \frac{\partial f}{\partial x}\bigg|_{p_0} + \frac{(y - y_0)}{1} \frac{\partial f}{\partial y}\bigg|_{p_0}$$
$$+ \frac{(x - x_0)^2}{2!} \frac{\partial^2 f}{\partial x^2}\bigg|_{p_0} + \frac{(x - x_0)}{1!} \frac{(y - y_0)}{1!} \frac{\partial^2 f}{\partial x\, \partial y}\bigg|_{p_0} + \frac{(y - y_0)^2}{2!} \frac{\partial^2 f}{\partial y^2}\bigg|_{p_0}$$
$$+ \cdots + \frac{(x - x_0)^n}{n!} \frac{\partial^n f}{\partial x^n}\bigg|_{p_0} + \frac{(x - x_0)^{n-1}}{(n-1)!} \frac{(y - x_0)}{1!} \frac{\partial^n f}{\partial x^{n-1}\, \partial y}\bigg|_{p_0}$$
$$+ \cdots + \frac{(y - y_0)^n}{n!} \frac{\partial^n f}{\partial y^n}\bigg|_{p_0} + R_n(x, y)$$

where

$$R_n(x, y) = \frac{(x - x_0)^{n+1}}{(n+1)!} \frac{\partial^{n+1} f}{\partial x^{n+1}}\bigg|_{p^*} + \cdots + \frac{(y - y_0)^{n+1}}{(n+1)!} \frac{\partial^{n+1} f}{\partial y^{n+1}}\bigg|_{p^*}$$

and p^* is a point on the line segment joining p_0 and p.

By adopting a special notation, this can be thrown into a simpler appearing form. Let $\Delta x = x - x_0$ and $\Delta y = y - y_0$, so that

$$\Delta p = (\Delta x, \Delta y) = p - p_0$$

Define a differential operator U by

$$U = \Delta x \frac{\partial}{\partial x} + \Delta y \frac{\partial}{\partial y}$$

so that, for instance, $Uf = \Delta x\, f_1 + \Delta y\, f_2$. Then, the Taylor expansion formula may be written as

$$f(p_0 + \Delta p) = f(p_0) + \frac{1}{1!} Uf(p_0) + \frac{1}{2!} U^2 f(p_0) + \cdots + \frac{1}{n!} U^n f(p_0) + \frac{1}{(n+1)!} U^{n+1} f(p^*) \tag{3-42}$$

For example,

$$U^2 f(p_0) = \left[(\Delta x)^2 \frac{\partial^2}{\partial x^2} + 2(\Delta x)(\Delta y) \frac{\partial^2}{\partial x\, \partial y} + (\Delta y)^2 \frac{\partial^2}{\partial y^2}\right](f)(p_0)$$
$$= (\Delta x)^2 \frac{\partial^2 f}{\partial x^2}\bigg|_{p_0} + 2(\Delta x)(\Delta y) \frac{\partial^2 f}{\partial x\, \partial y}\bigg|_{p_0} + (\Delta y)^2 \frac{\partial^2 f}{\partial y^2}\bigg|_{p_0}$$

This device makes it possible to state the n-variable form of Taylor's theorem without becoming overwhelmed with subscripts.
Set $\Delta p = (\Delta x_1, \Delta x_2, \ldots, \Delta x_n)$ and define the operator U by

(3-43) $$U = \Delta x_1 \frac{\partial}{\partial x_1} + \Delta x_2 \frac{\partial}{\partial x_2} + \cdots + \Delta x_n \frac{\partial}{\partial x_n}$$

so that $$Uf(q) = \Delta p \cdot \mathbf{D}f(q)$$

Then, the statement of Taylor's theorem appears the same for n variables as it does for two, namely (3-42). Of course, the powers of U will be considerably more complicated when they are translated into formulas involving the higher partial derivatives of f (see Exercise 17).

EXERCISES

1 Show that $\sin x$ can be approximated by $x - x^3/6$ within .01 on the interval $[-1, 1]$.

2 Determine the accuracy of the approximation

$$\cos x \sim 1 - \frac{x^2}{2} + \frac{x^4}{24}$$

on the interval $[-1, 1]$.

3 Determine the accuracy of the approximation

$$\log(1 + x) \sim x - \frac{x^2}{2} + \frac{x^3}{3} - \frac{x^4}{4}$$

on the interval $[-\frac{1}{2}, \frac{1}{2}]$.

4 Determine the accuracy of the approximation

$$\sqrt{x} \sim 1 + \frac{(x-1)}{2} - \frac{(x-1)^2}{8}$$

on the interval $[\frac{1}{2}, \frac{3}{2}]$.

5 How many terms of the Taylor expansion for $\sin x$ about a conveniently chosen point are needed to obtain a polynomial approximation accurate to .01 on the interval $[0, \pi]$? On the interval $[0, 2\pi]$?

6 Suppose that $f(0) = f(-1) = 0$, $f(1) = 1$, and $f'(0) = 0$. Assuming that f is of class C^3, show that there is a point c in $[-1, 1]$ where $f'''(c) \geq 3$.

7 Assume that $f \in C''$, that $|f''(x)| \leq M$, and that $f(x) \to 0$ as $x \uparrow \infty$. Prove that $f'(x) \to 0$ as $x \uparrow \infty$.

8 Let $$P(x) = f(c) + f'(c)(x - c) + \frac{f''(c)(x-c)^2}{2!} + \frac{f'''(c)(x-c)^3}{3!}$$

be a Taylor polynomial at c of degree 3. Let $g(x) = f(x) - P(x) - A(x - c)^4/4!$ Suppose that A is chosen so that $g(\bar{x}) = 0$. Prove that there is a point τ between c and $\bar{x}$ with $A = f^{(4)}(\tau)$. (*Hint:* Use Exercise 2, Sec. 3.2.)

9 Show that for all $x \geq 1$, $\log x \leq \sqrt{x} - 1/\sqrt{x}$.

10 Show that for all $x \geq 0$, $e^x \geq \frac{3}{2}x^2$. Can you replace $\frac{3}{2}$ by a larger number?

11 Let f obey the condition $|f^{(n)}(x)| \leq B^n$ for all x in an open interval I, and all n. Show that f is analytic on I.

*12 Let f be of class C'' on $[0, 1]$ with $f(0) = f(1) = 0$, and suppose that $|f''(x)| \le A$ for all x, $0 < x < 1$. Show that $|f'(\frac{1}{2})| \le A/4$ and that $|f'(x)| \le A/2$ for $0 < x \le 1$.

*13 If $f(0) = 0$ and $|f'(x)| \le M|f(x)|$ for $0 \le x \le L$, show that on that interval $f(x) \equiv 0$.

*14 Say that f is locally a polynomial on $-\infty, \infty$ if, given x_0, there is a neighborhood $\mathcal{N}_{x_0}$ and a polynomial $P(x)$, and on $\mathcal{N}_{x_0}$, $f(x) = P(x)$. Prove that f is a polynomial.

15 Sketch the graph of $y = 1/(1 + x^2)$ and then, for comparison, the graph of the polynomials

$$P_2(x) = 1 - x^2 \qquad \text{and} \qquad P_4(x) = 1 - x^2 + x^4$$

16 Sketch the graph of the function $f(x) = e^x$, and then the graph of its Taylor polynomials

$$P_2(x) = 1 + x + \tfrac{1}{2}x^2 \qquad \text{and} \qquad P_3(x) = 1 + x + x^2/2 + x^3/6$$

What is the contrast in behavior between this and the results in the preceding exercise?

17 Use (3-43) to obtain the coordinate form for Taylor's theorem in three variables, with a remainder of total degree 3.

18 Let $f \in C'$ in an open convex set S in $\mathbf{R}^n$. Show that f obeys a Lipschitz condition on any compact subset $E \subset S$.

3.6 EXTREMAL PROBLEMS

Any continuous function f defined on a closed bounded set D attains a maximum (and a minimum) value at some point of D. If f is of class C' in D, and p_0 is an interior point of D at which f attains such an extremal value, then (Theorem 11, Sec. 3.3) all the first-order partial derivatives of f are 0 at p_0. This suggests that we single out the points in the domain of a function which have the last property.

Definition 4 *A* ***critical point*** *for a function f is a point p where*

$$f_1(p) = f_2(p) = \cdots = 0$$

The discussion in the first paragraph can be rephrased as asserting that the extremal points for the function f, which lie in the set D but do not lie on the boundary of D, are among the critical points for f in D. A critical point need not yield a local maximum or minimum value of f. However, since such a point is one where the directional derivative of f is 0 in every direction, the point is a stationary point for the function; this is reflected in the fact that the tangent hyperplane to the graph of f will be horizontal there. Since such a point can be one where the surface rises in one direction and falls in another, as in Fig. 3-9, a critical point need not correspond to either a maximum or a minimum. Such points are often called **saddle points,** or **minimax** points.

Before proceeding further, let us recall the facts about functions of one variable. A critical point is a solution of the equation $f'(x) = 0$, and corresponds to a point on the curve with equation $y = f(x)$ at which the tangent line is horizontal. The critical point may be an extremal point for f, or it may yield

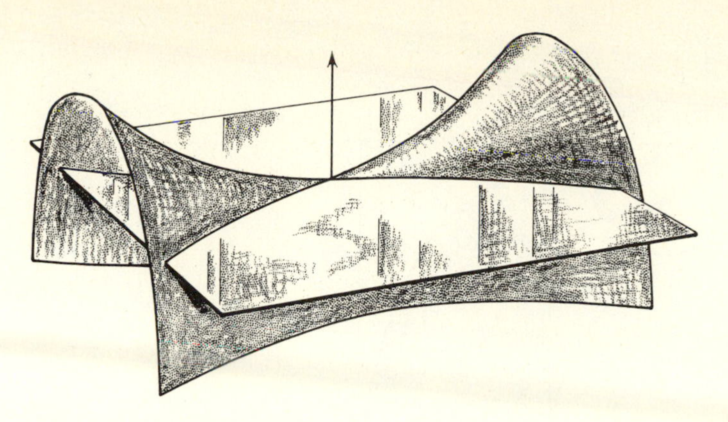

Figure 3-9 Horizontal tangent plane at saddle point.

a point of inflection on the curve. As an illustration, suppose we wish to find the maximum value of

$$f(x) = 4x^3 - 15x^2 + 18x$$

for x in the interval $I = [0, 2]$. The critical points for f in I are found to be 1 and $\frac{3}{2}$. The maximum value of f on I must therefore be attained either at one of these or on the boundary of I, that is, at one of the endpoints 0 and 2. Computing the value of f at each, we find $f(0) = 0$, $f(2) = 8$, $f(1) = 7$, $f(\frac{3}{2}) = \frac{27}{4}$. Hence, the maximum value of f on I is 8, and it is attained on the boundary. (However, f has a local maximum value of 7 at 1, and a local minimum value of $\frac{27}{4}$ at $\frac{3}{2}$.)

The same technique may be used for functions of several variables. As an illustration, consider the function f given by

$$f(x, y) = 4xy - 2x^2 - y^4$$

in the square $D = \{\text{all } (x, y) \text{ with } |x| \le 2, |y| \le 2\}$. The critical points for f are the simultaneous solutions of the equations

$$0 = f_1(x, y) = 4y - 4x$$

$$0 = f_2(x, y) = 4x - 4y^3$$

and are $(0, 0)$, $(1, 1)$, and $(-1, -1)$. The maximum value of f in D must be attained at one of these or on the boundary of D. We do not have to compute the values of f on all the edges of this square; since we are looking for a maximum value, we may discard the parts of ∂D lying in the second and fourth quadrants where the term xy which occurs in $f(x, y)$ is negative. Moreover, $f(-x, -y) = f(x, y)$, so that f takes the same values at symmetric points in the first and third quadrants. This reduces our work to an examination of the values of f on the line $x = 2$, $0 \le y \le 2$, and the line $y = 2$, $0 \le x \le 2$. On

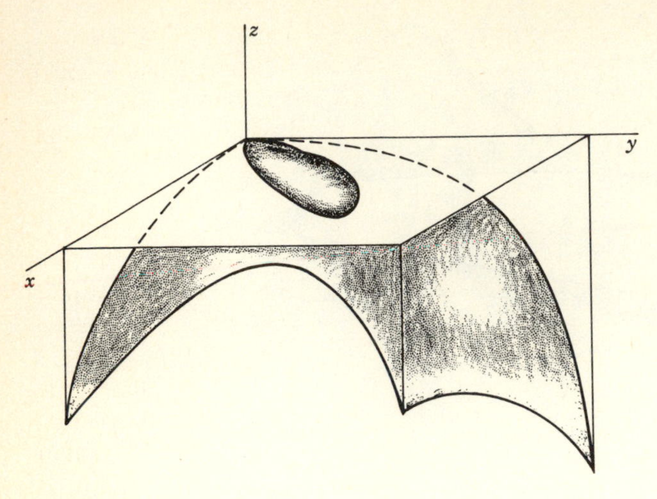

Figure 3-10 Graph of $z = 4xy - 2x^2 - y^4$ for $0 \le x \le 2$, $0 \le y \le 2$.

the former,

$$f(2, y) = 8y - 8 - y^4$$

Proceeding as in the previous illustration, we find that the largest value of this for y in $[0, 2]$ is $6(2)^{1/3} - 8 \sim -.44$, attained when $y = 2^{1/3}$. On the second part of ∂D, $f(x, 2) = 8x - 2x^2 - 16$, whose greatest value for x in $[0, 2]$ is -8. This shows that the maximum value of f on the boundary of D is approximately $-.44$. Comparing this with the values which f has at the three critical points, $f(0,0) = 0$, $f(1,1) = f(-1,-1) = 1$ we see that the (absolute) maximum value of f in D is 1, attained at the two points $(1, 1)$ and $(-1, -1)$. (A graph of f is given in Fig. 3-10.) In the same fashion, one may show that the minimum value of f in D is -40, attained at the two boundary points $(2, -2)$ and $(-2, 2)$.

In the more familiar case of a function of one variable, the second derivative may be used to test the nature of a critical point.

Theorem 18 *Let f be of class C'' in the interval $[a, b]$ and let c be an interior point of this interval with $f'(c) = 0$. Then, in order that c be a local maximum point for f, it is necessary that $f''(c) \le 0$, and sufficient that $f''(c) < 0$; for c to be a minimum point, the conditions are the same with the inequality signs reversed.*

The proof is an immediate deduction from Taylor's theorem. Write

$$f(c + h) = f(c) + f'(c)h + \frac{f''(\tau)h^2}{2}$$

$$= f(c) + \frac{f''(\tau)h^2}{2}$$

where τ is a point between c and $c + h$. If $f''(c) < 0$, then $f''(\tau) < 0$ whenever $|h|$ is sufficiently small, so that $f(c + h) < f(c)$ and f has a local maximum value at c. Conversely, if f has a local maximum value at c, then $f''(\tau)h^2 \le 0$ for all small h. Since $h^2 \ge 0$, and τ approaches c as $h \to 0$, we have $f''(c) \le 0$. ∎

What is the corresponding statement for functions of two variables? Clearly, if p_0 is a critical point for f lying interior to a set D, and if f has an extremal value at p_0, then this must also be extremal if we examine the values of f on any curve passing through p_0. In particular, approaching p_0 along the vertical and the horizontal directions, a necessary condition that p_0 be a maximum point for f is that $f_{11}(p_0) \le 0$ and $f_{22}(p_0) \le 0$. These conditions are not sufficient, nor are the conditions obtained by removing the equal signs (see Exercise 5). For example, the function given by $f(x, y) = xy$ has $(0, 0)$ for a critical point and

$$f_{11}(0, 0) = f_{22}(0, 0) = 0$$

but $(0, 0)$ is neither a maximum point nor a minimum point. The shape of the graph of f is again like the saddle shown in Fig. 3-9. The name "saddle point" or "minimax" is given to a critical point for a function which does not yield either a local maximum or a local minimum value for the function. A simple condition which is sufficient to ensure that a critical point p_0 be a *saddle point* is that $f_{11}(p_0)f_{22}(p_0)$ be strictly negative, since this implies that f has a local maximum at p_0 when p_0 is approached along one axis direction, and a local minimum when p_0 is approached along the other. A more general criterion can also be obtained.

Theorem 19 *Let f be of class C'' in a neighborhood of the critical point p_0, and let*

$$\Delta = (f_{12}(p_0))^2 - f_{11}(p_0)f_{22}(p_0)$$

Then,

(i) *If $\Delta > 0$, p_0 is a saddle point for f.*
(ii) *If $\Delta < 0$, p_0 is an extremal point for f, and is a maximum if $f_{11}(p_0) < 0$ and a minimum if $f_{11}(p_0) > 0$.*
(iii) *If $\Delta = 0$, the nature of p_0 is not determined by this test.*

In condition (ii), $f_{22}(p_0)$ may also be used to distinguish between maxima and minima. Let us apply the test to the function given by $f(x, y) = 4xy - 2x^2 - y^4$ which was used as an illustration prior to Theorem 18 and whose critical points are $(0, 0)$, $(1, 1)$, and $(-1, -1)$. We find that $f_{11}(x, y) = -4$, $f_{12}(x, y) = 4$, $f_{22}(x, y) = -12y^2$, so that

$$\Delta = 16 - 48y^2$$

At $(0, 0)$, $\Delta = 16 > 0$, so that $(0, 0)$ is a saddle point. At $(1, 1)$ and $(-1, -1)$,

$\Delta = -32 < 0$, so that each is extremal. Since

$$f_{11} = -4 < 0$$

both are local maxima.

The proof of the general test is based upon the following special case.

Lemma 2 *Let* $P(x, y) = Ax^2 + 2Bxy + Cy^2$, *and set*

$$\Delta = \frac{(P_{12})^2 - P_{11}P_{22}}{4} = B^2 - AC$$

Then,

(i) *If* $\Delta > 0$, *then there are two lines through the origin such that* $P(x, y) > 0$ *for all* (x, y) *on one, and* $P(x, y) < 0$ *for all* (x, y) *on the other, with the point* $(0, 0)$ *omitted.*
(ii) *If* $\Delta < 0$, *then* $P(x, y)$ *never changes sign, and* $P(x, y) > 0$ *for all points except* $(0, 0)$ *if* $A > 0$, *and* $P(x, y) < 0$ *for all points except* $(0, 0)$ *if* $A < 0$.

We prove (*ii*) first. If $\Delta < 0$, then $AC \neq 0$ so that $A \neq 0$. Write

$$\begin{aligned} AP(x, y) &= A^2x^2 + 2ABxy + ACy^2 \\ &= (Ax + By)^2 - \Delta y^2 \end{aligned}$$

Since Δ is negative, $AP(x, y) \geq 0$ for all (x, y), with equality only at $(0, 0)$. Thus, $P(x, y)$ always has the same sign as the number A.

To prove (*i*), we assume that $\Delta > 0$ and compute

$$P(B, -A) = AB^2 - 2B^2A + CA^2 = -A\Delta$$

$$P(C, -B) = AC^2 - 2B^2C + CB^2 = -C\Delta$$

Since P is homogeneous of degree 2, $P(\lambda x, \lambda y) = \lambda^2 P(x, y)$, it will be sufficient if we can find two points at which P has opposite signs, since the same will then hold on the entire line joining these to the origin. Let us suppose first that $A \neq 0$. Then, recalling that Δ is positive, $P(1, 0) = A$ and $P(B, -A) = -A\Delta$ have opposite signs. Similarly, if $C \neq 0$, then $P(0, 1)$ and $P(C, -B)$ have opposite signs. Finally, if $A = C = 0$, then $P(x, y) = 2Bxy$ and P takes opposite signs at $(1, 1)$ and $(-1, 1)$. ▌

To apply this lemma to the proof of Theorem 19, assume that p_0 is a critical point for f, and write the Taylor expansion of f near p_0

$$f(p_0 + \Delta p) = f(p_0) + \tfrac{1}{2}\{f_{11}(p^*)(\Delta x)^2 + 2f_{12}(p^*)(\Delta x)(\Delta y) + f_{22}(p^*)(\Delta y)^2\} \tag{3-44}$$

where p^* is a point lying on the line segment joining p_0 and $p_0 + \Delta p$. We shall again prove (*ii*) first. We note first that the expression in brackets in (3-44) has the form $P(\Delta x, \Delta y)$, where P is the quadratic polynomial with coefficients $A = f_{11}(p^*)$, $B = f_{12}(p^*)$, $C = f_{22}(p^*)$. If $\Delta < 0$, then $B^2 - AC < 0$ whenever $|\Delta p|$ is sufficiently small. Hence, $P(\Delta x, \Delta y)$ has, by the lemma, the same sign as A. Thus, if $f_{11}(p_0) < 0$, then $P(\Delta x, \Delta y) < 0$ for $0 < |\Delta p| < \delta$, and

$$f(p_0 + \Delta p) = f(p) \le f(p_0)$$

for all p in a neighborhood of p_0, with equality holding only at p_0. The point p_0 is then an external point for f which yields a local maximum. If $f_{11}(p_0) > 0$, then $P(\Delta x, \Delta y) > 0$, and $f(p) \ge f(p_0)$ so that p_0 yields a local minimum. To prove part (*i*), we must show that f takes values which are bigger than $f(p_0)$ and smaller than $f(p_0)$, in any neighborhood of p_0. Write $\Delta p = (\rho \cos \theta, \rho \sin \theta)$. Since P is homogeneous,

$$P(\Delta x, \Delta y) = \rho^2 P(\cos \theta, \sin \theta)$$

Since f is of class C'', we have

$$\text{(3-45)} \quad \lim_{\Delta p \to 0} P(\cos \theta, \sin \theta) = f_{11}(p_0)(\cos \theta)^2 + 2f_{12}(p_0)(\cos \theta \sin \theta) + f_{22}(p_0)(\sin \theta)^2 = P_0(\cos \theta, \sin \theta)$$

Since $\Delta > 0$, the lemma implies that $P_0(\cos \theta, \sin \theta)$ takes both positive and negative values for $0 \le \theta \le 2\pi$. In particular, we may choose θ' and θ'' so that $P_0(\cos \theta', \sin \theta') > 0$ and $P_0(\cos \theta'', \sin \theta'') < 0$. By (3-45) these relations are also true of P, when $|\Delta p|$ is sufficiently small. Thus $P(\Delta x, \Delta y) > 0$ when $\Delta p = (\rho \cos \theta', \rho \sin \theta')$ and $0 < \rho < \delta$, and $P(\Delta x, \Delta y) < 0$ when $\Delta p = (\rho \cos \theta'', \rho \sin \theta'')$ and $0 < \rho < \delta$. Since P takes values which are both positive and negative in a neighborhood of the origin, $f(p)$ is sometimes larger than $f(p_0)$ and sometimes smaller, and p_0 is indeed a saddle point for f. ▮

The reader may wonder about the source of the expression Δ that appears in Theorem 19. There is a simple explanation which we must leave incomplete for the present. In this chapter, we have introduced the first derivative of a function f of n variables as a vector-valued function $\mathbf{D}f$; however, although we have used the higher partial derivatives of f, we have not introduced the second (total) derivative of f, which would be the derivative of $\mathbf{D}f$. Derivatives of vector-valued functions are discussed in Chap. 7, where they can be regarded as matrix-valued functions. The number Δ in Theorem 19 is the negative of the determinant of the second derivative of f.

In the vicinity of a critical point for a function of class C'', the level curves form characteristic patterns which can also be used to determine the character of the critical point. For example, if p_0 is a simple maximum point for f so that $f(p) < f(p_0)$ for all $p \ne p_0$ near p_0, then the level lines of f have

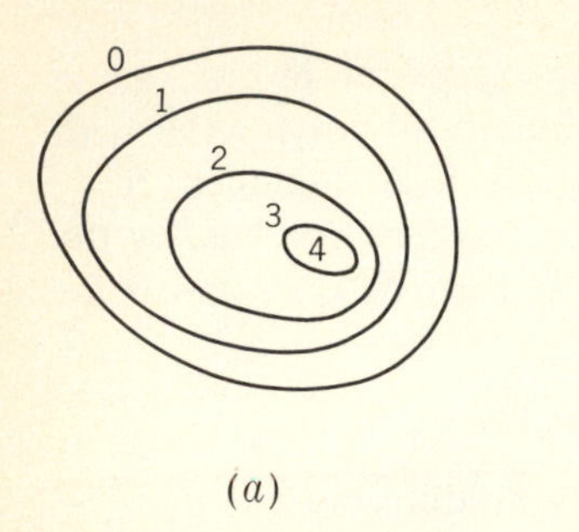

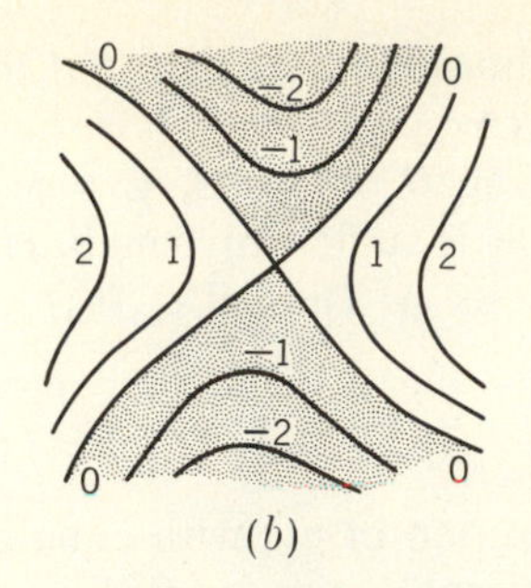

Figure 3-11 Level curves near a typical maximum point and near a saddle point.

an appearance similar to those shown in Fig. 3-11*a*. On the other hand, the pattern in Fig. 3-11*b* is typical of a simple saddle point corresponding to the pass between two peaks. A more complicated type of saddle point is illustrated by the so-called "monkey saddle" or triple peak pass (Fig. 3-12).

To give another simple illustration, let us study the function F given by $F(x, y) = x^2 + y^3 - 3xy$. Differentiating, we find

$$F_1(x, y) = 2x - 3y \qquad F_2(x, y) = 3y^2 - 3x$$

$$F_{11}(x, y) = 2 \qquad F_{12}(x, y) = -3 \qquad F_{22}(x, y) = 6y$$

The critical points of F are found to be $(0, 0)$ and $(\frac{9}{4}, \frac{3}{2})$. Computing $\Delta = 9 - 12y$ at each, the first gives $\Delta = 9$ and the second, $\Delta = -9$. Thus $(0, 0)$ is a saddle point, and $(\frac{9}{4}, \frac{3}{2})$ is an extremal point; since $F_{11} = 2$, it is a local minimum for F. (The level curves for F are sketched in Fig. 3-13.)

A more complicated example is the "Texas hat," whose level lines are shown in Fig. 3-14. From this diagram, it is evident that this surface has seven critical points, indicated on the diagram; three are saddle points, and

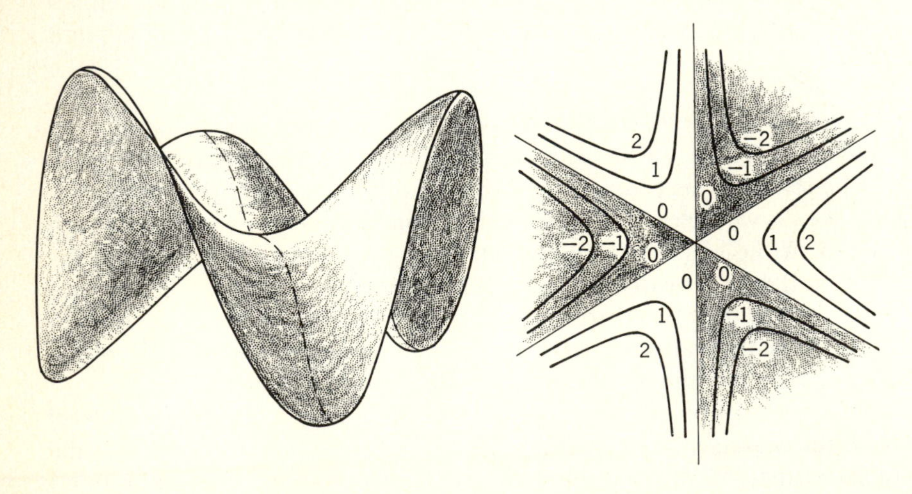

Figure 3-12 Monkey saddle.

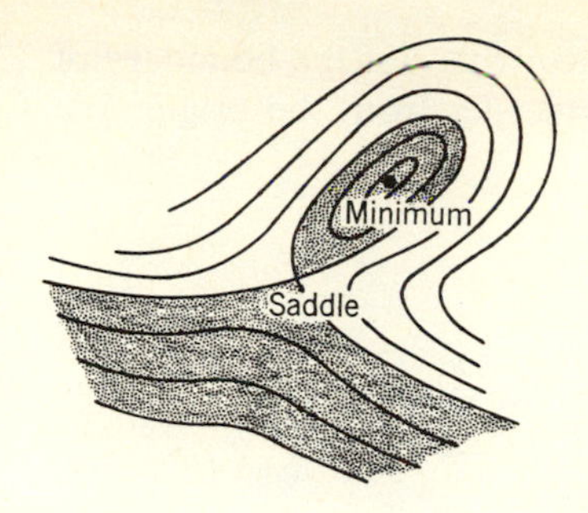

Figure 3-13

the remaining four are extremes. The contour chart does not have the numerical values of the level lines, so that it may not be evident that three are maxima and one a minimum. (The portion that is shaded in Fig. 3-14 is below level 0; the unshaded part is where the function is positive.) It is also instructive to attempt to visualize the graph of the function from this contour chart.

The technique can also be applied to the standard problem of obtaining the linear function which best fits a set of data, in the sense of **least squares.** Given n points $(x_1, y_1), \ldots, (x_n, y_n)$, not all the same, we wish to find the function F of the form $F(x) = ax + b$ for which

$$f(a, b) = \sum_{1}^{n} (F(x_j) - y_j)^2 = \sum_{1}^{n} (ax_j + b - y_j)^2 \tag{3-46}$$

is a minimum. The function f is of class C'' in the whole (a, b) plane and has a minimum value which is attained at some point, or points, $(a_0, b_0) = p_0$.

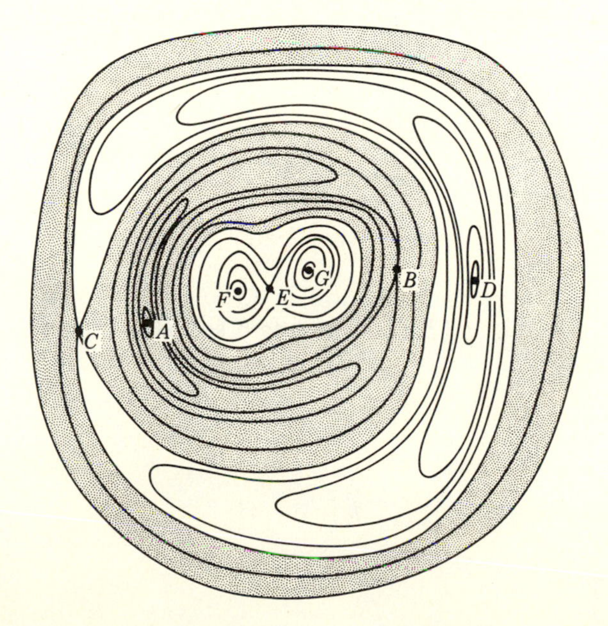

Figure 3-14

If we take the region D as a large disk, then we can rule out the boundary of D, for $f(a, b)$ becomes arbitrarily large as (a, b) recedes from the origin. We compute the derivatives of f.

$$f_1(a, b) = \sum_1^n 2(ax_j + b - y_j)x_j$$

$$= 2a\sum_1^n (x_j)^2 + 2b\sum_1^n x_j - 2\sum_1^n x_j y_j$$

$$f_2(a, b) = \sum_1^n 2(ax_j + b - y_j)$$

$$= 2a\sum_1^n x_j + 2nb - 2\sum_1^n y_j$$

$$f_{11}(a, b) = 2\sum_1^n (x_j)^2$$

$$f_{12}(a, b) = 2\sum_1^n x_j$$

$$f_{22}(a, b) = 2n$$

The critical points for f are the solutions of the following equations for a and b:

$$\begin{aligned} a\sum_1^n (x_j)^2 + b\sum_1^n x_j &= \sum_1^n x_j y_j \\ a\sum_1^n x_j + nb &= \sum_1^n y_j \end{aligned} \tag{3-47}$$

Introduce $\bar{x} = \left(\sum_1^n x_j\right)/n$ and $\bar{y} = \left(\sum_1^n y_j\right)/n$ so that $(\bar{x}, \bar{y})$ is the **center of gravity** of the points (x_j, y_j). The second equation in (3-47) can now be written $a\bar{x} + b = \bar{y}$, and asserts that a and b must be chosen so that $F(\bar{x}) = \bar{y}$, that is, the graph of F is a line passing through $(\bar{x}, \bar{y})$. A solution of (3-47) can be written as

$$a = \frac{\dfrac{1}{n}\displaystyle\sum_1^n x_j y_j - \bar{x}\bar{y}}{\dfrac{1}{n}\displaystyle\sum_1^n (x_j)^2 - (\bar{x})^2} \qquad b = \bar{y} - a\bar{x}$$

There is only one critical point. To check that it is a local minimum point for f, we compute

$$\Delta = (f_{12})^2 - f_{11}f_{22} = \left(2\sum_1^n x_j\right)^2 - (2n)\left(2\sum_1^n (x_j)^2\right)$$

$$= 4\left\{\left(\sum_1^n x_j\right)^2 - n\sum_1^n (x_j)^2\right\}$$

By the Schwarz inequality (Sec. 1.3),

$$\left(\sum_1^n x_j\right)^2 = \left(\sum_1^n 1 \cdot x_j\right)^2 \le \sum_1^n 1^2 \sum_1^n x_j^2 = n\sum_1^n x_j^2$$

and equality holds only if all the x_j are the same. Hence, $\Delta < 0$ and our critical point (a, b) is extremal. Since $f_{11} = 2\Sigma x_j^2 > 0$, it is indeed a local (and hence global) minimum point for f. (A different but related extremal problem is given in Exercise 17.)

In a numerical problem, especially one with many variables, it may be more efficient on a high-speed computer to use a simple search technique to find the maximum (or minimum) of a function F than it is to look for critical points and use Theorem 19. One such device is called the method of **steepest ascent.** Suppose that we wish to maximize the function F in a region D. The procedure is to select a starting point p_0, and then generate from it a sequence of points p_n which "go uphill" from p_0 and (hopefully) converge to the desired maximum point p^* for F in D. The algorithm for generating p_{n+1} from p_n is based on the fact that the gradient of F, $\mathbf{D}F(p)$, at a point p points in the direction of greatest increase. Choose a small positive number h, and define $\{p_n\}$ by

$$p_{n+1} = p_n + h\mathbf{D}F(p_n) \qquad n = 0, 1, 2, \ldots.$$

The distance between successive points, $|p_{n+1} - p_n|$, is determined by the size of h and also by the length of the vector $\mathbf{D}F$. At a critical point, $\mathbf{D}F = 0$, so the size of the successive steps will decrease as p_n approaches a critical point. The success of this ascent algorithm depends much upon the choice of a good starting point p_0, and upon the "landscape" of the function F. The surface shown in Fig. 3-15 illustrates some of the difficulties that can be met; this is the surface whose contour lines were shown in Fig. 3-14. A new branch of applied analysis called **optimization** has grown out of the need to find better ways for finding the extremal values of a function.

In searching for an extremal point for F on a set D, one must check both the boundary and the interior of D. Some functions, however, can be shown never to have *interior* extrema. For such a function F, it must then be the case that if $m \le F(p) \le M$ holds on the boundary of D, it holds inside D as well. This is true for the **harmonic** functions; *A function F of two variables is said to be harmonic in an open set D if $F \in C''$ and if $F_{xx} + F_{yy} = 0$ everywhere in D.*

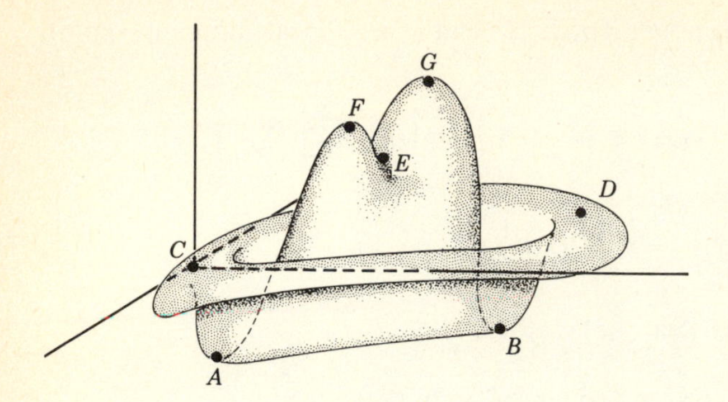

Figure 3-15 Surface with seven critical points.

Theorem 20 *Let D be a bounded open set, and F harmonic in D and continuous on the closure of D. If $F(p) \leq M$ for p in* bdy (D), *then $F(p) \leq M$ for p in D.*

Take any $\varepsilon > 0$ and set $f(x, y) = F(x, y) + \varepsilon x^2$. The function f is continuous on $\overline{D}$, which is compact, so the maximum of f on $\overline{D}$ is taken at some point p^* in $\overline{D}$. If p^* is interior to D, $f_{xx}(p^*) \leq 0$ and $f_{yy}(p^*) \leq 0$ and $f_{xx} + f_{yy} \leq 0$. Hence, $F_{xx} + F_{yy} + 2\varepsilon \leq 0$, which violates the assumption that F is harmonic in D. Accordingly, f must take its maximum on the boundary of D. Let A be the largest value of x for (x, y) in D. Then, for any point (x, y) in bdy (D),

$$f(p) = F(p) + \varepsilon x^2 \leq M + \varepsilon A^2$$

and therefore the same must hold for any point in D. This shows that if $p_0 \in D$, $p_0 = (x_0, y_0)$,

$$F(p_0) \leq M + \varepsilon(A^2 - x_0^2)$$

This estimate for $F(p_0)$ holds for any choice of ε, so let ε approach 0 and arrive at $F(p_0) \leq M$. ▮

Many problems arise in which it is known that the sought-for extremal point is not an interior point of the set D. A familiar example of this is the category of problems known as extremal problems with **constraints** or side conditions. As an illustration, we may be interested in the point $p = (x, y, z)$ at which $f(p)$ has a maximum value, where p is restricted to lie on a portion S of the surface described by $g(p) = 0$. As a subset of 3-space, S is a closed set which has no interior points; thus, the extremal point need not be among the critical points for f. The side condition $g(p) = 0$ which forces p to move on the surface piece S has the effect of decreasing the number of free variables from 3 to 2. This can be done explicitly if we can solve $g(x, y, z) = 0$ for one of the variables, in a neighborhood of the sought-for point. If $z = \phi(x, y)$, then

$$f(x, y, z) = f(x, y, \phi(x, y)) = F(x, y)$$

and we now look for the maximum value of F.

For example, let us find the point (x, y, z) obeying

$$g(x, y, z) = 2x + 3y + z - 12 = 0$$

for which $f(x, y, z) = 4x^2 + y^2 + z^2$ is minimum. We find that

$$z = 12 - 2x - 3y$$

so that

$$F(x, y) = 4x^2 + y^2 + (12 - 2x - 3y)^2$$

The critical points of F are found from the equations

$$0 = F_1(x, y) = 8x + 2(12 - 2x - 3y)(-2)$$

$$0 = F_2(x, y) = 2y + 2(12 - 2x - 3y)(-3).$$

These have only one solution, $(\frac{6}{11}, \frac{36}{11})$. Checking, we find that $F_{11} = 16$, $F_{12} = 12$, and $F_{22} = 20$, so that $\Delta = (12)^2 - (16)(20) < 0$ and this point yields a local minimum for F. Using the side condition to find z, we find that the solution to the original problem is the point $(\frac{6}{11}, \frac{36}{11}, \frac{12}{11})$.

A general approach to the solution of extremal problems with constraints, usually called the method of Lagrange multipliers, will be treated in Chap. 10. While of some theoretical interest, it has many practical limitations and is not often the most successful approach to a numerical solution.

EXERCISES

1 What are the maxima and minima of $f(x) = (2x^2 + 6x + 21)/(x^2 + 4x + 10)$?

2 (*a*) Does $P(x) = 1 - x + x^2/2 - x^3/3 + x^4/4$ have any real zeros?
(*b*) What about $Q(x) = P(x) - x^5/5 + x^6/6$?

3 Find the maximum and minimum value of $2x^2 - 3y^2 - 2x$ for $x^2 + y^2 \leq 1$.

4 Find the maximum and minimum value of $2x^2 + y^2 + 2x$ for $x^2 + y^2 \leq 1$.

5 Discuss the nature of the critical points of each of the functions described by:
(*a*) $f(x, y) = x^2 - y^2$ (*b*) $f(x, y) = 3xy - x^2 - y^2$
(*c*) $f(x, y) = 2x^4 + y^4 - x^2 - 2y^2$ (*d*) $f(x, y) = 4x^2 - 12xy + 9y^2$
(*e*) $f(x, y) = x^4 + y^4$ (*f*) $f(x, y) = x^4 - y^4$

6 Sketch the level curves of f for the functions given in Exercise 5, parts (*a*), (*b*), (*d*), and (*e*).

7 Given $f(x, y) = x^2 - 2xy + 3y^2 - x$ and the square $D = \{(x, y), 0 \leq x \leq 1, 0 \leq y \leq 1\}$. Find all critical points and find the maximum and minimum on D.

8 Show that $H(x, y) = x^2y^4 + x^4y^2 - 3x^2y^2 + 1 \geq 0$ for all (x, y).

9 Let $f(x, y) = (y - x^2)(y - 2x^2)$. Show that the origin is a critical point for f which is a saddle point, although on any line through the origin, f has a local minimum at $(0, 0)$.

10 Given n points in space, $P_1, P_2, \ldots, P_n$, find the point P for which

$$f(P) = \sum_1^n |P - P_j|^2$$

is a minimum.

11 Let $f \in C'$ in the open set Ω and have no critical points there. Let E be the set where $f(p) = 0$. Show that E has no interior points.

12 Find the point on the line through $(1, 0, 0)$ and $(0, 1, 0)$ which is closest to the line: $x = t$, $y = t$, $z = t$.

13 Find the maximum value of $x^2 + 12xy + 2y^2$, among the points (x, y) for which $4x^2 + y^2 = 25$.

14 Give a complete discussion of the problem of finding the right circular cone of greatest lateral area which may be inscribed upside down in the cone of radius 1 and altitude 3.

15 Given an equilateral triangular set, what location of P in the set will yield the maximum value of the product of the distances from P to the vertices? (*Hint:* Use the symmetry of the triangle.)

16 In the solution of the **normalized two-person game** whose **payoff matrix** is

$$\begin{bmatrix} 1 & 2 & -1 \\ -2 & 0 & 1 \\ 1 & -2 & 0 \end{bmatrix}$$

one is led to the problem of finding the saddle points of the function F described by

$$F(x_1, x_2, x_3, y_1, y_2, y_3) = (x_1 - 2x_2 + x_3)y_1 + (2x_1 - 2x_3)y_2 + (-x_1 + x_2)y_3$$

subject to the constraints $x_1 + x_2 + x_3 = 1$, $y_1 + y_2 + y_3 = 1$. Show that the saddle point is $x = (\frac{1}{3}, \frac{1}{3}, \frac{1}{3})$, $y = (\frac{2}{7}, \frac{1}{7}, \frac{4}{7})$.

***17** Let $(x_1, y_1) = P_1$, $(x_2, y_2) = P_2, \ldots, P_n$ be a set of points, not all the same. Find the line L which "fits" these points best in the sense that it minimizes $\sum_1^n d_j^2$, where d_j is the distance from P_j to L. [Note that this is not the same as the function f given in (3-46).]

18 Using Theorem 20, prove the following: If D is a closed bounded set, and if f and g are both harmonic in D, and if $f(p) = g(p)$ for all p on the boundary of D, then $f \equiv g$ in D.

19 For a function F of one variable, show that the method of steepest ascent leads to the algorithm

$$x_{n+1} = x_n + hF'(x_n)$$

If this is applied to find the maximum of $F(x) = x^4 - 6x^2 + 5$ for x in the interval $[-1, 1]$, show that the algorithm yields a convergent sequence $\{x_n\}$ if h is small, but that the method fails if h is too large and the starting point x_0 is unfortunately chosen.

20 Apply the method of steepest ascent to locate the maxima of the function $F(x, y) = x^3 - 3xy^2$ in the square

$$D = \{\text{all } (x, y) \text{ with } |x| \le 2, |y| \le 2\}$$

Examine the effect of the following three choices of initial point, $p_0 = (-1, 0)$, $(1, 0)$, $(0, 1)$, and the effect of the step size $h = .1$ and $h = .01$.

***21** Let $C_1 \ge C_2 \ge \cdots \ge C_n$ be a fixed set of positive numbers. Maximize the linear function $L(x_1, x_2, \ldots, x_n) = \sum_1^n C_j x_j$ in the closed set described by the inequalities $0 \le x_j \le 1$, $\sum_1^n x_j \le A$.

CHAPTER
FOUR

INTEGRATION

4.1 PREVIEW

In this chapter we start by discussing $\iint_D f$, the double integral of a function of two variables over a well-behaved set in the plane. This keeps the notation within reason, enables one to look at concrete cases easily, and can be generalized to the n-variable case or contracted to the familiar one-variable form. Too much experience with the latter in elementary calculus often leaves one with misleading impressions of both the nature of integration and the role of antidifferentiation in evaluating integrals, especially in connection with multiple integrals.

It is important to understand the difference between a multiple integral and an iterated integral, and how one moves from one to the other. Since not every integral can be evaluated in this manner, we give a brief treatment of certain numerical methods; more will be found in Chap. 10. Since high-dimensional integrals often seem artificial, we give several at the end of Sec. 4.3 that arise naturally in the study of urban transportation.

One topic that is usually slighted in elementary calculus is the treatment of improper integrals, especially those involving several variables. We cover this in Sec. 4.5.

Finally, we give a very brief discussion in Sec. 4.4 of the change-of-variables formula for multiple integrals, stating the general procedures and illustrating these with the familiar polar coordinate case. We have also chosen to present (as a piece of platform magic) the simple algorithm for carrying this out that is based on the algebraic manipulation of differential forms; a more complete explanation of this appears in Chap. 9.

4.2 THE DEFINITE INTEGRAL

In elementary calculus, integration is often equated to the notion of "area under a curve"; as a result, when the double integral is encountered, it is automatically viewed as the volume under a surface. However, this leaves one in some difficulty with triple integrals, and even more so with higher-order integrals. One major objective of this section is to gain another viewpoint that is extremely useful both in applications and in mathematical theory. It is that the definite integral involves integrating a function f over a set D, with the realization that the resulting numerical value depends both on f and on D. We begin with a simple case, the definition of the Riemann integral of a continuous function f over a rectangle R. By starting here, rather than with the one-variable case, we hope to overcome the tendency to connect integration too strongly with antidifferentiation.

By **grid,** we mean any finite set of horizontal and vertical lines which divide the rectangle R into a set of smaller rectangles, mutually disjoint except for their edges. If a particular grid is denoted by N and divides R into subrectangles R_{ij}, as shown in Fig. 4-1, the mesh size of the grid $d(N)$ is the largest diameter of the R_{ij}. The sides of the R_{ij} may be unequal. It is evident

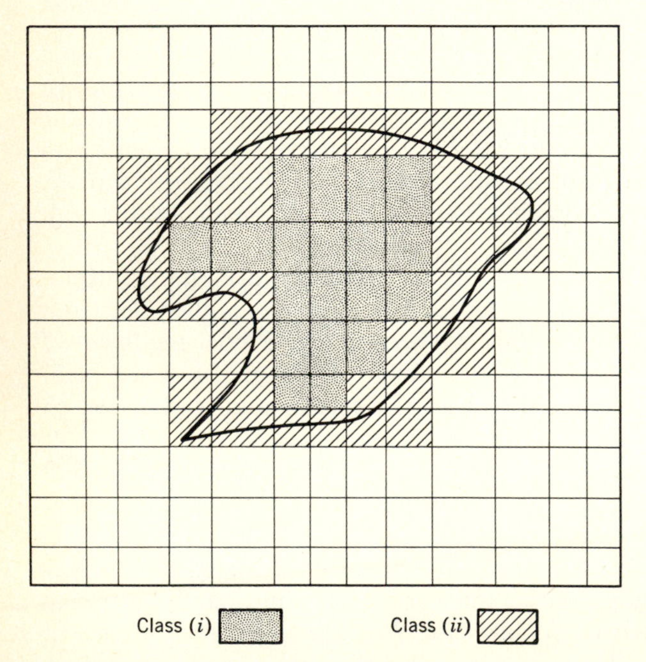

Figure 4-1

that if $d(N)$ is very small, the total number of subrectangles will be large, and each will have short sides. We take for granted the formula for the area $\mathrm{A}(R_{ij})$ of a rectangle. (However, see Exercise 1, which asks you to justify logically a simple property of this concept.)

In each R_{ij}, choose a point p_{ij}, and then form the associated Riemann sum:

$$S(f, N, \{p_{ij}\}) = \sum_{i,j} f(p_{ij})\mathrm{A}(R_{ij})$$

Since the sum on the right is finite, S is a number which depends on all the listed variables. Intuitively, it might be interpreted in various ways; for example, if the rectangle R were a metal plate of variable density, and $f(p)$ the density at the point p, then S would be a plausible estimate for the total mass of the plate.

Definition 1 *The double integral* $\iint_R f$ *exists and has value* v *if and only if for any* $\varepsilon > 0$ *there is a* $\delta > 0$ *such that*

$$|S(N, f, \{p_{ij}\}) - v| < \varepsilon$$

for any choice of $\{p_{ij}\}$ *and any grid* N *with* $d(N) < \delta$.

By analogy with preceding definitions, this is often written

$$\lim_{d(N)\to 0} S(N, f, \{p_{ij}\}) = \iint_R f$$

However, as a limit operation, it differs in many essentials from those that we have discussed earlier. The system of grids is not a sequence tending toward some limit, and the number $d(N)$ does not serve to identify the grid N. (The system of all grids is an example of what is called a "net" or "directed system" in more advanced courses, and is a generalization of the notion of sequence; we do not need the precise definition here, but refer to the article by McShane in [4] for anyone interested.) Having defined the integral, we need to know that it is valid for a wide class of functions f.

Theorem 1 *If* f *is continuous on* R, *then* $\iint_R f$ *exists.*

In the first stages of the proof, we assume only that f is bounded on R, since this will make the proof of later theorems slightly easier. Associated with each subrectangle R_{ij} determined by the grid N are two numbers

$$M_{ij} = \sup_{p \in R_{ij}} f(p)$$

$$m_{ij} = \inf_{p \in R_{ij}} f(p)$$

We set
$$\overline{S}(N) = \sum M_{ij}\,\mathrm{A}(R_{ij}) = \text{upper Riemann sum}$$
$$\underline{S}(N) = \sum m_{ij}\,\mathrm{A}(R_{ij}) = \text{lower Riemann sum}$$

It is clear that $\underline{S}(N) \le S(N, f, \{p_{ij}\}) \le \overline{S}(N)$ for any choice of the points p_{ij}, and if f is continuous, equality at either end can be obtained by proper choices. We separate the proof into several simple steps. In order to be able to compare two different grids and their upper and lower sums, we introduce the notion of a **refinement** of a grid.

Definition 2 *A grid N' is said to be a refinement of the grid N if N' is obtained by adding one or more lines to those which form N.*

We note that it is possible to have two grids, neither of which is a refinement of the other.

Lemma 1 *If N' is a refinement of N, then*
$$\underline{S}(N) \le \underline{S}(N') \le \overline{S}(N') \le \overline{S}(N)$$

Let us examine the effect of the refinement on a single term $M_{ij}\,\mathrm{A}(R_{ij})$ of the upper sum $\overline{S}(N)$. Under the new partition scheme, R_{ij} will be broken up into a collection of smaller rectangles $r_1, r_2, \ldots, r_m$. Thus, corresponding to this single term of $\overline{S}(N)$, there will be in $\overline{S}(N')$ a block of terms $M^{(1)}\mathrm{A}(r_1) + M^{(2)}\mathrm{A}(r_2) + \cdots + M^{(m)}\mathrm{A}(r_m)$, where $M^{(k)}$ is the least upper bound of f in r_k. Since each r_k lies in R_{ij}, $M^{(k)} \le M_{ij}$, and
$$\sum_1^m M^{(k)}\mathrm{A}(r_k) \le M_{ij} \sum_1^m \mathrm{A}(r_k) = M_{ij}\,\mathrm{A}(R_{ij})$$

Repeating this argument for each term of $\overline{S}(N)$, we find that $\overline{S}(N') \le \overline{S}(N)$. An analogous argument shows that $\underline{S}(N) \le \underline{S}(N')$. ▮

Every grid is a refinement of the empty grid which does not partition R at all, so that for any N,
$$m\mathrm{A}(R) \le \underline{S}(N) \le \overline{S}(N) \le M\mathrm{A}(R)$$

where $m = \inf_{p \in R} f(p)$ and $M = \sup_{p \in R} f(p)$. The set of lower sums $\underline{S}(N)$ forms a bounded set of numbers as N ranges over all possible grids N; let s be its least upper bound. The set of all upper sums $\overline{S}(N)$ is also bounded; let S be its greatest lower bound.

Lemma 2 $\quad$ *$s \le S$ and for any N, $S - s \le \overline{S}(N) - \underline{S}(N)$.*

This asserts that the four numbers mentioned have the relative position shown below:

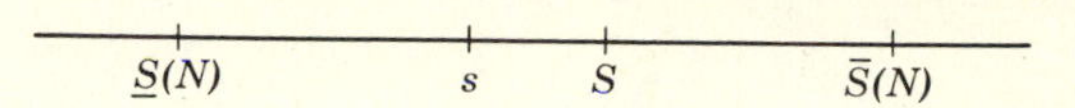

Since $\underline{S}(N)$ is always to the left of s, and $\overline{S}(N)$ to the right of S, it is always true that $S - s \leq \overline{S}(N) - \underline{S}(N)$, regardless of the relative positions of s and S. Let N_1 and N_2 be any two grids, and construct a third grid N by forming the union of N_1 and N_2. N will therefore have all the lines that make up N_1, and will be a refinement of N_1. It will also be a refinement of N_2. Applying Lemma 1,

$$\begin{matrix}\underline{S}(N_1)\\ \underline{S}(N_2)\end{matrix} \leq \underline{S}(N) \leq \overline{S}(N) \leq \begin{matrix}\overline{S}(N_1)\\ \overline{S}(N_2)\end{matrix}$$

In particular, $\underline{S}(N_1) \leq \overline{S}(N_2)$. Since N_1 and N_2 were arbitrary grids, this shows that every lower sum is smaller than each upper sum, $\overline{S}(N_2)$. Since s is the least upper bound for the set of lower sums, $s \leq \overline{S}(N_2)$. This holds for every N_2, so that $s \leq S$. ▮

Up to this point, we have not assumed that f was continuous. The numbers s and S are therefore defined for any bounded function f; they are called the **lower** and **upper** integrals of f over R.

Lemma 3 *If f is continuous on R, then* $\lim_{d(N)\downarrow 0} |\overline{S}(N) - \underline{S}(N)| = 0$.

Since R is closed and bounded, f is uniformly continuous on R. Given ε, we may choose $\delta > 0$ so that $|f(p) - f(q)| < \varepsilon$ whenever p and q lie in R and $|p - q| < \delta$. Let N be any grid which partitions R with mesh diameter $d(N) < \delta$. Since $M_{ij} = f(p)$ and $m_{ij} = f(q)$ for a particular choice of p and q in R_{ij}, and since R_{ij} has diameter less than δ, we have $M_{ij} - m_{ij} < \varepsilon$ for all i and j. This gives

$$\begin{aligned}0 \leq \overline{S}(N) - \underline{S}(N) &= \sum (M_{ij} - m_{ij})\mathrm{A}(R_{ij})\\ &\leq \varepsilon \sum \mathrm{A}(R_{ij}) = \varepsilon \mathrm{A}(R)\end{aligned}$$

This shows that $|\overline{S}(N) - \underline{S}(N)|$ can be made arbitrarily small merely by requiring that $d(N)$ be small. ▮

We now complete the proof of the theorem. Lemma 2 and Lemma 3 combined show that $s = S$. Call the common value v. Given ε, choose δ so that $\overline{S}(N) - \underline{S}(N) < \varepsilon$ whenever N obeys $d(N) < \delta$. The closed interval $[\underline{S}(N), \overline{S}(N)]$ contains v and also the value of the general Riemann sums

$S(N, f, \{p_{ij}\})$. Thus $|S(N, f, p_{ij}) - v| < \varepsilon$ whenever the grid N obeys $d(N) < \delta$, so that $\iint_R f$ exists. ∎

Two questions arise naturally: (i) What happens if f is not continuous on R? (ii) How can we define the integral of a function f over a set D in the plane that is not a rectangle? It turns out that the answer to the first provides the answer to the second.

Theorem 2 *Let R be a closed rectangle, and let f be bounded in R and continuous at all points of R except those in a set E of zero area. Then $\iint_R f$ exists.*

This introduces a new concept, **"sets of zero area,"** so that we must digress to discuss this before giving the proof of this theorem.

Historically, the concept of area preceded that of integral. Let D be any bounded set in the plane, and choose some rectangle R containing D, with its edges parallel to the coordinate axes. Any grid N on R will partition R into closed subrectangles R_{ij} which will together cover D.

We separate the rectangles R_{ij} into three classes (see Fig. 4-1): Class (i), all the R_{ij} which contain only interior points of D; class (ii), all the R_{ij} which contain at least one boundary point of D; class (iii), all R_{ij} which contain only exterior points of D. The union of the rectangles of class (i) is called the **inner** or inscribed set for D determined by the grid N, while the union of the rectangles of class (i) and class (ii) is called the **outer** or circumscribing set. Let $\overline{S}(N, D)$ be the total area of the outer set, and $\underline{S}(N, D)$ the total area of the inner set. Clearly, $0 \le \underline{S}(N, D) \le \overline{S}(N, D) \le$ area of R. As N ranges over all possible grids, the values $\overline{S}(N, D)$ generate a bounded set which is determined solely by the set D. Let $\overline{\mathrm{A}}(D)$ be the greatest lower bound (infimum) of this set. Similarly, the values $\underline{S}(N, D)$ generate another bounded set, and we let $\underline{\mathrm{A}}(D)$ be its least upper bound (supremum). We call $\overline{\mathrm{A}}(D)$ the **outer area** of D, and $\underline{\mathrm{A}}(D)$ the **inner area** of D. If they have the same value, we denote it by $\mathrm{A}(D)$ and call this **area** of D.

This process defines a notion of area for certain sets D. Not all sets have an area. For example, let D be the set of points (x, y) with $0 \le x \le 1$, $0 \le y \le 1$, and both x and y rational. Since D has no interior points, class (i) is always empty, so that $\underline{\mathrm{A}}(D) = 0$. On the other hand, every point of the unit square is a boundary point for D, so that every circumscribing set has area 1, and $\overline{\mathrm{A}}(D) = 1$. Since these are not equal, the set D does not have an area. (Note that this is distinctly different from saying that a set D has zero area, for this would mean $\underline{\mathrm{A}}(D) = \overline{\mathrm{A}}(D) = 0$.)

According to our definition, a set D will have zero area, $\mathrm{A}(D) = 0$, if and only if given $\varepsilon > 0$ we can cover D by a finite collection of rectangles R_k such that

$$A(R_1) + A(R_2) + \cdots + A(R_m) < \varepsilon$$

Any finite set D has zero area, as does any simple polygon. It is also easy to show (Exercise 14) that the graph of any continuous function $f(x)$ defined on an interval $[a, b]$ is a set of zero area; it can also be shown that the same is true for any sufficiently smooth curve in the plane (Exercise 15).

This can also be used to settle whether certain sets D are nice enough to have an area defined. (Return to Fig. 4-1.)

For any choice of N, the number $\overline{S}(N, D) - \underline{S}(N, D)$ is exactly the sum of the areas of the rectangles R_{ij} of the partition which are in class (ii). These are just the ones which form the circumscribing set for the boundary Γ of D. Thus, the number $\overline{A}(D) - \underline{A}(D)$ is exactly $\overline{A}(\Gamma)$, the outer area of the boundary of D. We have therefore shown that a set D is well behaved enough to *have* an area if and only if its boundary is a set with zero area. This can therefore be the case when the boundary is composed of polygonal curves, or more generally, piecewise smooth curves. However, an example has been constructed of a region D, bounded by a simple closed curve, which does not have an area.

PROOF We now return to the proof of Theorem 2. By assumption, the set of discontinuities of f is a set E of zero area. It is therefore possible to choose a grid on R such that those subrectangles that cover E have arbitrarily small total area. Given any $\varepsilon > 0$, we therefore assume that we have a finite union of R_0 of these subrectangles with $E \subset R_0$ and $A(R_0) < \varepsilon$. The union of the rectangles not needed to cover E forms a closed set R_1 containing no points of E, and in which f is continuous. Since f is then uniformly continuous in R_1, we may choose $\delta_1 > 0$, so that $|f(p) - f(q)| < \varepsilon$ whenever p and q lie in R_1 and $|p - q| < \delta_1$. Take any grid N, and form the difference

$$\overline{S}(N) - \underline{S}(N) = \sum (M_{ij} - m_{ij})A(R_{ij})$$

Divide the collection of rectangles R_{ij} into two classes. Into class $\mathscr{C}_1$ we put all R_{ij} which are subsets of R_1, and into the class $\mathscr{C}_2$ we put all that remain (see Fig. 4-2). We split the expression for $\overline{S}(N) - \underline{S}(N)$, in a corresponding way:

$$\overline{S}(N) - \underline{S}(N) = \sum_{R_{ij} \in \mathscr{C}_1} (M_{ij} - m_{ij})A(R_{ij}) + \sum_{R_{ij} \in \mathscr{C}_2} (M_{ij} - m_{ij})A(R_{ij})$$

If $d(N) < \delta_1$, we have $M_{ij} - m_{ij} < \varepsilon$ whenever $R_{ij} \subset R_1$, so that

$$\sum_{R_{ij} \in \mathscr{C}_1} (M_{ij} - m_{ij})A(R_{ij}) < \varepsilon \sum_{R_{ij} \in \mathscr{C}_1} A(R_{ij}) \leq 2BA(R_0')$$

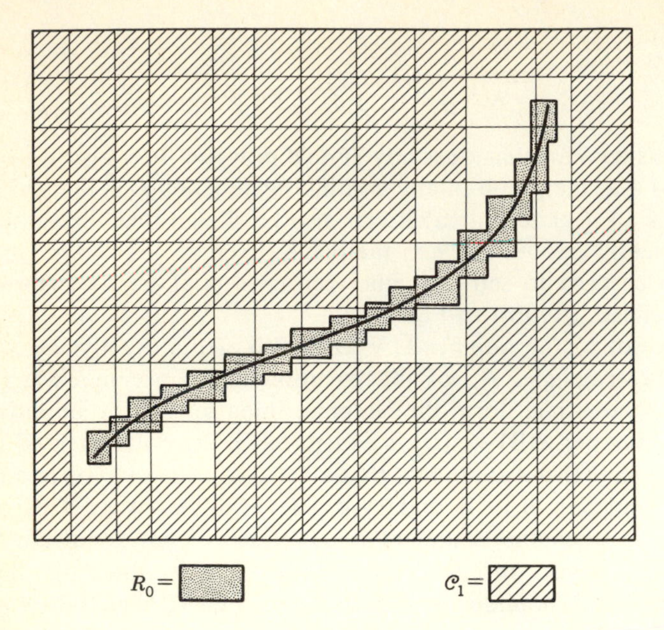

Figure 4-2

By assumption, f is bounded in R, so that $|f(p)| < B$ for all $p \in R$. Let R_0' be the union of the rectangles in $\mathscr{C}_2$. Then,

$$\sum_{R_{ij} \in \mathscr{C}_2} (M_{ij} - m_{ij})\text{A}(R_{ij}) \leq 2B \sum_{R_{ij} \in \mathscr{C}_2} \text{A}(R_{ij}) = 2B\text{A}(R_0')$$

The set R_0' is exactly the circumscribing set for R_0 in the partition determined by N. We can therefore choose δ_2 so that $\text{A}(R_0') \leq \text{A}(R_0) + \varepsilon < 2\varepsilon$ whenever $d(N) < \delta_2$. Letting δ be the smaller of δ_1 and δ_2, we have shown that whenever N is a grid which partitions R with $d(N) < \delta$, $\overline{S}(N) - \underline{S}(N) < \varepsilon\text{A}(R) + 4B\varepsilon = (4B + \text{A}(R))\varepsilon$. This reestablishes Lemma 3 with our weakened hypothesis, and the theorem follows as before. ▮

With Theorem 2, we have obtained an answer to our first question. We now use it to answer the second as well. We use a simple device to define the integral of f over D when D is not a rectangle. Suppose that D is any bounded set, and choose a rectangle R containing D. Define a new function F on R by

$$F(p) = \begin{cases} f(p) & \text{for } p \in D \\ 0 & \text{for } p \notin D \end{cases} \tag{4-1}$$

Suppose that f is continuous on the interior of D. What can be said about the points at which F is continuous? F will certainly be continuous at any interior point of D; if p_0 is a point not in D, then $F(p_0) = 0$, and if p_0 is exterior to D, all the points near p_0 are also points where F vanishes. Thus,

F is certainly continuous at every point p that is either interior to D or exterior to D. Hence, F is continuous off the boundary of D. If the set D is one that has an area (we shall call such sets **Jordan measurable**), then as seen above, bdy(D) will have zero area. By Theorem 2, $\iint_R F$ exists, and we use this to define the integral of f over D:

$$\iint_D f = \iint_R F \tag{4-2}$$

All that remains is to show that the answer does not depend on the choice of the containing rectangle R. Suppose R' is another closed rectangle containing D and F' is the corresponding function. Form $R'' = R \cap R'$, which is also a closed rectangle containing D. Then,

$$\iint_R F = \iint_{R''} F = \iint_{R''} F' = \iint_{R'} F'$$

and we obtain the same value for $\iint_D f$ from both R and R'. This shows that the definition for $\iint_D f$ is unambiguous, and we have a satisfactory definition. Suppose now that f is not continuous on D. If its discontinuities form a set E of zero area, then the function F will also be discontinuous on E, in addition to its discontinuities on the boundary of D. However, the union of two sets of zero area is again of zero area, and the preceding argument still applies. We have thus proved the following basic existence theorem for the definite integral.

Theorem 3 *Let D be a bounded Jordan-measurable set, and let f be bounded on D and continuous except for a set E of zero area. Then, $\iint_D f$ exists, as defined by* (4-1) *and* (4-2).

What we have said about the double integral is also valid for the general n-fold multiple integral. Because of the importance of the special case $n = 1$, we restate some of the treatment. A grid is now just a finite set of division points. If we impose a grid N on the closed interval $I = [a, b]$, we obtain a partition of I into closed intervals $I_k = [x_k, x_{k+1}]$, where $a = x_0 < x_1 < x_2 < \cdots < x_n = b$. The general Riemann sum becomes

$$S(N, f, \{p_k\}) = \sum f(p_k)\,\Delta x_k$$

where $p_k \in I_k$ and $\Delta x_k = x_{k+1} - x_k$ is the length of I_k. The number $d(N)$ is now the largest of the lengths Δx_k.

Definition 3 *The integral* $\int_I f$ *of* f *over the interval* I *exists and has the value* v *if and only if*

$$\lim_{d(N)\to 0} S(N, f, \{p_k\}) = v$$

Instead of $\int_I f$, it is customary to write $\int_a^b f$ or $\int_a^b f(x)\,dx$. Corresponding to Theorem 3 we have Theorem 3′:

Theorem 3′ *If* f *is bounded on* $[a, b]$ *and if* f *is continuous on* $[a, b]$ *except on a set of zero length, then* $\int_a^b f$ *exists.*

A set on the line has zero length if it can be covered by a finite collection of intervals of arbitrarily small total length. In particular, a finite set of points has zero length. As an instance of the theorem,

$$\int_{-1}^{1} \sin^2\left(\frac{1}{x}\right) dx$$

exists, since the integrand is bounded and continuous except at the single point $x = 0$.

Some of the familiar properties of the definite integral are set forth in the next theorem as they would be stated for double integrals. The sets D, D_1, and D_2 are assumed to have area.

Theorem 4 *Let* f *and* g *be continuous and bounded on* D. *Then,*

(i) $\iint_D (f + g)$ *exists and is* $\iint_D f + \iint_D g$.

(ii) *For any constant* C, $\iint_D Cf = C \iint_D f$.

(iii) *If* $f(p) \geq 0$ *for all* $p \in D$, $\iint_D f \geq 0$.

(iv) $\iint_D |f|$ *exists and* $\left|\iint_D f\right| \leq \iint_D |f|$.

(v) *If* $D = D_1 \cup D_2$ *and* D_1 *and* D_2 *intersect in a set with zero area, then*

$$\iint_D f = \iint_{D_1} f + \iint_{D_2} f$$

Most of these follow directly from corresponding properties of the Riemann sums. Take a rectangle R containing D and define f and g to vanish off D. Then, the relations

$$\sum [f(p_{ij}) + g(p_{ij})]\mathrm{A}(R_{ij}) = \sum f(p_{ij})\mathrm{A}(R_{ij}) + \sum g(p_{ij})\mathrm{A}(R_{ij})$$

$$\sum Cf(p_{ij})\mathrm{A}(R_{ij}) = C\sum f(p_{ij})\mathrm{A}(R_{ij})$$

lead at once to (i) and (ii). For (iii), we observe that if $f(p) \geq 0$ for all $p \in D$, then $S(N, f, \{p_{ij}\}) \geq 0$. In turn, (iii) leads to (iv); since $|f| + f$ and $|f| - f$ are nonnegative on D, and continuous, $\iint_D |f| + \iint_D f \geq 0$ and $\iint_D |f| - \iint_D f \geq 0$. To prove (v), we define a special function F by

$$F(p) = \begin{cases} f(p) & \text{for } p \in D_1 \\ 0 & \text{for } p \notin D_1 \end{cases}$$

Then
$$\iint_D f = \iint_D F + \iint_D (f - F) = \iint_{D_1} f + \iint_{D_2} (f - F)$$

$$= \iint_{D_1} f + \iint_{D_2} f \quad \blacksquare$$

With functions of one variable, the last property is usually stated differently. When $a \leq b$, $\int_a^b f$ is an alternative notation for $\int_{[a,\,b]} f$. When $a > b$, we define $\int_a^b f$ to mean $-\int_{[b,\,a]} f$.

Thus, for functions of one variable, if f is continuous on an interval I, and a, b, c are any three points in I, then no matter how they are arranged on the interval,

$$\int_a^b f + \int_b^c f = \int_a^c f$$

In this respect, the one-variable Riemann integral differs from the double or triple integral as they are usually presented. The single integral is an "oriented" integral, and in $\int_a^b f$ one speaks of integrating "from a to b." In the definition of $\iint_R f$, no notion of orientation of R occurred. We will need to add such a notion later in dealing with multiple integrals and with transformations of coordinates in n-space.

EXERCISES

Where nothing is said, assume that the sets D mentioned below have positive area and are bounded.

1 (*a*) Let R be a 3×5 rectangle divided as shown in Fig. 4-3, where the division points are arbitrary. Show that the sum of the areas of the subrectangles is always 15.

*(*b*) How can you prove that this is similarly true for a general rectangle and an arbitrary subdivision?

2 Show from Definition 1 that for any rectangle R,

(*a*) $\iint_R f = 0$ if $f \equiv 0$ on R.

(*b*) $\iint_R f = \text{A}(R)$ if $f = 1$ on R. (*Hint:* Use Exercise 1.)

3 If f is continuous on D and $m \le f(p) \le M$ for all $p \in D$, show that

$$m\text{A}(D) \le \iint_D f \le M\text{A}(D)$$

4 If f is defined on a rectangle R and $\iint_R f$ exists, as defined above, then f is necessarily bounded on R.

5 (*Mean Value Theorem*) Let D be compact, connected. Let f and g be continuous and bounded on D, with $g(p) \ge 0$ for all $p \in D$. Then, there is a point $\bar{p} \in D$ such that $\iint_D fg = f(\bar{p}) \iint_D g$.

6 If D is open, and if f is continuous, bounded, and obeys $f(p) \ge 0$ for all $p \in D$, then $\iint_D f = 0$ implies $f(p) = 0$ for all p.

7 Formulate corresponding definitions for the triple integral $\iiint_D f$ where D is a bounded set in 3-space.

8 If f is bounded and monotonic on $[a, b]$, show that $\int_a^b f$ exists, even if f is not continuous.

9 If $D_1 \subset D_2$, then $\text{A}(D_1) \le \text{A}(D_2)$.

10 Show that a finite set has zero area.

11 Let D be the set of all points $(1/n, 1/m)$ where n and m are positive integers. Does D have area?

12 Explain why the area of a Jordan-measurable region D is given by $\text{A}(D) = \iint_D 1$, using Theorem 3.

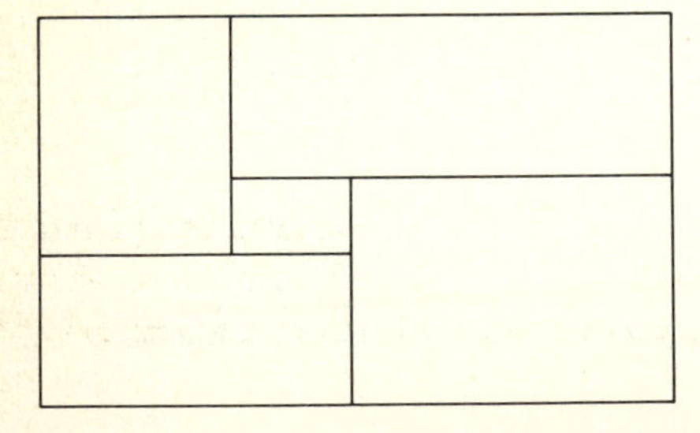

Figure 4-3

13 Let $f \in C^2$, $f(x) \geq 0$, and $f''(x) \leq 0$ for $a \leq x \leq b$. Prove that

$$\frac{1}{2}(b-a)(f(a)+f(b)) \leq \int_a^b f \leq (b-a)f\left(\frac{a+b}{2}\right)$$

14 Let $f(x)$ be continuous for $a \leq x \leq b$. Show that the graph of f has zero area.

***15** Let $f(t)$ be continuous for $0 \leq t \leq 1$. Let $g(t)$ be continuous for $0 \leq t \leq 1$ with $|g'(t)| \leq B$. Show that the set of points (x, y) with $x = f(t)$, $y = g(t)$, $0 \leq t \leq 1$, has zero area.

16 Let f be continuous and positive on $[a, b]$. Let D be the set of all points (x, y) with $a \leq x \leq b$ and $0 \leq y \leq f(x)$. Show that D has area $\text{A}(D) = \int_a^b f$.

***17** A general subdivision $\mathcal{S}$ of a closed rectangle R is a collection $D_1, D_2, \ldots, D_m$ of a finite number of domains D_j, each having area, which together cover R, and such that no pair have common interior points. The norm $d(\mathcal{S})$ of such a subdivision is the largest of the diameters of the set D_j. Let f be continuous on R. Show that for any ε there is a δ such that whenever $\mathcal{S}$ is a general subdivision of R, and

$$S(\mathcal{S}, f, \{p_i\}) = \sum f(p_j)\text{A}(D_j)$$

where $p_j \in D_j$, then

$$\left| \iint_R f - S(\mathcal{S}, f, \{p_j\}) \right| < \varepsilon$$

whenever $d(\mathcal{S}) < \delta$.

4.3 EVALUATION OF DEFINITE INTEGRALS

Several comments about notation are in order. It will be noticed that in most instances, we use $\int_a^b f$ and $\iint_D f$ rather than $\int_a^b f(x)\,dx$ and $\iint_D f(x, y)\,dx\,dy$. When the second form is used, it must be recalled that the occurrence of "x" in $\int_a^b f(x)\,dx$ or of "x" and "y" in $\iint_D f(x, y)\,dx\,dy$ is that of a dummy letter, and that one could equally well write $\int_a^b f(u)\,du$ or $\int_a^b f(t)\,dt$, or $\iint_D f(u, v)\,du\,dv$ or $\iint_D f(s, t)\,ds\,dt$. The same is true for the letter "j" in $\sum_{j=1}^n a_j$ and both "i" and "j" in $\sum_{i,j=1}^n i^2 j^3$. While the notation $\int_a^b f$ is thus preferable from some points of view, it also has some disadvantages. Without introducing additional notations, it is difficult to indicate $\int_0^1 (x^3 - 4x^2 - 1)\,dx$ except in this way; again, when f is a function of several variables, and we wish to indicate the result of integration with respect to only one, it is convenient to write $\int_a^b f(x, y)\,dy$. No simple and usable substitute which does not also have dummy letters suggests itself for these. It is also important to realize that in all of these, the letter "d" is irrelevant and serves merely to block off the letter which follows it from the function which is the integrand. In place of $\int_a^b f(x, y)\,dy$, we might write $\int_a^b f(x, y)\,\boxed{y}$ or even $\boxed{y}\int_a^b f(x, y)$ where the presence of "y" in the square is to inform us that in carrying out the integration, we must regard $f(x, y)$ as a function of the second variable alone. In the language of logic, "x" is free, while "y" is bound and "$\boxed{y}$" serves as a quantifier. There is a reason for the conventional choice of "dy" in place of the suggested "$\boxed{y}$"; this will be indicated when we discuss the rule for "change of variable" in integration.

If it is known that $\iint_D f$ exists, then its value can often be obtained by the use of special subdivision schemes. If $N_1, N_2, \ldots$ is a sequence of grids such that $\lim_{k\to\infty} d(N_k) = 0$, then the corresponding sequence of Riemann sums will converge to the value of the integral. To illustrate this, we first evaluate $\iint_R xy^2\,dx\,dy$, where R is the square: $0 \le x \le 1$, $0 \le y \le 1$. Let N_k be the grid which partitions R into k^2 equal squares, each of side $1/k$. Choosing the point p_{ij} in R_{ij} to be $(i/k\ j/k)$, we have

$$\bar{S}(N_k) = \sum f(p_{ij})\mathrm{A}(R_{ij}) = \sum_{i,\,j=1}^{k} \frac{i}{k}\left(\frac{j}{k}\right)^2 \frac{1}{k^2}$$

$$= \frac{1}{k^5}\sum_{i=1}^{k} i \sum_{j=1}^{k} j^2$$

Since $$\sum_1^k i = \frac{k(k+1)}{2} \quad \text{and} \quad \sum_1^k j^2 = \frac{k(k+1)(2k+1)}{6}$$

$$\bar{S}(N_k) = \frac{1}{k^5}\left(\frac{k(k+1)}{2}\right)\left(\frac{k(k+1)(2k+1)}{6}\right)$$

and $$\iint_R xy^2\,dx\,dy = \lim_{k\to\infty} \bar{S}(N_k) = \tfrac{1}{6}$$

As another illustration, let us compute $\int_1^2 \sqrt{x}\,dx$ from the definition. If we again use equal subdivisions, we arrive at the Riemann sums

$$\bar{S}(N_n) = \sum_{j=1}^{n} \sqrt{1 + \frac{j}{n}}\,\frac{1}{n} = \frac{1}{n\sqrt{n}}\sum_{j=1}^{n} \sqrt{j+n}$$

However, it is not easy to compute $\lim_{n\to\infty} \bar{S}(N_n)$. Instead, we use a different type of subdivision with unequal intervals. Choose

$$r = 2^{1/n} > 1$$

and let N_n be the grid with division points

$$1 = r^0 < r < r^2 < \cdots < r^{n-1} < r^n = 2$$

The longest of the intervals determined by this grid is the last, so that $d(N_n) = 2 - r^{n-1} = 2(1 - 1/r)$. Since $\lim_{n\to\infty} r = \lim_{n\to\infty} 2^{1/n} = 1$, $\lim_{n\to\infty} d(N_n) = 0$, and

$$\begin{aligned}\bar{S}(N_n) &= \sqrt{r}\,(r-1) + \sqrt{r^2}\,(r^2 - r) + \cdots + \sqrt{r^n}\,(r^n - r^{n-1})\\ &= \sqrt{r}\,(r-1)(1 + r\sqrt{r} + [r\sqrt{r}]^2 + \cdots + [r\sqrt{r}]^{n-1})\\ &= \sqrt{r}\,(r-1)\frac{r^{3n/2} - 1}{r^{3/2} - 1}\\ &= (2\sqrt{2} - 1)\sqrt{r}\,\frac{r-1}{r^{3/2} - 1}\end{aligned}$$

so that

$$\int_1^2 \sqrt{x}\,dx = \lim_{n\to\infty} \bar{S}(N_n) = (2\sqrt{2} - 1)\lim_{r\to 1} \sqrt{r}\,\frac{r-1}{r^{3/2} - 1} = \frac{(4\sqrt{2} - 2)}{3}$$

If this direct method were the only available procedure for computing the value of an integral, only the simplest of integrands could be used. We next prove the **fundamental theorem of integral calculus,** which justifies the process of evaluating the integral of a function of one variable by means of antidifferentiation.

Definition 4 *A function F is an antiderivative (or primitive or indefinite integral) of f on an interval I if $F'(x) = f(x)$ for all $x \in I$.*

Theorem 5 *If f is continuous on the interval $I = [a, b]$, then f has an antiderivative on I.*

Define a function F_0 on I by

$$F_0(x) = \int_a^x f \qquad \text{for } a \le x \le b$$

If x and $x + h$ both lie in I, then

$$\begin{aligned}F_0(x+h) - F_0(x) &= \int_a^{x+h} f - \int_a^x f\\ &= \int_x^{x+h} f = f(\bar{x})h\end{aligned}$$

where we use the mean value theorem for integrals (Exercise 5, Sec. 4.2) and $\bar{x}$ lies between x and $x + h$. Divide by h, and let h approach 0; since $\bar{x}$ must then approach x, and since f is continuous, we find $F_0'(x) = f(x)$, and F_0 is an antiderivative of f. (When x is an endpoint of I, the argument shows the appropriate one-sided derivative of F_0 has the correct value.) ▮

Some discontinuous functions have antiderivatives and some do not. When a function, continuous or not, has one, it has an infinite number, all differing by constants.

Theorem 6 *If F_1 and F_2 are both antiderivatives of the same function f on an interval I, then $F_1 - F_2$ is constant on I.*

For, $(F_1 - F_2)' = F_1' - F_2' = f - f = 0$ on I, and Exercise 1, Sec. 3.2, applies. ▮

Theorem 7 *If f is continuous on $[a, b]$ and F is any antiderivative of f, then*

$$\int_a^b f = F(b) - F(a)$$

Let F_0 be the particular antiderivative of f which was constructed in Theorem 5. By Theorem 6, we may write $F = F_0 + C$. Referring back to the definition of F_0, we have $F_0(a) = 0$, so that $C = F(a)$. Then,

$$\int_a^b f = F_0(b) = F(b) - C = F(b) - F(a) \quad ▮$$

One consequence of this is the familiar procedure for "change of variable" in integration.

Theorem 8 *Let ϕ' exist and be continuous on the interval $[\alpha, \beta]$ with $\phi(\alpha) = a$ and $\phi(\beta) = b$. Let f be continuous at all points $\phi(u)$ for $\alpha \le u \le \beta$. Then,*

$$\int_a^b f(x)\,dx = \int_\alpha^\beta f(\phi(u))\phi'(u)\,du$$

The rule may also be stated: To make the substitution $x = \phi(u)$ in an integral $\int_a^b f(x)\,dx$, replace $f(x)$ by $f(\phi(u))$, replace dx by $\phi'(u)\,du$, and replace the limits a and b by u values which correspond to them. It might seem that this is obviously true, since if $x = \phi(u)$,

$$dx = \frac{dx}{du}\,du = \phi'(u)\,du$$

However, we are misled by a matter of notation. Using the alternative square notation, the theorem asserts that

$$\int_a^b f(x)\,\boxed{x} = \int_\alpha^\beta f(\phi(u))\phi'(u)\,\boxed{u}$$

and the rule requires us to replace $\boxed{x}$ by $\phi'(u)\ \boxed{u}$; analogously, the differentiation rule $dy/dx = dy/ds\ ds/dx$ is not proved by canceling the two occurrences of "ds". It is a virtue of the notation used in differentiation and integration that such formalism is consistent with the truth, and serves as a guide in the correct application of theorems.

To give a valid proof of the theorem, let $F' = f$ and define G on $[\alpha, \beta]$ by $G(u) = F(\phi(u))$. Then, $G'(u) = F'(\phi(u))\phi'(u) = f(\phi(u))\phi'(u)$, so that

$$\begin{aligned}\int_\alpha^\beta f(\phi(u))\phi'(u)\,du &= \int_\alpha^\beta G'(u)\,du = \int_\alpha^\beta G' \\ &= G(\beta) - G(\alpha) = F(\phi(\beta)) - F(\phi(\alpha)) \\ &= F(b) - F(a) = \int_a^b f \quad \blacksquare\end{aligned}$$

For fixed a and b, there may be a number of possible choices of α and β; any of these may be used if ϕ is continuous and differentiable on the entire interval $[\alpha, \beta]$. For example, with $x = u^2 = \phi(u)$, we have

$$\int_1^4 f(x)\,dx = \int_1^2 f(u^2)2u\,du = \int_{-1}^{2} f(u^2)2u\,du = \int_1^{-2} f(u^2)2u\,du$$

This requires care in its use, as may be seen from the special choices $f(x) = x$ and $f(x) = \sqrt{x}$.

The following examples should be studied carefully. The first contains an incorrect application of Theorem 7, and the second an incorrect application of Theorem 8.

(i) To compute $\int_{-2}^{2} x^{-2}\,dx$, we observe that an antiderivative is $-x^{-1}$, so that

$$\int_{-2}^{2} x^{-2}\,dx = -x^{-1}\Big|_{-2}^{2} = -(\tfrac{1}{2}) = (\tfrac{1}{2}) = -1$$

(ii) Let $C = \int_{-1}^{1} [\sin 1/x]^2\,dx$. Since the integrand is bounded and continuous on $[-1, 1]$ except at the point $x = 0$, the integral exists. Moreover, since the integrand is never negative, $C > 0$. Put $u = 1/x$ so that $f(x)$ becomes $(\sin u)^2$ and $dx = -u^{-2}\,du$. When $x = 1$, $u = 1$, and when $x = -1$, $u = -1$.

Thus,

$$C = \int_{-1}^{1} (\sin u)^2(-u^{-2})\, du = -\int_{-1}^{1} \left[\frac{\sin u}{u}\right]^2 du$$

The integrand in this integral is continuous on $[-1, 1]$ since the discontinuity at $u = 0$ is removable; thus, the new integral exists, and is negative!

[In each of these examples, you should make sure that you understand the nature of the error involved, before reading further.]

We now take up the evaluation of multiple integrals. It is first important to understand the distinction between a multiple integral and an *iterated* integral. Here is an example of the latter.

$$\int_0^3 dx \int_1^{x^2} dy \int_{x-y}^{3y} (4x + yz)\, dz$$

This can be calculated using only elementary calculus, by starting with the inside integral and using antiderivatives and Theorem 7. Thus, the first step is to write

$$\int_{x-y}^{3y} (4x + yz)\, dz = 4xz + \frac{1}{2} yz^2 \Big|_{x-y}^{3y}$$

$$= 12xy + \frac{9}{2} y^3 - 4x(x - y) - \frac{1}{2} y(x - y)^2$$

This is then the integrand for the next integration, which will be done "with respect to y," and so on.

The standard evaluation process for multiple integrals replaces a multiple integral by one of a number of different iterated integrals.

Theorem 9 *Let R be the rectangle described by $a \le x \le b$, $c \le y \le d$, and let f be continuous on R. Then*

$$\iint_R f = \int_a^b dx \int_c^d f(x, y)\, dy$$

For fixed x, $f(x, y)$ is continuous in y, so that we may write

$$F(x) = \int_c^d f(x, y)\, dy$$

The theorem asserts that $\int_a^b F$ exists and is $\iint_R f$. We shall prove this by showing that any one-dimensional Riemann sum computed for a partition of the interval $[a, b]$ and the function F, has the same value as a special two-dimensional Riemann sum for R and f.

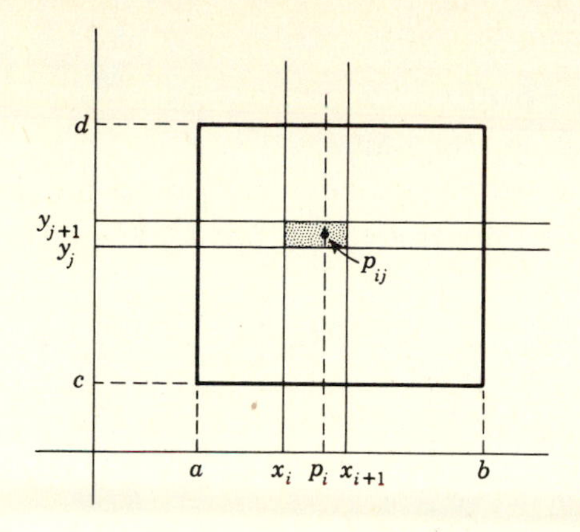

Figure 4-4

Let N be any grid which partitions $[a, b]$ with division points

$$a = x_0 < x_1 < \cdots < x_n = b$$

and choose any point p_i in the interval $[x_i, x_{i+1}]$. The corresponding sum is

$$S(N, F, \{p_i\}) = \sum_{i=0}^{n-1} F(p_i)\,\Delta x_i$$

Let $d(N) = \delta$ and choose any division points for the interval $[c, d]$, $c = y_0 < y_1 < \cdots y_m = d$ so that $\Delta y_j = y_{j+1} - y_j \leq \delta$ for each j (see Fig. 4-4). Then, for any x,

$$\begin{aligned} F(x) &= \int_c^d f(x, y)\,dy = \int_{y_0}^{y_m} f(x, y)\,dy \\ &= \int_{y_0}^{y_1} f(x, y)\,dy + \int_{y_1}^{y_2} f(x, y)\,dy + \cdots + \int_{y_{m-1}}^{y_m} f(x, y)\,dy \end{aligned}$$

To each of these, we apply the mean value theorem for integrals. A point $\bar{y}_j$ can be chosen in the interval $[y_j, y_{j+1}]$ so that

$$\begin{aligned} \int_{y_j}^{y_{i+1}} f(x, y)\,dy &= f(x, \bar{y}_j)[y_{j+1} - y_j] \\ &= f(x, \bar{y}_j)\,\Delta y_j \end{aligned}$$

In general, the choice of $\bar{y}_j$ depends upon the value of x, so that we should write $\bar{y}_j = Y_j(x)$. Summing for $j = 0, 1, \ldots, m-1$, we have

$$F(x) = f(x, Y_0(x))\,\Delta y_0 + \cdots + f(x, Y_{m-1}(x))\,\Delta y_{m-1}$$

When x is chosen as the particular point p_i, $(x, Y_j(x))$ becomes a point

$$p_{ij} = (p_i, Y_j(p_i))$$

in R, and

$$F(p_i) = \sum_{j=0}^{m-1} f(p_{ij})\,\Delta y_j$$

Returning to the original one-dimensional Riemann sum for F,

$$S(N, F, \{p_i\}) = \sum_{i=0}^{n-1} \left\{ \sum_{j=0}^{m-1} f(p_{ij})\,\Delta y_j \right\} \Delta x_i$$

$$= \sum_{i=0}^{n-1} \sum_{j=0}^{m-1} f(p_{ij})\,\Delta x_i\,\Delta y_j$$

The vertical lines $x = x_i$, $i = 0, 1, \ldots, n$, and the horizontal lines $y = y_j$, $j = 0, 1, \ldots, m$, define a grid N^* which partitions R into rectangles R_{ij} in such a fashion that $p_{ij} \in R_{ij}$ and $\mathrm{A}(R_{ij}) = \Delta x_i\,\Delta y_j$; moreover, since $\Delta x_i \le \delta$ and $\Delta y_j \le \delta$, $d(N^*) < 2\delta = 2d(N)$. We have therefore shown that corresponding to any grid N which partitions $[a, b]$ we can find a grid N^* which partitions R such that

$$S(N, F, \{p_i\}) = S(N^*, f, \{p_{ij}\})$$

Since $\iint_R f$ exists, the two-dimensional Riemann sums converge and $\int_a^b F$ exists and is equal to $\iint_R f$. ∎

We remark that in this proof, we could have obtained the existence of $\int_a^b F$ at the outset from the easily proven fact that F is continuous (see Exercise 18). However, if we extend the theorem to functions f which may have discontinuities in R, then F may be discontinuous. In fact, if f is bounded and continuous on R except at points of a set E of zero area, it may happen that for individual values of x, $\int_c^d f(x, y)\,dy$ does not exist, and the set of x for which this is true can fail to be a set of zero length.

One special case can be treated easily.

Theorem 10 *Let f be bounded in a closed rectangle R and continuous there except on a set E of zero area. Suppose there exists a k such that no vertical line meets E in more than k points. Then,*

$$\iint_R f = \int_a^b dx \int_c^d f(x, y)\,dy$$

The proof of this requires only slight modifications from that for

Theorem 9. As before, we consider F defined on $[a, b]$ by

$$F(x) = \int_c^d f(x, y)\, dy$$

Since $f(x, y)$ is continuous in y in $[c, d]$ except for at most k points, and since f is bounded, this integral exists and F is defined. Again take a general one-dimensional Riemann sum $S(N, F, \{p_i\}) = \sum_{i=0}^{n-1} F(p_i)\, \Delta x_i$ and choose division points y_j on $[c, d]$. These determine a grid N^* which partitions R into rectangles R_{ij}. Let R_0 be the union of those which contain points of E, and let R_1 be the union of the remaining rectangles. We may again write

$$F(p_i) = \int_{y_0}^{y_m} f(p_i, y)\, dy = \sum_{j=0}^{m-1} \int_{y_j}^{y_{j+1}} f(p_i, y)\, dy$$

Since $f(p_i, y)$ is continuous as a function of y in all but k of the intervals $[y_j, y_{j+1}]$, we may apply the mean value theorem as before, and replace these by terms of the form $f(p_{ij})\, \Delta y_j$. For the remaining k terms, we have bounds of the form $B\, \Delta y_j$, where B is an upper bound for $|f|$ in R. Adding these estimates for $F(p_i)\, \Delta x_i$, we arrive at

$$\left| S(N, F, \{p_i\}) - \sum_{R_{ij} \subset R_1} f(p_{ij}) \mathrm{A}(R_{ij}) \right| \le B \sum_{R_{ij} \subset R_0} \mathrm{A}(R_{ij}) \tag{4-3}$$

Since R_0 is a circumscribing set for E and E has zero area, the right side of (4-3) can be made arbitrarily small by requiring that $d(N^*)$ be small; the Riemann sum $S(N, F, \{p_i\})$ will converge to the integral of F, and the sum $\sum f(p_{ij}) \mathrm{A}(R_{ij})$ will converge to the integral of f on R, and we have shown that

$$\iint_R f = \int_a^b F(x)\, dx$$

which is the desired result. ▮

A standard special case of this is when we wish to evaluate $\iint_D f$, where D is the set described by $a \le x \le b$, $\phi(x) \le y \le \psi(x)$, where ϕ and ψ are continuous on $[a, b]$. If the function f is continuous on D and we extend it to a rectangle R containing D so as to be 0 off D, the set E of discontinuities of f will be the graphs of ϕ and ψ, and $\mathrm{A}(E) = 0$. Moreover, vertical lines cut E twice, so $k = 2$. Hence we have:

Corollary *If D is the region bounded by the lines $x = a$, $x = b$, and the graphs of ϕ and ψ, with $\phi(x) \le \psi(x)$, and if f is continuous on D, and ϕ and ψ*

continuous on $[a, b]$,

$$\text{(4-4)} \qquad \iint_D f = \int_a^b dx \int_{\phi(x)}^{\psi(x)} f(x, y)\, dy$$

The expression on the right is an iterated integral and may often be evaluated by antidifferentiation. As an example, let D be the region between the line $y = x$ and the parabola $y = x^2$, and take $f(x, y) = xy^2$; then,

$$\iint_D f = \int_0^1 dx \int_{x^2}^{x} xy^2\, dy = \int_0^1 dx \left[\frac{xy^3}{3}\right]_{x^2}^{x}$$

$$= \int_0^1 \frac{x^4 - x^7}{3}\, dx = \frac{1}{40}$$

Since this region D is such that every horizontal line cuts the boundary at most twice, $\iint_D f$ can be also be evaluated by an iterated integral in which the order of the variables is reversed.

$$\iint_D f = \int_0^1 dy \int_y^{\sqrt{y}} xy^2\, dx = \int_0^1 dy \left[\frac{x^2 y^2}{2}\right]_y^{\sqrt{y}}$$

$$= \tfrac{1}{2}\int_0^1 (y^3 - y^4)\, dy = \tfrac{1}{40}$$

In the cases which are usually encountered, the evaluation of a double

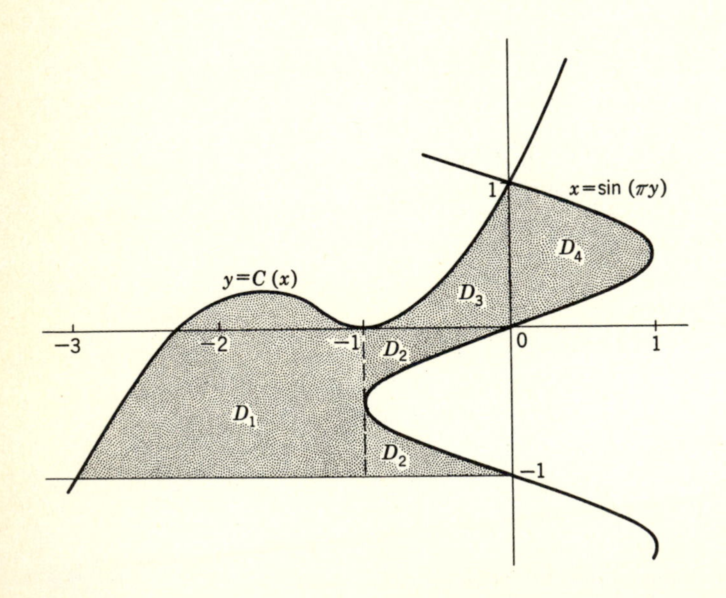

Figure 4-5

integral can be reduced to the computation of a number of such iterated integrals. As an example, consider the region D bounded by the line $y = -1$, and the curves

$$x = \sin(\pi y) \qquad \text{and} \qquad y = (x+1)^2(\tfrac{5}{12}x + 1) = C(x)$$

(see Fig. 4-5). To evaluate $\iint_D f$, it is convenient to split D into four regions as shown, so that

$$\iint_D f = \iint_{D_1} f + \iint_{D_2} f + \iint_{D_3} f + \iint_{D_4} f$$

where we have

$$\iint_{D_1} f = \int_{-3}^{-1} dx \int_{-1}^{C(x)} f(x, y)\, dy$$

$$\iint_{D_2} f = \int_{-1}^{0} dy \int_{-1}^{\sin(\pi y)} f(x, y)\, dx$$

$$\iint_{D_3} f = \int_{-1}^{0} dx \int_{0}^{C(x)} f(x, y)\, dy$$

$$\iint_{D_4} f = \int_{0}^{1} dy \int_{0}^{\sin(\pi y)} f(x, y)\, dx$$

The last two could be combined and written as

$$\iint_{D_3 \cup D_4} f = \int_{0}^{1} dy \int_{\lambda(y)}^{\sin(\pi y)} f(x, y)\, dx$$

where $x = \lambda(y)$ is the particular solution of the cubic equation $y = C(x)$ valid for $-1 \le x \le 0$.

We can also use the relation between double integrals and iterated integrals to prove the equality of mixed partial derivatives, promised in Sec. 3.3. The case of f_{xy} and f_{yx} is typical.

Theorem 11 *Let f be of class C'' in a rectangle R with vertices $P_1 = (a_1, b_1)$, $Q_1 = (a_2, b_1)$, $P_2 = (a_2, b_2)$, $Q_2 = (a_1, b_2)$, where $a_1 \le a_2$ and $b_1 \le b_2$. Then,*

$$\iint_R f_{12} = \iint_R \frac{\partial^2 f}{\partial y\, \partial x} dx\, dy = f(P_1) - f(Q_1) + f(P_2) - f(Q_2)$$

Writing the double integral as an iterated integral, we have

$$\iint_R f_{12} = \int_{a_1}^{a_2} dx \int_{b_1}^{b_2} \frac{\partial}{\partial y}\left(\frac{\partial f}{\partial x}\right) dy$$

$$= \int_{a_1}^{a_2} \left[\frac{\partial f}{\partial x}\right]_{y=b_1}^{y=b_2} dx$$

$$= \int_{a_1}^{a_2} f_1(x, b_2)\, dx - \int_{a_1}^{a_2} f_1(x, b_1)\, dx$$

$$= \left[f(x, b_2)\right]_{x=a_1}^{x=a_2} - \left[f(x, b_1)\right]_{x=a_1}^{x=a_2}$$

$$= f(a_2, b_2) - f(a_1, b_2) - [f(a_2, b_1) - f(a_1, b_1)]$$

$$= f(P_1) - f(Q_1) + f(P_2) - f(Q_2) \quad ▮$$

Corollary *If a function f is of class C'' in an open set D, then $f_{12} = f_{21}$ throughout D.*

If R is any rectangle lying in D, then the line of argument in the theorem shows that $\iint_R f_{12}$ and $\iint_R f_{21}$ are both equal to $f(P_1) - f(Q_1) + f(P_2) - f(Q_2)$, where P_1, Q_1, P_2, Q_2 are the vertices of R, in counterclockwise order. Thus, $\iint_R (f_{12} - f_{21}) = 0$ for every choice of R, and the integrand must be identically 0 in D. ▮

Although iterated integrals having a form like (4-4) thus appear in the course of evaluating double integrals, they often arise in other ways (e.g., estimations of a probability), and sometimes do not correspond immediately to a multiple integral. For example,

$$V = \int_0^2 dx \int_1^{x^2-1} f(x, y)\, dy \tag{4-5}$$

has the same general form as (4-4), and might seem to be the iterated integral arising from the double integral of f over a region D bounded by the curves $y = 1$, $y = x^2 - 1$, $x = 0$, $x = 2$. The corresponding picture (Fig. 4-6) suggests that this may not be so, since we have several regions. Moreover, we note that the upper limit of the inside integral is not always larger than the lower limit, as it should be if this iterated integral had the form (4-4).

Such an iterated integral, therefore, does *not* correspond to a double integral. However, it can be written as the difference of two double integrals.

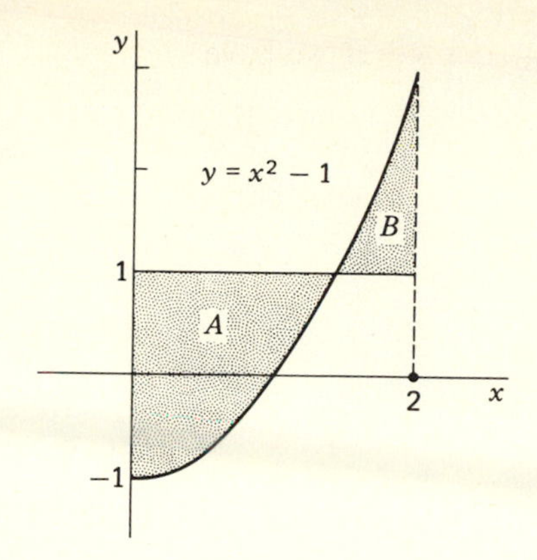

Figure 4-6

To see this, we use the fact that for any functions $\phi(x)$ and $\psi(x)$, and any c,

$$\int_{\phi(x)}^{\psi(x)} f\,dy = \int_{c}^{\psi(x)} f\,dy - \int_{c}^{\phi(x)} f\,dy$$

If we are dealing with an integral with limits $\phi(x)$, $\psi(x)$ where it is not always true that $\phi(x)$ is the smaller, we can choose c less than the minimum of either ϕ or ψ, and thus retain this desirable property by splitting up the integral.

Applying this to (4-5), we would proceed as follows: Take $c = -1$, which is the minimum of $x^2 - 1$ for $0 \le x \le 2$, and then rewrite the iterated integral as

$$\int_0^2 dx \int_{-1}^{x^2-1} f(x, y)\,dy - \int_0^2 dx \int_{-1}^{1} f(x, y)\,dy$$

Each of these is now the iterated integral corresponding to a double integral, and we may now rewrite (4-5) as

$$V = \iint_{D_1} f - \iint_{D_2} f \tag{4-6}$$

where the two regions are both shown in Fig. 4-7. D_1 is bounded above by $y = x^2 - 1$, and D_2 is a rectangle.

A second approach is sometimes used. Note that D_1 and D_2 overlap. This portion of each double integral in (4-6) is the same, and so can be cancelled. We thus arrive at an alternative form

$$V = \iint_{B} f - \iint_{A} f \tag{4-7}$$

where these are the two regions shown in Fig. 4-6. This interpretation of the original iterated integral could have been read from it immediately by observing that when x is between 0 and 1, y goes from 1 *downward* to $x^2 - 1$,

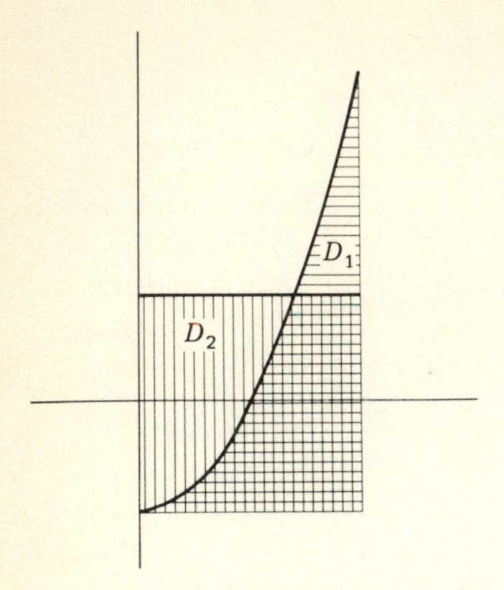

Figure 4-7

which suggests that the region A should be regarded as negative, rather than positive as is the case with B.

Why have we discussed the conversion of iterated integrals into combinations of double integrals? Consider the following.

$$V = \int_1^3 dx \int_x^2 e^{x/y}\, dy \tag{4-8}$$

We cannot carry out the first step in evaluating this, since we cannot find the function whose derivative, with respect to y, is $e^{x/y}$. (This is one of an infinite number of simple functions whose indefinite integral cannot be expressed in terms of any of the familiar functions.) Note, however, that we *could* integrate it with respect to x. If we could convert (4-8) into one or more double integrals, and then write these as iterated integrals in the opposite order of integration, with dy outside and dx inside, we could at least carry out the first integration step. If we apply the procedure we have outlined above, we find that (4-8) can be expressed as the difference of two double integrals over triangular regions, and writing each as an iterated integral, we arrive at

$$V = \int_1^2 dy \int_1^y e^{x/y}\, dx - \int_2^3 dy \int_y^3 e^{x/y}\, dx$$

which, in the usual way, leads to

$$V = 4e - \int_1^2 ye^{1/y}\, dy - \int_2^3 ye^{3/y}\, dy$$

and the remaining integrals can be done numerically.

The technique of reversal of order of integration, in iterated integrals, is a very useful one.

As another illustration, we have the useful formula

$$\int_a^b dx \int_a^x f(x, y)\, dy = \int_a^b dy \int_y^b f(x, y)\, dx \tag{4-9}$$

which follows at once from the observation that both sides reduce to the double integral of f over the triangle with vertices (a, a), (b, a), (b, b). The following is a special application of this which is often used. Consider the n-fold iterated integral

$$\int_0^b dx_1 \int_0^{x_1} dx_2 \int_0^{x_2} dx_3 \cdots \int_0^{x_{n-1}} f(x_n)\, dx_n$$

The last two inner integrals have the form

$$\int_0^{x_{n-2}} dx_{n-1} \int_0^{x_{n-1}} f(x_n)\, dx_n$$

to which the relation (4-9) may be applied, obtaining

$$\int_0^{x_{n-2}} dx_n \int_{x_n}^{x_{n-2}} f(x_n)\, dx_{n-1} = \int_0^{x_{n-2}} f(x_n)(x_{n-2} - x_n)\, dx_n$$

We have thus reduced the original integral to an $(n - 1)$-fold iterated integral. Repeating this process,

$$\int_0^{x_{n-3}} dx_{n-2} \int_0^{x_{n-2}} f(x_n)(x_{n-2} - x_n)\, dx_n$$

becomes

$$\int_0^{x_{n-3}} dx_n \int_{x_n}^{x_{n-3}} f(x_n)(x_{n-2} - x_n)\, dx_{n-2} = \int_0^{x_{n-3}} f(x_n) \frac{(x_{n-3} - x_n)^2}{2!}\, dx_n$$

and the original integral can finally be reduced to the single integral

$$\int_0^b f(x_n) \frac{(b - x_n)^{n-1}}{(n - 1)!}\, dx_n$$

As a final illustration of this technique, let us prove a theorem dealing with differentiation of a function defined by means of an integral. As we have mentioned, if $f(x, y)$ is continuous for $a \le x \le b$, and $c \le y \le d$, then

$$F(x) = \int_c^d f(x, y)\, dy \tag{4-10}$$

is continuous for x in $[a, b]$ (Exercise 18). It is reasonable to suppose that if f is differentiable as a function of x alone, with y fixed, so that the partial derivative

$$f_1(x, y) = \frac{\partial f}{\partial x} = \lim_{h \to 0} \frac{f(x + h, y) - f(x, y)}{h}$$

exists for each x in $[a, b]$ and y in $[c, d]$, then we may differentiate (4-10) under the integral sign, obtaining

$$F'(x) = \int_c^d \frac{\partial f}{\partial x}\, dy = \int_c^d f_1(x, y)\, dy \tag{4-11}$$

We justify this under the hypothesis that f_1 is continuous.

Theorem 12 *Let f and f_1 be defined and continuous for $x \in [a, b]$, $y \in [c, d]$, and F defined by* (4-10). *Then, $F'(x)$ exists on the interval $[a, b]$ and is given by* (4-11).

Since f_1 is continuous, Exercise 18 shows that $\phi(x) = \int_c^d f_1(x, y)\, dy$ is continuous for $x \in [a, b]$. Take any x_0 and consider

$$\int_a^{x_0} \phi = \int_a^{x_0} dx \int_c^d f_1(x, y)\, dy$$

Reverse the order of integration, obtaining

$$\int_a^{x_0} \phi = \int_c^d dy \int_a^{x_0} f_1(x, y)\, dx = \int_c^d dy \int_a^{x_0} \frac{\partial f}{\partial x}\, dx$$

$$= \int_c^d [f(x_0, y) - f(a, y)]\, dy = F(x_0) - F(a)$$

This shows that F is an antiderivative for ϕ, so that F' exists and is ϕ. ∎

For example, if

$$F(x) = \int_0^1 \frac{\sin (xy)}{y}\, dy$$

then

$$F'(x) = \int_0^1 \cos (xy)\, dy = \frac{\sin (xy)}{x}\bigg|_0^1 = \frac{\sin x}{x}$$

This technique is sometimes useful in the evaluation of special types of definite integrals. For example, let us show that for $x > 0$,

$$F(x) = \int_0^{\pi/2} \log (\sin^2 \theta + x^2 \cos^2 \theta)\, d\theta$$

$$= \pi \log \frac{x + 1}{2}$$

Differentiating the integral which defines F, we have

$$F'(x) = \int_0^{\pi/2} \frac{2x \cos^2 \theta}{\sin^2 \theta + x^2 \cos^2 \theta} d\theta$$

so that
$$\frac{F'(x)(x^2 - 1)}{2x} = \int_0^{\pi/2} \frac{(x^2 - 1) \cos^2 \theta}{\sin^2 \theta + x^2 \cos^2 \theta} d\theta$$

$$= \int_0^{\pi/2} \frac{x^2 \cos^2 \theta + \sin^2 \theta - 1}{\sin^2 \theta + x^2 \cos^2 \theta} d\theta$$

$$= \frac{\pi}{2} - \int_0^{\pi/2} \frac{d\theta}{\sin^2 \theta + x^2 \cos^2 \theta}$$

Now
$$\int \frac{d\theta}{\sin^2 \theta + x^2 \cos^2 \theta} = \int \frac{\sec^2 \theta \, d\theta}{\tan^2 \theta + x^2}$$

$$= x^{-1} \arctan [x^{-1} \tan \theta]$$

so that for $x > 0$ and $x \neq 1$,

$$F'(x) = \frac{2x}{x^2 - 1} \left\{ \frac{\pi}{2} - \frac{\pi}{2x} \right\}$$

$$= \frac{\pi}{x + 1}$$

We have established this for all $x > 0$ except $x = 1$. When $x = 1$, we have directly $F'(1) = 2 \int_0^{\pi/2} \cos^2 \theta \, d\theta = \pi/2$, or we may argue that F' is continuous, so that $F'(1) = \lim_{x \to 1} F'(x) = \pi/2$. Integrating F', we have $F(x) = \pi \log (x + 1) + C$. To determine C, we observe that

$$F(1) = \int_0^{\pi/2} \log (1) \, d\theta = 0$$

so that
$$C = -\pi \log 2, \text{ and } F(x) = \pi \log \frac{x + 1}{2}$$

At times, a more complicated rule for differentiating a function defined by an integral is needed. Consider the function given by

$$F(x) = \int_{\alpha(x)}^{\beta(x)} f(x, y) \, dy \tag{4-12}$$

To find $F'(x)$, we use the chain rule; let G be the function of three variables

$$G(x, u, v) = \int_u^v f(x, y) \, dy$$

We can compute the partial derivatives of G, using the fundamental theorem of calculus for $\partial/\partial u$ and $\partial/\partial v$.

$$G_1(x, u, v) = \int_u^v f_1(x, y)\, dy$$

$$G_2(x, u, v) = -f(x, u)$$

$$G_3(x, u, v) = f(x, v)$$

Then, since $F(x)$ in (4-12) is $G(x, \alpha(x), \beta(x))$, the chain rule of differentiation gives

$$F'(x) = \beta'(x) f(x, \beta(x)) - \alpha'(x) f(x, \alpha(x)) + \int_{\alpha(x)}^{\beta(x)} f_1(x, y)\, dy \tag{4-13}$$

For example, if

$$F(x) = \int_{x^2}^{e^x} \frac{\sin(xu)}{u}\, du$$

then

$$F'(x) = e^x \frac{\sin(xe^x)}{e^x} - 2x \frac{\sin(x^3)}{x^2} + \int_{x^2}^{e^x} \cos(xu)\, du$$

$$= (1 + x^{-1}) \sin(xe^x) - 3x^{-1} \sin(x^3)$$

In many cases, both with single and with multiple integrals, the best and most efficient procedure is to apply some method of approximate integration. Such an approach is forced on one if the indefinite integration steps are impossible, and may even be appropriate otherwise if the final step in getting an "exact" answer uses the approximate tabulated values of standard functions. One simple scheme for a double integral is to construct a convenient grid covering the region D, then compute the Riemann sums $\overline{S}(N)$ and $\underline{S}(N)$ for f. The exact value of $\iint_D f$ must then lie between these numbers, and their difference is a measure of the accuracy of the approximation. Any other Riemann sum $\sum_{i,j} f(p_{ij}) \mathrm{A}(R_{ij})$ may also be used as an approximation. In practice, a number of simple formulas are commonly used in the approximate evaluation of single integrals. The general Riemann sum for $\int_a^b f$ can be put into the form $\sum_0^{n-1} f_i\, \Delta x_i$, where $\Delta x_i = x_{i+1} - x_i$

$$a = x_0 < x_1 < x_2 < \cdots x_n = b$$

and where $f_i = f(p_i)$ is the value of f at some point of the interval $[x_i, x_{i+1}]$. By the intermediate value theorem, if f is continuous on such an interval and A and B are values of f there, then every number between A and B, and in particular $(A + B)/2$, is a value of f. More generally, if $A_1, A_2, \ldots, A_r$ are values of f on an interval, so is $c_1 A_1 + \cdots + c_r A_r$, where $c_j \geq 0$ and $\sum c_j = 1$. (This is merely a general weighted average of the numbers A_j.) Two special cases of this lead to the **trapezoidal rule** and **Simpson's rule.** For the first, we take $f_i = [f(x_i) + f(x_{i+1})]/2$, and for the latter, we take $f_i = [f(x_i) + 4f(\bar{x}) + f(x_{i+1})]/6$, where $\bar{x}$ is the midpoint $\bar{x} = (x_i + x_{i+1})/2$. The reason behind the second choice lies in the result of Exercise 26, which shows that Simpson's rule is exact whenever f is a polynomial of degree at most 3; application of the formula therefore amounts to approximating f on each of the intervals $[x_i, x_{i+1}]$ by such a polynomial, chosen to fit f at the endpoints and the midpoint. Other methods for estimating the value of an integral will be found in the exercises.

For single integrals, a good way to approximate the value of $\int_a^b f$ where f is a function that does not have an elementary indefinite integral is to approximate f by another function which can be integrated easily. Polynomials are the simplest choice here, and Taylor's theorem is helpful. Consider

$$C = \int_0^1 \sqrt{x}\, e^{\sqrt{x}}\, dx$$

Instead of applying the expansion process to the integrand directly, we first recall that

$$e^x \sim 1 + x + \frac{x^2}{2} + \frac{x^3}{6} + \frac{x^4}{24} + \frac{x^5}{120}$$

within .005 on $[-1, 1]$. On $[0, 1]$, we may write

$$\begin{aligned} \sqrt{x}\, e^{\sqrt{x}} &\sim \sqrt{x}\left[1 + x^{1/2} + \cdots + \frac{x^{5/2}}{120}\right] \\ &\sim x^{1/2} + x + \cdots + \frac{x^3}{120} \end{aligned}$$

also good to .005. Making this replacement in the integral,

$$C \sim \int_0^1 \left(x^{1/2} + x + \frac{x^{3/2}}{2} + \cdots + \frac{x^3}{120}\right) dx = 1.436\,210$$

While we can be sure that this answer has at most an error of $\int_0^1 (.005)\, dx = .005$, in fact the estimate we have obtained is far better than that. If one makes the substitution $u = \sqrt{x}$, the integral becomes

$$\begin{aligned} C &= \int_0^1 2u^2 e^u\, du = (4 - 4u + 2u^2)e^u \Big|_0^1 \\ &= 1.436\,563\,66 \end{aligned}$$

More about numerical integration will be found in Chap. 10.

Iterated integrals involving many integrations often arise in very natural problems. Consider first the following question. Let $I_1 = [a, b]$ and $I_2 = [c, d]$ be two disjoint intervals on the line. Choose a point in each, say $p_i \in I_i$. We ask: What is the expected distance between p_1 and p_2, in the sense of probability theory?

If we let x be the coordinate for p_1 and y that for p_2, and let L_i be the length of I_i, then the distance between the points is $|y - x| = f(x, y)$, and the problem is answered by asking for the value of the double integral

$$V = \frac{1}{L_1 L_2} \iint_R f \tag{4-14}$$

where R is the rectangle $I_1 \times I_2$ with vertices (a, c), (b, c), (a, d), (b, d). In this case, V can be calculated with ease, and the answer turns out to be the distance between the midpoints of the two intervals (Exercise 30). [This is not the answer if the intervals overlap!]

Suppose we now consider the corresponding problem for variable points in the plane. Consider two disjoint sets D_1 and D_2 and $p_i \in D_i$, and again ask for the expected value of the distance $|p_1 - p_2|$. Following the pattern above, let $p_i = (x_i, y_i)$, and

$$\begin{aligned} f(x_1, y_1, x_2, y_2) &= |p_1 - p_2| \\ &= \sqrt{(x_1 - x_2)^2 + (y_1 - y_2)^2} \end{aligned}$$

Then, the desired answer is

$$V = \frac{1}{\text{A}(D_1)\text{A}(D_2)} \iiiint_S f \tag{4-15}$$

where S is the set $D_1 \times D_2$ in 4 space.

For illustration, let D_1 be the square with opposite vertices at (0, 0) and (1, 1), and let D_2 be the square with vertices at (1, 0) and (2, 1). Then,

$$V = \int_0^1 dx_1 \int_0^1 dy_1 \int_1^2 dx_2 \int_0^1 f(x_1, y_1, x_2, y_2)\, dy_2$$

This is a challenge to carry out, and it is therefore more suitable as a candidate for numerical integration in the form (4-15).

EXERCISES

1 Calculate $\int_1^2 \sqrt[3]{x}\, dx$ from the definition of the integral. (*Hint:* Use a method similar to that of the text for $\sqrt{x}$.)

2 If possible, give an explicit formula for a function F such that for all x,

(*a*) $F'(x) = x + |x - 1|$ (*b*) $\log F'(x) = 2x + e^x$

3 Let $f(x) = \begin{cases} x & 0 \le x \le 1 \\ x - 1 & 1 < x \le 2 \\ 0 & 2 < x \le 3 \end{cases}$

(*a*) What is $F(x) = \int_0^x f$ on $[0, 3]$?

(*b*) Is F continuous?

(*c*) Is $F'(x) = f(x)$?

4 Explain the errors in illustrative examples (i) and (ii) on page 183.

5 Let F be defined by

$$F(x) = \int_1^x \exp\left(\frac{u^2+1}{u}\right)\frac{du}{u}$$

Show that $F(1/x) = -F(x)$.

6 Let f be continuous on $[0, 1]$, and suppose that for all x, $0 < x < 1$,

$$\int_0^x f = \int_x^1 f$$

Can you determine f?

***7** Find all continuous functions f such that for all $x \geq 0$

$$(f(x))^2 = \int_0^x f$$

8 Evaluate the double integral $\iint_D x^2y\,dx\,dy$ when D is the region bounded by (*a*) the line $y = x$ and the parabola $y = x^2$; (*b*) the line $y = x - 2$ and the parabola $x = 4 - y^2$.

9 Express the following iterated integral as a double integral, and then as an iterated integral with the order of integrations reversed.

$$\int_1^2 dx \int_0^x f(x, y)\,dy$$

10 Show that reversing the order of integration in the integral (4-5) yields

$$\int_1^3 dy \int_{\sqrt{y+1}}^2 f\,dx - \int_{-1}^1 dy \int_0^{\sqrt{y+1}} f\,dx$$

11 If the order of integration is reversed in

$$\int_0^1 dx \int_{2x^2}^{x+1} f(y)\,dy$$

the sum of two integrals of the form $\int_0^1 dy\,[\quad] + \int_1^2 dy\,[\quad]$ is obtained. Fill in the blanks [ ].

12 If D is a pyramid with vertices (1, 0, 0), (0, 1, 0), (0, 0, 1), (0, 0, 0), find

$$\iiint_D (xy + z)\,dx\,dy\,dz$$

13 Express the following iterated integral as a triple integral, and then rewrite it in several other orders of integration as iterated integrals.

$$\int_0^2 dx \int_1^{2-x/2} dy \int_x^2 f(x, y, z)\,dz$$

14 Evaluate the preceding integral with $f(x, y, z) = x + yz$.

15 The line $y = x$ divides the unit square with opposite vertices at (0, 0), (1, 1) into two triangular sets. On which set can the function $f(x, y) = y/x$ be integrated? Find the value of the integral over that triangle.

16 Evaluate:

(*a*) $\int_0^1 dy \int_y^1 e^{y/x}\, dx$ (*b*) $\int_1^2 dx \int_{1/x}^2 ye^{xy}\, dy$

17 Reverse the order of integration in the following iterated integral, and compute its value.

$$\int_{-6}^{8} dx \int_{x^{1/3}}^{(x+6)/7} xy\, dy$$

18 Let $f(x, y)$ be defined and continuous for $a \le x \le b$, $c \le y \le d$. Let

$$F(x) = \int_c^d f(x, y)\, dy$$

Prove that F is continuous on $[a, b]$.

19 Prove the following formula for "reversal of order of addition" in finite sums:

$$\sum_{i=r}^{n} \sum_{j=r}^{i} a_{ij} = \sum_{j=r}^{n} \sum_{i=j}^{n} a_{ij}$$

and compare with (4-9). *Hint:* Try this first with numerical values for r and n.

20 Let $f(x)$ be continuous for all x.

(*a*) Find the value of $\int_0^1 dx \int_x^{1-x} f(t)\, dt$.

(*b*) Can you explain the answer you obtain?

21 By choosing an appropriate grid and computing $\bar{S}(N)$ and $\underline{S}(N)$, estimate the value of $\int_0^1 dx/(1 + x^3)$ to within .05. Compare this with the work of computing the exact value by the usual process.

22 By choosing an appropriate grid, estimate the value of

$$\int_0^2 dy \int_0^1 \frac{dx}{x + y + 10}$$

to within .02, and again compare by computing the exact value.

23 Let f and g be continuous on $[a, b]$. Show that

$$\left[\int_a^b fg\right]^2 \le \int_a^b f^2 \int_a^b g^2$$

(This is the integral form for the Schwarz inequality, given in Sec. 1.3 for finite sums.)

24 Using Exercise 23, show that

$$\int_0^1 \sqrt{xe^{-x}}\, dx < .47$$

25 Use the same method to estimate the value of

(*a*) $\int_0^1 \sqrt{1 + x^3}\, dx$ (*b*) $\int_0^\pi \sqrt{\sin x}\, dx$

26 Show that

$$\int_a^b P = (b-a)\frac{P(a) + P(b) + 4P([a+b]/2)}{6}$$

whenever P is a polynomial of degree at most 3.

27 Verify the following statements:

(*a*) If $F(x) = \int_{-1}^{1} \log(1 + xe^u)\,du$,

then
$$F'(x) = x^{-1}\log\left(\frac{1+ex}{1+e^{-1}x}\right)$$

(*b*) If $F(x) = \int_1^x u^{-1}\cos(xu^2)\,du$,

then
$$F'(x) = \frac{3\cos(x^3)}{2x} - \frac{\cos x}{2x}$$

(*c*) If $F(x) = \int_0^\pi u^{-1}e^{xu}\sin u\,du$,

then
$$F'(x) = \frac{e^{\pi x}+1}{x^2+1}$$

28 Use formula (4-13) to find $F'(x)$ if:

(*a*) $F(x) = \int_x^{x^2} t^{-1}e^{xt}\,dt$ (*b*) $F(x) = \int_{2x}^{3x} \cos(4x)\,dx$

29 Let
$$F(x, y) = \begin{cases} 1 & \text{if } x \text{ is rational} \\ 2y & \text{if } x \text{ is irrational} \end{cases}$$

Show that
$$\int_0^1 dx \int_0^1 F(x, y)\,dy = 1$$

but that
$$\int_0^1 dy \int_0^1 F(x, y)\,dx \quad \text{does not exist.}$$

30 Let $I_1 = [a, b]$ and $I_2 = [c, d]$ with $b < c$. Verify that the value of (4-14) is $(c + d - a - b)/2$.

31 Two points are chosen at random from an interval of length L. Show that their expected distance apart is $L/3$.

32 (*a*) Apply Theorem 11 to evaluate the double integral

$$\int_2^5 dx \int_{-1}^3 (x^2y + 5xy^2)\,dy$$

(*b*) Formulate and prove a corresponding theorem for the evaluation of triple integrals.

4.4 SUBSTITUTION IN MULTIPLE INTEGRALS

In evaluating single integrals, we use either numerical methods or substitution and the fundamental theorem of calculus (Theorem 5). It is natural to expect that the situation would be similar for multiple integrals; in most instances, however, multiple integrals are evaluated numerically or by replacing them by equivalent iterated single integrals, and the technique of substitution is mainly used in certain standard special circumstances, or occasionally to simplify difficult problems. In this section, we merely quote the relevant formulas, and defer all proofs and theoretical discussions until Chap. 7, where the necessary machinery for dealing with transformations of coordinates has been obtained.

In one variable, we are accustomed to the following procedure for making a change of variable in an integral. Given

$$V = \int_a^b F(x)\,dx$$

suppose we wish to make the substitution $x = \phi(u)$. The three steps are:

(4-16) *Determine an interval $[\alpha, \beta]$ of u values which is mapped by $x = \phi(u)$ onto $[a, b]$.*

(4-17) *Replace $F(x)$ by $F(\phi(u))$, which is a function of u defined on the interval $[\alpha, \beta]$.*

(4-18) *Replace dx by $\phi'(u)\,du$.*

In Theorem 8 of the preceding section, we have shown that these steps convert the original integral into another with the same value

$$V = \int_\alpha^\beta F(\phi(u))\phi'(u)\,du$$

Consider now a double integral

$$V = \iint_D F(x, y)\,dx\,dy \tag{4-19}$$

and a substitution

$$\begin{cases} x = f(u, v) \\ y = g(u, v) \end{cases} \tag{4-20}$$

The three analogous steps are:

(4-21) *Determine a region D^* in the UV plane that is mapped by (4-20) onto the region D, 1-to-1.*

(4-22) *Replace $F(x, y)$ by $F(f(u, v), g(u, v))$, which is a function of (u, v) defined on D^*.*

(4-23) *Replace $dxdy$ by* XXXXX.

We have chosen to leave the third step incomplete for the present, since the correct formula here is somewhat mysterious. One special case is familiar from elementary calculus, namely, the formula arising from the coordinate substitutions of polar coordinates:

$$\begin{cases} x = r\cos\theta \\ y = r\sin\theta \end{cases} \tag{4-24}$$

In this case, the third step in the process is

$$\textit{Replace} \quad dxdy \quad \textit{by} \quad rdrd\theta \tag{4-25}$$

and a geometric explanation for this is often given.

In order to fill in the blank in (4-23), we must recall the expression given in (3-25) for a general Jacobian. Applied to (4-20), we write

$$\frac{\partial(x, y)}{\partial(u, v)} = \det \begin{bmatrix} \dfrac{\partial x}{\partial u} & \dfrac{\partial x}{\partial v} \\ \dfrac{\partial y}{\partial u} & \dfrac{\partial y}{\partial v} \end{bmatrix} = \begin{bmatrix} f_1 & f_2 \\ g_1 & g_2 \end{bmatrix} \tag{4-26}$$

$$= f_1 g_2 - f_2 g_1$$

Using this, the complete statement of the third step in the substitution procedure is as follows:

$$\textit{Replace} \quad dxdy \quad \textit{by} \quad \left| \frac{\partial(x, y)}{\partial(u, v)} \right| du\, dv \tag{4-27}$$

In the polar coordinates example, this becomes

$$\frac{\partial(x, y)}{\partial(r, \theta)} = \det \begin{bmatrix} \cos\theta & -r\sin\theta \\ \sin\theta & r\cos\theta \end{bmatrix}$$

$$= r\{(\cos\theta)^2 + (\sin\theta)^2\} = r$$

which agrees with (4-25) above.

While there is obvious parallelism between the three-step procedure for single integrals and that for double integrals, there is one visible difference. In that for multiple integrals, the second step mentions that the mapping shall be 1-to-1, and in the third step, the Jacobian (which replaces the derivative $\phi'(u)$) is inside an absolute value $|\quad|$. This is because it is easier to prove a better theorem (Theorem 8, Sec. 4.3) for one variable than for the case of several variables, which we shall take up in Chap. 7. Later, in Chap. 9, we will meet a version of the several-variables substitution more like that for one variable.

We end this brief discussion by describing another method for carrying out a variable substitution in a multiple integral which does not bring in the Jacobian symbol, and which is as simple and automatic as the familiar method used for single integrals. It will seem far more mysterious than the above since it introduces some new symbols without explanation, and depends on some strange algebraic rules that seem arbitrary and ad hoc.

First, as in (4-18), we agree that if $x = \phi(u)$, then

$$dx = \phi'(u)\, du$$

By analogy, we agree that if $x = f(u, v)$, then we write

$$dx = \frac{\partial x}{\partial u}\, du + \frac{\partial x}{\partial v}\, dv = f_1\, du + f_2\, dv \tag{4-28}$$

From (4-28), we would therefore also write $dy = g_1\, du + g_2\, dv$. Accordingly, turning to (4-23), we have

$$\begin{aligned} dx\, dy &= (f_1\, du + f_2\, dv)(g_1\, du + g_2\, dv) \\ &= f_1 g_1\, du\, du + f_1 g_2\, du\, dv + f_2 g_1 dv\, du + f_2 g_2\, dv\, dv \end{aligned} \tag{4-29}$$

We now invoke the following algebraic rules:

$$\begin{aligned} du\, du &= dv\, dv = 0 \\ dv\, du &= -du\, dv \end{aligned} \tag{4-30}$$

and using these in (4-29), we have

$$\begin{aligned} dx\, dy &= 0 + f_1 g_2\, du\, dv - f_2 g_1 du\, dv + 0 \\ &= (f_1 g_2 - f_2 g_1)\, du\, dv \end{aligned}$$

(We recognize that this is the same as (4-27), except that now we have dispensed with the absolute value of the Jacobian.)

This generalizes immediately to multiple integrals of any size, and can be used with success, provided that one is careful to multiply all factors in the order in which they appear.

Perhaps a word of explanation about this "black magic" method might be wise. In mathematical research, there seems to be a guiding platonic principle, most visible perhaps in many of the newer developments, which asserts that anything that leads to correct results, no matter how wild the method may look, may have some inner rationale. In the present case, a justification for the quasi-algebraic method for change of variable is found in the theory of differential forms. This is a fairly recent invention, growing out of earlier work (line and surface integrals, Green's theorem, and Stokes' theorem), and will be discussed in part in Chap. 9. Among other topics, this will contain a far-reaching generalization of the fundamental theorem of calculus.

EXERCISES

1 Let $f(x, y) = xy^2$, and let D be the region shown in Fig. 4-8. Use the substitution in (4-24) to evaluate $\iint_D f(x, y)\, dx\, dy$.

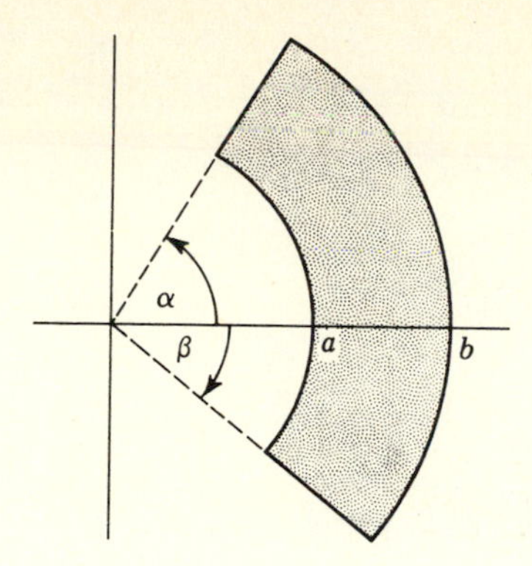

Figure 4-8

2 A standard transformation between spherical and cartesian coordinates in 3-space is

$$x = \rho \sin \phi \cos \theta$$
$$y = \rho \sin \phi \sin \theta$$
$$z = \rho \cos \phi$$

Show that a correct replacement formula is

$$dx\, dy\, dz = \rho^2 \sin \phi \, d\rho \, d\phi \, d\theta$$

3 In 4-space, "double" polar coordinates are defined by the equations

$$x = r \cos \theta, \qquad y = r \sin \theta, \qquad z = \rho \cos \phi, \qquad w = \rho \sin \phi$$

Obtain the correct substitution formula, and then show that the volume of the hypersphere $x^2 + y^2 + z^2 + w^2 \le R^2$ is $\pi^2 R^4/2$.

4.5 IMPROPER INTEGRALS

As outlined in Sec. 4-2, the notion of **area** applied only to bounded sets. If we seek to extend this to unbounded sets, a simple procedure suggests itself. Let $R_1 \subset R_2 \subset \cdots$ be an expanding sequence of closed rectangles whose union is the whole plane; for example, we may choose R_n as the square with center at the origin and one vertex at (n, n). Let D be an unbounded set whose area we wish to measure. We form the bounded set $D_n = R_n \cap D$, the part of D in R_n, and assume that the boundary of D is nice enough so that each of these has an area. Since the sets $\{D_n\}$ form an expanding sequence, the sequence of their areas $\{\text{A}(D_n)\}$ is a monotonic sequence. If it is bounded, it converges, and we write $\text{A}(D) = \lim_{n\to\infty} \text{A}(D_n)$; if it is unbounded, the sequence diverges, and we write $\text{A}(D) = \infty$.

To see how this process works, let D be the set of all points (x, y) with $0 \le y \le (1 + x^2)^{-1}$ (see Fig. 4-9). Choosing R_n as above, we have

$$\text{A}(D_n) = \int_{-n}^{n} dx \int_0^{1/(1+x^2)} dy = \int_{-n}^{n} \frac{dx}{1 + x^2}$$
$$= 2 \arctan (n)$$

and
$$\text{A}(D) = \lim_{n\to\infty} \text{A}(D_n) = \pi$$

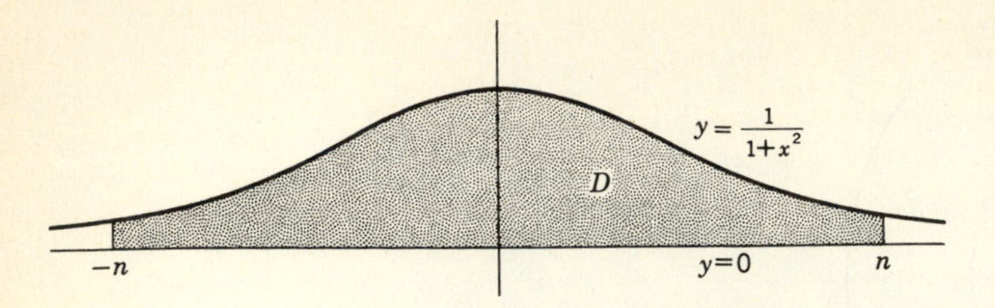

Figure 4-9

Again, let D be the set of all points (x, y) with $|x^2 - y^2| \leq 1$ (see Fig. 4-10). We have

$$\mathrm{A}(D_n) = 8\int_0^1 dx \int_0^x dy + 8\int_1^n dx \int_{\sqrt{x^2-1}}^x dy$$

$$= 4 + 8\int_1^n (x - \sqrt{x^2 - 1})\, dx$$

Since $x - \sqrt{x^2 - 1} = 1/(x + \sqrt{x^2 - 1}) \geq 1/(2x)$,

$$\mathrm{A}(D_n) \geq 4 + 8\int_1^n \frac{dx}{2x} = 4 + 4 \log n$$

and $\mathrm{A}(D) = \infty$.

If this example is modified slightly, a region of finite area results. Let D be those points (x, y) with $|x^4 - y^4| \leq 1$. The graph of D would look very

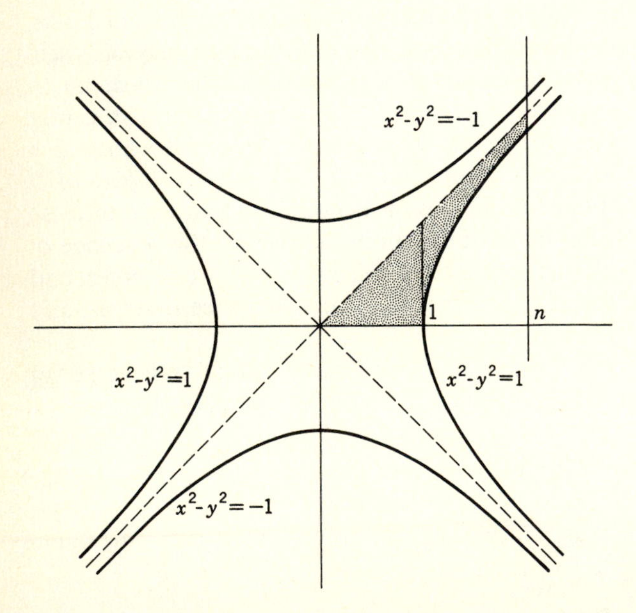

Figure 4-10

much the same as that in Fig. 4-10. As above, we have

$$\text{A}(D_n) = 4 + 8 \int_1^n (x - \sqrt[4]{x^4 - 1})\, dx$$

Now, it is not hard to see that, for any $x > 1$,

$$x - \sqrt[4]{x^4 - 1} \leq \frac{1}{x^3}$$

For example, this can be checked directly by showing that

$$\left(x - \frac{1}{x^3}\right)^4 = x^4 - 4 + 6x^{-4} - 4x^{-8} + x^{-12} \leq x^4 - 1$$

whenever $x \geq 1$. Hence,

$$\text{A}(D_n) \leq 4 + 8 \int_1^n x^{-3}\, dx = 4 + 8\left(\frac{1}{2} - \frac{1}{2n^2}\right) \leq 8$$

for all n. Since the D_n expand, $\{\text{A}(D_n)\}$ is monotone increasing and $\lim \text{A}(D_n)$ exists. We have shown that $\text{A}(D)$ is finite, and at most 8, although we have not found its exact value.

It is natural to ask whether or not the choice of the rectangles $\{R_n\}$ affects the final values of $\text{A}(D)$. If we require that a more general sequence share with the special sequence the property that their interiors cover the plane, so that any point of the plane is eventually interior to some R_n, then we can show that $\text{A}(D)$ is independent of this choice.

Theorem 13 *Let $\{R_n'\}$ and $\{R_n\}$ be two expanding sequences of closed and bounded rectangles whose interiors each cover the plane. Let D be a set such that the sets*

$$D_n = D \cap R_n \qquad \text{and} \qquad D_n' = D \cap R_n'$$

have area for each $n = 1, 2, 3, \ldots$. Then,

$$\lim_{n\to\infty} \text{A}(D_n) = \lim_{n\to\infty} \text{A}(D_n')$$

Let U_n be the set of interior points of R_n and set $E = R_j'$. The open sets U_n cover the plane, and thus E; by the Heine-Borel theorem, there is a k such that $E \subset U_k$. This implies that for each j, there is a corresponding k such that $R_j' \subset R_k$. Intersecting both with D, $D_j' \subset D_k$ and $\text{A}(D_j') \leq \text{A}(D_k) \leq \lim_{n\to\infty} \text{A}(D_n)$. Since this holds for any j, $\lim_{n\to\infty} \text{A}(D_n') \leq \lim_{n\to\infty} \text{A}(D_n)$. By an analogous argument, we obtain the reversed inequality, and thus equality of the limits. ▮

Having extended the notion of area, we may similarly seek to extend the notion of the definite integral

$$I(f, D) = \iint_D f$$

Regarded as a function of the pair (f, D), we have shown that it is defined when D is a bounded set having area, and f is bounded and continuous on D, except for a set of zero area. In both cases, the word "bounded" cannot be deleted. For example, if f were not bounded above on D, then for any grid N partitioning D, the upper Riemann sum $\bar{S}(N)$ would have the value ∞, since $\text{lub}_{p \in R_{ij}} f(p)$ would be infinite for some choice of i and j. In particular, the previous definition fails to give meaning to the integrals $\int_0^1 \log x \, dx$ and $\int_0^1 1/\sqrt{x} \, dx$. Guided by our discussion of area for unbounded sets, let us attempt to extend $I(f, D)$ so that it will be defined for some unbounded regions D and some unbounded functions f. To distinguish these from the original notion of integral, we call them **improper** integrals.

What meaning should be attached to $\int_c^\infty f$ when f is continuous on the unbounded interval $c \le x < \infty$? If f is positive-valued, we can fall back on the notion of area, and define the value of this to be the area of the plane region $D = \{\text{all } (x, y) \text{ with } c \le x < \infty \text{ and } 0 \le y \le f(x)\}$. Applying the previous discussion,

$$\int_c^\infty f = \text{A}(D) = \lim_{n \to \infty} \int_c^n dx \int_0^{f(x)} dy$$

$$= \lim_{n \to \infty} \int_c^n f(x) \, dx$$

Modifying this slightly, we adopt the following definition to be applied also when the integrand takes on positive and negative values.

Definition 5 *Let $f(x)$ be continuous for $c \le x \le \infty$. Then,*

$$\int_c^\infty f = \lim_{r \uparrow \infty} \int_c^r f$$

when this limit exists.

To illustrate this,

$$\int_0^\infty e^{-x} \, dx = \lim_{r \uparrow \infty} \int_0^r e^{-x} \, dx = \lim_{r \uparrow \infty} - e^{-x} \Big|_0^r = 1$$

$$\int_0^\infty \sin x \, dx = \lim_{r \uparrow \infty} - \cos x \Big|_0^r$$

which does not exist. Since the final step in the computation is the evaluation of a limit, it is customary to use the terms "convergent" and "divergent" in place of "exist" and "not exist." Thus $\int_0^\infty e^{-x}\,dx$ converges and $\int_0^\infty \sin x\,dx$ diverges. Following the same pattern, we shall understand $\int_{-\infty}^c f$ to mean $\lim_{r\uparrow\infty}\int_{-r}^c f$.

For the expression $\int_{-\infty}^\infty f$ two distinct definitions are used.

(4-31)
$$\int_{-\infty}^{\infty} f = \lim_{r\to\infty}\int_c^r f + \lim_{r\to\infty}\int_{-r}^c f$$
$$= \int_c^\infty f + \int_{-\infty}^c f$$

(4-32)
$$\int_{-\infty}^{\infty} f = \lim_{r\to\infty}\int_{-r}^r f$$

To distinguish these, we call the first the (ordinary) value of the improper integral, and the second the **Cauchy principal value.** These agree whenever both exist, but the Cauchy value may exist in some cases where the ordinary value does not. The reason for this lies in the fact that the limit calculations in the ordinary case must be computed separately, and if either diverges, so does the integral. However, in the Cauchy principal value, the limit operations are combined, and divergence of one may be offset by the other. For example, consider the improper integral

$$\int_{-\infty}^{\infty}\frac{(1+x)\,dx}{1+x^2}$$

We have
$$\int_0^r \frac{(1+x)\,dx}{1+x^2} = \left[\arctan x + \frac{1}{2}\log(1+x^2)\right]_0^r$$
$$= \arctan r + \frac{1}{2}\log(1+r^2)$$

and since $\lim_{r\to\infty}\log(1+r^2)=\infty$, the original integral is a divergent improper integral. However,

$$\int_{-r}^0 \frac{(1+x)\,dx}{1+x^2} = -\arctan(-r) - \frac{1}{2}\log(1+r^2)$$

so that
$$\int_{-r}^r \frac{(1+x)\,dx}{1+x^2} = 2\arctan r$$

and
$$(\text{CPV})\int_{-\infty}^{\infty}\frac{(1+x)\,dx}{1+x^2} = \pi$$

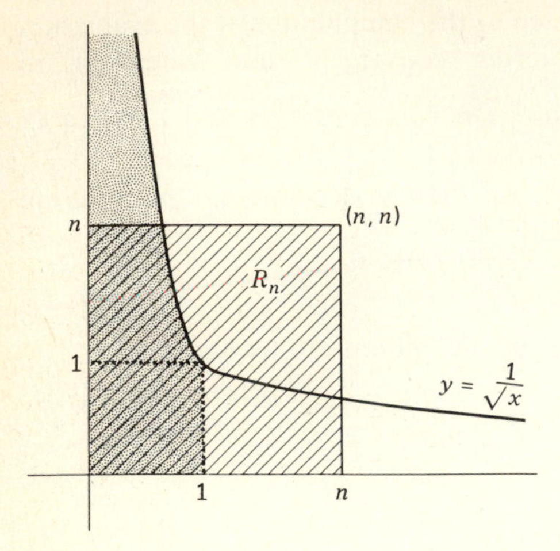

Figure 4-11

Unless there is some indication (such as the prefix (CPV)) that the Cauchy value is meant, the ordinary value is always to be understood, and is usually the one which is appropriate to the problem. The rather arbitrary nature of choice (ii) may be seen from the fact that

$$\lim_{r \to \infty} \int_{-r}^{2r} \frac{(1 + x)\,dx}{1 + x^2} = \pi + \log 2$$

One important case in which both choices agree is that in which the integrand is always positive, or always negative.

Let us consider now the case in which the range of integration is a bounded set, but the integrand is unbounded. For example, can we attach a meaning to $\int_0^1 1/\sqrt{x}\,dx$? We may again consider the set D bounded by the horizontal axis and the graph $y = x^{-1/2}$. This set is unbounded, and to compute its area, we use the rectangles R_n whose vertices are $(\pm n, \pm n)$ as before (see Fig. 4-11).

$$\begin{aligned} \text{A}(D \cap R_n) &= \int_0^1 dy \int_0^1 dx + \int_1^n dy \int_0^{1/y^2} dx \\ &= 1 + \int_1^n y^{-2}\,dy \to 2 \end{aligned}$$

which gives $\text{A}(D) = 2$, and we are led to write

$$\int_0^1 x^{-1/2}\,dx = 2$$

More generally, if f is continuous on $[a, b]$ except at a finite number of points, with $f(x) \geq 0$ for all x, let f_n be the function defined on $[a, b]$ by

$$f_n(x) = \text{minimum of } f(x) \text{ and } n$$

If D is the set of points lying above the horizontal axis, below the graph of f, and between the lines $x = a$, $x = b$, then the area of this unbounded set is given by

$$\mathrm{A}(D) = \lim_{n \to \infty} \int_a^b f_n$$

Were we to take this as our definition for $\int_a^b f$, we would find that it is not too easily applied, even when f is uncomplicated. Moreover, this would have to be modified to treat integrands taking both positive and negative values. Instead, we adopt a different definition which overcomes some of these objections, and which leads to the same results. We first isolate the discontinuities of f by splitting the interval of integration into subintervals in each of which f is continuous, except possibly for one endpoint. For example, if $f(x) = 1/(x^2 - 1)$ and the interval of integration is $[-1, 2]$, we would first write

$$\int_{-1}^{2} f(x)\,dx = \int_{-1}^{0} f(x)\,dx + \int_0^1 f(x)\,dx + \int_1^2 f(x)\,dx$$

since the discontinuities are at 1 and -1. We may therefore assume that we are considering a function f which is continuous on a half-open interval $a < x \le b$ and unbounded there; we do not require that f be positive.

Definition 6 *Let $f(x)$ be continuous for $a < x \le b$. Then,*

$$\int_a^b f = \lim_{r \downarrow a} \int_r^b f$$

whenever this limit exists.

Using this for the integral $\int_0^1 x^{-1/2}\,dx$, we obtain

$$\int_0^1 x^{-1/2}\,dx = \lim_{r \downarrow 0} \int_r^1 x^{-1/2}\,dx$$

$$= \lim_{r \downarrow 0} 2\sqrt{x}\,\Big|_r^1 = 2$$

in agreement with the former calculation. This agreement is of course not accidental.

Theorem 14 *Let f be continuous for $a < x \le b$ with $f(x) \ge 0$. Let D be the region bounded by the line $y = 0$, by $x = a$ and $x = b$, and the curve $y = f(x)$. Then, $\mathrm{A}(D) = \int_a^b f$.*

This asserts that $\lim_{n\to\infty} \int_a^b f_n$ and $\lim_{r\downarrow a} \int_r^b f$ are either both infinite, or both finite and equal. First, choose $r > a$ and let $M_r = \max_{x\in[r,b]} f(x)$. When $n > M_r$, $f_n(x) = f(x)$ on $[r, b]$ so that

$$\int_a^b f_n = \int_a^r f_n + \int_r^b f_n \geq \int_r^b f$$

using the fact that f is positive. Letting n increase, we find that for any $r > a$,

$$\lim_{n\to\infty} \int_a^b f_n \geq \int_r^b f$$

and

$$\lim_{n\to\infty} \int_a^b f_n \geq \lim_{r\downarrow a} \int_r^b f$$

To obtain the opposite inequality, choose any n; since $f(x) \geq f_n(x)$ for all x, while $f_n(x) \leq n$,

$$\int_r^b f_n \leq \int_r^b f \quad \text{and} \quad \int_a^r f_n \leq n(r-a)$$

Combining these, we have

$$\int_a^b f_n \leq n(r-a) + \int_r^b f$$

and letting r approach a, we see that for any choice of n,

$$\int_a^b f_n \leq \lim_{r\downarrow a} \int_r^b f$$

and thus

$$\lim_{n\to\infty} \int_a^b f_n \leq \lim_{r\downarrow a} \int_r^b f$$

Equality must hold, and we have shown that the alternative definition is consistent with the original area definition. ▮

The examples which follow show how the definition is to be modified if the discontinuity of the integrand occurs at the upper endpoint, and if there is more than one discontinuity in the interval of integration.

(i)
$$\int_0^1 \frac{dx}{\sqrt{1-x^2}} = \lim_{r\uparrow 1} \int_0^r \frac{dx}{\sqrt{1-x^2}} = \lim_{r\uparrow 1} \arcsin r$$
$$= \frac{\pi}{2}$$

(ii)
$$\int_0^1 \frac{dx}{\sqrt{x(1-x)}} = \int_0^{1/2} \frac{dx}{\sqrt{x(1-x)}} + \int_{1/2}^1 \frac{dx}{\sqrt{x(1-x)}}$$

Standard integration procedure shows that

$$\int \frac{dx}{\sqrt{x}\,(1-x)} = \log \frac{1+\sqrt{x}}{1-\sqrt{x}}$$

so that

$$\int_0^{1/2} \frac{dx}{\sqrt{x}\,(1-x)} = \lim_{r \downarrow 0} \log \frac{1+\sqrt{x}}{1-\sqrt{x}}\Bigg|_r^{1/2} = \log \frac{\sqrt{2}+1}{\sqrt{2}-1}$$

However,

$$\int_{1/2}^{1} \frac{dx}{\sqrt{x}(1-x)} = \lim_{r \uparrow 1} \log \frac{1+\sqrt{x}}{1-\sqrt{x}}\Bigg|_{1/2}^{r} = \infty$$

so that the original improper integral is divergent.

So far, we have extended the definite integral by relaxing separately the restriction that the interval of integration be bounded, and that the integrand be bounded. These may also be combined. We interpret

$$\int_0^\infty x^{-1/2}e^{-x}\,dx \qquad \text{to mean} \qquad \int_0^1 x^{-1/2}e^{-x}\,dx + \int_1^\infty x^{-1/2}e^{-x}\,dx$$

and speak of the original improper integral as convergent only when both of these integrals are convergent.

If care is exercised, the techniques applicable to ordinary proper integrals may also be used in evaluating improper integrals. If $\phi(u)$ is of class C' for $\alpha \le u \le \beta$, and if $\phi(\alpha) = a$, and $\lim_{u \uparrow \beta} \phi(u) = b$, then the change of variable $x = \phi(u)$ converts an improper integral $\int_a^b f(x)\,dx$ into $\int_\alpha^\beta f(\phi(u))\phi'(u)\,du$. If this is now a proper integral, the original integral was convergent; if the new integral is also improper, then both are convergent or both divergent. For example, the substitution $x = u^2$ in $\int_0^1 x^{-1/2}\,dx$ gives

$$\int_0^1 \frac{2u\,du}{u} = \int_0^1 2\,du = 2$$

This procedure is also valid for integrals with unbounded range of integration. The integral $\int_0^\infty dx/(1+x^2)^2$ becomes $\int_0^{\pi/2} (\cos\theta)^2\,d\theta$ under the substitution $x = \tan\theta$; likewise, $\int_1^\infty x^{-2}\sin x\,dx$ is convergent since the substitution $x = 1/u$ changes this into the proper integral

$$\int_1^0 -\sin\left(\frac{1}{u}\right) lu = \int_0^1 \sin\left(\frac{1}{u}\right) du \tag{4-33}$$

When the interval of integration is split into a union of subintervals, the original integral is represented as a sum of integrals, each having the same integrand, and it is divergent if any one of these diverges. A common error is to suppose that this is true whenever one integral is represented in *any* fashion as a sum of integrals. However, the integrals $\int_0^1 (1 + x)/x \, dx$ and $\int_0^1 (2x - 1)/x \, dx$ are each divergent, while their sum

$$\int_0^1 \frac{1+x}{x} dx + \int_0^1 \frac{2x-1}{x} dx = \int_0^1 \frac{1+x+2x-1}{x} dx = \int_0^1 3 \, dx$$

is convergent.

The convergence of an improper integral such as $\int_0^\infty f(x) \, dx$ depends upon the behavior of $f(x)$ when x is large. The following simple **comparison test** is often used to show convergence or divergence. The functions f and g are assumed continuous on the interval $a \le x < b$, and b may be a number or ∞.

Theorem 15 *Let $0 \le f(x) \le g(x)$ for $a \le x < b$. Then, if $\int_a^b g$ converges, so does $\int_a^b f$ and $\int_a^b f \le \int_a^b g$.*

Figure 4-12 will make this theorem plausible; if the area under the curve $y = g(x)$ is finite, so is the area under the curve $y = f(x)$. Of course, in this it is essential that f and g are both positive. To give a formal proof, we define F and G by $F(r) = \int_a^r f$, $G(r) = \int_a^r g$. Since $f(x) \le g(x)$, $F(r) \le G(r)$ for all $r < b$. Since the integral of g converges, $\lim_{r \to b} G(r)$ exists. Since f is positive, F is monotonic increasing. Since $F(r)$ is bounded above by $G(r)$, and therefore by $\lim_{r \to b} G(r)$, which is finite, $\lim_{r \to b} F(r)$ exists, and $\lim_{r \to b} F(r) = \int_a^b f \le \lim_{r \to b} G(r) = \int_a^b g$. ▮

A corollary of this comparison test is sometimes called the **ratio test** for improper integrals.

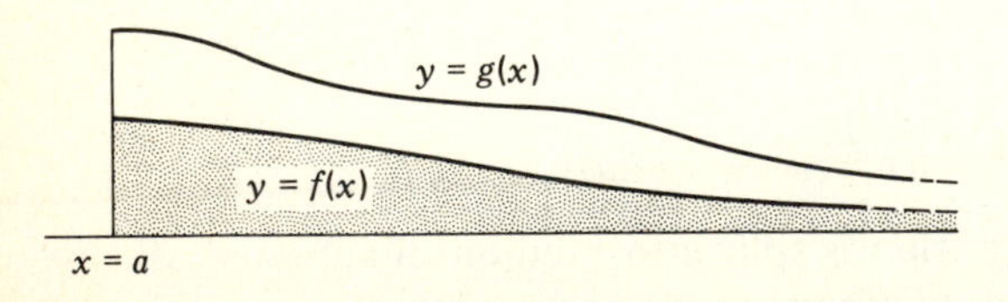

Figure 4-12

Corollary *Let $f(x) \geq 0$ and $g(x) \geq 0$ for $a \leq x < b$, and let*

$$\lim_{x \uparrow b} \frac{f(x)}{g(x)} = L$$

with $0 < L < \infty$. Then, the integrals $\int_a^b f$ and $\int_a^b g$ are either both convergent or both divergent.

If $\lim_{x \to b} f(x)/g(x) = L$, then there is a point x_0 between a and b such that whenever $x_0 < x < b$

$$\frac{L}{2} < \frac{f(x)}{g(x)} < 2L$$

Thus, $f(x) < 2Lg(x)$ and $g(x) < (2/L)f(x)$ for all x, $x_0 < x < b$. Applying the theorem, we see that if $\int_a^b g$ converges, so does $\int_a^b f$, and conversely. ∎

To apply these tests, one must have at hand a collection of known integrals for comparison. Most frequently used are

	$\displaystyle\int_1^\infty \frac{dx}{x^p}$	converges if and only if $p > 1$
(4-34)	$\displaystyle\int_0^1 \frac{dx}{x^p}$	converges if and only if $p < 1$
	$\displaystyle\int_a^c \frac{dx}{\lvert c - x \rvert^p}$	converges if and only if $p < 1$

For example, $\int_0^\infty dx/\sqrt{x + x^3}$ converges, since the integrand is dominated by $1/\sqrt{x}$ on the interval $[0, 1]$ and by $1/x^{3/2}$ on $[1, \infty]$. The integral $\int_0^\infty dx/\sqrt{x + x^2}$ is divergent, since $\lim_{x \uparrow \infty} x/\sqrt{x + x^2} = 1$ and $\int_1^\infty dx/x$ diverges.

These comparison tests apply directly only when the integrand is everywhere positive. However, the next result may often be used to reduce a general case to this special one.

Theorem 16 *$\int_a^b f$ always converges if $\int_a^b |f(x)|\,dx$ converges.*

Since $f(x)$ always lies between $-|f(x)|$ and $|f(x)|$,

$$0 \leq |f(x)| + f(x) \leq 2|f(x)|$$

If $\int_b^b |f|$ converges, then so does $\int_a^b [f(x) + |f(x)|]\, dx$, and subtracting the convergent integral $\int_a^b |f(x)|\, dx$, $\int_a^b f(x)\, dx$ must converge.

The following examples illustrate the use of this in combination with the preceding theorems.

(i) $\displaystyle\int_1^\infty \frac{\sin x}{x^2}\, dx$ converges, since $\displaystyle\left|\frac{\sin x}{x^2}\right| \le \frac{1}{x^2}$

(ii) $\displaystyle\int_0^1 \frac{\cos(1/x)}{\sqrt{x}}\, dx$ converges, since $\displaystyle\frac{|\cos(1/x)|}{\sqrt{x}} \le x^{-1/2}$

(iii) $\displaystyle\int_0^\infty \frac{\sin x}{x\sqrt{x}}\, dx$ converges, since $|x^{-3/2} \sin x| \le x^{-3/2}$ on the interval $[1, \infty]$ and $\displaystyle\lim_{x \downarrow 0} \sqrt{x}\, \frac{\sin x}{x\sqrt{x}} = 1$.

A convergent improper integral $\int_a^b f$ is said to be **absolutely convergent** if $\int_a^b |f|$ is convergent, and **conditionally convergent** if $\int_a^b |f|$ is divergent. All the convergent examples that we have discussed so far are absolutely convergent. A sample of an integral which is only conditionally convergent is $\int_1^\infty x^{-1} \sin x\, dx$. The type of argument which was used above fails here; $|(\sin x)/x|$ is dominated by $1/x$, but $\int_1^\infty x^{-1}\, dx$ is divergent, so that Theorem 16 does not apply, and this method gives no information about the convergence or divergence of either $\int_1^\infty x^{-1} \sin x\, dx$ or $\int_1^\infty x^{-1} |\sin x|\, dx$. To prove convergence of the first integral and divergence of the second, a different method must be used.

Recall the familiar formula for **integration by parts:**

$$\int_a^b f(x)\, dg(x) = f(x)g(x)\Big|_a^b - \int_a^b g(x)\, df(x) \tag{4-35}$$

We apply this to our integral, and have

$$\begin{aligned}
\int_1^r \frac{\sin x}{x}\, dx &= \int_1^r \frac{1}{x}\, d(-\cos x) \\
&= -\frac{\cos x}{x}\Big|_1^r + \int_1^r \cos x\, d\left(\frac{1}{x}\right) \\
&= -\frac{\cos r}{r} + \cos(1) - \int_1^r \frac{\cos x}{x^2}\, dx
\end{aligned}$$

Since $\lim_{r\to\infty} r^{-1}\cos r = 0$, we have shown that

$$\int_1^\infty \frac{\sin x}{x}\,dx = \cos(1) - \int_1^\infty \frac{\cos x}{x^2}\,dx$$

The technique of integration by parts has replaced the original improper integral with another; however, the new one is absolutely convergent, since $|x^{-2}\cos x| \le x^{-2}$. This shows that the original integral converges.

To show that $\int_1^\infty x^{-1}\sin x\,dx$ is itself not absolutely convergent, write

$$\int_1^{m\pi} \frac{|\sin x|}{x}\,dx = \int_1^\pi \frac{\sin x}{x}\,dx + \sum_1^{m-1}\int_{n\pi}^{(n+1)\pi}\frac{|\sin x|}{x}\,dx$$

Since the minimum value of $1/x$ on $[n\pi, (n+1)\pi]$ is $1/(n+1)\pi$,

$$\begin{aligned}\int_{n\pi}^{(n+1)\pi}\frac{|\sin x|}{x}\,dx &\ge \frac{1}{(n+1)\pi}\int_{n\pi}^{(n+1)\pi}|\sin x|\,dx\\ &\ge \frac{2}{(n+1)\pi} \ge \frac{2}{\pi}\int_{n+1}^{n+2}\frac{dx}{x}\end{aligned}$$

and

$$\begin{aligned}\sum_1^{m-1}\int_{n\pi}^{(n+1)\pi}\frac{|\sin x|}{x}\,dx &\ge \frac{2}{\pi}\sum_1^{m-1}\int_{n+1}^{n+2}\frac{dx}{x}\\ &\ge \frac{2}{\pi}\int_2^{m+1}\frac{dx}{x} = \frac{2}{\pi}\log\frac{m+1}{2}\end{aligned}$$

Hence,

$$\int_1^{m\pi}\frac{|\sin x|}{x}\,dx \ge \int_1^\pi \frac{\sin x}{x}\,dx + \frac{2}{\pi}\log\frac{m+1}{2}$$

and $\int_1^\infty |\sin x|/x\,dx$ diverges.

The device of integration by parts may be applied in other cases too. To study the improper integral $\int_2^\infty (\cos x)/(\log x)\,dx$, we first perform an integration by parts. (The use of "∞" is an abbreviation for the previous limit operations.)

$$\begin{aligned}\int_2^\infty \frac{\cos x}{\log x}\,dx &= \int_2^\infty \frac{1}{\log x}\,d(\sin x)\\ &= \frac{\sin x}{\log x}\Big|_2^\infty - \int_2^\infty \sin x\,d\frac{1}{\log x}\\ &= -\frac{\sin 2}{\log 2} + \int_2^\infty \frac{\sin x}{x(\log x)^2}\,dx\end{aligned}$$

Since $\int_2^\infty dx/x(\log x)^2$ converges, the new improper integral is (absolutely) convergent, and $\int_2^\infty (\cos x)/(\log x)\,dx$ converges.

The same procedure can be used to prove convergence for a general class of improper integrals.

Theorem 17 (Dirichlet Test) *Let f, g, and g' be continuous on the unbounded interval $c \le x \le \infty$. Then the integral $\int_c^\infty f(x)g(x)\,dx$ is convergent if f and g obey the following conditions:*

(i) $\lim_{x\to\infty} g(x) = 0$,

(ii) $\int_c^\infty |g'|$ *is convergent,*

(iii) $F(r) = \int_c^r f$ *is bounded for $c \le r < \infty$.*

Take
$$\int_c^r fg = \int_c^r g(x)\,dF(x) = F(x)g(x)\Big|_c^r - \int_c^r F(x)\,dg(x)$$
$$= F(r)g(r) - F(c)g(c) - \int_c^r F(x)g'(x)\,dx$$

By assumption, $|F(r)| \le M$ for all $r \ge c$, so that $|F(r)g(r)| \le M|g(r)|$ and $\lim_{r\to\infty} F(r)g(r) = 0$. Also, $|F(x)g'(x)| \le M|g'(x)|$ and by hypothesis, $\int_c^\infty |g'(x)|\,dx$ converges, so that by comparison $\int_c^\infty F(x)g'(x)\,dx$ converges This shows that $\lim_{r\to\infty} \int_c^r fg$ exists. ∎

Two special cases of this are used frequently.

Corollary 1 *$\int_c^\infty fg$ converges if f obeys condition* (iii) *and $g(x)$ decreases monotonically to 0 as $x \uparrow \infty$.*

For, $g'(x)$ is always negative, so that
$$\int_c^r |g'(x)|\,dx = -\int_c^r g'(x)\,dx = -g(x)\Big|_c^r = g(c) - g(r)$$
and $\lim_{r\to\infty} \int_c^r |g'(x)|\,dx$ exists and is $g(c)$.

Corollary 2 *If g is of class C' for $c \le x < \infty$ and $g(x)$ decreases monotonically to 0 as $x \uparrow \infty$, then the integrals $\int_c^\infty g(x) \sin x \, dx$ and $\int_c^\infty g(x) \cos x \, dx$ are convergent.*

This corollary covers both of the illustrations given above. It is often applied in combination with a change of variable.

Consider the integral

$$\int_0^\infty \cos (x^2) \, dx \tag{4-36}$$

We prove that this converges, thus showing that $\int_c^\infty f(x) \, dx$ can converge without having $f(x) \to 0$ as $x \uparrow$. Make the substitution $u = x^2$, and obtain instead the integral

$$\int_0^\infty \frac{\cos (u)}{2\sqrt{u}} \, du$$

For u near 0, the integrand behaves like $1/\sqrt{u}$. Corollary 2 takes care of the interval $1 \le u < \infty$, and the integral above converges.

Again, $\int_0^1 x^{-1} \sin (x^{-1}) \, dx$ becomes $\int_1^\infty (\sin u)/u \, du$ after the substitution $x = 1/u$, and is therefore convergent.

The discussion of improper double and triple integrals follows somewhat the same pattern, allowing for the change in dimension. The most significant difference is the absence of an analog for conditional convergence. The reason for this will appear later. *For the present, only positive integrands will be considered.* Let D be an unbounded plane set and f continuous and positive in D. If we agree that $\iint_D f$ is to measure the volume of the region V in 3-space lying above D and below the surface $z = f(x, y)$, then we are led to the following definition.

Definition 7 $\iint_D f$ *is* $\lim_{n\to\infty} \iint_{D_n} f$ *where* $D_n = D \cap R_n$ *and* $\{R_n\}$ *is an expanding sequence of closed rectangles whose interiors cover the plane.*

As in Theorem 13, the Heine-Borel theorem shows that the value obtained does not depend upon the choice of the sequence $\{R_n\}$. To illustrate this, let D be the first quadrant and $f(x, y) = xye^{-(x^2+y^2)}$. Choosing R_n as the square with center at the origin and vertices at $(\pm n, \pm n)$

$$\iint_{D_n} f = \int_0^n \int_0^n xye^{-(x^2+y^2)}\,dx\,dy$$

$$= \int_0^n dx \int_0^n xe^{-x^2}ye^{-y^2}\,dy$$

$$= \int_0^n xe^{-x^2}\,dx \int_0^n ye^{-y^2}\,dy = \left[\frac{1-e^{-n^2}}{2}\right]^2$$

and
$$\iint_D f = \lim_{n\to\infty} \iint_{D_n} f = \tfrac{1}{4}$$

When D is a bounded set, but f is unbounded, the region V above D and below the surface $z = f(x, y)$ is again an unbounded set in 3 space. Its volume is found by constructing the truncated functions f_n, where $f_n(p) = \min\{n, f(p)\}$. These are bounded and the volume of V is $\lim_{n\to\infty} \iint_D f_n$, which we may accept as the definition of $\iint_D f$, when the limit exists. Even in simple cases, this process can be somewhat complicated to carry out. For example, let D be the unit square with vertices at $(0, 0)$, $(0, 1)$, $(1, 0)$, $(1, 1)$, and $f(x, y) = yx^{-1/2}$. Cutting f off at height n, we see that $f(p) \le n$ when $y \le n\sqrt{x}$ and

$$f_n(p) = \begin{cases} yx^{-1/2} & \text{for } y \le n\sqrt{x} \\ n & \text{for } y > n\sqrt{x} \end{cases}$$

Thus (see Fig. 4-13)

$$\iint_D f_n = \int_0^{1/n^2} dx \int_{n\sqrt{x}}^1 n\,dy + \int_0^{1/n^2} dx \int_0^{n\sqrt{x}} yx^{-1/2}\,dy$$

$$+ \int_{1/n^2}^1 dx \int_0^1 yx^{-1/2}\,dy = \frac{1}{3n} + \frac{1}{3n} + \left(1 - \frac{1}{n}\right)$$

and $\iint_D f = 1$. As in the case of single integrals, an alternative procedure may be used. The integrand $yx^{-1/2}$ is continuous in D except on the left edge, $x = 0$. Let D_r be the rectangle bounded by the lines $x = 1$, $x = r$, $y = 0$, and $y = 1$. As $r \downarrow 0$, D_r approaches D. Since f is continuous in D_r, we may integrate f over D_r, and define $\iint_D f$ to be $\lim_{r\to 0} \iint_{D_r} f$. We have

$$\iint_{D_r} f = \int_r^1 dx \int_0^1 yx^{-1/2}\,dy = \int_r^1 dx \left[\frac{y^2}{2\sqrt{x}}\right]_0^1$$

$$= \int_r^1 \frac{dx}{2\sqrt{x}} = 1 - \sqrt{r}$$

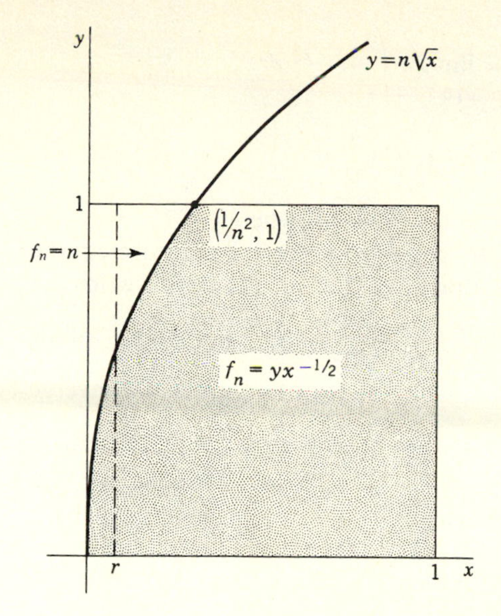

Figure 4-13

and
$$\iint_D f = 1$$

This suggests the following general definition.

Definition 8 *Let D be an open set whose boundary consists of a finite number of piecewise smooth curves and isolated points, and let f be continuous and positive-valued on D. Then, $\iint_D f$ converges and has value c if and only if there is an expanding sequence of closed sets $\{D_n\}$ which converges to D in the sense that every point of D is interior to some set D_k while each D_k lies in the closure of D, and such that f is bounded on each D_k and $c = \lim_{n\to\infty} \iint_{D_n} f$.*

As before, the existence of the limit and the value obtained are independent of the choice of the sequence $\{D_n\}$. To illustrate the definition, let us evaluate the integral of $(x^2 + y^2)^{-\lambda}$ over the unit disk. We take D to be the set of (x, y) with $0 < x^2 + y^2 < 1$ and D_n as the annulus $1/n = \rho \le \sqrt{x^2 + y^2} \le 1$, and compute $\iint_{D_n} (x^2 + y^2)^{-\lambda}\, dx\, dy$. To make this easier, we transform to polar coordinates, setting $x = r \cos \theta$, $y = r \sin \theta$, $dx\, dy = r\, dr\, d\theta$.

$$\iint_{D_n} (x^2 + y^2)^{-\lambda}\, dx\, dy = \int_\rho^1 dr \int_0^{2\pi} r^{-2\lambda} r\, d\theta$$

$$= 2\pi \int_\rho^1 r^{1-2\lambda}\, dr$$

so that
$$\iint_D (x^2+y^2)^{-\lambda}\,dx\,dy = \lim_{\rho\downarrow 0} 2\pi\int_\rho^1 r^{1-2\lambda}\,dr$$
$$= 2\pi\int_0^1 r^{1-2\lambda}\,dr$$

which diverges when $\lambda \geq 1$ and converges to $\pi/(1-\lambda)$ when $\lambda < 1$.

The following example may help to show why we have restricted the discussion thus far to integrands that are positive. Let us attempt to calculate $\iint_D f$ where D is the first quadrant, and $f(x, y) = \sin(x^2+y^2)$. We follow the pattern used before and choose a sequence of regions $\{D_n\}$ that converge to D, and examine $\lim_{n\to\infty}\iint_{D_n} f$. Suppose that D_n is the square with vertices at (0, 0), (0, n), (n, 0), (n, n). Then

$$\iint_{D_n} f = \int_0^n dx\int_0^n \sin(x^2+y^2)\,dy$$
$$= \int_0^n \sin(x^2)\,dx\int_0^n \cos(y^2)\,dy + \int_0^n \cos(x^2)\,dx\int_0^n \sin(y^2)\,dy$$
$$= 2\int_0^n \sin(x^2)\,dx\int_0^n \cos(x^2)\,dx$$

and
$$\lim_{n\to\infty}\iint_{D_n} f = 2\int_0^\infty \sin(x^2)\,dx\int_0^\infty \cos(x^2)\,dx$$

Each of these separate integrals can then be shown convergent as was done earlier for (4-36), be first making the variable change $u = x^2$, and then using Corollary 2 of Theorem 17. Using certain special methods in complex analysis involving contour integration in the complex plane, it is possible to show that

(4-37)
$$\int_0^\infty \sin(x^2)\,dx = \int_0^\infty \cos(x^2)\,dx = \sqrt{\frac{\pi}{8}}$$

[These are called the **Fresnel integrals** and arise in optics.] We are thus led to propose $2(\sqrt{\pi/8})^2 = \pi/4$ as the value of $\iint_D f$. However, let us examine the effect of a different choice for the sequence $\{D_n\}$. Take D_n to be the quarter circle $0 \leq x$, $0 \leq y$, $\sqrt{x^2+y^2} \leq n$. Using polar coordinates, we have

$$\iint_{D_n} f = \int_0^{\pi/2} d\theta\int_0^n r\sin(r^2)\,dr$$
$$= \frac{\pi}{4}[1-\cos(n^2)]$$

and $\lim_{n\to\infty}\iint_{D_n} f$ does not exist!

In this example, two equally natural choices of the sequence $\{D_n\}$ led to inconsistent results.

The key to this strange behavior is the fact that, in this example, the integral of $|f|$ over D is divergent. The following result supplies the clue.

Theorem 18 *Let f be continuous on D and $\iint_D |f|$ converge. Then, $\lim_{n\to\infty} \iint_{D_n} f$ exists and has the same value for any choice of the sequence $\{D_n\}$ converging to D.*

This is the analog for Theorem 16, and we prove it by means of the 2-space analog of the comparison test, Theorem 15 (Exercise 7). For any $p \in D$, $0 \le f(p) + |f(p)| \le 2|f(p)|$, and thus $\iint_D \{f + |f|\}$ converges. Let its value be A and the value of $\iint_D |f|$ be B. Then, for any sequence $\{D_n\}$ converging to D, $\lim \iint_{D_n} \{f + |f|\} = A$ and $\lim \iint_{D_n} |f| = B$ so that

$$\lim_{n\to\infty} \iint_{D_n} f = \lim_{n\to\infty} \left\{ \iint_{D_n} (f + |f|) - \iint_{D_n} |f| \right\}$$

exists and is $A - B$. ▌

This result enables us to prove convergence for a wide class of improper integrals whose integrands take both positive and negative values. Thus, we can immediately know that the integral of $\sin(x^2 + y^2)e^{-(x^2+y^2)}$ over the entire plane is convergent. However, there is no analog for the notion of a conditionally convergent improper integral or for Dirichlet's test, when it comes to multiple integrals. The reason lies in the definition for convergence of an improper double integral, and the freedom of choice of the sets $\{D_n\}$ that occur in that definition. We have stated that $\iint_D f$ will exist only when $\lim \iint_{D_n} f$ exists and is independent of the choice of the expanding regions D_n, which are restricted only by the requirement that every compact set interior to D shall be covered by the interior of D_n for all sufficiently large n. With this definition, it turns out that $\iint_D f$ cannot exist without $\iint_D |f|$ existing too! A sketch of the proof is as follows. Let $f_1 = (|f| + f)/2$ and $f_2 = (|f| - f)/2$. We may assume that the integrals $\iint_D f_i$ are each divergent. Note that $f_1 f_2 = 0$, so that the sets where f_1 and f_2 are positive are disjoint. It is then possible to choose $\{D_n\}$ which favor f_1 over f_2, so that $\iint_{D_n} f_1$ diverges faster than $\iint_{D_n} f_2$, with the result that $\iint_{D_n} f$, which is their difference, also diverges.

Another illustration may reinforce this. We will attempt to integrate the function $f(x, y) = (x^2 - y^2)/(x^2 + y^2)^2$ over the unit square S with opposite vertices (0, 0), (1, 1). Here are two different iterated integral calculations.

$$\int_0^1 dx \int_0^1 f(x, y)\, dy = \int_0^1 dx \left[\frac{y}{x^2 + y^2}\right|_{y=0}^1 = \int_0^1 \frac{dx}{1 + x^2} = \frac{\pi}{4} \tag{4-38}$$

$$\int_0^1 dy \int_0^1 f(x, y)\, dx = \int_0^1 dy \left[\frac{-x}{x^2 + y^2}\right|_{x=0}^1 = \int_0^1 \frac{-dy}{1 + y^2} = -\frac{\pi}{4} \tag{4-39}$$

Note that the integrand $f(x, y)$ is continuous everywhere in the square S except at (0, 0), where it becomes unbounded, both positively and negatively. If a third method is tried, integrating f over S with a small quarter disc of radius ρ at the origin removed, and then letting $\rho \to 0$, the value of the improper integral seems to be 0! Again, the key is simply that $\iint_S |f|$ is divergent, and we do not therefore *have* a convergent integral $\iint_S f$.

So far, we have said little about the evaluation of improper integrals. The methods of approximate integration apply as before. One must first replace the original interval or set over which the integration is to be carried out by a bounded set. For example, if we wish to find the approximate value of $\int_0^\infty e^{-t^2}\, dt$, accurate to .001, we first observe that

$$\left|\int_R^\infty e^{-t^2}\, dt\right| \le \int_R^\infty e^{-t}\, dt = e^{-R}$$

Choose R sufficiently large so that $e^{-R} < .0005$. Then, calculate the value of $\int_0^R e^{-t^2}\, dt$ by Simpson's rule, with an accuracy of .0005. This will be the desired answer. Many times, however, a special device may enable one to find an expression for the exact value in terms of known constants and functions.

As an illustration, let us find the exact value of the integral we have just discussed. We start by replacing the integration variable "t" by "x" and by "y", and consider the product of the resulting integrals, converting this to a double integral:

$$\int_0^R e^{-x^2}\,dx \int_0^R e^{-y^2}\,dy = \int_0^R \int_0^R e^{-x^2-y^2}\,dx\,dy$$

Let R increase, and letting I be the value we seek,

$$I^2 = \left(\int_0^\infty e^{-t^2}\,dt\right)^2 = \iint_D e^{-x^2-y^2}\,dx\,dy$$

where D is the first quadrant. The integrand is positive, and the integral converges. We evaluate the double improper integral by an expanding sequence of discs, changing to polar coordinates to simplify the calculations.

$$I^2 = \lim_{R\to\infty} \int_0^R dr \int_0^{\pi/2} e^{-r^2} r\,d\theta$$

$$= \frac{\pi}{2} \lim_{R\to\infty} \left[-\frac{1}{2} e^{-r^2} \right|_0^R = \frac{\pi}{4}$$

Thus,

(4-40)
$$I = \int_0^\infty e^{-t^2}\,dt = \frac{\sqrt{\pi}}{2}$$

Other special methods for evaluating integrals will be given in the next several chapters.

EXERCISES

1 What is the area of the region bounded by $y = e^x$, $y = 2\cosh x$, with $x \geq 0$?

2 Discuss the convergence of the following integrals:

(*a*) $\displaystyle\int_0^\infty \frac{dx}{x^2 - 1}$ (*b*) $\displaystyle\int_{-\infty}^\infty \frac{dx}{x^3 + 1}$

(*c*) $\displaystyle\int_0^\infty \sin 2x\,dx$ (*d*) $\displaystyle\int_{-\infty}^\infty \frac{e^x}{1 + e^{2x}}\,dx$

(*e*) $\displaystyle\int_0^1 \frac{x^2\,dx}{\sqrt{1 - x^4}}$ (*f*) $\displaystyle\int_0^\infty \sqrt{x}\,e^{-x}\,dx$

(*g*) $\displaystyle\int_0^\infty \frac{1 - \cos x}{x^2}\,dx$ (*h*) $\displaystyle\int_0^\infty \frac{dx}{\sqrt{x^3 + x^2}}$

3 Discuss the convergence of the following integrals:

(*a*) $\displaystyle\int_0^\infty e^x \sin e^x\,dx$ (*b*) $\displaystyle\int_0^\infty x \sin e^x\,dx$

(*c*) $\displaystyle\int_0^{\pi/2} \sin(\sec x)\,dx$ (*d*) $\displaystyle\int_0^1 \frac{\sin(1/x)}{x}\,dx$

(*e*) $\int_0^1 x \log x \, dx$

(*f*) $\int_0^{\pi/2} \sqrt{\tan \theta} \, d\theta$

(*g*) $\int_0^{\pi/2} \frac{d\theta}{\sqrt{1 - \sin \theta}}$

(*h*) $\int_0^{\pi/2} x \sqrt{\sec x} \, dx$

4 For what values of α and β does $\int_0^{\infty} \frac{dx}{x^\alpha + x^\beta}$ converge?

5 For what values of α and β does $\int_0^{\infty} x^\alpha |x - 1|^\beta \, dx$ converge?

6 Let D be the unbounded triangular region in the right half plane bounded by the lines $y = 0$ and $y = x$, and let $f(x, y) = x^{-3/2} e^{y-x}$. Does $\iint_D f$ converge?

7 Let $0 \le f(p) \le g(p)$ for all p in a set D, and suppose that $\iint_D g$ converges. Prove that $\iint_D f$ converges.

8 Discuss the convergence of the improper integral

$$\iint_D \frac{dx \, dy}{x^2 + y^2 + 1}$$

where D is the entire closed first quadrant.

9 Discuss convergence of the integral

$$\iiint \frac{dx \, dy \, dz}{(x^2 + y^2 + z^2 + 1)^2}$$

where the integration is over all 3-space.

10 Verify (4-38) and (4-39).

11 For which real numbers α is there a value c for which

$$\int_0^c \frac{dx}{1 + x^\alpha} = \int_c^\infty \frac{dx}{1 + x^\alpha}$$

12 Let $F(x, y) = y^2/\sqrt{(x^2 + y^2)^3}$ and let S be the unit square with opposite corners $(0, 0)$ and $(1, 1)$.

(*a*) Show why $\iint_S F$ exists.

(*b*) As an iterated integral, would one of the two orders of integration be preferable?

(*c*) Show that $\iint_S F = \log(1 + \sqrt{2})$

CHAPTER

FIVE

SERIES

5.1 PREVIEW

The topic of study is $\sum_1^\infty a_n = a_1 + a_2 + a_3 + \cdots$, an infinite series whose terms are numbers. For some readers, portions of this will be review. However, in addition to the standard ratio test, comparison test, and integral test, we also discuss the ratio comparison test, Raabe's test, and Dirichlet's test, which apply to series that escape the other tests for convergence and divergence. There is a brief discussion of the notion of conditional convergence, versus absolute convergence, and the effects that rearrangements have on the former.

We present a small portion of the theory of absolutely convergent double series, since this is directly related to the Cauchy product of series, and since simple geometric problems often lead to double series.

Finally, we describe and illustrate a number of different techniques for finding exact values or estimated values for the sum of specific series. Computers have not dispensed with the need for this; one does not blindly ask for the sum of a series, for one must know how many terms must be added to give the required accuracy, and any method which will decrease this number is worthwhile.

5.2 INFINITE SERIES

An infinite series is often defined to be "an expression of the form $\sum_1^\infty a_n$." It is recognized that this has many defects. In order to avoid some of these, we adopt the following formal definition.

Definition 1 *An infinite series of real numbers is a pair of real sequences* $\{a_n\}$ *and* $\{A_n\}$ *whose terms are connected by the relations:*

$$A_n = \sum_1^n a_k = a_1 + a_2 + \cdots + a_n \tag{5-1}$$

$$a_1 = A_1 \qquad a_n = A_n - A_{n-1} \qquad n \ge 2$$

The first sequence is called the **sequence of terms** of the series, and the second is called the **sequence of partial sums.** If either is given, the other can be found from the relations (5-1). To denote the series as a single entity, one might use the expression $\langle\{a_n\}, \{A_n\}\rangle$; it is more customary to use $\sum_1^\infty a_n$ or $a_1 + a_2 + \cdots$. Although the sequence of terms is given a dominant place in these expressions, the series itself is still the pair of sequences; we may speak of the sixth term of the series $\sum_1^\infty 1/(n^2 + n)$ (which is $\frac{1}{42}$) as well as the sixth partial sum (which is $\frac{6}{7}$). The index "n" is a dummy letter, and may be replaced by any other convenient choice. As with sequences, it is not necessary that the initial term of a series be labeled with index "1". It is often convenient to vary this, and use, for example, $a_0 + a_1 + a_2 + \cdots$.

There is a strong, but not perfect, analogy between infinite series $\sum_1^\infty a_n$ and improper integrals $\int_c^\infty f(x)\,dx$. The function $f(x)$ corresponds to the sequence of terms $\{a_n\}$, and the partial sums $\sum_1^n a_k = A_n$ to $F(r) = \int_c^r f$. This correspondence means that most of the theorems in Sec. 4.5 have their analogs in this chapter, with proofs that will seem very similar. For this reason, some of these proofs will be abbreviated.

Certain algebraic operations are defined for series.

Definition 2 *The product of the series* $\sum a_n$ *and the number c is the series*

$$\sum (ca_n) = ca_1 + ca_2 + ca_3 + \cdots \tag{5-2}$$

The sum of the series $\sum a_n$ *and* $\sum b_n$ *is the series*

$$\sum (a_n + b_n) = (a_1 + b_1) + (a_2 + b_2) + (a_3 + b_3) + \cdots \tag{5-3}$$

Thus, if $a_n = 1/n$ and $b_n = -1/(n+1)$, then

$$\begin{aligned}\sum a_n + \sum b_n &= (1 + \tfrac{1}{2} + \cdots) + (-\tfrac{1}{2} - \tfrac{1}{3} + \cdots) \\ &= (1 - \tfrac{1}{2}) + (\tfrac{1}{2} - \tfrac{1}{3}) + \cdots \\ &= \tfrac{1}{2} + \tfrac{1}{6} + \tfrac{1}{12} + \tfrac{1}{20} + \cdots\end{aligned}$$

which may be written as

$$\sum_{1}^{\infty} \frac{1}{n(n+1)} \tag{5-4}$$

Note that this series is not the same as the series

$$1 - \tfrac{1}{2} + \tfrac{1}{2} - \tfrac{1}{3} + \tfrac{1}{3} - \tfrac{1}{4} + \tfrac{1}{4} - \cdots \tag{5-5}$$

obtained from $\sum (a_n + b_n)$ by merely removing all the parentheses, for (5-4) and (5-5) have different terms.

Definition 3 *A series* $\sum a_n$ *is said to converge to the sum A whenever the sequence of partial sums* $\{A_n\}$ *converges to A. A series that does not converge is said to diverge.*

We note that the algebraic operations described above may be performed on either divergent or convergent series. We shall see that both $\sum_0^\infty a_n$ and $\sum_0^\infty b_n$ of the example above are divergent; however, their sum (5-4) is a convergent series, for the nth partial sum is

$$\left(1 - \frac{1}{2}\right) + \left(\frac{1}{2} - \frac{1}{3}\right) + \cdots + \left(\frac{1}{n} - \frac{1}{n+1}\right) = 1 - \frac{1}{n+1} \tag{5-6}$$

which converges to 1.

A frequent cause for confusion in discussions about series is the unfortunate habit mathematicians have of using the expression "$\sum a_n$" to stand both for the series and (when convergent) for its sum, letting the context distinguish between these meanings. In "$\sum a_n$ is divergent" or "$\sum a_n$ is alternating" it is clear that the series itself is intended, while in "$\sum a_n$ is larger than 3," the sum is meant. However, in "$\sum a_n$ is positive," either is possible, since it may be intended to mean that each of the terms is positive. A worse case is the statement: $\sum a_n + \sum b_n = \sum (a_n + b_n)$. If this is a statement about series, it is simply the definition for addition of two series; if this is a statement about sums, it is the theorem which asserts that the sum of two convergent series is itself convergent, and its sum is the sum of the numbers $\sum a_n$ and $\sum b_n$.

Another source for confusion stems from the fact that in English, the words "series" and "sequence" are used with almost identical meanings, whereas their mathematical meanings are quite distinct.

The series illustrated in (5-4) and (5-6) suggests a very simple way to construct lots of convergent series. From their nature, these are usually called "telescoping" series.

Theorem 1 *Let* $\{b_n\}$ *be any sequence with* $b_n \to L$, *and set* $a_n = b_n - b_{n+1}$. *Then,* $\sum_1^\infty a_n$ *converges to the sum* $A = b_1 - L$.

For, $A_n = (b_1 - b_2) + \cdots + (b_n - b_{n+1}) = b_1 - b_{n+1}$, which converges to $b_1 - L$.

For example, calculating $a_n = 1/n^2 - 1/(n+1)^2$, we see that

$$\sum_1^\infty \frac{2n+1}{n^2(n+1)^2} = \frac{3}{4} + \frac{5}{36} + \frac{7}{144} + \cdots \tag{5-7}$$

converges to 1.

A series cannot converge unless its terms approach 0.

Theorem 2 *If* $\sum a_n$ *converges, then* $\lim a_n = 0$.

For, $a_n = A_n - A_{n-1}$, and $\lim a_n = A - A = 0$.

The Cauchy criterion for convergence of a sequence immediately gives a similar criterion for series.

Theorem 3 *A series* $\sum a_n$ *converges if and only if*

$$\lim_{m,\, n \to \infty} (a_m + a_{m+1} + \cdots + a_n) = 0.$$

For, this sum of terms is exactly the difference $(a_1 + a_2 + \cdots + a_n) - (a_1 + a_2 + \cdots + a_{m-1}) = A_n - A_{m-1}$, and $\{A_k\}$ is convergent if and only if $\lim_{m,\, n \to \infty} |A_n - A_{m=1}| = 0$. ▮

One consequence of this result is both important and useful.

Theorem 4 *If* $\sum |a_n| = |a_1| + |a_2| + \cdots$ *is a convergent series, so is the series* $\sum a_n$.

This is true whether the terms of the series are real numbers or complex numbers, and it explains why the first step in working with a particular series is often to consider instead the positive series obtained by taking the absolute values of all the original terms. The proof of Theorem 4 is immediate, for

$$|a_m + a_{m+1} + \cdots + a_n| \le |a_m| + |a_{m+1}| + \cdots + |a_n|$$

and by hypothesis, the right side of this is approaching 0. ▮

Because of Theorem 4, many of the convergence tests for series are formulated for series with positive terms, but can be applied to the series $\sum |a_n|$ and thus provide information about $\sum a_n$. (Note, however, that if $\sum |a_n|$ diverges, one cannot therefore conclude that $\sum a_n$ diverges; this is discussed in the next section in more detail.)

The comparison test and the ratio test were extremely useful in dealing with improper integrals; they have their analogs for series.

Theorem 5 (Comparison Test) *If $0 \le a_n \le b_n$ for all sufficiently large n, and $\sum b_n$ converges, then $\sum a_n$ converges.*

The terms $\{a_n\}$ are positive from some index on, and at this point the sequence $\{A_n\}$ of partial sums becomes monotonic increasing. They are bounded above, so that $\lim_{n\to\infty} A_n$ exists. ▮

A corollary of this that is often easier to apply directly is:

Corollary *If $0 \le a_n$ and $0 \le b_n$ and $\lim_{n\to\infty} a_n/b_n = L$ where $0 < L < \infty$, then $\sum a_n$ and $\sum b_n$ are either both convergent or both divergent.*

The next result is sometimes called the **ratio comparison test.**

Theorem 6 *If $0 < a_n$, $0 < b_n$, and $\sum b_n$ converges, and if for all sufficiently large n, $a_{n+1}/a_n \le b_{n+1}/b_n$, then $\sum a_n$ converges.*

Writing the inequality as $a_{n+1}/b_{n+1} \le a_n/b_n$, we see that $\{a_n/b_n\}$ is an ultimately decreasing sequence, and is therefore bounded. Thus $a_n \le Mb_n$ for all n, and $\sum a_n$ converges by the simple comparison test. ▮

In order to apply either of these comparison tests, some known examples of divergent or convergent series must be at hand.

Theorem 7 *The geometric series $\sum_0^\infty x^n = 1 + x + x^2 + \cdots$ converges to $1/(1-x)$ for $|x| < 1$ and diverges when $|x| \ge 1$.*

The partial sums are given by $A_n = (1 - x^{n+1})/(1 - x)$ when $x \neq 1$ and by $A_n = n + 1$ when $x = 1$. ▮

Combining this with Theorem 3 gives the **ratio test.**

Theorem 8 *If* $0 < a_n$, *let* $L = \limsup_{n\to\infty} a_{n+1}/a_n$ *and* $l = \liminf_{n\to\infty} a_{n+1}/a_n$. *Then* $\sum a_n$ *converges if* $L < 1$, *and diverges if* $l > 1$; *if* $l \leq 1 \leq L$, *no conclusion can be reached about the behavior of* $\sum a_n$.

If $L < 1$, then a number x can be chosen so that $L < x < 1$ and $a_{n+1}/a_n \leq x$ for all but a finite number of indices n. Since $x = x^{n+1}/x^n$, this takes the form given in Theorem 6 with $b_n = x^n$, and since $\sum b_n = \sum x^n$ converges, so does $\sum a_n$. On the other hand, when $l > 1$, then for all sufficiently large n, $a_{n+1}/a_n \geq 1$, so that $\{a_n\}$ is an ultimately increasing sequence; as such, it cannot converge to 0, and $\sum a_n$ must diverge. ▮

In many cases, the sequence of ratios $\{a_{n+1}/a_n\}$ is convergent; when this happens, the statement of the theorem is simpler.

Corollary *If* $a_n > 0$ *and* $\lim a_{n+1}/a_n = r$, *then* $\sum a_n$ *converges if* $r < 1$ *and diverges if* $r > 1$. *If* $r = 1$, *no information about* $\sum a_n$ *results.*

The last part of the corollary is easily illustrated. We will see shortly that $\sum 1/n$ diverges and that $\sum 1/n^2$ converges; it is easily checked that $r = 1$ for both.

An example to illustrate the last part of Theorem 8 is also easily supplied. Consider the series

$$\frac{2}{3} + \frac{8}{9} + \frac{16}{27} + \frac{64}{81} + \frac{128}{243} + \frac{512}{729} + \cdots \tag{5-8}$$

Instead of calculating the ratios a_{n+1}/a_n, it is easier to observe that these are the numbers c_n such that $a_{n+1} = a_n c_n$. In (5-8) these are $c_1 = \frac{4}{3}$, $c_2 = \frac{2}{3}$, $c_3 = \frac{4}{3}$, etc., and we find that the numbers l and L in Theorem 8 are $l = \frac{2}{3}$ and $L = \frac{4}{3}$. Since $l < 1 < L$, we do not yet know if the series (5-8) is convergent or divergent. This can be settled by a different test called the **root test,** which also is derived by using the geometric series for comparison.

Theorem 9 *Let* $\limsup_{n\to\infty} |a_n|^{1/n} = r$. *Then,* $\sum a_n$ *converges if* $r < 1$ *and diverges if* $r > 1$; *when* $r = 1$, *no conclusion can be reached.*

If $r < 1$, choose x with $r < x < 1$ and have $|a_n|^{1/n} \leq x$ for all sufficiently large n. Thus $|a_n| \leq x^n$ for $n \geq N$, and since $\sum x^n$ converges, so does $\sum a_n$. If $r > 1$, then from the definition of limit superior, $|a_n|^{1/n} \geq 1$ for infinitely many indices n. We therefore have $|a_{k_1}| \geq 1$, $|a_{k_2}| \geq 1, \ldots, |a_{k_j}| \geq 1, \ldots,$ so that the sequence of terms $\{a_{n_j}\}$ is not convergent to 0; $\sum a_n$ must then diverge. ▌

These two tests are closely connected. The ratio test is often easier to apply; however, if a series can be shown convergent by the ratio test, it could also be treated by the root test (see Exercise 8).

If this test is applied to the previous example (5-8), the number r is seen to be $(\frac{2}{3})\sqrt{2} < 1$, and (5-8) converges (Exercise 4).

The next theorem shows the close connection between improper integrals and infinite series; it also provides a large class of useful series for comparison purposes. It is called the **integral test.**

Theorem 10 *If f is positive on the interval $1 \leq x < \infty$ and monotonic decreasing with $\lim_{x \uparrow \infty} f(x) = 0$, then the series $\sum_1^\infty f(n)$ and the improper integral $\int_1^\infty f$ are either both convergent or both divergent.*

Let $a_n = f(n)$ and $b_n = \int_n^{n+1} f$. Since f is monotonic,

$$f(n+1) \leq \int_n^{n+1} f(x)\,dx \leq f(n)$$

or $a_{n+1} \leq b_n \leq a_n$. By the comparison test, $\sum a_n$ converges if $\sum b_n$ converges, and $\sum b_n$ converges if $\sum a_n$ converges, so that $\sum a_n$ and $\sum b_n$ converge or diverge together. But, $\sum b_n$ converges exactly when the integral $\int_1^\infty f$ converges. ▌

Corollary *The series $\sum_0^\infty 1/n^p$ and $\sum_2^\infty 1/n(\log n)^p$ converge when $p > 1$ and diverge when $p \leq 1$.*

This follows immediately from the behavior of the corresponding improper integrals [see formula (4-34)]. If the first of these series is used as the comparison series $\sum b_n$ in Theorem 6, a test called **Raabe's test** is obtained.

Theorem 11 *Let* $0 < a_n$ *and* $p > 1$, *and suppose that* $a_{n+1}/a_n \le 1 - p/n$ *for all sufficiently large n. Then, the series* $\sum a_n$ *converges.*

For the proof, we need a simple lemma:

Lemma 1 *If* $p > 1$ *and* $0 < x < 1$, *then* $1 - px \le (1 - x)^p$.

Set $g(x) = px + (1 - x)^p$, observe that $g'(x) \ge 0$ and that $g(0) = 1$, and conclude that $g(x) \ge 1$.

To see that this implies Theorem 11, observe that, with $x = 1/n$,

$$\frac{a_{n+1}}{a_n} \le 1 - \frac{p}{n} \le \left(1 - \frac{1}{n}\right)^p = \frac{(n-1)^p}{n^p}$$

which is exactly b_{n+1}/b_n if $b_{n+1} = 1/n^p$, and Theorem 6 applies.

As an illustration of the last test, consider the series

$$\frac{1}{4} + \frac{1 \cdot 3}{4 \cdot 6} + \frac{1 \cdot 3 \cdot 5}{4 \cdot 6 \cdot 8} + \cdots + \frac{1 \cdot 3 \cdots\cdot (2n-1)}{4 \cdot 6 \cdots\cdot (2n+2)} + \cdots$$

The successive ratios a_{n+1}/a_n are $\frac{3}{6}$, $\frac{5}{8}$, $\cdots$, $(2n-1)/(2n+2)$ and $\lim_{n\to\infty} a_{n+1}/a_n = 1$, so that the simple ratio test fails. However,

$$\frac{2n-1}{2n+2} = \frac{2n+2-3}{2n+2} = 1 - \frac{3}{2n+2}$$

and since this has the form $1 - p/(n+1)$ with $p = \frac{3}{2} > 1$, the series is convergent by Raabe's test.

EXERCISES

1. Investigate the convergence of the following series:

(*a*) $\frac{1}{3} + \frac{2}{6} + \frac{3}{11} + \frac{4}{18} + \frac{5}{27} + \cdots$

(*b*) $\frac{1}{2} - \frac{2}{20} + \frac{3}{38} - \frac{4}{56} + \frac{5}{74} - \cdots$

(*c*) $\frac{1}{3} + \frac{1 \cdot 2}{3 \cdot 5} + \frac{1 \cdot 2 \cdot 3}{3 \cdot 5 \cdot 7} + \cdots$

(*d*) $\frac{1}{4} + \frac{1 \cdot 9}{4 \cdot 16} + \frac{1 \cdot 9 \cdot 25}{4 \cdot 16 \cdot 36} + \frac{1 \cdot 9 \cdot 25 \cdot 49}{4 \cdot 16 \cdot 36 \cdot 64} + \cdots$

2 Show that if $\sum a_n$ converges, then $\sum_N^\infty a_n \to 0$ as $N \to \infty$.

3 Investigate the convergence of $\sum a_n$ where

(*a*) $a_n = \dfrac{\sqrt{n+1} - \sqrt{n}}{n+1}$ (*b*) $a_n = \sqrt{\dfrac{\sqrt{n+1} - \sqrt{n}}{n+1}}$

4 Show that in (5-8)

$$a_n = \begin{cases} \dfrac{2^{3n/2}}{3^n} & \text{if } n \text{ is even} \\[2ex] \dfrac{1}{\sqrt{2}}\left(\dfrac{2^{3n/2}}{3^n}\right) & \text{if } n \text{ is odd} \end{cases}$$

and that $\lim (a_n)^{1/n} = 2\sqrt{2}/3 < 1$. Apply the root test to prove the convergence of (5-8).

5 (*a*) Let $0 < a_n$ and $0 < b_n$ and $a_{n+1}/a_n \geq b_{n+1}/b_n$ for all sufficiently large n. Show that if $\sum b_n$ diverges, so does $\sum a_n$.

(*b*) Prove: If $0 < a_n$ and $a_{n+1}/a_n \geq 1 - p/n$ for some $p \leq 1$ and all large n, then $\sum a_n$ diverges.

(*c*) Prove: If $0 < a_n$ and $a_{n+1}/a_n \geq 1 - 1/n - A/n^2$ for all large n, $A > 0$, then $\sum a_n$ diverges.

6 Let $\sum a_n$ and $\sum b_n$ converge, with $b_n > 0$ for all n. Suppose that $a_n/b_n \to L$. Prove that

$$\sum_N^\infty a_k \Big/ \sum_N^\infty b_k \quad \to \quad L$$

7 Let $\{a_n\} \downarrow 0$; show that $\sum_1^\infty a_n$ converges if and only if $\sum_1^\infty 2^n a_{2^n}$ converges.

8 Show that directly that if $\lim_{n\to\infty} |a_{n+1}/a_n| = L$, then $\lim_{n\to\infty} |a_n|^{1/n} = L$.

9 Some of the following statements are true and some are false; prove those that are true, and disprove those that are false.

(*a*) If $\sum a_n$ and $\sum b_n$ converge, so does $\sum (a_n + b_n)$.

(*b*) If $\sum a_n$ and $\sum b_n$ diverge, so does $\sum (a_n + b_n)$.

(*c*) If $\sum |a_n|$ is convergent, so is $\sum (a_n)^2$.

(*d*) If $\sum |a_n|$ and $\sum |b_n|$ converge so does $\sum a_n b_n$.

(*e*) If $\sum_1^\infty a_n^2$ converges, so does $\sum_1^\infty a_n/n$.

*(*f*) If $\{a_n\} \downarrow 0$ and $\sum a_n$ converges, then $\lim_{n\to\infty} na_n = 0$.

10 Show that if $f \geq 0$ and f is monotonically decreasing, and if $c_n = \sum_1^n f(k) - \int_1^n f(x)\,dx$, then $\lim_{n\to\infty} c_n$ exists.

11 Show that if $c_n \geq 0$ and $\sum c_n$ converges, then $\sum \sqrt{c_n}/n$ also converges.

12 Show that if $a_n > 0$, $\sum a_n$ diverges, and $S_n = a_1 + \cdots + a_n$, then $\sum a_n/S_n$ also diverges, but more slowly.

***13** Let f and f' be continuous on the interval $1 \leq x < \infty$ with $f(x) > 0$ and

$$\int_1^\infty |f'(x)|\,dx$$

convergent. Show that the series $\sum_1^\infty f(k)$ and the improper integral $\int_1^\infty f(x)\,dx$ are either both convergent or both divergent.

***14** Show that if $\sum a_n^2/n$ converges, then $1/N \sum_1^N a_k \to 0$.

5.3 CONDITIONALLY CONVERGENT SERIES

A convergent series $\sum a_n$ for which the series $\sum |a_n|$ is divergent is said to be **conditionally convergent;** if $\sum |a_n|$ is convergent, $\sum a_n$ is said to be **absolutely convergent.** The preceding theorems show that each of the following series is absolutely convergent.

$$1 - \tfrac{1}{4} + \tfrac{1}{9} - \tfrac{1}{16} + \tfrac{1}{25} - \cdots = \sum_1^\infty (-1)^{n+1} \frac{1}{n^2}$$

$$1 + \tfrac{1}{2} - \tfrac{1}{4} - \tfrac{1}{8} + \tfrac{1}{16} + \tfrac{1}{32} - \cdots = \sum_0^\infty (-1)^{n(n-1)/2} \frac{1}{2^n}$$

$$1 + \tfrac{1}{2} + \tfrac{1}{5} - \tfrac{1}{10} + \tfrac{1}{17} + \tfrac{1}{26} + \tfrac{1}{37} - \tfrac{1}{50} + \cdots = \sum_0^\infty (-1)^{n(n-1)(n-2)/2} \frac{1}{n^2+1}$$

One may also use Theorem 2 to prove that a series having both positive and negative terms may be divergent. For example, the series $1 - \frac{2}{5} + \frac{3}{9} - \frac{4}{13} + \frac{5}{17} - \cdots$ which has the general term

$$a_n = (-1)^{n+1} \frac{n}{4n-3}$$

is divergent since it is not true that $\lim_{n \to \infty} a_n = 0$.

The methods given so far for testing a series do not apply to the (convergent) alternating harmonic series

$$1 - \tfrac{1}{2} + \tfrac{1}{3} - \tfrac{1}{4} + \tfrac{1}{5} - \cdots$$

Since $\sum_1^\infty 1/n$ diverges, this series is not absolutely convergent, while $\lim_{n\to\infty} (-1)^{n+1} 1/n = 0$, so that Theorem 2 cannot be used to show divergence. The next result contains as a special case the alternating series test, and is the analog for series of Theorem 17, Sec. 4.5, dealing with improper integrals with integrands that change sign. It is usually called the **Dirichlet test for series.**

Theorem 12 *Let $\{a_n\}$ and $\{b_n\}$ be sequences (real or complex) which obey the following:*

(5-9) $$\lim_{n\to\infty} a_n = 0$$

(5-10) $$\sum |a_{n+1} - a_n| \text{ converges}$$

(5-11) *The series $\sum b_n$ is such that its partial sums are uniformly bounded.*

Then, $\sum a_n b_n$ converges.

The last condition, (5-11), means that if $B_n = \sum_1^n b_k$, there is a number M such that $|B_n| \le M$ for all n. The proof of this follows a pattern similar to that of Theorem 17 cited above; we use a discrete analog of integration by parts, sometimes called **"partial summation."**

$$\begin{aligned}\sum_1^n a_k b_k &= a_1 b_1 + a_2 b_2 + \cdots + a_n b_n \\ &= a_1 B_1 + a_2(B_2 - B_1) + \cdots + a_n(B_n - B_{n-1}) \\ &= (a_1 - a_2)B_1 + (a_2 - a_3)B_2 + \cdots + (a_{n-1} - a_n)B_{n-1} + a_n B_n \\ &= a_n B_n - \sum_1^{n-1} (a_{k+1} - a_k)B_k\end{aligned}$$

Because of (5-9) and (5-11), $\lim_{n\to\infty} a_n B_n = 0$. Since

$$|(a_{k+1} - a_k)B_k| \le M|a_{k+1} - a_k|$$

and (5-10) holds, the series $\sum_1^\infty (a_{k+1} - a_k)B_k$ is convergent. These together prove that $\lim_{n\to\infty} \sum_1^n a_k b_k$ exists and $\sum_1^\infty a_n b_n$ converges. ▮

This theorem takes a simpler form when $\{a_n\}$ is monotonic.

Corollary 1 *If the sequence $\{a_n\}$ is monotonic decreasing with $\lim_{n\to\infty} a_n = 0$, and the partial sums of $\sum b_n$ are bounded, then $\sum a_n b_n$ converges.*

$$\begin{aligned}\sum_1^n |a_{k+1} - a_k| &= (a_1 - a_2) + (a_2 - a_3) + \cdots + (a_n - a_{n+1}) \\ &= a_1 - a_{n+1}\end{aligned}$$

and $\lim_{n\to\infty} \sum_1^n |a_{k+1} - a_k| = \lim_{n\to\infty} (a_1 - a_{n+1}) = a_1$, so that hypothesis is satisfied. ▮

A special choice of the sequence $\{b_n\}$ yields the usual **alternating series test.**

Corollary 2 *If $\{a_n\}$ is monotonic decreasing with $\lim_{n\to\infty} a_n = 0$, then $\sum_1^\infty (-1)^{n+1} a_n$ converges.*

With $b_n = (-1)^{n+1}$ the partial sums of $\sum_1^\infty b_n$ are always either 1 or 0, and are therefore bounded. ▮

The following examples illustrate the use of these tests; we note that none of the series is absolutely convergent.

The series $1 - \frac{1}{2} + \frac{1}{3} - \frac{1}{4} + \cdots$ is a simple alternating series and therefore converges. It should be noted that alternation of signs and $\lim a_n = 0$ are not alone sufficient; the partial sums of the series $1/3 - 1/2 + 1/5 - 1/2^2 + 1/7 - 1/2^3 + 1/9 - 1/2^4 + \cdots$ are unbounded, since the positive terms form the divergent series $1/3 + 1/5 + 1/7 + \cdots$, while the negative terms form a convergent geometric series $-1/2 - 1/2^2 - 1/2^3 - \cdots$.

The series $1 + 1/2 - 2/3 + 1/4 + 1/5 - 2/6 + 1/7 + \cdots$ converges by appeal to Corollary 1; take $a_n = 1/n$ and let $\sum b_n$ be $1 + 1 - 2 + 1 + 1 - 2 + 1 + 1 - 2 + \cdots$, which has a bounded sequence of partial sums.

The conditionally convergent series $\sum_1^\infty (-1)^{n+1}(1/n)$ is often used to illustrate a property which is shared by all conditionally convergent series, namely, that rearrangement of the order in which the terms appear may change the sum or even render the series divergent. Denoting the sum of this series by S (approximately .693), we write

$$S = 1 - \tfrac{1}{2} + \tfrac{1}{3} - \tfrac{1}{4} + \tfrac{1}{5} - \tfrac{1}{6} + \tfrac{1}{7} - \cdots \tag{5-12}$$

Then

$$\tfrac{1}{2}S = \tfrac{1}{2} - \tfrac{1}{4} + \tfrac{1}{6} - \tfrac{1}{8} + \tfrac{1}{10} - \tfrac{1}{12} + \tfrac{1}{14} - \cdots$$

Neither the convergence of a series nor the value of its sum is altered by the insertion or deletion of zero terms, so that

$$\tfrac{1}{2}S = 0 + \tfrac{1}{2} + 0 - \tfrac{1}{4} + 0 + \tfrac{1}{6} + 0 - \tfrac{1}{8} + \cdots$$

Adding this series to the first one, we have

$$\tfrac{3}{2}S = 1 + 0 + \tfrac{1}{3} - \tfrac{1}{2} + \tfrac{1}{5} + 0 + \tfrac{1}{7} - \tfrac{1}{4} + \tfrac{1}{9} + 0 + \cdots$$

or, dropping the zero terms,

$$\tfrac{3}{2}S = 1 + \tfrac{1}{3} - \tfrac{1}{2} + \tfrac{1}{5} + \tfrac{1}{7} - \tfrac{1}{4} + \tfrac{1}{9} + \tfrac{1}{11} - \tfrac{1}{6} + \cdots$$

If the terms of this series are compared with those of the original series whose sum was S, it will be seen that these series are rearrangements of each other; each term of one appears exactly once somewhere among the terms of the other series. This emphasizes the fact that an infinite series is not merely the "sum" of an infinite set of numbers; if we return to the view that a series is a pair of related sequences, then we see that the two series are quite different, having entirely different *sequences* of terms, and that it should not be surprising, therefore, that they converge to different sums.

It is not difficult to see that more drastic rearrangements can convert the alternating harmonic series (5-12) into a divergent series, or in fact into a convergent series with any preassigned sum. This depends on three facts:

i. $\lim_{n\to\infty} 1/n = 0$.
ii. The positive terms of (5-12) form a divergent series $1 + \frac{1}{3} + \frac{1}{5} + \frac{1}{7} + \cdots$.
iii. The negative terms in (5-12) form a divergent series $-\frac{1}{2} - \frac{1}{4} - \frac{1}{6} - \frac{1}{8} - \cdots$.

Suppose we wish to rearrange (5-12) to converge to $A = 10$. Choose positive terms in order of occurrence in (5-12) until their sum first exceeds A. Follow these by negative terms, in order, until the cumulative sum drops below A. Continue then with unused positive terms until the cumulative sum exceeds A again. Follow these by more negative terms until one drops below A. This process will succeed in producing a rearrangement of (5-12) whose partial sums shift back and forth across the assigned value A; because of (i) above, the oscillations will decrease and the partial sums will converge to A.

Our next theorem shows that this property does not hold for a series that is absolutely convergent.

Theorem 13 *If* $\sum_1^\infty a_n$ *is an absolutely convergent series with sum* A, *then every rearrangement of* $\sum_1^\infty a_n$ *converges to* A.

Let the series $\sum_1^\infty a_n'$ result from an arbitrary rearrangement of the series $\sum_1^\infty a_n$. This means that $a_n' = a_{r_n}$ where the sequence $\{r_n\}$ is some ordering of the sequence of positive integers 1, 2, ... Given ε, choose N so that $\sum_{k>N} |a_k| < \varepsilon$. This is possible since $\sum |a_k|$ converges. Each of the integers 1, 2, ..., N appears once somewhere among the integers $r_1, r_2, \ldots$. Choose n_0 so that all are contained in the set $\{r_1, r_2, \ldots, r_{n_0}\}$. Write

$$\left|A - \sum_1^n a_k'\right| = \left|A - \sum_1^N a_k + \sum_1^N a_k - \sum_1^n a_k'\right|$$

$$\leq \left|A - \sum_1^N a_k\right| + \left|\sum_1^n a_k' - \sum_1^N a_k\right|$$

The first is dominated by $\sum_{k>N} |a_k|$; if $n \geq n_0$, then $\sum_1^n a_k' - \sum_1^N a_k$ can be written as a sum of terms a_j with $j > N$, since each term a_k with $k = 1, 2, \ldots, N$ already appears in the sum $\sum_1^n a_k'$. Thus, for $n > n_0$,

$$\left|A - \sum_1^n a_k'\right| \leq \sum_{k>N} |a_k| + \sum_{j>N} |a_j| < 2\varepsilon$$

and $\sum_1^\infty a_k'$ converges to A. ▮

Another operation on series which can sometimes alter sums is the removal of brackets. If the series $\sum a_n$ has partial sums $\{A_n\}$, then the partial sums of the series $(a_1 + a_2) + (a_3 + a_4 + a_5) + (a_6 + a_7) + (a_8 + a_9 + a_{10}) + \cdots$ are $A_2, A_5, A_7, A_{10}, \ldots$, a subsequence of the original sequence A_n. Any subsequence of a convergent sequence converges to the same limit, so that any convergent series may have its terms grouped in brackets without altering the sum. If the original series is divergent, grouping terms may render it convergent. The series $\sum_1^\infty (-1)^{n+1}$ diverges, while $(1-1)+(1-1)+\cdots$ converges to 0, and $1-(1-1)-(1-1)-\cdots$ converges to 1.

The techniques we have discussed for investigating the convergence of a specific series whose terms are numbers are equally applicable if the terms have variables in them; in this case, we seek the set of values which these variables must have if the resulting series is to converge. An instance of this is Theorem 7, where we showed that $\sum x^n$ converges exactly for those x with $|x| < 1$. Technically, what we have described is called pointwise convergence, since we are examining the convergence of the series for each individual choice of the variable points. (In the next chapter, we will study uniform convergence of series of functions, which looks at such matters differently.)

Power series form the most common example of series that involve variables. A general power series in x has the form $\sum_0^\infty a_n x^n$, and a power series in $x - c$ (or about $x = c$) has the form $\sum_0^\infty a_n(x-c)^n$. The behavior of power series with respect to pointwise convergence is especially simple and valid both for real and complex $\{a_n\}$.

Theorem 14 *With any power series $\sum_0^\infty a_n x^n$ is associated a radius of convergence R, $0 \le R \le \infty$, such that the series converges (absolutely) for all x with $|x| < R$, and diverges for all x with $|x| > R$. Moreover, R may be calculated from the relation*

$$\frac{1}{R} = \limsup_{n\to\infty} |a_n|^{1/n}$$

or

$$\frac{1}{R} = \lim_{n\to\infty} \left|\frac{a_{n+1}}{a_n}\right|$$

when the latter exists.

We have only to apply the root test for convergence. Let

$$L = \limsup |a_n|^{1/n}$$

Then, $\limsup |a_n x^n|^{1/n} = L|x|$, so that $\sum a_n x^n$ converges whenever $L|x| < 1$ and diverges whenever $L|x| > 1$. If $L = 0$, we see that the series converges for all x; if $L = \infty$, then it converges only for $x = 0$. Setting $R = 1/L$, and interpreting $L = 0$ to correspond to $R = \infty$, and $L = \infty$ to $R = 0$, we see that $\sum a_n x^n$ converges for all x with $|x| < R$. The final statement of the theorem comes from the fact that when $\lim |a_{n+1}/a_n|$ exists, its value is always the same as $\lim |a_n|^{1/n}$. (See Exercise 8 Sec. 5.2.) ∎

The corresponding facts for more general power series can be obtained from this by substitution. The series $\sum_0^\infty a_n[g(x)]^n$ may be called a power series in $g(x)$. If we set $g(x) = y$, this becomes $\sum_0^\infty a_n y^n$. If this power series in y has radius of convergence R, then $\sum_0^\infty a_n[g(x)]^n$ converges for all x with $|g(x)| < R$. In particular, this gives the corollary.

Corollary *If* $1/R = \limsup |a_n|^{1/n}$, *then* $\sum_0^\infty a_n(x-c)^n$ *converges for all* x *with* $|x - c| < R$, *and diverges when* $|x - c| > R$.

The convergence set is thus an interval of length $2R$ centered at the point c, with endpoints $c - R$ and $c + R$, and the series may or may not converge at either endpoint.

As an illustration, consider the power series

$$\sum_0^\infty \frac{(x+2)^n}{3^n\sqrt{2n+1}} = 1 + \frac{x+2}{3\sqrt{3}} + \frac{(x+2)^2}{9\sqrt{5}} + \cdots \tag{5-13}$$

If we take absolute values of the terms, then apply the ratio test, we arrive at

$$\begin{aligned}\lim_{n\to\infty} \frac{a_{n+1}}{a_n} &= \lim_{n\to\infty} \left|\frac{(x+2)^{n+1}}{3^{n+1}\sqrt{2n+3}}\right|\left|\frac{3^n\sqrt{2n+1}}{(x+2)^n}\right| \\ &= \lim_{n\to\infty} \frac{|x+2|}{3}\frac{\sqrt{2n+1}}{\sqrt{2n+3}} = \frac{|x+2|}{3}\end{aligned}$$

Thus, we conclude that the series converges if $|x+2|/3 < 1$ and diverges if $|x+2|/3 > 1$, and we do not yet know what happens if $|x+2|/3 = 1$. Rewriting these, the first becomes $|x+2| < 3$, which, for real x, is the same as $-3 < x + 2 < 3$, or $-5 < x < 1$. It remains to test each of the endpoints separately. When $x = 1$, (5-13) becomes the series

$$1 + \frac{1}{\sqrt{3}} + \frac{1}{\sqrt{5}} + \frac{1}{\sqrt{7}} + \cdots + \frac{1}{\sqrt{2n+1}} + \cdots$$

which diverges (since the terms behave like $n^{-1/2}$). When $x = -5$, the series (5-13) becomes

$$1 - \frac{1}{\sqrt{3}} + \frac{1}{\sqrt{5}} - \frac{1}{\sqrt{7}} + \cdots + \frac{(-1)^n}{\sqrt{2n+1}} + \cdots$$

which is a convergent alternating series. We have now found the exact convergence set for (5-13); it converges for all x with $-5 \le x < 1$, and diverges for all other (real) values of x.

This is typical of the behavior of many power series, except that it may be much more difficult to decide on convergence or divergence at the endpoints. Examples of this sort will be found in the exercises.

Sometimes a simple substitution will simplify the work. Consider the series

$$\sum_{1}^{\infty} \frac{1}{n^2} \frac{(x+1)^n}{(x-3)^n} \tag{5-14}$$

If we set $y = (x+1)/(x-3)$, this becomes $\sum y^n/n^2$, which is easily seen to converge for all y with $|y| \leq 1$. Hence, (5-14) converges exactly for those x with

$$\left| \frac{x+1}{x-3} \right| \leq 1$$

This, in turn, can be written (if $x \neq 3$) as

$$|x+1| \leq |x-3| \tag{5-15}$$

and since in general, $|x - b|$ is the distance between x and b, the relation (5-15) is equivalent to

$$\text{dist } (x \text{ to } -1) \leq \text{dist } (x \text{ to } 3)$$

we conclude that (5-15) converges exactly for those x with $x \leq 1$.

The same techniques that work for power series are also used for other types of series involving variable parameters. For example, let us prove the following statement about one form of the so-called **hypergeometric series:**

The series

$$\frac{\alpha}{\beta} + \frac{\alpha(\alpha+1)}{\beta(\beta+1)} + \frac{\alpha(\alpha+1)(\alpha+2)}{\beta(\beta+1)(\beta+2)} + \cdots \qquad \alpha > 0, \beta > 0 \tag{5-16}$$

converges when $\beta > 1 + \alpha$ and diverges when $\beta \leq 1 + \alpha$.

The term ratios are $(\alpha+1)/(\beta+1)$, $(\alpha+2)/(\beta+2)$, and in general $(\alpha+n)/(\beta+n)$ so that

$$\lim_{n \to \infty} \frac{a_{n+1}}{a_n} = 1$$

which indicates nothing about convergence or divergence. Raabe's test is applicable, however, and writing $(\alpha+n)/(\beta+n) = 1 - (\beta-\alpha)/(n+\beta)$, we see that the series converges when $\beta - \alpha > 1$. If $\beta - \alpha = 1$, the series becomes

$$\frac{\alpha}{\alpha+1} + \frac{\alpha(\alpha+1)}{(\alpha+1)(\alpha+2)} + \frac{\alpha(\alpha+1)(\alpha+2)}{(\alpha+1)(\alpha+2)(\alpha+3)} + \cdots$$

$$= \frac{\alpha}{\alpha+1} + \frac{\alpha}{\alpha+2} + \frac{\alpha}{\alpha+3} + \cdots + \frac{\alpha}{\alpha+n} + \cdots$$

which is divergent. If $\beta < 1 + \alpha$, then the terms become even larger and the series is also divergent.

A somewhat more complicated example is the series

$$\sum_{1}^{\infty} \frac{\sin(nx)}{\sqrt{n}} = \sin x + \frac{\sin 2x}{\sqrt{2}} + \frac{\sin 3x}{\sqrt{3}} + \cdots \tag{5-17}$$

We shall show that it converges for every value of the parameter x. Apply Corollary 1 of Theorem 12 with $a_n = 1/\sqrt{n}$ and $b_n = \sin nx$; it is only necessary to show that the partial sums of the series $\sum_1^\infty \sin nx$ are bounded.

Lemma 2
$$\sum_{1}^{n} \sin kx = \frac{\cos(x/2) - \cos(n + \frac{1}{2})x}{2\sin(x/2)}$$

for all x with $\sin(x/2) \neq 0$.

We have

$$\sin\left(\frac{x}{2}\right)\sum_{1}^{n} \sin kx = \sin\left(\frac{x}{2}\right)\sin x + \sin\left(\frac{x}{2}\right)\sin 2x + \cdots + \sin\left(\frac{x}{2}\right)\sin nx$$

Using the identity: $2\sin A \sin B = \cos(B - A) - \cos(B + A)$, this may be written as

$$2\sin\left(\frac{x}{2}\right)\sum_{1}^{n} \sin kx = \left(\cos\frac{x}{2} - \cos\frac{3x}{2}\right) + \left(\cos\frac{3x}{2} + \cos\frac{5x}{2}\right)$$

$$+ \cdots + \left(\cos\left[n - \frac{1}{2}\right]x - \cos\left[n + \frac{1}{2}\right]x\right)$$

$$= \cos\left(\frac{x}{2}\right) - \cos\left(n + \frac{1}{2}\right)x$$

and the required relation follows. ∎

It is now clear that the partial sums of $\sum_1^\infty \sin(nx)$ are bounded by $1/|\sin(x/2)|$, so that Corollary 1 of Theorem 12 applies and $\sum_1^\infty (\sin nx)/\sqrt{n}$ converges for all x except possibly those for which $\sin(x/2) = 0$. However, these are the values $x = 0, \pm 2\pi, \ldots$, and the series is clearly convergent for these also.

EXERCISES

1 Investigate the convergence of the following series:

(*a*) $\frac{2}{1} - \frac{3}{2} + \frac{2}{3} - \frac{1}{4} + \frac{2}{5} - \frac{3}{6} + \frac{2}{7} - \frac{1}{8} + \cdots$

(*b*) $1+\frac{1}{\sqrt{3}}-\frac{1}{\sqrt{2}}+\frac{1}{\sqrt{5}}+\frac{1}{\sqrt{7}}-\frac{1}{\sqrt{4}}+\frac{1}{\sqrt{9}}+\frac{1}{\sqrt{11}}-\frac{1}{\sqrt{6}}+\frac{1}{\sqrt{13}}+\cdots$

2 Determine the values of the parameters for which the following series converge.

(*a*) $\frac{r}{2}+\frac{4r^2}{9}+\frac{9r^3}{28}+\frac{16r^4}{65}+\cdots$

(*b*) $1+\frac{x}{3}+\frac{2x^2}{9}+\frac{(2\cdot 3)x^3}{27}+\frac{(2\cdot 3\cdot 4)x^4}{81}+\cdots$

(*c*) $\sum_1^\infty \frac{(x+2)^n}{n\sqrt{n+1}}$ (*d*) $\sum_1^\infty \frac{(2n)!x^n}{n(n!)^2}$

(*e*) $\sum_1^\infty \frac{(x-1)^{2n}}{n^2 3^n}$ (*f*) $\sum_1^\infty ne^{-ns}$

(*g*) $\sum_1^\infty \frac{(\beta n)^n}{n!}$

(*h*) $\frac{\alpha\beta}{\gamma}+\frac{\alpha(\alpha+1)\beta(\beta+1)}{2!\,\gamma(\gamma+1)}+\frac{\alpha(\alpha+1)(\alpha+2)\beta(\beta+1)(\beta+2)}{3!\,\gamma(\gamma+1)(\gamma+2)}+\cdots$

(*i*) $\sum_1^\infty \frac{x^n(1-x^n)}{n}$ (*j*) $\sum_1^\infty \frac{nx^n}{n^3+x^{2n}}$

(*k*) $\sum_1^\infty \frac{\sqrt{n}}{(n+1)(2x+3)^n}$ (*l*) $\sum_1^\infty \frac{1}{\sqrt{n}}\left[\frac{x+1}{2x+1}\right]^n$

(*m*) $\sum_1^\infty \sin\left(\frac{x}{n^2}\right)$

3 Show why the following two statements are false.

(*a*) If $\sum a_n$ converges, and $\lim c_n = 0$, then $\sum a_n c_n$ converges.

(*b*) If $\sum b_n$ converges, and $\lim_{n\to\infty} a_n/b_n = 1$, then $\sum a_n$ converges.

4 Show that the sum of an alternating series lies between any pair of successive partial sums, so that the error made in stopping at the nth term does not exceed the absolute value of the next term.

5 Determine the radius of convergence of the following series:

(*a*) $\sum_0^\infty n!\,x^n$ (*b*) $\sum_1^\infty \frac{n!}{n^n}x^n$

(*c*) $\sum_0^\infty c^{n^2}x^n$ where $0 < x < \infty$

6 Determine the radius of convergence of each of these power series:

(*a*) $\sum_1^\infty n(x-1)^n$ (*b*) $\sum_1^\infty \frac{n}{n^2+1}x^n$

(c) $\sum_{1}^{\infty} \frac{n-1}{n+1}(x+2)^n$

7 Formal algebra yields the expression:

$$(1+x)^p = 1 + px + \frac{p(p-1)}{2}x^2 + \cdots + \binom{p}{n}x^n + \cdots$$

Show that this series converges for any x, $|x| < 1$, and any $p > 0$.

8 Discuss the convergence of the series: $\sum_{n=1}^{\infty} (1/\sqrt{n})(x+1)^n/(x-3)^n$.

9 Suppose that there is a sequence of points $\{p_n\}$ such that $|p_{n+1} - p_n| \le c_n$, where $\sum_{n=1}^{\infty} c_n$ is convergent. Show that $\lim_{n\to\infty} p_n$ exists.

10 It was pointed out in Sec. 5.2 that there is an analogy between improper integrals $\int_c^{\infty} f(x)\,dx$ and the infinite series $\sum_{n=1}^{\infty} a_n$ in which $f(x)$ corresponds to a_n. The analog of $f'(x)$ is $a_{n+1} - a_n$. Justify this statement by comparing the Dirichlet test for series (Theorem 12) with the Dirichlet test for integrals (Theorem 17, Sec. 4.5).

11 Using Exercise 10, what is the analog for improper integrals of Theorem 1 on telescoping series?

12 Is there an analog for improper integrals of Theorem 2 for series?

5.4 DOUBLE SERIES

The analog of an improper double integral is a **double series** $\sum\sum a_{ij}$. There are several possible and acceptable definitions for convergence of double series. We select a simple one which is often used.

Definition 4 *The double series* $\sum\sum a_{ij}$ *converges to the sum A if and only if for any* $\varepsilon > 0$ *there is a number N such that*

$$\left| A - \sum_{i=1}^{i=n} \sum_{j=1}^{j=m} a_{ij} \right| < \varepsilon$$

whenever $n \ge N$ *and* $m \ge N$.

If we arrange the terms a_{ij} in a square array with a_{ij} in the ith row and jth column,

$$[a_{ij}] = \begin{bmatrix} a_{11} & a_{12} & a_{13} & \cdots & a_{1N} & \cdots & a_{1m} & \cdots \\ a_{21} & a_{22} & a_{23} & \cdots & a_{2N} & \cdots & a_{2m} & \cdots \\ \cdots & \cdots & \cdots & \cdots & \cdots & \cdots & \cdots & \cdots \\ a_{N1} & a_{N2} & a_{N3} & \cdots & a_{NN} & \cdots & a_{Nm} & \cdots \\ \cdots & \cdots & \cdots & \cdots & \cdots & \cdots & \cdots & \cdots \\ a_{n1} & a_{n2} & a_{n3} & \cdots & a_{nN} & \cdots & a_{nm} & \cdots \\ \cdots & \cdots & \cdots & \cdots & \cdots & \cdots & \cdots & \cdots \end{bmatrix}$$

we see that this definition amounts to summing $\sum\sum a_{ij}$ by rectangles. As with single series, the behavior of series with positive terms is particularly simple, and comparison theorems may be proved. Again, if $\sum\sum |a_{ij}|$ converges, so does $\sum\sum a_{ij}$; such a series is said to be absolutely convergent, and it can be shown that an absolutely convergent series can be arbitrarily rearranged without altering its convergence or its sum. In particular, any absolutely convergent double series $\sum\sum a_{ij}$ can be rearranged as a convergent single series

$$a_{11} + a_{12} + a_{21} + a_{13} + a_{22} + a_{31} + a_{14} + a_{23} + \cdots$$

One immediate application of this observation is the following important theorem dealing with multiplication of absolutely convergent single series.

Definition 5 *The Cauchy product of the series* $\sum_0^\infty a_n$ *and* $\sum_0^\infty b_n$ *is the series* $\sum_0^\infty c_n$, *where*

$$c_n = a_0 b_n + a_1 b_{n-1} + a_2 b_{n-2} + \cdots + a_n b_0 \tag{5-18}$$

The motivation for this definition comes from the study of power series. If we treat power series simply as infinite-degree polynomials and use the same algebraic rules for them, then the product of $\sum_0^\infty a_n x^n$ and $\sum_0^\infty b_n x^n$ will produce a new power series, which will turn out to be $\sum_0^\infty c_n x^n$, where c_n is given by (5-18).

$$\begin{array}{l} a_0 + a_1 x + a_2 x^2 + a_3 x^3 + a_4 x^4 + \cdots \\ \underline{b_b + b_1 x + b_2 x^2 + b_3 x^3 + b_4 x^4 + \cdots} \\ a_0 b_0 + (a_0 b_1 + a_1 b_0)x + (a_0 b_2 + a_1 b_1 + a_2 b_0)x^2 + \cdots \end{array} \tag{5-19}$$

If we leave the parentheses in and set $x = 1$, we arrive at the definition (5-18) introduced by Cauchy.

The following result justifies this definition.

Theorem 15 *Let* $\sum_0^\infty a_n$ *and* $\sum_0^\infty b_n$ *be absolutely convergent, with sums A and B. Then, their product series*

$$\sum_0^\infty c_n = \sum_0^\infty (a_0 b_n + a_1 b_{n-1} + \cdots + a_n b_0)$$

is absolutely convergent with sum AB.

Put $a_{ij} = a_i b_j$. Then,

$$\sum_{i=0}^{i=n} \sum_{j=0}^{j=m} |a_{ij}| = \sum_{0}^{n} |a_i| \sum_{0}^{m} |b_j|$$

so that $\sum_0^\infty \sum_0^\infty |a_{ij}|$ is a convergent double series. The series $\sum_0^\infty \sum_0^\infty a_{ij}$ is then absolutely convergent, and its sum is

$$\lim_{\substack{n\to\infty \\ m\to\infty}} \sum_{i=0}^{i=n} \sum_{j=0}^{j=m} a_{ij} = \lim_{n\to\infty} \sum_{0}^{n} a_i \lim_{m\to\infty} \sum_{0}^{m} b_j$$
$$= AB$$

Rewriting the double series as a simple series, and inserting brackets, we have

$$AB = (a_0 b_0) + (a_0 b_1 + a_1 b_0) + (a_0 b_2 + a_1 b_1 + a_2 b_0) + \cdots$$
$$= \sum_{0}^{\infty} c_n \quad \blacksquare$$

This theorem is also true if one of the series $\sum a_n$ or $\sum b_n$ is absolutely convergent and the other is conditionally convergent. A proof of this, together with many other refinements and additional results, may be found in special treatises on the theory of infinite series. Mere convergence is not enough, as the following example shows.

Consider the convergent alternating series (for $n \geq 0$)

$$1 - \frac{1}{\sqrt{2}} + \frac{1}{\sqrt{3}} - \frac{1}{\sqrt{4}} + \cdots + \frac{(-1)^n}{\sqrt{n+1}} + \cdots \tag{5-20}$$

and form its Cauchy product with itself, according to the scheme (5-18); call the result $\sum c_n$. We find that the signs of this new series alternate and that

$$|c_n| = \frac{1}{\sqrt{n+1}} + \frac{1}{\sqrt{2}\sqrt{n}} + \frac{1}{\sqrt{3}\sqrt{n-1}} + \cdots + \frac{1}{\sqrt{n}\sqrt{2}} + \frac{1}{\sqrt{n+1}} \tag{5-21}$$

However, calculation suggests that c_n does not approach 0. In fact, an easy estimate shows that $|c_n| > \frac{1}{2}$ (Exercise 1). Thus, (5-20) is a convergent series whose Cauchy product with itself is divergent!

The series analog for iterated integration is iterated summation, such as in $\sum_{n=1}^\infty \left(\sum_{k=1}^\infty a_{nk}\right)$, which is interpreted to mean

$$\sum_{n=1}^{\infty} \sum_{k=1}^{\infty} a_{nk} = \sum_{n=1}^{\infty} A_n \qquad \text{where } A_n = \sum_{k=1}^{\infty} a_{nk} \tag{5-22}$$

When the terms a_{nk} are positive, the following result on reversal of order of summation is often used.

Theorem 16 *If* $a_{nk} \geq 0$ *and* $\sum_{n=1}^{\infty} \{\sum_{k=1}^{\infty} a_{nk}\}$ *converges to* S, *then* $\sum_{k=1}^{\infty} \{\sum_{n=1}^{\infty} a_{nk}\}$ *also converges to* S.

The hypothesis asserts that each of the series $\sum_{k=1}^{\infty} a_{nk} = \alpha_n$ is convergent, and that $S = \sum_1^{\infty} \alpha_n$. For any k, $a_{nk} \leq \alpha_n$; invoking the comparison theorem, $\sum_{n=1}^{\infty} a_{nk}$ converges to a sum β_k for $k = 1, 2, \ldots$. Take any integer N and write

$$\sum_1^N \beta_k = \beta_1 + \beta_2 + \cdots + \beta_N = \sum_{n=1}^{\infty} a_{n,1} + \cdots + \sum_{n=1}^{\infty} a_{n,N}$$

$$= \sum_{n=1}^{\infty} \{a_{n,1} + a_{n,2} + \cdots + a_{n,N}\}$$

$$\leq \sum_{n=1}^{\infty} \alpha_n = S$$

Since this bound is independent of N, $\sum_1^{\infty} \beta_k$ converges with sum less than or equal to S, and we have proved that

$$\sum_{k=1}^{\infty} \left\{ \sum_{n=1}^{\infty} a_{nk} \right\} \leq \sum_{n=1}^{\infty} \left\{ \sum_{k=1}^{\infty} a_{nk} \right\}$$

We could now start over again with the series on the left, and obtain the opposite inequality; this shows that the two sides are actually equal. ▌

This result need not hold if the term a_{nk} are not all positive. In the array given below, the sum of the nth row is $1/2^n$, so that

$$\sum_{n=1}^{\infty} \left\{ \sum_{k=1}^{\infty} a_{nk} \right\} = \sum_{n=1}^{\infty} 1/2^n = 1$$

However, the sum of the kth column is $1/2^{k-1}$, so that

$$\sum_{k=1}^{\infty} \left\{ \sum_{n=1}^{\infty} a_{nk} \right\} = 2$$

This shows that the bookkeeper's check may fail for infinite matrices of numbers, unless all the entries are nonnegative. The array is:

$$[a_{nk}] = \begin{bmatrix} 1 & -\frac{1}{2} & 0 & 0 & 0 & 0 & \cdots & \cdots \\ 0 & 1 & -\frac{3}{4} & 0 & 0 & 0 & 0 & \cdots \\ 0 & 0 & 1 & -\frac{7}{8} & 0 & 0 & 0 & \cdots \\ 0 & 0 & 0 & 1 & -\frac{15}{16} & 0 & 0 & \cdots \\ 0 & 0 & 0 & 0 & 1 & -\frac{31}{32} & 0 & \cdots \\ \cdots & \cdots & \cdots & \cdots & \cdots & \cdots & \cdots & \cdots \end{bmatrix}$$

EXERCISES

1 Verify (5-21), and show that $|c_n| > \frac{1}{2}$.

2 Form the Cauchy product of the following series:

$$1 + 2 + 4 + 8 + 16 + 32 + \cdots$$

$$1 - 1 + 1 - 1 + 1 - 1 + 1 - 1 + \cdots$$

and find a formula for the coefficients of the resulting series.

3 Find a formula for the coefficients of the Cauchy product of the series $\sum_0^\infty A^n$ and $\sum_0^\infty B^n$.

4 If $\sum_0^\infty a_n x^n = \left(\sum_0^\infty x^n\right)\left(\sum_0^\infty x^{2n}\right)$, what is a_n?

5 Define the sine and cosine functions by

$$\sin(x) = S(x) = \sum_0^\infty (-1)^n x^{2n+1}/(2n+1)!$$

$$\cos(x) = C(x) = 1 + \sum_0^\infty (-1)^n x^{2n}/(2n)!$$

Show that $\sin(2x) = 2 \sin x \cos x$ directly by multiplying power series.

6 For which values of r and s does the double series $\sum\sum_{m,n=1}^\infty r^m s^n$ converge?

7 For which values of x does the following double series converge?

$$\sum\sum_{m,n=1}^\infty \frac{(2+x)^m(1-x)^n}{3^{m+n}}$$

8 Investigate the convergence of the series $\sum\sum_{k,n=1}^\infty 1/(n+3)^{2k}$.

9 Consider $\sum\sum a_{ij}$ and $\iint f(x, y)\,dx\,dy$, where $f(i, j) = a_{ij}$. Formulate a correct version of the integral test for double series.

10 Investigate the convergence of $\sum\sum_{m,n=1}^\infty 1/(m^2 + n^2)$.

5.5 SOME SUMS

While it is enough in some cases to know that a series $\sum a_n$ converges, it is often more important to know something about the value of its sum. We know that $\sum_1^\infty 1/n^2$ converges. Denote its sum by C; convergence implies that C is almost equal to $\sum_1^N 1/n^2$ if N is large enough, but how large must we take N to calculate C to a specified precision?

A reexamination of the integral test will help here. If $a_n \geq 0$ and f is a decreasing positive function with $f(n) = a_n$, and $b_n = \int_n^{n+1} f(x)\,dx$, then as in the proof of Theorem 10, $a_{n+1} \leq b_n$ and $\sum_1^\infty b_n = \int_1^\infty f$, and therefore

$$\sum_1^\infty a_n \leq a_1 + \int_1^\infty f(x)\,dx \tag{5-23}$$

and

$$\sum_{n \geq N+1} a_n \leq \int_N^\infty f(x)\,dx \tag{5-24}$$

Applied to $\sum 1/n^p$, $p > 1$, these estimates are

$$\sum_1^\infty \frac{1}{n^p} \leq 1 + \frac{1}{p-1} \tag{5-25}$$

$$\sum_{n>N} \frac{1}{n^p} \leq \frac{1}{p-1}\frac{1}{N^{p-1}} \tag{5-26}$$

These tell us that the number C obeys $C \leq 2$, and that the finite sum $\sum_1^{1000} 1/n^2$ will differ from the exact value of C by not more than .001. (We will find out later how accurate these estimates are.)

Similar remarks apply to divergent series. While it may be enough sometimes to know that $\sum a_n$ diverges, at other times one may need to know how fast it diverges—e.g., how rapidly $A_N = \sum_1^N a_k$ approaches infinity, in a case where $a_n \geq 0$. For example, the **harmonic series** $\sum 1/n$ diverges. Calculation shows that

$$\sum_1^{50} \frac{1}{n} = 4.499\,205 \cdots$$

$$\sum_1^{100} \frac{1}{n} = 5.187\,377 \cdots$$

$$\sum_1^{500} \frac{1}{n} = 6.792\,823 \cdots$$

Since the series diverges and has positive terms, we know that the partial sums must exceed 100 eventually. How many terms must we use to achieve $A_N > 100$? Again, the proof of the integral test suggests a way to estimate this. As before, if f is continuous for $1 \le x < \infty$ and is positive and decreasing, then

$$f(m+1) \le \int_m^{m+1} f(x)\,dx \le f(m)$$

so that
$$\sum_2^n f(k) \le \int_1^n f(x)\,dx \le \sum_1^{n-1} f(k) \qquad \text{and}$$

(5-27)
$$f(n) \le \sum_1^n f(k) - \int_1^n f(x)\,dx \le f(1)$$

Applying this, for example, to the series $\sum_1^\infty 1/n$, we see that

$$\sum_1^n \frac{1}{k} = \log n + C_n$$

where $0 < C_n \le 1$. (In fact, $\lim_{n\to\infty} C_n$ exists, by Exercise 10, Sec. 5.2.)

Applied to our case of the harmonic series, we see that $A_N > 100$ if $\log N > 100$, or $N > e^{100}$, which is about 2.69×10^{43}. (R. P. Boas has informed us that the first value of N for which $A_N > 100$ is

1509 26886 22113 78832 36935 63264 53810 14498 59497

which is smaller, being 1.509×10^{43}.)

If we analyze the relationship between partial sums and related integrals more carefully, we can obtain a result that is useful in estimations.

Theorem 17 *Let $f \in C''$ with $f(x) \geq 0, f''(x) \leq 0$ for $1 \leq x < \infty$, Let $a_n = f(n)$ for $n \geq 1$, and set*

$$S_n = \sum_1^n a_k - \int_1^n f(x)\,dx - \tfrac{1}{2}f(n)$$

Then, $\{S_n\}$ is a bounded sequence; specifically,

$$f(1) - \tfrac{1}{2}f(2) \leq S_n \leq \tfrac{1}{2}f(1) \tag{5-28}$$

Before proving this, let us use it to obtain a standard estimate. We will find an approximation for the value of $n!$ which is useful when n is large. Observe that $\log(n!) = \sum_1^n \log k$. Apply the theorem with $f(x) = \log x$, and we find

$$\begin{aligned} S_n &= \log(n!) - \int_1^n \log(x)\,dx - \tfrac{1}{2}\log n \\ &= \log(n!) - n\log n + n - 1 - \tfrac{1}{2}\log n \end{aligned} \tag{5-29}$$

where $-\frac{1}{2}\log 2 \leq S_n \leq 0$. We can rewrite (5-29) as

$$n! = n^n e^{-n}\sqrt{n}C_n \tag{5-30}$$

where $C_n = \exp(1 + S_n)$, and where we have

$$1.922 < \frac{e}{\sqrt{2}} \leq C_n \leq e = 2.718\cdots \tag{5-31}$$

(In Sec. 6.5, it will be found from a different estimate that the sequence $\{C_n\}$ converges to $\sqrt{2\pi}$; this fact, together with (5-30), is called **Stirling's formula** for factorials.)

This formula is often useful in working with series whose coefficients involve factorials. Consider

$$1 + \frac{(1)(1\cdot 2)}{(1\cdot 2\cdot 3)}x + \frac{(1\cdot 2)(1\cdot 2\cdot 3\cdot 4)}{6!}x^2 + \cdots + \frac{n!(2n)!}{(3n)!}x^n + \cdots$$

We can use (5-30) to estimate the size and behavior of the coefficient of x^n, writing

$$\begin{aligned} \frac{n!\,(2n)!}{(3n)!} &= \frac{(n^n e^{-n}\sqrt{n})[(2n)^{2n}e^{-2n}\sqrt{2n}]}{(3n)^{3n}e^{-3n}\sqrt{3n}}\cdot\frac{C_n C_{2n}}{C_{3n}} \\ &= \left(\frac{n^{3n}e^{-3n}n2^{2n}}{n^{3n}e^{-3n}\sqrt{n}\,3^{3n}}\right)\left(\frac{\sqrt{2}C_n C_{2n}}{\sqrt{3}C_{3n}}\right) \approx \sqrt{n}\left(\frac{4}{27}\right)^n \end{aligned}$$

Returning to the series, this tells us that it converges for $|x| < \frac{27}{4}$ and that it diverges at each endpoint, since the series there becomes one whose nth term is like $\sqrt{n}$ or $\pm\sqrt{n}$, and thus does not approach 0.

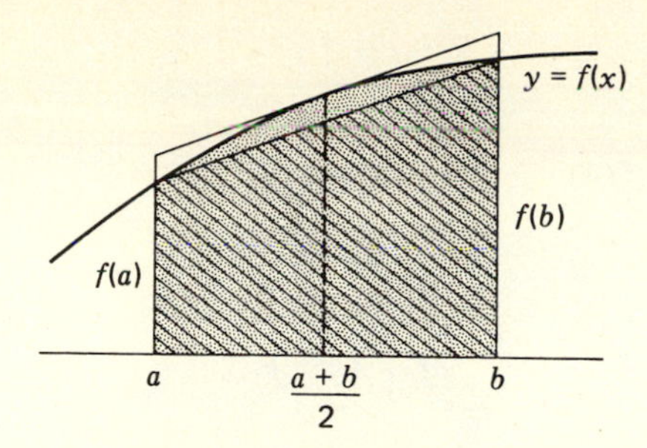

Figure 5-1

The proof of Theorem 17 starts with several simple geometric observations. Since f'' is negative, f' is decreasing for all $x > 1$. Since f remains positive, f' can never be strictly negative; thus, $f'(x) \geq 0$ and f itself is increasing (Exercise 31, Sec. 3.2). Also, Fig. 5-1 makes plausible the following (which was proved as Exercise 16, Sec. 4.2.):

$$\frac{1}{2}(b-a)(f(a)+f(b)) \leq \int_a^b f \leq (b-a)f\left(\frac{a+b}{2}\right) \tag{5-32}$$

For any $n \geq 1$, we have

$$\int_1^n f = \int_1^2 f + \int_2^3 f + \cdots + \int_{n-1}^n f$$

The left side of (5-32) gives

$$\frac{1}{2}(f(k)+f(k+1)) \leq \int_k^{k+1} f$$

so that adding these,

$$\frac{1}{2}f(1)+f(2)+f(3)+\cdots+f(n-1)+\frac{1}{2}f(n) \leq \int_1^n f$$

and

$$\sum_1^n f(k) \leq \int_1^n f + \frac{1}{2}f(n) + \frac{1}{2}f(1)$$

This proves the right side of (5-28).

To complete the proof of the theorem, two cases must be considered. Suppose first that n is odd. Write

$$\int_1^n f = \int_1^3 f + \int_3^5 f + \cdots + \int_{n-2}^n f$$

From the right side of (5-32), we have

$$\int_{k-2}^k f \leq 2f(k-1)$$

which yields at once

$$\int_1^n f \le (2)(f(2) + f(4) + \cdots + f(n-1)) \tag{5-33}$$

Alternatively, we can write

$$\int_1^n f = \int_1^2 f + \int_2^4 f + \int_4^6 f + \cdots + \int_{n-3}^{n-1} f + \int_{n-1}^n f$$
$$\le (2)(f(3) + f(5) + \cdots + f(n-2)) + \int_1^2 f + \int_{n-1}^n f$$

Adding this to (5-33), we arrive at

$$2\int_1^n f \le 2\sum_2^{n-1} f(k) + \int_1^2 f + \int_{n-1}^n f$$

Using the fact that f is increasing to estimate the integrals, and including the terms $f(1)$ and $f(n)$ in the sum, we obtain

$$2\int_1^n f \le 2\sum_1^n f(k) - 2f(1) - 2f(n) + f(n) + f(2)$$

which yields

$$\int_1^n f + \frac{1}{2} f(n) - \sum_1^n f(k) \le \frac{1}{2} f(2) - f(1)$$

and we have proved the left half of (5-28) when n is odd. When n is even, the argument is almost the same, using

$$\int_1^n f = \int_1^2 f + \int_2^4 f + \cdots + \int_{n-2}^n f$$

and

$$\int_1^n f = \int_1^3 f + \int_3^5 f + \cdots + \int_{n-3}^{n-1} f + \int_{n-1}^n f$$

to end with the same estimate. ▮

A very similar estimation theorem can be proved if $f''(x)$ is positive (see Exercise 8). Something similar can also be done with double series (see Exercise 11).

At times, a very different process is used to obtain estimates for the sum of a convergent series. It is usually referred to as "acceleration of convergence," since it is a method which replaces the given series by another whose terms tend to 0 faster, and for which a numerical estimate can be obtained by a partial sum involving fewer terms. The method depends on finding another series whose sum is known and whose terms are similar in size to those of the given series.

As an illustration, consider the series $\sum_1^\infty 1/n^2$, whose sum we denoted earlier by C. Write

$$\frac{1}{n^2} = \frac{n+1}{n^2(n+1)} = \frac{n}{n^2(n+1)} + \frac{1}{n^2(n+1)}$$
$$= \frac{1}{n(n+1)} + \frac{1}{n^2(n+1)}$$

Accordingly,

$$\sum_1^\infty \frac{1}{n^2} = \sum_1^\infty \frac{1}{n(n+1)} + \sum_1^\infty \frac{1}{n^2(n+1)}$$

The first series on the right is recognized as a telescoping series $\sum_1^\infty [1/n - 1/(n+1)]$, convergent to 1. Hence, we have replaced the original series by one whose terms, being cubic in n, decrease more rapidly:

$$\text{(5-34)} \quad C = \sum_1^\infty \frac{1}{n^2} = 1 + \sum_1^\infty \frac{1}{n^2(n+1)} = 1 + \frac{1}{2} + \frac{1}{12} + \frac{1}{36} + \frac{1}{80} + \cdots$$

This process can be repeated again, to accelerate the convergence of the new series. We have

$$\frac{1}{n^2(n+1)} = \frac{n+1}{n^2(n+1)^2} = \frac{1}{2}\frac{2n+1}{n^2(n+1)^2} + \frac{1}{2}\frac{1}{n^2(n+1)^2}$$

and noting that

$$\frac{2n+1}{n^2(n+1)^2} = \frac{1}{n^2} - \frac{1}{(n+1)^2}$$

we find

$$\sum_1^\infty \frac{1}{n^2} = 1 + \frac{1}{2}\sum_1^\infty \left(\frac{1}{n^2} - \frac{1}{(n+1)^2}\right) + \frac{1}{2}\sum_1^\infty \frac{1}{n^2(n+1)^2}$$
$$= 1 + \left(\frac{1}{2}\right)(1) + \frac{1}{2}\sum_1^\infty \frac{1}{n^2(n+1)^2}$$

(Since the first series telescopes.) The last series converges more rapidly than the original series, since the terms decrease like n^{-4}. The number $C = \sum_1^\infty 1/n^2$

has now been shown to be

$$C = \frac{3}{2} + \frac{1}{2}\left(\frac{1}{4} + \frac{1}{36} + \frac{1}{144} + \frac{1}{400} + \frac{1}{900} + \frac{1}{1764} + \cdots\right) \tag{5-35}$$

and nine terms of this gives $C \approx 1.644\,767\,731$.

Kummer replaced this ad hoc procedure by a systematic technique, described in the following statement.

Theorem 18 *Let* $\sum a_n$ *converge, and let* $\sum b_n$ *converge with sum* B; *suppose also that* $\lim a_n/b_n = L$. *Then,* $\sum_1^\infty a_n = BL + \sum_1^\infty u_k a_k$, *where* $u_k = 1 - Lb_k/a_k$.

The point of this result, whose proof is left to the reader, is that $u_k \to 0$; thus the new series converges more rapidly than the original series. The chief difficulty in using this result is in finding a convenient series $\sum b_n$ with a known sum B whose terms $\{b_n\}$ behave enough like those of the original series $\{a_n\}$ so that $\lim a_n/b_n$ exists. As illustrated in the examples, $\sum b_n$ is often chosen as a telescoping series for this reason.

Simple algebraic ingenuity is often the key in evaluating the sum of a series. Here is a sample from the early years of the subject. We will show the following:

$$\sum_{n=3}^{\infty} \frac{1}{n^2 - 4} = \frac{25}{48} \tag{5-36}$$

Our first step is to write out the first half dozen terms in factored form.

$$\frac{1}{1\cdot 5} + \frac{1}{2\cdot 6} + \frac{1}{3\cdot 7} + \frac{1}{4\cdot 8} + \frac{1}{5\cdot 9} + \frac{1}{6\cdot 10} + \frac{1}{7\cdot 11} + \frac{1}{8\cdot 12} + \cdots$$

We next observe the following:

$$1 = \left(\frac{1}{1} - \frac{1}{5}\right) + \left(\frac{1}{5} - \frac{1}{9}\right) + \left(\frac{1}{9} - \frac{1}{13}\right) + \cdots$$

$$= \frac{4}{1\cdot 5} + \frac{4}{5\cdot 9} + \frac{4}{9\cdot 13} + \frac{4}{13\cdot 17} + \frac{4}{17\cdot 21} + \cdots$$

This suggests the next observation:

$$\frac{1}{2} = \left(\frac{1}{2} - \frac{1}{6}\right) + \left(\frac{1}{6} - \frac{1}{10}\right) + \left(\frac{1}{10} - \frac{1}{14}\right) + \left(\frac{1}{14} - \cdots\right.$$

$$= \frac{4}{2\cdot 6} + \frac{4}{6\cdot 10} + \frac{4}{10\cdot 14} + \frac{4}{14\cdot 18} + \cdots$$

Comparing these two series with the terms in (5-36), we try two more telescoping series, $(\frac{1}{3} - \frac{1}{7}) + (\frac{1}{7} - \frac{1}{11}) + \cdots$ and $(\frac{1}{4} - \frac{1}{8}) + (\frac{1}{8} - \frac{1}{12}) + \cdots$, and thus obtain

$$\sum_{3}^{\infty} \frac{1}{n^2 - 4} = \frac{1}{4}\left(1 + \frac{1}{2} + \frac{1}{3} + \frac{1}{4}\right) = \frac{25}{48}$$

Several other equally simple procedures for estimating the sum of a series will be found in the exercises.

We end this chapter with a computation of the exact value of the number C, showing that $C = \pi^2/6 = 1.644\,934\,06\ldots$; however, this computation is done in the spirit of the eighteenth century, rather than of today. We leave it to the reader to evaluate the rigor of the method. (That the answer is indeed correct will be shown rigorously in the next chapter, Sec. 6.6.)

We borrow a result which may be familiar from elementary calculus, but if not, will be shown in the section on power series in the next chapter, namely that

$$\cos x = 1 - \frac{x^2}{2} + \frac{x^4}{4!} - \frac{x^6}{6!} + \cdots \tag{5-37}$$

We also record a fact dealing with the roots of polynomials: If $P(x)$ has roots $\beta_1, \beta_2, \ldots, \beta_m$, none 0, and we write

$$\begin{aligned} P(x) &= (\beta_1 - x)(\beta_2 - x)\cdots(\beta_m - x) \\ &= a_0 + a_1 x + a_2 x^2 + \cdots \end{aligned}$$

then

$$a_0 = \beta_1 \beta_2 \beta_3 \cdots \beta_m$$

and

$$\begin{aligned} -a_1 = {} & \beta_2 \beta_3 \cdots \beta_m + \beta_1 \beta_3 \beta_4 \beta_5 \cdots \beta_m + \beta_1 \beta_2 \beta_4 \cdots \beta_m \\ & + \cdots + \beta_1 \beta_2 \cdots \beta_{m-1} \end{aligned}$$

Thus,

$$-\frac{a_1}{a_0} = \sum \frac{1}{\beta_k} \tag{5-38}$$

We next observe that $\cos x = 0$ if and only if $x = \pm\pi/2, \pm 3\pi/2, \pm 5\pi/2, \ldots$. Hence, the roots β_k of $\cos\sqrt{x} = 0$ are exactly $\pi^2/4, 9\pi^2/4, 25\pi^2/4, \ldots$. From (5-37), we have

$$\cos\sqrt{x} = 1 - \frac{1}{2}x + \frac{1}{24}x^2 - \cdots$$

Comparing this with $P(x)$, we see $a_0 = 1$ and $a_1 = -\frac{1}{2}$. Hence, by (5-38),

$$-\frac{a_1}{a_0} = \frac{1}{2} = \frac{4}{\pi^2} + \frac{4}{9\pi^2} + \frac{4}{25\pi^2} + \frac{4}{49\pi^2} + \cdots$$

from which we conclude that

$$\frac{\pi^2}{8} = 1 + \frac{1}{9} + \frac{1}{25} + \frac{1}{49} + \cdots$$

However,
$$\begin{aligned} C &= 1 + \frac{1}{4} + \frac{1}{9} + \frac{1}{16} + \frac{1}{25} + \frac{1}{36} + \frac{1}{49} + \cdots \\ &= \frac{\pi^2}{8} + \frac{1}{4} + \frac{1}{16} + \frac{1}{36} + \frac{1}{64} + \cdots \\ &= \frac{\pi^2}{8} + \frac{1}{4}C \end{aligned}$$

from which we find $\frac{3}{4}C = \pi^2/8$ and $C = \pi^2/6$. Thus, we are led by the "calculations" to write

$$\frac{\pi^2}{6} = 1 + \frac{1}{4} + \frac{1}{9} + \frac{1}{16} + \cdots \tag{5-39}$$

(although perhaps this argument did not convince you!)

EXERCISES

1 Since $\sum_2^\infty 1/(n \log n)$ diverges, $\lim_{n\to\infty} \sum_2^n 1/(k \log k) = \infty$. How many terms must be taken before the partial sums exceed 10?

2 Estimate:

(a) $\sum_1^n \sqrt{k}$ (b) $\sum_1^n \frac{\log k}{k}$ (c) $\sum_1^n (\log k)^2$

3 A mobile is to be made from 50 uniform sticks of length L by hanging each by a thread 1 inch long and of negligible mass from the end of the stick above it (see Fig. 5-2). When all

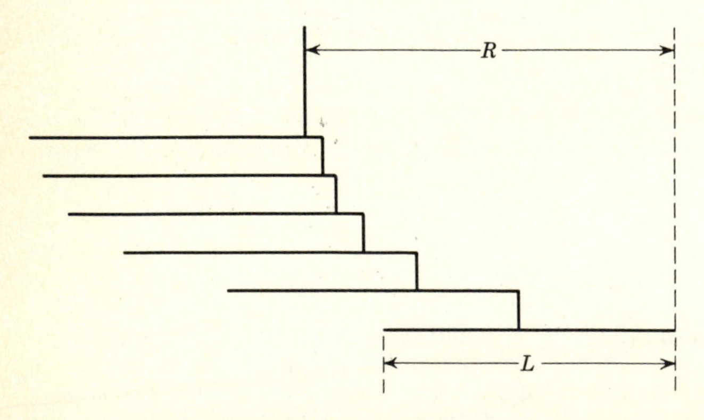

Figure 5-2 A balanced mobile.

are balanced in a horizontal position, the whole is supported by a thread from the top stick to the ceiling. How much space must be allowed for the rotation of the mobile?

4 Apply Exercise 4, Sec. 5.3, to show that the number $S = \sum_0^\infty (-1)^n/(n+1)^2$ lies between .818 and .828. (*Remark:* The average of two successive partial sums of an alternating series is often very accurate; in this example, two such sums are .8179 and .8279, and their average is .8229, while the true value of S is $\pi^2/12 = .82246703\cdots$.)

5 Estimate the sum of each of the following series, accurate to .005:

(*a*) $\displaystyle\sum_1^\infty (-1)^n n/10^n$ (*b*) $\displaystyle\sum_1^\infty (-1)^{n+1}\frac{1}{n^3}$

6 Decide how many terms of the series $\sum_1^\infty (-1)^{n+1}(1/\sqrt{n})$ would have to be used to be sure of an estimate for the sum, accurate to .005.

7 Let $f(x) \ge 0, f'(x) \ge 0, f''(x) \ge 0$ for $1 \le x < \infty$. Show that

$$0 \le \sum_1^n f(k) - \int_1^n f - \tfrac{1}{2}f(n) - \tfrac{1}{2}f(1) \le \tfrac{1}{4}f'(n) \text{ for } n \ge 1.$$

8 Let f be of class C'' on $1 \le x < \infty$, with $f(x) > 0, f'(x) < 0, f''(x) > 0$. Suppose that the series $\sum_1^\infty f(n)$ converges. Using the same ideas as those of Theorem 17, prove that

$$0 \le \sum_N^\infty f(n) - \int_N^\infty f(x)\,dx - \tfrac{1}{2}f(N) \le \tfrac{1}{2}f'(N)$$

9 Apply Exercise 7 to estimate the sums $\sum_1^N k^2$ and $\sum_1^N k^3$, then compare these estimates with the exact values. (The latter can be found by induction.)

10 Show that $\sum_2^\infty 1/(n^2 + 3n - 4) = 137/300$.

11 Apply Exercise 8 to estimate the sum of the series $\sum_1^\infty (-1)^{n+1}(1/\sqrt{n})$ within .005 by grouping terms in pairs, choosing

$$f(x) = \frac{1}{\sqrt{2x-1}} - \frac{1}{\sqrt{2x}}$$

[You will have to select N, apply Exercise 8, and make a separate calculation for $\sum_1^{N-1} f(n)$.]

12 Discuss the convergence of $\displaystyle\sum_{n=1}^\infty \frac{(3n)!}{(n!)^3} x^n$

13 In Exercise 1(*c*) and (*d*), Sec. 5.2, rewrite the terms of the series using factorials and use (5-30) to determine convergence or divergence of the series.

14 Show that the sum of the double series in Exercise 8 Sec. 5.4 is exactly 7/24.

15 Approximately how many points $p = (m, n)$, with m and n integers larger than 0, obey $|p| \le R$?

16 Show that a reasonable value for this unending product is 1/2.

$$\left(1 - \frac{1}{4}\right)\left(1 - \frac{1}{9}\right)\left(1 - \frac{1}{16}\right)\left(1 - \frac{1}{25}\right)\left(1 - \frac{1}{36}\right)\cdots$$

17 What is a reasonable value to assign to the unending expression:

$$\sqrt{2 + \sqrt{2 + \sqrt{2 + \sqrt{2 + \sqrt{2 + \cdots}}}}}$$

CHAPTER

SIX

UNIFORM CONVERGENCE

6.1 PREVIEW

This chapter and the one to follow involve a noticeable increase in mathematical sophistication and rigor. As motivation for this, the chapter begins with a number of paradoxical examples in which a reasonable procedure is shown to yield an incorrect answer. Each time, the procedure involves the interchange of two limiting operations.

The main theme of most of the chapter is "uniformity," treated first for series and sequences, and later, in Sec. 6.4, for improper integrals. A series of functions that is uniformly convergent turns out to behave far more rationally than one that is merely pointwise convergent. Thus, one can integrate termwise, take limits under the summation sign, etc. The power of this is shown by sketching a proof of the Tietze extension theorem (for sets in n-space) and by constructing a standard example of a continuous function that is everywhere nondifferentiable.

An important part of Secs. 6.3 and 6.4 are the illustrations of useful techniques for evaluating integrals and summing power series. In the same spirit, Sec. 6.5 (which is not essential for later sections) treats the gamma and beta functions, and contains a derivation of Stirling's formula to illustrate how one may estimate the asymptotic behavior of a typical improper integral.

Finally, Sec. 6.6 contains an introduction to the study of general orthogonal expansions and Fourier series, presented as a study of the space of continuous functions with a notion of convergence different from pointwise or uniform.

6.2 SERIES AND SEQUENCES OF FUNCTIONS

In the last chapter, we discussed the convergence of series whose terms involved one or more parameters or variables. In essence, we were there dealing with a particular notion of convergence of a series of functions which can be formalized as follows.

Definition 1 *Let each of the functions u_n be defined for points of a set D. Then the series $\sum u_n$ is said to converge pointwise on a set $E \subset D$ if and only if $\sum u_n(p)$ converges for each $p \in E$.*

If we denote the sum of $\sum u_n(p)$ by $F(p)$, then we say that $\sum u_n$ converges pointwise to F on E. For sequences, a similar definition is used; $\{f_n\}$ converges pointwise to F on E if for each point $p \in E$, $\lim_{n\to\infty} f_n(p) = F(p)$.

There are also a number of other important and useful notions of convergence for series and sequences of functions; before introducing these, let us observe some of the shortcomings of pointwise convergence. Each of the following examples represents a plausible, but unfortunately invalid, argument dealing with series or sequences of functions.

Consider the series

(6-1) $$\sum_{1}^{\infty} x(1-x)^n = x(1-x) + x(1-x)^2 + \cdots$$

Standard techniques from the last chapter show that this converges for each x with $0 \le x < 2$. For these values we denote the sum of the series by $F(x)$. We also note that (6-1) converges for $x = 0$, and thus $F(0) = 0$. We pose the following question:

(6-2) What is the value of $\lim_{x \downarrow 0} F(x)$?

Examining (6-1), we note that $\lim_{x\to 0} x(1-x)^n = 0$. This means that if we were to let x approach 0 in the series in each term, separately, we would obtain the series $0 + 0 + 0 + \cdots$. Is this argument valid? Can we find the limit of a function defined by a series by taking the limit termwise?

In this example, the answer is "no." To explain this, we calculate $F(x)$. We have, from (6-1),

$$\begin{aligned} F(x) &= x(1-x) + x(1-x)^2 + x(1-x)^3 + \cdots \\ &= x(1-x)\{1 + (1-x) + (1-x)^2 + \cdots\} \end{aligned}$$

and summing the geometric series, we have

$$F(x) = x(1-x)\frac{1}{1-(1-x)} = 1-x$$

so that the correct value for $\lim_{x \downarrow 0} F(x) = 1$, not 0.

Our second example is the following. Consider the series

$$\sum_{1}^{\infty} \frac{nx^2}{n^3+x^3} = \frac{x^2}{1+x^3} + \frac{2x^2}{8+x^3} + \cdots \tag{6-3}$$

This converges for every x, $0 \le x < \infty$, since

$$\frac{nx^2}{n^3+x^3} \le \frac{nx^2}{n^3} = \frac{x^2}{n^2}$$

and $\sum x^2/n^2$ converges. Again, let $F(x)$ denote the sum of (6-3). This time, we ask:

$$\text{What is } \lim_{x\to\infty} F(x)? \tag{6-4}$$

If we look at the behavior of each term of the series separately, we find $\lim_{n\to\infty} nx^2/(n^3+x^3) = 0$. Thus, termwise, the series for $F(x)$ approaches 0, suggesting that the answer to (6-4) is the number 0. This again is false, as seen thus.

Take any $x > 0$. Then, for any n with $x/2 < n < 2x$,

$$\frac{nx^2}{n^3+x^3} \ge \frac{(x/2)x^2}{(2x)^3+x^3} = \frac{1}{18}$$

In the series for $F(x)$, there are therefore $(2x) - (x/2)$ terms, each larger than $\frac{1}{18}$, so that since all the terms of the series are positive, $F(x) \ge (3x/2)(\frac{1}{18}) = x/12$. Since this is true for any x,

$$\lim_{x\uparrow\infty} F(x) = \infty$$

For the third example, consider

$$\sum_{1}^{\infty} [(n+1)x - n]x^n = (2x-1)x + (3x-2)x^2 + \cdots \tag{6-5}$$

Calling the sum of this $F(x)$, what is $\lim_{x\to 1} F(x)$? Termwise, we obtain $1+1+1+1+\cdots$, which strongly suggests that $\lim_{x\to1} F(x) = \infty$. However, if we expand each term of the original series, we obtain

$$(2x^2 - x) + (3x^3 - 2x^2) + (4x^4 - 3x^2) + \cdots$$

which is a telescoping series whose partial sum is $-x + nx^n$, and which therefore converges to $F(x) = -x$ for all x, $|x| < 1$. Accordingly,

$$\lim_{x\to1} F(x) = -1.$$

The next two examples involve differentiation and integration. Consider the series

$$\sum_{1}^{\infty} \frac{\sin(n^2x)}{n^2} = F(x) \tag{6-6}$$

which converges for all x (absolutely). If we differentiate termwise, we obtain the supposed equality

$$\sum_{1}^{\infty} \cos(n^2x) = F'(x)$$

However, the truth of this is dubious, since the series can be seen to be divergent for every choice of x. (The terms do not approach 0.)

Finally, we propose to calculate the value of

$$\lim_{n\to\infty} \int_0^1 n^2 x e^{-nx}\,dx \tag{6-7}$$

For any choice of x, n^2xe^{-nx} approaches 0 as n increases, since e^{nx} grows so much faster than n^2. Thus, the integrand approaches 0 pointwise for all x, $0 \le x \le 1$, and one might conclude that the value of the limit (6-7) must be 0. However, if we put $u = nx$, the integral becomes

$$\int_0^n ue^{-u}\,du = 1 - (n+1)e^{-n} \to 1$$

All these paradoxical examples involve the reversal of two limit processes. In the first, for instance, we were concerned with the possible equality of

$$\lim_{x\to 0} \lim_{n\to\infty} \sum_{1}^{n} u_k(x) \quad \text{and} \quad \lim_{n\to\infty} \lim_{x\to 0} \sum_{1}^{n} u_k(x) \tag{6-8}$$

In the last two examples, this may not be so apparent, but integration and differentiation both involve hidden limit operations.

In circumstances such as these, the notion of **uniform convergence** of series or sequences of functions is especially useful. To simplify the discussion, we introduce a special notation. If f is a function which is defined on a set E, then $\|f\|_E$ will denote the least upper bound of the set of values $|f(p)|$ for $p \in E$; when f is continuous on E and E is closed and bounded, this is the maximum value of $|f(p)|$ on E. If f and g are both defined on E, then $\|f - g\|_E$ is a measure of the distance between f and g over the set E. If $\|f - g\|_E < \varepsilon$, then $|f(p) - g(p)| < \varepsilon$ for all $p \in E$ so that f approximates g within ε uniformly on E (see Fig. 2.5).

Definition 2 *A sequence of functions $\{f_n\}$ is uniformly convergent to a function F on a set E if and only if* $\lim_{n\to\infty} \|F - f_n\|_E = 0$.

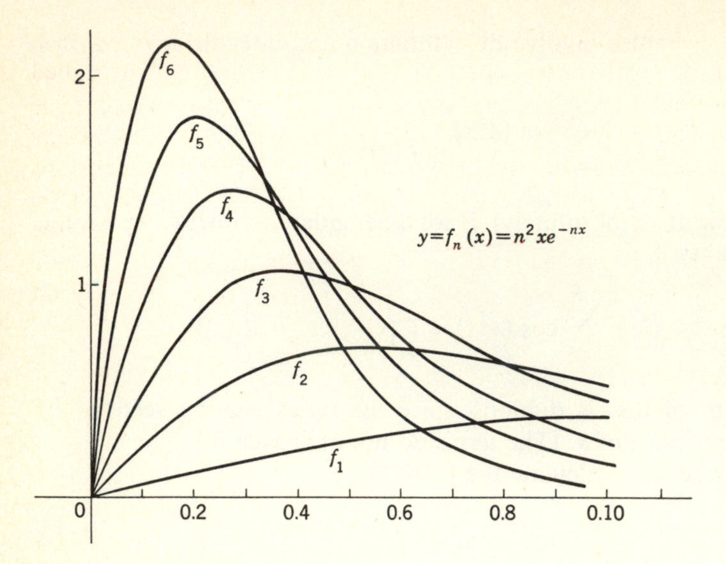

Figure 6-1 Nonuniform convergence.

If we restate this without using the special notation, it becomes: $\{f_n\}$ *converges to F uniformly on E if and only if for any* ε *there is an N such that for any* $n \geq N$ *and any* $p \in E$, $|F(p) - f_n(p)| < \varepsilon$. If a sequence $\{f_n\}$ is uniformly convergent on a set E, then it is certainly pointwise convergent for at least all points of E. However, it may converge pointwise on E and not uniformly on E. Examine again the sequence

$$f_n(x) = n^2xe^{-nx}$$

As we saw, this sequence converges pointwise to 0 on the interval $E = [0, 1]$. The convergence is not uniform on E. The maximum of f_n on E occurs for $x = 1/n$ and is $f_n(1/n) = n/e$, so that $\lim_{n\to\infty} \|f_n\|_E$ is not 0. This behavior is also evident from the graphs of the functions $f_1, f_2, \ldots$ (see Fig. 6-1). The definition of pointwise convergence can also be given as follows: $\{f_n\}$ *converges to F pointwise on E if and only if for any* $\varepsilon > 0$ *and any point* $p \in E$, *there is an N such that whenever* $n \geq N$, $|F(p) - f_n(p)| < \varepsilon$. Comparing this with the corresponding definition of uniform convergence, we see that the essential difference lies in the fact that in uniform convergence, N depends only upon ε, while in pointwise convergence, N depends upon both ε and p.

Uniform convergence of series is defined by throwing it back onto the sequence of partial sums: $\sum_1^\infty u_n$ *converges to F uniformly on E if and only if* $\{f_n\}$ *converges to F uniformly on E, where* $f_n = \sum_1^n u_k$. An alternative statement is: $\sum u_n$ *converges uniformly on E if and only if* $\sum u_n$ *converges pointwise on E and* $\lim_{n\to\infty} \left\|\sum_n^\infty u_k\right\|_E = 0$.

It is often convenient to adopt the viewpoint that functions defined on a common set E can be regarded as geometric points in what is usually called **function space,** and that the **metric,** or measure of distance between points, is given by $\|f - g\|_E$. This enables us to use geometric terminology and analogy to help motivate concepts and calculations. For example, a sequence of functions f_n is said to have the **Cauchy property** uniformly on a set E if for any $\varepsilon > 0$ there is an N such that $\|f_n - f_m\|_E < \varepsilon$ whenever $n \geq N$ and $m \geq N$ (see Sec. 1-6). As before, this is often written $\lim_{n, m \to \infty} \|f_n - f_m\|_E = 0$. Any uniformly convergent sequence has the Cauchy property; for, if $\{f_n\} \to F$ uniformly on E, then for any point $p \in E$,

$$\begin{aligned} |f_n(p) - f_m(p)| &= |f_n(p) - F(p) + F(p) - f_m(p)| \\ &\leq \|f_n - F\|_E + \|F - f_m\|_E \end{aligned}$$

so that $\lim_{n, m \to \infty} \|f_n - f_m\|_E = 0$. The converse also holds.

Theorem 1 *If* $\lim_{n, m \to \infty} \|f_n - f_m\|_E = 0$, *then there is a function* F *to which the sequence* $\{f_n\}$ *converges uniformly on* E.

Since $|f_n(p) - f_m(p)| \leq \|f_n - f_m\|_E$ for each point $p \in E$, $\{f_n(p)\}$ is a Cauchy sequence of numbers and is therefore convergent. Define F by $F(p) = \lim_{n \to \infty} f_n(p)$. F is then the pointwise limit of f_n on E. To show that the convergence is actually uniform, take any $p \in E$ and write

$$\begin{aligned} |F(p) - f_n(p)| &= |F(p) - f_k(p) + f_k(p) - f_n(p)| \\ &\leq |F(p) - f_k(p)| + |f_k(p) - f_n(p)| \\ &\leq |F(p) - f_k(p)| + \|f_k - f_n\|_E \end{aligned}$$

Given $\varepsilon > 0$, choose N so that $\|f_k - f_n\|_E < \varepsilon$ whenever $n \geq N$, $k \geq N$. For each $p \in E$, $\lim_{k \to \infty} f_k(p) = F(p)$; we may then choose k larger than N and dependent upon p and ε, so that $|F(p) - f_k(p)| < \varepsilon$. Making this choice of k in the inequality above, we have

$$|F(p) - f_n(p)| < \varepsilon + \varepsilon = 2\varepsilon$$

holding now for each $n \geq N$ and each point $p \in E$. Hence,

$$\|F - f_n\|_E < 2\varepsilon$$

for all $n \geq N$, and $\{f_n\}$ is uniformly convergent to F on E. ▌

In the language of Sec. 1.7, the space of functions is **complete** with respect to uniform convergence; for series, the corresponding statement is the following.

Corollary *If* $\lim_{n, m \to \infty} \left\|\sum_n^m u_k\right\|_E = 0$, *then* $\sum u_k$ *is uniformly convergent on* E.

The simplest and most useful test for uniform convergence is the following comparison test, also called the M test.

Theorem 2 (Weierstrass Comparison Test) *If* $\|u_k\|_E \le M_k$ *for all* k, *and* $\sum_1^\infty M_k$ *converges, then* $\sum_1^\infty u_k$ *converges uniformly on* E.

For any $p \in E$, $|u_k(p)| \le M_k$, so that by the simple comparison test (Theorem 5, Sec. 5.2), $\sum_1^\infty u_k(p)$ converges pointwise on E. Estimating the tail of this series, uniformly, we have $\left|\sum_n^\infty u_k(p)\right| \le \sum_n^\infty M_k$ for all $p \in E$, and thus $\lim_{n\to\infty} \left\|\sum_n^\infty u_k\right\|_E \le \lim_{n\to\infty} \sum_n^\infty M_k = 0$, proving uniform convergence of $\sum u_k$ on E. ▮

As an illustration, $\sum_1^\infty \sin(nx)/n^2$ converges uniformly for all x, $-\infty < x < \infty$, since $|\sin(nx)/n^2| \le 1/n^2$ and $\sum 1/n^2$ converges.

We take up next a very important property of uniform convergence.

Theorem 3 *If* $\{f_n\}$ *converges to* F, *uniformly on* E, *and each function* f_n *is continuous on* E, *then* F *is continuous on* E.

We prove that F is continuous at an arbitrary point $p_0 \in E$. Given $\varepsilon > 0$, first choose N so that $\|F - f_N\|_E < \varepsilon$. For any point p in E we may write

$$\begin{aligned}
|F(p) - F(p_0)| &= |F(p) - f_N(p) + f_N(p) - f_N(p_0) + f_N(p_0) - F(p_0)| \\
&\le |F(p) - f_N(p)| + |f_N(p) - f_N(p_0)| + |f_N(p_0) - F(p_0)| \\
&\le 2\|F - f_N\|_E + |f_N(p) - f_N(p_0)| \\
&\le 2\varepsilon + |f_N(p) - f_N(p_0)|
\end{aligned}$$

Since f_N is continuous on E, we may now choose δ so that $|f_N(p) - f_N(p_0)| < \varepsilon$ whenever $p \in E$ and $|p - p_0| < \delta$. We thus obtain

$$|F(p) - F(p_0)| < 3\varepsilon$$

whenever $p \in E$ and $|p - p_0| < \delta$, proving that F is continuous at p_0. ▮

Stated for series, this becomes:

Corollary *If each of the functions* u_n *is continuous on a set* E, *and* $\sum u_n$ *converges to* F *uniformly on* E, *then* F *is continuous on* E.

Theorem 3 and its corollary can also be described in topological terms. Let $\mathscr{C}$ be the class of functions f that are continuous on E. Then, Theorem 3 states that the set $\mathscr{C}$ is a closed set in the space of all functions on E.

The corollary also has another interpretation that is related to the examples with which we started this section, for an equivalent statement is

$$\lim_{p \to p_0} \sum_1^\infty u_n(p) = \sum_1^\infty u_n(p_0)$$

and this has the same meaning as (6-8), dealing with the interchange of limit operations.

Returning to the examples which opened this section, the series in (6-1) was a series of continuous functions, but its sum is the discontinuous function described by $F(x) = 1, 0 < x \le 1$, $F(0) = 0$. The series is therefore not uniformly convergent on [0, 1]. In fact, it cannot be uniformly convergent on the open interval $0 < x < 1$, even though F is continuous there. This is shown by the following general theorem.

Theorem 4 *Let E be the closure of an open set. Let $\{f_n\}$ converge uniformly in the interior of E. Suppose that each function f_n is continuous on E. Then $\{f_n\}$ is uniformly convergent on E.*

Given $\varepsilon > 0$, choose N so that $|f_n(p) - f_m(p)| < \varepsilon$ whenever $n \ge N$, $m \ge N$, and p is an *interior* point of E. Since f_n and f_m are both continuous on E, so is $\phi(p) = |f_n(p) - f_m(p)|$, and since ϕ is bounded by ε on the interior of E, it is bounded by ε on all of E. This shows that $\|f_n - f_m\|_E \le \varepsilon$ for all n and m with $n \ge N$, $m \ge N$, and $\{f_n\}$ converges uniformly on E, by Theorem 1. ▌

A useful application of this to series of continuous functions is:

Corollary *Let $\sum_1^\infty u_n(x)$ converge to $F(x)$, uniformly for all x with $c \le x < \infty$. Let $\lim_{x \uparrow \infty} u_n(x) = b_n < \infty$ for $n = 1, 2, \ldots$. Then, $\sum_1^\infty b_n$ converges, and $\lim_{x \uparrow \infty} F(x) = \sum_1^\infty b_n$.*

Setting $x = 1/t$, we obtain a series of functions which converges uniformly for $0 < t \le a$. Since $\lim_{t \to 0} u_n(1/t) = b_n$, we can define the terms so as to be continuous at $t = 0$. Applying the theorem, the series is uniformly convergent for $0 \le t \le a$, and we can evaluate

$$\lim_{t \downarrow 0} F\left(\frac{1}{t}\right) = \lim_{x \uparrow \infty} F(x)$$

termwise. ▌

Returning to Example (6-3), the series $F(x) = \sum_1^\infty nx^2/(n^3 + x^3)$ converges uniformly on each of the intervals $[0, R]$, since $|nx^2/(n^3 + x^3)| \le nR^2/n^3 = R^2/n^2$ and $\sum_1^\infty 1/n^2$ converges. However, it cannot converge *uniformly* on the whole unbounded interval $0 \le x < \infty$, since $\lim_{x \uparrow \infty} F(x) = \infty$, although $\lim_{x \uparrow \infty} nx^2/(n^3 + x^3) = 0$ for each n.

The next theorem is the fundamental result dealing with integration of uniformly convergent series or sequences. We state it in a two-dimensional form.

Theorem 5 *Let D be a closed bounded set in the plane which has area, and let the functions f_n be continuous on D. Then, if $\{f_n\}$ converges to F uniformly on D, $\lim_{n\to\infty} \iint_D f_n = \iint_D F$.*

$\iint_D F$ exists since F is continuous on D. For any n, we have

$$\left| \iint_D F - \iint_D f_n \right| = \left| \iint_D (F - f_n) \right| \le \iint_D |F - f_n| \le \|F - f_n\|_D \, A(D)$$

since $\|F - f_n\|_D$ is the maximum of $|F(p) - f_n(p)|$ for $p \in D$. Since $\lim_{n\to\infty} \|F - f_n\|_D = 0$, $\lim_{n\to\infty} \iint_D f_n = \iint_D F$. ▮

If $\{f_n\}$ is the sequence of partial sums of a series $\sum_1^\infty u_k$, then

$$\iint_D f_n = \iint_D (u_1 + u_2 + \cdots + u_n) = \iint_D u_1 + \iint_D u_2 + \cdots + \iint_D u_n$$

leading to the following corollary.

Corollary *If each of the functions u_n is continuous on D and $\sum_1^\infty u_n$ converges to F uniformly on D, then $\iint_D F = \sum_1^\infty \iint_D u_n$.*

This is usually abbreviated to the statement that a uniformly convergent series may be integrated termwise. As an illustration, consider the series $\sum_0^\infty (-t)^n$. This converges to $1/(1 + t)$ uniformly on any interval $-r \le t \le r$, for $r < 1$. Integrating the series termwise between 0 and x, $|x| < 1$,

$$\int_0^x \frac{dt}{1+t} = \log(1 + x) = \sum_0^\infty (-1)^n \int_0^x t^n \, dt = \sum_0^\infty (-1)^n \frac{x^{n+1}}{n+1}$$

Thus, for any x with $|x| < 1$,

$$\log(1 + x) = x - \frac{x^2}{2} + \frac{x^3}{3} - \cdots + (-1)^{n-1} \frac{x^n}{n} + \cdots$$

We can also extend this to the endpoint $x = 1$ by a special argument. If $x > 0$, the series $\sum_1^\infty (-1)^{n-1}(x^n/n)$ is an alternating series and converges for $0 \le x \le 1$. Using the fact (see Exercise 4, Sec. 5.3) that the partial sums of an alternating series constantly approach the sum and alternatively lie above and below, so that the error at any stage does not exceed the next term, we may write

$$\left|\log(1+x) - \sum_1^n (-1)^{k-1}\frac{x^k}{k}\right| \le \frac{x^{n+1}}{n+1} \le \frac{1}{n+1}$$

for each x with $0 \le x < 1$. However, since all the functions involved in this are continuous at $x = 1$, the same inequality holds when $x = 1$. This shows that the series is uniformly convergent on the *closed* interval $[0, 1]$, and termwise integration is valid for all x with $0 \le x \le 1$. In particular, setting $x = 1$, we have

$$\log 2 = 1 - \tfrac{1}{2} + \tfrac{1}{3} - \tfrac{1}{4} + \tfrac{1}{5} - \tfrac{1}{6} + \cdots$$

The functions $f_n(x) = n^2xe^{-nx}$ of Example (6-7) converge to 0 pointwise on $[0, 1]$ but do not converge uniformly on $[0, 1]$, and their integrals $\int_0^1 f_n$ converge to 1, not 0. Examining their graphs (Fig. 6-1), we see that f_n has a peak near the origin which becomes narrower but higher as n increases, leaving the total area underneath the curve about constant.

For contrast, consider the function $g_n(x) = nxe^{-nx}$. These functions likewise converge to 0 pointwise, but not uniformly, on $[0, 1]$. However, it is easily seen from the earlier calculations with (6-7) that $\int_0^1 g_n \to 0$. The difference is that this time, the functions g_n are uniformly bounded on $[0, 1]$. None of the peaks of the functions g_n reaches higher than $1/e$.

This example is an instance of a general convergence theorem for bounded functions. While it is very useful, its proof requires techniques that go beyond this text, involving portions of the theory of Lebesgue measure. We therefore state the result here without proof, but include a sketch of the argument in Appendix 6.

Theorem 6 *If the functions f_n and F are integrable on a bounded closed set E, and $\{f_n\} \to F$ pointwise on E, and if $\|f_n\|_E \le M$ for some M and all $n = 1, 2, \ldots$, then $\lim_{n\to\infty} \int_E f_n = \int_E F$.*

This may often be used when uniform convergence does not occur. Let g be continuous (and therefore bounded) on $[-1, 1]$. Then, the sequence $\{f_n\}$ with $f_n(x) = e^{-nx^2}g(x)$ is uniformly bounded on $[-1, 1]$ and converges pointwise to the function F

$$F(x) = \begin{cases} 0 & x \ne 0 \\ g(0) & x = 0 \end{cases}$$

This is not continuous if $g(0) \neq 0$, so that $\{f_n\}$ does not in general converge uniformly. However, Theorem 6 applies, so that

$$\lim_{n\to\infty} \int_{-1}^{1} \exp(-nx^2)g(x)\,dx = 0$$

As we have seen in illustrative example (6-6), the process of differentiation of series is not well behaved. There we saw that a series such as $\sum_1^\infty n^{-2} \sin(n^2 x)$ could be uniformly convergent everywhere, and yet not allow termwise differentiation. Termwise integration is much better. This can be used to obtain a valid result for differentiation, showing that termwise differentiation is justified if the resulting series of derivatives is itself uniformly convergent.

Theorem 7 *Let $\sum_1^\infty u_n(x)$ converge to $F(x)$ for each x in $[a, b]$. Let $u_n'(x)$ exist and be continuous for $a \le x \le b$, and let $\sum u_n'(x)$ converge uniformly on $[a, b]$. Then, $\sum_1^\infty u_n'(x) = F'(x)$.*

Setting $g(x) = \sum u_n'(x)$, integrate termwise between a and x, so that

$$\int_a^x g = \sum_1^\infty \int_a^x u_n' = \sum_1^\infty [u_n(x) - u_n(a)]$$

$$= F(x) - F(a)$$

This shows that F is an antiderivative (indefinite integral) of g, so that $F'(x) = g(x)$ for all x in $[a, b]$. ▮

As an illustration of this, consider the series $F(x) = \sum_1^\infty e^{-n^2 x}$, which converges for all $x > 0$. We shall show that F is continuous and of class C^∞ on this interval, that is, $F^{(k)}(x)$ exists for all $x > 0$. The termwise derivative of the series is $(-1)\sum_1^\infty n^2 \exp(-n^2 x)$. If $\delta > 0$, then for all $x \ge \delta$, $|n^2 \exp(-n^2 x)| \le n^2 \exp(-n^2 \delta)$, so that the derived series is uniformly convergent for all x, $\delta \le x < \infty$. This shows that $F'(x)$ exists and is given by $(-1)\sum_1^\infty n^2 \exp(-n^2 x)$ for all $x > 0$. Repetition of this process leads to

$$F^{(k)}(x) = (-1)^k \sum_1^\infty n^{2k} \exp(-n^2 x) \qquad k = 1, 2, \ldots$$

where the series is uniformly convergent for $\delta \le x < \infty$ and any $\delta > 0$.

We conclude this section with a number of special examples.

Let

$$F(x) = \sum_1^\infty x/n(x + n) \tag{6-9}$$

Since $|x/n(x+n)| \le 1/n^2$ for all x in $[0, 1]$, this series is uniformly convergent there, and may be integrated termwise. The resulting series,

$$\sum_1^\infty \int_0^1 \frac{x\,dx}{n(x+n)},$$

must converge. Denote its sum by γ, so that

$$\begin{aligned}\gamma &= \sum_1^\infty \int_0^1 \left\{\frac{1}{n} - \frac{1}{x+n}\right\} dx \\ &= \sum_1^\infty \left\{\frac{1}{n} - \log\frac{n+1}{n}\right\} \\ &= \lim_{N\to\infty} \left\{\sum_1^N \frac{1}{n} - \log(N+1)\right\}\end{aligned}$$

Since $\lim_{N\to\infty} [\log(N+1) - \log N] = 0$, we have thus shown that there is a positive number γ such that

$$\sum_1^N \frac{1}{n} = \log N + \gamma + \sigma_N$$

where $\lim \sigma_N = 0$. The number γ is called **Euler's constant,** and is approximately .577 21 $\cdots$.

Let

$$\text{(6-10)} \qquad F(x) = \sum_1^\infty 1/(1+n^2x^2) \qquad \text{converging for all } x \ne 0$$

For any $\delta > 0$, and $x \ge \delta$, we see that $(1+n^2x^2)^{-1} \le 1/n^2\delta^2$, so that the series is uniformly convergent for all x, $x \ge \delta$. Appealing to the corollary to Theorem 4, $\lim_{x\uparrow\infty} F(x) = 0$. How does F behave near the origin? When $x = 0$, the series becomes $1 + 1 + 1 + \cdots$, which suggests that

$$\lim_{x\downarrow 0} F(x) = \infty.$$

This conjecture is correct. For any N,

$$F(x) \ge \sum_1^N \frac{1}{1+n^2x^2} = g(x) \qquad \text{and} \qquad g(0) = N$$

so that

$$\liminf_{x\downarrow 0} F(x) \ge N$$

and letting N increase, we see that $\lim_{x\downarrow 0} F(x) = \infty$. This approach cannot always be used, and depends upon the fact that the terms of the present series are positive. (For comparison, recall illustrative example (6-5), where we had a series whose terms also approached $1 + 1 + 1 + \cdots$, but for which this did not mean that the function in question approached infinity.)

Continuing our study of the function F, consider now

$$x^2F(x) = \sum_1^\infty \frac{x^2}{1 + n^2x^2}$$

Since $x^2/(1 + n^2x^2) \le 1/n^2$ for all x, $-\infty < x < \infty$, this series is uniformly convergent on the whole axis. In particular, $\lim_{x\to 0} x^2F(x) = \sum 0 = 0$, and

$$\lim_{x\uparrow\infty} x^2F(x) = \sum_1^\infty \lim_{x\uparrow\infty} \frac{x^2}{1 + n^2x^2} = \sum_1^\infty \frac{1}{n^2}$$

Finally, consider $xF(x) = \sum_1^\infty x/(1 + n^2x^2)$. This converges uniformly on the intervals $\delta \le x < \infty$, for any $\delta > 0$. Does it converge uniformly for $0 \le x \le \delta$? If so, then $\lim_{x\downarrow 0} xF(x)$ would have to be 0; we shall show that instead, $\lim_{x\downarrow 0} xF(x) = \pi/2$. For any n,

$$\frac{x}{1 + (n + 1)^2x^2} \le \int_n^{n+1} \frac{x}{1 + t^2x^2}\, dt \le \frac{x}{1 + n^2x^2}$$

and adding,

$$\sum_{n=0}^\infty \frac{x}{1 + (n + 1)^2x^2} \le \int_0^\infty \frac{x}{1 + t^2x^2}\, dt \le \sum_{n=0}^\infty \frac{x}{1 + n^2x^2}$$

or

$$xF(x) \le \int_0^\infty \frac{x}{1 + x^2t^2}\, dt \le x + xF(x)$$

However,

$$\int_0^\infty \frac{x\, dt}{1 + x^2t^2} = \int_0^\infty \frac{du}{1 + u^2} = \frac{1}{2}\pi$$

so that $\frac{1}{2}\pi - x \le xF(x) \le \frac{1}{2}\pi$, for all $x > 0$. Letting x approach 0, we have $\lim_{x\downarrow 0} xF(x) = \frac{1}{2}\pi$. Summarizing what we have found, the function F behaves near the origin like $\pi/(2x)$, and for large x like Ax^{-2}, where

$$A = \sum 1/n^2 \qquad (= \pi^2/6).$$

Theorem 3, on the continuity of the uniformly convergent sum of a series of continuous functions, is often used to construct continuous functions that have strange and unlikely properties. For example, $(1 + x^2)^{-1}x^2 \sin(1/x) = g(x)$ is a continuous function that is differentiable everywhere, even at $x = 0$, where $g'(0) = 0$. Note also that $|g(x)| \le 1$ for all x. The function g is not monotonic on any neighborhood of $x = 0$ because of the rapid oscillations of the sine function near there. Displacing g by forming $g_c(x) = g(x - c)$, we have a function that has the same sort of behavior at $x = c$. If c_n is any

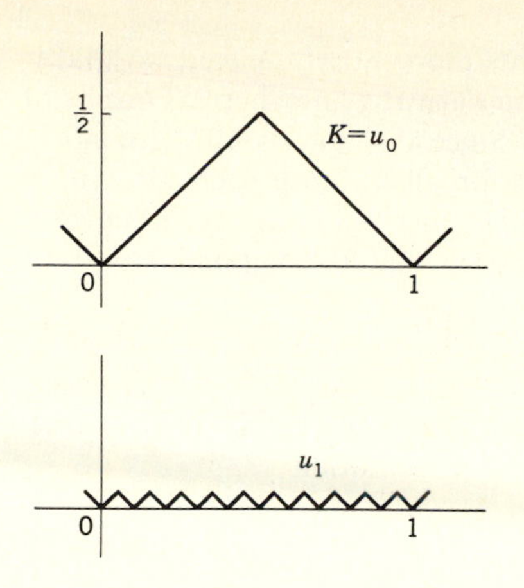

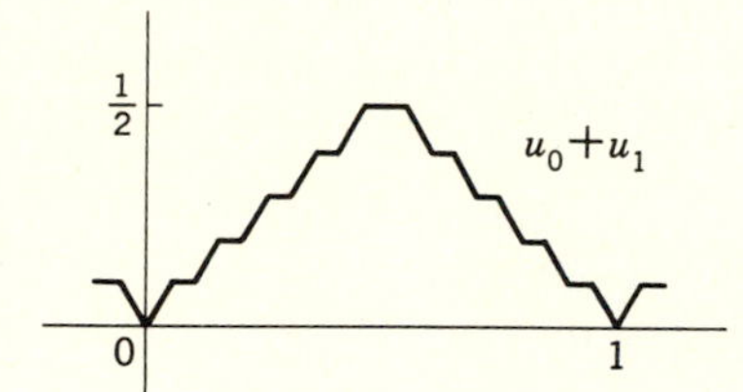

Figure 6-2

sequence of real numbers, then

$$G(x) = \sum_{n=1}^{\infty} \frac{1}{2^n} g_{c_n}(x) = \sum_{n=1}^{\infty} \frac{1}{2^n} g(x - c_n)$$

is a continuous function that is not monotonic on any neighborhood of the points $c_1, c_2, \ldots$. These points c_n can be everywhere dense on the line.

The same technique makes it possible to construct a simple example of a nowhere differentiable continuous function. Let K be the special function defined by saying that $K(x)$ is the distance from x to the nearest integer; K is continuous everywhere, and periodic with period 1 (see Fig. 6-2). We note that K also has the property that

$$|K(b) - K(a)| = |b - a|$$

whenever a and b both lie in the same half of any interval $[m, m + 1]$. Set $u_n(x) = 10^{-n}K(10^n x)$ for $n = 0, 1, \ldots$, and then define the sought-for function $H(x)$ for all x by

$$H(x) = \sum_{0}^{\infty} u_j(x) \tag{6-11}$$

The function K has corners where the derivative fails to exist, and the function u_n inherit a similar behavior at points more closely spaced, so that it is intuitively plausible that $H(x)$ might have a derivative nowhere. (Graphs of u_0, u_1, and $u_0 + u_1$ are shown in Fig. 6-2.) Since $0 \le u_j(x) \le 10^{-j}$ for all x, the series for $H(x)$ is uniformly convergent for all x. Since each term u_j is everywhere continuous and satisfies the relation $u_j(x+1) = u_j(x)$, H is everywhere continuous, and has period 1. We shall show that for any point b in $[0, 1]$,

$$H'(b) = \lim_{x \to b} \frac{H(x) - H(b)}{x - b}$$

fails to exist. We choose a special sequence $\{x_n\}$ approaching b for which this is easy to prove. Let b have the decimal representation

$$.b_1 b_2 b_3 \cdots = \sum_1^\infty \frac{b_j}{10^j}$$

To achieve uniqueness, we adopt the convention of using terminations $\cdots 0000000 \cdots$ rather than $\cdots 999999 \cdots$, so that, for example, we write $.241\,000\,00 \cdots$ in place of the equivalent $.240\,999\,999 \cdots$. Given any positive integer $n = 1, 2, \ldots$, we define a real number x_n near b by

$$x_n = \begin{cases} b + 10^{-n} & \text{if } b_n \text{ is different from 4 and 9} \\ b - 10^{-n} & \text{if } b_n \text{ is either 4 or 9} \end{cases}$$

So chosen, the pairs x_n and b, $10x_n$ and $10b$, $\ldots$, $10^{n-1}x_n$ and $10^{n-1}b$, will always lie in the same half of any interval $[m, m+1]$ which contains one of the pair. Moreover, for $j \ge n$, $10^j x_n$ and $10^j b$ will differ by some integer. The special properties of $K(x)$ then show that

$$|u_j(x_n) - u_j(b)| = \begin{cases} 10^{-n} = |x_n - b| & \text{for } j = 0, 1, \ldots, n-1 \\ 0 & \text{for } j \ge n \end{cases}$$

and adding,
$$\frac{H(x_n) - H(b)}{x_n - b} = \overbrace{\pm 1 \pm 1 \cdots \pm 1 \pm 1}^{n \text{ terms}}$$

where the signs are determined by the particular digits in the decimal representation of b. However, regardless of the signs, we see that this quotient is an *even* integer for $n = 2, 4, 6, 8, \ldots$ and an *odd* integer for $n = 1, 3, 5, \ldots$. This shows that $\lim_{n\to\infty} (H(x_n) - H(b))/(x_n - b)$ does not exist, and that H does not have a derivative at b.

Our final illustration of the power of these techniques is directed not toward the construction of strange functions, but rather toward showing a very useful extension property for continuous functions defined on closed sets.

Theorem 8 (Tietze Extension) *Let E be a closed set in n space, and let f be a function that is continuous on E and obeys $|f(p)| \le M$ for all $p \in E$. Then, there is a continuous function F, defined on all of n space and bounded there by M, such that F coincides with f on E.*

By virtue of this result, any bounded continuous function on a closed set may always be regarded as bounded and continuous on all of n space.

The proof of this result depends on the use of a special auxiliary function which is defined geometrically.

Lemma *If C is a closed set, and $\phi(p) = d(p, C)$, the distance from p to C, then ϕ is everywhere continuous, and strictly positive off C.*

Let p, q, and c be three points with $c \in C$. The triangle property for distances shows that $|p - c| \leq |p - q| + |q - c|$. Since

$$d(p, C) = \inf_{c \in C} |p - c|$$

we have $\phi(p) \leq |p - q| + |q - c|$. This holds in turn for every $c \in C$, so that $\phi(p) \leq |p - q| + \phi(q)$. By symmetry, $\phi(q) \leq |q - p| + \phi(p)$, so that, putting these together, $|\phi(p) - \phi(q)| \leq |p - q|$. This shows that ϕ is everywhere (uniformly) continuous. If $\phi(p) = 0$, then

$$p = \lim c_n$$

for a sequence $\{c_n\}$ of points of C. Since C is closed, $p \in C$. ▮

To prove the theorem, we produce a series $\sum_1^\infty F_n$ of continuous functions which converges uniformly in the whole plane, and whose sum on the set E is f. Suppose that $|f(p)| \leq M$ for $p \in E$. Divide E into three sets

$$A = \left\{\text{all } p \in E \quad \text{where } \frac{M}{3} \leq f(p) \leq M\right\}$$

$$C = \left\{\text{all } p \in E \quad \text{where } \frac{-M}{3} < f(p) < \frac{M}{3}\right\}$$

$$B = \left\{\text{all } p \in E \quad \text{where } -M \leq f(p) \leq \frac{-M}{3}\right\}$$

and construct a function F_1 by the definition

$$F_1(p) = \left(\frac{M}{3}\right)\frac{d(p, B) - d(p, A)}{d(p, B) + d(p, A)}$$

Since A and B are disjoint closed sets, F_1 is everywhere defined and is everywhere continuous. For any point p in the plane, $|F_1(p)| \leq M/3$. On E, F_1 behaves as follows:

If $p \in A$, $d(p, A) = 0$ and $F_1(p) = M/3$.
If $p \in B$, $d(p, B) = 0$ and $F_1(p) = -M/3$.
If $p \in C$, then $-M/3 \leq F_1(p) \leq M/3$.

An examination of the values of f on E shows that

$$|f(p) - F_1(p)| \le \tfrac{2}{3}M$$

for all $p \in E$. Repeat this argument with $f - F_1$ playing the role of f. On E, $f - F_1$ is bounded by $(2M)/3$. We can therefore construct a function F_2 which is everywhere continuous and such that $|F_2(p)| \le (\tfrac{1}{3})(2M/3)$ for all p, while $|[f(p) - F_1(p)] - F_2(p)| \le (\tfrac{2}{3})(2M/3)$ for $p \in E$. Continuing this, we arrive at a sequence of continuous functions $\{F_n\}$ which obey the two conditions:

$$|F_n(p)| \le \tfrac{1}{3}(\tfrac{2}{3})^{n-1}M \qquad \text{all } p$$

$$|f(p) - \{F_1(p) + \cdots + F_n(p)\}| \le (\tfrac{2}{3})^n M \qquad \text{all } p \in E$$

The first condition assures us that the series $\sum_1^\infty F_n$ is uniformly convergent for all p. The sum F is then, by Theorem 3, continuous everywhere in the plane. The second condition shows that

$$F(p) = \sum_1^\infty F_n(p) = f(p)$$

for all $p \in E$. Finally, for any p

$$|F(p)| \le \sum_1^\infty |F_n(p)| \le \frac{M}{3}\left(1 + \frac{2}{3} + \left(\frac{2}{3}\right)^2 + \cdots\right) \le \frac{M}{3}\frac{1}{1 - (\frac{2}{3})} = M$$

so that F is bounded on the whole plane by the same bound that applied to f on E. ∎

EXERCISES

1 Show that if f_n converges pointwise to f on E, then $f_n \to f$ uniformly on every finite subset of E.

2 Exhibit a sequence $\{f_n\}$ which converges uniformly on every interval $[0, L]$ for every $L > 0$, but not uniformly on the interval $0 \le x < \infty$.

3 Let $f_n(x) = x^n$ for $0 \le x \le 1$. Does $\{f_n\}$ converge pointwise on $[0, 1]$? Does it converge uniformly on $[0, 1]$? Does it converge uniformly on $[0, \tfrac{1}{2}]$?

4 Let $f_n(x) = nx^n(1 - x)$ for $0 \le x \le 1$. Show that $\{f_n\}$ converges pointwise, but not uniformly, on $[0, 1]$. Does $\lim_{n\to\infty} \int_0^1 f_n = \int_0^1 \lim_{n\to\infty} f_n$?

5 Let $F(x) = \sum_1^\infty x^2/(x^2 + n^2)$. Study the uniform convergence of this series, and investigate the existence of $\lim_{x\downarrow 0} F(x)$, $\lim_{x\downarrow 0} F(x)/x$, $\lim_{x\downarrow 0} F(x)/x^2$, $\lim_{x\uparrow\infty} F(x)$, $\lim_{x\uparrow\infty} F(x)/x^2$, $\lim_{x\uparrow\infty} F(x)/x$.

6 Let f be continuous on the interval $0 \le x < \infty$, and let $\lim_{x\to\infty} f(x) = L$. What can you say about $\lim_{n\to\infty} \int_0^2 f(nx)\,dx$?

7 Let g be continuous on $[0, 1]$ with $g(1) = 0$. Show that $\{g(x)x^n\}$ converges uniformly for x in $[0, 1]$.

8 Prove the corollary to Theorem 4 without making the change of variable $x = 1/t$.

9 Extend Theorem 5 to improper integrals as follows: Suppose $\{f_n\}$ is a sequence of functions continuous on $0 \le x < \infty$ and such that $|f_n(x)| \le g(x)$ for all $x \ge 0$. Suppose that $\int_0^\infty g$ converges and that $f_n \to f$ where convergence is uniform on every interval $[0, L]$ for any $L > 0$. Prove $\int_0^\infty f_n \to \int_0^\infty f$.

***10** Let $\phi_n(x)$ be positive-valued and continuous for all x in $[-1, 1]$ with

$$\lim_{n\to\infty} \int_{-1}^{1} \phi_n = 1$$

Suppose further that $\{\phi_n\}$ converges to 0 uniformly on the intervals $[-1, -c]$ and $[c, 1]$ for any $c > 0$. Let g be any function which is continuous on $[-1, 1]$. Show that

$$\lim_{n\to\infty} \int_{-1}^{1} g(x)\phi_n(x)\, dx = g(0).$$

***11** Let f_n be continuous on a closed and bounded set E, for each n, and let $\{f_n\}$ converge pointwise to a continuous function F. Suppose that for any $p \in E$, the sequence $\{f_n(p)\}$ is an increasing sequence of real numbers. Prove that $\{f_n\}$ in fact converges uniformly on E. (*Hint:* For a given ε, consider the set

$$C_n = \{\text{all } p \in E \text{ with } F(p) - f_n(p) \ge \varepsilon\}$$

and apply the nested set property.)

12 Apply Exercise 11 to prove the following result: Let $\{u_n\}$ be a sequence of nonnegative continuous functions defined on the interval $[a, b]$, and suppose that the series $\sum_1^\infty u_n(x)$ converges pointwise to a continuous function $F(x)$. Then, termwise integration is allowed; $\int_a^b F(x)\, dx = \sum_1^\infty \int_a^b u_n(x)\, dx$.

13 Use the Tietze extension theorem to show that if C is a compact set in n space, $\mathcal{O}$ an open set containing C, and f a continuous real-valued function defined on C, then f has a continuous extension to all of n space such that f is 0 everywhere in the complement of $\mathcal{O}$.

14 Show that the special operator $\|\ \|_E$, defined in this section by

$$\|f\|_E = \sup_{x\in E} |f(x)|$$

has the following properties of the Euclidean norm in n space:

(*a*) $\|f + g\|_E \le \|f\|_E + \|g\|_E$.

(*b*) $\|f - g\|_E = 0$ if and only if $f = g$.

15 Interpreting $\|f - g\|_E$ as a distance between the functions f and g, show that $\frac{1}{2}(f + g)$ is equally far from f and from g.

16 If $\mathcal{M}$ is a collection of functions f, then the distance from a function F to the collection is defined to be

$$d(F, \mathcal{M}) = \inf_{f\in\mathcal{M}} \|F - f\|_E$$

Find the distance between F, where $F(x) = x^2$, and the set $\mathcal{M}$ of all linear functions of the form $f(x) = Ax$. Choose E as the set $[0, 1]$. (You will have found the best uniform approximation on this interval to the function F by functions in the class $\mathcal{M}$.)

6.3 POWER SERIES

The basic facts about pointwise convergence of power series were presented in Sec. 5.3. We now examine power series as functions defined by series whose terms are polynomials. We must therefore study the uniform convergence properties of power series.

Theorem 9 *If* $\sum_0^\infty a_n x^n$ *has radius of convergence R, then it is uniformly convergent on every compact subset of the open interval* $-R < x < R$.

Any such compact set lies in a closed interval $[-b, b]$, where $b < R$. The power series converges when $x = b$, and in fact converges absolutely. Thus, $\sum_0^\infty |a_n| b^n$ converges. If $|x| \le b$, then $|a_n x^n| \le |a_n| b^n$, so the Weierstrass test applies, and the series converges uniformly on $[-b, b]$. ▮

If this is combined with Theorem 7, it follows that a power series can always be differentiated termwise within its interval of convergence.

Theorem 10 *Let* $f(x) = \sum_0^\infty a_n x^n$ *converge for* $|x| < R$, *Then,* f' *exists, and* $f'(x) = \sum_1^\infty n a_n x^{n-1}$ *for all x with* $|x| < R$.

Consider $x \sum_1^\infty n a_n x^{n-1} = \sum_1^\infty n a_n x^n$. Since multiplication by x does not affect convergence properties, the differentiated series has radius of convergence $1/L$, where $L = \limsup_{n\to\infty} |na_n|^{1/n} = \limsup_{n\to\infty} |a_n|^{1/n}$, using the fact that $\lim_{n\to\infty} n^{1/n} = 1$. Thus, the original power series and its termwise derivative always have the same radius of convergence. The derived series is then uniformly convergent for $|x| \le b$, and any $b < R$, and, by Theorem 7, the differentiation was justified. ▮

Repeating this argument, we see that a function given by a convergent power series may be differentiated as many times as desired, and the derivatives computed termwise, within the interval of convergence.

Corollary 1 *If* $f(x) = \sum_0^\infty a_n(x - c)^n$, *convergent for some interval about c, then* $a_n = f^{(n)}(c)/n!$.

For we have

$$f^{(n)}(x) = n!\,a_n + 2 \cdot 3 \cdots (n+1)a_{n+1}(x - c) + 3 \cdot 4 \cdots (n+2)a_{n+2}(x - c)^2 + \cdots$$

in an interval about c, and setting $x = c$, we have $f^{(n)}(c) = n!\,a_n$. ▮

Corollary 2 *If* $\sum_0^\infty a_n(x-c)^n = \sum_0^\infty b_n(x-c)^n$ *for all* x *in a neighborhood of* c, *then* $a_n = b_n$ *for* $a = 0, 1, \ldots$.

If the common value is $f(x)$, then a_n and b_n are of necessity both given by $f^{(n)}(c)/n!$. ∎

It should be observed that this shows that a function that is analytic on a neighborhood of c can have only one power series expansion of the form $\sum a_n(x-c)^n$, and that this one is the Taylor series which is obtained as the limit of the Taylor polynomials. Conversely, any function which is given by a convergent power series is analytic within the interval of convergence.

Differentiation of a power series does not change the radius of convergence, but can destroy convergence at the endpoints. Consider the series

$$f(x) = \sum_1^\infty \frac{x^n}{n^2}$$

which has radius of convergence $R = 1$. This converges in fact for all x with $-1 \le x \le 1$. If we differentiate this, we obtain

$$f'(x) = \sum_1^\infty \frac{x^{n-1}}{n}$$

which converges only for $-1 \le x < 1$. If we again differentiate it, we obtain

$$f''(x) = \sum_2^\infty \frac{n-1}{n} x^{n-2}$$

which now converges only for $-1 < x < 1$. The first of the three series converges uniformly for all x with $-1 \le x \le 1$ since its terms are dominated there by $1/n^2$. Theorem 9 shows that the second series converges uniformly in intervals $|x| \le b$ with $b < 1$; since it is (pointwise) convergent on the larger interval $-1 \le x \le b$, it might be conjectured that it is *uniformly* convergent there as well. The truth of this follows from a general result, due to Abel.

Theorem 11 *Let* $\sum_0^\infty a_n x_n$ *have radius of convergence* R, *and let it also converge for* $x = R$ [*for* $x = -R$]. *Then, it is uniformly convergent on the interval* $0 \le x \le R$ [*the interval* $-R \le x \le 0$].

Without loss of generality, we may assume that $R = 1$, and that $\sum a_n x^n$ converges when $x = 1$. Put $B_n = \sum_n^\infty a_k$ so that $\lim_{n\to\infty} B_n = 0$.

Then, for any x, $0 \le x < 1$.

$$\begin{aligned}
\sum_{n}^{\infty} a_k x^k &= a_n x^n + a_{n+1} x^{n+1} + \cdots \\
&= (B_n - B_{n+1})x^n + (B_{n+1} - B_{n+2})x^{n+1} + \cdots \\
&= B_n x^n + B_{n+1}(x^{n+1} - x^n) + B_{n+2}(x^{n+2} - x^{n+1}) + \cdots \\
&= B_n x^n + (x - 1)x^n\{B_{n+1} + B_{n+2}x + \cdots\}
\end{aligned}$$

Given ε, choose N so that $|B_j| < \varepsilon$ whenever $j \ge N$. Then, for $0 \le x < 1$, and $n \ge N$,

$$\begin{aligned}
\left|\sum_{n}^{\infty} a_k x^k\right| &\le \varepsilon x^n + (1 - x)x^n\{\varepsilon + \varepsilon x + \varepsilon x^2 + \cdots\} \\
&\le \varepsilon x^n + \varepsilon x^n(1 - x)\{1 + x + x^2 + \cdots\} \\
&\le 2\varepsilon x^n < 2\varepsilon
\end{aligned}$$

This also holds when $x = 1$, since $|B_n| < \varepsilon < 2\varepsilon$. Thus, $\left|\sum_n^\infty a_k x^k\right| < 2\varepsilon$ uniformly for all x with $0 \le x \le 1$, and all $n \ge N$, proving uniform convergence. ▌

Corollary *If a power series converges at an endpoint of the interval of convergence, then the function defined by the power series is continuous at that endpoint: If* $\sum_0^\infty a_n r^n$ *converges, with* $r > 0$, *then* $\lim_{x \uparrow r} \sum_0^\infty a_n x^n = \sum_0^\infty a_n r^n$.

[We remark that while the results on pointwise convergence, and the general results on uniform convergence, apply equally to series of real-valued and series of complex-valued functions, the theorem of Abel (Theorem 11) on uniform convergence at endpoints does not generalize directly. Thus, a power series $\sum_0^\infty a_n z^n$ can converge for all complex numbers z with $|z| = R$, but need not converge uniformly in the closed disc $|z| \le R$; it will, however, converge uniformly in every closed polygon with a finite number of vertices inscribed in the disc.]

Since power series are always uniformly convergent in closed intervals which lie in the interval of convergence, they can be integrated termwise over any such interval. The process of integration, or differentiation, is often combined with algebraic operations and with substitution to obtain the power series expansion of special functions; although the coefficients of the expansion of a function f can always be obtained from the formula $a_n = f^{(n)}(c)/n!$, this is often a long and complicated task. The following examples will illustrate these remarks.

Let us start from the simple geometric series

$$\frac{1}{1-x} = 1 + x + x^2 + x^3 + \cdots \tag{6-12}$$

which converges for $-1 < x < 1$. Repeated differentiation of this yields

$$\frac{1}{(1-x)^2} = 1 + 2x + 3x^2 + 4x^3 + \cdots + (n+1)x^n + \cdots$$

$$\frac{2}{(1-x)^3} = 2 + 6x + 12x^2 + \cdots + (n+1)(n+2)x^n + \cdots$$

and in general

$$\frac{k!}{(1-x)^{k+1}} = k! + \frac{(k+1)!}{1!}x + \frac{(k+2)!}{2!}x^2 + \cdots$$

with convergence for $-1 < x < 1$. If we choose instead to integrate the first series, we obtain

$$-\log(1-x) = x + \frac{x^2}{2} + \frac{x^3}{3} + \cdots \tag{6-13}$$

convergent for $-1 < x < 1$. Since the series is also convergent for $x = -1$, the corollary to Theorem 11 shows that (6-13) holds also for $x = -1$, with uniform convergence for $-1 \le x \le b$, $b < 1$. In (6-13) replace x by $-x$, obtaining

$$\log(1+x) = x - \frac{x^2}{2} + \frac{x^3}{3} - \frac{x^4}{4} + \cdots \tag{6-14}$$

Divide by x, and integrate again on the interval $[0, x]$, $x \le 1$

$$\int_0^x \frac{\log(1+t)}{t}\,dt = x - \frac{x^2}{4} + \frac{x^3}{9} - \cdots (-)^{n+1}\frac{x^n}{n^2} + \cdots$$

In the first series (6-12), replace x by $-x^2$, getting

$$\frac{1}{1+x^2} = 1 - x^2 + x^4 - x^6 + x^8 - \cdots$$

convergent for $-1 < x < 1$. Integrate from 0 to x, $|x| < 1$, obtaining

$$\arctan x = x - \frac{x^3}{3} + \frac{x^5}{5} - \cdots \tag{6-15}$$

Since this converges for $x = 1$ and $x = -1$, it converges uniformly for $-1 \le x \le 1$. In particular, setting $x = 1$,

$$\frac{\pi}{4} = 1 - \frac{1}{3} + \frac{1}{5} - \frac{1}{7} + \cdots \tag{6-16}$$

where we have again used the corollary of Theorem 11.

Let us define the exponential function E by the power series

$$E(x) = \sum_{0}^{\infty} \frac{x^n}{n!} = 1 + x + \frac{x^2}{2!} + \cdots \tag{6-17}$$

Then, since this converges for all x, it converges uniformly in every interval $|x| \le R$, $R < \infty$. Differentiating, we obtain

$$E'(x) = \sum_{1}^{\infty} \frac{nx^{n-1}}{n!} = \sum_{1}^{\infty} \frac{x^{n-1}}{(n-1)!} = E(x)$$

This shows that the function E is a solution of the differential equation $y' = y$. The other properties of the exponential function can also be obtained from the series definition. To verify the relation

$$E(a)E(b) = E(a + b)$$

we form the Cauchy product of the series for $E(a)$ and $E(b)$, obtaining

$$\begin{aligned} E(a)E(b) &= \sum_{0}^{\infty} \frac{a^n}{n!} \sum_{0}^{\infty} \frac{b^n}{n!} = \sum_{n=0}^{\infty} \left\{ \frac{a^n b^0}{n!\,0!} + \frac{a^{n-1}b}{(n-1)!\,1!} + \cdots + \frac{a^0 b^n}{0!\,n!} \right\} \\ &= \sum_{n=0}^{\infty} \frac{1}{n!} \sum_{k=0}^{n} \frac{n!}{(n-k)!\,k!} a^{n-k} b^k \\ &= \sum_{0}^{\infty} \frac{1}{n!} (a + b)^n = E(a + b) \end{aligned}$$

With $e = E(1)$, $e^n = E(n)$, and we may define e^x for general real exponents as $E(x)$. Suppose we wish to expand e^x in a power series in powers of $x - c$. Write

$$\begin{aligned} e^x &= E(x - c + c) = E(x - c)E(c) \\ &= e^c \left\{ 1 + \frac{x - c}{1} + \frac{(x - c)^2}{2!} + \frac{(x - c)^3}{3!} + \cdots \right\} \end{aligned}$$

Replacing x by $-x^2$ in the series for E, we have

$$e^{-x^2} = 1 - x^2 + \frac{x^4}{2!} - \frac{x^6}{3!} + \cdots$$

convergent for all x. Integrating this, we have

$$\int_0^t e^{-x^2}\,dx = t - \frac{t^3}{3} + \frac{t^5}{10} - \frac{t^7}{42} + \cdots \tag{6-18}$$

We note that this expansion does not enable us to evaluate the probability integral $\int_0^\infty e^{-x^2}\,dx$ merely by letting t increase, since we cannot take the termwise limit of the right side of (6-18). [Recall that in Sec. 4-5 we used a method to show that this improper integral has the exact value $\sqrt{\pi}/2$.]

The series which defines the exponential function, (6-17), is also convergent if x is replaced by any complex number $z = x + iy$, and the identities derived by multiplying power series still hold. Thus, writing e^z for $E(z)$, we have

$$e^{x+iy} = e^x e^{iy} \tag{6-19}$$

Using (6-17), we have

$$e^{iy} = 1 - \frac{y^2}{2!} + \frac{y^4}{4!} + \cdots + i\left(y - \frac{y^3}{3!} + \frac{y^5}{5!} + - \cdots\right)$$

Adopting the definitions suggested in Exercise 5, Sec. 5.4, we may now write this relation as

$$e^{x+iy} = e^x \cos y + ie^x \sin y \tag{6-20}$$

which permits us the standard exponential definitions

$$\sin\theta = \frac{1}{2i}(e^{i\theta} - e^{-i\theta}) = \sum_0^\infty (-1)^n \frac{\theta^{2n+1}}{(2n+1)!} \tag{6-21}$$

$$\cos\theta = \frac{1}{2}(e^{i\theta} + e^{-i\theta}) = \sum_0^\infty (-1)^n \frac{\theta^{2n}}{(2n)!}$$

$$e^{i\theta} = \cos\theta + i\sin\theta$$

It is an instructive exercise to see how much of the standard trigonometric lore can be recovered from these formulas.

Division of power series is possible, but difficult to carry out except in relatively simple cases. The test is simple in theory. For example, one writes

$$\frac{\sum_0^\infty a_n x^n}{\sum_0^\infty b_n x^n} = \sum_0^\infty c_n x^n$$

exactly when there is a common open interval I of convergence for all the series, and when, on I, it is true that the product of $\sum c_n x^n$ and $\sum b_n x^n$ is $\sum a_n x^n$.

EXERCISES

1 Find the power series representations for the following functions which converge in some interval containing the indicated point.

(*a*) $\sin(x^2)$ near $x = 0$ (*b*) $1/x$ near $x = 1$

(*c*) $\log(1 + x^2)$ near $x = 0$ (*d*) $\cosh(x)$ near $x = 0$

2 By integration, differentiation, or any other valid operation, find the functions which are given by the following power series.

(*a*) $\sum_1^\infty n^2 x^n$ (*b*) $\sum_0^\infty \frac{x^n}{(2n)!}$

(*c*) $\sum_0^\infty \frac{x^{2n+1}}{2n+1}$ (*d*) $x + x^4 + x^7 + x^{10} + x^{13} + \cdots$

3 Can the following functions be expressed as power series in x which converge in a neighborhood of 0?

(*a*) $f(x) = |x|$ (*b*) $f(x) = \cos\sqrt{x}$

4 The Bessel function of zero order may be defined by

$$J_0(x) = \sum_0^\infty \frac{(-1)^n x^{2n}}{4^n (n!)^2}$$

Find its radius of convergence, and show that $y = J_0(x)$ is a solution of the differential equation: $xy'' + y' + xy = 0$.

***5** Let $\lim_{n\to\infty} a_n = L$ and $f(x) = \sum_0^\infty a_n x^n$. Show that $\lim_{x\uparrow 1} (1 - x)f(x) = L$.

6 Find power series expansions for the following functions about the indicated points:

(*a*) $f(x) = xe^{-x}$ near $x = 0$ (*b*) $f(x) = e^{2x} - e^x$ near $x = 0$

(*c*) $f(x) = \frac{1}{1 + 2x}$ near $x = 0$ (*d*) $f(x) = \frac{1}{1 - 2x}$ near $x = 1$

(*e*) $f(x) = \frac{1}{1 - x^2}$ near $x = 0$

7 Show that $(\sin x)^2 + (\cos x)^2 = 1$ by showing directly that $|e^{i\theta}| = 1$.

8 From the fact that $e^{i\alpha}e^{i\beta} = e^{i(\alpha+\beta)}$, deduce the addition formulas for $\sin(\alpha + \beta)$ and $\cos(\alpha + \beta)$.

9 Show that $F(x) = 1 + x^3/3! + x^6/6! + x^9/9! + \cdots$ can be expressed in terms of e^x and the number $-\frac{1}{2} + i\sqrt{3}/2$.

10 Let $F(x) = 1 + 2x + x^2 + 2x^3 + x^4 + 2x^5 + \cdots$. By division, find a power series expansion near $x = 0$ for $1/F(x)$.

11 Find a power series for $1/F(x)$ where $F(x) = \sum_0^\infty (n + 1)x^n$.

12 If $f(x) = a_1 x + a_2 x^2 + \cdots$, then $1/f(x) = 1 - a_1 x + \{(a_1)^2 - a_2\}x^2 + \{2a_1 a_2 - (a_1)^3 - a_3\}x^3 + Ax^4 + \cdots$. Find A.

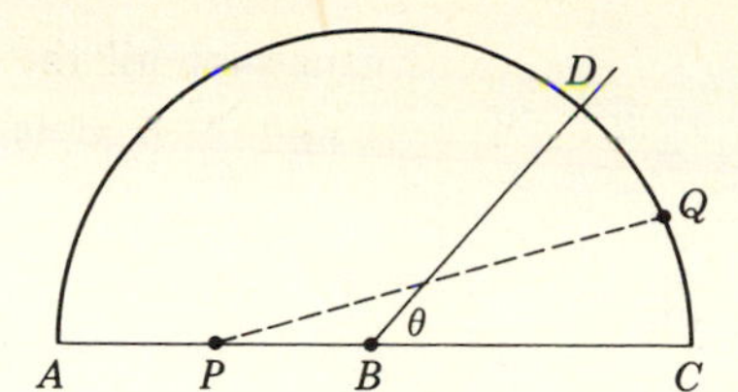

Figure 6-3

13 There is an approximate angle trisection method due to d'Ocagne. Given an angle θ in a unit semicircle (see Fig. 6-3), let P be the midpoint of the segment AB and Q the midpoint of the arc CD. Show that angle $QPC \approx \theta/3$.

14 Let $y = f(x)$ be a solution of the differential equation

$$x^2 \frac{dy}{dx} - xy = \sin x$$

obeying $f(0) = C$. Find a power series expansion for f near $x = 0$.

15 Let $y = f(x)$ be a solution of the differential equation

$$\frac{dy}{dx} = x^2 + y^2$$

By differentiating this equation repeatedly, find $f^{(n)}(0)$ for $n = 0, 1, \ldots, 5$, and thus find the first few terms of a power series for f about the point $x = 0$. Can you decide whether this series converges?

6.4 IMPROPER INTEGRALS WITH PARAMETERS

The notion of uniform convergence is not limited to sequences and series of functions. For example, instead of a sequence $\{f_n(p)\}$, we may consider $f(p, t)$ and study the behavior of this as $t \to t_0$ or $t \to \infty$. If $\lim_{t \to t_0} f(p, t) = F(p)$ for each individual choice of p in E, then one says that the limit is "pointwise in E." Based on the analogy with sequences, we choose the following definition for uniform convergence.

Definition 3 $\lim_{t \to t_0} f(p, t) = F(p)$ *uniformly for* $p \in E$ *if, given* $\varepsilon > 0$, *there is a deleted neighborhood* $\mathcal{N}$ *of* t_0 *such that* $|F(p) - f(p, t)| < \varepsilon$ *for all* t *in* $\mathcal{N}$ *and all* $p \in E$.

As an illustration, take $f(x, t) = \sin(xt)/t(1 + x^2)$. Then, letting $t \downarrow 0$, and evaluating the (pointwise) limit by L'Hospital's rule, we have

$$\lim_{t \downarrow 0} f(x, t) = \frac{x}{1 + x^2}$$

for all x, $-\infty < x < \infty$. Let us prove the convergence is uniform. For any

x and any $t \neq 0 = t_0$, we have

$$|f(x, t) - F(x)| = \left|\frac{\sin xt}{t(1 + x^2)} - \frac{x}{1 + x^2}\right|$$
$$= \frac{|x|}{1 + x^2}\left|\frac{\sin xt}{xt} - 1\right|$$

Since $\lim_{\theta \to 0} (\sin \theta)/\theta = 1$, we can choose β so that $|(\sin \theta)/\theta - 1| < \varepsilon$ whenever $|\theta| < \beta$. For any R let $\delta = \beta/R$. Then, if $|x| \leq R$ and $|t| < \delta$,

$$|xt| < R\left(\frac{\beta}{R}\right) = \beta$$

so that
$$|f(x, t) - F(x)| \leq R\left|\frac{\sin xt}{xt} - 1\right| \leq R\varepsilon$$

This shows that $\lim_{t \to 0} f(x, t) = x/(1 + x^2)$ uniformly on each of the intervals $[-R, R]$. To prove uniform convergence on the whole axis, we need an additional estimate. Since $|(\sin \theta)/\theta| \leq 1$ for all θ, we see that for any x and t,

$$|f(x, t) - F(x)| \leq \frac{2|x|}{1 + x^2}$$

Since $\lim_{|x| \to \infty} 2x/(1 + x^2) = 0$, we can choose R_0 so that for any t

$$|f(x, t) - F(x)| < \varepsilon$$

whenever $|x| \geq R_0$. Combining these inequalities, we have

$$|f(x, t) - F(x)| < \varepsilon$$

for all x and any t with $|t| < \beta/R_0$.

The continuous analog of an infinite series $\sum_1^\infty u_n(p)$ of functions is an improper integral $\int_c^\infty f(p, t)\, dt$ in which p is a parameter, and one speaks of this being pointwise convergent for $p \in E$ if it is a convergent improper integral for each individual choice of $p \in E$.

Definition 4 *The integral* $\int_c^\infty f(p, t)\, dt$ *converges to* $F(p)$, *uniformly for* $p \in E$ *if, given any* $\varepsilon > 0$, *there is an* r_0 *that depends on* ε *but not* p, *such that*

$$\left|F(p) - \int_c^r f(p, t)\, dt\right| < \varepsilon$$

for all $r > r_0$, *and all* $p \in E$.

Most of the theorems in Sec. 6.2 dealing with sequences and series have analogs for continuous limits and improper integrals. Some of the proofs follow the same pattern, but others have essential differences brought in by the contrast between finite sums and integrals. We give abbreviated proofs of the former.

Theorem 12 *Let* $\lim_{t\to t_0} f(p, t) = F(p)$, *uniformly for* $p \in E$, *and suppose that for each* t, $f(p, t)$ *is continuous for all* $p \in E$. *Then,* F *is continuous on* E.

(Refer to Theorem 3, Sec. 6.2.)

The next result is the analog of the corollary to Theorem 3.

Theorem 13 *Let* $\int_c^\infty f(p, t)\,dt$ *converge to* $F(p)$ *uniformly for all* p *in an open set* E, *and suppose that* $f(p, t)$ *is continuous for all* (p, t) *with* $p \in E$ *and* $t \geq c$. *Then* F *is continuous in* E.

The key to the truth of this is the fact that continuity is a local property, and that every point in E has a neighborhood that is compact. Given $p_0 \in E$ and $\varepsilon > 0$, choose $r = r(\varepsilon)$ so that $\left| F(p) - \int_c^r f(p, t)\,dt \right| < \varepsilon$ for all $p \in E$. Then, one arrives at

$$|F(p) - F(p_0)| < 2\varepsilon + \left| \int_c^r f(p, t)\,dt - \int_c^r f(p_0, t)\,dt \right|$$

However, $f(p, t)$ is uniformly continuous for all (p, t) with $|p - p_0| \leq \delta$, $c \leq t \leq r$, so that by Exercise 18, Sec. 4.3, $\lim_{p\to p_0} \int_c^r f(p, t)\,dt = \int_c^r f(p_0, t)\,dt$, and it follows that F is continuous at p_0. ▮

The simplest standard test of the uniform convergence of an improper integral with parameters is again the **Weierstrass comparison test.** The proof is similar to that of Theorem 2.

Theorem 14 *Suppose that* f *is continuous and obeys* $|f(p, t)| \leq g(t)$ *for all* $t \geq c$ *and all* $p \in E$. *Suppose that* $\int_c^\infty g(t)\,dt$ *converges. Then,* $\int_c^\infty f(p, t)\,dt$ *converges uniformly for all* $p \in E$.

We now examine the first of a number of illustrations designed to show that what seems plausible is not always true when dealing with improper integrals. Consider

$$F(x) = \int_1^\infty \frac{x^2}{1 + x^2t^2}\,dt \tag{6-22}$$

Since the integrand obeys $|f(x, t)| \leq 1/t^2$ for all x, and since $\int_1^\infty t^{-2}\,dt$ converges, we know that (6-22) converges uniformly for all x. We may therefore conclude by Theorem 13 that F is continuous for all x; in particular, since $F(0) = 0$, we know that the integral in (6-22) is such that $\lim_{x\to 0} F(x) = 0$.

Return to (6-22), and observe that

$$\lim_{x\to\infty} \frac{x^2}{1 + x^2t^2} = \lim_{x\to\infty} \frac{1}{x^{-2} + t^2} = \frac{1}{t^2}$$

Can we also conclude immediately that

$$\lim_{x\to\infty} F(x) = \int_1^\infty t^{-2}\,dt = 1$$

That is, can we also evaluate the limit "at infinity" by simply taking the limit inside the integral? In general, the answer must be "no." (However, see Exercise 3.)

What we are doing here is extending a statement from the set E on which the uniformity was established to a boundary point of E—since ∞ plays this role for the unbounded interval $0 \leq x < \infty$. (If you are uncomfortable with this, swap ∞ for 0 by the substitution $s = 1/x$.)

This is a situation where the analogy between series and integrals is not perfect. What we are asking for is a direct analogy for the result stated in the corollary to Theorem 4; the analogy for the theorem *does* hold, but that for the corollary fails. Here is an instance of this.

Consider the integral

$$F(x) = \int_0^\infty x^2te^{-xt}\,dt \qquad x \geq 1 \tag{6-23}$$

For any $t \geq 0$, $\lim_{x\to\infty} f(x, t) = 0$, since e^{xt} grows faster than x^2 for any $t > 0$. Thus, one might expect that $\lim_{x\to\infty} F(x) = 0$. However, if we put $s = xt$ into (6-23), then the integral becomes

$$F(x) = \int_0^\infty se^{-s}\,ds = 1 \tag{6-24}$$

so that in fact, $F(x) = 1$ for all $x \geq 1$ and $\lim_{x\to\infty} F(x) = 1$.

We can also check that (6-23) converges uniformly for all $x \geq 1$ in the same way.

$$\int_0^r x^2te^{-xt}\,dt = \int_0^{rx} se^{-s}\,ds = -(1 + s)e^{-s}\Big|_0^{rx}$$
$$= 1 - (1 + rx)e^{-rx}$$

so that

$$\left|F(x) - \int_0^r f(x, t)\,dt\right| = (1 + rx)e^{-rx}$$
$$\leq (1 + r)e^{-r}$$

for all $x \geq 1$. Since $\lim_{r\to\infty} (1 + r)e^{-r} = 0$, we have proved uniform convergence for (6-23) directly. [We remark that this example is not amenable to the Weierstrass test; see Exercise 10.]

Thus, in (6-23) we have an instance of an improper integral that converges uniformly for all $x \geq 1$ while $\lim_{x\to\infty} F(x)$ cannot be calculated by carrying the limit operation inside the integral; this is in contrast to the behavior of series in the corollary to Theorem 4. The analogy breaks down, and it is natural to wonder why. The answer lies in the behavior of finite sums versus integrals. In the series case, we use the fact that $\sum_1^N u_k(x)$ is continuous in the entire set where each of the functions u_k is continuous. The analog would be the assertion that $\int_c^r f(x, t)\, dt$ is continuous for all those x where $f(x, t)$ is continuous. The proof of this statement (which was Exercise 18, Sec. 4.3) depended upon having $f(x, t)$ uniformly continuous, which in turn depended on having x in a compact set. However, in our present context, the set of x that concerns us is $1 \leq x < \infty$, which is certainly not compact.

Indeed, this problem cannot be resolved merely by a different mode of proof, for the substitution $s = xt$ shows that

$$\lim_{x\to\infty} \int_0^r x^2te^{-xt}\, dt = \lim_{x\to\infty} \int_0^{rx} se^{-s} = 1$$

which is clearly not the same as

$$\int_0^r \lim_{x\to\infty} x^2te^{-xt}\, dt = \int_0^r 0\, dt = 0$$

One convenient way around this is to require a stronger limit behavior of the integrand as $x \to \infty$, namely that $\lim_{x\to\infty} f(x, t)$ is *uniform* in t on every interval $[c, L]$.

Theorem 15 *Let $f(x, t)$ be continuous for $x \geq b$, $t \geq c$, and suppose that $\int_c^\infty f(x, t)\, dt$ converges to $F(x)$, uniformly for all $x \geq b$. Suppose also that $\lim_{x\to\infty} f(x, t) = g(t)$, where this convergence is uniform in t on every bounded interval $c \leq t \leq L$, for any L. Then,*

$$\lim_{x\to\infty} F(x) = \int_c^\infty g(t)\, dt \tag{6-25}$$

Given $\varepsilon > 0$, suppose $r_0 = r_0(\varepsilon)$ so that

$$\left| F(x) - \int_c^{r_0} f(x, t)\, dt \right| < \varepsilon$$

for all $x \geq b$. Then, choose $x_0 = x_0(r_0, \varepsilon) = x_0(\varepsilon)$ so that $|f(x, t) - g(t)| < \varepsilon/(r_0 - c)$ for all $x \geq x_0$ and all t, $c \leq t \leq r_0$. Take any points x_1 and

x_2, with $x_i > x_0$, and an easy argument yields

$$|F(x_1) - F(x_2)| \le 2\varepsilon + \int_c^{r_0} |f(x_1, t) - g(t)|\, dt + \int_c^{r_0} |f(x_2, t) - g(t)|\, dt$$

$$\le 2\varepsilon + 2(r_0 - c)\frac{\varepsilon}{r_0 - c} = 4\varepsilon$$

This shows that $F(x)$ has the Cauchy property as $x \to \infty$, and that $\lim_{x\to\infty} F(x)$ exists, from which the conclusion of the theorem readily follows. ▮

There are other circumstances in which we can be sure that

$$\lim_{x\to\infty} \int_c^r f(x, t)\, dt = \int_c^r \lim_{x\to\infty} f(x, t)\, dt,$$

such as a form of the Lebesgue bounded convergence theorem. Further refinements of this nature are left to texts on advanced real analysis.

According to the corollary of Theorem 5, Sec. 6.2, a series of functions that converges uniformly on a compact set can be integrated termwise on that set. This has a direct analogy for improper integrals which can also be regarded as a statement about the interchange of orders of integration when one is improper. (In connection with Theorem 16, we note that uniform convergence is not enough to justify such an interchange when *both* integrals are improper, as is demonstrated in Exercise 16. This is also connected with the absence of a suitable theory of conditionally convergent improper double integrals, as was explained at the end of Sec. 4.5.)

Theorem 16 $\int_a^b dx \int_c^\infty f(x, u)\, du = \int_c^\infty du \int_a^b f(x, u)\, dx$ *if* $f(x, u)$ *is continuous for* $a \le x \le b$, $c \le u < \infty$, *and* $\int_c^\infty f(x, u)\, du$ *converges uniformly for* x *on* $[a, b]$.

Using Theorem 9, Sec. 4.3, to reverse the order of integration, we have

$$\int_c^r du \int_a^b f(x, u)\, dx = \int_a^b dx \int_c^r f(x, u)\, du$$

so that
$$\int_c^\infty du \int_a^b f(x, u)\, dx = \lim_{r\uparrow\infty} \int_a^b dx \int_c^r f(x, u)\, du$$

On the other hand,

$$\int_a^b dx \int_c^\infty f(x, u)\, du = \int_a^b dx \int_c^r f(x, u)\, du + \int_a^b dx \int_r^\infty f(x, u)\, du$$

Since $\int_c^\infty f(x, u)\, du$ converges uniformly,

$$\lim_{r \uparrow \infty} \int_r^\infty f(x, u)\, du = 0$$

uniformly for $x \in [a, b]$ and

$$\lim_{r \uparrow \infty} \int_a^b dx \int_r^\infty f(x, u)\, du = 0 \quad \blacksquare$$

Without some restriction on the integrand, and on the mode of convergence of an improper integral, reversal of the order of integration is not valid. Consider, for example, a function F defined by

$$F(x) = \int_0^\infty (2xu - x^2u^2)e^{-xu}\, du \tag{6-26}$$

For $x = 0$, $F(0) = 0$. With $x > 0$, we may evaluate F directly, obtaining

$$\begin{aligned} F(x) &= \lim_{R\to\infty} \Big[xu^2e^{-xu} \Big]_{u=0}^{u=R} \\ &= \lim_{R\to\infty} xR^2e^{-xR} = 0 \end{aligned}$$

Thus, $F(x) = 0$ for all $x \geq 0$. In particular, then,

$$\int_0^1 F(x)\, dx = \int_0^1 dx \int_0^\infty (2xu - x^2u^2)e^{-xu}\, du = 0$$

Consider the effect of reversing the order of integration:

$$\begin{aligned} \int_0^\infty du \int_0^1 (2xu - x^2u^2)e^{-xu}\, dx &= \int_0^\infty du \Big[x^2ue^{-xu} \Big]_{x=0}^{x=1} \\ &= \int_0^\infty ue^{-u}\, du = 1 \end{aligned}$$

This discrepancy is explained by the fact that the original improper integral (6-26) is not uniformly convergent for x in the interval $[0, 1]$ over which we wish to integrate. We have

$$F(x) = \lim_{R\to\infty} xR^2e^{-xR} = \lim_{R\to\infty} g(x, R)$$

and the convergence is not uniform since $g(1/R, R) = R/e$, which is unbounded. (The graphs of these functions for several values of R are given in Fig. 6-1.)

When certain conditions are fulfilled, an improper integral containing a parameter can be differentiated with respect to that parameter underneath the integral sign.

Theorem 17 *If* $\int_c^\infty f(x, u)\,du$ *converges to* $F(x)$ *for all* x, $a \le x \le b$, *and if* f *and* $f_1 = \partial f/\partial x$ *are continuous for* $a \le x \le b$, $c \le u < \infty$, *and if* $\int_c^\infty f_1(x, u)\,du$ *is uniformly convergent for* x *in* $[a, b]$, *then for any* x *in* $[a, b]$,

$$F'(x) = \frac{d}{dx}\int_c^\infty f(x, u)\,du = \int_c^\infty f_1(x, u)\,du$$

The proof is essentially the same as that of Theorem 12, Sec. 4.3. Set

$$g(x) = \int_c^\infty f_1(x, u)\,du$$

Since this is uniformly convergent and f_1 is continuous, g is continuous, and for any $\bar{x}$, $a \le \bar{x} \le b$,

$$\int_a^{\bar{x}} g = \int_a^{\bar{x}} g(x)\,dx = \int_a^{\bar{x}} dx \int_c^\infty f_1(x, u)\,du$$

$$= \int_c^\infty du \int_a^{\bar{x}} f_1(x, u)\,dx$$

But
$$\int_a^{\bar{x}} f_1(x, u)\,dx = \int_a^{\bar{x}} \frac{\partial f}{\partial x}(x, u)\,dx = f(\bar{x}, u) - f(a, u)$$

and
$$\int_a^{\bar{x}} g = \int_c^\infty [f(\bar{x}, u) - f(a, u)]\,du$$

$$= F(\bar{x}) - F(a)$$

Since g is continuous, this shows that F is differentiable, and that $F' = g$. ∎

We shall give a number of examples which illustrate these theorems and the manner in which they may be used in the evaluation of certain special definite integrals.

Let us start with the formula

$$\frac{1}{x} = \int_0^\infty e^{-xu}\,du \tag{6-27}$$

valid for all $x > 0$. If we differentiate this, we obtain

$$\frac{1}{x^2} = \int_0^\infty ue^{-xu}\,du \tag{6-28}$$

To check the validity of this process, we observe that $|ue^{-xu}| \le ue^{-\delta u}$ for all x with $x \ge \delta > 0$; since $\int_0^\infty ue^{-\delta u}\,du$ converges, the integral in (6-28) is uniformly convergent for all x with $\delta \le x < \infty$. This justifies the differentiation, and the formula (6-28) holds for every point x which can be included in one of these intervals, that is, for every x with $x > 0$.

More generally, the same argument may be used to show that for any $n = 1, 2, 3, \ldots,$

$$\frac{n!}{x^{n+1}} = \int_0^\infty u^n e^{-xu}\, du \tag{6-29}$$

As another illustration, consider the improper integral

$$\int_0^\infty \frac{e^{-x} - e^{-2x}}{x}\, dx \tag{6-30}$$

Since the integrand may be expressed as $\int_1^2 e^{-xu}\, du$, we may write (6-30) as an iterated integral

$$\int_0^\infty dx \int_1^2 e^{-xu}\, du$$

Reversing the order of integration, we obtain

$$\int_1^2 du \int_0^\infty e^{-xu}\, dx \tag{6-31}$$

This does not alter the value, since the inner integral is uniformly convergent for all u with $1 \le u \le 2$. Using (6-27), we find that the exact value of the original integral (6-30) is $\int_1^2 u^{-1}\, du = \log 2$.

Sometimes the operations of differentiation and integration are combined, as in the following illustration. On page 217 we saw that the improper integral $\int_0^\infty x^{-1} \sin x\, dx$ was conditionally convergent; we shall now show that its value is $\frac{1}{2}\pi$. Let

$$F(u) = \int_0^\infty e^{-xu} \frac{\sin x}{x}\, dx \tag{6-32}$$

This converges for all $u \ge 0$, and $F(0)$ is the value we seek. If we differentiate (6-32), we obtain

$$F'(u) = -\int_0^\infty e^{-xu} \sin x\, dx \tag{6-33}$$

which may be integrated exactly, yielding $F'(u) = -(1 + u^2)^{-1}$. This is valid for all u with $u > 0$, since the integrand in (6-33) is dominated by e^{-xu} and, as we have seen, the integral of this is uniformly convergent for all u with $\delta \le u < \infty$ and any $\delta > 0$. Integrating, we find that $F(u) = C - \arctan u$, for all $u > 0$. Now, suppose we let u increase; from (6-32), it would seem that $\lim_{u\to\infty} F(u) = 0$, so that

$$0 = C - \lim_{u\to\infty} \arctan u = C - \tfrac{1}{2}\pi$$

and $C = \frac{1}{2}\pi$. We then have $F(u) = \pi/2 - \arctan u$, so that

$$F(0) = \int_0^\infty x^{-1} \sin x \, dx = \tfrac{1}{2}\pi$$

There are two gaps in this argument. We have assumed that

$$\lim_{u\to\infty} \int_0^\infty e^{-xu} \frac{\sin x}{x} \, dx = 0 \tag{6-34}$$

and

$$\lim_{u\downarrow 0} \int_0^\infty e^{-xu} \frac{\sin x}{x} \, dx = \int_0^\infty \frac{\sin x}{x} \, dx \tag{6-35}$$

Neither of these is immediately obvious, although both are plausible. The first can be treated quickly. Since

$$\frac{\sin x}{x} = \int_0^1 \cos (xs) \, ds$$

it follows that $|x^{-1} \sin x| \le 1$ for all x. Thus, for any u,

$$\left| \int_0^\infty e^{-xu} \frac{\sin x}{x} \, dx \right| \le \int_0^\infty e^{-ux} \, dx = \frac{1}{u}$$

and therefore $\lim_{u\to\infty} F(u) = 0$, which is (6-34).

To verify (6-35), we show that the integral involved which defines $F(u)$ is uniformly convergent for all $u \ge 0$, for this will, by Theorem 13, permit us to evaluate (6-35) as shown, taking the limit inside the integral. The Weierstrass test will not show uniform convergence in this example, since the integrand in (6-35) gives

$$\left| e^{-xu} \frac{\sin x}{x} \right| \le \frac{|\sin x|}{x}$$

and $\int_0^\infty x^{-1} |\sin x| \, dx$ diverges. If we use the trick behind the Dirichlet test (Theorem 17, Sec. 4.5) and integrate by parts once, the situation improves. We have

$$\int_r^\infty e^{-xu} \frac{\sin x}{x} \, dx = \Big[-e^{-xu} x^{-1} \cos x \Big|_{x=r}^{x=\infty} - \int_r^\infty \frac{(1 + xu)e^{-xu} \cos x}{x^2} \, dx$$

and (estimating the integrand in the second integral),

$$\left|\int_r^\infty e^{-xu}x^{-1}\sin x\,dx\right| \le \frac{e^{-ur}|\cos r|}{r} + \int_r^\infty x^{-2}\,dx$$
$$\le \frac{2}{r}$$

for any $u \ge 0$. Since $\lim_{r\to\infty} 2/r = 0$, (6-32) converges uniformly for this set of values of u. Putting this with all the rest of the discussion, we have proved that the value of $\int_0^\infty x^{-1}\sin x\,dx$ is exactly $\pi/2$.

Another specialized integral that can be treated by the same method is

$$\int_0^\infty \exp(-x^2 - x^{-2})\,dx \tag{6-36}$$

Consider the related integral

$$F(u) = \int_0^\infty e^{-x^2}e^{-u^2/x^2}\,dx \tag{6-37}$$

As partial motivation for this choice, we observe that $F(1)$ is the desired integral, while $F(0)$ is the familiar integral $\int_0^\infty e^{-x^2}\,dx$ which we have shown (Sec. 4.5) to have the value $\sqrt{\pi}/2$. Differentiate (6-37), getting

$$F'(u) = -2\int_0^\infty \left(\frac{u}{x^2}\right)e^{-x^2}e^{-u^2/x^2}\,dx \tag{6-38}$$

To justify this, we must show that this integral converges uniformly. Writing the integrand in (6-38) as $u^{-1}(u/x)^2e^{-(u/x)^2}e^{-x^2}$ and using the fact that $s^2e^{-s^2}$ has maximum value e^{-1}, we see that the integrand is dominated by $(1/eu)e^{-x^2}$, so that the integral is uniformly convergent for all $u \ge \delta > 0$. We cannot easily integrate (6-38); however, let us make the substitution $t = u/x$, obtaining for any $u > 0$,

$$\begin{aligned} F'(u) &= -2\int_\infty^0 u^{-1}t^2e^{-t^2}e^{-(u/t)^2}\left(\frac{-u}{t^2}\right)dt \\ &= -2\int_0^\infty e^{-t^2}e^{-u^2/t^2}\,dt \\ &= -2F(u) \end{aligned}$$

Solving this differential equation, we find that $F(u) = Ce^{-2u}$, valid for all $u > 0$. However, the integral (6-37) which defines F is uniformly convergent for all u, since $|e^{-x^2}e^{-u^2/x^2}| \le e^{-x^2}$ for all u. In particular, F is then continuous at 0, and $\lim_{u\downarrow 0} F(u) = F(0)$. We know that $F(0)$ has the value $\sqrt{\pi}/2$, so that

$$C = \sqrt{\pi}/2 \qquad \text{and} \qquad F(1) = \int_0^\infty e^{-(x^2+1/x^2)}\,dx = \tfrac{1}{2}\sqrt{\pi}\,e^{-2}$$

EXERCISES

1 Investigate the existence and uniformity of

$$\lim_{t\to 0} \frac{x \sin (xt)}{1 + x^2}$$

2 Let f be continuous on the interval $0 \le x < \infty$ with $|f(x)| \le M$. Set

$$F(u) = \frac{2}{\pi}\int_0^\infty \frac{uf(x)\,dx}{u^2 + x^2}$$

Show that $\lim_{u \downarrow 0} F(u) = f(0)$.

3 Show that, in fact, $\lim_{x\to\infty} F(x) = 1$ where $F(x)$ is given by (6-22).

4 Evaluate $\displaystyle\lim_{r\to\infty} \int_0^{\pi/2} e^{-r\sin\theta}\,d\theta$.

5 Evaluate $\displaystyle\lim_{x\to\infty} \int_0^\infty \frac{x^2u\,du}{x^3 + u^3}$.

6 Evaluate $\displaystyle\int_0^\infty \frac{\sin^2(xu)\,du}{u^2}$.

7 Evaluate $\displaystyle\int_0^\infty \frac{1-\cos x}{x^2}\,dx$.

8 Evaluate $\displaystyle\int_0^\infty e^{-u^2}\cos(xu)\,du$.

9 Evaluate $\displaystyle\int_0^\infty e^{-(x-1/x)^2}\,dx$.

10 Show that there is no function $g(t)$ such that $\int_0^\infty g(t)\,dt$ converges but such that $x^2te^{-xt} \le g(t)$ for all $x \ge 1$, $t \ge 0$.

11 Investigate the existence and uniformity of the limit

$$\lim_{t\to 0} \frac{e^{xt} - 1}{x} \qquad 0 < x < \infty$$

12 Evaluate:

(*a*) $\displaystyle\lim_{x\to 0} \int_0^1 \frac{dt}{\sqrt{x^2 + t^2}}$ (*b*) $\displaystyle\lim_{x\to 0} \int_0^1 \frac{x\,dt}{\sqrt{x^2 + t^2}}$

13 Evaluate:

(*a*) $\displaystyle\lim_{x\to\infty} \int_0^\infty \sin(e^{xt})\,dt$ (*b*) $\displaystyle\lim_{x\to\infty} \int_0^\infty \frac{x\,dt}{1 + x^2t^2}$

14 Write out a detailed argument for the proof of Theorem 13.

15 Write out a detailed argument for the proof of Theorem 14.

***16** Show that $\int_1^\infty dx \int_1^\infty G(x, y)\,dy \ne \int_1^\infty dy \int_1^\infty G(x, y)\,dx$ if $G(x, y) = (x - y)/(x + y)^3$.

6.5 THE GAMMA FUNCTION

In Sec. 6.4, we discussed certain properties of functions which have been expressed as integrals. It is the exception when such functions can be expressed in terms of elementary functions; when this is not possible, the integral itself is often taken as the definition of the function, and additional properties of the function must be deduced from the integral representation. A simple and familiar instance of this procedure is the definition of the logarithm function by the formula

$$L(x) = \int_1^x \frac{dt}{t} \qquad x > 0 \tag{6-39}$$

By making changes of variable, one may verify all the algebraic properties. For example, to prove that $L(1/x)$ is $-L(x)$, we set $t = 1/u$, and have

$$L\left(\frac{1}{x}\right) = \int_1^{1/x} \frac{dt}{t} = \int_1^x u(-u^{-2})\,du$$

$$= -\int_1^x \frac{du}{u} = -L(x)$$

A less familiar example is the approach to the trigonometric functions which starts with the definition

$$A(x) = \int_0^x \frac{dt}{1+t^2} \tag{6-40}$$

As an example of what may be done with this, let us make the variable change $t = 1/u$, and obtain

$$A(x) = \int_\infty^{1/x} \frac{-u^{-2}}{1+u^{-2}}\,du = \int_{1/x}^\infty \frac{du}{1+u^2}$$

Let $K = \int_0^\infty dt/(1+t^2) = \lim_{x\to\infty} A(x)$. Then, the last integral is

$$K - \int_0^{1/x} \frac{du}{1+u^2} = K - A\left(\frac{1}{x}\right)$$

so that we have shown the identity $A(x) + A(1/x) = K$. In particular, $A(1) = K/2$; π may be defined as $2K$. (6-40) can be taken as the definition of $\tan\theta$ by the equation $\theta = A(x)$. Exercise 2 provides a proof of the familiar identity for $\tan(2\theta)$.

We devote this section to the study of some of the simpler properties of the **gamma function,** as defined by the improper integral

$$\Gamma(x) = \int_0^\infty u^{x-1}e^{-u}\,du \tag{6-41}$$

This converges for all $x > 0$, and converges uniformly for all x in the interval $[\delta, L]$, for any $\delta > 0$ and $L < \infty$, so that $\Gamma(x)$ is continuous for all $x > 0$. If x is chosen as an integer, (6-41) becomes an integral which we can evaluate exactly, so that comparing this with Eq. (6-29) of Sec. 6.4,

$$\Gamma(n+1) = \int_0^\infty u^n e^{-u}\,du = n!$$

If we write $x! = \Gamma(x+1)$, we therefore obtain a definition of "factorial" applying to nonintegral values of x which agrees with the customary definition when x is an integer. The gamma function also obeys the identity $(x+1)! = (x+1)(x!)$.

Theorem 18 $\Gamma(x+1) = x\Gamma(x)$ *for any* $x > 0$.

Integrating by parts, we have

$$\Gamma(x+1) = \int_0^\infty u^x e^{-u}\,du = \Big[-u^x e^{-u}\Big]_{u=0}^{u=\infty} + \int_0^\infty e^{-u}\,d(u^x)$$

$$= 0 + \int_0^\infty x e^{-u} u^{x-1}\,du = x\Gamma(x) \quad \blacksquare$$

By making appropriate changes of variable, many useful alternative definitions of the gamma function may be obtained. We list several below, together with the necessary substitution:

(6-42) $$\Gamma(x) = 2\int_0^\infty t^{2x-1} e^{-t^2}\,dt \qquad \{u = t^2\}$$

$$\Gamma(x) = \int_0^1 \left[\log\left(\frac{1}{t}\right)\right]^{x-1} dt \qquad \{u = -\log t\}$$

(6-43) $$\Gamma(x) = c^x \int_0^\infty t^{x-1} e^{-ct}\,dt \qquad \{u = ct\}$$

(6-44) $$\Gamma(x) = \int_{-\infty}^\infty e^{xt} \exp(-e^t)\,dt \qquad \{u = e^t\}$$

The first of these, (6-42), makes it possible to determine the value of $\Gamma(x)$ when x is any odd multiple of $\frac{1}{2}$.

Theorem 19 $\Gamma(\frac{1}{2}) = \sqrt{\pi}$, $\Gamma(\frac{3}{2}) = \sqrt{\pi}/2$ *and in general,*

$$\Gamma\left(n + \frac{1}{2}\right) = \frac{(2n)!\sqrt{\pi}}{4^n n!}$$

Putting $x = \frac{1}{2}$ in (6-42), we have $\Gamma(\frac{1}{2}) = 2\int_0^\infty e^{-t^2}\,dt = \sqrt{\pi}$. Using the functional equation $\Gamma(x+1) = x\Gamma(x)$, we have

$$\Gamma\left(\frac{3}{2}\right) = \left(\frac{1}{2}\right)\Gamma\left(\frac{1}{2}\right) = \frac{\sqrt{\pi}}{2}$$

The general formula may be verified by an inductive argument. $\blacksquare$

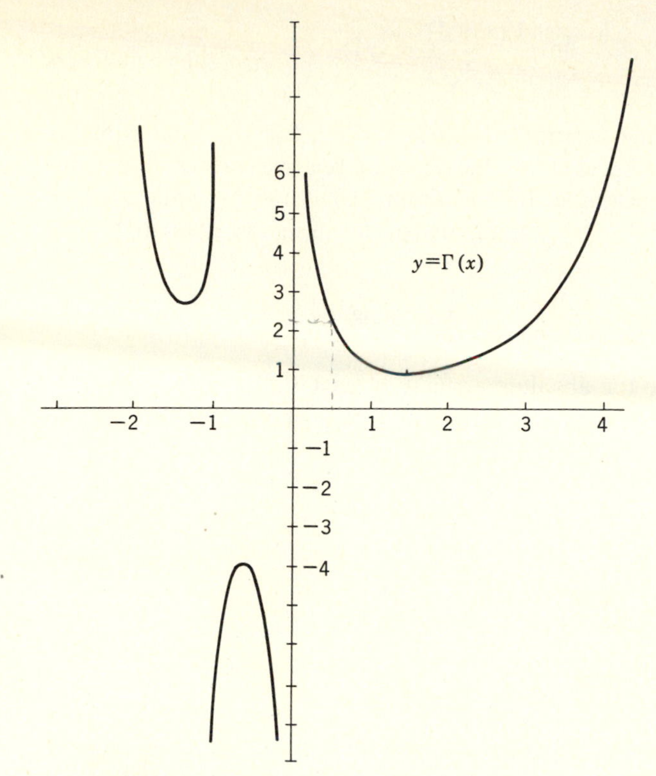

Figure 6-4 The gamma function.

If we know the values of the gamma function on one interval $[k, k+1]$, we can use the functional equation to compute the values on the adjacent intervals, and thus tabulate the function. Turning the formula around, we may write $\Gamma(x) = x^{-1}\Gamma(x+1)$; since $\Gamma(1) = 1$, this shows that $\Gamma(x) \sim x^{-1}$ as x approaches 0. In this form, we can also use the functional equation to extend the definition of the gamma function to negative nonintegral values of x. For example,

$$\Gamma(-\tfrac{1}{2}) = (-2)\Gamma(\tfrac{1}{2}) = -2\sqrt{\pi}$$

and $\Gamma(-\frac{3}{2}) = (-\frac{2}{3})\Gamma(-\frac{1}{2}) = (\frac{4}{3})\sqrt{\pi}$. $\Gamma(x)$ becomes unbounded as x approaches a negative integer since this is the case near 0. (An approximate graph of Γ is given in Fig. 6-4.)

As an illustration of the use of certain techniques for estimating the behavior of functions defined by an integral, we shall obtain a standard asymptotic formula for $\Gamma(x)$. When x is an integer n, this gives a form of Stirling's approximation for $n!$.

Recall that in (5-30) we showed that $n! = n^n e^{-n}\sqrt{n}\,C_n$, where the sequence $\{C_n\}$ is a bounded sequence whose terms lie between 1.9 and 2.8. Our next result will show that, in fact, $\lim_{n\to\infty} C_n = \sqrt{2\pi}$, and that n can be replaced by x, so that we obtain an asymptotic formula for $x! = \Gamma(x+1)$.

Theorem 20 $$\lim_{x\to\infty} \frac{\Gamma(x+1)}{x^x e^{-x}\sqrt{x}} = \sqrt{2\pi}$$

This result is often written $x! \sim x^x e^{-x}\sqrt{2\pi x}$, but the approximation symbol is to be interpreted in the sense of relative rather than small absolute error. For example, 100! is about $(9.3326)10^{157}$, while Stirling's formula gives $100^{100}e^{-100}\sqrt{200\pi}$, which is about $(9.3248)10^{157}$. The relative error is thus

$$\frac{9.3326}{9.3248} - 1 = .0008$$

or .08 percent, while the absolute error is at least 10^{155}.

Starting from the original formula for $\Gamma(x+1)$, we make the change of variable $u = xt$.

$$\Gamma(x+1) = \int_0^\infty u^x e^{-u}\, du = x^{x+1}\int_0^\infty t^x e^{-xt}\, dt$$

so that $$\frac{\Gamma(x+1)}{x^x e^{-x}\sqrt{x}} = \sqrt{x}\, e^x \int_0^\infty [te^{-t}]^x\, dt$$

$$= \sqrt{x}\int_0^\infty [te^{1-t}]^x\, dt$$

The function $g(t) = te^{1-t}$ has its maximum on $0 \le t < \infty$ at $t = 1$, where $g(1) = 1$. Thus, $0 \le g(t) < 1$ for all $t \ne 1$, $t \ge 0$. This suggests splitting the interval of integration into three portions to emphasize the contribution near $t = 1$, for if x is large and t is not 1, $[g(t)]^x$ will be very small. Choose an interval $[a, b]$ about $t = 1$, and write

$$C(x) = \frac{\Gamma(x+1)}{x^x e^{-x}\sqrt{x}} = \sqrt{x}\int_0^a [g(t)]^x\, dt + \sqrt{x}\int_a^b [g(t)]^x\, dt + \sqrt{x}\int_b^\infty [g(t)]^x\, dt$$

where $0 < a < 1 < b < \infty$ (see Fig. 6-5). Our objective is to show that $\lim_{x\to\infty} C(x) = \sqrt{2\pi}$. We first look at the first and third integrals in this sum. On the interval $[0, a]$, $g(t) \le g(a) < 1$, and

$$\sqrt{x}\int_0^a [g(t)]^x\, dt \le a\sqrt{x}\,[g(a)]^x \to 0$$

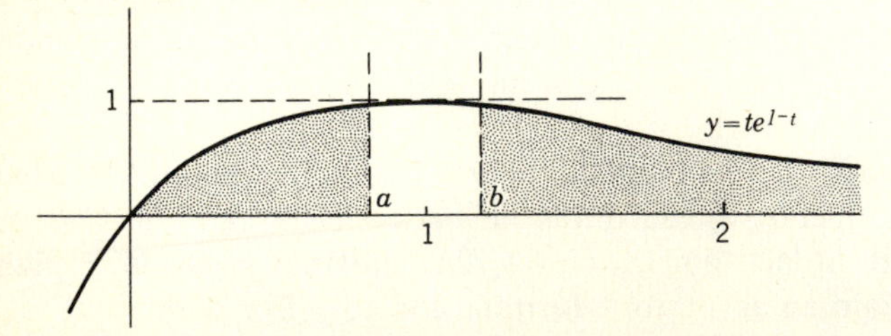

Figure 6-5

as x increases. On the interval $b \le t < \infty$, $g(t) \le g(b) < 1$. Thus,

$$\sqrt{x}\int_b^\infty [g(t)]^x\,dt = \sqrt{x}\int_b^\infty [g(t)]^{x-1}g(t)\,dt$$

$$\le \sqrt{x}\,[g(b)]^{x-1}\int_0^\infty g(t)\,dt$$

which again approaches 0 as x increases.

This leaves us with the middle integral. We choose $a = 1 - \delta$ and $b = 1 + \delta$, where δ has yet to be chosen. Then the middle integral, after the substitution $s = t - 1$, becomes

(6-45) $$I(x) = \sqrt{x}\int_{-\delta}^{\delta} [(1+s)e^{-s}]^x\,ds$$

Lemma 1 *For s near 0, $(1+s)e^{-s} = e^{-s^2h(s)}$, where $\lim_{s\to 0} h(s) = \frac{1}{2}$.*

This is seen by applying L'Hospital's rule to

$$h(s) = \frac{-\log(1+s) + s}{s^2} \quad ∎$$

The integrand in (6-45) is therefore $e^{-xs^2h(s)}$. We now choose δ. Given $\varepsilon > 0$, choose δ so that $|h(s) - \frac{1}{2}| < \varepsilon$ for all s, $|s| \le \delta$. Then, for any s, $-\delta \le s \le \delta$,

(6-46) $$e^{-xs^2(1/2+\varepsilon)} \le e^{-xs^2h(s)} \le e^{-xs^2(1/2-\varepsilon)}$$

We will use this to estimate $I(x)$. First, we need a simple calculation.

Lemma 2 *For any $\delta > 0$ and any $c > 0$,*

$$\lim_{x\to\infty} \sqrt{x}\int_{-\delta}^{\delta} e^{-cxs^2}\,ds = \sqrt{\frac{\pi}{c}}$$

Put $u = \sqrt{cx}\,s$, which turns the given integral into

(6-47) $$\frac{1}{\sqrt{c}}\int_{-\delta\sqrt{cx}}^{\delta\sqrt{cx}} e^{-u^2}\,du$$

and as x increases, this converges to $(1/\sqrt{c})\int_{-\infty}^{\infty} e^{-u^2}\,du$, which by (4-40), is $\sqrt{\pi/c}$. ∎

Returning to (6-46) and integrating between $-\delta$ and δ, we see that $I(x)$, given by (6-45), is sandwiched between two integrals that converge to $\sqrt{\pi/(\frac{1}{2} - \varepsilon)}$ and $\sqrt{\pi/(\frac{1}{2} + \varepsilon)}$ as x increases. Since ε is arbitrarily small, we may conclude that $\lim_{x\to\infty} I(x) = \sqrt{\pi/\frac{1}{2}} = \sqrt{2\pi}$. ∎

If we admit the gamma function to the collection of functions which we may use, many otherwise intractable definite integrals can be evaluated exactly.

Theorem 21 $B(p, q) = \int_0^1 x^{p-1}(1-x)^{q-1}\,dx = \Gamma(p)\Gamma(q)/\Gamma(p+q)$, *where p and q are positive real numbers.*

The integral defines B as a function of two real variables; it is known as the **beta function.** When p and q are positive integers, the integrand is a polynomial, and $B(p, q)$ can be easily evaluated; the theorem makes the evaluation possible for all positive p and q.

We follow a procedure similar to that which we used in evaluating $\int_0^\infty e^{-t^2}\,dt$ in (4-40). Using the alternative formula (6-42) for the gamma function, and introducing different dummy letters for the variables of integration, we have

$$\Gamma(p) = 2\int_0^\infty y^{2p-1}e^{-y^2}\,dy$$

$$\Gamma(q) = 2\int_0^\infty x^{2q-1}e^{-x^2}\,dx$$

The product of these can then be expressed as

$$\begin{aligned}\Gamma(p)\Gamma(q) &= \lim_{R\to\infty} 4\int_0^R y^{2p-1}e^{-y^2}\,dy \int_0^R x^{2q-1}e^{-x^2}\,dx \\ &= \int_0^\infty \int_0^\infty y^{2p-1}x^{2q-1}e^{-(x^2+y^2)}\,dx\,dy\end{aligned}$$

Since the integrand of this improper double integral is positive, we can also integrate over the first quadrant by using quarter circles, and polar coordinates. Replace x by $r\cos\theta$, y by $r\sin\theta$, and $dx\,dy$ by $r\,dr\,d\theta$. Carrying this out, we have

$$\begin{aligned}\Gamma(p)\Gamma(q) &= \lim_{R\to\infty} 4\int_0^R dr \int_0^{\pi/2} (r\sin\theta)^{2p-1}(r\cos\theta)^{2q-1}e^{-r^2}r\,d\theta \\ &= 4\int_0^\infty r^{2p+2q-1}e^{-r^2}\,dr \int_0^{\pi/2} (\cos\theta)^{2q-1}(\sin\theta)^{2p-1}\,d\theta\end{aligned}$$

In the first, put $u = r^2$ so that $dr = du/(2r)$; in the second, put $v = \sin^2\theta$ so that $d\theta = dv/(2\sin\theta\cos\theta)$. Then

$$\begin{aligned}\Gamma(p)\Gamma(q) &= 4\int_0^\infty \frac{u^{p+q+1/2}e^{-u}}{2u^{1/2}}\,du \int_0^1 \frac{(1-v)^{q-1/2}v^{p-1/2}}{2v^{1/2}(1-v)^{1/2}}\,dv \\ &= \int_0^\infty u^{p+q-1}e^{-u}\,du \int_0^1 v^{p-1}(1-v)^{q-1}\,dv \\ &= \Gamma(p+q)B(p, q) \quad \blacksquare\end{aligned}$$

Many integrals which are not originally in the form of a beta function can be converted into such a form, and thus evaluated. We give two samples of this.

$$\int_0^1 \frac{dx}{(1-x^4)^{1/2}} = \int_0^1 \frac{u^{-3/4}\,du}{4(1-u)^{1/2}} = \tfrac{1}{4}B(\tfrac{1}{4}, \tfrac{1}{2})$$

$$= \frac{\Gamma(\frac{1}{4})\Gamma(\frac{1}{2})}{4\Gamma(\frac{3}{4})}$$

$$\int_0^{\pi/2} \sqrt{\sin\theta}\,d\theta = \int_0^1 \frac{v^{1/4}\,dv}{2v^{1/2}(1-v)^{1/2}} = \tfrac{1}{2}B(\tfrac{3}{4}, \tfrac{1}{2})$$

$$= \frac{\Gamma(\frac{3}{4})\Gamma(\frac{1}{2})}{2\Gamma(\frac{5}{4})} = \frac{2\Gamma(\frac{3}{4})\Gamma(\frac{1}{2})}{\Gamma(\frac{1}{4})}$$

(In the second, we have used the substitution $v = \sin^2\theta$, and in the last step, the functional equation for the gamma function.)

EXERCISES

1 Show directly from the integral definition in Eq. (6-39) that the function L has the property $L(xy) = L(x) + L(y)$.

2 (*a*) Show that the function A, defined in (6-40), has the property that

$$A\left(\frac{2x}{1-x^2}\right) = 2A(x) \qquad \text{for any } x,\ |x| < 1$$

(*b*) Show how this leads to the identity $\tan(2\theta) = (2\tan\theta)/(1-\tan^2\theta)$.

3 Complete the induction argument in Theorem 19.

4 In terms of the gamma function, evaluate the following integrals:

(*a*) $\displaystyle\int_0^1 \frac{x^3\,dx}{\sqrt{1-x^3}}$ (*b*) $\displaystyle\int_0^1 \frac{dx}{\sqrt{x\log(1/x)}}$

(*c*) $\displaystyle\int_0^1 [1-1/x]^{1/3}\,dx$ (*d*) $\displaystyle\int_0^{\pi/2} \sqrt{\tan\theta}\,d\theta$

5 Show that $\Gamma(x)\Gamma(1-x) = \displaystyle\int_0^\infty u^{x-1}/(1+u)\,du$.

6 Evaluate $\displaystyle\int_0^\infty u^p e^{-u^q}\,du$.

7 Evaluate $\displaystyle\int_0^1 x^r[\log(1/x)]^s\,dx$.

8 The error function is defined by

$$\operatorname{erf}(x) = \frac{2}{\sqrt{\pi}}\int_0^x e^{-t^2}\,dt$$

Use this to express (evaluate) the following integrals:

(*a*) $\displaystyle\int_0^L e^{-1/s^2}\,ds$ (*b*) $\displaystyle\int_0^1 x^2 e^{-x^2}\,dx$

(c) $F(x) = \int_0^\infty e^{-xt^2}\, dt$

9 (a) Show that $\int_0^1 dx/(1 - x^{1/4}) = 128/35$.

(b) Show that $\int_0^1 (1 - x^{2/3})^{3/2}\, dx = 3\pi/32$.

10 Show that $B(p, q) = \int_0^\infty u^{p-1}/(1 + u)^{p+q}\, du$. (Hint: $x = u/(1 + u)$.)

11 Using Exercise 10, show that $I = \int_{-\infty}^\infty dt/(1 + t^2)^4 = 5\pi/16$.

12 Express $\int_1^\infty (x - 1)^{2/3} x^{-2}\, dx$ in terms of the gamma function.

13 The following identity was stated by Wallis (c. 1650)

$$\frac{\pi}{2} = \left(\frac{2}{1}\right)\left(\frac{2}{3}\right)\left(\frac{4}{3}\right)\left(\frac{4}{5}\right)\left(\frac{6}{5}\right)\left(\frac{6}{7}\right)\left(\frac{8}{7}\right)\cdots$$

Use Stirling's formula to show this is correct.

6.6 FOURIER SERIES

It is often useful to borrow the terms "point" and "distance" from elementary geometry and apply them in quite different contexts in order to help motivate and explain mathematical techniques. We have already done this in treating n space, and have indicated in Sec. 6.2 an analog that is useful in working with functions. We now formalize this.

A set $\mathscr{M}$, whose members shall also be called points, is said to be a **metric space** if for every pair p, q in $\mathscr{M}$, there is a real number $d(p, q)$ which we call the distance from p to q, such that:

(6-48) $$0 < d(p, q) < \infty \qquad \text{unless } p = q \text{ when } d(p, q) = 0$$

(6-49) $$d(p, q) = d(q, p)$$

(6-50) $$d(p, q) \le d(p, r) + d(r, q) \qquad \text{for any } r \in \mathscr{M}$$

The last of these is again called the triangle law. Clearly, with $d(p, q) = |p - q|$, $\mathbf{R}^n$ is a metric space. All of Section 6.2 dealt with the metric space $\mathscr{C}[I]$ whose "points" were the continuous real-valued functions defined on a fixed compact set I, using the metric (Exercise 14, Sec. 6.2)

$$d(f, g) = \|f - g\|_I = \max_{p \in I} |f(p) - g(p)|$$

In any metric space, one can introduce all the topological concepts described in Sec. 1.5. For example, a sequence of "points" $\{p_n\}$ is said to converge to a point p if

(6-51) $$\lim_{n \to \infty} d(p_n, p) = 0$$

In the metric space $\mathscr{C}[I]$, (6-51) is the same as saying that the functions $\{f_n\}$ converge to a function f uniformly on I.

In this section, we look at two other metric spaces, both of which have been of central importance in mathematics and its applications, particularly to physics. The first of these is ℓ^2 (usually called "little L two"), which is the natural infinite-dimensional generalization of ordinary n space. The points of ℓ^2 are infinite real sequences, written as though the terms were coordinates:

$$p = (a_1, a_2, a_3, \ldots)$$
$$q = (b_1, b_2, b_3, \ldots)$$

The metric is defined by

$$d(p, q) = \|p - q\| = \sqrt{\sum_1^\infty |a_k - b_k|^2} = \sqrt{|a_1 - b_1|^2 + |a_2 - b_2|^2 + \cdots} \tag{6-52}$$

and in order to use it, the series inside the square root must converge. The origin of the space ℓ^2 is $\mathbf{0}$, the sequence all of whose terms are 0, so that if we want $d(p, \mathbf{0})$ to exist, we must have

$$\sum_1^\infty |a_k|^2 = \|p\|^2 < \infty \tag{6-53}$$

This imposes a requirement on all the points of ℓ^2. For example, the choice $a_n = 1/n$ yields a point of ℓ^2, since $\sum_1^\infty 1/n^2$ converges, but the choice $a_n = 1/\sqrt{n}$ is not allowed, and does not yield a point in the space ℓ^2. (This explains the "2" in ℓ^2; there is also a metric space ℓ^p whose points are the sequences $\{a_n\}$ for which $\sum_1^\infty |a_n|^p$ converges. The space ℓ^2 is also called real separable **Hilbert space.**)

Imitating n space, we also introduce a scalar or dot product in ℓ^2, defining

$$p \cdot q = \sum_1^\infty a_k b_k \tag{6-54}$$

The fact that this series is convergent for every pair of points p and q in ℓ^2, and that the metric satisfies the desired laws, especially the triangle law (6-50), follows from an appropriate form of the Schwarz inequality,

$$|p \cdot q| \le \|p\| \, \|q\| \tag{6-55}$$

whose proof comes at once from the corresponding finite-dimensional version (Theorem 1, Sec. 1.3). For any n, we have

$$\left(\sum_1^n |a_k||b_k|\right)^2 \le \sum_1^n |a_k|^2 \sum_1^n |b_k|^2$$

and the right side is less than $\|p\|^2\|q\|^2$, which is finite by (6-53), so that the series in (6-54) is absolutely convergent and is bounded as indicated in (6-55).

The space ℓ^2 has a special set of points $\{\mathbf{e}_n\}$ which we call its **standard basis.** These are

$$\mathbf{e}_1 = (1, 0, 0, \ldots, 0, \ldots)$$
$$\mathbf{e}_2 = (0, 1, 0, 0, \ldots, 0, \ldots)$$
$$\mathbf{e}_n = (0, 0, 0, 0, \ldots, 1, 0, \ldots)$$

We note that each has length 1, meaning that $\|\mathbf{e}_n\| = 1$, and that they are pairwise orthogonal, meaning that $\mathbf{e}_i \cdot \mathbf{e}_j = 0$ if $i \neq j$. It is customary to write both of these facts in the single equation

(6-56) $$\mathbf{e}_i \cdot \mathbf{e}_j = \delta_{ij}$$

where δ_{ij} is a special function of i and j that is 0 unless $i = j$, when it is 1.

The reason for the term "basis" lies in the following result.

Theorem 22 *For any* $p \in \ell^2$, $p = \sum_1^\infty c_n \mathbf{e}_n$, *where the numbers* c_n *are defined by*

(6-57) $$c_n = p \cdot \mathbf{e}_n$$

The proof of this is not quite as trivial as it looks. If we start with $p = (a_1, a_2, \ldots)$, then (6-57) immediately gives $a_n = c_n$ for all n, and all that remains is to show that

(6-58) $$p = (c_1, c_2, c_3, \ldots) = \sum_1^\infty c_n \mathbf{e}_n$$

The only question lies in knowing what the right side means. As with series of numbers, we interpret the right side as $\lim_{N\to\infty} S_N$ where $S_N = \sum_1^N c_n \mathbf{e}_n$, but "limit" must now be understood in the sense of the metric of ℓ^2, meaning that $\lim_{N\to\infty} \|p - S_N\| = 0$. However, since S_N is a finite sum, $S_N = (c_1, c_2, \ldots, c_N, 0, 0, 0, \ldots)$, and

$$\|p - S_N\|^2 = |c_{N+1}|^2 + |c_{N+2}|^2 + \cdots$$

Since this is the tail of the convergent series $\sum_1^\infty |c_k|^2$, we see that, indeed, $\lim_{N\to\infty} \|p - S_N\| = 0$, as required. ▌

We state one more theorem about ℓ^2 which will be of use to us later in this section. We need a preliminary definition.

Definition 5 *A metric space $\mathscr{M}$ is called* **complete** *if every Cauchy sequence of points in $\mathscr{M}$ is convergent in $\mathscr{M}$.*

Thus, in a complete space $\mathscr{M}$, any sequence $\{p_n\}$ whose terms p_n are points of $\mathscr{M}$ and which is such that $\lim_{m,n\to\infty} \|p_n - p_m\| = 0$ will converge to some point $p \in \mathscr{M}$. The function space $\mathscr{C}[E]$ and $\mathbf{R}^n$ are each complete, the former because of Theorem 1 in Sec. 6.2, and the latter by Theorem 23 of Sec. 1.7. However, there are many metric spaces that are not complete (see Exercise 3).

Theorem 23 *ℓ^2 is complete.*

Since the proof of this is not needed for what we do later, and since it is long and complicated, we have left this as a challenge in Exercise 14, where you are asked to fill in a skeleton proof.

While the space ℓ^2 may seem to be a transparent generalization of n space, it is not. The understanding of its geometric and analytic properties is far from complete, even after a century of active research. By the end of this section, some of the reasons for this will have become plain.

We now introduce another example of a metric space. The points in this space will be continuous functions, for simplicity functions defined on a fixed interval $I = [a, b]$. We use a different metric from that in the space $\mathscr{C}[I]$, and therefore use different symbols to denote both. We denote the new space by $^*\mathscr{C}[I]$, and the new metric is given by

$$d(f, g) = {}^*\|f - g\| = \left\{\int_a^b |f(t) - g(t)|^2\, dt\right\}^{1/2} \tag{6-59}$$

We will use the elevated * in connection with operations in the new space, to distinguish them from similar operations in the function space $\mathscr{C}[I]$. Thus, we write

$$^*\lim_{n\to\infty} f_n = g$$

to mean

$$\lim_{n\to\infty} {}^*\|f_n - g\| = 0$$

which in turn (after squaring the result) translates into

$$\lim_{n\to\infty} \int_a^b |f_n(t) - g(t)|^2\, dt = 0 \tag{6-60}$$

In much of the earlier work with Fourier series, the type of convergence described by the formula in (6-60) was called **mean square convergence,** or convergence in the mean; in some papers, it was denoted by l.i.m. $f_n(t) = g(t)$, where the letters were read "limit in the mean."

Mean convergence and uniform convergence are different, just as the two metrics $\|\ \|_I$ and $^*\|\ \|$ are different, even though both are being used on the same collection of functions. Thus, if $f(t) = t^2$ and $g(t) = t^3$ and $I = [0, 1]$, then computation easily shows that $\|f-g\|_I = \frac{4}{27}$, while

$$^*\|f-g\| = \sqrt{\frac{1}{5} - \frac{1}{3} + \frac{1}{7}} = \frac{1}{\sqrt{105}}$$

However, there is a universal inequality between these two metrics, as shown by the following result (Exercise 4).

Theorem 24 *If $I = [a, b]$ and f and g are continuous on I, then*

$$^*\|f-g\| \le \sqrt{b-a}\,\|f-g\|_I$$

Accordingly, if $\{f_n\}$ converges to g uniformly on I, then $^*\lim_{n\to\infty} f_n = g$.

The converse of this does not hold. If $f_n(t) = n\sqrt{t}\exp(-n^2 t)$ and $I = [0, 1]$, then $^*\lim_{n\to\infty} f_n = 0$, since

$$\int_0^1 |f_n(t)|^2\,dt = \int_0^1 n^2 t e^{-2n^2 t}\,dt$$

$$= \frac{1}{n^2}\int_0^{n^2} s e^{-2s}\,ds \to 0$$

but $\{f_n\}$ does not converge to 0 uniformly on I, since

$$\|f_n\| = \max_{0\le t\le 1} |f_n(t)| = |f_n(1/n^2)| = \frac{1}{e}$$

for all n.

Although we have called $^*\|\ \ \|$ a metric on the space $^*\mathscr{C}$, we have not yet verified the three requirements (6-48), (6-49), and (6-50). The first two are easily checked. For the third, we are able to use the same approach adopted in treating the spaces ℓ^2 and $\mathbf{R}^n$. (Indeed, as will be seen later in this section, there is an innate similarity between the space ℓ^2 and the space $^*\mathscr{C}$.) We do this by introducing an inner product in $^*\mathscr{C}$ by writing

$$\langle f, g\rangle = \int_a^b f(t)g(t)\,dt \tag{6-61}$$

We use this notation, rather than $f \cdot g$, in order to avoid confusion with ordinary multiplication of functions. Notice that we are again making use of the analogy between sums and integrals in choosing definitions (6-59) and (6-61), which should be compared with (6-52) and (6-54). The properties of $\langle\ ,\ \rangle$ are described in Exercise 2, and the same procedure that was used to prove the Schwarz inequality in $\mathbf{R}^n$ (Theorem 1, Sec. 1.3) now proves the analog

$$|\langle f, g\rangle| \le {}^*\|f\|\,{}^*\|g\| \tag{6-62}$$

and from this, the triangle law (6-50) for the metric $^*\|\ \ \|$ readily follows, using the fact that $^*\|f+g\|^2 = \langle f+g, f+g\rangle$.

Following the pattern set by ℓ^2, we say that two functions f and g in $^*\mathscr{C}$ are **orthogonal** when neither is 0, but $\langle f, g\rangle = 0$. An infinite set of functions $\{\phi_n\}$ is said to be an **orthogonal set** or **orthogonal system** if every distinct pair is orthogonal. This can be stated as

(6-63) $$\langle \phi_n, \phi_m \rangle = 0 \qquad n \neq m$$

and in integral form as

$$\int_a^b \phi_n(t)\phi_m(t)\, dt = 0 \qquad n \neq m$$

If the ϕ_n have the additional property that $^*\|\phi_n\| = 1$ for every n, then the set $\{\phi_n\}$ is said to be **orthonormal** on the interval I. Using the special function δ_{nm}, this can be written in the abbreviated form $\langle \phi_n, \phi_m \rangle = \delta_{nm}$. We note that a set that is merely orthogonal can be converted into an orthonormal set by multiplying each ϕ_n by an appropriately chosen constant.

Unlike ℓ^2, there is no immediately obvious standard orthonormal set for the space $^*\mathscr{C}$. Instead, one has many to choose from; indeed, one can be constructed from any infinite linearly independent set of functions by taking appropriate linear combinations to achieve the orthogonality. For example, starting from the collection of all polynomials, we pick ϕ_1 to be the constant function 1, then choose $\phi_2(t) = a + bt$ to be orthogonal to ϕ_1, and then $\phi_3(t) = a + bt + ct^2$ so that ϕ_3 is orthogonal to both ϕ_1 and ϕ_2, etc., each time solving for the required coefficients. For example, if I is the interval $[0, 1]$, then one might arrive at the following polynomials:

(6-64)

$$\phi_1 = 1 \qquad \phi_2 = 2t - 1 \qquad \phi_3 = 6t^2 - 6t + 1 \qquad \phi_4 = 20t^3 - 30t^2 + 12t - 1$$

These are not normalized; in order to make them orthonormal, each must be multiplied by the appropriate factor, leading to the set $\phi_1 = 1$, $\phi_2 = \sqrt{3}\,(2t - 1)$, $\phi_3 = \sqrt{5}\,(6t^2 - 6t + 1)$, etc. Another common set, orthogonal on the interval $[-1, 1]$, is the **Legendre polynomials,** $P_n(x)$, defined by the formula of Rodrigues

(6-65) $$P_n(x) = \frac{1}{2^n n!}\left(\frac{d}{dx}\right)^n (x^2 - 1)^n$$

so that $P_0 = 1$, $P_1 = x$, $P_2 = \frac{3}{2}x^2 - \frac{1}{2}$, etc. (These can be seen to be closely related to the polynomials ϕ_n given above in connection with the interval $[0, 1]$; see Exercise 12.)

While there are many obvious analogies between the space $^*\mathscr{C}$ and the sequence space ℓ^2, there are also many important differences. Perhaps the most significant one is that the space $^*\mathscr{C}$ is not complete. This means that there are sequences $\{f_n\}$ in $^*\mathscr{C}$ which obey the Cauchy criterion $\lim_{n,\, m\to\infty} {}^*\|f_n - f_m\| = 0$ but which are not convergent. An example of such a sequence $\{f_n\}$ is given

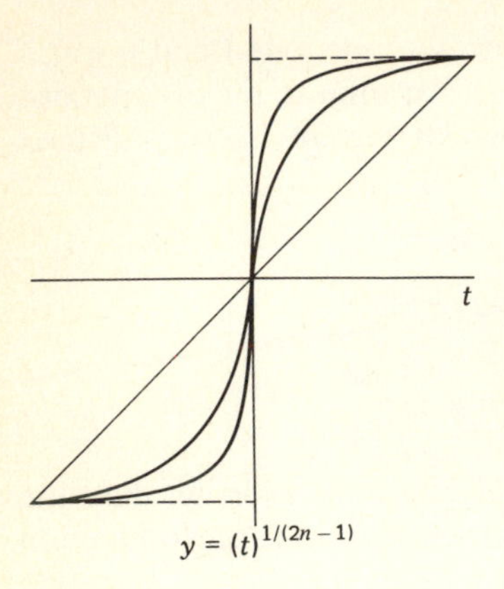

Figure 6-6

on the interval $[-1, 1]$ by

$$f_1(t) = t \qquad f_2(t) = \sqrt[3]{t} \qquad f_3(t) = \sqrt[5]{t} \qquad f_n(t) = t^{1/(2n-1)}$$

whose graphs are shown in Fig. 6-6. Computation shows easily that this sequence is Cauchy (see Exercise 3). However, there is no function $g \in {}^*\mathscr{C}$ with $g = {}^*\lim f_n$.

The reason for this is in fact quite simple: the space $^*\mathscr{C}$ is too small. As can be conjectured from Fig. 6-6, the sequence $\{f_n\}$ *does* in fact converge to a function g, in the special metric defined in (6-59), but this function turns out to be the *discontinuous* function

$$g(t) = \begin{cases} 1 & \text{if } t > 0 \\ 0 & \text{if } t = 0 \\ -1 & \text{if } t < 0 \end{cases}$$

which is not in the space $^*\mathscr{C}$. The situation is similar to what would happen if we were to try to build analysis on the rational numbers alone; here, we would find that a sequence of rationals could be a Cauchy sequence but still fail to converge if its limit (in the reals) were a number such as $\sqrt{2}$, which is not rational.

It would therefore seem plausible that we could overcome this defect in the function space $^*\mathscr{C}$ by enlarging it to include many more functions, all discontinuous, so that any sequence $\{f_n\}$ that is Cauchy in the metric $^*\| \ \|$ will now converge in that metric to one of the functions in the new space. This is (almost) what is in fact done in a thorough and complete treatment of the topic of Fourier series. However, certain technical difficulties must be overcome. As a first step, $^*\mathscr{C}$ is enlarged by including all the piecewise continuous functions on the interval $I = [a, b]$. However, the resulting space is still not large enough, and one must include functions having an infinite number of discontinuities. The basic definitions of the metric and inner product given in (6-59) and (6-61) remain the same, but Riemann integration must be replaced by Lebesgue integration to handle the new functions that appear. Finally, while the metric still obeys the triangle law, one can now have $^*\|f - g\| = 0$ without having f and g identical; for example, suppose f and g agree except at one point of the interval I. Since a treatment of all this would take us far afield, we will instead continue to work with the smaller function space $^*\mathscr{C}$ of continuous functions, and accept the fact that it is not metrically complete.

Having defined the notion of an orthonormal set $\{\phi_n\}$, it is natural to ask if the pattern followed in the space ℓ^2 still holds, so that one can choose a natural orthonormal set which forms a basis for the space $^*\mathscr{C}$. Let us formulate a general definition.

Definition 6 *An orthonormal set $\{\phi_n\}$ in $^*\mathscr{C}$ is called a Fourier* **basis** *for $^*\mathscr{C}$ if every $f \in {}^*\mathscr{C}$ has the unique expansion*

(6-66) $$f = {}^*\sum_1^\infty c_n \phi_n$$

(*convergent in the metric of $^*\mathscr{C}$*), *where the coefficients $\{c_n\}$ are given by*

(6-67) $$c_n = \langle f, \phi_n \rangle \qquad \text{for } n = 1, 2, 3, \ldots$$

Here, the elevated * in (6-66) indicates that the convergence of the series is in the topology of the space $^*\mathscr{C}$, meaning that $^*\lim_{N\to\infty} S_N = f$ where $S_N = \sum_1^N c_n \phi_n$, which in turn is equivalent to $\lim_{N\to\infty} {}^*\|f - S_N\| = 0$. The sequence $\{c_n\}$ is usually called the sequence of **Fourier coefficients** of f with respect to the set $\{\phi_n\}$; it is important to note that $\{c_n\}$ can be constructed for any orthonormal set $\{\phi_n\}$, whether they form a basis or not. We shall usually write $f \sim \{c_n\}$ to indicate the association between the function f and the sequences of Fourier coefficients. In the same way, it is always possible to form the formal series

(6-68) $$\sum_1^\infty c_n \phi_n = c_1\phi_1 + c_2\phi_2 + \cdots$$

where the coefficients c_n are given by (6-67); this is called the **Fourier series** for f with respect to the orthonormal set $\{\phi_n\}$. The word "formal" is used here to emphasize that nothing is implied about the convergence of the series (6-68) in any of the possible meanings.

Because of their importance, we rewrite some of these formulas in integral form. Thus, given any orthonormal set $\{\phi_n\}$, and a function $f \in {}^*\mathscr{C}$, we have $f \sim \{c_n\}$, where

(6-69) $$c_n = \int_a^b f(t)\phi_n(t)\,dt$$

and the association of f with its Fourier series is often written

(6-70) $$f(t) \sim \sum_{1}^{\infty} c_n \phi_n(t)$$

(Note that "$\sim$" is used here instead of "$=$" to emphasize that this is a formal series, rather than one that is known to converge.) The set $\{\phi_n\}$ is a basis if and only if

$$\lim_{N\to\infty} \int_a^b |f(t) - S_N(t)|^2 \, dt = 0$$

for every $f \in {}^*\mathscr{C}$. [In the older literature, one says that the series (6-70) converges to f in the mean.]

From the above discussion, it is clear that a basis for ${}^*\mathscr{C}$ will play the same role there as that played by the standard basis $\{\mathbf{e}_k\}$ for the space ℓ^2. However, we have not yet shown that such a basis exists for ${}^*\mathscr{C}$. Unlike in ℓ^2, none is immediately obvious, and indeed, much of the research in general Fourier theory deals with the study of different orthonormal sets, the development of methods to decide which form a basis for ${}^*\mathscr{C}$, and the study of the various modes of convergence of the series (6-70) and the degree to which the Fourier coefficients $\{c_n\}$ characterize f.

The systematic development of classicial Fourier series started with the work of Fourier about 1820 in connection with the study of wave motion and the flow of heat. Fourier asserted that the system

(6-71) $$\frac{1}{\sqrt{2\pi}}, \quad \frac{\cos(nx)}{\sqrt{\pi}}, \quad \frac{\sin(nx)}{\sqrt{\pi}} \qquad \text{for } n = 1, 2, 3, \ldots$$

formed a basis for ${}^*\mathscr{C}[-\pi, \pi]$. (We indicate a proof of this later on.) It is easy to verify that this system is orthonormal by using the trigonometric identities

(6-72) $$\begin{aligned} 2 \sin A \sin B &= \cos(A - B) - \cos(A + B) \\ 2 \sin A \cos B &= \sin(A + B) + \sin(A - B) \\ 2 \cos A \cos B &= \cos(A - B) + \cos(A + B) \end{aligned}$$

For example, if $n \neq m$,

$$\begin{aligned} \int_{-\pi}^{\pi} \frac{\sin nx}{\sqrt{\pi}} \frac{\sin mx}{\sqrt{\pi}} \, dx &= \frac{1}{2\pi} \int_{-\pi}^{\pi} \{\cos(n - m)x - \cos(n + m)x\} \, dx \\ &= \frac{1}{2\pi} \left[\frac{\sin(n - m)x}{n - m} - \frac{\sin(n + m)x}{n + m} \right|_{-\pi}^{\pi} \\ &= 0 \end{aligned}$$

while $$\int_{-\pi}^{\pi} \frac{\sin nx}{\sqrt{\pi}} \frac{\sin nx}{\sqrt{\pi}}\, dx = \frac{1}{2\pi} \int_{-\pi}^{\pi} (1 - \cos 2nx)\, dx$$

$$= \frac{\pi}{\pi} = 1$$

The Fourier coefficients of $f(x)$ on the interval $[-\pi, \pi]$ with respect to the functions (6-71) will lead to a Fourier series having the form

$$f(x) \sim A_0 \frac{1}{\sqrt{2\pi}} + \sum_1^\infty A_n \frac{\cos nx}{\sqrt{\pi}} + \sum_1^\infty B_n \frac{\sin nx}{\sqrt{\pi}} \tag{6-73}$$

where the A_k and B_k are calculated from (6-67). However, it is easier to absorb the terms involving $\sqrt{\pi}$ into the coefficients, arriving at the more traditional series

$$f(x) \sim \frac{1}{2} a_0 + \sum_1^\infty (a_n \cos nx + b_n \sin nx) \tag{6-74}$$

where the coefficients are now given by the formulas

$$a_n = \frac{1}{\pi} \int_{-\pi}^{\pi} f(x) \cos nx\, dx$$

$$b_n = \frac{1}{\pi} \int_{-\pi}^{\pi} f(x) \sin nx\, dx \tag{6-75}$$

For example, if $f(x) = x$, then we find $a_n = 0$ for all n, and $b_n = (2/n)(-1)^{n+1}$. (In calculating these, it is helpful to observe that any odd function, obeying $g(-x) = -g(x)$, has a vanishing integral over any symmetric interval $[-c, c]$; since $f(x) = x$ is odd and $\cos nx$ is even, $a_n = 0$.) After dividing by 2, we therefore obtain the following Fourier series:

$$\frac{x}{2} \sim \sin x - \frac{1}{2} \sin 2x + \frac{1}{3} \sin 3x - \frac{1}{4} \sin 4x + \cdots \tag{6-76}$$

Following the treatment of (5-17), we can apply Dirichlet's test to show that the series is convergent pointwise for every x. This shows that (6-76) cannot be taken as equality for all x on $[-\pi, \pi]$, for the right side is 0 if $x = \pi$ or $-\pi$. However, if we set $x = \pi/2$, (6-76) becomes

$$\frac{\pi}{4} \sim 1 - \frac{1}{3} + \frac{1}{5} - \frac{1}{7} + \frac{1}{9} - \cdots$$

which agrees with the result in (6-16), found from the power series for arctan x.

Again, if $f(x) = x^2/4$, then the same process (simplified by observing that f is an even function) yields the Fourier series

$$\frac{x^2}{4} \sim \frac{\pi^2}{12} - \cos x + \frac{1}{4} \cos 2x - \frac{1}{9} \cos 3x + \cdots \tag{6-77}$$

Since $\sum_1^\infty 1/n^2$ converges, we see that the series converges uniformly for all x. However, we do not as yet know that the function to which it converges must be $x^2/4$. We can check that this is plausible by setting $x = \pi$ in (6-77) to obtain

$$\frac{\pi^2}{4} \sim \frac{\pi^2}{12} + 1 + \frac{1}{4} + \frac{1}{9} + \frac{1}{16} + \cdots$$

which is confirmed by the "computation" in Sec. 5.5 that led to (5-39), although at this stage we certainly cannot say we have proved $\sum 1/n^2 = \pi^2/6$.

Additional "confirmation" of the consistency of all this can be seen by looking at the formal derivative of (6-77) and comparing the result with (6-76). However, legitimate doubts should arise, since if we differentiate (6-74) in the same way, we then obtain the relation

$$\text{(6-78)} \qquad \frac{1}{2} \sim \cos x - \cos 2x + \cos 3x - \cos 4x + \cdots$$

and this series diverges for every choice of x. The mixture of sense and nonsense which these examples illustrate make it necessary to develop more theory.

We suppose that $\{\phi_n\}$ is an arbitrary orthonormal set, not necessarily a basis, and we propose to study the relationship between a function f and its Fourier series; there are three main questions which we examine. The first deals with the Fourier coefficients themselves. If $f \sim \{c_n\}$, then we ask if the sequence $\{c_n\}$ uniquely characterizes the function f among all other functions in $*\mathscr{C}$, and if so, how properties of f are reflected in $\{c_n\}$. The first half of this question can be rephrased usefully. If f and g both have the same Fourier sequence, then $\langle f, \phi_n\rangle = \langle g, \phi_n\rangle$ for all n, and if $h = f - g$, $\langle h, \phi_n\rangle = 0$ for all n. Since $f = g$ if and only if $h = 0$, we are led to ask if 0 is the only function in $*\mathscr{C}$ that is orthogonal to all the functions ϕ_n. The uniqueness question for a given orthonormal set $\{\phi_n\}$ is therefore the same as asking if the set is maximal in $*\mathscr{C}$, meaning that no additional function can be found in $*\mathscr{C}$ which is orthogonal to all the ϕ_n already chosen and which is not identically 0. If an orthonormal set $\{\phi_n\}$ is maximal in $*\mathscr{C}$, then every f in $*\mathscr{C}$ is characterized uniquely by its Fourier coefficients. (Instead of "maximal," some writers on Fourier theory use the more ambiguous terms "complete" or "closed.") It should be noted that a given set might be maximal with respect to $*\mathscr{C}$ but not maximal with respect to a larger space of functions, since it might be possible to find a *discontinuous* function that is orthogonal to all the original ϕ_n which could then be adjoined to the set to form a larger orthonormal set.

Once we have found the Fourier coefficients $\{c_n\}$ for f, we can form the Fourier series $\sum_1^\infty c_n \phi_n$. The remaining questions deal with the convergence of this series, and in particular, whether it converges to f, either in the mean (i.e., in the topology of the space $*\mathscr{C}$) or in more familiar ways such as pointwise or uniformly on I.

We start by looking at finite linear combinations (sums), such as

$P = \sum_1^N a_n \phi_n$. We call these **Fourier polynomials,** by analogy with elementary algebra; in the classical trigonometric case, these are called trigonometric polynomials. We prove several useful identities.

Lemma 1 *If* $P = \sum_1^N a_n \phi_n$, *and* $f \in {}^*\mathscr{C}$, *then*

(6-79) $$\langle f, P \rangle = \sum_1^N a_n \langle f, \phi_n \rangle = \sum_1^N a_n c_n$$

(6-80) $$\langle \phi_k, P \rangle = \begin{cases} a_k & \text{if } k \le N \\ 0 & \text{if } k > N \end{cases}$$

For,
$$\begin{aligned} \langle f, P \rangle &= \langle f, a_1\phi_1 + a_2\phi_2 + \cdots + a_N\phi_N \rangle \\ &= a_1\langle f, \phi_1 \rangle + a_2\langle f, \phi_2 \rangle + \cdots\ a_N\langle f, \phi_N \rangle \\ &= a_1c_1 + a_2c_2 + \cdots + a_Nc_N \end{aligned}$$

Lemma 2 *If* $P = \sum_1^N a_n \phi_n$, then

(6-81) $$^*\|P\| = \left\{ \sum_1^N |a_n|^2 \right\}^{1/2}$$

For,
$$\begin{aligned} ^*\|P\|^2 = \langle P, P \rangle &= \sum_1^N a_n \langle P, \phi_n \rangle \\ &= \sum_1^N a_n \langle \phi_n, P \rangle = \sum_1^N |a_n|^2 \quad \blacksquare \end{aligned}$$

We next use these to show that the partial sums of the Fourier series for a function f provide optimal approximations to f, among all Fourier polynomials.

Theorem 25 *If* $f \sim \{c_n\}$ *and* $S_N = \sum_1^N c_n \phi_n$, *then*

(6-82) $$^*\|f - S_N\| \le {}^*\|f - P\|$$

where P may be any Fourier polynomial of the form $\sum_1^N a_n \phi_n$.

We have

$$\begin{aligned} ^*\|f - P\|^2 &= \langle f - P, f - P \rangle \\ &= \langle f, f \rangle - \langle f, P \rangle - \langle P, f \rangle + \langle P, P \rangle \end{aligned}$$

Then, by (6-79) and (6-81), this becomes

$$*\|f-P\|^2 = \langle f,f\rangle - 2\sum_1^N a_n c_n + \sum_1^N |a_n|^2$$

and then, with minor adjustments,

$$*\|f-P\|^2 = *\|f\|^2 - \sum_1^N |c_n|^2 + \sum_1^N |a_n - c_n|^2 \tag{6-83}$$

Since the numbers c_n are determined by f, the minimum of the right-hand side of (6-83), as the a_n vary, is clearly achieved uniquely by the choice $a_1 = c_1$, $a_2 = c_2, \ldots$, and the optimal polynomial P is S_N itself. ▮

From this simple result, we obtain a large number of very useful implications.

Corollary 1 *If $f \sim \{c_n\}$ then*

$$\sum_1^\infty |c_n|^2 \quad \textit{converges} \tag{6-84}$$

$$\lim_{n\to\infty} c_n = 0 \tag{6-85}$$

$$\sum_1^\infty |c_n|^2 \le *\|f\|^2 \quad \textbf{(Bessel's Inequality)} \tag{6-86}$$

These all arise from the identity

$$*\|f-S_N\|^2 = *\|f\|^2 - \sum_1^N |c_n|^2 \tag{6-87}$$

which comes from (6-83) by setting $a_n = c_n$. Since the left side of (6-87) cannot be negative,

$$\sum_1^N |c_n|^2 \le *\|f\|^2$$

for any N, and (6-84), (6-85), and (6-86) follow.

Corollary 2 *If $f \sim \{c_n\}$ and S_N is the Nth partial sum of the Fourier series $\sum_1^\infty c_n \phi_n$, then $*\|f - S_N\|$ is monotonic decreasing as $N \to \infty$, and*

$$f = *\sum_1^\infty c_n \phi_n = *\lim_{N\to\infty} S_N \tag{6-88}$$

if and only if

(6-89) $$*\|f\| = \left\{\sum_{1}^{\infty} |c_n|^2\right\}^{1/2}$$

This follows directly from (6-84); the relation (6-89) is sometimes called **Parseval's formula.**

At this stage, we can now apply these results to the study of any specific orthonormal set, such as the Legendre polynomials or the classical trigonometric functions, (6-71). For example, because of (6-85) we know immediately that the trigonometric series (6-78) cannot be the Fourier series of a function in $*\mathscr{C}$, since the coefficients do not converge to 0. Again, the following trigonometric series is not a Fourier series

(6-90) $$\sin x + \frac{1}{\sqrt{2}} \sin (2x) + \frac{1}{\sqrt{3}} \sin (3x) + \cdots$$

since $\sum_1^\infty (1/\sqrt{n})^2$ diverges. We note that this series in fact converges pointwise for all x, by the Dirichlet test. This leads to the reasonable conjecture that (6-90) ought to be the Fourier series (in some sense) for something, even if it is not the Fourier series for a well-behaved function. We return to this at the end of the present section.

Turning now to the classical trigonometric Fourier series, based on the orthogonal functions (6-71) and the formulas (6-74) and (6-75), the Parseval relation becomes

(6-91) $$\frac{1}{2}|a_0|^2 + \sum_1^\infty |a_n|^2 + \sum_1^\infty |b_n|^2 = \frac{1}{\pi}\int_{-\pi}^{\pi} |f(x)|^2\, dx$$

which, by Corollary 2, is a necessary and sufficient condition for the Fourier series (6-74) to be convergent "in the mean" to $f(x)$, in the metric of $*\mathscr{C}$.

If we apply this to the series (6-76), Parseval's relation becomes the assertion

$$0 + 0 + \sum_1^\infty \frac{1}{n^2} = \frac{1}{\pi}\int_{-\pi}^{\pi} \frac{x^2}{4}\, dx = \frac{\pi^2}{6}$$

(which is by now an old friend). Thus, we see that we can conclude that the Fourier series (6-76) is convergent in the mean to $x/2$ if we can finally establish the series identity (5-39), and vice versa.

As we have seen from (6-78) and (6-90), not every series in the orthogonal functions ϕ_n is a Fourier series, including some that converge pointwise as series of functions. This is not the case if we use the notion of convergence in the space $*\mathscr{C}$, for then every convergent series is a Fourier series.

Theorem 26 *Let* $\sum_1^\infty a_n \phi_n$ *converge in* $^*\mathscr{C}$ *to a function* $g \in {}^*\mathscr{C}$. *Then*

i. $\sum_1^\infty |a_n|^2 < \infty$.
ii. $g \sim \{a_n\}$.
iii. *If f is any function in* $^*\mathscr{C}$ *and* $f \sim \{c_n\}$, *then*

$$\langle f, g \rangle = \sum_1^\infty a_n c_n \tag{6-92}$$

The proof uses the Schwarz inequality, (6-62). Let $F_N = \sum_1^N a_n \phi_n$, so that we have, by hypothesis, $^*\lim_{N\to\infty} F_N = g$. By Lemma 1, for any $f \in {}^*\mathscr{C}$,

$$\langle f, F_N \rangle = \sum_1^N a_n c_n$$

so that
$$\langle f, g \rangle - \sum_1^N a_n c_n = \langle f, g \rangle - \langle f, F_N \rangle$$
$$= \langle f, g - F_N \rangle$$

and thus by (6-62),

$$\left| \langle f, g \rangle - \sum_1^N a_n c_n \right| = |\langle f, g - F_N \rangle|$$
$$\le {}^*\|f\|^*\|g - F_N\|$$

But, $\lim_{N\to\infty} {}^*\|g - F_N\| = 0$, and we have proved (iii). To obtain (ii), we take f as ϕ_k, and (6-92) gives us $\langle \phi_k, g \rangle = a_k$, so that $g \sim \{a_n\}$, from which (i) follows. ▌

It is natural to ask if the condition $\sum_1^\infty |a_n|^2 < \infty$ is *sufficient* to imply that the series $\sum_1^\infty a_n \phi_n$ converges. The complicated nature of this question can be seen from the following simple observation.

Theorem 27 *If* $\sum_1^\infty |a_n|^2$ *converges, then the series* $\sum_1^\infty a_n \phi_n$ *is a Cauchy series, meaning that the sequence of partial sums is a Cauchy sequence in the space* $^*\mathscr{C}$.

For, if $F_N = \sum_1^N a_n \phi_n$, then, applying Lemma 2 to the Fourier polynomial

$$F_M - F_N = a_{N+1}\phi_{N+1} + \cdots + a_M \phi_M$$

we have
$$*\|F_M - F_N\|^2 = \sum_{N+1}^{M} |a_n|^2$$

and since $\sum |a_n|^2$ converges, the right side of this approaches 0 as N and M become large, proving that $\{F_k\}$ is a Cauchy sequence in the metric of $*\mathscr{C}$.

However, as we have seen, the space $*\mathscr{C}$ is not complete, and not every Cauchy sequence converges. As mentioned earlier, this can be overcome by enlarging the space $*\mathscr{C}$ by including a great many discontinuous functions. When this is done, arriving at a function space $\mathscr{F}$ that *is* complete, then every series $\sum_1^\infty a_n \phi_n$ with $\sum_1^\infty |a_n|^2 < \infty$ is convergent, and is therefore the Fourier series of a member of $\mathscr{F}$. (As noted before, this process requires a much more thorough discussion of integration and Lebesgue measure theory, and is best treated as a subject in itself.)

Let us return to the initial question: How can one tell if an orthonormal set $\{\phi_n\}$ is a Fourier basis for the space $*\mathscr{C}$?

Theorem 28 *The set $\{\phi_n\}$ is a basis for $*\mathscr{C}$ if and only if the collection of all Fourier polynomials P is dense in $*\mathscr{C}$.*

By "dense," we mean that given any $f \in *\mathscr{C}$, and any $\varepsilon > 0$, there is a P with $*\|f - P\| < \varepsilon$. This clearly holds if $\{\phi_n\}$ is a basis, since P could be chosen as one of the partial sums of the Fourier series for f, which by assumption converges to f. Conversely, suppose the polynomials are dense; we must then prove that the Fourier series for any f converges to f. Choose any ε, and then P with $*\|f - P\| < \varepsilon$. Suppose that P has the form $\sum_1^N a_n \phi_n$. Then, combining the minimality property of the partial sums $\{S_k\}$ of the Fourier series $\sum c_n \phi_n$ for f and their monotonic property (Theorem 25), we have

$$*\|f - S_k\| \le *\|f - S_N\| \le *\|f - P\| < \varepsilon$$

for all $k > N$, and $*\lim_{k \to \infty} S_k = f$, and $f = *\sum_1^\infty c_n \phi_n$. ∎

At this point, we leave the general theory and turn instead to some results that deal specifically with the classical trigonometric Fourier series. Our first theorem shows that the Fourier coefficients of a function $f \in *\mathscr{C}$ characterize the function.

Theorem 29 *The trigonometric functions* (6-71) *form a maximal orthogonal set on the interval* $[-\pi, \pi]$.

What we must prove is that if f is continuous in the interval $I = [-\pi, \pi]$ and if

$$\int_{-\pi}^{\pi} f(x)\cos(nx)\,dx = \int_{-\pi}^{\pi} f(x)\sin(nx)\,dx = 0 \tag{6-93}$$

for all $n = 0, 1, 2, \ldots$, then f is identically 0 on I. We first note that because of (6-93),

$$\int_{-\pi}^{\pi} f(x)P(x)\,dx = 0 \tag{6-94}$$

for any trigonometric polynomial of the form

$$P(x) = \sum_{0}^{N} (\alpha_k \cos(kx) + \beta_k \sin(kx))$$

We will use this to show that $f(x) = 0$ for every $x \in I$ by choosing a very special trigonometric polynomial $P(x)$, and do this first for $x = 0$. Suppose that $f(0) \neq 0$, and assume that $f(0) > 0$. Since f is continuous, there is an $\varepsilon > 0$ and a neighborhood about 0 in I on which $f(x) > \varepsilon$. We next wish to choose $P(x)$ so that it is extremely small outside this neighborhood of 0, but very large for the points x near 0; such a function clearly cannot satisfy relation (6-94). To construct $P(x)$, we need two elementary results about the cosine function.

Lemma 3 *Each of the functions* $(\cos x)^m$ *is a trigonometric polynomial.*

The proof is by induction. We have $(\cos x)^2 = \frac{1}{2} + (\frac{1}{2})\cos 2x$. Suppose that

$$(\cos x)^N = c_0 + c_1 \cos x + c_2 \cos(2x) + \cdots + c_N \cos(Nx)$$

Then, using the identity (6-72),

$$\begin{aligned}(\cos x)^{N+1} &= \sum_{0}^{N} c_k \cos x \cos(kx) \\ &= \sum_{0}^{N} \tfrac{1}{2}c_k(\cos(k+1)x + \cos(k-1)x)\end{aligned}$$

Lemma 4 *If* $|x| \leq \pi$, *then*

$$1 - \tfrac{1}{2}x^2 \leq \cos x \leq 1 - \tfrac{1}{5}x^2 \tag{6-95}$$

This is a routine application of Taylor's theorem with remainder.

Using these, we now construct the needed polynomial $P(x)$. We suppose that $f(x) > \varepsilon$ for all x with $|x| < \delta$. Then, if $|x| < \delta/3$, we have by (6-95)

$$\tfrac{1}{6}\delta^2 + \cos x > \tfrac{1}{6}\delta^2 + 1 - \tfrac{1}{2}(\tfrac{1}{9}\delta^2) = 1 + \tfrac{1}{9}\delta^2$$

while if $\delta < |x| \le \pi$,

$$-1 + \tfrac{1}{6}\delta^2 \le \tfrac{1}{6}\delta^2 + \cos x < \tfrac{1}{6}\delta^2 + 1 - \tfrac{1}{5}\delta^2 = 1 - \tfrac{1}{30}\delta^2$$

Choose n very large, and set

$$P(x) = (\tfrac{1}{6}\delta^2 + \cos x)^{2n}$$

Then, expanding this in powers of $\cos x$ and using Lemma 3, we see that $P(x)$ is a trigonometric polynomial which is uniformly arbitrarily large for $|x| < \delta/3$ and uniformly arbitrarily small for $\delta < |x| \le \pi$, and for it (6-94) must fail. This shows that $f(0) = 0$. A similar argument can be made at any point $x = c$ on the interval I by using the polynomial

$$P(x) = (\tfrac{1}{6}\delta^2 + \cos(x - c))^{2n}$$

where $|f(x)| > \varepsilon$ for all x with $|x - c| < \delta$. ∎

We next use this to prove that any sufficiently smooth periodic continuous function has a classical Fourier series that converges to it uniformly on the interval I. We prove this in two steps.

Theorem 30 *Let f be of class C' on $[-\pi, \pi]$, and suppose that $f(\pi) = f(-\pi)$. Then the Fourier coefficients $\{a_n\}$, $\{b_n\}$ for f obey $\sum_0^\infty |a_n| < \infty$, $\sum_1^\infty |b_n| < \infty$.*

By hypothesis, f' is continuous on $[-\pi, \pi]$. Using formulas (6-75), we find the Fourier coefficients of f' in terms of those of f.

$$\begin{aligned}
\alpha_n &= \frac{1}{\pi}\int_{-\pi}^{\pi} f'(x)\cos(nx)\,dx \\
&= \frac{1}{\pi} f(x)\cos(nx)\Big|_{-\pi}^{\pi} + \frac{n}{\pi}\int_{-\pi}^{\pi} f(x)\sin(nx)\,dx \\
&= (-1)^n(f(\pi) - f(-\pi)) + nb_n \\
&= 0 + nb_n \\
\beta_n &= \frac{1}{\pi}\int_{-\pi}^{\pi} f'(x)\sin(nx)\,dx \\
&= \frac{1}{\pi} f(x)\sin(nx)\Big|_{-\pi}^{\pi} - \frac{n}{\pi}\int_{-\pi}^{\pi} f(x)\cos(nx)\,dx \\
&= 0 - na_n
\end{aligned}$$

Since α_n and β_n are Fourier coefficients of a function in $*\mathscr{C}$, Theorem 26 applies and $\sum |\alpha_n|^2$ and $\sum |\beta_n|^2$ both converge; accordingly, $\sum n^2|a_n|^2$ and $\sum n^2|b_n|^2$ both converge. However, a simple argument with Schwarz' inequality shows that in general, if $\sum n^2|c_n|^2$ converges, so must the series $\sum |c_n|$, for

$$\left(\sum_1^N |c_n|\right)^2 = \left(\sum_1^N (n|c_n|)\left(\frac{1}{n}\right)\right)^2$$
$$\leq \sum_1^N n^2|c_n|^2 \sum_1^N \frac{1}{n^2}$$
$$\leq \sum_1^\infty n^2|c_n|^2 \sum_1^\infty \frac{1}{n^2}$$

for all N. Applying this separately, we have $\sum |a_n|$ and $\sum |b_n|$ both convergent. ▌

The next result depends upon the fact that the trigonometric orthogonal set is uniformly bounded.

Theorem 31 *If f is continuous on $[-\pi, \pi]$ and has Fourier coefficients $\{a_n\}$, $\{b_n\}$, and $\sum |a_n|$ and $\sum |b_n|$ both converge, then the Fourier series for f*

$$\tfrac{1}{2}a_0 + \sum_1^\infty (a_n \cos(nx) + b_n \sin(nx)) \tag{6-96}$$

converges uniformly to $f(x)$ on $[-\pi, \pi]$.

First, we observe that the hypothesis on the coefficients ensures that the series (6-96) converges uniformly on I to some function $g(x)$, necessarily continuous. The relation between the norm $*\|\ \|$ and $\|\ \|_I$ (Theorem 24) implies that the series (6-96) converges to g in the metric of $*\mathscr{C}$, and Theorem 26 implies that the sequences $\{a_n\}$, $\{b_n\}$ must be the Fourier coefficients of g. Accordingly, f and g are both continuous on I and have the same Fourier coefficients. By Theorem 29, $f = g$. ▌

Combining the last two results yields a number of useful corollaries.

Corollary 1 *Any periodic function f of class C' on $[-\pi, \pi]$ has a uniformly convergent Fourier series.*

In particular, this applies to the function $x^2/4$, so that we have at last verified (6-77) and the formula for the sum of the numerical series $1/n^2$. We next obtain one form of the Weierstrass approximation theorem.

Corollary 2 *Any periodic continuous function on* $[-\pi, \pi]$ *can be uniformly approximated there by trigonometric polynomials.*

In Sec. 2.3, Theorem 7, we saw that f could be approximated uniformly by continuous piecewise linear functions, formed from a finite number of line segments. By rounding off the corners, we see that such a piecewise linear function can be uniformly approximated by a function g of class C'. If f is periodic, so that $f(\pi) = f(-\pi)$, then g can be chosen to obey $g(\pi) = g(-\pi)$, and if we now approximate g on $[-\pi, \pi]$ by a partial sum of its uniformly convergent Fourier series, the resulting trigonometric polynomial will be the desired uniform approximation to the function f.

If we now apply Theorem 28, we obtain the following.

Corollary 3 *The trigonometric orthonormal set* (6-71) *is a basis for* $^*\mathscr{C}$, *and every continuous function f has a Fourier series which converges to it "in the mean," that is, in the metric of the space* $^*\mathscr{C}$.

We must prove that the trigonometric polynomials are dense in $^*\mathscr{C}$. By Corollary 2, they are uniformly dense in the space of periodic continuous functions on $[-\pi, \pi]$, and thus by Theorem 24, they are also dense in the metric $^*\| \quad \|$. We need now only observe that the periodic functions are dense in $^*\mathscr{C}$. For, if f is continuous, but $f(\pi) \neq f(-\pi)$, then a continuous function g can be chosen which coincides with f exactly, except in a small neighborhood of $x = \pi$, and which is such that $g(-\pi) = g(\pi)$ and $^*\|f - g\|$ is as small as desired. (The same argument shows considerably more; the trigonometric polynomials are dense in the metric $^*\| \quad \|$ in a much larger space than $^*\mathscr{C}$, including, for example, all Riemann-integrable discontinuous functions. Thus, any piecewise continuous function on $[-\pi, \pi]$ is the limit in the mean of the partial sums of its Fourier series.)

We conclude this brief introduction to the theory of general Fourier series with two remarks. The first deals with the sequence space ℓ^2. Perhaps it has been apparent that this space has played a central role. Given any infinite orthonormal set ϕ_n in $^*\mathscr{C}$, the correspondence $f \sim \{c_n\}$, where $c_n = \langle f, \phi_n \rangle$, defines a mapping from $^*\mathscr{C}$ into ℓ^2, because of the Bessel inequality. If the set $\{\phi_n\}$ is maximal, this mapping is 1-to-1. However, even if $\{\phi_n\}$ is a basis, this mapping need not be onto, so that the complete image of $^*\mathscr{C}$ will not fill up ℓ^2. However, the mapping has several very nice properties. For example, it preserves distance and angle; this is the geometric meaning of Parseval's relation, (6-89), and (6-92). For, if $f \sim \{c_n\}$ and $g \sim \{b_n\}$, and p and q

are the points in ℓ^2 identified with the sequences $\{c_n\}$ and $\{b_n\}$, then

$$*\|f - g\| = \|p - q\|$$

and
$$\langle f, g\rangle = p \cdot q$$

Thus, each basis for $*\mathscr{C}$ yields a different congruent embedding of $*\mathscr{C}$ as a subset of ℓ^2. The seemingly simple space ℓ^2 contains all the complexity associated with any basis for the space of continuous functions, as well as that for the enlarged function spaces mentioned earlier in this section.

Our second remark returns to several examples given earlier, namely the series

$$\sin x + \frac{1}{\sqrt{2}} \sin (2x) + \frac{1}{\sqrt{3}} \sin (3x) + \cdots \tag{6-97}$$

and

$$\cos x - \cos (2x) + \cos (3x) - \cos (4x) + \cdots \tag{6-98}$$

The first converges for all x, by Dirichlet's test. The second converges for no x. Neither is a Fourier series, for the coefficients do not satisfy (6-84). The series can be obtained by formal differentiation of the following series:

$$-\cos x - \frac{1}{2\sqrt{2}} \cos (2x) - \frac{1}{3\sqrt{3}} \cos (3x) - \cdots \tag{6-99}$$

$$\sin x - \tfrac{1}{2} \sin (2x) + \tfrac{1}{3} \sin (3x) - \cdots \tag{6-100}$$

which *are* in fact Fourier series [again by reference to (6-84)]. This suggests that (6-97) and (6-98) ought to be (in some new sense) the Fourier series for the derivatives of the functions associated with the Fourier series (6-99) and (6-100).

Such a mathematical theory has been developed, based on what is called the theory of distributions. For it, one must first invent a new concept of differentiation; it then turns out that the "derivative" of a function which does not have an ordinary derivative can exist but may be something quite different from the conventional notion of function, and it is these objects that have Fourier series such as (6-97). (Some aspects of this are discussed in Appendix 6.)

EXERCISES

1 Find the angle between the following points in ℓ^2.

$$p = \left(1, \frac{1}{2}, \frac{1}{3}, \frac{1}{4}, \frac{1}{5}, \ldots\right) \qquad q = \left(\frac{1}{2}, \frac{1}{3}, \frac{1}{4}, \frac{1}{5}, \ldots\right)$$

2 Verify the following properties of the inner product and norm in the space $*\mathscr{C}$.

(*a*) $d(f, g) = d(g, f)$ and $d(f, g) = 0$ if and only if $f = g$

(*b*) $\langle f, g_1 + g_2\rangle = \langle f, g_1\rangle + \langle f, g_2\rangle$

$\langle f, \alpha g\rangle = \alpha\langle f, g\rangle = \langle \alpha f, g\rangle$

$*\|f\| = \sqrt{\langle f, f\rangle}$

3 Verify that the functions $f_n(t) = t^{1/(2n-1)}$ are *Cauchy in the space $^*\mathscr{C}[-1, 1]$.

4 Prove Theorem 24 by estimating the integral in (6-59).

5 Use (6-72) to finish verifying that the set (6-71) is orthonormal on $[-\pi, \pi]$.

6 Verify formulas (6-74) and (6-75).

7 Verify (6-76).

8 Verify (6-77).

9 Find the Fourier series of e^x on $[-\pi, \pi]$ with respect to the orthonormal functions in (6-71).

10 The space of complex-valued continuous functions on the interval $I = [0,1]$ can be made an inner product metric space by means of the definition: $\langle f, g\rangle = \int_0^1 f(t)\overline{g(t)}\, dt$, $\|f\| = \sqrt{\langle f, f\rangle}$ and $d(f, g) = \|f - g\|$ where $\overline{g}$ is the complex conjugate of g. Verify the following properties:

(*a*) $\langle f, g\rangle = \langle \overline{g}, \overline{f}\rangle = \overline{\langle g, f\rangle}$

(*b*) $\beta\langle f, g\rangle = \langle \beta f, g\rangle = \langle f, \overline{\beta} g\rangle$

(*c*) $\|\alpha f\| = |\alpha|\,\|f\|$

(*d*) $\|f - g\| = 0$ implies that $f = g$

(*e*) $\langle f, g\rangle + \langle g, f\rangle \le 2\|f\|\,\|g\|$

(*f*) $|\langle f, g\rangle| \le \|f\|\,\|g\|$

(*g*) $\|f + g\| \le \|f\| + \|g\|$

(*h*) Complex ℓ^2 has points $p = (a_1, a_2, \ldots)$ where the a_n are complex numbers obeying $\sum |a_n|^2 < \infty$. What is an appropriate definition for the inner product $p \cdot q$?

11 (*a*) Verify that the set $\{\phi_n\} = \{e^{inx}\}$, for $n = 0, \pm 1, \pm 2, \pm 3, \ldots$, forms an orthogonal set on $[0, 2\pi]$. Find the normalizing constants that convert this set into an orthonormal system, and find the formula for the Fourier coefficients of a function $f(x)$ with respect to these functions $\{\phi_n\}$.

(*b*) Show how this set is related to the set (6-71).

12 (*a*) Use (6-65) to obtain the Legendre polynomials for $n \le 5$.

(*b*) Verify directly that P_3 is orthogonal on $[-1, 1]$ to P_1 and to P_2.

(*c*) Make a change of variable: $t = (x + 1)/2$, and compare the result with the special orthogonal polynomials $\phi_n(t)$ in (6-64).

13 The Chebyshev polynomials $T_n(x)$ are defined by the recursion

$$T_0(x) = 1$$
$$T_1(x) = x$$
$$T_{n+1}(x) = 2xT_n(x) - T_{n-1}(x)$$

(*a*) Find $T_n(x)$ for $n \le 5$.

(*b*) Verify that $T_n(\cos\theta) = \cos(n\theta)$.

(*c*) Verify that $|T_n(x)| \le 1$ for $-1 \le x \le 1$.

(*d*) Verify that if $n \ne m$

$$\int_{-1}^{1} T_n(x)T_m(x)\,\frac{dx}{\sqrt{1 - x^2}} = 0.$$

14 Complete the following proof that ℓ^2 is complete.

(*a*) If $p = (b_1, b_2, b_3, \ldots) \in \ell^2$, define for any k,

$$U_k(p) = \sum_{n \le k} b_n e_n, \qquad V_k(p) = \sum_{n > k} b_n e_n$$

(*b*) Verify that $\|U_k(p)\| \le \|p\|$, $\|V_k(p)\| \le \|p\|$, $\lim_{k\to\infty} \|U_k(p)\| = \|p\|$, and $\lim_{k\to\infty} \|V_k(p)\| = 0$.

(*c*) Let $p_n = (a_1^n, a_2^n, a_3^n, \ldots)$ be a Cauchy sequence of points in ℓ^2. Use $U_k(p_n)$ to prove that $\{a_k^n\}$ is a Cauchy sequence for each k.

(*d*) Let $c_k = \lim_{n\to\infty} a_k^n$ and prove that $q = (c_1, c_2, c_3, \ldots)$ belongs to ℓ^2.

(*e*) Show that for any $\varepsilon > 0$, there is a k_0 and an N such that $\|V_{k_0}(p_n)\| < 2\varepsilon$ for all $n \geq N$.

(*f*) Prove that $\lim_{n\to\infty} \|p_n - q\| = 0$.

CHAPTER

SEVEN

DIFFERENTIATION OF TRANSFORMATIONS

7.1 PREVIEW

This chapter deals with the central core of multidimensional analysis, the study of transformations from n space into m space. A key tool in this is the differential (or derivative) of such a transformation T, regarded either as a matrix-valued function or as a function whose value at a point is a linear transformation. Thus, one has $T'(p_0) = L$, where L is the unique linear transformation such that

$$\lim_{\Delta p \to 0} \frac{|T(p_0 + \Delta p) - T(p_0) - L(\Delta p)|}{|\Delta p|} = 0$$

a formula that resembles the corresponding one for the derivative of a function of one variable.

In terms of this general concept of differentiation, which extends the previous concepts defined for real-valued functions, we then obtain the general chain rule in the simple form $d(ST) = dS\, dT$, and study the existence of local and global inverses for transformations and the properties of nonsingular mappings, defined as those T having a nonvanishing Jacobian. This leads immediately to the implicit function theorems, which give sufficient conditions for the existence of a solution of a system of nonlinear equations, and to the study of functional dependence.

7.2 TRANSFORMATIONS

In Chaps. 1 and 2, we discussed many general properties of functions that were defined on one set and took values in another set, both possibly being subsets of a space of high dimension. However, most of the material discussed since then has concentrated on the study of scalar-valued (i.e., numerical-valued) functions. In the present chapter, we will be working almost entirely with functions whose range of values will be points in m space, for some specific choice of m such as 2 or 3. These were called **vector-valued** functions, to distinguish them from scalar-valued functions. To emphasize a different viewpoint which lays stress on their role as mappings, we shall now use the special term **transformation**. Thus, if A is a set in n space and B a set in m space, a transformation T of A onto B is a function whose domain is A and whose range is B. We shall say that T carries a point p into the point $T(p) = q$ and call q the **image** of p under the transformation T. If p is a point in the plane and q a point in space, we may write $p = (x, y)$ and $q = (u, v, w)$ so that $T(x, y) = (u, v, w)$. In this case, we may also describe the transformation by specifying three coordinate functions, and writing

$$T: \begin{cases} u = f(x, y) \\ v = g(x, y) \\ w = h(x, y) \end{cases}$$

For brevity, we may also refer to such a transformation as a **mapping** from $\mathbf{R}^2$ into $\mathbf{R}^3$, even though its domain may not be all of the plane.

To illustrate these notions, let us begin with a particular transformation of the plane into itself:

$$T: \begin{cases} u = x^2 + y^2 \\ v = x + y \end{cases} \tag{7-1}$$

Under T, the point (x, y) is sent into the point (u, v). For example, the image of $(1, 2)$ is $(5, 3)$, the image of $(0, 2)$ is $(4, 2)$, and the image of $(2, 0)$ is also $(4, 2)$. Any set of points in the XY plane is carried into a corresponding set of points in the UV plane. It often helps to compute the image of a number of selected curves and regions. To determine the image of the horizontal line $y = c$, make this substitution in the equations for T, obtaining

$$\begin{cases} u = x^2 + c^2 \\ v = x + c \end{cases}$$

This can be regarded as parametric equations for the image curve in the UV plane, or x may be eliminated, resulting in the equation:

$$u = (v - c)^2 + c^2$$

(In Fig. 7-1 these curves are shown for several choices of c.) Turning to regions, we first observe that the point (a, b) and the point (b, a) both have the

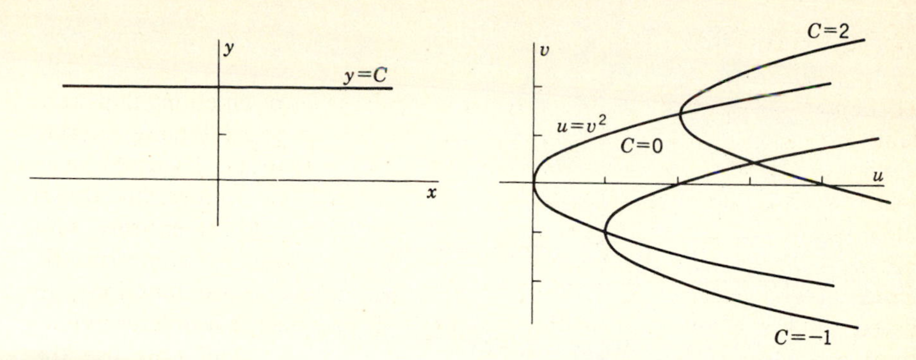

Figure 7-1

same image. Thus, the line $y = x$ divides the XY plane into two half planes which are mapped by T onto the same set in the UV plane. To determine this set, we first find the image of the line $y = x$. Substituting into (7-1), we obtain $u = 2x^2$, $v = 2x$, which are parametric equations for the parabola $v^2 = 2u$. The image of any point (x, y) lies within this curve, for

$$\begin{aligned} 2u - v^2 &= 2(x^2 + y^2) - (x + y)^2 \\ &= x^2 + y^2 - 2xy = (x - y)^2 \geq 0 \end{aligned}$$

Conversely, it is easy to see that every point (u, v) on or within this parabola is in turn the image of a point (x, y) (see Fig. 7-2). Accordingly, one may

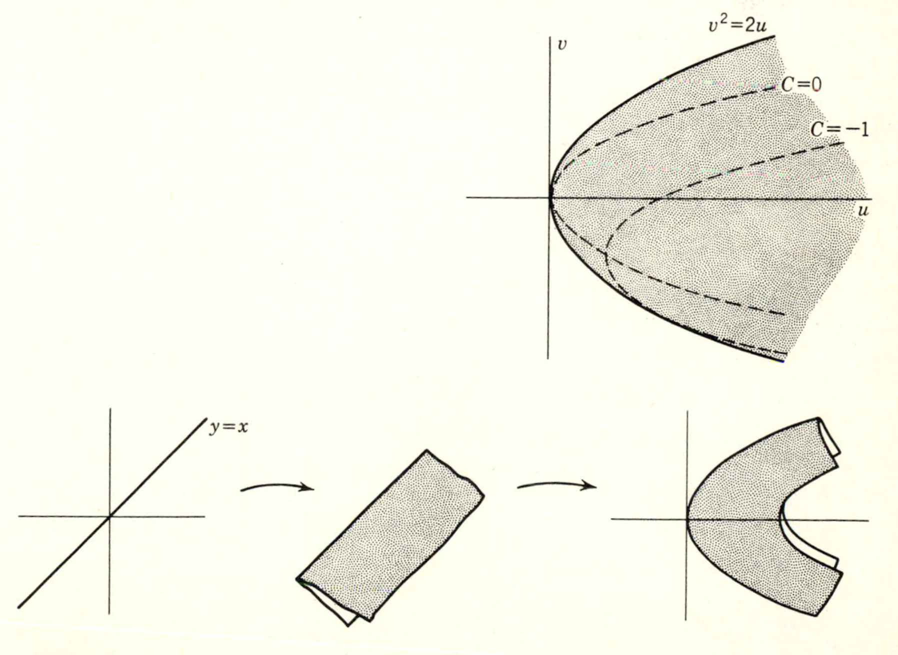

Figure 7-2

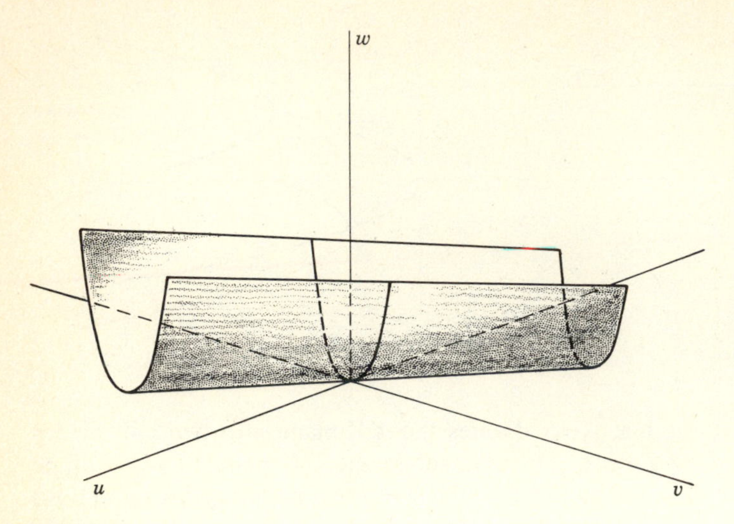

Figure 7-3

picture the effect of T approximately as follows: first, fold the XY plane along the line $y = x$, and then fit the folded edge along the parabola $v^2 = 2u$, and flatten out the rest to cover the inside of the parabola smoothly. (To permit the necessary distortion, think of the XY plane as a sheet of rubber.)

As another illustration, consider the following transformation of 2-space into 3-space.

$$(7\text{-}2) \qquad S: \begin{cases} u = x + y \\ v = x - y \\ w = x^2 \end{cases}$$

The image of $(1, 2)$ under S is $(3, -1, 1)$. The image of the line $y = x$ is the curve given in parametric form by

$$\begin{cases} u = 2x \\ v = 0 \\ w = x^2 \end{cases}$$

which is a parabola lying in the UW plane. The image of the whole XY plane is a parabolic cylinder resting on the UV plane (see Fig. 7-3).

Since transformations belong to the general category of functions, one may also consider their graphs. Except in the simplest cases, this is usually difficult and of little help; for example, the graph of the transformation T described by (7-1) is the set of all points (x, y, u, v) in $\mathbf{R}^4$ for which (7-1) holds, that is, the set of all points

$$(x, y, x^2 + y^2, x + y) \qquad \text{for } -\infty < x < \infty \quad \text{and} \quad -\infty < y < \infty$$

Therefore, we usually adopt other devices in studying transformations. The general transformation from $\mathbf{R}^n$ to $\mathbf{R}^m$ can be described by means of the equations:

$$(7\text{-}3) \qquad T: \begin{cases} y_1 = f_1(x_1, x_2, \ldots, x_n) \\ y_2 = f_2(x_1, x_2, \ldots, x_n) \\ \cdots\cdots\cdots\cdots\cdots\cdots\cdots \\ y_m = f_m(x_1, x_2, \ldots, x_n) \end{cases}$$

where we have written $p = (x_1, x_2, \ldots, x_n)$ for the general point in n space, and $q = (y_1, y_2, \ldots, y_m)$ for the general point in m space. (Note that the subscripts on the functions f_i do *not* indicate partial differentiation.) The labeling of the coordinates of p and q is quite arbitrary; convenience and habit are the guiding criteria. For example, the general transformation from 1-space to 3-space can be described by

$$T: \begin{cases} x = f(t) \\ y = g(t) \\ z = h(t) \end{cases}$$

where we have put $p = t$ and $q = (x, y, z)$. In this case, it often helps in studying T to regard these equations as a particular set of parametric equations for a curve in $\mathbf{R}^3$. Similarly, when $n = 2$ and $m = 3$, we may write

$$\begin{cases} x = f(u, v) \\ y = g(u, v) \\ z = h(u, v) \end{cases}$$

and regard these as a set of parametric equations for a surface in $\mathbf{R}^3$.

Transformations can be combined by substitution to yield new transformations, provided the dimensions of the domains and ranges are compatible. If T is a transformation from $\mathbf{R}^n$ to $\mathbf{R}^m$, and S is a transformation from $\mathbf{R}^m$ to $\mathbf{R}^k$ the transformation R defined by $R(p) = S(T(p))$ is a mapping from $\mathbf{R}^n$ into $\mathbf{R}^k$; R is often called the product or composite of S and T. For example, let T be the mapping which sends (x, y) into $(xy, 2x, -y)$ and S the mapping which sends (x, y, z) into $(x - y, yz)$. Then, the mapping $R = ST$ sends (x, y) into $(xy - 2x, -2xy)$. T is a mapping of $\mathbf{R}^2$ into $\mathbf{R}^3$, and S a mapping of $\mathbf{R}^3$ into $\mathbf{R}^2$, so that R is a mapping from $\mathbf{R}^2$ into $\mathbf{R}^2$.

In order to calculate the product $R = ST$, given equations for S and T, it is convenient to use the same coordinate labels for points in the middle space in both transformations. Thus, if T sends (x, y) into (r, s, t) and S sends (r, s, t) into (u, v), as shown in Fig. 7-4, and if the equations for T and S are

$$(7\text{-}4) \qquad T: \begin{cases} r = xy \\ s = 2x \\ t = -y \end{cases} \qquad S: \begin{cases} u = r - s \\ v = \phantom{r-{}} st \end{cases}$$

then R sends (x, y) into (u, v) and we can read off the equation for $R = ST$

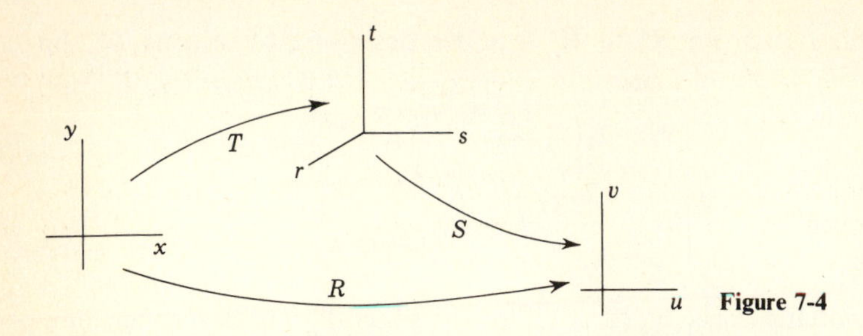

Figure 7-4

directly from (7-4), obtaining

$$R:\begin{cases} u = xy - 2x \\ v = -2xy \end{cases}$$

In this particular example, since T maps 2-space into 3-space, and S maps 3-space into 2-space, we can also form the product of S and T in the opposite order, obtaining $H = TS$, which will be a mapping from 3-space into itself. To calculate H, one might restate the equations for S and T with new coordinate labels for the points. Suppose we say that S maps (x, y, z) into (u, v), and T maps (u, v) into (r, s, t). Then, (7-4) would appear instead in the form

(7-5)
$$S:\begin{cases} u = x - y \\ v = yz \end{cases} \qquad T:\begin{cases} r = uv \\ s = 2u \\ t = -v \end{cases}$$

and $H = TS$ is seen to be

$$H:\begin{cases} r = (x - y)yz \\ s = 2(x - y) \\ t = -yz \end{cases}$$

Note that $ST \neq TS$.

The definition of continuity for a function as stated in Chap. 2 applies to transformations, as do many of the theorems in Secs. 2.2, 2.3, and 2.4 which were proved there for functions whose values could either be numbers or points. For convenience of reference, we restate some of these here, with references to the earlier discussion.

Definition 1 *A transformation T defined on a set D is said to be continuous at a point $p_0 \in D$ if and only if for any $\varepsilon > 0$ there is a $\delta > 0$ such that $|T(p) - T(p_0)| < \varepsilon$ whenever $|p - p_0| < \delta$ and $p \in D$.*

We remark that a transformation described by equations such as (7-3) is continuous on a set D exactly when each of the component functions f_i is continuous on D.

Theorem 1 *Let T be a transformation defined on a set D in n space and taking values in m space. Then, T is continuous on D if and only if $T^{-1}(\mathcal{O})$ is open, relative to D, for every open set $\mathcal{O}$ in m space.*

This was Theorem 3, Sec. 2.2. We observe that the word "open" could be replaced by "closed," since the complement of any open set is closed, and vice versa.

Theorem 2 *Let T be a transformation defined on a compact set C in n space and taking values in m space. Then, T is continuous on C if and only if the graph of T is a compact set.*

This was Theorem 12, Sec. 2.4, and the proof there was written to apply to this general case, although the accompanying figure illustrated the one-variable form.

Theorem 3 *Let the transformation S be continuous on a set A and T be continuous on a set B, and let $p_0 \in A$ and $S(p_0) = q_0 \in B$. Then, the product transformation TS, defined by $TS(p) = T(S(p))$, is continuous at p_0.*

This is Theorem 5, Sec. 2.2; in general, the product or composite of continuous transformations is continuous where it is defined.

Theorem 4 *Let T be continuous on a set D. Then, any compact set $C \subset D$ is carried by T into a compact set $T(C)$, and any connected set $S \subset D$ is carried into a connected set $T(S)$.*

This combines Theorems 13 and 15, Sec. 2.4, and generalizes the more elementary properties of real-valued functions usually described as the intermediate value theorem and the fact that continuous functions always achieve a bounded maximum on any closed and bounded set.

Many of these statements follow in fact from the simpler statements about scalar-valued functions because one can throw the argument back upon the coordinate functions f_i that describe a general transformation T, as in (7-3). For example, Theorem 4 above implies that the continuous image of any compact set C must be a bounded set. This follows from the fact that each of the scalar functions f_i is bounded on C.

EXERCISES

1 Discuss the nature of the transformation T of $\mathbf{R}^2$ into $\mathbf{R}^2$ which:

(*a*) Sends (x, y) into $(x + 3, y - 1)$. (*b*) Sends (x, y) into (y, x).

(*c*) Sends (x, y) into $(x - y, x + y)$. (*d*) Sends (x, y) into (x^2, y^2).

2 Verify the fact that every (u, v) inside the parabola $v^2 = 2u$ is an image under the transformation T described by (7-1) of a point (x, y).

3 What is the image of the line $x = 0$ under the transformation S described by (7-2)?

4 Discuss the nature of the transformation T which sends (x, y) into $(x^2 - y^2, 2xy)$.

5 Find the products in each order of each pair of transformations given in Exercise 1.

6 When a transformation T of $\mathbf{R}^n$ into $\mathbf{R}^m$ is described by means of coordinate functions, as in (7-3), show that T is continuous in a set D if and only if each coordinate function f_i is continuous in D.

7 Construct a transformation of $\mathbf{R}^2$ into $\mathbf{R}^2$ which maps the curve $y = x^2$ onto the horizontal axis, and the line $y = 3$ onto the vertical axis.

8 (*a*) Formulate a definition of uniform continuity for transformations.

(*b*) Show that a transformation T which is continuous in a closed and bounded set D is uniformly continuous there, by representing T in coordinate form.

9 A transformation T is said to be distance preserving if $|T(p) - T(q)| = |p - q|$ for all points p and q in the domain of T. Show that the transformation of the plane into itself which sends (x, y) into $((x + y)/\sqrt{2}, (x - y)/\sqrt{2})$ is distance preserving.

10 Discuss the nature of the transformations T of $\mathbf{R}^2$ into $\mathbf{R}^2$ described by:

(*a*) $(x, y) \xrightarrow{T} (\sin x, \cos y)$ (*b*) $(x, y) \xrightarrow{T} (x^2 + y^2, x^2 - y^2)$

(*c*) $(x, y) \xrightarrow{T} (xy, y)$ (*d*) $T(x, y) = (xy, x^2y^2)$

(*e*) $T(x, y) = (x \cos y, x \sin y)$

11 A transformation T on $\mathbf{R}^n$ to $\mathbf{R}^n$ is said to be **distance decreasing** if there is a constant r, $r < 1$, such that

$$|T(p) - T(q)| \le r|p - q|$$

Show that the transformation defined by $T(x, y) = (3 - \frac{1}{2}x, \frac{3}{2} - \frac{1}{2}y)$ is distance decreasing.

***12** Let T be any distance-decreasing transformation of the plane into itself. Prove that T leaves exactly one point of the plane fixed; that is, $T(p) = p$ has one and only one solution p^*.

13 Write a set of equations for the transformation T which reflects the plane

(*a*) Across the line $x = 0$ (*b*) Across the line $x = 2$

(*c*) Across the line $y = x$ (*d*) Across the line $y = 2x - 1$

7.3 LINEAR FUNCTIONS AND TRANSFORMATIONS

The section which follows this one is devoted to the theory of differentiation for general transformations. The key to this is the use of linear and affine transformations as local approximations to a general transformation. For this, only the most elementary aspects of matrix theory and linear algebra are needed; an abbreviated summary of the latter is given in Appendix 3, although the present section is intended to be sufficient without requiring a review of this.

Definition 2 *A transformation T from $\mathbf{R}^n$ to $\mathbf{R}^m$ is said to be linear if and only if it has the following two properties:*

$$\text{(7-6)} \qquad \begin{aligned} T(p + q) &= T(p) + T(q) && \textit{for all } p \textit{ and } q \\ T(\lambda p) &= \lambda T(p) && \textit{for any real number } \lambda \end{aligned}$$

These defining properties immediately prescribe the form of linear transformations.

Theorem 5 *The general linear transformation L from $\mathbf{R}^n$ to $\mathbf{R}^m$ has the coordinate form:*

$$(7\text{-}7)\qquad \begin{cases} y_1 = a_{11}x_1 + a_{12}x_2 + \cdots + a_{1n}x_n \\ y_2 = a_{21}x_1 + a_{22}x_2 + \cdots + a_{2n}x_n \\ \cdots\cdots\cdots\cdots\cdots\cdots\cdots\cdots \\ y_m = a_{m1}x_1 + a_{m2}x_2 + \cdots + a_{mn}x_n \end{cases}$$

where the coefficients a_{ij} are (real) constants.

For, if we use the standard basis vectors $\mathbf{e}_k$ for $\mathbf{R}^n$, so that $p = (x_1, x_2, \ldots, x_n)$ is written as

$$p = \sum_1^n x_k \mathbf{e}_k = x_1\mathbf{e}_1 + \cdots + x_n\mathbf{e}_n$$

then $L(p) = L\left(\sum_1^n x_k \mathbf{e}_k\right)$, which by (7-6) can be written

$$q = L(p) = \sum_1^n L(\mathbf{e}_k)x_k$$

and if $L(\mathbf{e}_k)$ is the point $(a_{1k}, a_{2k}, a_{3k}, \ldots, a_{mk})$ in m space, and $q = (y_1, y_2, \ldots, y_m)$, then we have obtained the formula (7-7).

The linear transformation L described by (7-7) is also completely specified by giving the **coefficient matrix**

$$(7\text{-}8)\qquad A = [a_{ij}] = \begin{bmatrix} a_{11} & a_{12} & a_{13} & \cdots & a_{1n} \\ a_{21} & a_{22} & a_{23} & \cdots & a_{2n} \\ \cdots & \cdots & \cdots & \cdots & \cdots \\ a_{m1} & a_{m2} & a_{m3} & \cdots & a_{mn} \end{bmatrix}$$

whose columns are the points $L(\mathbf{e}_k)$; the formula in (7-8) has the abbreviated equation

$$(7\text{-}9)\qquad y = Ax$$

where x and y are column vectors and we use the standard way of multiplying matrices. Note that we could also write formula (7-7) as $y_j = v_j \cdot p$, using the ordinary dot product of vectors, where v_j is the jth row of the matrix A.

Scalar-valued linear transformations form an important special case. These are merely real-valued functions that are also linear according to definition. As linear transformations from n space into 1-space, the general formula (7-7) yields

$$(7\text{-}10)\qquad L(p) = a_1x_1 + a_2x_2 + \cdots + a_nx_n$$

and the corresponding matrix has only a single row. Restated, every linear

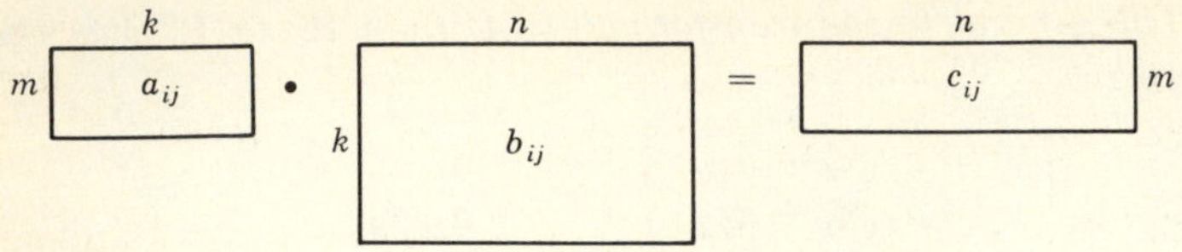

$$\underset{m}{\overset{k}{\boxed{a_{ij}}}} \cdot \underset{k}{\overset{n}{\boxed{b_{ij}}}} = \overset{n}{\boxed{c_{ij}}}\ m$$

Figure 7-5 Size restrictions for multiplication of matrices.

function on n space has the form $L(p) = v \cdot p$ where v is the specific vector $(a_1, a_2, \ldots, a_n)$ associated with L.

For linear transformations, the product of two transformations can be calculated by multiplying their representing matrices. If T maps n space into k space, and S maps k space into m space,

$$\mathbf{R}^n \underset{T}{\rightarrow} \mathbf{R}^k \underset{S}{\rightarrow} \mathbf{R}^m \tag{7-11}$$

then the product ST maps n space into m space, and its matrix is $[S][T]$, where $[S]$ is the matrix for S and $[T]$ the matrix for T. (Multiplication of matrices is done row-by-column, calculating the dot product to obtain the appropriate entry in the product matrix; see Fig. 7-5.)

As an illustration, consider the following linear transformations of the plane into itself; we give both the equations and the representing matrices.

$$S: \begin{array}{l} u = 2x - 3y \\ v = x + y \end{array} \qquad [S] = \begin{bmatrix} 2 & -3 \\ 1 & 1 \end{bmatrix}$$

$$T: \begin{array}{l} u = x + y \\ v = 3x + y \end{array} \qquad [T] = \begin{bmatrix} 1 & 1 \\ 3 & 1 \end{bmatrix}$$

Then,
$$[S][T] = \begin{bmatrix} 2 & -3 \\ 1 & 1 \end{bmatrix} \begin{bmatrix} 1 & 1 \\ 3 & 1 \end{bmatrix} = \begin{bmatrix} -7 & -1 \\ 4 & 2 \end{bmatrix}$$

so that ST has the equation

$$u = -7x - y$$
$$v = 4x + 2y$$

Note also that

$$[T][S] = \begin{bmatrix} 1 & 1 \\ 3 & 1 \end{bmatrix} \begin{bmatrix} 2 & -3 \\ 1 & 1 \end{bmatrix} = \begin{bmatrix} 3 & -2 \\ 7 & -8 \end{bmatrix}$$

which is not the same as $[S][T]$.

The geometric character of a linear transformation is determined by its matrix. The discussion is simplest for square matrices, representing mappings of a space into itself.

Theorem 6 *Let L be a linear transformation of n space into itself, represented by an n-by-n matrix $A = [a_{ij}]$. Then, L is a* 1-*to*-1 *mapping if and* *only if the matrix A is* **nonsingular,** *meaning any of the following equivalent statements: (i) the determinant* det (A) *is not* 0, *(ii) the rows of A are linearly independent, (iii) the columns of A are linearly independent.*

A review of the terms and concepts used in this result may be found in Appendix 3, together with a sketch of the proof. The statements in the theorem have immediate translation into statements about the solution of the system (7-7) of linear equations in the special case when $n = m$ and there are the same number of equations as unknowns. If det $(A) \neq 0$, then for any given set of y_j, there is one and only one solution for the x_i. In particular, when $y_1 = y_2 = \cdots = y_n = 0$ then the only solution of the system is $x_1 = x_2 = \cdots = x_n = 0$. Conversely, if det $(A) = 0$, then L is not 1-to-1, and there exists a point $p = (x_1, x_2, \ldots, x_n) \neq \mathbf{0}$ with $L(p) = \mathbf{0}$. Accordingly, when det $(A) = 0$, the system of linear equations obtained by setting $y_j = 0$ for all j has a solution x_i not all 0.

A linear transformation L that is nonsingular has an **inverse,** L^{-1}, and the matrix representing it is called the **inverse of the matrix** A that represents L. Thus, if det $(A) \neq 0$, there is a unique matrix A^{-1} such that $AA^{-1} = A^{-1}A = I$, the n-by-n identity matrix. There are a number of computational ways to construct A^{-1} from A. If $n = 2$ and

$$A = \begin{bmatrix} a & b \\ c & d \end{bmatrix} \qquad \text{then} \qquad A^{-1} = \begin{bmatrix} \dfrac{d}{\Delta} & -\dfrac{b}{\Delta} \\ -\dfrac{c}{\Delta} & \dfrac{a}{\Delta} \end{bmatrix}$$

where $\Delta = \det(A) = ad - bc \neq 0$

At times it is economical of notation to use the same symbol for a linear transformation and for a matrix representation of it, where this will not lead to confusion. Thus, one might write det (L) rather than det (A).

The behavior of a singular transformation L is further characterized by the rank of its matrix. The **rank** of A is defined as the number of rows (equivalently, the number of columns) that are linearly independent; rank (A) is an integer between 0 and n. One can have rank $(A) = 0$ only if all the entries in A are 0, and rank $(A) = 1$ only if all the rows are multiples of one of the rows, itself not $(0, 0, 0, \ldots, 0) = \mathbf{0}$.

Theorem 7 *Let L be a linear transformation represented by a square n-by-n matrix A; then the following statements are equivalent: (i)* rank $(A) = r$*; (ii) L maps all of n space onto an r-dimensional subspace of n space; (iii) the null space of L, which is the set of all points p with $L(p) = \mathbf{0}$, is a subspace of n space of dimension $n - r$.*

When $r = n$, this result reduces to that of the preceding theorem. For a review of the concepts that appear in these two results and a sketch of their proof, we refer again to Appendix 3.

A linear transformation is continuous everywhere, since this is clearly true for a linear function. More than this is true; any linear transformation from $\mathbf{R}^n$ into $\mathbf{R}^m$ is everywhere uniformly continuous. This is equivalent to the next theorem.

Theorem 8 *Let L be a linear transformation from $\mathbf{R}^n$ into $\mathbf{R}^m$ represented by the matrix $[a_{ij}]$. Then, there is a constant B such that $|L(p)| \le B|p|$ for all points p.*

We shall find that the number $\left\{\sum\sum |a_{ij}|^2\right\}^{1/2}$ will serve for B. Put $p = (x_1, x_2, \ldots, x_n)$ and $q = L(p) = (y_1, y_2, \ldots, y_m)$, so that

$$y_i = \sum_{j=1}^{n} a_{ij} x_j \qquad i = 1, 2, \ldots, m$$

We have $|p|^2 = \sum_{j=1}^{n} |x_j|^2$ and $|q|^2 = \sum_{i=1}^{m} |y_i|^2$. Accordingly,

$$|y_i|^2 \le \left\{\sum_{j=1}^{n} |a_{ij}|\,|x_j|\right\}^2 \le \sum_{j=1}^{n} |a_{ij}|^2 \sum_{j=1}^{n} |x_j|^2 = |p|^2 \sum_{j=1}^{n} |a_{ij}|^2$$

where we have used the Schwarz inequality (Sec. 1.3):

$$\left(\sum a_k b_k\right)^2 \le \sum |a_k|^2 \sum |b_k|^2$$

Adding these for $i = 1, 2, \ldots, m$, we obtain

$$|q|^2 = \sum_{i=1}^{m} |y_i|^2 \le |p|^2 \sum_{i=1}^{m} \sum_{j=1}^{n} |a_{ij}|^2$$

and $|L(p)| = |q| \le B|p|$ where $B = \left\{\sum_{i=1}^{m} \sum_{j=1}^{n} |a_{ij}|^2\right\}^{1/2}$. ∎

It should be remarked that the number B which we have found is not the smallest number with this property. For example, the transformation L specified by the identity matrix

$$\begin{bmatrix} 1 & 0 \\ 0 & 1 \end{bmatrix}$$

is such that $|L(p)| = |p|$, while the theorem provides the number

$$B = \sqrt{2} > 1$$

However, this is not the case for linear *functions*. Let L be specified by the row matrix $[c_1, c_2, \ldots, c_n]$. Then, according to the theorem,

$$|L(p)| \le \left(\sum_{i}^{n} |c_j|^2\right)^{1/2} |p|$$

This is best possible, for taking $p = (c_1, c_2, \ldots, c_n)$, we have

$$L(p) = c_1 c_1 + \cdots + c_n c_n$$
$$= \sum_1^n |c_j|^2 = \left(\sum_1^n |c_j|^2\right)^{1/2} |p|$$

In the study of general transformations on n space. it is convenient to introduce a class of special transformations that is somewhat larger than the linear transformations. These are called **affine,** and have the simple form

$$T(p) = A(p) + \beta$$

where A is linear and β is a fixed vector. In coordinate form, an affine transformation from $\mathbf{R}^n$ to $\mathbf{R}^m$ is described by writing $T(p) = q$, where $p = (x_1, x_2, \ldots, x_n)$, $q = (y_1, y_2, \ldots, y_m)$, and

$$y_1 = a_{11}x_1 + a_{12}x_2 + \cdots + a_{1n}x_n + b_1$$
$$y_2 = a_{21}x_1 + a_{22}x_2 + \cdots + a_{2n}x_n + b_2$$
$$\cdots\cdots\cdots\cdots\cdots\cdots\cdots\cdots\cdots\cdots\cdots$$
$$y_m = a_{m1}x_1 + a_{m2}x_2 + \cdots + a_{mn}x_n + b_n$$

These can be regarded as the generalization of straight lines, since the equation for an affine map from $\mathbf{R}$ to $\mathbf{R}$ is merely $y = ax + b$.

Since the effect of the added vector β is to translate m space by a shift in a certain direction, the geometry of affine transformations as mappings is determined again by the nature of the linear transformation A (see Exercise 11).

EXERCISES

1 Let L be the linear function specified by the coefficient matrix $[2, 0, -1, 3]$. What is $L(1, 1, -1, -1)$? What is $L(2, 0, 0, 1)$?

2 Find the linear function L such that $L(1, 0, 0) = 2$, $L(0, 1, 0) = -1$, $L(0, 0, 1) = 3$.

3 Find the linear function L such that $L(1, 0, -1) = 3$, $L(2, -1, 0) = 0$, $L(0, 1, 0) = 2$.

4 Let T be the linear transformation of $\mathbf{R}^2$ into $\mathbf{R}^2$ specified by the matrix

$$\begin{bmatrix} 2 & -1 \\ -3 & 0 \end{bmatrix}$$

Find the images of the points $(1, 2)$, $(-2, 1)$, $(1, 0)$, $(0, 1)$.

5 Let T be the linear transformation specified by

$$\begin{bmatrix} 2 & 0 & -1 \\ -1 & 3 & 1 \end{bmatrix}$$

Find the images of $(1, 2, 1)$, $(1, 0, 0)$, $(0, 1, 0)$.

6 Find the matrix representation for the linear transformation T which

(*a*) Maps $(1, 0, 0)$ into $(0, 1, 1)$, $(0, 1, 0)$ into $(1, 4, 0)$, and $(0, 0, 1)$ into $(2, 3, 1)$.

(*b*) Maps $(1, 1, 0)$ into $(0, 0, 1)$, $(0, 1, 1)$ into $(1, 0, 0)$, and $(1, 0, 1)$ into $(1, 0, 0)$.

7 Compute the ranks of the following matrices:

$$(a)\begin{bmatrix} 1 & 2 & -3 \\ 2 & -1 & 4 \\ 3 & 1 & 1 \end{bmatrix} \qquad (b)\begin{bmatrix} 0 & 1 & -1 \\ 2 & -1 & 3 \\ 1 & 0 & 1 \end{bmatrix}$$

$$(c)\begin{bmatrix} 3 & -6 & 9 \\ 2 & -4 & 6 \\ -2 & 4 & -12 \end{bmatrix} \qquad (d)\begin{bmatrix} 1 & -1 & 0 & 1 \\ 2 & 3 & 1 & 0 \\ 4 & 1 & 1 & -1 \end{bmatrix}$$

8 By computing ranks, discuss the nature of the image in UVW space of all of XYZ space. If either transformation is nonsingular, find the equations for its inverse.

$$(a)\begin{cases} u = x + 2y - 3z \\ v = 2x - y + 4z \\ w = 3x + y + 4z \end{cases} \qquad (b)\begin{cases} u = y - z \\ v = 3x - y + 3z \\ w = x + z \end{cases}$$

9 Compute the indicated matrix products:

$$(a)\begin{bmatrix} 2 & -3 \\ 0 & 1 \end{bmatrix}\begin{bmatrix} -1 & 4 \\ 1 & 3 \end{bmatrix} \qquad (b)\begin{bmatrix} 2 & -7 & 1 \\ -1 & 0 & 2 \end{bmatrix}\begin{bmatrix} 2 & -1 \\ 0 & 2 \\ 4 & 1 \end{bmatrix}$$

$$(c)\begin{bmatrix} 1 & -1 \\ -1 & 0 \\ 0 & 1 \end{bmatrix}\begin{bmatrix} 2 & 3 \\ -1 & 2 \end{bmatrix} \qquad (d)\begin{bmatrix} -1 & -1 & 0 \\ 2 & 0 & 1 \\ -1 & 1 & 1 \end{bmatrix}\begin{bmatrix} -1 & 1 \\ 3 & 2 \\ -2 & 0 \end{bmatrix}$$

10 Find the product ST of the transformations given by

$$S: (x, y, z) \to (6x - y + 2z, 2x + 4z)$$

$$T: (x, y) \quad \to (x - y, 2x, -x + 2y)$$

Can you form the product TS?

11 (*a*) If T is an affine transformation, show that for any points p and q and all real λ,

$$T(\lambda p + (1 - \lambda)q) = \lambda T(p) + (1 - \lambda)T(q)$$

(*b*) What is the geometric meaning of this condition?
(*c*) Does this condition characterize affine transformations?

***12** Given any function f of one variable, infinitely differentiable, define

$$M(f) = \begin{bmatrix} f & 0 & 0 \\ f' & f & 0 \\ f'' & 2f' & f \end{bmatrix}$$

Verify the following facts:

(*a*) $\dfrac{d}{dx}M(f) = M(f')$ (*b*) $M(f)M(g) = M(fg)$

(*c*) $M(P(f)) = P(M(f))$ where P is any polynomial

(*d*) $e^{M(f)} = M(e^f)$ (*e*) $M\left(\dfrac{1}{f}\right) = M(f)^{-1}$

7.4 DIFFERENTIALS OF TRANSFORMATIONS

In Chap. 3, we defined the (total) derivative of a scalar-valued function f defined on an open set S to be a vector-valued function $\mathbf{D}f$ defined on S and constructed from the partial derivatives of f. Specifically, if $y = f(x_1, x_2, \ldots, x_n)$, then

$$\mathbf{D}f = \left(\frac{\partial y}{\partial x_1}, \frac{\partial y}{\partial x_2}, \frac{\partial y}{\partial x_3}, \ldots, \frac{\partial y}{\partial x_n}\right) \tag{7-12}$$

$$= (f_1, f_2, f_3, \ldots, f_n)$$

At any point $p \in S$, $\mathbf{D}f(p)$ is a numerical vector with n components, the values $f_i(p)$ of the partial derivatives of f at p; this was also called the gradient of f at p.

In this section we take the next step and define the derivative of a vector-valued function T; this will turn out to be a matrix-valued function whose entries are the partial derivatives of the coordinate functions describing T. Note that we prefer to treat T at present as a transformation.

We adopt a slightly different notation and terminology from that used for he simpler case discussed in Chap. 3. The derivative of a transformation T will now be called the **differential** of T, written dT. This is because our viewpoint must now be somewhat more abstract in order to develop a general theory. If T defined on a set $S \subset \mathbf{R}^n$, and is of class C' there, then dT will be a function that is also defined on S, but whose value at a point $p \in S$ is a linear transformation L on n space. Since linear transformations can be represented by matrices, dT can also be regarded as a function on S whose value at each point of S is a matrix with numerical entries. Computationally, dT is therefore a matrix-valued function on S whose entries are the first partial derivatives of the component functions that describe the transformation T. The dual role of dT, as a function whose values are linear transformations, and as a function whose values are numerical matrices, is chiefly a notational problem and will become clearer as we develop more of the theory.

Let us start by considering a transformation T of $\mathbf{R}^3$ into itself which is given by a set of equations such as

$$T: \begin{cases} u = f(x, y, z) \\ v = g(x, y, z) \\ w = h(x, y, z) \end{cases} \tag{7-13}$$

We shall say that T is of class $C^{(n)}$ in a region D whenever each of the coordinate functions f, g, and h is of this class in D. In particular, T is of class C' in D if all the partial derivatives $\partial u/\partial x, \partial u/\partial y, \ldots, \partial w/\partial y, \partial w/\partial z$ exist and are continuous in D.

For transformations of class C', we define dT for such a T as that in (7-13) by

$$dT = \begin{bmatrix} f_1 & f_2 & f_3 \\ g_1 & g_2 & g_3 \\ h_1 & h_2 & h_3 \end{bmatrix} = \begin{bmatrix} \dfrac{\partial u}{\partial x} & \dfrac{\partial u}{\partial y} & \dfrac{\partial u}{\partial z} \\ \dfrac{\partial v}{\partial x} & \dfrac{\partial v}{\partial y} & \dfrac{\partial v}{\partial z} \\ \dfrac{\partial w}{\partial x} & \dfrac{\partial w}{\partial y} & \dfrac{\partial w}{\partial z} \end{bmatrix}$$

If this is evaluated at a point $p \in D$, the resulting matrix of numbers specifies a linear transformation of $\mathbf{R}^3$ into itself which is called the differential of T at p. (This can be denoted by $dT|_p$; we shall occasionally use merely dT when the context makes clear at which point we are computing the differential of T.)

To illustrate this, let T be given by

$$\text{(7-14)} \qquad \begin{cases} u = x^2 + y - z \\ v = xyz^2 \\ w = 2xy - y^2 z \end{cases}$$

The differential of T at (x, y, z) is

$$\text{(7-15)} \qquad dT = \begin{bmatrix} 2x & 1 & -1 \\ yz^2 & xz^2 & 2xyz \\ 2y & 2x - 2yz & -y^2 \end{bmatrix}$$

so that the differential of T at $p_0 = (1, 1, 1)$ is

$$\text{(7-16)} \qquad dT|_{p_0} = \begin{bmatrix} 2 & 1 & -1 \\ 1 & 1 & 2 \\ 2 & 0 & -1 \end{bmatrix}$$

The differential of a general transformation is obtained in the same fashion. If T is a transformation from $\mathbf{R}^n$ into $\mathbf{R}^m$ and

$$\text{(7-17)} \qquad T(x_1, x_2, \ldots, x_n) = (y_1, y_2, \ldots, y_m)$$

then

$$\text{(7-18)} \qquad dT = \begin{bmatrix} \dfrac{\partial y_1}{\partial x_1} & \dfrac{\partial y_1}{\partial x_2} & \cdots & \dfrac{\partial y_1}{\partial x_n} \\ \cdots & \cdots & \cdots & \cdots \\ \dfrac{\partial y_m}{\partial x_1} & \dfrac{\partial y_m}{\partial x_2} & \cdots & \dfrac{\partial y_m}{\partial x_n} \end{bmatrix}$$

If T is of class C' in a region D, we have thus associated with each point of D a linear transformation.

It may be observed that if (7-17) is written in the explicit coordinate form

$$T: \begin{cases} y_1 = f_1(x_1, x_2, \ldots, x_n) \\ y_2 = f_2(x_1, x_2, \ldots, x_n) \\ \cdots\cdots\cdots\cdots\cdots\cdots\cdots \\ y_m = f_m(x_1, x_2, \ldots, x_n) \end{cases} \tag{7-19}$$

then the rows of the matrix (7-18) are exactly the total derivatives, as in (7-12), of the coordinate functions in (7-19). In our new notation, we write these as df_1, df_2, etc.

Before examining the significance of the differential dT and explaining its role in obtaining local approximations to T, we will try to clarify the dual nature of dT, using for this purpose the more familiar case of a scalar function g whose differentiation theory was discussed in Chap. 3. As preparation for the next paragraph, please review the connection between linear functions L and their representation in terms of vectors and dot products as outlined in Sec. 7.3.

Suppose that g is a real-valued function of class C' defined on an open set S in 3-space. Writing $g(x, y, z)$, we recall that its vector-valued derivative in S is the function $\mathbf{D}g$ defined by (7-12) as

$$\mathbf{D}g(p) = (g_x(p), g_y(p), g_z(p)) \qquad \text{for } p \in S \tag{7-20}$$

On the other hand, the differential dg is a function defined on S whose value at $p \in S$ is the linear transformation that is represented by the 1-by-3 matrix $[g_x(p), g_y(p), g_z(p)]$. Such a matrix with only one row represents a linear function L on $\mathbf{R}^3$, so that we now have $dg(p) = L$; like any linear function, L is defined everywhere in $\mathbf{R}^3$, and if $u = (a, b, c)$, then by (7-10),

$$L(u) = g_x(p)a + g_y(p)b + g_z(p)c$$

However, we can also use the alternative dot product formulation of this and have

$$L(u) = [g_x(p), g_y(p), g_z(p)] \cdot (a, b, c)$$

which, from (7-20), can also be written

$$L(u) = \mathbf{D}g(p) \cdot u$$

Since $L = dg|_p$, we have proved the following basic identity connecting the differential dg of the scalar function g with the vector-valued derivative $\mathbf{D}g$ treated in Chap. 3.

Theorem 9 *For any $p \in S$ and any $u \in \mathbf{R}^3$,*

$$dg|_p(u) = \mathbf{D}g(p) \cdot u \tag{7-21}$$

The obvious similarity of the notation on both sides of this equation emphasizes the identity itself and shows that our new viewpoint gives nothing basically new when we apply it to scalar-valued functions.

Turning now to transformations T, which in the present context can be regarded merely as vector-valued functions defined on an open set S, we again will refer to the differential of T in two ways. The primary meaning of dT is a function defined on S whose value at any point $p_0 \in S$ is a linear transformation L; the secondary meaning is a function defined on S whose value at any point $p_0 \in S$ is the matrix that represents the corresponding linear transformation L. The latter is what we calculate with when we work with derivatives of transformations; the former is what we use when we prove theorems. The significance of the differential lies in the next result, which shows that these linear transformations provide **local approximations** for T.

Theorem 10 *Let T be of class C' in an open region D, and let E be a closed bounded subset of D. Let $dT|_{p_0}$ be the differential of T at a point $p_0 \in E$. Then,*

$$T(p_0 + \Delta p) = T(p_0) + dT|_{p_0}(\Delta p) + R(\Delta p) \tag{7-22}$$

where

$$\lim_{\Delta p \to 0} \frac{|R(\Delta p)|}{|\Delta p|} = 0 \tag{7-23}$$

uniformly for $p_0 \in E$.

Let T be given by $T(x_1, x_2, \ldots, x_n) = (y_1, y_2, \ldots, y_m)$ where $y_i = \phi_i(p) = \phi_i(x_1, x_2, \ldots, x_n)$. (The subscript on ϕ_i does not indicate differentiation.) Then

$$T(p_0 + \Delta p) - T(p_0) = (\Delta y_1, \Delta y_2, \ldots, \Delta y_m)$$

where
$$\Delta y_i = \phi_i(p_0 + \Delta p) - \phi_i(p_0)$$

To each of these, we apply the approximation theorem for functions, in the stronger form outlined at the end of the proof of Theorem 8, Sec. 3.3, to write [using (7-21)]

$$\Delta y_i = d\phi_i(\Delta p) + R_i(\Delta p)$$

Combining these, we have

$$T(p_0 + \Delta p) = T(p_0) + L(\Delta p) + R(\Delta p)$$

where
$$L(\Delta p) = (d\phi_1(\Delta p), d\phi_2(\Delta p), \ldots, d\phi_m(\Delta p))$$

and
$$R(\Delta p) = (R_1(\Delta p), R_2(\Delta p), \ldots, R_m(\Delta p))$$

If we now set $\Delta p = (\Delta x_1, \Delta x_2, \ldots, \Delta x_n)$, then

$$d\phi_i(\Delta p) = \frac{\partial y_i}{\partial x_1}\Delta x_1 + \frac{\partial y_i}{\partial x_2}\Delta x_2 + \cdots + \frac{\partial y_i}{\partial x_n}\Delta x_n$$

where the partial derivatives are evaluated at p_0. Thus,

$$L(\Delta p) = dT|_{p_0}(\Delta p)$$

where $dT|_{p_0}$ is the linear transformation with matrix (7-18). Again we know that for each i, $i = 1, 2, \ldots, m$, $\lim_{\Delta p \to 0} |R_i(\Delta p)|/|\Delta p| = 0$, uniformly for all $p_0 \in E$. Since

$$|R(\Delta p)| \le |R_1(\Delta p)| + \cdots + |R_m(\Delta p)|$$

it follows at once that

$$\lim_{\Delta p \to 0} \frac{|R(\Delta p)|}{|\Delta p|} = 0$$

uniformly for $p_0 \in E$. ∎

We can restate the property described in this theorem in a form which relates it directly to the familiar notion of derivative from elementary calculus; (7-22) and (7-23) together imply that

$$\lim_{\Delta p \to 0} \frac{|T(p_0 + \Delta p) - T(p_0) - L(\Delta p)|}{|\Delta p|} = 0 \tag{7-24}$$

For comparison, we note that a function of one variable is said to be differentiable at x_0 if there is a number A such that

$$\lim_{\Delta x \to 0} \frac{f(x_0 + \Delta x) - f(x_0) - A\,\Delta x}{\Delta x} = 0 \tag{7-25}$$

Guided by this observation, we are led to formulate a general definition of differentiability that does not make reference to coordinate functions, partial derivatives, or matrices, and which can therefore be regarded as the primary definition from which all others follow.

Definition 3 *Let T be defined on an open set $S \subset \mathbf{R}^n$, and taking values in $\mathbf{R}^m$. Then, T is said to be* **differentiable** *at $p_0 \in S$ if there is a linear transformation L on $\mathbf{R}^n$ to $\mathbf{R}^m$ such that* (7-24) *holds. In this case, we write $dT|_{p_0} = L$, and call L the differential* (*or derivative*) *of T at p_0. T is said to be differentiable in S if T has a differential at each point of S.*

Now that we have adopted this definition, a number of facts can be checked easily. First, T cannot have two different differentials at the same point p_0 (Exercise 10); second, if dT exists at p_0, and T is described by $T(x_1, x_2, \ldots, x_n) = (y_1, y_2, \ldots, y_m)$, where each y_j is a specified function of the x_i, then all the partial derivatives $\partial y_j/\partial x_i$ exist at p_0, and the linear transformation $L = dT|_{p_0}$ has (7-18) for its representing matrix. (This is verified in Exercise 5 in a typical case.) Finally, Theorem 10 now tells us that any transformation of class C' in an open set S is differentiable everywhere in S.

The differential of a transformation T is also seen to correspond to the initial terms in a power series representation for T. Suppose that $T(x, y) = (u, v)$, where

$$T: \begin{cases} u = f(x, y) \\ v = g(x, y) \end{cases}$$

and suppose that we can expand f and g in double power series at the origin.

$$u = a_{00} + a_{10}x + a_{01}y + a_{20}x^2 + a_{11}xy + a_{02}y^2 + \cdots$$

$$v = b_{00} + b_{10}x + b_{01}y + b_{20}x^2 + b_{11}xy + b_{02}y^2 + \cdots$$

Then, this can be written

$$T(x, y) = (a_{00}, b_{00}) + \begin{bmatrix} a_{10} & a_{01} \\ b_{10} & b_{01} \end{bmatrix} \begin{bmatrix} x \\ y \end{bmatrix} + R(x, y)$$

where $R = (R_1, R_2)$ and

$$R_1 = a_{20}x^2 + a_{11}xy + a_{02}y^2 + \cdots$$

$$R_2 = b_{20}x^2 + b_{11}xy + b_{02}y^2 + \cdots$$

Note that
$$dT|_0 = \begin{bmatrix} a_{10} & a_{01} \\ b_{10} & b_{01} \end{bmatrix}$$

and that the expression for $T(x, y)$ given above corresponds exactly to the form in (7-22), with $p_0 = (0, 0)$ and $\Delta p = (x, y)$. This process of cutting off all terms in a power series higher than the first order is called **linearization,** and is a frequently used technique for gaining understanding about the local behavior of a transformation.

Our next result is the general chain rule of differentiation for products of transformations. It includes the earlier forms of the chain rule that were discussed in Sec. 3.4 in connection with differentiation of composite functions. Consider two transformations T and S, with T defined from $\mathbf{R}^n$ into $\mathbf{R}^k$ and S from $\mathbf{R}^k$ into $\mathbf{R}^m$. Their product ST is the composite mapping from $\mathbf{R}^n$ into $\mathbf{R}^m$. We show that $d(ST) = dS\, dT$.

Theorem 11 *Let T be differentiable on an open set D, and let S be differentiable on an open set containing $T(D)$. Then, ST is differentiable on D, and if $p \in D$ and $q = T(p)$, then*

$$d(ST)|_p = dS|_q\, dT|_p \tag{7-26}$$

Before proving this, we illustrate it in several typical cases similar to some treated in Sec. 3.4; the calculations are reduced to multiplication of matrices.

Let S and T be given by

$$S:\begin{cases} u = f(x, y, z) \\ v = g(x, y, z) \\ w = h(x, y, z) \end{cases} \qquad T:\begin{cases} x = F(s, t) \\ y = G(s, t) \\ z = H(s, t) \end{cases}$$

Their differentials are

$$dS = \begin{bmatrix} \dfrac{\partial u}{\partial x} & \dfrac{\partial u}{\partial y} & \dfrac{\partial u}{\partial z} \\ \dfrac{\partial v}{\partial x} & \dfrac{\partial v}{\partial y} & \dfrac{\partial v}{\partial z} \\ \dfrac{\partial w}{\partial x} & \dfrac{\partial w}{\partial y} & \dfrac{\partial w}{\partial z} \end{bmatrix} \qquad dT = \begin{bmatrix} \dfrac{\partial x}{\partial s} & \dfrac{\partial x}{\partial t} \\ \dfrac{\partial y}{\partial s} & \dfrac{\partial y}{\partial t} \\ \dfrac{\partial z}{\partial s} & \dfrac{\partial z}{\partial t} \end{bmatrix}$$

The transformation ST is given by

$$ST:\begin{cases} u = f(F(s, t), G(s, t), H(s, t)) \\ v = g(F(s, t), G(s, t), H(s, t)) \\ w = h(F(s, t), G(s, t), H(s, t)) \end{cases}$$

and its differential is

$$d(ST) = \begin{bmatrix} \dfrac{\partial u}{\partial s} & \dfrac{\partial u}{\partial t} \\ \dfrac{\partial v}{\partial s} & \dfrac{\partial v}{\partial t} \\ \dfrac{\partial w}{\partial s} & \dfrac{\partial w}{\partial t} \end{bmatrix}$$

Multiplying the matrices, we obtain

$$dS\,dT = \begin{bmatrix} \dfrac{\partial u}{\partial x}\dfrac{\partial x}{\partial s} + \dfrac{\partial u}{\partial y}\dfrac{\partial y}{\partial s} + \dfrac{\partial u}{\partial z}\dfrac{\partial z}{\partial s} & \dfrac{\partial u}{\partial x}\dfrac{\partial x}{\partial t} + \dfrac{\partial u}{\partial y}\dfrac{\partial y}{\partial t} + \dfrac{\partial u}{\partial z}\dfrac{\partial z}{\partial t} \\ \dfrac{\partial v}{\partial x}\dfrac{\partial x}{\partial s} + \dfrac{\partial v}{\partial y}\dfrac{\partial y}{\partial s} + \dfrac{\partial v}{\partial z}\dfrac{\partial z}{\partial s} & \dfrac{\partial v}{\partial x}\dfrac{\partial x}{\partial t} + \dfrac{\partial v}{\partial y}\dfrac{\partial y}{\partial t} + \dfrac{\partial v}{\partial z}\dfrac{\partial z}{\partial t} \\ \dfrac{\partial w}{\partial x}\dfrac{\partial x}{\partial s} + \dfrac{\partial w}{\partial y}\dfrac{\partial y}{\partial s} + \dfrac{\partial w}{\partial z}\dfrac{\partial z}{\partial s} & \dfrac{\partial w}{\partial x}\dfrac{\partial x}{\partial t} + \dfrac{\partial w}{\partial y}\dfrac{\partial y}{\partial t} + \dfrac{\partial w}{\partial z}\dfrac{\partial z}{\partial t} \end{bmatrix}$$

However, this is exactly the matrix which results from computing the partial derivatives by the chain rule and substituting into $d(ST)$.

Other instances of this general formula for computing the differential of a composite transformation will be found in the exercises. The more complicated examples of Sec. 3.4 can also be obtained in this fashion. Consider, for

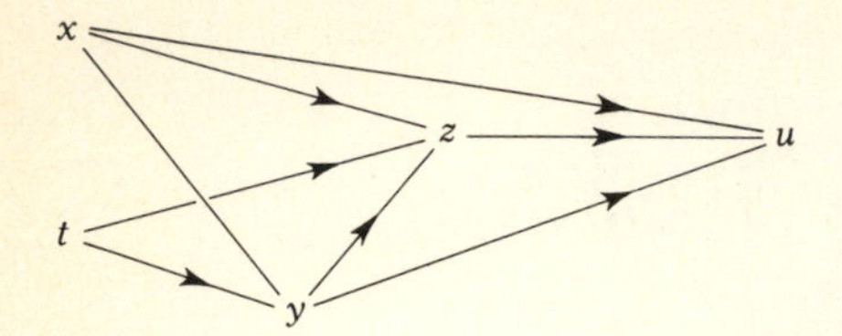

Figure 7-6

example, the following set of equations:

$$u = f(x, y, z) \qquad \begin{array}{l} z = g(x, y, t) \\ y = h(x, t) \end{array}$$

These may be used to express u in terms of x and t, and the corresponding diagram is Fig. 7-6. We introduce three transformations, R, S, and T, such that $u = (STR)(x, t)$. R is the mapping of $\mathbf{R}^2$ into $\mathbf{R}^3$ given by

$$R: \begin{cases} x = x \\ y = h(x, t) \\ t = t \end{cases}$$

T is the mapping of $\mathbf{R}^3$ into $\mathbf{R}^3$ given by

$$T: \begin{cases} x = x \\ y = y \\ z = g(x, y, t) \end{cases}$$

S is the mapping of $\mathbf{R}^3$ into $\mathbf{R}^1$ given by

$$S: \quad u = f(x, y, z)$$

In order to find the partial derivatives $\partial u/\partial x$ and $\partial u/\partial t$, we shall find the differential

$$d(STR) = \left[\frac{\partial u}{\partial x}, \frac{\partial u}{\partial t}\right]$$

By Theorem 11, $d(STR) = dS\, dT\, dR$, where we have

$$dS = [f_1, f_2, f_3] \qquad dT = \begin{bmatrix} 1 & 0 & 0 \\ 0 & 1 & 0 \\ g_1 & g_2 & g_3 \end{bmatrix} \qquad dR = \begin{bmatrix} 1 & 0 \\ h_1 & h_2 \\ 0 & 1 \end{bmatrix}$$

Computing these products, we have

$$dS\, dT = [f_1 + f_3 g_1, f_2 + f_3 g_2, f_3 g_3]$$

and $$dS\, dT\, dR = [f_1 + f_3 g_1 + (f_2 + f_3 g_2)h_1, (f_2 + f_3 g_2)h_2 + f_3 g_3]$$

so that we may read off the correct expressions for the desired partial derivatives

$$\frac{\partial u}{\partial x} = f_1 + f_3 g_1 + f_2 h_1 + f_3 g_2 h_1$$

$$\frac{\partial u}{\partial t} = f_2 h_2 + f_3 g_2 h_2 + f_3 g_3$$

We now return to the *proof of Theorem 11*. We have $T(p) = q$. Take Δp small, and write $\Delta q = T(p + \Delta p) - T(p)$. Thus, $q + \Delta q = T(p + \Delta p)$, and $(ST)(p + \Delta p) = S(q + \Delta p)$. Apply the local approximation theorem (Theorem 10):

$$S(q + \Delta q) = S(q) + dS|_q(\Delta q) + R_1(\Delta q)$$

The same theorem applied to $T(p + \Delta p)$ yields

$$\Delta q = T(p + \Delta p) - T(p)$$

$$= dT|_p(\Delta p) + R_2(\Delta p) \tag{7-27}$$

and combining these, we have

$$(ST)(p + \Delta p) = ST(p) + dS|_q dT|_p(\Delta p) + R(\Delta p) \tag{7-28}$$

where

$$R(\Delta p) = dS|_q(R_2(\Delta p)) + R_1(\Delta q) \tag{7-29}$$

To prove ST differentiable at p, we must now show that $\lim_{\Delta p \to 0} |R(\Delta p)|/|\Delta p| = 0$. We use the fact that S and T are each differentiable, which implies that $R_1(\Delta q) < \varepsilon|\Delta q|$ for small $|\Delta q|$ and that $R_2(\Delta p) < \varepsilon|\Delta p|$ for small $|\Delta p|$. We also use the standard boundedness property of linear transformations, stated in Theorem 8. This enables us to know that there is a number M such that

$$\left| dT|_p(\Delta p) \right| \le M|\Delta p|$$

$$\left| dS|_q(R_2(\Delta p)) \right| \le M|R_2(\Delta p)|$$

for all Δp. From (7-27), we next have

$$|\Delta q| \le M|\Delta p| + \varepsilon|\Delta p| < (M + 1)|\Delta p|$$

Then, using this in (7-29), we have

$$\begin{aligned}|R(\Delta p)| &\le M|R_2(\Delta p)| + \varepsilon|\Delta q|\\ &\le M\varepsilon|\Delta p| + \varepsilon(M+1)|\Delta p|\\ &\le (2M+1)\varepsilon|\Delta p|\end{aligned}$$

and since ε is arbitrarily small, $\lim_{\Delta p \to 0} |R(\Delta p)|/|\Delta p| = 0$ ∎

Before leaving the chain rule, we note that if S and T are of class C', the correctness of the formula $d(ST) = dS\,dT$ can be verified as illustrated above by falling back upon the earlier versions of the chain rule. The advantage of the proof given above is that it merely requires existence of the differentials, and therefore gives a sharper theorem.

We conclude this section with the **mean value theorem** for general transformations. To avoid making the notation too complicated, we state it only for transformations from 3-space into 3-space; the general formulation of the result is evident from this. Note also that this mean value theorem differs from the corresponding result for scalar functions, given in Theorem 16, Sec. 3.5, in that we cannot now expect to have a *single* auxiliary point p^*.

Theorem 12 *Let T be a transformation of class C' defined for all points $p = (x, y, z)$ in an open set D by*

$$T:\begin{cases} u = f(x, y, z)\\ v = g(x, y, z)\\ w = h(x, y, z)\end{cases}$$

Let D contain the points p' and p'' and the line segment which joins them. Then, there are three points p_1^*, p_2^*, and p_3^* lying on this line segment such that

$$T(p'') - T(p') = L(p'' - p')$$

where L is the linear transformation represented by the matrix

$$\begin{bmatrix} f_1(p_1^*) & f_2(p_1^*) & f_3(p_1^*)\\ g_1(p_2^*) & g_2(p_2^*) & g_3(p_2^*)\\ h_1(p_3^*) & h_2(p_3^*) & h_3(p_3^*)\end{bmatrix} \tag{7-30}$$

This may be proved by applying the mean value theorem (Theorem 16 of Sec. 3.5) to each of the functions f, g, and h. Setting

$$p'' - p' = \Delta p = (\Delta x, \Delta y, \Delta z)$$

we have

$$f(p'') - f(p') = f_1(p_1^*)\,\Delta x + f_2(p_1^*)\,\Delta y + f_3(p_1^*)\,\Delta z$$

where p_1^* is some point on the line segment joining p' and p''. Similarly,

$$g(p'') - g(p') = g_1(p_2^*)\,\Delta x + g_2(p_2^*)\,\Delta y + g_3(p_2^*)\,\Delta z$$
$$h(p'') - h(p') = h_1(p_3^*)\,\Delta x + h_2(p_3^*)\,\Delta y + h_3(p_3^*)\,\Delta z$$

and the result follows. ▮

The following consequence of this theorem is sometimes handier to use.

Corollary *Let T be a transformation of class C' on an open set D in n space, and let E be a compact set in D. Then, there are numbers M and $\delta > 0$ such that $|T(p) - T(q)| \le M|p - q|$ for all p and q in E with $|p - q| < \delta$. If E is convex, $\delta = \text{diam}\,(E)$.*

To show this, let $\delta_1 = \text{dist}\,(E, \text{bdy}\,(D))$; $\delta_1 > 0$ since E is compact. Let E_1 be the set of all points p whose distance from some point of E is not more than $\delta_1/2$. Then E_1 is also compact, and $E \subset E_1 \subset D$. Moreover, if $\delta < \delta_1/2$ and p and q lie in E_1 with $q \in E$ and $|p - q| < \delta$, then the entire segment from p to q lies in E_1. Applying Theorem 12, $T(p) - T(q) = L(p - q)$; the functions in the matrix for L are continuous on E_1 and thus bounded there, and the auxiliary points $p_1^*, p_2^*, \ldots, p_n^*$ which are used in the definitions of L and which lie on the segment from p to q also lie in E_1. Thus, the entries of L are uniformly bounded, and Theorem 8 provides a number M such that $|L(p - q)| \le M|p - q|$. If E itself is convex, then the segment joining p and q lies in E whenever p and q do, so that we do not need to construct the set E_1 and do not need to have p and q close together. As before, the bound M on L is obtained from uniform bounds on E of all the functions in the matrix for L, and we again have $|L(p - q)| \le M|p - q|$, and therefore

$$|T(p) - T(q)| \le M|p - q| \tag{7-31}$$

EXERCISES

1 Compute the differentials of the following transformations at the indicated points.

(*a*) $\begin{cases} u = xy^2 - 3x^3 \\ v = 3x - 5y^2 \end{cases}$ at $(1, -1)$ and $(1, 3)$

(*b*) $\begin{cases} u = xyz^2 - 4y^2 \\ v = 3xy^2 - y^2z \end{cases}$ at $(1, -2, 3)$

(*c*) $\begin{cases} u = x + 6y \\ v = 3xy \\ w = x^2 - 3y^2 \end{cases}$ at $(1, 1)$

2 If L is a linear transformation, show that $dL|_p = L$ at every point p.

3 Using the methods of Sec. 3.4, verify Theorem 11 for the following transformations:

(*a*) $S\colon u = F(x, y) \qquad T\colon\begin{cases} x = \phi(t) \\ y = \psi(t) \end{cases}$

(*b*) $S\colon\begin{cases} u = F(x, y) \\ v = G(x, y) \end{cases} \qquad T\colon\begin{cases} x = \phi(t) \\ y = \psi(t) \end{cases}$

4 Let $w = F(x, y, t)$, $x = \phi(t)$, $y = \psi(t)$. Show that Theorem 11 can be applied to yield Eq. (3-22) for dw/dt.

5 Let T be a transformation from $\mathbf{R}^2$ into $\mathbf{R}^2$ given by $u = f(x, y)$, $v = g(x, y)$. Let L be a linear transformation

$$L = \begin{bmatrix} A & B \\ C & D \end{bmatrix}$$

such that $T(p_0 + \Delta p) - T(p_0) = L(\Delta p) + R(\Delta p)$, where $\lim_{\Delta p \to 0} R(\Delta p)/|\Delta p| = \mathbf{0}$. Prove that $L = dT|_{p_0}$.

6 Prove: If T is of class C'' in an open connected set D, and $dT = 0$ at each point of D, then T is constant in D.

7 Is the a transformation T of the plane into itself whose differential at (x, y) is given by

$$\begin{bmatrix} 3x^2y & x^3 \\ y & x \end{bmatrix}$$

8 Is there a transformation whose differential is

$$\begin{bmatrix} y & x \\ xy & x + y \end{bmatrix}$$

9 Explain the difference between (7-24) as applied to a scalar function of one variable and (7-25).

10 Show that a differentiable transformation (Definition 3) cannot have two different differentials at the same point.

11 Given the transformation

$$T\colon\begin{cases} u = 3x^2y^3 \\ v = x^3 - y^2 \end{cases}$$

on E, the unit square with (0, 0) and (1, 1) as diagonal corners, show that an estimate for the Lipschitz constant M for E in (7-31) is $\sqrt{130}$.

12 Consider the transformation $T(x, y) = (u, v)$ defined on the unit square $E\colon 0 \le x \le 1, 0 \le y \le 1$, where

$$u = 2x^2 + 6xy - 4x^3/3 - 3xy^2 \qquad \text{and} \qquad v = x^3 - y^2$$

Show that an estimate for the Lipschitz constant M for E in (7-31) is $M = \sqrt{65}$.

7.5 INVERSES OF TRANSFORMATIONS

Much of the discussion in Sec. 2.7, dealing with inverses of functions of one variable, applies with few changes to the study of inverses of transformations from n space to n space. If T is a transformation of the plane into itself, and maps a set D 1-to-1 onto a set D', then this defines a transformation T^{-1} which maps D' onto D, reversing the action of T. If $p \in D$, then $T^{-1}T(p) = p$, and if $q \in D'$, $TT^{-1}(q) = q$. A transformation which is not 1-to-1 may have a number of partial inverses. For example, the transformation

$$\text{(7-32)} \qquad T: \begin{cases} u = 2xy \\ v = x^2 - y^2 \end{cases}$$

maps the whole XY plane onto the UV plane. It is not 1-to-1 in the whole plane, since T sends both $(1, 1)$ and $(-1, -1)$ into $(2, 0)$. More generally, $T(p) = T(-p)$ for any point p. However, if we take D to be the open half plane {all (x, y) with $x > 0$}, then T is 1-to-1 in D. To see this, let $T(x, y) = T(a, b)$. Then $2xy = 2ab$ and

$$x^2 - y^2 = a^2 - b^2$$

so that

$$\begin{aligned} 0 &= x^2(x^2 - y^2 - a^2 + b^2) \\ &= x^4 - x^2y^2 - a^2x^2 + b^2x^2 \\ &= x^4 - a^2b^2 - a^2x^2 + b^2x^2 \\ &= (x^2 + b^2)(x^2 - a^2) \end{aligned}$$

Hence, $x^2 = a^2$, and since $x > 0$ and $a > 0$, $x = a$ and $y = b$. T thus maps D onto a set D' in the UV plane in a 1-to-1 fashion. This mapping has an inverse which maps D' onto D. Solving Eqs. (7-32), we obtain

$$S_1: \begin{cases} x = \left[\dfrac{v + \sqrt{u^2 + v^2}}{2}\right]^{1/2} \\ y = u[2v + 2\sqrt{u^2 + v^2}]^{-1/2} \end{cases}$$

The set D' is the set of points (u, v) for which $v + \sqrt{u^2 + v^2} > 0$, that is, all points (u, v) except those of the form $(0, c)$ with $c \le 0$.

The graphical approach used in Sec. 2.7 can still be used here, and the proof given there for Theorem 25 also proves the following theorem.

Theorem 13 *If T is continuous and* 1-*to*-1 *on a compact set D, then T has a unique inverse T^{-1} which maps $T(D) = D^*$* 1-*to*-1 *onto D, and T^{-1} is continuous on D^**,

For, the graph of T is itself a compact set by Theorem 12, Sec. 2.4, and the graph of T^{-1} is just the reflection of the graph of T and is also compact, so that the transformation T^{-1} must also be continuous. ▌

(Another equally brief proof of this theorem is outlined in Exercise 14.)

It is easy to tell if a real-valued continuous function f of one variable is 1-to-1 on an interval, for it must be monotonic there; if f is differentiable, and $f'(x) \neq 0$ on an interval I, then f is 1-to-1 on I and has an inverse that is also differentiable (Theorem 7, Sec. 3.2). We seek a similar criterion for transformations. The special case of linear transformations gives the clue. If L is a linear map on $\mathbf{R}^n$ into itself, it is 1-to-1 if and only if it is nonsingular, and this is equivalent to the condition $\det(L) \neq 0$. This would lead one to conjecture that a general transformation T is 1-to-1 on an open set D if dT is nonsingular at every point of D. (Note that this reduces to the requirement $f'(x) \neq 0$ if T is a function f on **R** to **R**.) We shall see that this conjecture is almost correct. We introduce the following:

Definition 4 *If T is a transformation from $\mathbf{R}^n$ into $\mathbf{R}^n$ which is of class C' in a set D, then the Jacobian of T is the function J defined in D by*

$$J(p) = \det(dT|_p)$$

For example, if T is given by

$$\begin{cases} u = f(x, y, z) \\ v = g(x, y, z) \\ w = h(x, y, z) \end{cases}$$

then
$$J(p) = \begin{vmatrix} f_1(p) & f_2(p) & f_3(p) \\ g_1(p) & g_2(p) & g_3(p) \\ h_1(p) & h_2(p) & h_3(p) \end{vmatrix} = \left.\frac{\partial(u, v, w)}{\partial(x, y, z)}\right|_p$$

The Jacobian of the transformation given in (7-32) is

$$\begin{vmatrix} 2y & 2x \\ 2x & -2y \end{vmatrix} = -4(x^2 + y^2)$$

One is thus led to the conjecture that if $J(p) \neq 0$ throughout a region D, then T is 1-to-1 in D. This conjecture is false. A simple counterexample is supplied by the transformation

$$T: \begin{cases} u = x \cos y \\ v = x \sin y \end{cases} \tag{7-33}$$

whose Jacobian is

$$J(x, y) = \begin{vmatrix} \cos y & -x \sin y \\ \sin y & x \cos y \end{vmatrix} = x$$

In the right half plane $D = \{\text{all } (x, y) \text{ with } x > 0\}$, J is never 0. However, T is not 1-to-1 in D, for (a, b) and $(a, b + 2\pi)$ always have the same image. The effect of the transformation may be seen from Fig. 7-7, where a set of $S \subset D$ is shown, together with its image $T(S)$. We notice that although T is not 1-to-1 in S, two distinct points of S which have the same image must be widely separated; thus, in S (and in fact in D), T has the property of being **locally** 1-to-1. We give this a formal definition.

Definition 5 *A transformation which is defined in an open set D is said to be locally* 1-*to*-1 *(or locally univalent) in D if about any point $p \in D$ there is a neighborhood in which T is* 1-*to*-1.

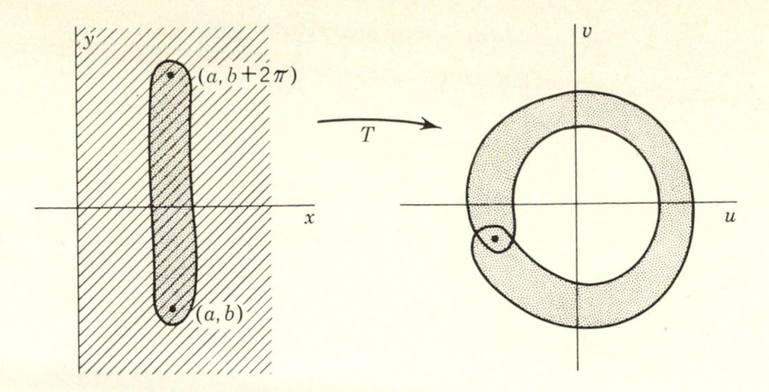

Figure 7-7

The transformation in (7-33) illustrates the fact that a transformation can be locally 1-to-1 in a region D without being globally 1-to-1 in D. (This is not true for functions of one variable; see Exercise 15.)

We are now ready to state the fundamental theorem on the existence of inverses for transformations.

Theorem 14 *Let T be a transformation from $\mathbf{R}^n$ into $\mathbf{R}^n$ which is of class C' in an open set D, and suppose that $J(p) \neq 0$ for each $p \in D$. Then, T is locally 1-to-1 in D.*

The proof of this depends upon the mean value theorem for transformations (Theorem 12, Sec. 7.4). Again, although our present result is true regardless of the value of n, we shall write out a proof only for the case $n = 3$; the general case requires no change in method. Let us suppose that T is described by

$$\begin{cases} u = f(x, y, z) \\ v = g(x, y, z) \\ w = h(x, y, z) \end{cases}$$

Given a point $p \in D$, we shall produce a neighborhood of p in which T is 1-to-1. Let p' and p'' be two points near p such that the line segment joining p' and p'' lies in D. By Theorem 12, we may then choose three points p_1^*, p_2^*, and p_3^* on this line segment such that

$$T(p'') - T(p') = L(p'' - p') \tag{7-34}$$

where L is the linear transformation represented by

$$L = \begin{bmatrix} f_1(p_1^*) & f_2(p_1^*) & f_3(p_1^*) \\ g_1(p_2^*) & g_2(p_2^*) & g_3(p_2^*) \\ h_1(p_3^*) & h_2(p_3^*) & h_3(p_3^*) \end{bmatrix}$$

Introduce a special function F defined for any triple of points of D by

$$F(p_1, p_2, p_3) = \det \begin{bmatrix} f_1(p_1) & f_2(p_1) & f_3(p_1) \\ g_1(p_2) & g_2(p_2) & g_3(p_2) \\ h_1(p_3) & h_2(p_3) & h_3(p_3) \end{bmatrix}$$

Thus, $F(p_1^*, p_2^*, p_3^*) = \det(L)$, and $F(p, p, p) = J(p)$. Moreover, since $T \in C'$, F is continuous, so that, since $J(p) \neq 0$, there is a spherical neighborhood N about p and lying in D such that $F(p_1, p_2, p_3) \neq 0$ for all choices of the points p_1, p_2, and p_3 in N. We shall show that T is 1-to-1 in N. We must therefore show that if p' and p'' are points of N for which $T(p') = T(p'')$, then $p' = p''$. Since p' and p'' lie in N and N is convex, the entire line segment joining p' to p'' also lies in N; in particular, each of the points p_1^*, p_2^*, and p_3^* is a point of N. Using the characteristic property of N, we have

$$F(p_1^*, p_2^*, p_3^*) = \det(L) \neq 0$$

The linear transformation L is therefore nonsingular. Returning to Eq. (7-34) and using the assumption that $T(p') = T(p'')$, we have $L(p'' - p') = 0$ or $L(p') = L(p'')$. But, since L is nonsingular, L is 1-to-1, and $p' = p''$. ▌

Corollary *If T is a transformation from $\mathbf{R}^n$ into $\mathbf{R}^n$ which is of class C' in a neighborhood of a point p_0, and $J(p_0) \neq 0$, then T is 1-to-1 on a (usually smaller) neighborhood of p_0, and has there an inverse T^{-1}.*

Thus, a transformation T defined on an open set D whose Jacobian never vanishes on D is very well behaved there; it has continuous local inverses everywhere, although it need not have a single inverse defined on the total image set $T(D)$. The next result describes an important property of such maps.

Theorem 15 *Let T be of class C' on an open set D in n space, taking values in n space; suppose that $J(p) \neq 0$ for all $p \in D$. Then, $T(D)$ is an open set; thus, T carries every open set in D into an open set.*

Take any point $q_0 \in T(D)$. We must show that q_0 is surrounded by a neighborhood which is composed entirely of image points of D. Let p_0 be any point in D with $T(p_0) = q_0$. Since D is open and $J(p_0) \neq 0$, we can choose a closed neighborhood N about p_0 (a closed disk if $n = 2$, a closed ball if $n = 3$) which lies in D and on which T is a 1-to-1 transformation. Let C be the closed set which is the boundary of N (a circle for $n = 2$, a spherical surface for $n = 3$). Since T is continuous and 1-to-1 in N, Theorem 4 shows that the image $T(C)$ of this set is a closed and bounded set which does not contain q_0, the image of p_0. Let d be the distance from q_0 to the nearest point of $T(C)$ (see Fig. 7-8). We shall show that any point within $d/3$ of q_0 is in $T(D)$. Let q_1 be any such point, so that $|q_1 - q_0| < d/3$. As p wanders throughout N, how

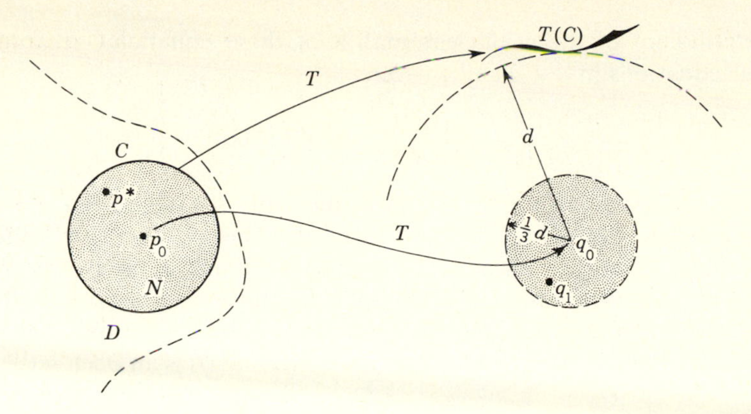

Figure 7-8

close does $T(p)$ come to q_1? The square of the distance from $T(p)$ to q_1 is

$$\phi(p) = |T(p) - q_1|^2$$

which is continuous for $p \in N$. Let p^* be a point in N for which $\phi(p)$ is minimum; this is then a point whose image $T(p^*)$ is as close as possible to the point q_1. Can p^* lie on C, the boundary of N? When p is on C, $|T(p) - q_0| \geq d$, so that $|T(p) - q_1| \geq d - \frac{1}{3}d = \frac{2}{3}d$. Thus, the closest that $T(p)$ can get to q_1 for p on C is $\frac{2}{3}d$. However, the point p_0 itself has image q_0, which is only $\frac{1}{3}d$ away from q_1. Thus, the point p^* will not lie on C, and is therefore an interior point of N. Since p^* minimizes $\phi(p)$, the partial derivatives of ϕ must all vanish at p^*. To see what this implies, we return to a coordinate description of T, and obtain a formula for ϕ. Let us suppose that $n = 2$, so that T may be given by

$$\begin{cases} u = f(x, y) \\ v = g(x, y) \end{cases}$$

Let $q_1 = (a, b)$. Then, the formula for the distance between q_1 and $(u, v) = T(p)$ gives

$$\phi(p) = (u - a)^2 + (v - b)^2$$

The partial derivatives of ϕ are given by

$$\phi_1(p) = 2(u - a)\frac{\partial u}{\partial x} + 2(v - b)\frac{\partial v}{\partial x}$$

$$\phi_2(p) = 2(u - a)\frac{\partial u}{\partial y} + 2(v - b)\frac{\partial v}{\partial y}$$

At the extremal point p^*, both are 0, so that

$$\begin{aligned} 0 &= (u - a)f_1(p^*) + (v - b)g_1(p^*) \\ 0 &= (u - a)f_2(p^*) + (v - b)g_2(p^*) \end{aligned} \tag{7-35}$$

The determinant of the coefficient matrix of these equations, regarded as linear equations in $u - a$ and $v - b$, is

$$\begin{vmatrix} f_1(p^*) & g_1(p^*) \\ f_2(p^*) & g_2(p^*) \end{vmatrix} = J(p^*) \neq 0$$

The only solutions of (7-35) are then the null solutions, $u - a = 0$, $v - b = 0$. Thus, $T(p^*) = (u, v) = (a, b)$, and $T(p^*) = q_1$. This shows that the minimum value of ϕ is actually 0, and the point q_1 is always the image point of some point in N. This holds for any point q_1 lying in the neighborhood of radius $d/3$ about q_0, so that q_0 is an interior point of the image set $T(D)$. Since q_0 was any point of $T(D)$, $T(D)$ is an open set. ▌

One fact about mappings with nonvanishing Jacobians remains to be proved. We must prove that the local inverses are themselves differentiable transformations, and find a formula for their differentials.

Theorem 16 *Let T be of class C' in an open set D, with $J(p) \neq 0$ for all $p \in D$. Suppose also that T is globally 1-to-1 in D, so that there is an inverse transformation T^{-1} defined on the open set $T(D) = D^*$. Then, T^{-1} is of class C' on D^*, and $d(T^{-1})\big|_q = (dT\big|_p)^{-1}$, where $q = T(p)$.*

Let q_0 and $q_0 + \Delta q$ be nearby points of $T(D)$. We shall show that T^{-1} is of class C' by exhibiting a linear transformation which has the characteristic approximating property of the differential, and whose entries are continuous functions. By Exercise 5, Sec. 7.4, this transformation must be the differential of T^{-1} and the entries are the required partial derivatives.

Let $p_0 = T^{-1}(q_0)$ and $p = T^{-1}(q_0 + \Delta q)$, and set $\Delta p = p - p_0$. Thus,

$$\Delta q = T(p) - T(p_0) = T(p_0 + \Delta p) - T(p_0) \tag{7-36}$$

$$\Delta p = T^{-1}(q_0 + \Delta q) - T^{-1}(q_0) \tag{7-37}$$

Let dT be the differential of T at p_0. By the approximation property (Theorem 10), applied to (7-36), we have

$$\Delta q = dT(\Delta p) + R(\Delta p) \tag{7-38}$$

where

$$\lim_{\Delta p \to 0} \frac{|R(\Delta p)|}{|\Delta p|} = 0 \tag{7-39}$$

Since $J(p_0) \neq 0$, dT is a nonsingular linear transformation and has an inverse $(dT)^{-1}$. Applying this linear transformation to both sides of (7-38),

$$\begin{aligned} (dT)^{-1}(\Delta q) &= (dT)^{-1}dT(\Delta p) + (dT)^{-1}(R(\Delta p)) \\ &= \Delta p + (dT)^{-1}(R(\Delta p)) \end{aligned} \tag{7-40}$$

so that, starting from (7-37),

$$\begin{aligned}T^{-1}(q_0 + \Delta q) - T^{-1}(q_0) &= \Delta p \\ &= (dT)^{-1}(\Delta q) - (dT)^{-1}(R(\Delta p)) \\ &= (dT)^{-1}(\Delta q) + R^*(\Delta q)\end{aligned}$$

where

(7-41) $$R^*(\Delta q) = -(dT)^{-1}(R(\Delta p))$$

If we show that

(7-42) $$\lim_{\Delta q \to 0} \frac{|R^*(\Delta q)|}{|\Delta q|} = 0$$

then we will have shown that T^{-1} satisfies the requirements of Definition 3 and is therefore differentiable at q_0, and that its differential at that point is $(dT)^{-1}$. To prove (7-41) we use the boundedness theorem for linear transformations to choose a number M such that $|(dT)^{-1}(u)| \le M|u|$ for all u. We apply this twice to obtain from (7-41)

(7-43) $$|R^*(\Delta q)| \le M|R(\Delta p)|$$

and from (7-40),

(7-44) $$|\Delta p| \le M|\Delta q| + M|R(\Delta p)|$$

Because of (7-39), we can assume $|R(\Delta p)| \le \varepsilon|\Delta p|$, and putting this in (7-44), we obtain $(1 - \varepsilon M)|\Delta p| \le M|\Delta q|$, and then

(7-45) $$|\Delta p| \le \frac{M}{1 - \varepsilon M}|\Delta q|$$

Returning to (7-43), we have

$$|R^*(\Delta q)| \le M\varepsilon|\Delta p| \le \varepsilon\frac{M^2}{1 - \varepsilon M}|\Delta q|$$

and since ε is arbitrarily small, (7-42) is proved.

The proof of the rest of Theorem 16 is easy. Now that we know that T^{-1} is differentiable, and that $dT^{-1} = (dT)^{-1}$ at corresponding points, then since the entries of $(dT)^{-1}$ are rational functions of the entries of dT, and since the latter are all continuous in D, so are the entries of $(dT)^{-1}$, and since T^{-1} is continuous, these become continuous function on the set $D^* = T(D)$. Hence, T^{-1} is of class C' in D^*. ▮

We may illustrate all of this with the transformation used earlier:

(7-46) $$T: \begin{cases} u = x \cos y \\ v = x \sin y \end{cases}$$

whose Jacobian is $J(x, y) = x$. Since this is 0 only on the vertical axis, T has local inverses about any point $p_0 = (x_0, y_0)$ with $x_0 \neq 0$. To find these explicitly, we must solve (7-46). We have $u^2 + v^2 = x^2$, and $2uv = 2x^2 \sin y \cos y = x^2 \sin(2y)$, so that $\sin(2y) = 2uv/(u^2 + v^2)$. If p_0 is a point in the right half plane, so that $x_0 > 0$, then the desired solution for x is $x = \sqrt{u^2 + v^2}$. If $x_0 < 0$, then $x = -\sqrt{u^2 + v^2}$. If $\sin(2y_0) \neq \pm 1$, then one of the inverses for the sine function may be chosen so that in an interval about y_0, $y = \frac{1}{2} \arcsin(2uv/(u^2 + v^2))$. If $\sin(2y_0) = 1$ or -1, then we proceed differently; we have

$$u^2 - v^2 = x^2(\cos^2 y - \sin^2 y) = x^2 \cos(2y)$$

so that $\cos(2y) = (u^2 - v^2)/(u^2 + v^2)$. Choosing an appropriate inverse for the cosine function, we obtain $y = \frac{1}{2} \arccos([u^2 - v^2]/[u^2 + v^2])$ in an interval about y_0. For example, with $p_0 = (1, 0)$ we have as equations for the desired inverse

$$S: \begin{cases} x = \sqrt{u^2 + v^2} \\ y = \frac{1}{2} \arcsin \left[\dfrac{2uv}{u^2 + v^2} \right] \end{cases}$$

This transformation is defined and of class C' at all points $(u, v) \neq (0, 0)$. If we also rule out $u = 0$, the simpler equivalent formula $y = \arctan(v/u)$ may also be used. Computing the differential of S, one finds

$$dS = \begin{bmatrix} \dfrac{u}{(u^2 + v^2)^{1/2}} & \dfrac{v}{(u^2 + v^2)^{1/2}} \\ \dfrac{-v}{u^2 + v^2} & \dfrac{u}{u^2 + v^2} \end{bmatrix}$$

so that the Jacobian of S at (u, v) is $(u^2 + v^2)^{-1/2}$. It will be noticed that this is x^{-1}, in agreement with Theorem 16 and the fact that the Jacobian of T is x. Also, changing back to x and y, the differential of S may be written as

$$dS = \begin{bmatrix} \cos y & \sin y \\ -x^{-1} \sin y & x^{-1} \cos y \end{bmatrix}$$

so that

$$dS\, dT = \begin{bmatrix} \cos y & \sin y \\ -x^{-1} \sin y & x^{-1} \cos y \end{bmatrix} \begin{bmatrix} \cos y & -x \sin y \\ \sin y & x \cos y \end{bmatrix} = \begin{bmatrix} 1 & 0 \\ 0 & 1 \end{bmatrix} = I$$

As we have seen from (7-33), a transformation T can be locally 1-to-1 on an open set D without being globally 1-to-1 on D. In fact, it is possible to have such a transformation which maps an open disk D onto itself and is locally 1-to-1 but not globally 1-to-1 on D. The sequence of pictures in Fig. 7-9 is intended to suggest the stages in a construction of such a mapping; note that T is at most 3-to-1 and that T is certainly not 1-to-1 on the boundary of D.

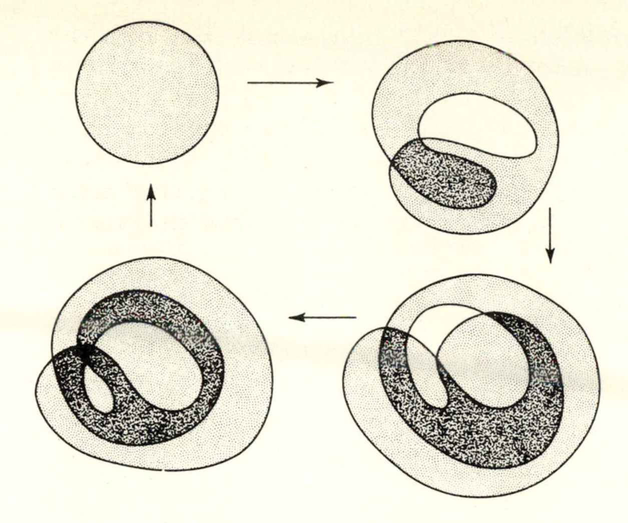

Figure 7-9

On the other hand, it can be shown that if T maps a convex set D onto another convex set D^* in such a way that the boundary of D is mapped exactly onto the boundary of D^*, and if T is locally 1-to-1 in D, then T must be globally 1-to-1 in D.

Other criteria have been obtained that look at the nature of the differential dT. For example, if T is defined on a rectangular region D in the plane, and if neither the Jacobian of T nor any of the diagonal entries of dT has a 0 value in D, then T must be globally 1-to-1 in D. (For a reference to this and other results, see an article by Gale and Nikaido in *Mathematische Annalen*, vol. 159, pp. 81–93, 1965.)

EXERCISES

1 Find the Jacobians of each of the transformations described in Exercise 12, Sec. 7.2.

2 Compute the Jacobians of the following transformations:

(a) $\begin{cases} u = e^x \cos y \\ v = e^x \sin y \end{cases}$ (b) $\begin{cases} u = x^2 \\ v = y/x \end{cases}$

(c) $\begin{cases} u = x^2 + 2xy + y^2 \\ v = 2x + 2y \end{cases}$ *(d) $\begin{cases} u = x + y \\ v = 2xy^2 \end{cases}$

3 Discuss the local behavior of the transformations in Exercise 2.

4 Where it is possible, find formulas for the local inverses of the transformations in Exercise 2.

5 Find the image under each of the transformations of Exercise 2 of the open set $D = [\text{all } (x, y),\ 0 < x < 1,\ 0 < y < 1\}$. For which is the image an open set?

6 The second half of the proof of Theorem 15 assumed that T was a transformation from 2-space into 2-space. Carry out the corresponding discussion with $n = 3$.

7 Let T be the transformation sending (x, y) into $(2x + 4y, x - 3y)$. Find T^{-1} and verify directly that the differential of T^{-1} is $(dT)^{-1}$.

8 Let T be a transformation from $\mathbf{R}^3$ into $\mathbf{R}^3$ which is of class C' in an open set D, and let $J(p) = \det (dT|_p)$. Show that J is continuous throughout D. Is the rank of dT a continuous function of p?

9 Let $T(x, y) = (u, v)$. Show that

$$\frac{\partial(u, v)}{\partial(x, y)} \frac{\partial(x, y)}{\partial(u, v)} = 1$$

10 (*a*) Find an example of a function f that is infinitely differentiable on an interval I, maps I onto itself 1-to-1, has a continuous inverse f^{-1}, but for which f^{-1} fails to be differentiable at some point of I.

(*b*) Show by an example that a transformation of the plane can be 1-to-1, have an everywhere-defined inverse, and still have the Jacobian vanishing somewhere.

11 Let T be defined by

$$T: \begin{cases} u = \sin x \cos y + \sin y \cos x \\ v = \cos x \cos y - \sin x \sin y \end{cases}$$

Find the Jacobian of T. Is there anywhere that T is locally 1-to-1?

12 The following transformation is continuous everywhere in the plane and differentiable there except on the lines $y = \pm x$.

$$T: \begin{cases} u = x^2 + y^2 - |x^2 - y^2| \\ v = x^2 + y^2 + |x^2 - y^2| \end{cases}$$

(*a*) Find dT where it exists.

(*b*) Discuss the local and global mapping behavior of T.

(*c*) Is T differentiable at $(0, 0)$ according to definition 3?

13 Let T be the transformation sending (x, y) into $(2xy, x^2 + y^2)$ and S the transformation sending (x, y) into $(x - y, x + y)$.

(*a*) Using the Jacobians, discuss the local and global mapping behavior of T and S.

(*b*) Obtain formulas for the two product transformations, ST and TS, and then repeat part (*a*) for these new transformations.

14 Prove Theorem 13 by filling in the details in the following argument: T^{-1} is continuous if its inverse carries closed sets into closed sets; however, the inverse of T^{-1} is T and any closed subset of D is compact.

15 Let f be a real-valued continuous function defined for $-\infty < x < \infty$. Suppose that f is locally 1-to-1 everywhere. Prove that f is globally 1-to-1. [*Note*: Do not assume that f' exists.]

16 Let $J_T(p)$ denote the Jacobian of a transformation T at the point p. Show that if S and T are transformations from n space into itself, and $T(p) = q$, then $J_{ST}(p) = J_S(q)J_T(p)$.

17 Let D be the unit disk, $x^2 + y^2 \le 1$. Consider a transformation T of class C' on an open set containing D,

$$T: \begin{cases} u = f(x, y) \\ v = g(x, y) \end{cases}$$

whose Jacobian is never 0 in D. Suppose that T is near the identity map in the sense that $|T(p) - p| \le \frac{1}{3}$ for all $p \in D$. Prove that there is a point p_0 with $T(p_0) = (0, 0)$.

7.6 THE IMPLICIT FUNCTION THEOREMS

The theorems of the present section have to do with the solution of one or more equations in several unknowns. They are existence theorems in that they give assurance that there are solutions, but do not give directions for obtaining them. In the simplest of these situations, suppose we are concerned with the solution of an equation $F(x, y, z) = 0$ for one of the variables as a function of the others. To "solve for z," for example, means to find a function ϕ such that $F(x, y, \phi(x, y)) = 0$ for all x and y in some open set. This is not always possible. In general, there are two reasons for this. One reason is illustrated by the equation

$$x^2 - z^2 + (z - y)(z + y) = 0$$

This cannot be solved for z, for z is not "really" present. The second reason is illustrated by the equation

$$x^2 + y^2 + z^2 + 10 = 0$$

This cannot be solved for z, and in fact, there is no real triple (x, y, z) which satisfies the equation. Our first result is a theorem which gives sufficient conditions for such an equation to be solvable for z.

Theorem 17 *Let F be a function of three variables which is of class C' in an open set D, and let $p_0 = (x_0, y_0, z_0)$ be a point of D for which $F(p_0) = 0$. Suppose that $F_3(p_0) \neq 0$. Then, there is a function ϕ of class C' in a neighborhood N of (x_0, y_0) such that $z = \phi(x, y)$ is a solution of $F(x, y, z) = 0$ for (x, y) in N, and such that $\phi(x_0, y_0) = z_0$.*

Consider the following special transformation from $\mathbf{R}^3$ into $\mathbf{R}^3$:

$$T: \begin{cases} u = x \\ v = y \\ w = F(x, y, z) \end{cases}$$

This is of class C' in D, and its Jacobian is

$$J(p) = \det \begin{bmatrix} 1 & 0 & 0 \\ 0 & 1 & 0 \\ F_1(p) & F_2(p) & F_3(p) \end{bmatrix} = F_3(p)$$

Since $J(p_0) = F_3(p_0) \neq 0$, we may apply the principal result of Sec. 7.5 (Theorem 16) and conclude that there is a transformation T^{-1} which is of class C' in a neighborhood of $q_0 = T(p_0) = (x_0, y_0, 0)$, and which is an inverse for T. Moreover, from the nature of the equations which describe T, we have for T^{-1}

$$T^{-1}: \begin{cases} x = u \\ y = v \\ z = f(u, v, w) \end{cases}$$

where f is of class C' in a neighborhood of q_0. Since T and T^{-1} are inverses, we must have

$$w = F(x, y, z) = F(u, v, f(u, v, w)) \tag{7-47}$$

holding identically for all (u, v, w) near $(x_0, y_0, 0) = q_0$. Set $w = 0$, replace u by x and v by y, and define ϕ by $\phi(x, y) = f(x, y, 0)$. Then, (7-47) becomes

$$0 = F(x, y, \phi(x, y))$$

holding for all points (x, y) in a neighborhood of (x_0, y_0). This shows that $z = \phi(x, y)$ is a solution of $F(x, y, z) = 0$. Since f is of class C', so is ϕ. ▌

By the same device, we can find sufficient conditions for the solution of an equation $F(x, y, z, u, v, \dots) = 0$ for any one of the variables as a function of the remaining. In addition to the differentiability conditions, the two requirements are: (1) there is a point p_0 that satisfies the equation, and (2) the partial of F with respect to the sought-for variable does not vanish at p_0.

Turning to the more complicated cases in which more than one equation is involved, the following will amply illustrate the general procedure.

Theorem 18 *Let F and G be of class C' in an open set $D \subset \mathbf{R}^5$. Let $p_0 = (x_0, y_0, z_0, u_0, v_0)$ be a point of D at which both of the equations*

$$\begin{aligned} F(x, y, z, u, v) &= 0 \\ G(x, y, z, u, v) &= 0 \end{aligned} \tag{7-48}$$

are satisfied. Suppose also that $\partial(F, G)/\partial(u, v) \neq 0$ at p_0. Then, there are two functions ϕ and ψ of class C' in a neighborhood N of (x_0, y_0, z_0) such that

$$\begin{cases} u = \phi(x, y, z) \\ v = \psi(x, y, z) \end{cases}$$

is a solution of (7-48) *in N giving u_0 and v_0 at (x_0, y_0, z_0).*

To prove this, we construct a special transformation from $\mathbf{R}^5$ into $\mathbf{R}^5$

$$T: \begin{cases} t_1 = x \\ t_2 = y \\ t_3 = z \\ t_4 = F(x, y, z, u, v) \\ t_5 = G(x, y, z, u, v) \end{cases}$$

The Jacobian of T is

$$\begin{aligned} J &= \det \begin{bmatrix} 1 & 0 & 0 & 0 & 0 \\ 0 & 1 & 0 & 0 & 0 \\ 0 & 0 & 1 & 0 & 0 \\ F_1 & F_2 & F_3 & F_4 & F_5 \\ G_1 & G_2 & G_3 & G_4 & G_5 \end{bmatrix} = \begin{vmatrix} F_4 & F_5 \\ G_4 & G_5 \end{vmatrix} \\ &= \frac{\partial(F, G)}{\partial(u, v)} \end{aligned}$$

By our hypotheses, $J(p_0) \neq 0$, so that T has a local inverse there. The form of this must be

$$\begin{cases} x = t_1 \\ y = t_2 \\ z = t_3 \\ u = f(t_1, t_2, t_3, t_4, t_5) \\ v = g(t_1, t_2, t_3, t_4, t_5) \end{cases}$$

where f and g are of class C'. Setting $t_4 = 0$ and $t_5 = 0$ to correspond to the original equations $F(x, y, z, u, v) = 0$ and $G(x, y, z, u, v) = 0$, and replacing t_1, t_2, t_3 by x, y, and z, and defining

$$\phi(x, y, z) = f(x, y, z, 0, 0)$$

$$\psi(x, y, z) = g(x, y, z, 0, 0)$$

we have

$$0 = F(x, y, z, \phi(x, y, z), \psi(x, y, z))$$

$$0 = G(x, y, z, \phi(x, y, z), \psi(x, y, z))$$

holding for all (x, y, z) in a neighborhood of (x_0, y_0, z_0). Thus, $u = \phi(x, y, z)$ and $v = \psi(x, y, z)$ are the desired solutions. ▮

To give a simple illustration of this theorem, let us discuss the solution of the following equations for u and v.

$$\begin{cases} x^2 - yu = 0 \\ xy + uv = 0 \end{cases} \tag{7-49}$$

Put $F(x, y, u, v) = x^2 - yu$ and $G(x, y, u, v) = xy + uv$. We find

$$\frac{\partial(F, G)}{\partial(u, v)} = \begin{vmatrix} -y & 0 \\ v & u \end{vmatrix} = -yu$$

Thus, if x_0, y_0, u_0, v_0 satisfy Eqs. (7-49) and $y_0 u_0 \neq 0$, then there are continuous solutions for u and v around the point (x_0, y_0). The limitation $u_0 \neq 0$, $y_0 \neq 0$ is needed, and both imply $x_0 \neq 0$. In this example, it is possible to solve explicitly and obtain

$$u = \frac{x^2}{y}$$

$$v = \frac{-y^2}{x}$$

which is valid at all points (x, y) except those on the axes.

The general case follows the same pattern as Theorem 17. Given a system of m simultaneous equations

$$\begin{aligned} \phi_1(x_1, x_2, \ldots, x_n) &= 0 \\ \phi_2(x_1, x_2, \ldots, x_n) &= 0 \\ &\cdots\cdots \\ \phi_m(x_1, x_2, \ldots, x_n) &= 0 \end{aligned}$$

in n variables, and a point $p = (\bar{x}_1, \bar{x}_2, \ldots, \bar{x}_n)$ that satisfies the system, we can (in theory) solve for a specific set of m of the variables, say $x_{i_1}, x_{i_2}, \ldots, x_{i_m}$, in terms of the rest in a neighborhood of p if the Jacobian

$$\frac{\partial(\phi_1, \phi_2, \ldots, \phi_m)}{\partial(x_{i_1}, x_{i_2}, \ldots, x_{i_m})} \neq 0$$

at p.

EXERCISES

1 Can the curve whose equation is $x^2 + y + \sin(xy) = 0$ be described by an equation of the form $y = f(x)$ in a neighborhood of the point (0, 0)? Can it be described by an equation of the form $x = g(y)$?

2 Can the surface whose equation is $xy - z \log y + e^{xz} = 1$ be represented in the form $z = f(x, y)$ in a neighborhood of (0, 1, 1)? In the form $y = g(x, z)$?

3 The point $(1, -1, 2)$ lies on both of the surfaces described by the equations $x^2(y^2 + z^2) = 5$ and $(x - z)^2 + y^2 = 2$. Show that in a neighborhood of this point, the curve of intersection of the surfaces can be described by a pair of equations of the form $z = f(x)$, $y = g(x)$.

4 Study the corresponding question for the surfaces with equations $x^2 + y^2 = 4$ and $2x^2 + y^2 - 8z^2 = 8$ and the point (2, 0, 0) which lies on both.

5 The pair of equations

$$\begin{cases} xy + 2yz = 3xz \\ xyz + x - y = 1 \end{cases}$$

is satisfied by the choice $x = y = z = 1$. Study the problem of solving (either in theory or in practice) this pair of equations for two of the unknowns as a function of the third, in the vicinity of the (1, 1, 1) solution.

6 (*a*) Let f be a function of one variable for which $f(1) = 0$. What additional conditions on f will allow the equation

$$2f(xy) = f(x) + f(y)$$

to be solved for y in a neighborhood of (1, 1)?

(*b*) Obtain the explicit solution for the choice $f(t) = t^2 - 1$.

***7** With f again a function of one variable obeying $f(1) = 0$, discuss the problem of solving the equation $f(xy) = f(x) + f(y)$ for y near the point (1, 1).

8 Using the method of Theorem 18, state and prove a theorem which gives sufficient conditions for the equations

$$F(x, y, z, t) = 0 \qquad G(x, y, z, t) = 0 \qquad \text{and} \quad H(x, y, z, t) = 0$$

to be solvable for x, y, and z as functions of t.

9 Apply Theorem 18 to decide if it is possible to solve the equations

$$xy^2 + xzu + yv^2 = 3 \qquad \text{and} \qquad u^3yz + 2xv - u^2v^2 = 2$$

for u and v as functions of (x, y, z) in a neighborhood of the points $(x, y, z) = (1, 1, 1)$, $(u, v) = (1, 1)$.

10 Find the conditions on the function F which allow you to solve the equation

$$F(F(x, y), y) = 0$$

for y as a function of x near $(0, 0)$. Assume $F(0, 0) = 0$.

11 Find conditions on the functions f and g which permit you to solve the equations

$$f(xy) + g(yz) = 0 \qquad \text{and} \qquad g(xy) + f(yz) = 0$$

for y and z as functions of x, near the point where $x = y = z = 1$; assume that $f(1) = g(1) = 0$.

7.7 FUNCTIONAL DEPENDENCE

In Sec. 7.5, we studied at some length the properties of transformations of class C' whose Jacobian is never 0 in an open set. We found that they map open sets onto open sets of the same dimension, are locally 1-to-1, and therefore have local inverses. In this section, we examine the behavior of a transformation T whose Jacobian vanishes everywhere in an open set.

We illustrate this first with a simple example. Consider the transformation described by

$$T: \begin{cases} u = \cos(x + y^2) \\ v = \sin(x + y^2) \end{cases}$$

At (x, y), the Jacobian of T is

$$\begin{aligned} J(x, y) &= \det \begin{bmatrix} -\sin(x + y^2) & -2y \sin(x + y^2) \\ \cos(x + y^2) & 2y \cos(x + y^2) \end{bmatrix} \\ &= -2y \sin(x + y^2) \cos(x + y^2) + 2y \sin(x + y^2) \cos(x + y^2) \\ &= 0 \end{aligned}$$

This transformation fails to have many of the properties which were shown to hold for those with nonvanishing Jacobian. For example, although it is continuous and in fact of class C^∞, it does not map open sets in the XY plane into open sets in the UV plane. Since $u^2 + v^2 = 1$ for any choice of (x, y), T maps the entire XY plane onto the set of points on this circle of radius 1. Furthermore, it is not locally 1-to-1. All the points on the parabola $x + y^2 = c$ map into the same point $(\cos c, \sin c)$, and as c changes, these parabolas cover the entire XY plane. Thus, any disk, no matter how small, contains points having the same image. Speaking on the intuitive level for the moment, T might be called a dimension-reducing transformation; if we regard open sets in the plane as two-dimensional, and curves as one-dimensional, then T takes a two-dimensional set into a one-dimensional set.

All this degenerate behavior of T stems from the fact that the functions which we chose for the coordinates u and v were not independent, but were functionally related. There was a function F such that

$$F(u, v) = 0$$

for all x and y, namely

$$F(u, v) = u^2 + v^2 - 1$$

Definition 6 *Two functions, f and g, are said to be functionally dependent in a set D if there is a function F of two variables, which itself is not identically* 0 *in any open set, such that $F(f(p), g(p)) = 0$ for all $p \in D$.*

A similar definition may be formulated to describe functional dependence for any finite set of functions. As a special case of this, we say that a function g is **functionally dependent** in D upon the functions $f_1, f_2, \ldots, f_m$ if there is a function F of m variables such that

$$g(p) = F(f_1(p), f_2(p), \ldots, f_m(p))$$

for all $p \in D$. When the function F is a *linear* function, g is said to be **linearly dependent** upon $f_1, f_2, \ldots, f_m$. In this case, there are m numbers $C_1, C_2, \ldots, C_m$ such that g can be expressed in D as a linear combination of the functions f_j:

$$g = C_1 f_1 + C_2 f_2 + \cdots + C_m f_m$$

Linear dependence is thus a special case of the general notion of functional dependence. The sine and cosine functions are *linearly* independent, since neither is a constant multiple of the other; however, they are *functionally* dependent, since

$$\cos x = \sqrt{1 - (\sin x)^2}$$

for x in the interval $[0, \frac{1}{2}\pi]$.

We return to the study of a general transformation from $\mathbf{R}^3$ into $\mathbf{R}^3$. Let us recall the effect on a linear transformation T of the vanishing of its Jacobian. If T is represented by a matrix $A = [a_{ij}]$, then the Jacobian of T is $\det(A)$; if this is 0, then T maps all of $\mathbf{R}^3$ onto a (two-dimensional) plane, or onto a (one-dimensional) line, or onto a single point. Which it does is determined by the rank of A. If rank $(A) = 2$, the image is a plane, and if rank $(A) = 1$, it is a line. Using this as a guide, one is led to guess the correct generalization of this for transformations T which are not linear. When the Jacobian of T vanishes throughout the open domain D, we expect T to be a dimension-reducing transformation. It will map D onto something like a surface, or a curve, or a single point; which it is will depend upon the rank of dT, the differential of T. In the statements of the next two theorems, we shall use the terms "surface" and "curve" in their intuitive meanings; we postpone formal discussion of these notions until Chap. 8.

Theorem 19 *Let T be a transformation from $\mathbf{R}^3$ into $\mathbf{R}^3$ described by*

$$u = f(x, y, z)$$
$$(7\text{-}50) \qquad v = g(x, y, z)$$
$$w = h(x, y, z)$$

which is of class C' in an open set D, and suppose that at each point $p \in D$ the differential dT has rank 2. Then, T maps D onto a surface in UVW space, and the functions f, g, and h are functionally dependent in D.

Theorem 20 *If T is given by (7-50), and dT has rank 1 at each point of D, then T maps D onto a curve in UVW space, that is, f, g, and h satisfy two independent functional relations in D. In particular, about any point of D there is a neighborhood in which one of the functions can be used to express each of the others.*

The conclusion of the first theorem says that near any point of D, one may write either $u = \phi(v, w)$ or $v = \psi(w, u)$ or $w = \gamma(u, v)$. As p moves about in D, one may be forced to change from one type of relation to another. Thus, all the points (u, v, w) which arise as image points $T(p)$ lie on the graph of a surface. Similarly, the conclusion of the second theorem says that near any point of D, one may write either $u = \phi(v)$ and $w = \psi(v)$, or $v = \alpha(w)$ and $u = \beta(w)$, or $w = \gamma(u)$ and $v = \eta(u)$; again, it may not be possible to adhere to one of these relationships throughout D. Thus, in this case, the points (u, v, w) lie on the graph of a curve in UVW space. Stated in terms of Jacobians, we have the following simple condition for functional dependence.

Corollary *If u, v, and w are C' functions of x, y, and z in D, and if $\partial(u, v, w)/\partial(x, y, z) = 0$ at all points of D, then u, v, and w are functionally related in D.*

We prove both theorems together. The differential of T is given by

$$dT = \begin{bmatrix} f_1 & f_2 & f_3 \\ g_1 & g_2 & g_3 \\ h_1 & h_2 & h_3 \end{bmatrix}$$

Let us first suppose that dT has rank 1 at all points of D. This means that every 2-by-2 submatrix of dT has 0 determinant, but at least one entry of dT is not 0. Let us suppose that at a point $p_0 \in D$, $f_1(p_0) \neq 0$. Writing the first line of (7-50) as $f(x, y, z) - u = 0$ and regarding the left side as $F(x, y, z, u)$, we may apply the implicit function theorem (see Theorem 17); since $F_1 = f_1$, which does not vanish at p_0, we can solve the equation $f(x, y, z) - u = 0$ for x, getting

$$x = K(y, z, u)$$

Making this substitution in the remaining equations of (7-50), we obtain

$$(7\text{-}51) \qquad \begin{aligned} v &= g(K(y, z, u), y, z) = G(y, z, u) \\ w &= h(K(y, z, u), y, z) = H(y, z, u) \end{aligned}$$

We shall next show that the variables y and z are not really present, that is, that G and H do not depend upon y and z. To show that y is absent, let us return to Eqs. (7-50), and differentiate the first and second with respect to y while holding z and u constants. (This may be done since Eq. (7-51) allow us to regard y, z, and u as the independent variables.) Doing so, we obtain

$$0 = f_1 \frac{\partial x}{\partial y} + f_2$$

$$\frac{\partial v}{\partial y} = g_1 \frac{\partial x}{\partial y} + g_2$$

Solving the first, we have $\partial x/\partial y = -f_2/f_1$ and

$$\frac{\partial v}{\partial y} = G_1(y, z, u) = \frac{f_1 g_2 - g_1 f_2}{f_1}$$

However

$$f_1 g_2 - g_1 f_2 = \begin{vmatrix} f_1 & f_2 \\ g_1 & g_2 \end{vmatrix}$$

is the determinant of one of the 2-by-2 submatrices in dT, and by assumption, it has the value 0. Thus, in a neighborhood of p_0, we find that $G_1(y, z, u) = 0$. By a previous result (Theorem 13, Sec. 3.3), this shows that G does not depend upon y. Similar computations show that G and H are both independent of y and of z. We may therefore write Eqs. (7-51) in the simpler form

$$\begin{aligned} v &= \phi(u) \\ w &= \psi(u) \end{aligned}$$

This proves Theorem 20. ▮

To prove Theorem 19, let us assume that dT has rank 2 throughout D. This means that the Jacobian of T is 0 at each point of D, but that about any point $p_0 \in D$ is a neighborhood in which one of the 2-by-2 submatrices is nonsingular. We may suppose that it is the upper left-hand submatrix. Thus, $\partial(f, g)/\partial(x, y) \neq 0$. By the implicit function theorem, we can solve the equations

$$f(x, y, z) - u = 0$$

$$g(x, y, z) - v = 0$$

for x and y as functions of u, v, and z, obtaining

$$\text{(7-52)} \qquad \begin{aligned} x &= F(u, v, z) \\ y &= G(u, v, z) \end{aligned}$$

Substituting these into the last equation in (7-50), we have

$$\text{(7-53)} \qquad \begin{aligned} w &= h(F(u, v, z), G(u, v, z), z) \\ &= H(u, v, z) \end{aligned}$$

Now, because $J = 0$, the variable z plays no role here and H does not really depend upon its third variable. To see this, return to Eqs. (7-50), and differentiate each with respect to z, holding u and v constant. (This may be done since Eqs. (7-52) and (7-53) allow us to regard u, v, and z as the independent variables.) Doing this, we obtain

$$0 = f_1 \frac{\partial x}{\partial z} + f_2 \frac{\partial y}{\partial z} + f_3$$

$$0 = g_1 \frac{\partial x}{\partial z} + g_2 \frac{\partial y}{\partial z} + g_3$$

$$\frac{\partial w}{\partial z} = h_1 \frac{\partial x}{\partial z} + h_2 \frac{\partial y}{\partial z} + h_3$$

Solving for $\partial w/\partial z$, we obtain

$$\frac{\partial w}{\partial z} = \frac{\begin{vmatrix} f_1 & f_2 & -f_3 \\ g_1 & g_2 & -g_3 \\ h_1 & h_2 & -h_3 \end{vmatrix}}{\begin{vmatrix} f_1 & f_2 & 0 \\ g_1 & g_2 & 0 \\ h_1 & h_2 & -1 \end{vmatrix}} = \frac{J}{\begin{vmatrix} f_1 & f_2 \\ g_1 & g_2 \end{vmatrix}}$$

Since $J = 0$, we have $\partial w/\partial z = H_3(u, v, z) = 0$ in a neighborhood, and w must be expressible as $w = \phi(u, v)$; z is a ghost in H. This proves the functional dependence of w on u and v (Theorem 19). ▮

So far, we have discussed only transformations whose domain and whose range lay in spaces of the same dimension, e.g., transformations from $\mathbf{R}^2$ into $\mathbf{R}^2$, from $\mathbf{R}^3$ into $\mathbf{R}^3$, etc. The techniques that have been developed will also apply to other types of transformations. Several samples will be sufficient to illustrate the manner in which this is done.

Theorem 21 *Let T be a transformation which is of class C' in an open set D in $\mathbf{R}^3$ and mapping this into $\mathbf{R}^2$. Let T be described by*

$$\begin{aligned} u &= f(x, y, z) \\ v &= g(x, y, z) \end{aligned}$$

Then, if the differential dT has rank 2 throughout D, f and g are functionally independent in D, and T maps D onto an open set in the UV plane; while if dT has rank 1 throughout D, then f and g are functionally dependent in D, and the image of D under T is a curve in the UV plane.

Corollary *If*

$$\frac{\partial(u, v)}{\partial(x, y)} = \frac{\partial(u, v)}{\partial(y, z)} = \frac{\partial(u, v)}{\partial(z, x)} = 0$$

throughout D, then about every point of D is a neighborhood in which $u = \phi(v)$ or $v = \psi(u)$.

Theorem 22 *Let T be a transformation from $\mathbf{R}^2$ into $\mathbf{R}^3$ which is of class C' in an open set D, and is given there by the equations*

$$\begin{aligned} x &= f(u, v) \\ y &= g(u, v) \\ z &= h(u, v) \end{aligned} \tag{7-54}$$

If the rank of dT is 2 throughout D, then T maps D onto a surface in XYZ space, while if the rank of dT is 1 throughout D, T maps D onto a curve.

Both of these may be reduced to special cases of Theorems 19 and 20. To prove Theorem 21, we adjoin an equation $w = 0$ to make T a transformation from $\mathbf{R}^3$ into $\mathbf{R}^3$. Its differential is now

$$\begin{bmatrix} f_1 & f_2 & f_3 \\ g_1 & g_2 & g_3 \\ 0 & 0 & 0 \end{bmatrix}$$

whose rank is always the same as that of dT. ▮

To prove Theorem 22, we introduce a dummy variable w into Eqs. (7-54) which describe T to again make T a transformation from $\mathbf{R}^3$ into $\mathbf{R}^3$. (For instance, one might write $x = f(u, v) + w - w$.) The differential of the resulting transformation is

$$\begin{bmatrix} f_1 & f_2 & 0 \\ g_1 & g_2 & 0 \\ h_1 & h_2 & 0 \end{bmatrix}$$

whose rank is again the same as that of dT. ▮

For future use, we note that the condition that the rank of this matrix be 2 may also be expressed in the equivalent form

$$\left[\frac{\partial(x, y)}{\partial(u, v)}\right]^2 + \left[\frac{\partial(y, z)}{\partial(u, v)}\right]^2 + \left[\frac{\partial(z, x)}{\partial(u, v)}\right]^2 > 0 \tag{7-55}$$

The reader will have noticed that the hypotheses of recent theorems have been of two types. It was assumed either that the Jacobian was nonzero everywhere in a region D, or that it was identically 0 in the region; it was assumed that a matrix either had rank 2 everywhere in D, or had rank 1 everywhere in D. For a general transformation $T: \mathbf{R}^3 \to \mathbf{R}^3$, the following behavior would be much more typical: The Jacobian would be nonzero everywhere in space, except on certain surfaces. On these surfaces, dT would have rank 2, except on certain curves. On these curves, dT would have rank 1, except for certain points where the rank is 0.

We shall say that a critical point for T is any point where the Jacobian is 0; more generally, if T is a transformation from $\mathbf{R}^n$ into $\mathbf{R}^m$, then a point p is a **critical point** for T if the rank of dT at p is less than optimal, i.e., less than the smaller of n and m. (Note that for a function f of n variables, values in $\mathbf{R}^1$, this would require that $df = 0$.)

What sort of behavior should one expect for a transformation in the neighborhood of a critical point? Even in the simplest cases, one is led into serious difficulties; one simple example may serve to illustrate the possibilities.

Consider the transformation T from $\mathbf{R}^2$ into $\mathbf{R}^2$ given by

$$u = x^2, \; v = y^2$$

The differential is
$$dT = \begin{bmatrix} 2x & 0 \\ 0 & 2y \end{bmatrix}$$

and $J(x, y) = 4xy$. The rank of dT is 2 everywhere except on the lines $x = 0$ and $y = 0$. On these, it has rank 1, except for the origin where the rank is 0. The effect of T may be indicated crudely by Fig. 7-10. We see

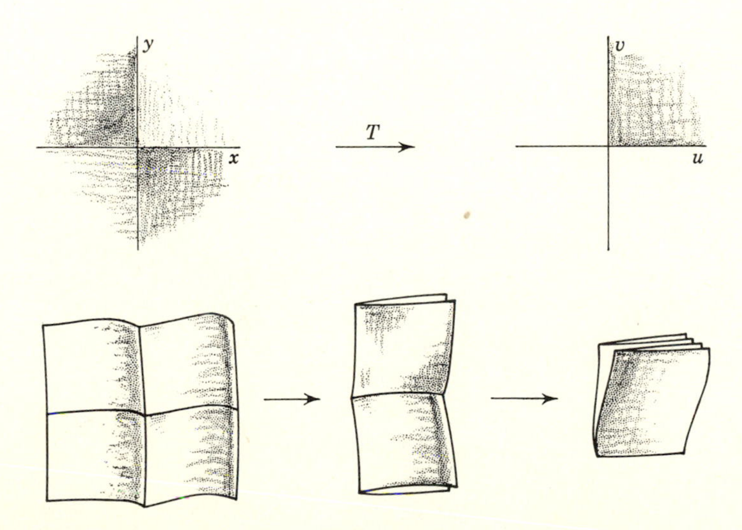

Figure 7-10

that the lines $x = 0$ and $y = 0$ are the creases along which T folds the XY plane, and where T is locally 2-to-1, while the origin corresponds to the point of the final fold, and at which T is locally 4-to-1.

EXERCISES

1 By consideration of their differentials, discuss the nature of the following transformations:

(*a*) $\begin{cases} u = x^2 \\ v = y \end{cases}$ (*b*) $\begin{cases} u = x^2y^2 \\ v = 2xy \end{cases}$

(*c*) $\begin{cases} u = x - y \\ v = x^2 + y^2 - 2xy \end{cases}$ (*d*) $\begin{cases} u = x + y \\ v = x + z \\ w = y^2 + z^2 - 2yz \end{cases}$

2 Show that the following sets of functions are functionally dependent and find the functional relationship.

(*a*) $\begin{cases} u = \log x - \log y \\ v = \dfrac{x^2 + 3y^2}{2xy} \end{cases}$ (*b*) $\begin{cases} u = x + y + z \\ v = xy + zx \\ w = x^2 + y^2 + z^2 + 2yz \end{cases}$

3 Find the functional relation of the following pairs if such a relation exists.

(*a*) $u = \log (x + y)$ $\quad v = x^2 + y^2 + 2xy + 1$
(*b*) $u = (x + y)/x$ $\quad v = (x + y)/y$
(*c*) $u = x + y$ $\quad v = x^2 + y^2$

4 For the transformation

$$u = \frac{x + y}{1 - xy} \qquad v = \frac{(x + y)(1 - xy)}{(1 + x^2)(1 + y^2)}$$

(*a*) Verify that $\dfrac{\partial(u, v)}{\partial(x, y)} \equiv 0$.

(*b*) Express v in terms of u.

5 In the proof of Theorem 20, complete the argument to show that the function G in Eqs. (7-51) is independent of z.

6 Discuss the nature of a transformation T whose differential is of rank 0 throughout an open set D.

7 Let $u = F(x, y)$ and $v = G(x, y)$, and suppose that F and G are functionally dependent in a set D. Show that $\partial(u, v)/\partial(x, y) = 0$ in D.

8 Show that $x - y$, xy, and xe^y are functionally dependent.

9 Let $u = f(x, y, z)$ and $v = g(x, y, z)$ and suppose that f and g are functionally dependent in an open set D, with $F(u, v) = 0$, where $F_1(u, v)$ and $F_2(u, v)$ are never both 0. Show that

$$\frac{\partial(u, v)}{\partial(x, y)} = \frac{\partial(u, v)}{\partial(y, z)} = \frac{\partial(u, v)}{\partial(z, x)} = 0$$

in D.

CHAPTER

EIGHT

APPLICATIONS TO GEOMETRY AND ANALYSIS

8.1 PREVIEW

This chapter deals with a number of geometric applications of integration and differentiation. Building on the techniques of the previous section, we treat the change-of-variable formula for multiple integrals in Sec. 8.3, using an approach that emphasizes the role of the differential of the transformation T that describes the coordinate change. The key idea is to look at the integral of a function f over the image set $T(D)$ as though one had integrated some other function over the set D, then use differentiation to evaluate this new integrand. Thus, one needs a multiple-integral form of the fundamental theorem of calculus. This is supplied in Sec. 8.2, which gives an elementary treatment of set functions and a weak version of the Radon-Nikodym theorem, which is the key also to the physical notions of density and mass and to some of the basic aspects of probability theory.

The remainder of the chapter deals with curves and surfaces, discussing such topics as tangents, normals, curvature, arc length and surface area, and integration of functions along a curve or over a surface. There is also a brief intuitive introduction to the notion of manifold, as an extension of "surface," in order to present the contrasting concepts of orientability and nonorientability.

While it might seem somewhat out of place, Sec. 8.5 turned out to be the appropriate place to discuss the total second derivative of a function or transformation, since this characterizes geometric aspects of the associated graphs; as the Hessian, it also discriminates between saddle points and extreme points.

8.2 SET FUNCTIONS

By the term **set function** we mean a function F which assigns a number $F(S)$ to each set S in some specified class of sets. By contrast, the term point functions might be used for the functions f which have been discussed up until now, and which assign a number $f(p)$ to each point p in a specified set of points. Examples of set functions are common in pure and in applied mathematics, embracing such diverse notions as area, force, mass, moment of inertia, and probability. To illustrate this, let $\mathscr{A}$ be the class of compact sets in the plane which have area. Then, the area function A, which assigns to each set $S \in \mathscr{A}$ its area $\text{A}(S)$, is a set function defined on the class $\mathscr{A}$. Taking an example from physics, we suppose we are given a specific distribution of matter throughout space, which in some places may be continuous, and in others, discrete. With this, we may associate a particular set function m by taking $\text{m}(S)$ to be the total mass of the matter lying within the set S. In particular, if the distribution of matter is taken to be only a single particle of mass 1 located at the origin, then $\text{m}(S) = 0$ whenever S fails to contain the origin, and $\text{m}(S) = 1$ if S contains the origin. Again, if a horizontal plate is subjected to a variable load distribution, we may obtain a set function F by choosing $F(S)$ to be the total force pressing down on each region S of the plate. Finally, if an experiment involves dropping shot from a height onto a target board, then to each region S of the target we can associate a number $P(S)$ which is the probability that an individual shot will land in S.

All the examples presuppose that one starts with a prior knowledge (e.g., the meaning of area, mass, force, or probability). Reversing the procedure, one might postulate the existence of the set functions, together with certain appropriate properties, and then use them to develop the corresponding physical or mathematical notions. One property that a set function may possess, shared by all those mentioned above, is particularly important.

Definition 1 *Let $\mathcal{S}$ be a collection of sets such that if S_1 and S_2 belong to $\mathcal{S}$, so do $S_1 \cup S_2$, $S_1 \cap S_2$, and $S_1 - S_2$. Then, a set function F defined on $\mathcal{S}$ is* **finitely additive** *if $F(S_1 \cup S_2) = F(S_1) + F(S_2)$ whenever S_1 and S_2 are disjoint members of $\mathcal{S}$.*

A finitely additive function F must automatically satisfy certain other properties. For example, $F(\varnothing) = 0$, since S and the empty set $\varnothing$ are necessarily disjoint, and hence $F(S \cup \varnothing) = F(S) = F(S) + F(\varnothing)$. Again, if the sets $S_k \in \mathcal{S}$

are pairwise disjoint, then

$$F\left(\bigcup_1^N S_k\right) = \sum_1^N F(S_k) \tag{8-1}$$

We note also that there are corresponding addition formulas which apply when the sets are not disjoint (see Exercise 2).

Another property shared by many set functions is positivity; F is said to be **positive** if $F(S) \geq 0$ for all $S \in \mathcal{S}$. This is equivalent to the property of being **monotone**, which means that $F(S_1) \leq F(S_2)$ whenever $S_1 \subset S_2$ (Exercise 3).

An extensive category of finitely additive set functions can be constructed by means of integration. Let ϕ be a real-valued point function defined and continuous on the whole plane. Then ϕ yields a set function defined on the class $\mathscr{A}$ by the equation

$$F(S) = \iint_S \phi \tag{8-2}$$

The fact that F is finitely additive is merely a restatement of one of the familiar properties of integration. Note also that F will be a positive set function if and only if $\phi(p) \geq 0$ for all points p (Exercise 5).

The central result of this section will show that every finitely additive set function obeying certain simple conditions must have the form (8-2), where the appropriate point function ϕ is obtained from the set function F by a process analogous to differentiation. Indeed, the entire result can be regarded as the multiple integration version of the fundamental theorem of calculus that connects integration and differentiation in one variable. For concreteness, we work with set functions defined on the class $\mathscr{A}$ of Jordan-measurable subsets of the plane—i.e., bounded sets B that have an area $\mathrm{A}(B)$. The same analysis could be done for n space.

We introduce a special type of limit operation.

Definition 2 *Given an arbitrary set function F, not necessarily additive, and defined at least for all rectangles R, we write $\lim_{R \downarrow p_0} F(R) = c$ if and only if, given $\varepsilon > 0$, there is a $\delta > 0$ such that*

$$|F(R) - c| < \varepsilon \tag{8-3}$$

whenever R is a rectangle containing the point p_0 and with diameter $\operatorname{diam}(R) < \delta$.

The rectangles R can be thought of as "closing down" on the point p_0, and the limit value c, when it exists, can be regarded as the value which F should assign to the set consisting of p_0 alone. (The restriction to rectangles makes this notion of limit rather special, but it is all we need for simple applications.)

If the set function F has such a limit for each point p in a region D, then this process defines a point function f on D by writing, for each $p \in D$,

$$f(p) = \lim_{R \downarrow p} F(R) \tag{8-4}$$

Definition 3 *We say that the limit* (8-4) *holds uniformly for all p in a set E if and only if, given* $\varepsilon > 0$, *there is a* $\delta > 0$ *such that*

$$|F(R) - f(p)| < \varepsilon$$

for every $p \in E$ *and every rectangle R containing p with* diam $(R) < \delta$.

Using these notions of limits, we now introduce a process of differentiation of set functions defined on the class $\mathscr{A}$, with respect to the area function A. When the derivative of F exists, it will be a point function f; not all set functions have derivatives.

Definition 4 *An arbitrary set function F, defined on* $\mathscr{A}$, *is said to be* **differentiable** *on a set D if*

$$\lim_{R \downarrow p} \frac{F(R)}{\text{A}(R)} \tag{8-5}$$

exists for each $p \in D$, *and to be* **uniformly differentiable** *on D if* (8-5) *exists, uniformly for all* $p \in D$.

It is to be understood in the definition that $\text{A}(R) > 0$.

This is not a new concept, especially in the case of the examples of set functions which we have given above. Taking F as the mass function m, the value of the quotient $\text{m}(S)/\text{A}(S)$ is the average mass per unit area (or volume, if we take the corresponding three-dimensional derivative) in the region S. As S closes down on p, the limit can be interpreted as the *density* of matter at p. Similarly, the derivative of the set function *force* is the point function *pressure* (= force per unit area). Not all set functions have a derivative; the mass function produced by an isolated unit point mass at the origin fails to have a derivative there since $\text{m}(S) = 1$ for any S containing $\mathbf{0}$, while $\text{A}(S) \to 0$ (see also Exercise 8).

Our first result shows that the special additive set function defined by (8-2) is differentiable everywhere, and that the integrand ϕ can be recovered from F itself.

Theorem 1 *The set function F given in* (8-2) *is uniformly differentiable on any compact set E, and its derivative is the point function* ϕ.

Given E, we can choose a small number δ_1 and a compact set E_1 containing E such that if $p \in E$ and $|q - p| < \delta_1$, then $q \in E_1$. The function

ϕ is uniformly continuous in E_1. Given ε, we can choose $\delta < \delta_1$ so that if $p \in E$ and $|p - q| < \delta$, then $|\phi(p) - \phi(q)| < \varepsilon$. Let R be any rectangle containing p and with diam $(R) < \delta$; note that R lies entirely in E_1. By the mean value theorem for integrals,

$$F(R) = \iint_R \phi = \phi(q)\mathrm{A}(R)$$

where q is some point in R. Thus, $F(R)/\mathrm{A}(R) = \phi(q)$, and we have shown that

$$\left| \frac{F(R)}{\mathrm{A}(R)} - \phi(p) \right| < \varepsilon$$

holding now for any choice of R, diam $(R) < \delta$, and any $p \in E$. ∎

The main result of this section is the converse of this, showing that a simple set of properties characterizes the set functions obtained by integration as in (8-2). In addition to finite additivity, we will use another property that is obviously true of those functions F defined by (8-2).

Definition 5 *A set function F is* **a.c.** (*area continuous*) *if $F(S) = 0$ for every set S with $\mathrm{A}(S) = 0$ with $S \in \mathscr{A}$.*

In fact, set functions F defined by (8-2) even obey the following more restrictive condition: *For any compact set E, there is a number M such that $|F(S)| \le M\mathrm{A}(S)$ for all sets $S \in \mathscr{A}$, $S \subset E$* (Exercise 6).

Theorem 2 *Let F be an additive set function, defined on $\mathscr{A}$ and* a.c. *Suppose also that F is differentiable everywhere, and uniformly differentiable on compact sets, with the derivative a point function f. Then, f is continuous everywhere, and*

$$F(S) = \iint_S f \tag{8-6}$$

holds for every rectangle S. If F is positive on an open set D (*meaning that $F(S) \ge 0$ for any set $S \subset D$, $S \in \mathscr{A}$*), *then* (8-6) *also holds for all sets $S \in \mathscr{A}$ contained in D.*

The pattern of the proof of this is similar to that of the fundamental theorem of calculus for functions of one variable, as given in Sec. 4.3. We first prove that the derivative f is continuous. Let E be any closed disk. Since F is uniformly differentiable in E, given ε we can choose δ so that

$$\left| \frac{F(R)}{\mathrm{A}(R)} - f(p) \right| < \varepsilon$$

for any $p \in E$ and any rectangle R containing p, and with diam $(R) < \delta$. Let p_1 and p_2 lie in E, with $|p_1 - p_2| < \delta/2$, and choose R, diam $(R) < \delta$, so that $p_1, p_2 \in R$. Then,

$$|f(p_1) - f(p_2)| = \left| f(p_1) - \frac{F(R)}{A(R)} + \frac{F(R)}{A(R)} - f(p_2) \right|$$
$$\leq \varepsilon + \varepsilon = 2\varepsilon$$

proving that f is (uniformly) continuous in E. Next, define a set function F_0 by using (8-2) with $\phi = f$. By Theorem 1, F_0 is also uniformly differentiable on compact sets, and its derivative is also the point function f. Construct a third set function H by $H(S) = F(S) - F_0(S)$. This is again finitely additive, a.c., defined on $\mathscr{A}$, and uniformly differentiable on compact sets, and its derivative is everywhere $f - f = 0$. We will now show that any such set function is necessarily 0, at least on rectangles. Let E be a large closed disk. Then, since the uniform derivative of H in E is the point function 0, we know that for any $\varepsilon > 0$ there is a $\delta > 0$ such that if R is a rectangle contained in E with diam $(R) < \delta$, then

$$\left| \frac{H(R)}{A(R)} \right| = \left| \frac{H(R)}{A(R)} - 0 \right| < \varepsilon$$

or, $|H(R)| \leq \varepsilon A(R)$. Now take any rectangle S in E, and express it as the union of rectangles $R_1, R_2, \ldots, R_n$, each of diameter smaller than δ and mutually disjoint except for edges. Since each edge has area 0 and H is both a.c. and additive, $H(S) = \sum_1^n H(R_i)$. Applying the general estimate above, we then have $|H(R_i)| \leq \varepsilon A(R_i)$ and

$$|H(S)| \leq \varepsilon \sum_1^n A(R_i) = \varepsilon A(S)$$

Since this holds for every ε, $H(S) = 0$, and it follows that $F(S) = F_0(S)$ for any rectangle, and hence for finite unions.

To extend this now to general sets $S \in \mathscr{A}$ contained in an open set D where F is a positive set function, we use the fact that F is monotonic, and approximate S from inside and outside by finite unions of rectangles. If $S_n \subset S$ with $A(S_n) \to A(S)$, then $F(S_n) \leq F(S)$, and since S_n is a finite union of rectangles,

$$F(S_n) = F_0(S_n) = \iint_{S_n} f$$

which converges to $\iint_S f = F_0(S)$ by properties of integration. This argument shows $F_0(S) \leq F(S)$ for any $S \subset D$. A similar argument using circumscribing sets S_n that are unions of rectangles yields $F(S) \leq F_0(S)$, and we have completed the proof. ∎

This result provides the justification for obtaining the mass of a body by integrating a density function over the corresponding region, for computing the force on a wall by integrating a pressure function, or for finding the probability of an event by integrating a probability density function. You will meet it again in courses in measure theory, where it is generalized to the Radon-Nikodym theorem.

In the next section, we apply Theorem 2 to derive the formula for change of variables in a multiple integral, which was discussed without proof in Sec. 4.4.

EXERCISES

1 Do the following describe set functions that are finitely additive?

(*a*) $F(S)$ is the square of the area of the plane set S.

(*b*) $F(S)$ is the area of the smallest closed circular disk which contains the closed set S.

(*c*) $F(S)$ is the moment of inertia of the plane set S about an axis through the origin, and perpendicular to the plane.

(*d*) $F(S)$ is the diameter of the closed set S.

2 (*a*) If F is additive on $\mathscr{A}$, then prove that

$$F(S_1 \cup S_2) = F(S_1) + F(S_2) - F(S_1 \cap S_2)$$

(*b*) Is there a similar expression for $F(S_1 \cup S_2 \cup S_3)$?

3 Show that any additive set function F defined on $\mathscr{A}$ is positive on $\mathscr{A}$ if and only if F is monotonic.

4 If F is positive and additive on $\mathscr{A}$, show that for any finite collection of sets $S_i \in \mathscr{A}$, $F(\bigcup_1^n S_i) \le \sum_1^n F(S_i)$.

5 Show that if F is defined by (8-2), where ϕ is continuous, then $F \ge 0$ if and only if $\phi \ge 0$.

6 Let F be defined by (8-2), with ϕ continuous on the whole plane. Show that for every compact set E there is a number M such that $|F(S)| \le M\text{A}(S)$ for all $S \in \mathscr{A}$ with $S \subset E$.

7 By considering the one-dimensional case, show why Theorems 1 and 2 are analogous to the fundamental theorem of calculus (see Theorems 5, 6, and 7, Sec. 4.3).

8 Let f be a bounded function which is continuous everywhere in the plane except at the origin. Define F by $F(S) = \iint_S f$. Is F differentiable at the origin? (Your answer will depend upon the nature of the discontinuity of f.)

9 Let m be the set function described in the first paragraph of this section, based upon a single particle of unit mass, located at the origin. Show that $\lim_{S \downarrow p_0} \text{m}(S)$ exists for each point p_0 but that m is not a differentiable set function on a square containing the origin.

10 Let f be a function of one variable, defined on the interval

$$-\infty < x < \infty$$

Define a set function on intervals by $F(I) = f(b) - f(a)$, where $I = [a, b]$, and assign the same value to $F(I)$ whether the endpoints of I are included or not.

(*a*) Show that F is an additive set function, for intervals.

(*b*) Show that $\lim_{I \downarrow x_0} F(I)$ exists if f is continuous at x_0.

(*c*) When is F differentiable?

***11** Let f be a continuous function of two variables, defined on the whole plane and of class C''. For any rectangle R, whose vertices are $P_0 = (x_0, y_0)$, $P_1 = (x_0, y_1)$, $P_2 = (x_1, y_1)$, $P_3 = (x_1, y_0)$, and $x_0 < x_1$, $y_0 < y_1$, let

$$F(R) = f(P_0) - f(P_1) + f(P_2) - f(P_3)$$

(*a*) Show that F is an additive set function on rectangles, in the sense that, if a rectangle R is the union of rectangles $R_1, R_2, \ldots, R_m$ which have disjoint interiors, then $F(R) = \sum_1^m F(R_k)$.

(*b*) Show that the derivative of F exists and is the point function $g = f_{12} = \partial^2 f/(\partial x\, \partial y)$.

8.3 TRANSFORMATIONS OF MULTIPLE INTEGRALS

In this section we shall discuss certain additional properties of transformations, in particular, their effect on volume and area. This is a subject of considerable technical complexity, and one that is still near the frontier of research.

To gain some insight into what may be expected, let us start with a linear transformation L from $\mathbf{R}^n$ into $\mathbf{R}^n$. If D is a bounded set in $\mathbf{R}^n$ whose n-dimensional volume is $\mathrm{v}(D)$, what can be said about the n-dimensional volume of the image $L(D)$?

Theorem 3 *The volume of $L(D)$ is $k\mathrm{v}(D)$, where $k = |\det(L)|$.*

This asserts that the effect of L on volumes is simply to multiply by a fixed numerical factor k, which is independent of the set D. Moreover, if L is singular, this factor is 0, so that the image of any set D is a set of zero volume; this agrees with the fact that in this case, the image of all n space will be a set of lower dimension. To make the geometry easier, let us take $n = 3$ so that L may be represented by a matrix of the form

$$A = \begin{bmatrix} a_{11} & a_{12} & a_{13} \\ a_{21} & a_{22} & a_{23} \\ a_{31} & a_{32} & a_{33} \end{bmatrix}$$

A general set D in 3-space which has volume can be approximated from inside and outside by finite unions of cubes; thus, the general result will follow if it can be shown to hold when D is restricted to be a cube. In this case, the theorem can be interpreted as the equivalent of a familiar statement about determinants, and the fact that the general 3-by-3 matrix A can be factored as a product of simpler matrices. The proof for the general n-by-n case can be found in any of the standard references quoted in Appendix 3. Rather than digress here, we merely show that the formula $\mathrm{v}(L(D)) = k\mathrm{v}(D)$ holds whenever D is a tetrahedron with one of its vertices at the origin. Let the other vertices be $P_j = (x_j, y_j, z_j)$, for $j = 1, 2, 3$. The volume of such a tetrahedron can be expressed by the formula

$$\mathrm{v}(D) = (\tfrac{1}{6})|\det(U)| \tag{8-7}$$

where U is the matrix

(8-8)
$$U = \begin{bmatrix} x_1 & x_2 & x_3 \\ y_1 & y_2 & y_3 \\ z_1 & z_2 & z_3 \end{bmatrix}$$

The transformation L carries D into another tetrahedron D' with vertices $(0, 0, 0)$ and $P'_j = (x'_j, y'_j, z'_j)$, for $j = 1, 2, 3$. Form the corresponding matrix U' from these points. The fact that $P'_j = L(P_j)$ can also be expressed by the matrix equation $U' = AU$. Since the determinant of the product of two square matrices is the product of their determinants, we also have the equation $\det(U') = \det(A)\det(U)$. Since the volume of the tetrahedron D' is $(\frac{1}{6})|\det(U')|$, this shows that

$$\mathrm{v}(L(D)) = k\mathrm{v}(D)$$

where $k = |\det(A)|$.

L carries the triangular faces of the tetrahedron D into the corresponding faces of D'. It is natural to ask if there is an equally simple relation holding between the areas of these triangles. However, such is not the case; the area of the image of a triangle under a linear transformation of 3-space into itself will depend not only upon the area of the original triangle, but also upon its position. Congruent triangles may have images of different area. [Consider, for example, the effect of the simple projection $(x, y, z) \to (x, y, 0)$.] Of course, this cannot happen if the transformation maps 2-space into itself, for here Theorem 3 tells us that the area of any set D in the plane, and that of its image $L(D)$, are connected by the formula $\mathrm{A}(L(D)) = k\mathrm{A}(D)$, where k depends only upon L and not upon D.

If, however, we consider linear transformations from 2-space into 3-space (or more generally, into n space), then we can obtain a result of comparable simplicity. Let L be the transformation represented by the matrix

$$\begin{bmatrix} a_{11} & a_{12} \\ a_{21} & a_{22} \\ a_{31} & a_{32} \end{bmatrix}$$

If R is a rectangle with vertices $(0, 0)$, $(a, 0)$, $(0, b)$, and (a, b), then L carries R into a parallelogram $L(R)$ with vertices $(0, 0, 0)$, aP_1, bP_2, and $aP_1 + bP_2$, where

$$P_1 = (a_{11}, a_{21}, a_{31}) \qquad P_2 = (a_{12}, a_{22}, a_{32})$$

The area of R is $|ab|$; the area of $L(R)$ is

$$|aP_1||bP_2| \sin\theta = |ab||P_1||P_2| \sin\theta$$

where θ is the angle between the lines OP_1 and OP_2 (see Fig. 8-1). Thus, we again have a relation of the form $\mathrm{A}(L(D)) = k\mathrm{A}(D)$, where

$$k = |P_1||P_2| \sin\theta$$

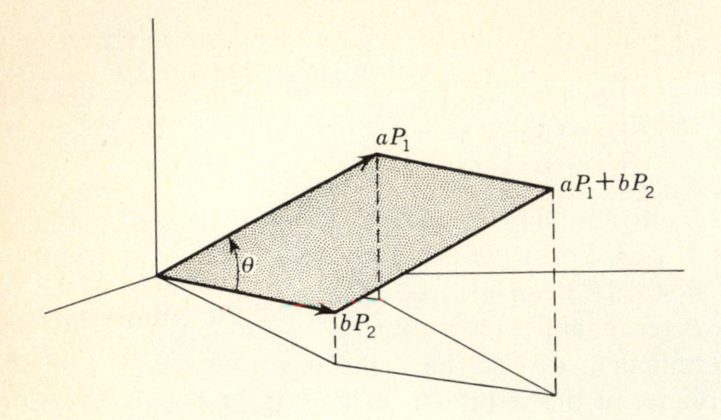

Figure 8-1

is a constant which is determined solely by the matrix of L. To obtain an explicit formula for k, we recall that the cosine of the angle between two vectors (points) may be found by means of their inner product

$$\cos\theta = \frac{(P_1 \cdot P_2)}{|P_1||P_2|}$$

Using this, we find that

$$\begin{aligned} k &= |P_1||P_2| \sin\theta \\ &= |P_1||P_2|\left\{1 - \left(\frac{P_1 \cdot P_2}{|P_1||P_2|}\right)^2\right\}^{1/2} \\ &= \{|P_1|^2|P_2|^2 - (P_1 \cdot P_2)^2\}^{1/2} \\ &= \left\{\sum_{i=1}^{3} (a_{i1})^2 \sum_{i=1}^{3} (a_{i2})^2 - \left(\sum_{i=1}^{3} a_{i1}a_{i2}\right)^2\right\}^{1/2} \end{aligned}$$

In the general case where L is a linear transformation from 2-space into n space, the result is the same, with

$$k = \left\{\sum_{i=1}^{n} (a_{i1})^2 \sum_{i=1}^{n} (a_{i2})^2 - \left(\sum_{i=1}^{n} a_{i1}a_{i2}\right)^2\right\}^{1/2} \tag{8-9}$$

For example, if L is the transformation

$$\begin{bmatrix} 1 & -1 \\ 2 & 0 \\ -1 & 1 \\ 0 & 3 \end{bmatrix}$$

then

$$k^2 = (1 + 4 + 1 + 0)(1 + 0 + 1 + 9) - (-1 + 0 - 1 - 0)^2 = 62$$

and $k = \sqrt{62}$.

There is a special identity involving determinants which gives an alternative expression for k. For example, direct computation shows that for $n = 3$,

$$k^2 = (a_{11}^2 + a_{21}^2 + a_{31}^2)(a_{12}^2 + a_{22}^2 + a_{32}^2) - (a_{11}a_{12} + a_{21}a_{22} + a_{31}a_{32})^2$$

$$= \begin{vmatrix} a_{11} & a_{12} \\ a_{21} & a_{22} \end{vmatrix}^2 + \begin{vmatrix} a_{21} & a_{22} \\ a_{31} & a_{32} \end{vmatrix}^2 + \begin{vmatrix} a_{31} & a_{32} \\ a_{11} & a_{12} \end{vmatrix}^2$$

In the general case, k^2 is equal to the sum of the squares of the determinants of all the 2-by-2 submatrices of the matrix for L. Thus, in the numerical example given above, we would have

$$k^2 = \begin{vmatrix} 1 & -1 \\ 2 & 0 \end{vmatrix}^2 + \begin{vmatrix} 1 & -1 \\ -1 & 1 \end{vmatrix}^2 + \begin{vmatrix} 1 & -1 \\ 0 & 3 \end{vmatrix}^2 + \begin{vmatrix} 2 & 0 \\ -1 & 1 \end{vmatrix}^2 + \begin{vmatrix} 2 & 0 \\ 0 & 3 \end{vmatrix}^2 + \begin{vmatrix} -1 & 1 \\ 0 & 3 \end{vmatrix}^2$$

$$= 4 + 0 + 9 + 4 + 36 + 9$$

$$= 62$$

Let us turn now to the study of a general (nonlinear) transformation T. If T is of class C' in an open set Ω, then the approximation theorem allows us to write as in Theorem 10, Sec. 7.4,

$$T(p + \Delta p) = T(p) + dT|_p(\Delta p) + R(\Delta p) \tag{8-10}$$

(8-11) where
$$\lim_{\Delta p \to 0} \frac{|R(\Delta p)|}{|\Delta p|} = 0$$

uniformly for all points p in any closed bounded subset of Ω. If T is a transformation from $\mathbf{R}^n$ into $\mathbf{R}^n$, then $dT|_p$ is a linear transformation from $\mathbf{R}^n$ into $\mathbf{R}^n$. As such, it alters volumes by the factor

$$k = \left|\det(dT|_p)\right| = |J(p)|$$

the absolute value of the Jacobian of T at p. If it is true that the local behavior of T is the same as the behavior of dT, then we would expect the volume of $T(D)$ to be about $J(p)\mathrm{v}(D)$ if D is a sufficiently small set surrounding p. As a first step in proving this, we have Theorem 4.

Theorem 4 *If E is a closed bounded subset of Ω of zero volume, then $T(E)$ has zero volume.*

Since E is compact and $\mathrm{v}(E) = 0$, we can enclose E in a finite union of balls $B_1, B_2, \ldots, B_N$, each of diameter less than any preassigned δ, and such that $\sum_1^N \mathrm{v}(B_k) < \varepsilon$. Also, by the corollary to Theorem 12,

Sec. 7.4, we know that there is a constant M such that if $p_0 \in E$ and $|p_0 - p| < \delta$, then $|T(p) - T(p_0)| < M|p - p_0|$. Accordingly, the image under T of any ball B of radius $r < \delta$ will be a subset of another ball of radius Mr, whose n-dimensional volume is $M^n \mathrm{v}(B)$. Thus, $T(E)$ is a set that can be covered by the union of balls of total volume $\sum_1^N M^n \mathrm{v}(B_k) < M^n \varepsilon$. Since M depends only on E, and ε is arbitrarily small, $T(E)$ is a set of zero volume. ▌

As an application of this, we see that any nonsingular C' transformation takes Jordan-measurable sets (i.e., sets that have an area or a volume) into Jordan-measurable sets.

Corollary *Let T be a C' transformation on an open set Ω on which its Jacobian $J(p)$ never vanishes. Then if D is a compact subset of Ω which has a boundary that has zero area (volume), the same is true for $T(D)$.*

For T carries open sets into open sets. Thus, the image of the interior of D is an open set contained in the compact set $T(D)$, and hence contained in the interior of $T(D)$. Accordingly, every boundary point of $T(D)$ is the image of a boundary point of D; bdy $(T(D)) \subset T(\text{bdy } (D))$. Since $\mathrm{v}(\text{bdy } (D)) = 0$, $\mathrm{v}(T(\text{bdy } (D))) = 0$, and the boundary of $T(D)$ is a subset of a set of zero volume, and therefore of zero volume itself. This is the criterion for $T(D)$ to be a set having a measure in the Jordan sense. ▌

What can be said about the value of $\mathrm{v}(T(D))$, in comparison with the value of $\mathrm{v}(D)$? Suppose that for concreteness we consider a transformation from 3-space into itself, given by a formula such as

$$T: \begin{cases} x = f(u, v, w) \\ y = g(u, v, w) \\ w = h(u, v, w) \end{cases} \tag{8-12}$$

which we may abbreviate by $T(u, v, w) = (x, y, z)$. With an appropriate hypothesis on T, we can find a formula for $\mathrm{v}(T(D))$.

Theorem 5 *Let T be a transformation from 3-space into 3-space, with $T(u, v, w) = (x, y, z)$, which is of class C' and 1-to-1 in an open set Ω, with $J(p) \neq 0$ throughout Ω. If D is a closed bounded subset of Ω in UVW space, then the volume of its image is given by*

$$\mathrm{v}(T(D)) = \iiint_{T(D)} dx\, dy\, dz = \iiint_D \left| \frac{\partial(x, y, z)}{\partial(u, v, w)} \right| du\, dv\, dw = \iiint_D |J|$$

In proving this, we shall use the theory of differentiation of set functions, as developed in Sec. 8.2. Define a set function F on the class of sets D

possessing volume by the equation

$$F(D) = \mathrm{v}(T(D))$$

Since T is 1-to-1, disjoint sets have disjoint images, and F is an additive set function. If F has a derivative f, that is, if the limit

$$\lim_{D \downarrow p} \frac{F(D)}{\mathrm{v}(D)} = f(p)$$

exists, then by Theorem 2, Sec. 8.2, extended to 3-space,

$$F(D) = \mathrm{v}(T(D)) = \iiint_D f$$

In the light of the remarks made earlier about the local behavior of T, it is easy to conjecture that this will hold, and that $f(p) = J(p)$. Since this result is important in itself, we single it out.

Lemma 1 *With Ω and T as above, let E be a compact subset of Ω. Then*

(8-13) $$\lim_{C \downarrow p} \frac{\mathrm{v}(T(C))}{\mathrm{v}(C)} = |J(p)|$$

where C ranges over the family of cubes lying in Ω with center $p \in E$, and the limit is uniform for all $p \in E$.

We start from the more precise statement of the approximation property of the differential of a transformation, as given in Theorem 10, Sec. 7.4. Given ε, there is a $\delta > 0$ such that for any point $p_0 \in E$ and $|\Delta p| < \delta$, $p = p_0 + \Delta p$,

(8-14) $$T(p) - T(p_0) = dT|_{p_0}(\Delta p) + R(\Delta p)$$

where $|R(\Delta p)| < \varepsilon |\Delta p|$.

In applying this, we first consider the case in which $dT|_{p_0} = I$, the identity transformation. Since $I(\Delta p) = \Delta p$, (8-14) becomes

(8-15) $$T(p) = T(p_0) + \Delta p + R(\Delta p)$$

If the remainder term $R(\Delta p)$ were absent, then this equation would assert that the transformation T is nothing more than a translation; it would shift a cube with center at p_0 so that its center would become $T(p_0)$, without altering lengths or direction. The term $R(\Delta p)$ causes a slight alteration of this picture. To estimate its effect, we see from (8-15) that whenever $|\Delta p| < \delta$,

(8-16) $$|T(p) - T(p_0)| < |\Delta p| + \varepsilon |\Delta p| = (1 + \varepsilon)|\Delta p|$$

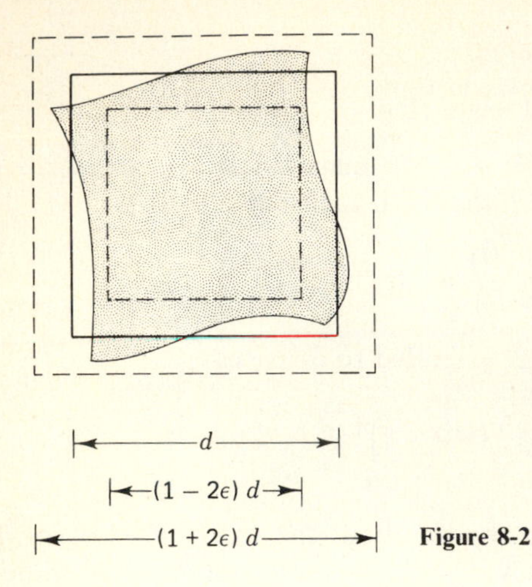

Figure 8-2

and that

$$|T(p) - T(p_0)| > |\Delta p| - \varepsilon|\Delta p| = (1 - \varepsilon)|\Delta p| \tag{8-17}$$

Intuitively, this means that the actual image of each point in C is only slightly displaced, and that the image of the entire cube C must cover a smaller concentric cube completely, and lie inside another (see Fig. 8-2, which shows the analogous situation for the plane, rather than 3-space). Accordingly, the volume of the image set $T(C)$ will lie between $(1 - 2\varepsilon)^3 \mathrm{v}(C)$ and $(1 + 2\varepsilon)^3 \mathrm{v}(C)$, which yields

$$(1 - 2\varepsilon)^3 \leq \frac{v(T(C))}{\mathrm{v}(C)} \leq (1 + 2\varepsilon)^3$$

and letting C close down, ε can become arbitrarily small and we arrive at

$$\lim_{C \downarrow p} \frac{\mathrm{v}(T(C))}{\mathrm{v}(C)} = 1$$

This is (8-13) for our present special case, since we have assumed that $dT|_p = I$ and $\det(I) = J_T(p) = 1$.

This plausible geometric argument about the way a transformation T which obeys (8-15) must map small cubes requires a more rigorous proof. To show that this is possible, we give an argument for the corresponding assertion about mappings of spheres, which happens to be slightly easier. Since the result itself is of some use, we make this a separate statement. (By an appropriate translation of space, we can assume that we are studying the given transformation T at the origin, $\mathbf{0}$, and that $dT|_{\mathbf{0}} = I$ and $T(\mathbf{0}) = \mathbf{0}$; in fact, we assume slightly less than this.)

Lemma 2 *Let B be the closed ball in n space, center $\mathbf{0}$, radius r. Let T be a C' transformation defined on an open set containing B on which* *its Jacobian $J(p)$ never vanishes. Suppose also that T is close to the identity map, meaning that there is a number ρ such that $0 < \rho < \frac{1}{2}$ and*

$$|T(p) - p| \le \rho r \qquad \text{for all } p \in B \tag{8-18}$$

Then, T maps B onto a set $T(B)$ that contains all the points in the open ball centered at $\mathbf{0}$ of radius $(1 - 2\rho)r$.

We give a proof that is quite similar to that for the open mapping theorem (Theorem 15, Sec. 7.5). Take any point q_0 with $|q_0| < (1 - 2\rho)r$; we wish to show that there is a point $p \in B$ with $T(p) = q_0$. Consider the real-valued function defined on B by

$$\psi(p) = |T(p) - q_0|$$

and let $m = \min_{p \in B} \psi(p)$, which we assume is achieved at a point p_0 in B. We note that $\psi(q_0) = |T(q_0) - q_0| \le \rho r$ by (8-18); thus, $m \le \rho r$. We next show that the minimum cannot be achieved on the boundary of B. For, if $p \in \text{bdy}\,(B)$,

$$\begin{aligned}\psi(p) &= |T(p) - p + p - q_0| \\ &\ge |p - q_0| - |T(p) - p| \\ &\ge |p - q_0| - \rho r \ge |p| - |q_0| - \rho r\end{aligned}$$

However, $|q_0| < (1 - 2\rho)r$ and $|p| = r$, so that

$$\psi(p) > r - (1 - 2\rho)r - \rho r = \rho r$$

and $\psi(p) > m$ everywhere on $|p| = r$.

We now know that the minimum of $\psi(p)$, and thus the minimum of $\psi(p)^2$, must occur at a point p_0 interior to B. The function $\psi(p)^2$ is of class C', and its differential must vanish at p_0. The 3-space case of the next step is typical of the general case, so we may suppose that T is described by (8-12). Then, if $q_0 = (a, b, c)$,

$$\psi(p)^2 = (x - a)^2 + (y - b)^2 + (z - c)^2$$

and at p_0, we must have

$$(x - a)\frac{\partial x}{\partial u} + (y - b)\frac{\partial y}{\partial u} + (z - c)\frac{\partial z}{\partial u} = 0$$

$$(x - a)\frac{\partial x}{\partial v} + (y - b)\frac{\partial y}{\partial v} + (z - c)\frac{\partial z}{\partial v} = 0$$

$$(x - a)\frac{\partial x}{\partial w} + (y - b)\frac{\partial y}{\partial w} + (z - c)\frac{\partial z}{\partial w} = 0$$

Since the coefficient matrix is exactly $J(p_0)$, which is not 0, the only solution of this system is $x - a = y - b = z - c = 0$, and we have shown that $T(p_0) = q_0$.

A similar argument can be made for cubical sets, using a different function ψ; this takes care of the proof of Lemma 1 when $dT|_{p_0}$ is the identity matrix.

Suppose now that $dT|_{p_0} = L$. Since det $(L) = J(p_0) \neq 0$, L is non-singular and has an inverse L^{-1}. Since $T \in C'$ and the entries of L^{-1} are continuous functions of p_0 and E is compact, we can choose a number M_0 (again by the boundedness theorem for linear transformations) such that $|L^{-1}(s)| \leq M_0|s|$ for all points s and all $p_0 \in E$. Return to (8-14) and apply L^{-1} to both sides, obtaining

$$L^{-1}\{T(p) - T(p_0)\} = L^{-1}\{L(\Delta p) + R(\Delta p)\}$$

or
$$\begin{aligned} L^{-1}T(p) - L^{-1}T(p_0) &= L^{-1}L(\Delta p) + L^{-1}(R(\Delta p)) \\ &= \Delta p + L^{-1}(R(\Delta p)) \end{aligned}$$

In this, $|L^{-1}(R(\Delta p))| \leq M_0|R(\Delta p)| \leq M_0\varepsilon|\Delta p|$. Setting $T^* = L^{-1}T$, we have now shown that

$$T^*(p) = T^*(p_0) + \Delta p + R^*(\Delta p) \tag{8-19}$$

where $|R^*(\Delta p)| \leq M_0\varepsilon|\Delta p|$. This differs from (8-15) only in that we have a different transformation T^* replacing T and we have replaced ε by $M_0\varepsilon$ in the estimate of $|R(\Delta p)|$. Using (8-19) in place of (8-15), the previous argument applies, and we find

$$(1 - 2M_0\varepsilon)^3 < \frac{\mathrm{v}(T^*(C))}{\mathrm{v}(C)} < (1 + 2M_0\varepsilon)^3$$

so that

$$\lim_{C \downarrow p_0} \frac{\mathrm{v}(T^*(C))}{\mathrm{v}(C)} = 1$$

However, $T^* = L^{-1}T$, so that

$$\mathrm{v}(T^*(C)) = \mathrm{v}(L^{-1}(T(C))) = |\det(L^{-1})|\mathrm{v}(T(C)),$$

since L^{-1} is linear. But, det $(L^{-1}) = 1/\det(L) = 1/J(p_0)$, and we arrive at

$$\lim_{C \downarrow p_0} \frac{1}{|J(p_0)|} \frac{\mathrm{v}(T(C))}{\mathrm{v}(C)} = 1$$

which (finally) proves (8-13). ▮

The proof of Theorem 5 itself is immediate. Since the set function F is defined on cubes by $F(C) = \mathrm{v}(T(C))$, the lemma shows that F is uniformly differentiable on E, with derivative at p the point function $|J(p)|$. By Theorem 2, F can be obtained by integrating its derivative, and for any set D having volume,

$$F(D) = \mathrm{v}(T(D)) = \iiint_D |J| = \iiint_D \left|\frac{\partial(x, y, z)}{\partial(u, v, w)}\right| du\, dv\, dw \quad ▮$$

With a slight modification of the argument, we also obtain a general theorem dealing with transformation of multiple integrals.

Theorem 6 *Let T be a transformation from 3-space into 3-space, with $T(u, v, w) = (x, y, z)$, which is of class C' and 1-to-1 in an open set Ω with $J(p) \neq 0$ throughout Ω. Let D^* be a closed bounded set in XYZ space which is the image under T of a set $D \subset \Omega$ in UVW space. Let f be a continuous function defined on D^*. Then,*

$$(8\text{-}20) \quad \iiint\limits_{D^*} f(x, y, z)\, dx\, dy\, dz = \iiint\limits_{D} f(T(u, v, w))|J(u, v, w)|\; du\, dv\, dw$$

The set D and D^* are assumed to possess volume. If f is a constant function, the result is covered by Theorem 5. We may suppose that f is everywhere positive by adding, if necessary, a suitable constant. We may also suppose that f is continuous throughout Ω. [This is a typical application of the Tietze extension theorem (Theorem 8, Sec. 6.2).] Define a set function F on subsets of Ω having volume by

$$F(S) = \iiint\limits_{T(S)} f$$

Let E be a closed bounded subset of Ω which contains D in its interior. Let C be a cube lying in E. Applying the mean value theorem for integrals, we can write $F(C) = f(q^*)\mathrm{v}(T(C))$, where $q^* \in T(C)$. As the cube C closes down on a point p_0, $T(C)$ closes down on $T(p_0)$, and q^* approaches $T(p_0)$. Thus,

$$\begin{aligned}
\lim_{C \downarrow p_0} \frac{F(C)}{\mathrm{v}(C)} &= \lim_{C \downarrow p_0} f(q^*)\frac{\mathrm{v}(T(C))}{\mathrm{v}(C)} \\
&= f(T(p_0)) \lim_{C \downarrow p_0} \frac{\mathrm{v}(T(C))}{\mathrm{v}(C)} \\
&= f(T(p_0))\; |J(p_0)|
\end{aligned}$$

using Lemma 1 for the last step. Moreover, since f is uniformly continuous on $T(E)$, we have uniform convergence of this limit for all $p_0 \in E$. Having thus computed the derivative of the set function F, application of Theorem 2 yields

$$\begin{aligned}
F(D) = \iiint\limits_{T(D)} f &= \iiint\limits_{D^*} f(x, y, z)\, dx\, dy\, dz \\
&= \iiint\limits_{D} f(T(u, v, w))|J(u, v, w)|\; du\, dv\, dw \\
&= \iiint\limits_{D} f(T(u, v, w)) \left| \frac{\partial(x, y, z)}{\partial(u, v, w)} \right| du\, dv\, dw \quad \blacksquare
\end{aligned}$$

It is enlightening to compare this result with the corresponding theorem dealing with transformation of (substitution into) integrals of functions of one variable. As stated in Theorem 8, Sec. 4.3, the substitution $x = \phi(u)$ in the integral $\int_a^b f(x)\,dx$ results in the integral $\int_\alpha^\beta f(\phi(u))\phi'(u)\,du$, where $\phi(\alpha) = a$ and $\phi(\beta) = b$. Comparing these, we see that the factor $\phi'(u) = dx/du$ corresponds to the Jacobian $\partial(u, v, w)/\partial(x, y, z)$ and the new limits of integration $[\alpha, \beta]$ to the set D. However, the one-variable theorem was considerably stronger than the present form of the several-variable theorem. For example, there is no need to have $\phi'(u) \neq 0$, nor must ϕ be a 1-to-1 mapping from the interval $[\alpha, \beta]$ onto the interval $[a, b]$. Another contrast is that $\phi'(u)$ rather than $|\phi'(u)|$ appears; however, the explanation for this lies in the fact that the simple one-variable integral is a directed or oriented integral, while we have so far not oriented the multiple integral. This suggests, correctly, that similar improvements can be made in Theorem 6. In Chap. 9, we shall obtain such a result, using a different approach which assumes that T is of class C''.

Returning to the result given in Theorem 6, we recall that the symbols "$dx\,dy\,dz$" or "$du\,dv\,dw$" serve to show what integration is to be performed, but have no meaning out of context; nor does the order $dx\,dy\,dz$ have as yet any different meaning from the order $dy\,dx\,dz$. As in Sec. 4.3, we might emphasize this by writing (8-20) in the form

$$\iiint_{D^*} f(x, y, z)\boxed{x\ y\ z} = \iiint_D f(T(u, v, w))|J(u, v, w)|\boxed{u\ v\ w}$$

The rule for change of variable in a triple integral can be given thus: To make the substitution $x = \phi(u, v, w)$, $y = \psi(u, v, w)$, $z = \theta(u, v, w)$ in the triple integral $\iiint_{D^*} f(x, y, z)\,dx\,dy\,dz$, replace the region D^* by the region D in the UVW plane which corresponds to D^* under the transformation, replace the integrand $f(x, y, z)$ by

$$f(\phi(u, v, w), \psi(u, v, w), \theta(u, v, w))$$

and replace $dx\,dy\,dz$ by $|\partial(x, y, z)/\partial(u, v, w)|\,du\,dv\,dw$. In the one-variable case, the notation assisted the application of the rule. When the substitution was $x = \phi(u)$, the formula $dx/du = \phi'(u)$ made the replacement of dx by $\phi'(u)\,du$ a routine operation. It is natural to ask if a similar formalism can be used for multiple integrals. In Sec. 4.4, where we gave a preliminary discussion of change-of-variable techniques for multiple integrals, we gave a preview of such a routine, using differential forms such as

$$dx = \frac{\partial x}{\partial u}\,du + \frac{\partial x}{\partial v}\,dv + \frac{\partial x}{\partial w}\,dw$$

and a multiplication process to obtain the (correct) replacement for $dx\,dy\,dz$.

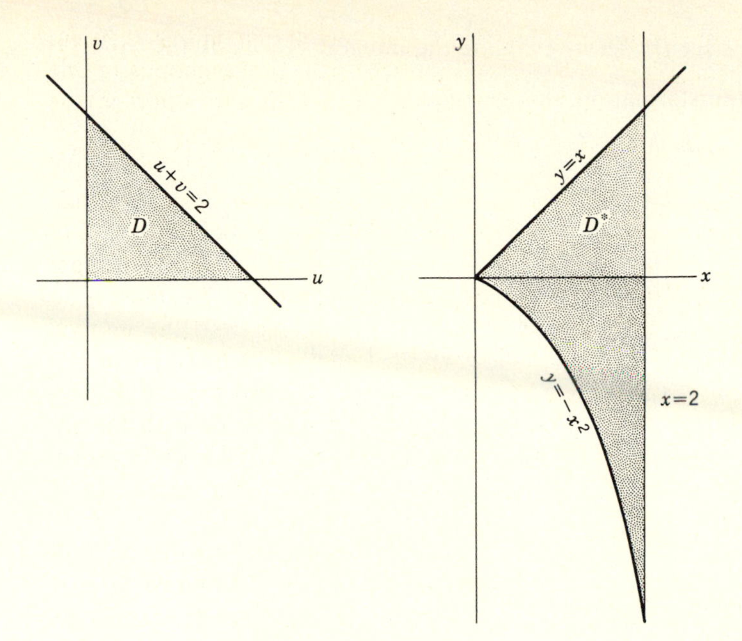

Figure 8-3

We conclude this section by giving a number of illustrations of the use of these formulas.

Consider first the transformation described by

$$T: \begin{cases} x = u + v \\ y = v - u^2 \end{cases}$$

and let D be the set in the UV plane bounded by the lines $u = 0$, $v = 0$, and $u + v = 2$. The image of D is the set D^* bounded by $x = 2$, $y = x$, and $y = -x^2$ (see Fig. 8-3). Computing the Jacobian of T,

$$J(u, v) = \frac{\partial(x, y)}{\partial(u, v)} = \det \begin{bmatrix} 1 & 1 \\ -2u & 1 \end{bmatrix} = 1 + 2u$$

By Theorem 5, the area of D^* is

$$\iint_D J = \iint_D (1 + 2u)\, du\, dv = \int_0^2 dv \int_0^{2-v} (1 + 2u) du$$

$$= \int_0^2 (v^2 - 5v + 6)\, dv = \tfrac{14}{3}$$

For comparison, we may calculate the area of D^* directly:

$$A(D^*) = \int_0^2 dx \int_{-x^2}^{x} dy = \int_0^2 (x + x^2)\, dx = \tfrac{14}{3}$$

With the same set D^*, let us evaluate the integral $\iint_{D^*} dx\, dy/(x - y + 1)^2$. Using the same transformation, this becomes

$$\iint_D \frac{(1 + 2u)\, du\, dv}{(u^2 + u + 1)^2} = \int_0^2 dv \int_0^{2-v} \frac{(1 + 2u)\, du}{(u^2 + u + 1)^2}$$

$$= \int_0^2 dv \left[-(u^2 + u + 1)^{-1} \right]_{u=0}^{u=2-v}$$

$$= \int_0^2 \{1 - (v^2 - 5v + 7)^{-1}\}\, dv$$

$$= 2 - \frac{2}{\sqrt{3}} \left[\arctan\left(\frac{5}{\sqrt{3}}\right) - \arctan\left(\frac{1}{\sqrt{3}}\right) \right]$$

$$= 2 - \frac{2}{\sqrt{3}} \arctan\left(\frac{\sqrt{3}}{2}\right)$$

For another example, consider the integral

$$\iint_{D^*} \exp\left(\frac{x - y}{x + y}\right) dx\, dy$$

where D^* is the region bounded by the lines $x = 0$, $y = 0$, $x + y = 1$. The form of the integrand suggests the use of the linear transformation

$$T: \begin{cases} u = x - y \\ v = x + y \end{cases}$$

This maps D^* onto a triangular region D (see Fig. 8-4). Since T is linear, it is sufficient to observe that T sends $(0, 1)$ into $(-1, 1)$, $(1, 0)$ into $(1, 1)$, and $(0, 0)$ into $(0, 0)$. The inverse of T is the (linear) transformation

$$T^{-1}: \begin{cases} x = \frac{1}{2}u + \frac{1}{2}v \\ y = -\frac{1}{2}u + \frac{1}{2}v \end{cases}$$

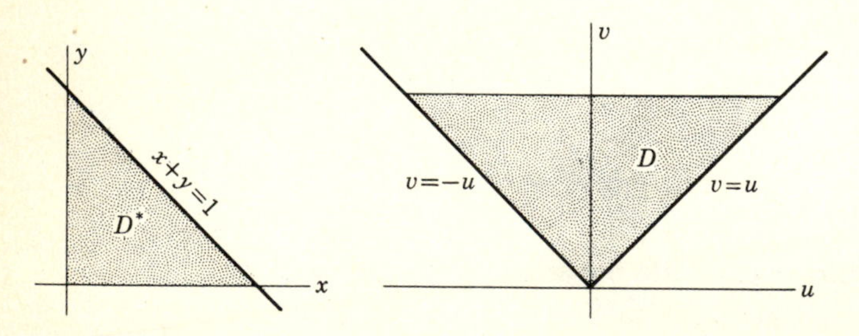

Figure 8-4

so that
$$\frac{\partial(x, y)}{\partial(u, v)} = \det \begin{bmatrix} \frac{1}{2} & \frac{1}{2} \\ -\frac{1}{2} & \frac{1}{2} \end{bmatrix} = \frac{1}{2}$$

Accordingly,

$$\iint_{D^*} \exp\left(\frac{x-y}{x+y}\right) dx\, dy = \iint_{D} \tfrac{1}{2}(e^{u/v})\, du\, dv$$

$$= \tfrac{1}{2}\int_0^1 dv \int_{-v}^{v} e^{u/v}\, du = \tfrac{1}{2}\int_0^1 dv \left[ve^{u/v}\right]_{-v}^{v}$$

$$= \tfrac{1}{2}\int_0^1 (e - e^{-1})v\, dv = \frac{e - e^{-1}}{4}$$

Another very familiar transformation is $x = r\cos\theta$, $y = r\sin\theta$, which we regard as a mapping from the (r, θ) plane into the (x, y) plane. The Jacobian of this transformation is

$$\frac{\partial(x, y)}{\partial(r, \theta)} = \det \begin{bmatrix} \cos\theta & -r\sin\theta \\ \sin\theta & r\cos\theta \end{bmatrix} = r$$

so that $dx\, dy$ is to be replaced by $r\, dr\, d\theta$ when transforming a double integral. The condition $J(p) \neq 0$ requires that we avoid the line $r = 0$ in the (r, θ) plane; this corresponds to the origin in the (x, y) plane. If D^* is a region in the XY plane which does not contain the origin, then the inverse transformation $r = (x^2 + y^2)^{1/2}$, $\theta = \arctan(y/x)$ determines a region D in the (r, θ) plane which is mapped onto D^* 1-to-1, and in which $J(p) \neq 0$ (see Fig. 8-5). Theorem 6 then gives the familiar formula

(8-21) $$\iint_{D^*} f(x, y)\, dx\, dy = \iint_{D} f(r\cos\theta, r\sin\theta) r\, dr\, d\theta$$

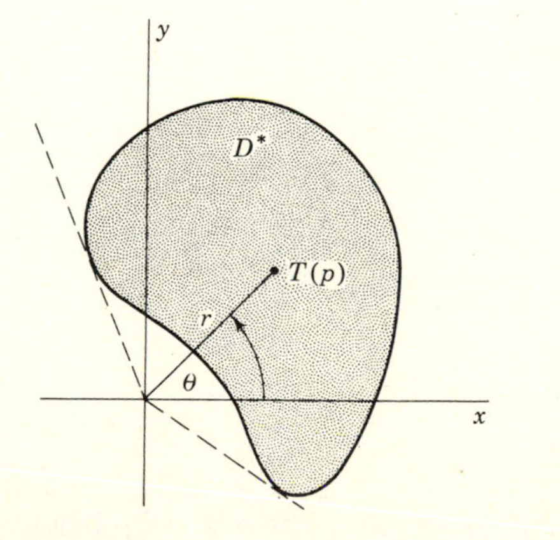

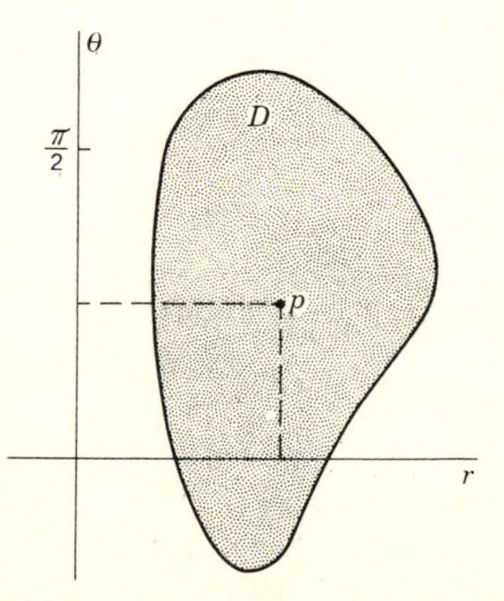

Figure 8-5

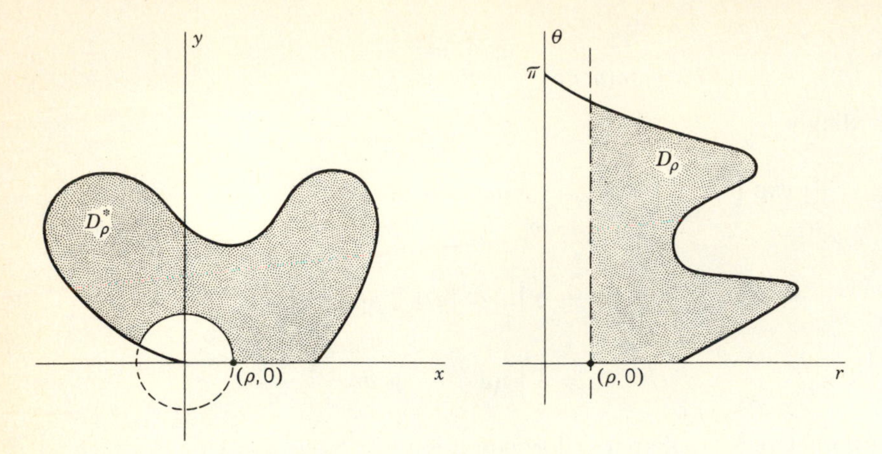

Figure 8-6 Polar coordinate mapping.

If D^* is a region containing the origin, then the corresponding set D will contain points on the line $r = 0$ where $J(p) = 0$. Theorem 6, as stated, does not apply. However, for this familiar transformation, it is not difficult to extend it. If D^* has the origin for an interior point, we can divide it into two pieces by the horizontal axis; each of these will have the origin as a boundary point, and if we can prove formula (8-21) for such a set, addition will give it for a general set D^*. Assuming therefore that D^* lies in the half plane $y \geq 0$, we choose a region D in the (r, θ) plane which is mapped onto D^*, 1-to-1, except for the points of D on $r = 0$, all of which map into $(0, 0)$ (see Fig. 8-6). Let D_ρ^* be the set which is obtained from D^* by removing the points within the open disk of radius ρ, and center $\mathbf{0}$. The corresponding set D_ρ is obtained by removing all the points (r, θ) in D with $0 \leq r < \rho$. In D_ρ, $J(p) = r \neq 0$, so that Theorem 6 applies, and

$$\iint\limits_{D_\rho^*} f(x, y)\, dx\, dy = \iint\limits_{D_\rho} f(r \cos \theta, r \sin \theta) r\, dr\, d\theta$$

This holds for all $\rho > 0$, so that we can obtain (8-21) by allowing ρ to approach 0.

A similar treatment can be applied to the equations

$$\text{(8-22)} \qquad \begin{cases} x = \rho \sin \phi \cos \theta \\ y = \rho \sin \phi \sin \theta \\ z = \rho \cos \phi \end{cases}$$

which serve to define the system of spherical coordinates (see Fig. 8-7). If we regard (8-22) as describing a transformation T mapping (ρ, ϕ, θ) into (x, y, z),

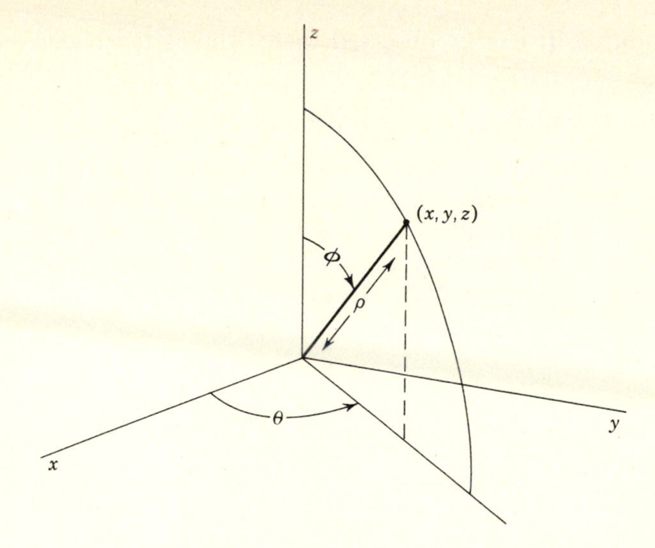

Figure 8-7 Spherical coordinates.

then T has for its Jacobian

$$\frac{\partial(x, y, z)}{\partial(\rho, \phi, \theta)} = \begin{vmatrix} \sin\phi\cos\theta & \rho\cos\phi\cos\theta & -\rho\sin\phi\sin\theta \\ \sin\phi\sin\theta & \rho\cos\phi\sin\theta & \rho\sin\phi\cos\theta \\ \cos\phi & -\rho\sin\phi & 0 \end{vmatrix}$$

$$= \rho^2 \sin\phi$$

Thus, $dx\,dy\,dz$ is to be replaced by $\rho^2 \sin\phi\,d\rho\,d\phi\,d\theta$ in transforming a triple

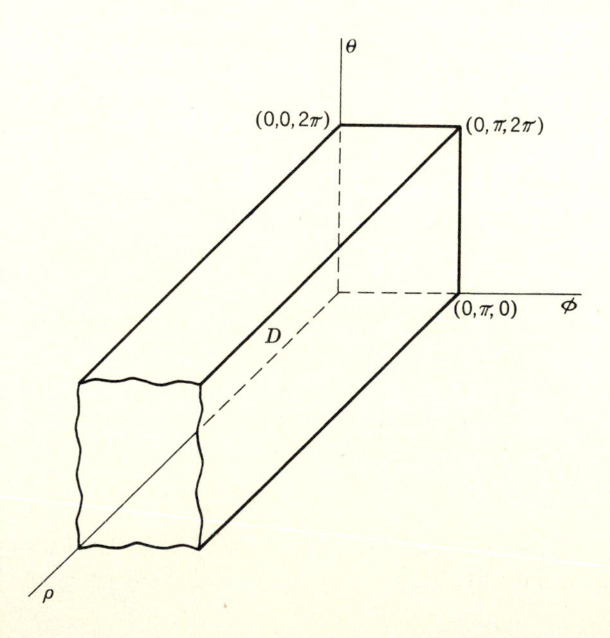

Figure 8-8 Spherical coordinate mapping.

integral with this substitution. It will be observed that T maps the region (Fig. 8-8)

$$D = \{\text{all } (\rho, \phi, \theta) \text{ with } 0 \le \rho, 0 \le \phi \le \pi, 0 \le \theta \le 2\pi\}$$

onto all of XYZ in a fashion which is 1-to-1 at all interior points of D. It is not 1-to-1 on the boundary of D; for example, the entire face $\rho = 0$ maps onto the single point (0, 0, 0). Moreover, the Jacobian is also 0 at points on portions of the boundary. However, by a limiting argument similar to that used above, the transformation formula can also be shown to hold for bounded subsets of D.

EXERCISES

1 Show that the linear transformation which sends (1, 1) into (2, 5) and (1, −1) into (0, −1) is an area-preserving transformation.

2 Let $A = [a_{ij}]$ be a matrix with two columns and four rows. Show that the number $k^2 = \sum_{i=1}^{4} (a_{i1})^2 \sum_{i=1}^{4} (a_{i2})^2 - \left(\sum_{i=1}^{4} a_{i1}a_{i2}\right)^2$ is in fact the same as the sum of the squares of the determinants of all of the 2-by-2 submatrices.

3 Make the indicated change of variables in the following integrals, and evaluate the result.

(a) $\int_0^1 dx \int_0^x xy\,dy, \qquad x = u + v,\ y = u - v$

(b) $\int_0^1 dx \int_{1-x}^{1+x} xy\,dy, \qquad x = u,\ y = u + v$

(c) $\iint_D xy\,dx\,dy, \qquad x = u^2 - v^2,\ y = 2uv$

where D is the unit disk, $x^2 + y^2 \le 1$.

4 Let D^* be the parallelogram bounded by the lines $y = \frac{1}{2}x$, $y = \frac{1}{2}x + 2$, $y = 3x$, $y = 3x - 4$. Make an appropriate substitution, and evaluate $\iint_{D^*} xy\,dx\,dy$.

5 Let D be the region in the first quadrant which is bounded by the curves $xy = 1$, $xy = 3$, $x^2 - y^2 = 1$, and $x^2 - y^2 = 4$. Make an appropriate substitution, and evaluate $\iint_D (x^2 + y^2)\,dx\,dy$.

6 The equation of a curve in polar coordinates is $r = \sin(\theta/2)$ for $0 \le \theta \le 2\pi$. Find the area of the region which is bounded by this curve.

7 In 4-space, "double" polar coordinates are defined by the equations

$$x = r\cos\theta \qquad y = r\sin\theta \qquad z = \rho\cos\phi \qquad w = \rho\sin\phi$$

Obtain the correct formula for making this substitution in a fourfold multiple integral, and use this formula to show that the volume of the spherical region $x^2 + y^2 + z^2 + w^2 \le R^2$ is $\frac{1}{2}\pi^2 R^4$.

8 The linear transformation L whose matrix is

$$\begin{bmatrix} 1 & 0 & -1 \\ 2 & 1 & 1 \\ 0 & -1 & 2 \end{bmatrix}$$

maps the unit cube with vertices (0, 0, 0), (1, 0, 0), (0, 1, 0), (0, 0, 1), (1, 1, 0), (0, 1, 1), (1, 0, 1), (1, 1, 1) into a parallelepiped R. Find the area of each of the faces of R, and find its volume.

9 Let $P = (-1, 0)$ and $Q = (1, 0)$ and let $p = (x, y)$ be any point with $y \geq 0$. Set $s = |P - p|^2$ and $t = |Q - p|^2$. Show that the correspondence $p \to (s, t)$ provides a coordinate system for the upper half plane, and discuss this as a transformation between (x, y) and (s, t). What is the image in the (x, y) plane on lines in the (s, t) plane? Where is $\partial(s, t)/\partial(x, y)$ zero?

10 (Alternative approach to Theorem 6.) Let T be the transformation described by $x = u$, $y = \psi(u, v)$. Let D be the region bounded by the lines $u = a$, $u = b$, and the smooth curves $v = \beta(u)$ and $v = \alpha(u)$ with $\beta(u) > \alpha(u)$ for all $u \in [a, b]$. Let D^* be the image of D under T, and suppose that $\psi_2(u, v) > 0$ for all $(u, v) \in D$. Show that for any function f which is continuous in D^*,

$$\iint_{D^*} f(x, y)\, dx\, dy = \iint_{D} f(T(u, v))\psi_2(u, v)\, du\, dv$$

(*Hint:* Write each side as an iterated integral, and use the substitution formula for single integrals.)

***11** (*Continuation*) Let T be a transformation of class C' in an open set Ω which is described by

$$x = \phi(u, v)$$

$$y = \psi(u, v)$$

Assume also that $\partial(x, y)/\partial(u, v) > 0$ throughout Ω. Show that in a sufficiently small neighborhood of any point $p \in \Omega$, the transformation T can be factored in one of the following two ways:

$$\begin{cases} x = s \\ y = G(s, t) \end{cases} \qquad \begin{cases} s = \phi(u, v) \\ t = v \end{cases}$$

or

$$\begin{cases} x = s \\ y = G(s, t) \end{cases} \qquad \begin{cases} s = \phi(u, v) \\ t = u \end{cases}$$

where G is a function of class C' which is determined by the functions ϕ and ψ.

12 (*Continuation*) When such a factoring is made, show that Exercise 10 can be applied to each successively to obtain the result stated in Lemma 1.

8.4 CURVES AND ARC LENGTH

In our treatment of these topics, we shall deviate from the older traditional approach and adopt one which is of greater significance for applications in analysis. The terms "curve" and "surface" have several meanings in mathematics. For example, in one usage, "curve" means a set of points which has certain topological properties associated with the intuitive notion of "thinness." The word is also used to refer to a set of equations. We adopt the following formal definition.

Definition 6 *A curve γ in n space is a mapping or transformation from* $\mathbf{R}^1$ *into* $\mathbf{R}^n$.

For example, the general continuous curve in 3-space is a continuous transformation of the form

$$\begin{cases} x = \phi(t) \\ y = \psi(t) \\ z = \theta(t) \end{cases} \tag{8-23}$$

defined for t in one or more intervals $[a, b]$. In more familiar terminology, we have identified the notion of curve with what is often called a parametric representation. A point p in $\mathbf{R}^3$ is said to **lie on** the curve when there is a t for which $p = \gamma(t)$. The set of all points which lie on γ is called the **trace** of γ. It is important to keep in mind the fact that the curve is the transformation (8-23) and not the set of points lying on γ. Many different curves can have the same trace. For example, each of the following curves has for its trace the unit circle $C = \{\text{all } (x, y) \text{ with } x^2 + y^2 = 1\}$.

$$\begin{cases} x = \cos t \\ y = \sin t \end{cases} \qquad \begin{cases} x = \cos(2t) \\ y = \sin(2t) \end{cases} \qquad \begin{cases} x = \sin t \\ y = \cos t \end{cases}$$

$$0 \le t \le 2\pi \qquad 0 \le t \le 2\pi \qquad 0 \le t \le 2\pi$$

However, the three curves are quite different. The first and third have length 2π, while the second (which goes around C twice) has length 4π. Again, the first curve starts and ends at $(1, 0)$ and goes around C in the counterclockwise direction, while the third begins and ends at $(0, 1)$ and goes in the opposite direction. Such distinctions are of great importance in the analytical theory of curves. In this section, we shall take up some of the simpler portions of this theory.

We begin with some convenient terminology. If a curve γ is defined for the interval $a \le t \le b$, then the **endpoints** of γ are $\gamma(a)$ and $\gamma(b)$; the former is called the first point on γ, and the latter the last point on γ. A curve γ is **closed** if its endpoints coincide, so that $\gamma(a) = \gamma(b)$. The three curves given above are examples of closed curves. A point p which lies on a curve γ is said to be a multiple point if there is more than one value of t for which $\gamma(t) = p$. A curve is said to be **simple** if it has no multiple points; the mapping γ is then 1-to-1. A continuous closed curve is said to be simple if the only multiple points are the coincident endpoints. (The terms "Jordan arc" and "Jordan curve" are also used.) In the examples given above, the first and third are simple closed curves, but the second is not.

The trace of a continuous curve can fill an entire region of the plane (see Appendix 2). The existence of such space-filling curves is one reason why we do not attempt to define a curve as a particular kind of point set but concentrate instead upon the mapping itself. (It can be shown, however, that a *simple* curve cannot be space-filling.)

It is also convenient to impose certain differentiability requirements.

Definition 7 *A curve γ is said to be smooth on an interval I if γ is of class C' and the differential $d\gamma$ is always of rank 1 on I.*

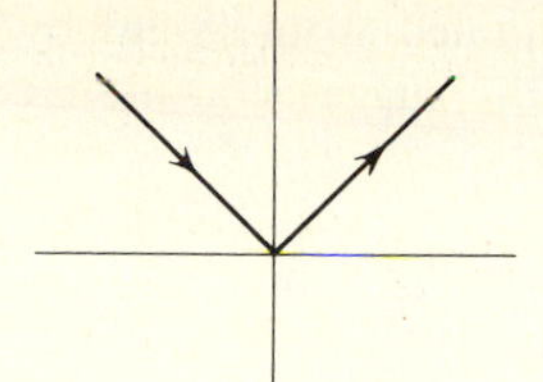

Figure 8-9 $x = t^3$, $y = |t^3|$.

When γ is given in the form (8-23), this asserts that ϕ', ψ', and θ' exist and are continuous on $[a, b]$ and that at no point of this interval do all of them become 0. This latter condition can also be written as

$$\left(\frac{dx}{dt}\right)^2 + \left(\frac{dy}{dt}\right)^2 + \left(\frac{dz}{dt}\right)^2 > 0 \tag{8-24}$$

The word "smooth" is used to suggest that the motion of a point which traces the curve has no abrupt changes of direction. The need for a condition such as (8-24) to ensure this is shown by the following example of a curve of class C' which violates condition (8-24) when $t = 0$:

$$\gamma\colon \begin{cases} x = t^3 \\ y = |t^3| \end{cases} \qquad -\infty < t < \infty$$

We note that $dx/dt = 3t^2$, and that $dy/dt = 3t^2$ for $t \geq 0$ and $-3t^2$ for $t \leq 0$; both are continuous, and both become zero for $t = 0$. However, the trace of γ is the set of points (x, y) with $y = |x|$ (Fig. 8-9).

A line is the simplest smooth curve. Recall from (1-20) that this is given by

$$\gamma(t) = p_0 + \mathbf{v}t \qquad -\infty < t < \infty \tag{8-25}$$

where the point (vector) $\mathbf{v}$ obeys the restriction $|\mathbf{v}| \neq 0$. For example, in 3-space, the general straight line is

$$\begin{cases} x = x_0 + at \\ y = y_0 + bt \\ z = z_0 + ct \end{cases} \tag{8-26}$$

where $a^2 + b^2 + c^2 > 0$. Since $\gamma(0) = p_0$ and $\gamma(1) = p_0 + \mathbf{v}$, the line (8-25) may be graphed as the line which goes through p_0 toward $p_0 + \mathbf{v}$, in a direction which is specified by $\mathbf{v}$. Since we have defined the notion of direction in n space by means of the points on the unit sphere, we form the unit vector $\beta = \mathbf{v}/|\mathbf{v}|$, which has length 1 and specifies the same direction. In the customary language of analytical geometry, $\mathbf{v}$ is a set of **direction components** for the line, while β is a set of **direction cosines**; either may be used. A line through $(1, 2, -1)$ toward $(3, 1, 1)$ has the equation

$$\begin{aligned} \gamma(t) &= (1, 2, -1) + \{(3, 1, 1) - (1, 2, -1)\}t \\ &= (1, 2, -1) + (2, -1, 2)t \end{aligned}$$

or $x = 1 + 2t, y = 2 - t, z = -1 + 2t$. The direction of this line is $\beta = (\frac{2}{3}, -\frac{1}{3}, \frac{2}{3})$.

We observe that the coordinates of $\mathbf{v}$ can be obtained from (8-26) by differentiation; $a = dx/dt$, $b = dy/dt$, and $c = dz/dt$. This suggests a similar procedure for the general smooth curve.

Definition 8 *If γ is a smooth curve, then the direction of γ at a point p corresponding to the value t is $\beta = \mathbf{v}/|\mathbf{v}|$, where*

$$\mathbf{v} = \gamma'(t) = \left(\frac{dx}{dt}, \frac{dy}{dt}, \frac{dz}{dt}\right)$$

When the parameter t is interpreted as time, $\mathbf{v}$ is interpreted as the velocity vector of the point whose position at time t is $\gamma(t)$, and the speed of the point along its path is

$$|\mathbf{v}| = \left\{\left(\frac{dx}{dt}\right)^2 + \left(\frac{dy}{dt}\right)^2 + \left(\frac{dz}{dt}\right)^2\right\}^{1/2}$$

Because of the restriction that $d\gamma$ always have rank 1, there is no t where $\gamma'(t) = (0, 0, 0)$, so the speed is never 0.

For the same reason, the unit vector β is always well defined and a direction exists at each point p on a smooth curve. At a multiple point, γ can have several different directions corresponding to the value of $\gamma'(t)$ for each t at which γ passes through the point. Two curves with the same direction at a common point p are said to be **tangent** there. The straight line α:

$$\alpha(t) = p_0 + \mathbf{v}t \qquad -\infty < t < \infty$$

where $p_0 = \gamma(t_0)$ and $\mathbf{v} = \gamma'(t_0)$, is tangent to γ at p_0.

In discussing a particular curve, it is helpful to plot the trace of the curve, being sure to record points in the order assigned by t. Consider, for example, the plane curve

$$\text{(8-27)} \qquad \gamma: \begin{cases} x = t - t^3 \\ y = t^2 - t \end{cases} \qquad -\infty < t < \infty$$

which has the trace shown in Fig. (8-10). The origin is a double point corresponding to both $t = 0$ and $t = 1$. Differentiating, we have $\gamma'(t) = (1 - 3t^2, 2t - 1)$, so that

$$\gamma'(0) = (1, -1)$$

and

$$\gamma'(1) = (-2, 1)$$

At the origin, γ has two tangent lines. The first is $x = t$, $y = -t$ with slope -1, and the second is $x = -2t$, $y = t$ with slope $-\frac{1}{2}$. The direction of γ for a general value of t is

$$\beta = \frac{\gamma'(t)}{|\gamma'(t)|} = \frac{(1 - 3t^2, 2t - 1)}{(9t^4 - 2t^2 - 4t + 2)^{1/2}}$$

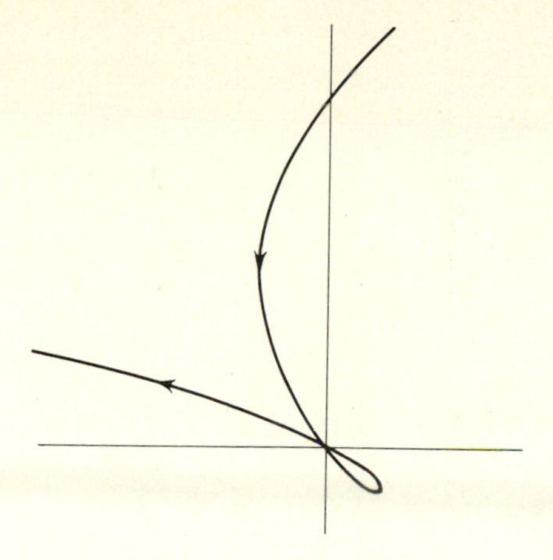

Figure 8-10

Thus, γ is horizontal $[\beta = (1, 0)]$ when $2t - 1 = 0$, and it is vertical $[\beta = (0, 1)]$ when $1 - 3t^2 = 0$.

Taking a curve in 3-space as a second illustration, consider

$$\gamma: \begin{cases} x = \sin(2\pi t) \\ y = \cos(2\pi t) \qquad 0 \le t \le 2 \\ z = 2t - t^2 \end{cases}$$

In graphing the trace of this curve, it is helpful to observe that

$$x^2 + y^2 = 1$$

for all t (see Fig. 8-11). γ is a closed curve, but is not simple since $t = \frac{1}{2}$ and $t = \frac{3}{2}$ both correspond to the point $(0, -1, \frac{3}{4})$. We, therefore, have $\gamma'(t) = (2\pi \cos(2\pi t), -2\pi \sin(2\pi t), 2 - 2t)$. At $t = 0$, this becomes $(2\pi, 0, 2)$, and at $t = 2$, $(2\pi, 0, -2)$, so that the curve has two tangent lines at the point $(0, 1, 0)$, which is at once the first and last point on γ.

We next take up the important subject of **arc length.** If γ is a smooth curve defined for $a \le t \le b$, then we define the length of γ by the formula

$$L(\gamma) = \int_a^b |\gamma'(t)|\, dt \tag{8-28}$$

In 3-space this takes the form

$$L(\gamma) = \int_a^b \left\{ \left(\frac{dx}{dt}\right)^2 + \left(\frac{dy}{dt}\right)^2 + \left(\frac{dz}{dt}\right)^2 \right\}^{1/2} dt$$

Alternatively, this formula can be obtained by starting with a geometrical definition for the length of a curve. Given γ, let us choose a subdivision of the interval $[a, b]$ by points t_j, $a = t_0 < t_1 \cdots < t_n = b$, and set $P_j = \gamma(t_j)$. The consecutive line segments joining P_0 to P_1, P_1 to $P_2, \ldots,$ form a polygonal path C. Since each point P_j lies on γ, we speak of C as inscribed in γ. For such

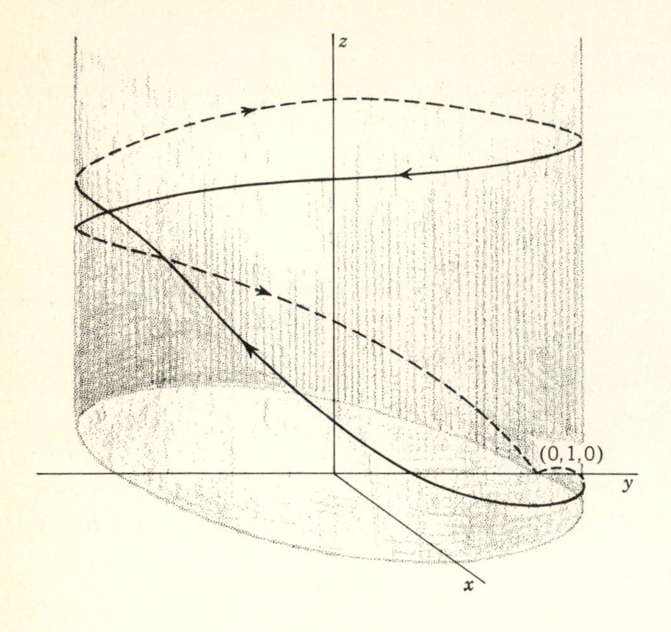

Figure 8-11

a polygon, we define length by

$$L(C) = |P_0 - P_1| + |P_1 - P_2| + \cdots$$

$$(8\text{-}29) \qquad = \sum_{j=0}^{n-1} |\gamma(t_{j+1}) - \gamma(t_j)|$$

Intuitively, the length of γ itself will exceed that of any of these inscribed polygons. This suggests that we adopt the following general definition, whether γ is smooth or not.

Definition 9 *The length of a continuous curve γ is defined to be the least upper bound of the numbers $L(C)$, where C ranges over all polygons inscribed in γ.*

When the set of numbers $L(C)$ has no finite upper bound, then we write $L(\gamma) = \infty$, and say that γ has infinite length. If $L(\gamma) < \infty$, then γ is said to be **rectifiable.**

Theorem 7 *If γ is a smooth curve whose domain is the interval $[a, b]$, then γ is rectifiable, and $L(\gamma)$ is given by formula* (8-28).

We shall prove this for curves in the plane; the proof of the general case follows the same method. Consider a general inscribed polygon C. The length of the jth segment is

$$|\gamma(t_{j+1}) - \gamma(t_j)| = \{[X(t_{j+1}) - X(t_j)]^2 + [Y(t_{j+1}) - Y(t_j)]^2\}^{1/2}$$

where we have written the curve as $x = X(t)$, $y = Y(t)$, $a \le t \le b$. Since X and Y are differentiable functions, we may use the mean value theorem and write the right side as

$$\{[X'(\tau_j')(t_{j+1} - t_j)]^2 + [Y'(\tau_j'')(t_{j+1} - t_j)]^2\}^{1/2}$$
$$= \{[X'(\tau_j')]^2 + [Y'(\tau_j'')]^2\}^{1/2} \Delta t_j$$

where $\Delta t_j = t_{j+1} - t_j$ and τ_j' and τ_j'' are two points in the interval $[t_j, t_{j+1}]$. Thus,

$$L(C) = \sum_{j=0}^{n-1} \{[X'(\tau_j')]^2 + [Y'(\tau_j'')]^2\}^{1/2} \Delta t_j$$

This closely resembles a Riemann sum for the definite integral $\int_a^b f(t)\,dt$, where $f(t) = \{[X'(t)]^2 + [Y'(t)]^2\}^{1/2}$. In fact, it would be one if the points in each pair τ_j', τ_j'' were to coincide. Introduce a special function F by

$$F(t', t'') = \{[X'(t')]^2 + [Y'(t'')]^2\}^{1/2}$$

and observe that $F(t, t) = f(t)$ while

$$L(C) = \sum_{j=0}^{n-1} F(\tau_j', \tau_j'') \Delta t_j$$

Since γ is smooth, F is continuous in the closed rectangle R consisting of the points (t', t'') with $a \le t' \le b$, $a \le t'' \le b$. Accordingly, F is uniformly continuous in R; given ε, we may choose δ so that $|F(p) - F(q)| < \varepsilon$ whenever p and q lie in R and $|p - q| < \delta$. In particular, we shall have

$$|F(\tau_j', \tau_j'') - F(t_j, t_j)| < \varepsilon$$

or equivalently,

$$|F(\tau_j', \tau_j'') - f(t_j)| < \varepsilon$$

whenever τ_j' and τ_j'' lie in an interval $[t_j, t_{j+1}]$ with $\Delta t_j < \delta$. Assuming that the subdivision of $[a, b]$ has mesh less than δ, we obtain

$$\left| \sum_{j=0}^{n-1} F(\tau_j', \tau_j'') \Delta t_j - \sum_{j=0}^{n-1} f(t_j) \Delta t_j \right| \le \sum_0^{n-1} |F(\tau_j', \tau_j'') - f(t_j)| \Delta t_j$$
$$\le \varepsilon \sum_0^{n-1} \Delta t_j = \varepsilon(b - a)$$

However, $\sum_0^{n-1} F(\tau_j', \tau_j'') \Delta t_j = L(C)$, while $\sum_0^{n-1} f(t_j) \Delta t_j$ is a Riemann sum for the integral $\int_a^b f$, so that we find that $L(C)$ approaches this integral as the mesh size of the subdivision decreases. Since $L(\gamma)$ is the least upper bound of the numbers $L(C)$, and since $L(C)$ increases (or at worst, remains the same) when an additional subdivision point is introduced, we also have $L(\gamma) = \int_a^b f$, the desired formula. ∎

For example, let us find the length of the helical curve $x = \cos t$, $y = \sin t$, $z = t^{3/2}$ for $0 \le t \le 4$. The function f is

$$f(t) = |\gamma'(t)| = \{(-\sin t)^2 + (\cos t)^2 + ((\tfrac{3}{2})t^{1/2})^2\}^{1/2}$$
$$= \sqrt{1 + (\tfrac{9}{4})t}$$

so that its length is $\int_0^4 \sqrt{1 + (\frac{9}{4})t}\, dt = (80\sqrt{10} - 8)/27$.

It is interesting that formula (8-29), which was motivated by the geometric definition of arc length, also arises in a different context (see Exercise 12). A real-valued function f of one variable defined on $[a, b]$ is said to be of **bounded variation** there if there is a number M such that

$$\sum_0^{m-1} |f(t_{k+1}) - f(t_k)| \le M \tag{8-30}$$

for every m and every set of points t_j with $a = t_0 < t_1 < t_2 < \cdots < t_m = b$. It is easily seen that any monotonic function on $[a, b]$ is of bounded variation, even if it is not continuous, and the same holds for any continuous function f for which f' exists everywhere and is bounded on $[a, b]$ (Exercise 13). The class of functions that are b.v. is rather large, since (Exercise 12) it contains any function that is the difference of two monotonic functions. What is also true (but not obvious) is that this characterizes the class; every function f that is b.v. on $[a, b]$ has the form $f = g_1 - g_2$, where g_1 and g_2 are monotonic on $[a, b]$. It then follows that not every continuous function is of bounded variation. The connection between these concepts and arc length is still present; a continuous function f is b.v. if and only if the graph of f is a rectifiable curve (Exercise 15).

As we have seen, different curves may have the same trace. Some of these curves are closely related and have many properties in common.

Definition 10 *Two curves γ and γ^* are said to be* **parametrically equivalent** *when the following conditions hold:* (i) *γ is a continuous mapping from an interval $[a, b]$ into $\mathbf{R}^n$;* (ii) *γ^* is a continuous mapping from an interval $[\alpha, \beta]$ into $\mathbf{R}^n$;* (iii) *there is a continuous function f which maps $[\alpha, \beta]$ onto $[a, b]$* 1-*to*-1 *with $f(\alpha) = a$, $f(\beta) = b$ and with $\gamma^*(t) = \gamma(f(t))$ for all t in $[\alpha, \beta]$.*

Briefly, this means that γ and γ^* are connected by a reversible change of parameter. For example, the following three curves are equivalent.

$$\begin{cases} x = t \\ y = t^2 \end{cases} \qquad \begin{cases} x = t^2 \\ y = t^4 \end{cases} \qquad \begin{cases} x = \frac{1}{2}(1 + t^{1/3}) \\ y = \frac{1}{4}(1 + 2t^{1/3} + t^{2/3}) \end{cases}$$

$$0 \le t \le 1 \qquad 0 \le t \le 1 \qquad -1 \le t \le 1$$

It should be noticed that the second of these is not smooth, due to the fact that $dx/dt = dy/dt = 0$ when $t = 0$, while the third is not even of class C' at $t = 0$.

When we consider parametric equivalence of smooth curves, more is true of the function f. If γ and γ^* are smooth and are parametrically equivalent, with $\gamma^*(t) = \gamma(f(t))$, then f itself will be of class C', with $f'(t) > 0$ for $\alpha < t < \beta$. For, at a point t_0, find $c = f(t_0)$, and since $\gamma'(c) \neq \mathbf{0}$, there is a component of γ whose derivative is not 0 at c, say γ_k. Since γ is C', $\gamma_k'(t) \neq 0$ on a neighborhood of c, and γ_k^{-1} exists locally and is of class C'. Thus, $f(t) = \gamma_k^{-1}\gamma_k^*(t)$ locally at t_0, and f is of class C' everywhere on the interior of $[\alpha, \beta]$. From the chain rule, we then have $d\gamma^* = (d\gamma)f'(t)$, so that if $f'(t_0) = 0$, $d\gamma^*|_{t_0} = \mathbf{0}$, and γ^* would fail to be smooth.

The relation of equivalence separates the class of all smooth curves into classes of mutually equivalent curves. All the curves in any one equivalence class have the same trace, and also share other geometric properties. We cite the next two results as examples.

Theorem 8 *Let γ_1 and γ_2 be smoothly equivalent smooth curves, and let p be a simple point on their trace. Then, γ_1 and γ_2 have the same direction at p.*

Let $\gamma_2(t) = \gamma_1(f(t))$, where $f'(t) > 0$ for all t. If $p = \gamma_2(c)$, then $p = \gamma_1(c^*)$, where $c^* = f(c)$. To find the direction of γ_2 at p, we write

$$\gamma_2'(t) = \frac{d}{dt}\gamma_1(f(t)) = \gamma_1'(f(t))f'(t)$$

so that

$$\gamma_2' = \gamma_1'(f(c))f'(c) = \gamma_1'(c^*)f'(c)$$

The direction of γ_2 at p is therefore

$$\frac{\gamma_2'(c)}{|\gamma_2'(c)|} = \frac{\gamma_1'(c^*)f'(c)}{|\gamma_1'(c^*)f'(c)|} = \frac{\gamma_1'(c^*)}{|\gamma_1'(c^*)|}$$

which is thus the same as the direction of γ_1 at p. ▮

Equivalent curves also have the same length.

Theorem 9 *If γ_1 and γ_2 are smoothly equivalent curves, then $L(\gamma_1) = L(\gamma_2)$.*

Assume that $\gamma_2(t) = \gamma_1(f(t))$ for $a \le t \le b$. Then, as before,

$$\gamma_2'(t) = \gamma_1'(f(t))f'(t)$$

Since $f'(t)$ is never negative, $|\gamma_2'(t)| = |\gamma_1'(f(t))| f'(t)$ and

$$L(\gamma_2) = \int_a^b |\gamma_2'(t)|\, dt = \int_a^b |\gamma_1'(f(t))| f'(t)\, dt$$

Putting $s = f(t)$ and $\alpha = f(a)$, $\beta = f(b)$, this becomes

$$L(\gamma_2) = \int_\alpha^\beta |\gamma_1'(s)|\, ds = L(\gamma_1) \quad \blacksquare$$

In the equivalence class containing a smooth γ, one curve has a special role. This is obtained by using arc length as a parameter. If γ is defined on the interval $[a, b]$, define a function g by

$$g(t) = \int_a^t |\gamma'|$$

Then, $g'(t) = |\gamma'(t)|$ for all t, $a \le t \le b$ and $g(a) = 0$, $g(b) = L(\gamma) = l$, so that g is a C' 1-to-1 map of $[a, b]$ onto $[0, l]$. Its inverse, which we call f, maps $[0, l]$ onto $[a, b]$ and is also of class C'. Let $\gamma^*(t) = \gamma(f(t))$; γ^* is then a smooth curve which is equivalent to γ, and is distinguished among all the curves equivalent to γ by the fact that $|(d/dt)\gamma^*(t)| = 1$ for all t. Many of the formulas involving curves take on a simpler form if arc length is the parameter. The geometric concept called curvature is one example. We first need some additional terminology.

As with general transformations, a curve γ is said to be of **class $\mathbf{C}^m$** if the coordinate functions which describe the curve are of class $\mathbf{C}^m$, and to be **analytic** if the coordinate functions are analytic, that is, representable by power series. All the examples that we have discussed are analytic curves, and may be represented in the form

$$\gamma(t) = \gamma(c) + \gamma'(c)(t - c) + \frac{\gamma''(c)(t - c)^2}{2!} + \cdots$$

convergent for all t in a neighborhood of c. In this Taylor series representation of γ, the coefficient of $t - c$ determines the direction of the curve when $t = c$. It is natural to ask for similar geometric interpretations of the remaining coefficients.

Definition 11 *If γ is a curve of class C'' with arc length as the parameter, then the curvature of γ at the point corresponding to $t = c$ is*

$$k = |\gamma''(c)| \tag{8-31}$$

Several illustrations will help motivate this. First, a straight line obviously should have zero curvature. Since a line is given by $\gamma(t) = p_0 + \mathbf{v}t$, $\gamma''(t) = \mathbf{0}$ as it should. Next, the curvature of a circle is the same everywhere, and should be small if the radius is large. The equation of a circle of radius R, with arc length as parameter, is

$$\gamma(t) = (R \cos(t/R), R \sin(t/R))$$

for

$$\gamma'(t) = (-\sin(t/R), \cos(t/R))$$

which clearly satisfies $|\gamma'(t)| = 1$ as required. Then,

$$\gamma''(t) = (-R^{-1} \cos(t/R), -R^{-1} \sin(t/R))$$

and $|\gamma''(t)| = 1/R$ for any t; thus, a circle of radius R has curvature $k = 1/R$ everywhere.

We can use this to give a geometric approach to curvature. The idea is to define the curvature of a general curve at a point p_0 to be the same as that of the best-fitting tangent circle at p_0. For simplicity, we assume that p_0 is the origin $\mathbf{0}$ and also assume that γ is an analytic curve given by a power series

$$p = \gamma(t) = \mathbf{a}t + \mathbf{b}t^2 + \mathbf{c}t^3 + \cdots \tag{8-32}$$

where $\mathbf{a}, \mathbf{b}, \mathbf{c}, \ldots$ are points (vectors) in n space. Note that $\gamma(0) = \mathbf{0}$, $\gamma'(0) = \mathbf{a}$, $\gamma''(0) = 2\mathbf{b}$.

Construct the circle which is tangent to γ at $\mathbf{0}$ and passes through p (see Fig. 8-12). Its radius is $R = R(t)$, and its curvature is $1/R$. As $t \to 0$, p moves along the curve γ toward $\mathbf{0}$, and we define the curvature of γ at the point $\mathbf{0}$ to be $1/\mathrm{R}_0$, where $R_0 = \lim_{t\to 0} R(t)$. We have $|p| = 2R \sin\theta$ and $\cos\theta = (\mathbf{a} \cdot p)/|\mathbf{a}|\,|p|$. This pair of equations can be solved for R by eliminating θ, to obtain

$$R^2 = \left(\frac{|\mathbf{a}|^2}{4}\right) \frac{|p|^2|p|^2}{|\mathbf{a}|^2|p|^2 - (\mathbf{a} \cdot p)^2} \tag{8-33}$$

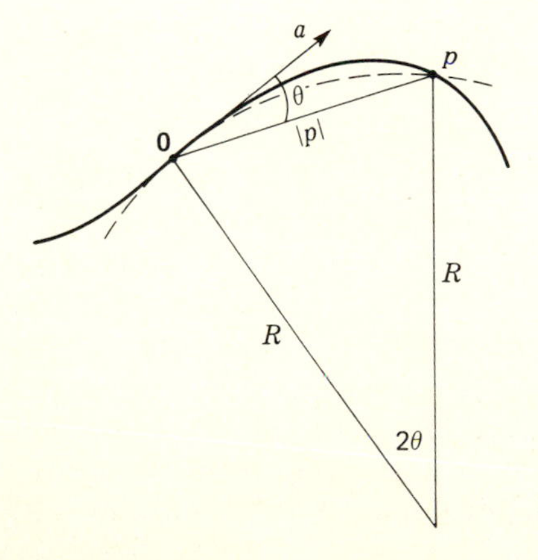

Figure 8-12

Now, from (8-32) we have

$$\mathbf{a} \cdot p = |\mathbf{a}|^2 t + (\mathbf{a} \cdot \mathbf{b})t^2 + (\mathbf{a} \cdot \mathbf{c})t^3 + \cdots$$

and squaring this, in the usual way one multiplies power series,

$$(\mathbf{a} \cdot p)^2 = |\mathbf{a}|^4 t^2 + 2(\mathbf{a} \cdot \mathbf{b})|\mathbf{a}|^2 t^3 + \{(\mathbf{a} \cdot \mathbf{b})^2 + 2(\mathbf{a} \cdot \mathbf{c})|\mathbf{a}|^2\}t^4 + \cdots$$

In the same way, if we take the dot product of (8-32) by itself, we obtain

$$(p \cdot p) = |p|^2 = |\mathbf{a}|^2 t^2 + 2(\mathbf{a} \cdot \mathbf{b})t^3 + \{|\mathbf{b}|^2 + 2(\mathbf{a} \cdot \mathbf{c})\}t^4 + \cdots$$

and then

$$|\mathbf{a}|^2|p|^2 = |\mathbf{a}|^4 t^2 + 2(\mathbf{a} \cdot \mathbf{b})|\mathbf{a}|^2 t^3 + \{|\mathbf{a}|^2|\mathbf{b}|^2 + 2(\mathbf{a} \cdot \mathbf{c})|\mathbf{a}|^2\}t^4 + \cdots$$

$$|p|^2|p|^2 = |\mathbf{a}|^4 t^4 + 4(\mathbf{a} \cdot \mathbf{b})|\mathbf{a}|^2 t^5 + \cdots$$

Substituting these into (8-33), we arrive at

$$R^2 = \left(\frac{|\mathbf{a}|^2}{4}\right) \frac{|\mathbf{a}|^4 t^4 + \cdots}{\{|\mathbf{a}|^2|\mathbf{b}|^2 - (\mathbf{a} \cdot \mathbf{b})^2\}t^4 + \cdots}$$

and thus obtain

$$R_0 = \lim_{t \to 0} R(t) = \left[\frac{|\mathbf{a}|^2|\mathbf{a}|^4}{4\{|\mathbf{a}|^2|\mathbf{b}|^2 - (\mathbf{a} \cdot \mathbf{b})^2\}}\right]^{1/2}$$

Substituting for **a** and **b** the values $\gamma'(0)$ and $\gamma''(0)/2$, and writing k for $1/R_0$, the curvature, we have a general formula for k, valid even if arc length is not the parameter and p_0 is not $\mathbf{0}$,

$$k = \frac{\sqrt{|\gamma'|^2|\gamma''|^2 - (\gamma' \cdot \gamma'')^2}}{|\gamma'|^3} \tag{8-34}$$

Suppose now that the curve γ does have arc length for the parameter. Then, $|\gamma'| = 1$, so that (8-34) reduces to

$$k = \sqrt{|\gamma''|^2 - (\gamma' \cdot \gamma'')^2} \tag{8-35}$$

However, this can be reduced still further. The characteristic property of curves with arc length as the parameter is $|\gamma'| = 1$, which we can write as

$$\gamma'(t) \cdot y'(t) = 1 \qquad \text{all } t \tag{8-36}$$

Differentiate this, and obtain

$$2\gamma'(t) \cdot \gamma''(t) = 0 \qquad \text{all } t \tag{8-37}$$

and use this in (8-35) to find the simple formula $k = |\gamma''|$, as given in our original definition, (8-31).

As an illustration of the use of the general formula (8-34), return to the curve given in (8-27), for which the origin is a multiple point. We have $\gamma'(t) = (1 - 3t^2, 2t - 1)$ and $\gamma''(t) = (-6t, 2)$. The point p is at $(0, 0)$ when $t = 0$

and when $t = 1$, from which we find the respective curvatures to be

$$k_1 = \frac{\sqrt{(2)(4) - (-2)^2}}{(\sqrt{2})^3} = \frac{1}{\sqrt{2}} \qquad k_2 = \frac{\sqrt{(5)(40) - (14)^2}}{(\sqrt{5})^3} = \frac{2}{5\sqrt{5}}$$

Curvature is a geometric property of curves; by this we mean that it is preserved under appropriate parameter changes, as was shown in Theorems 8 and 9 for length and direction. Thus, if γ_1 and γ_2 are curves of class C'' that are smoothly equivalent under a parameter change f, also of class C'', and p is a simple point that is on both curves, then γ_1 and γ_2 have the same curvature at p. Since the curvature reflects only the *magnitude* of the vector function $\gamma''(t)$ and not its direction, it might be expected that there are other geometric aspects of curves which can be described in terms of γ'' and the higher derivatives of γ. This is the classical differential geometry of curves, and we leave this to more specialized treatments of the subject.

Equivalence classes of curves have been introduced in topology and differential geometry to obtain alternative definitions for the notion of "curve." In this approach, a curve is no longer an individual mapping γ, but is, instead, an entire **equivalence class** of such mappings, under a prescribed collection of permissible parameter changes. Depending upon the nature of these, special designations are used, such as "Frechet curve," "Lebesgue curve," etc., when one wishes to refer to the individual equivalence classes. As an example, let us consider briefly the notion of an **algebraic curve.** One of the simplest ways in which a curve in the plane can be given is by an equation of the form $y = f(x)$; this can be thrown at once into a standard parametric form $x = t$, $y = f(t)$. One also says that an equation of the form $F(x, y) = 0$ specifies a curve. In a sense, one is again speaking of equivalence classes here, the class of all curves $x = \phi(t)$, $y = \psi(t)$ which satisfy the equation $F(\phi(t), \psi(t)) = 0$. If $F(x, y)$ is a polynomial, then one might restrict ϕ and ψ to be rational functions of t, and thus discuss the class of curves which are equivalent to a given curve under 1-to-1 birational correspondences. For example, the circle $x^2 + y^2 = 1$ has the rational parametrization $x = (1 - t^2)/(1 + t^2)$, $y = 2t/(1 + t^2)$. The situation becomes considerably more complicated when we turn to space curves. Here, one deals with pairs of algebraic equations, $F(x, y, z) = 0$, and $G(x, y, z) = 0$; again, with this one may associate a class of curves $x = \phi(t)$, $y = \psi(t)$, $z = \theta(t)$ which satisfy both equations for all t and whose coordinate functions ϕ, ψ, θ are of specified sort, for example, rational functions.

One may also consider the effect on a curve of a transformation which is applied to the space containing its trace. If γ is a curve whose trace lies in a region D and T is a continuous transformation which maps D onto a region D^*, then T carries γ into a curve γ^* lying in D^* defined by $\gamma^*(t) = T(\gamma(t))$. If T is 1-to-1 in a neighborhood N of p_0 and γ is a simple closed curve lying in N, then γ^* is a simple closed curve lying in $T(N)$. Suppose that D and D^* are each sets in the plane. As p moves along the trace of γ, $T(p)$ will

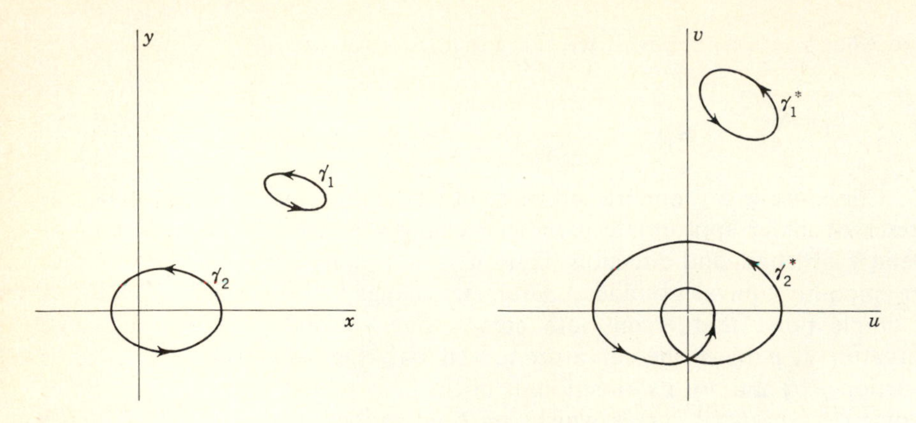

Figure 8-13

move along the trace of γ^*. They may move in the same direction—i.e., both clockwise or both counterclockwise—or they may move in opposite directions. When the first holds for all simple closed curves γ in N, T is said to be **orientation preserving;** when the second holds, T is said to be **orientation reversing.** When T is a nonsingular linear transformation, this may be determined by the sign of det (T); if $\det(T)$ is positive, T is orientation preserving, and if $\det(T)$ is negative, T is orientation reversing. For a general transformation of the plane, the same role is played by the sign of the Jacobian. For example, the transformation

$$T: \begin{cases} u = x^2 - y^2 \\ v = 2xy \end{cases} \tag{8-38}$$

has Jacobian $4(x^2 + y^2)$, which is never negative. T is therefore orientation preserving at all points of the plane, with the possible exception of the origin. Here, the Jacobian is 0; examining the image of a general curve γ which loops around the origin, one sees that T preserves the orientation of γ, but not its winding number. In Fig. 8-13 the curve γ_2 which loops the origin once has an image γ_2^* which loops the origin twice. There is a similar theory for transformations from $\mathbf{R}^n$ into $\mathbf{R}^n$; again, T is orientation preserving in a region D if the Jacobian of T is everywhere positive in D, the orientation reversing in D if the Jacobian is everywhere negative. The behavior of T near points where $J(P) = 0$ is indeterminate. In 3-space, the notion of "orientation preserving" can be visualized in terms of a spherical surface enclosing a point p_0 and its image under T (see Fig. 8-14).

We may also examine the effect which a transformation has upon angles at a point. If two smooth curves γ_1 and γ_2 pass through a point p_0, their images under T will pass through $T(p_0)$. Let θ be the angle between γ_1 and γ_2 at p_0. (More precisely, θ is the angle between the tangent lines to γ_1 and γ_2 at p_0.) If T is of class C', then the image curves will be smooth, and we

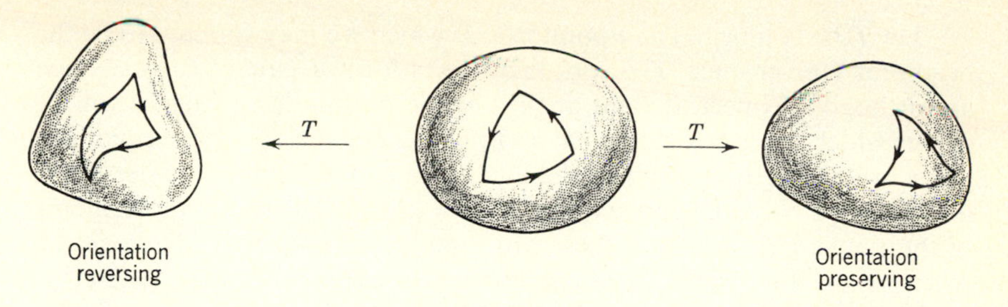

Figure 8-14

can speak of the angle θ^* which they form at $T(p_0)$ (see Fig. 8-15). T is said to be a **conformal** transformation if it is always true that $\theta^* = \theta$, and **directly conformal** if T is also orientation preserving.

Theorem 10 *Let T be a transformation from $\mathbf{R}^2$ into $\mathbf{R}^2$ which is of class C' in an open region D. Furthermore, let T be conformal and have a strictly positive Jacobian throughout D. Then, at each point of D, the differential of T has a matrix representation of the form*

$$\begin{bmatrix} A & B \\ -B & A \end{bmatrix}$$

Equivalently, if T is described by

$$u = f(x, y) \qquad v = g(x, y) \tag{8-39}$$

then Theorem 10 states that u and v must obey the **Cauchy-Riemann** differential equations: $u_x = v_y$, $u_y = -v_x$ in D.

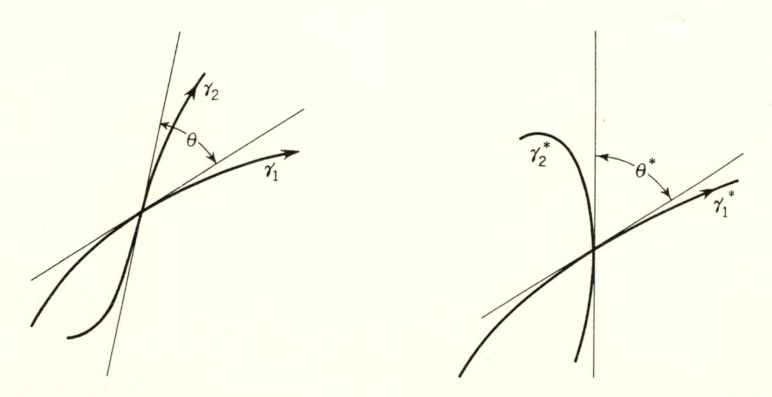

Figure 8-15

Let T be conformal at a point $p_0 \in D$ which we may suppose to be the origin for convenience. The coordinate axes form a pair of curves which are orthogonal (meet at right angles) at the origin. Their images will be curves which must be orthogonal at $T(0, 0)$. Since the axes have equations $x = t$, $y = 0$ and $x = 0$, $y = t$, the equations of their images under T are $u = f(t, 0)$, $v = g(t, 0)$ and $u = f(0, t)$, $v = g(0, t)$. The direction numbers of the tangent to either of these are found by computing $(du/dt, dv/dt)$; when $t = 0$, this results in (f_1, g_1) and (f_2, g_2), respectively, where the partial derivatives are evaluated at $(0, 0)$. These directions are perpendicular if their inner product is 0; this gives us one relation connecting the functions f and g, namely,

$$f_1 f_2 + g_1 g_2 = 0 \tag{8-40}$$

To obtain a second relation, we consider another pair of orthogonal lines through the origin; we choose the lines of slopes 1 and -1 with equations $x = t$, $y = t$ and $x = t$, $y = -t$, Their images are the curves $u = f(t, t)$, $v = g(t, t)$ and $u = f(t, -t)$, $v = g(t, -t)$. At $t = 0$, the tangents to these have direction numbers $(f_1 + f_2, g_1 + g_2)$ and $(f_1 - f_2, g_1 - g_2)$; since these must be perpendicular,

$$0 = (f_1{}^2 - f_2{}^2) + (g_1{}^2 - g_2{}^2) \tag{8-41}$$

Multiplying this by $f_2{}^2$ and making use of (8-40),

$$\begin{aligned} 0 &= f_1{}^2 f_2{}^2 + g_1{}^2 f_2{}^2 - (f_2{}^2 + g_2{}^2) f_2{}^2 \\ &= g_1{}^2 g_2{}^2 + g_1{}^2 f_2{}^2 - (f_2{}^2 + g_2{}^2) f_2{}^2 \\ &= (f_2{}^2 + g_2{}^2)(g_1{}^2 - f_2{}^2) \end{aligned}$$

If the first factor were 0, then $f_2 = g_2 = 0$ and the Jacobian

$$\frac{\partial(u, v)}{\partial(x, y)} = f_1 g_2 - f_2 g_1$$

would be 0. Thus, $g_1{}^2 = f_2{}^2$. Returning to (8-41), we must also have $f_1{}^2 = g_2{}^2$, so that $g_1 = \pm f_2$ and $g_2 = \pm f_1$. If the choice of signs were the same in each, then (8-40) would hold only if all the quantities were 0, and the Jacobian of T would again be 0; thus, we either have $f_2 = g_1$ and $f_1 = -g_2$ or we have $f_2 = -g_1$ and $f_1 = g_2$. In the first event, dT, which is

$$\begin{bmatrix} f_1 & f_2 \\ g_1 & g_2 \end{bmatrix}$$

would have the form

$$\begin{bmatrix} A & B \\ B & -A \end{bmatrix}$$

This can be ruled out, since a matrix of this form always has a negative determinant, while T has a positive Jacobian. When the second possibility holds, dT has the form

$$\begin{bmatrix} A & B \\ -B & A \end{bmatrix}$$

as stated in the theorem. (The entries, of course, need not be constant; we are only describing the form of the matrix.) ∎

The converse of this theorem is also true; if T is a transformation of class C' whose differential is represented by a matrix of the form

$$\begin{bmatrix} A & B \\ -B & A \end{bmatrix}$$

at each point of a region D, then T is conformal at all points of D, except those where the Jacobian is 0 (see Exercise 7). An example of a conformal transformation is supplied by that given in (8-38). The differential of T at (x, y) is

$$\begin{bmatrix} 2x & -2y \\ 2y & 2x \end{bmatrix}$$

which has the required form; T is therefore conformal everywhere in the plane, except at the origin. It fails to be conformal there; two lines through the origin which form the angle θ have as images two lines forming the angle 2θ (see Fig. 8-16).

It may be shown that there is no similar extensive class of transformations of $\mathbf{R}^n$ into itself, with $n > 2$. If such a transformation T is required to be conformal throughout an open set, then it may be shown that T is necessarily a linear transformation plus a translation, where the linear part is represented by an orthogonal matrix; thus, the only conformal transformations of 3-space into itself are the ordinary rigid motions.

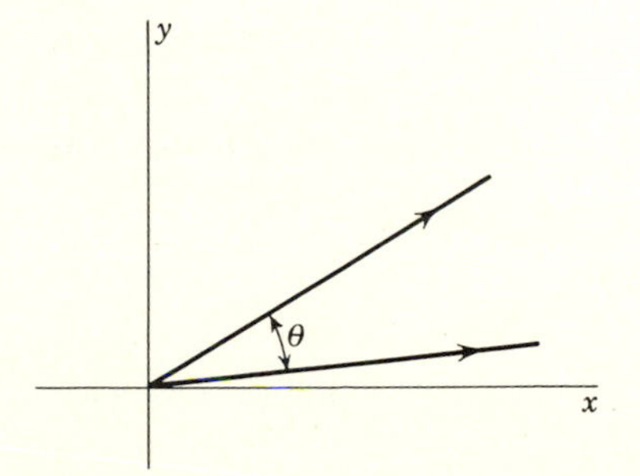

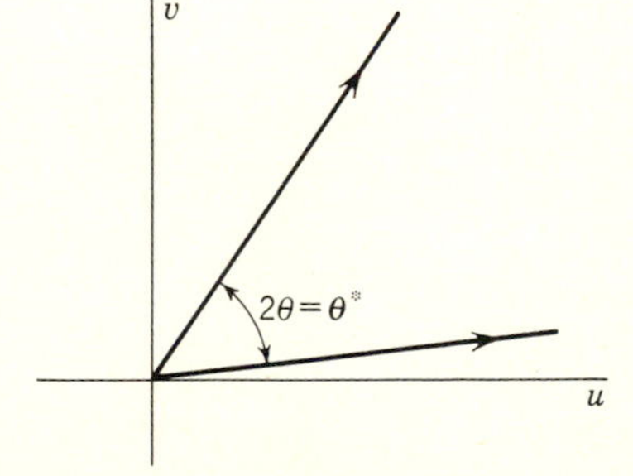

Figure 8-16 $u = x^2 - y^2 = \rho^2 \cos 2\phi$
$v = 2xy = \rho^2 \sin 2\phi$

EXERCISES

1 Plot the trace of the plane curve $x = t^2$, $y = t^3 - 4t$ for $-\infty < t < \infty$. Find the tangents at the double point.

2 Find the curvature of the helical curve of Fig. 8-11 at the double point.

3 Express the length of the loop of the curve (8-27) as a definite integral.

4 Find the curvature of the curve $x = t^2 - 2t$, $y = 3t$, $z = -t^3$, $w = t - t^2$ at the origin.

5 If γ is the curve given by $y = f(x)$, show that the curvature is given by

$$k = \frac{|f''(x)|}{(1 + f'(x)^2)^{3/2}}$$

and the length of γ between $x = a$ and $x = b$ by $\int_a^b \{1 + [f'(x)]^2\}^{1/2}\, dx$.

6 If γ is a curve which satisfies the equation $F(x, y) = 0$, show that the curvature at a point on γ is given by

$$k = \frac{|F_{11}F_2{}^2 + F_{22}F_1{}^2 - 2F_1F_2F_{12}|}{|F_1{}^2 + F_2{}^2|^{3/2}}$$

7 Using the converse of Theorem 10, show that the transformation

$$u = x^3 - 3xy^2 \qquad v = 3x^2y - y^3$$

is directly conformal everywhere in the plane except at the origin.

8 Find the length of one arch of the cycloid $x = a(t - \sin t)$, $y = a(1 - \cos t)$.

9 Find the angle between the curves where they intersect:

$$\begin{cases} x = t \\ y = 2t \\ z = t^2 \end{cases} \qquad \begin{cases} x = s^2 \\ y = 1 - s \\ z = 2 - s^2 \end{cases}$$

10 Let γ be a curve in 3-space which satisfies both of the equations $f(x, y, z) = 0$, $g(x, y, z) = 0$. Show that the direction of γ at a point on it has direction components

$$\left(\begin{vmatrix} f_2 & f_3 \\ g_2 & g_3 \end{vmatrix}, \begin{vmatrix} f_3 & f_1 \\ g_3 & g_1 \end{vmatrix}, \begin{vmatrix} f_1 & f_2 \\ g_1 & g_2 \end{vmatrix} \right)$$

11 If γ_1 and γ_2 are smoothly equivalent curves of class C'' and if the parameter change is effected by a function f which is also of class C'', show that γ_1 and γ_2 have the same curvature at corresponding points.

12 Let $f = g_1 - g_2$ where g_i is monotonic on [0, 1] and not necessarily continuous. Prove that f is of bounded variation on [0, 1], and that $\int_0^1 f$ exists.

13 If f is continuous, and f' exists and is bounded on [0, 1], show that f is of bounded variation on [0, 1].

14 Let f be continuous on [0, 1], f' exist for all x, $0 < x < 1$, and $\int_0^1 |f'| < \infty$. Show that f is of bounded variation on [0, 1].

15 Suppose γ is given by $x = \phi(t)$, $y = \psi(t)$, $0 \le t \le 1$. Prove that γ is a rectifiable curve if and only if ϕ and ψ are of bounded variation.

16 Let $f \in C'$ on [0, 1]. Prove that $f = g_1 - g_2$ where g_1 and g_2 are monotonic.

17 Find a rational parametrization for the hyperbola $x^2 - y^2 = 1$.

18 The graph of the equation $x^3 + y^3 = 3xy$ is known as the folium of Descartes. Show that a rational parametrization of this is $x = 3t/(1 + t^3)$, $y = 3t^2/(1 + t^3)$, and graph this curve.

8.5 SURFACES AND SURFACE AREA

The most elementary portions of the theory of curves in n space, outlined in Sec. 8.4, can be presented in a way that is relatively free of topological difficulties. This is much more difficult to do for the theory of surfaces; for this reason, certain topics in the present section will be presented on the intuitive level. Many of the basic definitions are direct analogs of those for curves.

Definition 12 *A surface Σ in n space is a transformation or mapping from* $\mathbf{R}^2$ *into* $\mathbf{R}^n$.

For example, the general continuous surface in 3-space can be expressed as a continuous transformation of the form

$$\Sigma: \begin{cases} x = \phi(u, v) \\ y = \psi(u, v) \\ z = \theta(u, v) \end{cases} \tag{8-42}$$

whose domain is a set D in the UV plane. A point p is said to **lie on** the surface Σ if $p = \Sigma(u, v)$ for some $(u, v) \in D$, and the set of all points that lie on Σ is called the **trace** of the surface Σ. As with curves, may different surfaces can have the same trace. A point p lying on Σ is said to be a **multiple** point if it is the image of more than one point in the domain of Σ. A surface is **simple** if it has no multiple points; the mapping Σ is then 1-to-1 in D.

Any set D which lies on one of the coordinate XYZ planes is the trace of a simple surface which is obtained from the identity mapping; if D is contained in the XY plane, then the surface may be defined by $x = x$, $y = y$, $z = 0$, using x and y as parameters rather than u and v.

Let Σ be a continuous surface whose domain D is such that its boundary is the trace of a simple closed curve γ defined for $a \le t \le b$. The image of γ under Σ is a closed curve Γ defined by $\Gamma(t) = \Sigma(\gamma(t))$ which is called the **boundary** or **edge** of Σ (see Fig. 8-17). We shall use $\partial\Sigma$ as a notation for Γ, rather than bdy (Σ), to emphasize the fact that we are dealing with both curves and surfaces as mappings rather than sets of points. The trace of $\partial\Sigma$ is the image, under Σ, of the boundary of the parameter domain D. If the boundary of D is composed of several simple closed curves, as indicated in Fig. 8-18, $\partial\Sigma$ will also be made up of several closed curves.

Any simple closed curve γ will be the edge $\partial\Sigma$ of an infinite number of different surfaces. In Fig. 8-19, we show a closed curve γ whose trace is a simple trefoil knot, and in Fig. 8-20 we show the trace of a surface whose edge is γ. To see how this can be done in general, let γ be a continuous mapping from the unit circle $u^2 + v^2 = 1$ into 3-space. We can extend this function γ in an infinite number of ways so that it becomes a continuous function Σ defined on the unit disk D into 3-space. For example, we can set

$$\Sigma(u, v) = \Sigma(p) = |p|^2\gamma(p/|p|)$$

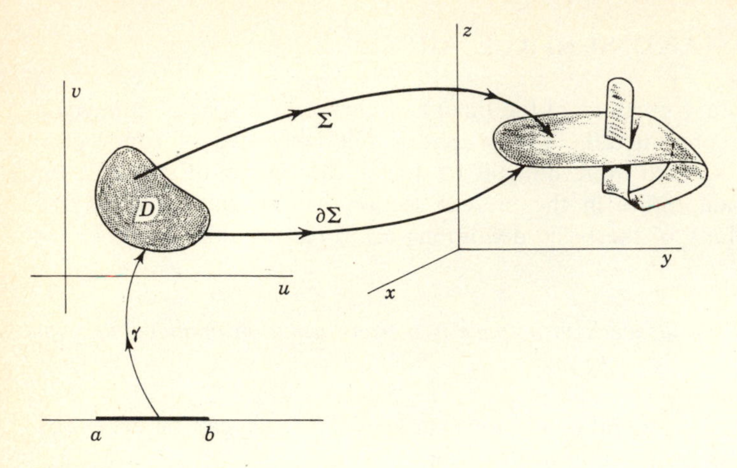

Figure 8-17

The mapping Σ is then a surface whose edge $\partial\Sigma$ is γ.

To obtain a nice theory, it is again convenient to impose certain differentiability requirements.

Definition 13 *A surface Σ is said to be smooth if Σ is of class C' and $d\Sigma$ is of rank 2 throughout D.*

When Σ is described by (8-42), then

$$d\Sigma = \begin{bmatrix} \dfrac{\partial x}{\partial u} & \dfrac{\partial x}{\partial v} \\[2ex] \dfrac{\partial y}{\partial u} & \dfrac{\partial y}{\partial v} \\[2ex] \dfrac{\partial z}{\partial u} & \dfrac{\partial z}{\partial v} \end{bmatrix}$$

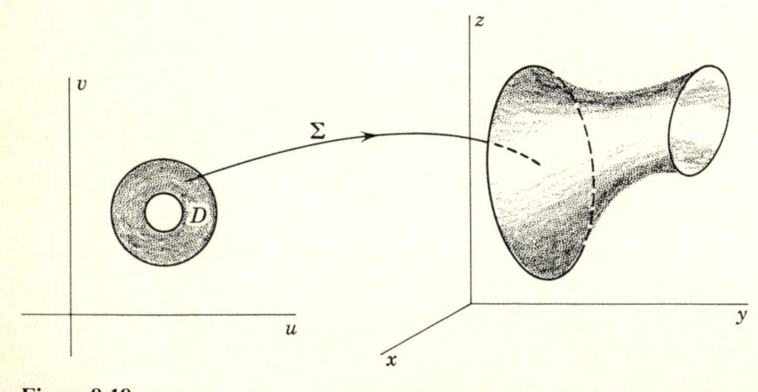

Figure 8-18

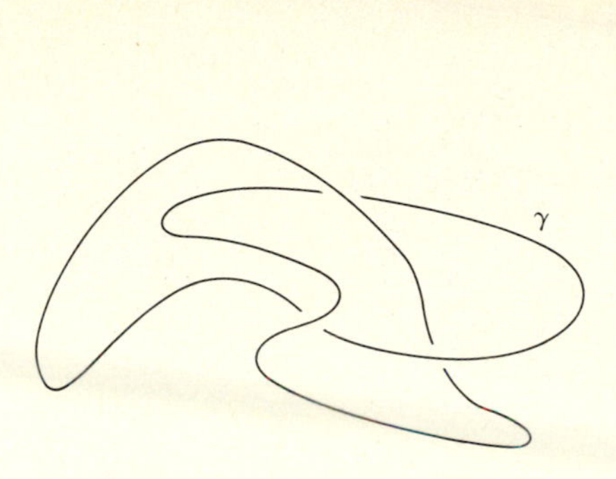

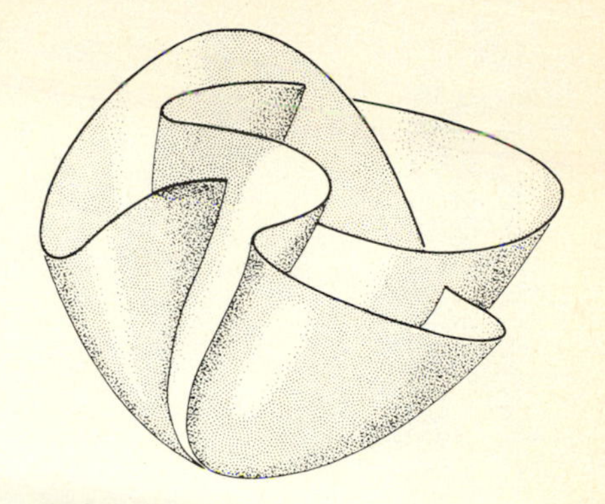

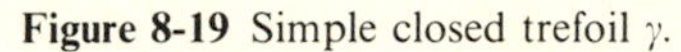
Figure 8-19 Simple closed trefoil γ.

Figure 8-20 A surface Σ with $\partial\Sigma = \gamma$.

The requirement that this have rank 2 is equivalent to the assertion that the 2-by-2 determinants formed from this matrix do not all vanish at any point of D. This can also be given in the form:

$$\left|\frac{\partial(x, y)}{\partial(u, v)}\right|^2 + \left|\frac{\partial(y, z)}{\partial(u, v)}\right|^2 + \left|\frac{\partial(z, x)}{\partial(u, v)}\right|^2 > 0 \tag{8-43}$$

The significance of this condition is to be found in the fact that if (8-43) holds, Σ is locally 1-to-1 in D (see Theorem 21, Sec. 7.7). It also corresponds to the intuitive notion of smoothness in the sense of "lack of sharp corners."

Consider, for example, the surfaces given by the following sets of equations:

$$\Sigma_1: \begin{cases} x = u \\ y = v \\ z = \sqrt{u^2 + v^2} \end{cases} \qquad \Sigma_2: \begin{cases} x = u \cos v \\ y = u \sin v \\ z = u \end{cases}$$

$$D_1: u^2 + v^2 \le 1 \qquad D_2: \begin{matrix} 0 \le u \le 1 \\ 0 \le v \le 2\pi \end{matrix}$$

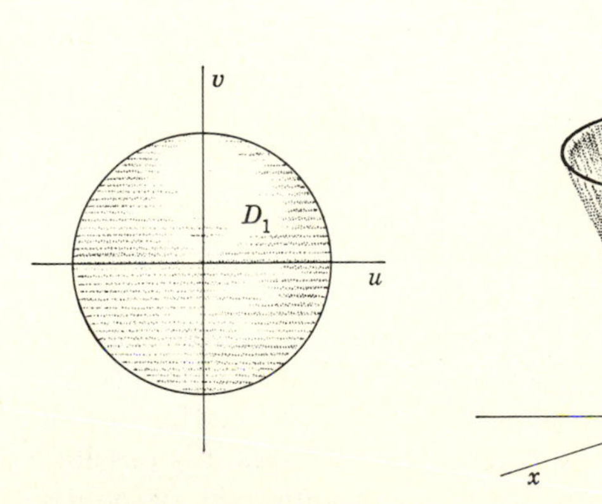

Figure 8-21 Graph of Σ_1 and $\partial\Sigma_1$.

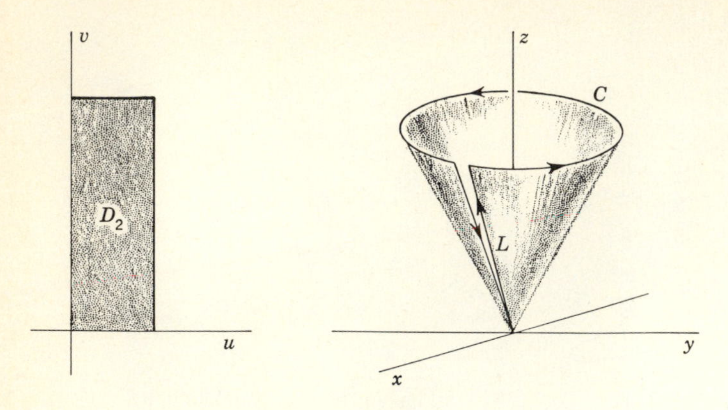

Figure 8-22 Graph of Σ_2 and $\partial\Sigma_2$.

The trace of each is a portion of the cone whose equation is $x^2 + y^2 = z^2$ (see Figs. 8-21 and 8-22). The edge of Σ_1 is the curve $\partial\Sigma_1$ whose trace is the circle C. The edge of Σ_2 is the curve $\partial\Sigma_2$ whose trace is the curve consisting of C and the line segment L taken twice, once in each direction. Σ_1 is not a smooth surface, since $\partial z/\partial u = u/(u^2 + v^2)^{1/2}$ is not continuous at $u = 0$, $v = 0$; geometrically, this corresponds to the presence of the point of the cone as an interior point of the surface. Σ_2 is a smooth surface on the interior of D_2, since

$$d\Sigma_2 = \begin{bmatrix} \cos v & -u \sin v \\ \sin v & u \cos v \\ 1 & 0 \end{bmatrix}$$

and

$$\begin{vmatrix} \cos v & -u \sin v \\ \sin v & u \cos v \end{vmatrix}^2 + \begin{vmatrix} \sin v & u \cos v \\ 1 & 0 \end{vmatrix}^2 + \begin{vmatrix} 1 & 0 \\ \cos v & -u \sin v \end{vmatrix}^2$$

$$= u^2 + u^2(\cos v)^2 + u^2(\sin v)^2 = 2u^2$$

which is 0 only on the edge $u = 0$ of the rectangle D_2. (The fact that the rank of dT is less than 2 when $u = 0$ is matched by the fact that the curve $\partial\Sigma_2$ sends this entire edge of D_2 into the point at the apex of the cone.)

Any equation of the form $z = f(x, y)$ will serve to define a surface, either by the equations $x = u$, $y = v$, $z = f(u, v)$, or simply $x = x$, $y = y$, $z = f(x, y)$. An equation of the form $F(x, y, z) = 0$ which can be solved locally for one of the variables leads similarly to one or more surfaces. Sometimes it is possible to obtain a single smooth surface whose trace is the entire set of points (x, y, z) satisfying the equation $F(x, y, z) = 0$. For example, a suitable choice

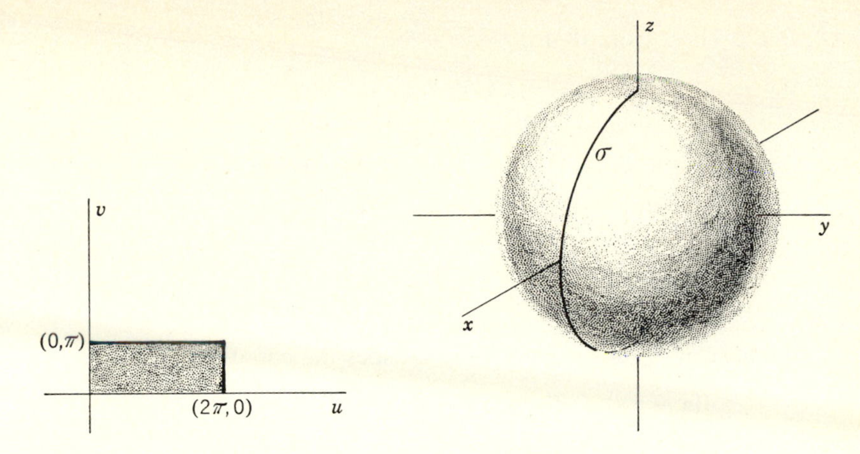

Figure 8-23

for the unit sphere $x^2 + y^2 + z^2 = 1$ is

$$(8\text{-}44) \qquad \begin{cases} x = \cos u \sin v \\ y = \sin u \sin v \\ z = \cos v \end{cases}$$

where D is the set of (u, v) obeying $0 \leq u \leq 2\pi$, $0 \leq v \leq \pi$. The edge of this surface is the closed curve whose trace is the semicircular arc σ taken twice in opposite directions (see Fig. 8-23).

By the term "plane" we shall mean any surface of the form

$$(8\text{-}45) \qquad p = \Sigma(u, v) = p_0 + \alpha u + \beta v$$

defined for all (u, v), where α and β are assumed nonparallel in order to satisfy condition (8-43). Since p_0, $p_0 + \alpha$, and $p_0 + \beta$ lie on Σ, the trace of this surface contains these three noncollinear points. If the surface lies in 3-space and $\alpha = (a_1, a_2, a_3)$, $\beta = (b_1, b_2, b_3)$,

$$p_0 = (x_0, y_0, z_0)$$

then (8-45) becomes

$$\Sigma: \begin{cases} x = x_0 + a_1 u + b_1 v \\ y = y_0 + a_2 u + b_2 v \\ z = z_0 + a_3 u + b_3 v \end{cases}$$

Put

$$(8\text{-}46) \qquad A = \begin{vmatrix} a_2 & b_2 \\ a_3 & b_3 \end{vmatrix} \qquad B = \begin{vmatrix} a_3 & b_3 \\ a_1 & b_1 \end{vmatrix} \qquad C = \begin{vmatrix} a_1 & b_1 \\ a_2 & b_2 \end{vmatrix}$$

and $\mathbf{n} = (A, B, C)$. Then, from (8-46),

$$\begin{aligned} &A(x - x_0) + B(y - y_0) + C(z - z_0) \\ &\quad = A(a_1 u + b_1 v) + B(a_2 u + b_2 v) + C(a_3 u + b_3 v) \\ &\quad = (Aa_1 + Ba_2 + Ca_3)u + (Ab_1 + Bb_2 + Cb_3)v \\ &\quad = \begin{vmatrix} a_1 & a_1 & b_1 \\ a_2 & a_2 & b_2 \\ a_3 & a_3 & b_3 \end{vmatrix} u + \begin{vmatrix} b_1 & a_1 & b_1 \\ b_2 & a_2 & b_2 \\ b_3 & a_3 & b_3 \end{vmatrix} v \\ &\quad = 0 + 0 = 0 \end{aligned}$$

Thus, every point (x, y, z) on the trace of Σ satisfies the equation

$$A(x - x_0) + B(y - y_0) + C(z - z_0) = 0$$

which justifies our use of the word "plane" for the surface Σ. The requirement that Σ be smooth is equivalent to the condition $A^2 + B^2 + C^2 > 0$, and may thus be stated in the form $\mathbf{n} \neq \mathbf{0}$. In the analytical geometry of 3-space, the vector $\mathbf{n}$ is a set of direction numbers called the **normal** to the plane

$$A(x - x_0) + B(y - y_0) + C(z - z_0) = 0$$

Since $(x - x_0, y - y_0, z - z_0) = \mathbf{v}$ is a vector from the point p_0 to the general point (x, y, z) on the plane, this equation asserts that $\mathbf{n} \cdot \mathbf{v} = 0$, and $\mathbf{n}$ and $\mathbf{v}$ are always orthogonal. By analogy, we may introduce a normal vector for a general smooth surface.

Definition 14 *If Σ is a smooth surface in 3-space described by the standard equations* (8-42), *then the normal to Σ at a point p is*

$$\mathbf{n} = \left(\frac{\partial(y, z)}{\partial(u, v)}, \frac{\partial(z, x)}{\partial(u, v)}, \frac{\partial(x, y)}{\partial(u, v)} \right) \tag{8-47}$$

Comparing this with (8-46), we see that this gives the correct answer when Σ is a plane. Moreover, the next theorem shows that this is in agreement with the intuitive notion of the normal as a "direction which is orthogonal to the surface" (see also Exercise 19).

Theorem 11 *Let Σ be a smooth surface and p a point lying on Σ. Then, the normal to Σ at p is orthogonal to any smooth curve which lies on Σ and passes through p.*

Let Σ have domain D, and let γ be any smooth curve whose trace lies in D. The mapping Σ carries γ into a curve Γ lying on the surface Σ (see Fig. 8-24). The equation of Γ is $\Gamma(t) = \Sigma(\gamma(t))$. Let us suppose that Γ passes through p when $t = 0$. The direction of Γ is determined by the vector $\mathbf{v} = \Gamma'(0)$. Since $\Gamma(t) = (x, y, z)$ where $(x, y, z) = \Sigma(u, v)$ and $(u, v) = \gamma(t)$, we apply the chain rule of differentiation to find that

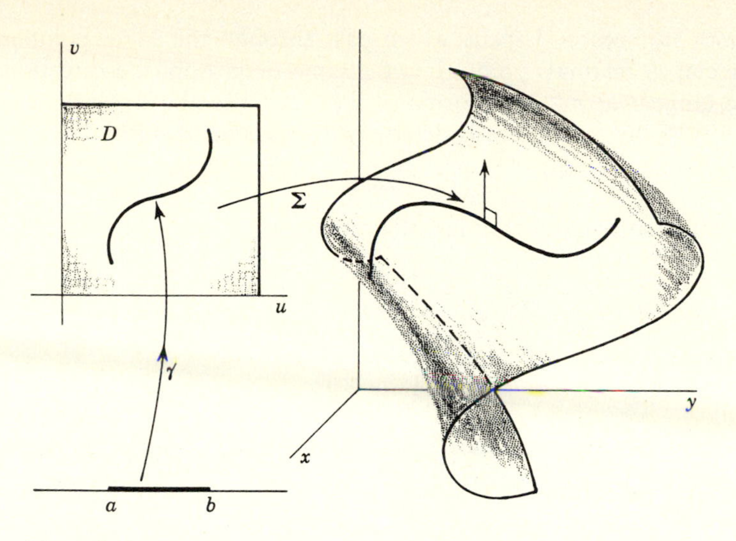

Figure 8-24

$$\mathbf{v} = (d\Sigma)(d\gamma) = \left(\frac{\partial x}{\partial u}\frac{du}{dt} + \frac{\partial x}{\partial v}\frac{dv}{dt}, \frac{\partial y}{\partial u}\frac{\partial u}{dt} + \frac{\partial y}{\partial v}\frac{dv}{dt}, \frac{\partial z}{\partial u}\frac{du}{dt} + \frac{dz}{\partial v}\frac{\partial v}{dt}\right)$$

The vectors $\mathbf{v}$ and $\mathbf{n}$ are orthogonal if $\mathbf{v} \cdot \mathbf{n} = 0$.

$$\mathbf{v} \cdot \mathbf{n} = \left(\frac{\partial x}{\partial u}\frac{du}{dt} + \frac{\partial x}{\partial v}\frac{dv}{dt}\right)\frac{\partial(y, z)}{\partial(u, v)} + \left(\frac{\partial y}{\partial u}\frac{du}{dt} + \frac{\partial y}{\partial v}\frac{dv}{dt}\right)\frac{\partial(z, x)}{\partial(u, v)}$$

$$+ \left(\frac{\partial z}{\partial u}\frac{du}{dt} + \frac{\partial z}{\partial v}\frac{dv}{dt}\right)\frac{\partial(x, y)}{\partial(u, v)}$$

$$= \left(\frac{\partial x}{\partial u}\frac{\partial(y, z)}{\partial(u, v)} + \frac{\partial y}{\partial u}\frac{\partial(z, x)}{\partial(u, v)} + \frac{\partial z}{\partial u}\frac{\partial(x, y)}{(\partial u, v)}\right)\frac{du}{dt}$$

$$+ \left(\frac{\partial x}{\partial v}\frac{\partial(y, z)}{\partial(u, v)} + \frac{\partial y}{\partial v}\frac{\partial(z, x)}{\partial(u, v)} + \frac{\partial z}{\partial v}\frac{\partial(x, y)}{\partial(u, v)}\right)\frac{dv}{dt}$$

$$= \begin{vmatrix} \frac{\partial x}{\partial u} & \frac{\partial x}{\partial u} & \frac{\partial x}{\partial v} \\ \frac{\partial y}{\partial u} & \frac{\partial y}{\partial u} & \frac{\partial y}{\partial v} \\ \frac{\partial z}{\partial u} & \frac{\partial z}{\partial u} & \frac{\partial z}{\partial v} \end{vmatrix} \frac{du}{dt} + \begin{vmatrix} \frac{\partial x}{\partial v} & \frac{\partial x}{\partial u} & \frac{\partial x}{\partial v} \\ \frac{\partial y}{\partial v} & \frac{\partial y}{\partial u} & \frac{\partial y}{\partial v} \\ \frac{\partial z}{\partial v} & \frac{\partial z}{\partial u} & \frac{\partial z}{\partial v} \end{vmatrix} \frac{dv}{dt}$$

$$= 0\frac{du}{dt} + 0\frac{dv}{dt} = 0$$

since a matrix with two columns alike has zero determinant. ∎

Two smooth surfaces in 3-space which pass through the same point p_0 and which have at p_0 normals pointing in the same or in opposite directions are said to be tangent at p_0. The normal $\mathbf{n} = (A, B, C)$ to the general plane (8-45) is given by (8-46) and hence is determined by α and β; since

$$\Sigma(u, v) = p_0 + \alpha u + \beta v$$

these may be obtained by partial differentiation, with

$$\alpha = \Sigma_u = \frac{\partial \Sigma}{\partial u} = \left(\frac{\partial x}{\partial u}, \frac{\partial y}{\partial u}, \frac{\partial z}{\partial u}\right)$$

and
$$\beta = \Sigma_v = \frac{\partial \Sigma}{\partial v} = \left(\frac{\partial x}{\partial v}, \frac{\partial y}{\partial v}, \frac{\partial z}{\partial v}\right)$$

The tangent plane to a smooth surface Σ at a point p_0 is therefore given by the equation

$$p = p_0 + \left(\frac{\partial \Sigma}{\partial u}\right)_{p_0} u + \left(\frac{\partial \Sigma}{\partial v}\right)_{p_0} v$$

We may also obtain a formula for the normal to a surface Σ when it is described by means of an equation $F(x, y, z) = 0$.

Theorem 12 *If Σ is a surface which satisfies the equation*

$$F(x, y, z) = 0$$

where F is of class C', then the normal to Σ at a simple point p_0 lying on Σ is a scalar multiple of $dF\big|_{p_0} = (F_1(p_0), F_2(p_0), F_3(p_0))$, unless this is $(0, 0, 0)$.

Assume that Σ has the form $x = \phi(u, v)$, $y = \psi(u, v)$, $z = \theta(u, v)$ for $(u, v) \in D$. Then,

$$F(\phi(u, v), \psi(u, v), \theta(u, v)) = \theta$$

for all $(u, v) \in D$. Differentiating, and setting $(u, v) = (u_0, v_0)$ so that $(x, y, z) = p_0$, we obtain

$$0 = F_1(p_0)\frac{\partial x}{\partial u} + F_2(p_0)\frac{\partial y}{\partial u} + F_3(p_0)\frac{\partial z}{\partial u}$$

$$0 = F_1(p_0)\frac{\partial x}{\partial v} + F_2(p_0)\frac{\partial y}{\partial v} + F_3(p_0)\frac{\partial z}{\partial v}$$

If we set $\beta = (F_1(p_0), F_2(p_0), F_3(p_0)) = dF\big|_{p_0}$, these equations assert that β is orthogonal to both Σ_u and Σ_v at p_0; but, the normal vector $\mathbf{n}$ is also orthogonal to these, so that β is a multiple of $\mathbf{n}$. ▮

As with curves, we say that a surface Σ is of class C^k if the coordinate functions that defined the mapping Σ are all of class C^k on the parameter domain, and is analytic if the coordinate functions are analytic in u and v. This means that an analytic surface can be represented locally as an absolutely convergent double Taylor series

$$\Sigma(u, v) = \Sigma(u_0, v_0) + (u - u_0)\Sigma_u + (v - v_0)\Sigma_v + \tfrac{1}{2}\{(u - u_0)^2\Sigma_{uu} + 2(u - u_0)(v - v_0)\Sigma_{uv} + (v - v_0)^2\Sigma_{vv}\} + \cdots$$

where the coefficients Σ_u, Σ_v, Σ_{uu}, Σ_{uv}, etc., are vector-valued functions.

Noting that Σ_u and Σ_v are the columns of $d\Sigma$, and thus together determine, for example, the normal to the surface at any point by (8-47), one might reasonably expect that the other coefficients in the series expansion of $\Sigma(u, v)$ would also have geometric significance. While we will leave this for more advanced works dealing with the classical differential geometry of surfaces, a few additional remarks may be helpful.

Up until now, we have spoken only of the *first* total derivative $\mathbf{D}g$ of a scalar function g and the first differential dT of a transformation T. However, $\mathbf{D}g$ (or dg) is itself a vector-valued function, and thus can be viewed as a transformation, so that one may automatically apply the definitions of Chap. 7 to obtain its derivative or differential $d(dg) = d^2g$. Suppose we try this with $w = g(x, y, z)$. We have $dg = (g_x, g_y, g_z)$, and then

$$d^2g = \begin{bmatrix} g_{xx} & g_{xy} & g_{xz} \\ g_{yx} & g_{yy} & g_{yz} \\ g_{zx} & g_{zy} & g_{zz} \end{bmatrix}$$

Thus, the second (total) derivative of a scalar function turns out to be a matrix-valued function. Note that if g is of class C'', the matrix d^2g is **symmetric** because $g_{yx} = g_{xy}$, $g_{xz} = g_{zx}$, etc.

For the general case, suppose we have a real-valued function $g(x_1, x_2, \ldots, x_n)$ of class C'' in an open set D. Then, d^2g is defined in D and is the matrix-valued function

$$d^2g = \begin{bmatrix} g_{11} & g_{12} & g_{13} & \cdots & g_{1n} \\ g_{21} & g_{22} & g_{23} & \cdots & g_{2n} \\ \cdots & \cdots & \cdots & \cdots & \cdots \\ g_{n1} & g_{n2} & g_{n3} & \cdots & g_{nn} \end{bmatrix} \tag{8-48}$$

[This is sometimes called the **Hessian** matrix of g, and its determinant the **Hessian** of g.]

While we have not met d^2g before in earlier topics, it has been implicit in several. For example, in Theorem 19 of Sec. 3.6, the nature of a critical point p_0 for a function $f(x, y)$ was determined by examining the sign of the expression $(f_{12})^2 - f_{11}f_{22}$. Note that this is exactly the negative of the determinant of d^2f. Thus, we could now restate Theorem 19 as follows: A critical point p_0 for $f(x, y)$ is an extreme point if $d^2f\,|_{p_0}$ has a strictly positive determinant, is a saddle point if the determinant is strictly negative, and is undetermined if $d^2f\,|_{p_0}$ is singular.

This result for two variables suggests that there is a similar theorem for functions of n variables, characterizing the nature of critical points of f by properties of the second derivative d^2f. This is correct, but it is not sufficient merely to look at the sign of the determinant of the matrix (8-48)—i.e., the sign of the Hessian of f. Instead, one must analyze the behavior of the quadratic form associated with the symmetric matrix $d^2f|_{p_0} = [a_{ij}] = A$. This arises by the same approach used in the proof of the two-variable version in Sec. 3.6. If p_0 is a critical point for f, then Taylor's theorem shows that on a neighborhood of p_0,

$$f(p_0 + \Delta p) = f(p_0) + 0 + Q(\Delta p) + {}^*R(\Delta p)$$

where $\Delta p = (\Delta x_1, \Delta x_2, \ldots, \Delta x_n)$ and

$$Q(\Delta p) = \sum_{i,\, j=1}^{n} a_{ij}(\Delta x_i)(\Delta x_j) \tag{8-49}$$

and where $|{}^*R(\Delta p)| \ll |\Delta p|^2$. The local behavior of f at p_0 is therefore determined by the behavior of $Q(\Delta p)$, and there are three cases.

Definition 15 *A nonsingular symmetric matrix A is:*

i. *Positive definite if*

$$Q(u) = \sum_{i,\, j=1}^{n} a_{ij} u_i u_j > 0$$

for all $u = (u_1, u_2, \ldots, u_n) \neq 0$.

ii. *Negative definite if* $Q(u) < 0$ *for all* $u \neq 0$.

iii. *Indefinite if neither* (i) *nor* (ii) *holds.*

Correspondingly, one arrives at the following characterization theorem for critical points.

Theorem 13 *Let f be of class C'' in an open set D, and let $p_0 \in D$ with $df|_{p_0} = 0$, so that p_0 is a critical point for f. Let $A = d^2 f|_{p_0}$ and assume* $\det(A) \neq 0$. *Then, p_0 is a local minimum for f if A is positive definite, is a local maximum if A is negative definite, and is a saddle point if A is indefinite.*

A simple sufficient criterion for A to be indefinite is that two of the diagonal entries in A have opposite signs. Algebraic criteria which use the determinants of diagonal submatrices of A can be given, and are mentioned in Appendix 3.

Another example of the role of d^2f may be given. Suppose $w = f(x, y)$, where $x = \phi(t)$, $y = \psi(t)$. Write this as $w = f(\gamma(t))$ with $\gamma(t) = (x, y)$. Then, by the usual chain rule,

$$\frac{dw}{dt} = (df)\Big|_{\gamma(t)} \gamma'(t) = [f_x \quad f_y]\begin{bmatrix} dx/dt \\ dy/dt \end{bmatrix}$$

$$= f_x \frac{dx}{dt} + f_y \frac{dy}{dt}$$

In the usual way, differentiate again, and obtain

$$\frac{d^2w}{dt^2} = f_{xx}\left(\frac{dx}{dt}\right)^2 + 2f_{xy}\frac{dx}{dt}\frac{dy}{dt} + f_{yy}\left(\frac{dy}{dt}\right)^2 + f_x \frac{d^2x}{dt^2} + f_y \frac{d^2y}{dt^2}$$

We now recognize this result as

$$\frac{d^2w}{dt^2} = Q(\gamma'(t)) + (df)\gamma''(t) \tag{8-50}$$

where Q is the quadratic form associated with d^2f. However, if we choose to write $Q(u)$ as $A(u)(u)$, (8-50) becomes

$$\frac{d^2w}{dt^2} = d^2f\,\gamma'(t)\gamma'(t) + df\,\gamma''(t)$$

and we see that we have returned to the type of notational simplicity of the one-variable calculus; if $w = f(\gamma(t))$, then $dw/dt = f'(\gamma(t))\gamma'(t)$, and $d^2w/dt^2 = f''(\gamma(t))\gamma'(t)\gamma'(t) + f'(\gamma(t))\gamma''(t)$.

It is natural now to ask for the second derivative of vector-valued functions (transformations), and to ask if they too have useful geometric interpretations. Suppose we have

$$\Sigma: \begin{cases} x = f(u, v) \\ y = g(u, v) \\ w = h(u, v) \end{cases}$$

Then, we may write

$$d\Sigma = \begin{bmatrix} f_u & f_v \\ g_u & g_v \\ h_u & h_v \end{bmatrix} = [\Sigma_u, \Sigma_v] = \begin{bmatrix} df \\ dg \\ dh \end{bmatrix}$$

and thus are led to write

$$d^2\Sigma = \begin{bmatrix} d^2f \\ d^2g \\ d^2h \end{bmatrix}$$

This allows many different descriptions. It is a 3-vector whose entries are 2-by-2 matrices, or the linear transformations that they represent. Or it is a solid array, obtained by stacking three 2-by-2 matrices on top of one another (see Fig. 8-25). Note that the vector-valued partial derivatives Σ_{uu}, $\Sigma_{uv} = \Sigma_{vu}$,

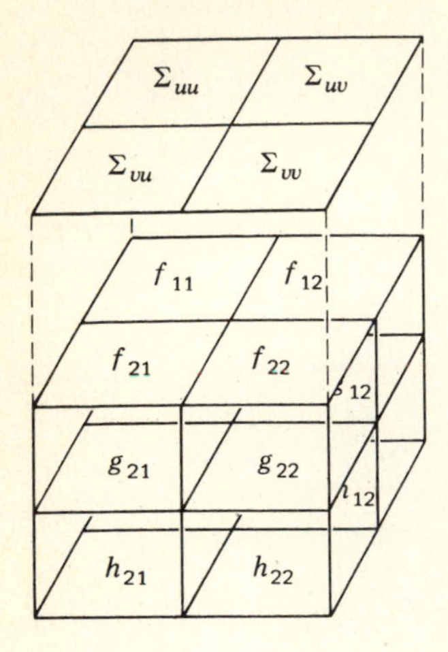

Figure 8-25 $d^2\Sigma$

and Σ_{vv} are visible in this array as the vertical columns, forming in fact the Hessian matrix (with vector entries)

$$d^2\Sigma = \begin{bmatrix} \Sigma_{uu} & \Sigma_{uv} \\ \Sigma_{vu} & \Sigma_{vv} \end{bmatrix}$$

In the differential geometry of surfaces, these are used to discuss the various aspects of curvature for surfaces, just as $d^2\gamma = \gamma''$ was used to discuss this for curves.

We turn next to the topic of surface area.

Definition 16 *The area of a smooth surface Σ with domain D is defined to be*

$$(8\text{-}51) \qquad A(\Sigma) = \iint_D |\mathbf{n}(u, v)|\, du\, dv$$

$$= \iint \left\{ \left| \frac{\partial(x, y)}{\partial(u, v)} \right|^2 + \left| \frac{\partial(y, z)}{\partial(u, v)} \right|^2 + \left| \frac{\partial(z, x)}{\partial(u, v)} \right|^2 \right\}^{1/2} du\, dv$$

This rather arbitrary definition can be motivated by geometrical considerations. First, it agrees with our previous notion for area when Σ is merely a region embedded in one of the coordinate planes. For example, if Σ is the surface defined by $x = u$, $y = v$, $z = 0$ for $(u, v) \in D$, then $\Sigma_u = (1, 0, 0)$ and $\Sigma_v = (0, 1, 0)$ and the normal $\mathbf{n}$ is $(0, 0, 1)$. The integral for the area of Σ becomes $\iint_D 1\, du\, dv$, which is precisely $\mathrm{A}(D)$, the area of the set D. Again, when Σ is the portion of a plane defined by

$$\Sigma: \begin{cases} x = a_1 u + b_1 v \\ y = a_2 u + b_2 v \\ z = a_3 u + b_3 v \end{cases} \qquad (u, v) \in D$$

then
$$d\Sigma = \begin{bmatrix} a_1 & b_1 \\ a_2 & b_2 \\ a_3 & b_3 \end{bmatrix}$$

and (8-51) yields

$$A(\Sigma) = \iint_D k\, du\, dv = k\mathrm{A}(D)$$

where
$$k^2 = \begin{vmatrix} a_1 & b_1 \\ a_2 & b_2 \end{vmatrix}^2 + \begin{vmatrix} a_2 & b_2 \\ a_3 & b_3 \end{vmatrix}^2 + \begin{vmatrix} a_3 & b_3 \\ a_1 & b_1 \end{vmatrix}^2$$

Since Σ, in this case, is a linear transformation, this result is in agreement with that obtained earlier (8-9). We may also use this to justify (not prove!) the general formula. Suppose that the domain of Σ is a rectangle R. If we subdivide R by a net into small rectangles R_j, then in any one of these, the transformation Σ might be expected to behave very much like its differential

$$d\Sigma = \begin{bmatrix} \dfrac{\partial x}{\partial u} & \dfrac{\partial x}{\partial v} \\[2ex] \dfrac{\partial y}{\partial u} & \dfrac{\partial y}{\partial v} \\[2ex] \dfrac{\partial z}{\partial u} & \dfrac{\partial z}{\partial v} \end{bmatrix}$$

As a linear transformation from $\mathbf{R}^2$ into $\mathbf{R}^3$, this multiplies area by the factor k, where

$$\begin{aligned} k^2 &= \begin{vmatrix} \dfrac{\partial x}{\partial u} & \dfrac{\partial x}{\partial v} \\[2ex] \dfrac{\partial y}{\partial u} & \dfrac{\partial y}{\partial v} \end{vmatrix}^2 + \begin{vmatrix} \dfrac{\partial y}{\partial u} & \dfrac{\partial y}{\partial v} \\[2ex] \dfrac{\partial z}{\partial u} & \dfrac{\partial z}{\partial v} \end{vmatrix}^2 + \begin{vmatrix} \dfrac{\partial z}{\partial u} & \dfrac{\partial z}{\partial v} \\[2ex] \dfrac{\partial x}{\partial u} & \dfrac{\partial x}{\partial v} \end{vmatrix}^2 \\ &= \left|\frac{\partial(x, y)}{\partial(u, v)}\right|^2 + \left|\frac{\partial(y, z)}{\partial(u, v)}\right|^2 + \left|\frac{\partial(z, x)}{\partial(u, v)}\right|^2 \end{aligned}$$

Referring to Fig. 8-26, we may expect that the area of the surface Σ should be approximately $\Sigma k(p_j)A(R_j)$, where p_j is an appropriate point in R_j at which the Jacobians are evaluated. Since this has the form of a Riemann sum for the integral $\iint_D k$, we are led to the formula (8-51) as a reasonable definition for $A(\Sigma)$.

We remark that there are many ways to generalize this approach to surface area. The analog of the method used for arc length does not work here, since there is no simple inequality between the area of an inscribed polyhedron and the area of the smooth surface itself, as there was between the length of a curve and the length of an inscribed polygon. For more information about the

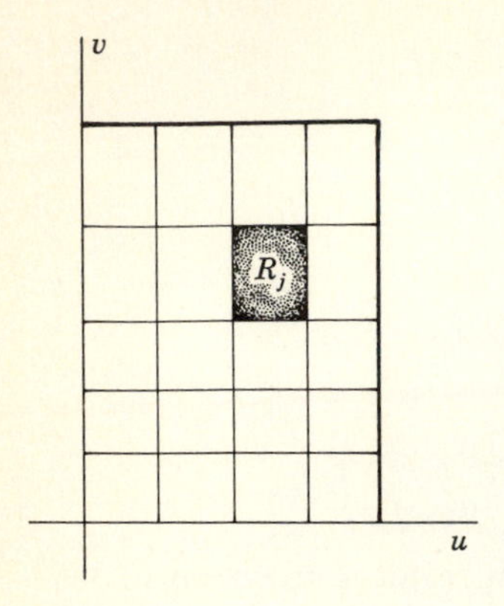

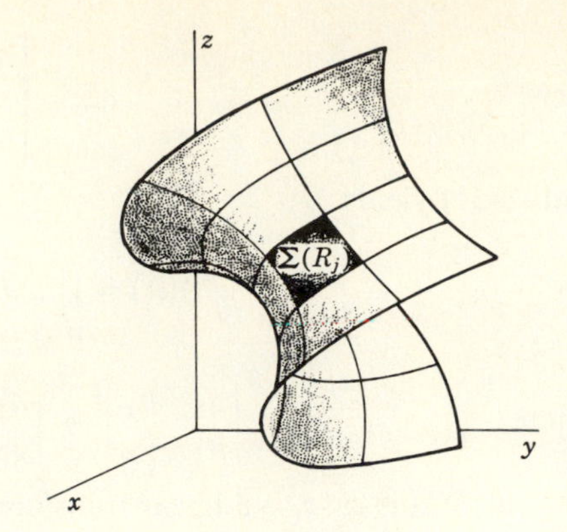

Figure 8-26

problem of an adequate general definition of surface area, we refer to the article by Radó given in the Reading List.

As an illustration of the use of formula (8-51), let us find the surface area of the torus (doughnut) in Fig. (8-27), whose equation is

$$\Sigma: \begin{cases} x = (R - \cos v) \cos u & -\pi \le u \le \pi \\ y = (R - \cos v) \sin u & -\pi \le v \le \pi \\ z = \sin v & R > 1 \end{cases} \tag{8-52}$$

We have
$$d\Sigma = \begin{bmatrix} -(R - \cos v) \sin u & \sin v \cos u \\ (R - \cos v) \cos u & \sin v \sin u \\ 0 & \cos v \end{bmatrix}$$

so that

$$\begin{aligned} |\mathbf{n}(u, v)|^2 &= |-(R - \cos v)(\sin^2 u \sin v + \cos^2 u \sin v)|^2 \\ &\quad + |(R - \cos v) \cos u \cos v|^2 + |(R - \cos v) \sin u \cos v|^2 \\ &= (R - \cos v)^2(\sin^2 v + \cos^2 u \cos^2 v + \sin^2 u \cos^2 v) \\ &= (R - \cos v)^2 \end{aligned}$$

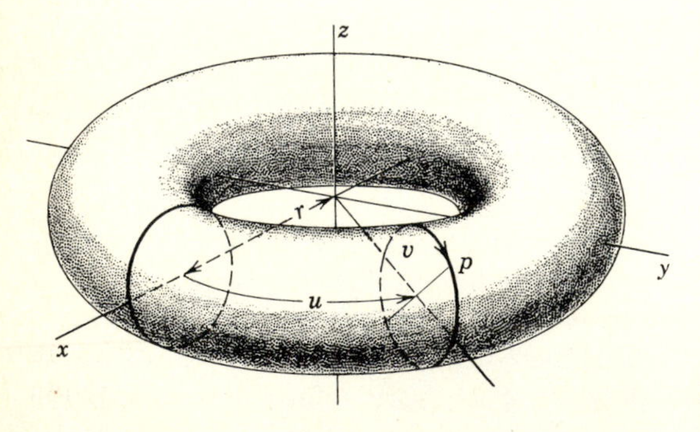

Figure 8-27

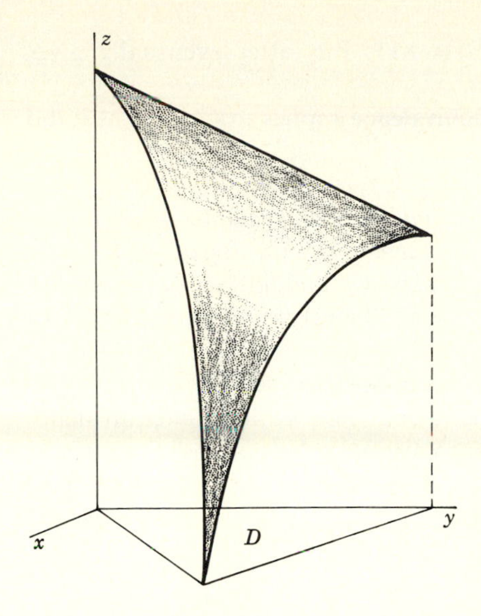

Figure 8-28

Accordingly,

$$A(\Sigma) = \int_{-\pi}^{\pi} du \int_{-\pi}^{\pi} (R - \cos v)\, dv = (2\pi)^2 R$$

Again, consider the surface described by the equation $z = 2 - x^2 - y$, with (x, y) restricted to lie in the triangle D bounded by the lines $x = 0$, $y = 1$, $y = x$ (see Fig. 8-28). Using x and y in place of u and v, we find that

$$\begin{aligned} |\mathbf{n}(x, y)|^2 &= \left|\frac{\partial(x, y)}{\partial(x, y)}\right|^2 + \left|\frac{\partial(y, z)}{\partial(x, y)}\right|^2 + \left|\frac{\partial(z, x)}{\partial(x, y)}\right|^2 \\ &= 1 + \left(\frac{\partial z}{\partial x}\right)^2 + \left(\frac{\partial z}{\partial y}\right)^2 \\ &= 1 + 4x^2 + 1 = 2 + 4x^2 \end{aligned}$$

so that the area of this surface is

$$\begin{aligned} A(\Sigma) &= \iint_D \sqrt{2 + 4x^2}\, dx\, dy \\ &= \int_0^1 dx \sqrt{2 + 4x^2} \int_x^1 dy \\ &= \int_0^1 (1 - x)\sqrt{2 + 4x^2}\, dx \end{aligned}$$

This either can be evaluated approximately or may be transformed into an elementary form by the substitution $u = 2^{-1/2} \tan \theta$. The former, with a division of [0, 1] into 10 subintervals, gives $A(\Sigma) = .811$; the latter eventually gives $A(\Sigma) = \frac{1}{2}\log(\sqrt{2}+\sqrt{3}) + \sqrt{2}/6 = .809$.

The important notion of **parametric equivalence** applies to surfaces as it did to curves, with appropriate modifications.

Definition 17 *Two smooth surfaces, Σ and Σ^*, are said to be smoothly equivalent if there is a transformation T mapping D^*, the domain of Σ^*, onto D, the domain of Σ, which is 1-to-1, is of class C', and has a strictly positive Jacobian in D^*, and such that $\Sigma^*(u, v) = \Sigma(T(u, v))$ for all $(u, v) \in D^*$.*

As with curves, it can be shown that if two smooth surfaces are continuously equivalent, then they are smoothly equivalent. The notion of equivalence separates the class of all smooth surfaces into distinct subclasses, all the surfaces in a subclass being mutually equivalent. All the surfaces in a subclass have the same trace and the same normal direction at any point (see Exercise 5). They also have the same area.

Theorem 14 *If Σ and Σ^* are smoothly equivalent surfaces, then $A(\Sigma) = A(\Sigma^*)$.*

We may assume that $\Sigma^*(u, v) = \Sigma(r, s) = \Sigma(T(u, v))$, where T maps the set of points (u, v) forming the domain D^* of Σ^* onto the set of points (r, s) forming the domain D of Σ. The area of Σ is given by

$$A(\Sigma) = \iint_D \left\{ \left| \frac{\partial(x, y)}{\partial(r, s)} \right|^2 + \left| \frac{\partial(y, z)}{\partial(r, s)} \right|^2 + \left| \frac{\partial(z, x)}{\partial(r, s)} \right|^2 \right\}^{1/2} dr\, ds$$

To this double integral, apply Theorem 6, and make the change of variable produced by the transformation T. D will be replaced by D^*, r and s will be replaced by their expressions in terms of u and v in the integrand, and $dr\, ds$ will be replaced by $|\partial(r, s)/\partial(u, v)|\, du\, dv$. This results in the formula

$$(8\text{-}53)\quad A(\Sigma) = \iint_{D^*} \left\{ \left| \frac{\partial(x, y)}{\partial(r, s)} \right|^2 + \left| \frac{\partial(y, z)}{\partial(r, s)} \right|^2 + \left| \frac{\partial(z, x)}{\partial(r, s)} \right|^2 \right\}^{1/2} \frac{\partial(r, s)}{\partial(u, v)}\, du\, dv$$

However, since $\Sigma^*(u, v) = \Sigma(T(u, v))$, the chain rule of differentiation (Theorem 11, Sec. 7.4) asserts that $d\Sigma^* = d\Sigma\, dT$, so that

$$\frac{\partial(x, y)}{\partial(u, v)} = \frac{\partial(x, y)}{\partial(r, s)} \frac{\partial(r, s)}{\partial(u, v)}$$

$$\frac{\partial(y, z)}{\partial(u, v)} = \frac{\partial(y, z)}{\partial(r, s)} \frac{\partial(r, s)}{\partial(u, v)}$$

$$\frac{\partial(z, x)}{\partial(u, v)} = \frac{\partial(z, x)}{\partial(r, s)} \frac{\partial(r, s)}{\partial(u, v)}$$

If we use these relations in (8-53), we obtain

$$A(\Sigma) = \iint_{D^*} \left\{ \left| \frac{\partial(x, y)}{\partial(u, v)} \right|^2 + \left| \frac{\partial(y, z)}{\partial(u, v)} \right|^2 + \left| \frac{\partial(z, x)}{\partial(u, v)} \right|^2 \right\}^{1/2} du\, dv$$

$$= A(\Sigma^*)$$

proving equality of the areas of the surfaces. ▮

From a more sophisticated point of view, it is now natural to alter the terminology so that the word "surface" no longer refers to an individual mapping Σ but rather to the entire equivalence class that contains Σ. Depending upon the notion of equivalence that is used, one speaks of Frechet surfaces, Lebesgue surfaces, etc. It is in the guise of equivalence classes, perhaps, that one is best able to speak of "the" surface defined by an equation $F(x, y, z) = 0$; this may be understood to be the equivalence class of smooth surfaces Σ which satisfy this equation.

So far, we have been chiefly concerned with the local theory of surfaces, that is, the behavior of a surface in the neighborhood of a point lying on it. Such considerations have to do with the existence and direction of a normal, with the curvature of the surface at the point, or with the relationship between the normal vector and curves which lie on the surface and pass through its base. Much of the modern theory of surfaces deals instead with properties "in the large," which require consideration of the surfaces as a whole. For example, all smooth surfaces might be said to be approximately locally planar; thus, the essential distinction between a sphere and a torus is a property "in the large." Similarly, the property of being orientable is one which cannot be settled by local considerations.

To discuss such aspects as these, it is necessary to deal with the more general notion of a two-dimensional **manifold.** This can be thought of as a generalization of our previous notion of surface, and is motivated by considering the union of a number of pieces of surfaces to form a larger object.

Definition 18 *A 2-manifold M is a set of mappings* $\Sigma_1, \Sigma_2, \ldots$ *such that* (i) *the domain of each is an open disk D in the* (u, v) *plane and* (ii) *each* Σ_j *is a continuous* 1-*to*-1 *mapping having a continuous inverse.*

If we denote the trace of Σ_j by S_j, then the **trace** of M is defined to be the union of the sets S_j. It is usual to require that the trace of M be a connected set of points; a manifold that is not connected is therefore the union of a number of connected manifolds. We may say that the manifold M is the union of the surface pieces (or surface elements) Σ_j. This may be visualized as shown in Fig. 8-29. It is not necessary that all the mappings Σ_j have the same domain. Any smooth surface which is simple is a manifold with just one surface element; any smooth surface Σ with multiple points is a manifold with a number of

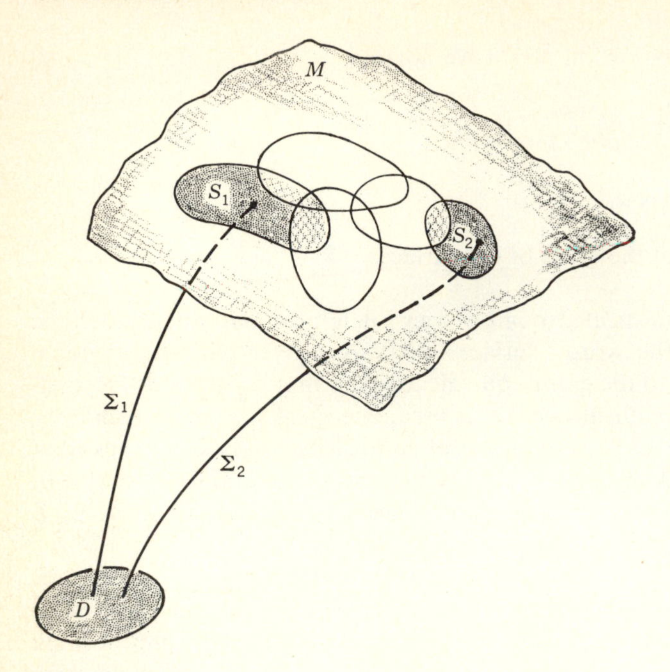

Figure 8-29

surface elements, for the mapping Σ is locally 1-to-1, and its domain can be covered with open disks D_j in each of which Σ is 1-to-1. Each of the mappings Σ_j can be thought of as carrying the coordinate system in D (or in D_j) up into the trace of M, where it creates a local coordinate system in the image set S_j. For this reason, the mappings Σ_j are sometimes called **charts,** and the entire collection an **atlas.**

In order to speak of differentiability properties of a manifold, it is convenient to impose restrictions on the way the surface elements fit together. Let Σ_i and Σ_j be any two pieces of M, and let $S = S_i \cap S_j$, the intersection of their traces (see Fig. 8-30). Since Σ_i maps D onto S_i, it maps a part of it, D_i, onto S. In the same way, Σ_j maps a part, D_j, also onto S, 1-to-1. (In Fig. 8-30 we have shown D_i and D_j disjoint, although this may not be the case in general.) Consider the transformation T_{ij} from D into D which is defined on D_j by

$$T_{ij}(p) = \Sigma_i^{-1}\Sigma_j(p)$$

This is a 1-to-1 mapping of D_j onto D_i. Such a mapping is defined for every pair of overlapping sets S_i and S_j. (Note that T_{ij} and T_{ji} are inverses, and that T_{ii} is the identity map of D onto itself.) The manifold M is said to be a **differentiable manifold** of class C' if each of the transformations T_{ij} is of class C' and has a nonvanishing Jacobian. More generally, M is of class C^k (or C^∞) if all the maps T_{ij} are of class C^k (or C^∞). In terms of the concept of local coordinate systems, these requirements are interpreted as meaning that the

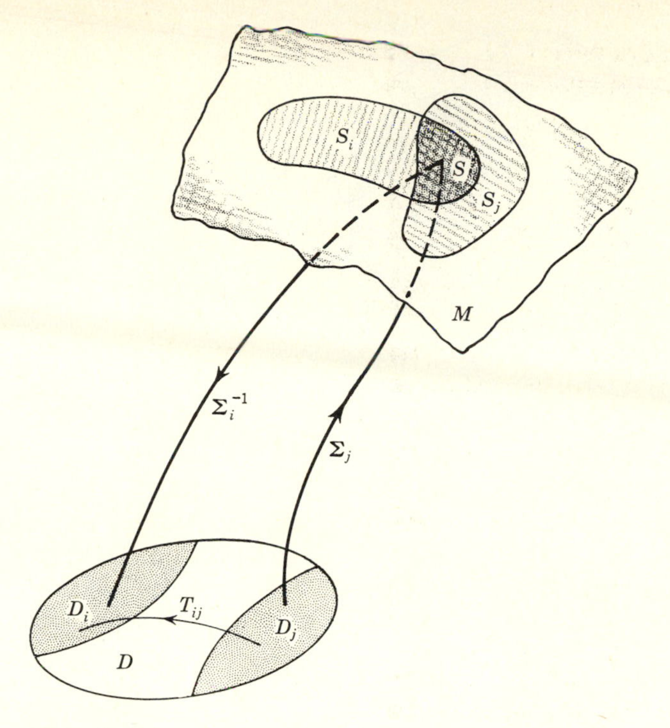

Figure 8-30

change-of-variable equations which relate the competing coordinates of a point which lies in two local coordinate patches must themselves be reversible and of class C^k (or C^∞).

We are now ready to say what is meant by an **orientable** manifold.

Definition 19 *A differentiable manifold M is orientable if each of the connection transformations T_{ij} has a positive Jacobian. Otherwise, M is said to be nonorientable.*

An orientable manifold can be given a consistent orientation by orienting D. Choose, once and for all, a positive "side" for D. This may be done by choosing which way a normal to D shall point. This in turn determines a direction of positive rotation for simple closed curves lying in D, counterclockwise about the normal. The mappings Σ_i carry these over to the individual surface elements. Since the Jacobian of the transformation T_{ij} is positive, T_{ij} is an orientation-preserving mapping of D_j onto D_i; this guarantees that the orientations produced by Σ_i and Σ_j are consistent in the overlapping set (see Fig. 8-31). Thus, we arrive at a consistent sense of "outward" normal on M, or equivalently, a consistent sense of positive rotation for neighborhoods of points on M. The Möbius strip is a familiar example of a nonorientable

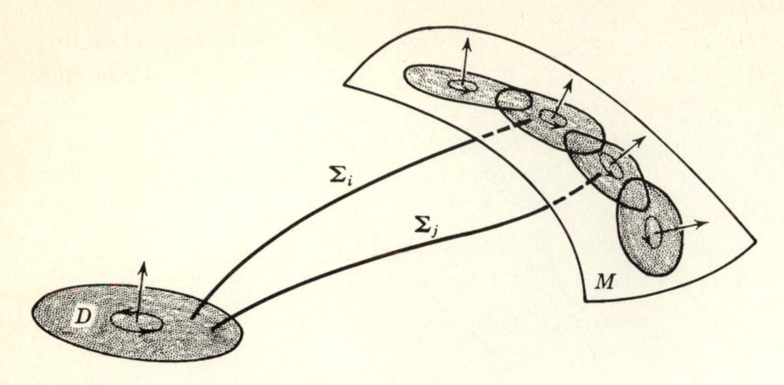

Figure 8-31

manifold (Fig. 8-32). If one starts with an "outward" normal at one point of the strip, and attempts to carry this around the strip, one finds that it will have reversed direction when the starting point is again reached. To verify that this manifold is nonorientable directly from the definition, we express M as the union of two surface elements, defined by the same mapping Σ but with differing domains. Let

$$\text{(8-54)} \qquad \Sigma: \begin{cases} x = \left(2 - v \sin\left(\dfrac{u}{2}\right)\right) \cos u \\ y = \left(2 - v \sin\left(\dfrac{u}{2}\right)\right) \sin u \\ z = v \cos\left(\dfrac{u}{2}\right) \end{cases}$$

and let Σ_1 be the surface obtained by restricting (u, v) to the rectangle $|v| < 1$, $0 < u < 2\pi$, and Σ_2 the surface obtained by restricting (u, v) to the rectangle $|v| < 1$, $|u| < \pi/2$. The traces of these overlap, and together cover

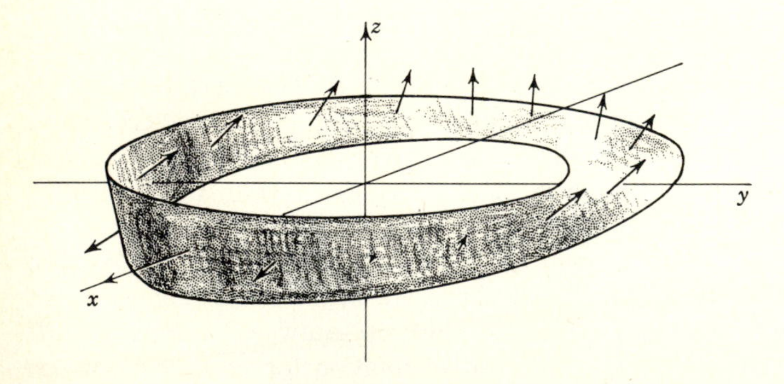

Figure 8-32 Möbius strip (nonorientable).

the twisted strip completely. The sets D_1 and D_2 are not connected; examining the equations for Σ and Fig. 8-30, we see that D_1 consists of the two rectangles

$$D_1' = \left\{\text{all } (u, v) \text{ with } |v| < 1 \text{ and } 0 < u < \frac{\pi}{2}\right\}$$

$$D_1'' = \{\text{all } (u, v) \text{ with } |v| < 1 \text{ and } \tfrac{3}{2}\pi < u < 2\pi\}$$

The transformation T_{21} is therefore given by

$$T_{21}(u, v) = \begin{cases} (u, v) \text{ in } D_1' \\ (u - 2\pi, -v) \text{ in } D_1'' \end{cases}$$

and has negative Jacobian in D_1''.

Another property "in the large" is the classification of manifolds as open and closed; this has little to do with the previous use of these words. A manifold is **closed** if it is compact, that is, if it has the property that any sequence of points lying on it has a subsequence which converges to a point lying on it. A manifold that is not closed is called **open.** The surface of a sphere defines a closed manifold; if we punch out a closed disk, the resulting manifold is open, since a sequence of points which converges to a point on the deleted rim of the disk will have no subsequence converging to a point on the manifold. A closed manifold has no edge or boundary; to define a notion of boundary for an open manifold, it is convenient to modify the construction by contracting the open domains of the surface elements Σ_{ij} so that we have overlapping of the sets S_j only along their edges. If this is done, then the manifold can be regarded as obtained by gluing these surface elements together along specified edges. When there are only a finite number of such pieces, the edge of the manifold is defined to be the curve or set of curves which is made up of the unmatched edges. When a manifold is orientable, the surface elements can be oriented so that each matched edge occurs once in one direction and once in the other; these may be said to cancel out, and the remaining unmatched curves constitute the oriented boundary of the manifold (see Fig. 8-33). (A detailed treatment of these subjects must be left to a course in topology or modern differential geometry.)

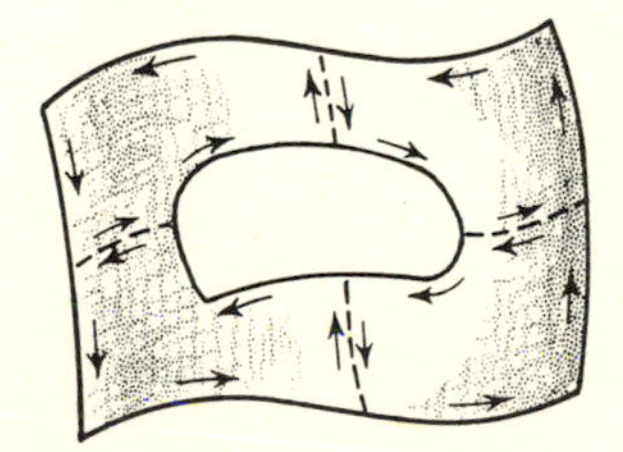

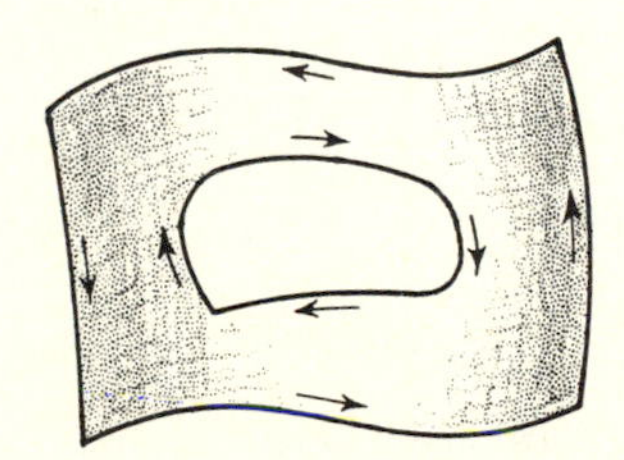

Figure 8-33 Manifold with two boundary curves formed from simple elements.

EXERCISES

1 Plot the trace of the surface $x = u^2 - v^2$, $y = u + v$, $z = u^2 + 4v$, $|u| \le 1$, $0 \le v \le 1$, and find the tangent plane at $(-\frac{1}{4}, \frac{1}{2}, 2)$.

2 Show that all the normal lines to the sphere (8-44) pass through the origin.

3 Let Σ be the surface described by $z = f(x, y)$, $(x, y) \in D$, with $f \in C'$. Show that the normal to Σ at p_0 is $(-f_1(p_0), -f_2(p_0), 1)$, and that the area of Σ is

$$A(\Sigma) = \iint_D \sqrt{1 + f_1^2 + f_2^2}$$

4 The curve γ given by $x = t$, $y = t^2$, $z = 2t^3$, $-\infty < t < \infty$, has a point in common with the surface Σ given by $z = x^2 + 3y^2 - 2xy$. What is the angle between γ and the normal to Σ at this point?

5 Show that two surfaces which are smoothly equivalent have the same normal directions at corresponding points.

6 Find the surface area of the sphere (8-44).

7 Set up a definite integral for the surface area of the ellipsoid

$$\left(\frac{x}{a}\right)^2 + \left(\frac{y}{b}\right)^2 + \left(\frac{z}{c}\right)^2 = 1$$

8 Set up an integral for the area of the Möbius strip as described by (8-54).

9 Show that the normal to a surface Σ is always orthogonal to Σ_u and Σ_v.

10 Find the surface area of the portion of the paraboloid $z = x^2 + y^2$ which is cut out by the region between the cylinder $x^2 + y^2 = 2$ and the cylinder $x^2 + y^2 = 6$.

11 Show that an alternative expression for the area of a surface Σ with domain D is

$$A(\Sigma) = \iint_D \{|\Sigma_u|^2 |\Sigma_v|^2 - (\Sigma_u \cdot E_v)^2\}^{1/2}\, du\, dv$$

[In the study of the differential geometry of surfaces the notation $E = |\Sigma_u|^2$, $F = \Sigma_u \cdot \Sigma_v$, and $G = |\Sigma_v|^2$ is customary.]

12 Let

$$E = \left(\frac{\partial x}{\partial u}\right)^2 + \left(\frac{\partial y}{\partial u}\right)^2 + \left(\frac{\partial z}{\partial u}\right)^2$$

$$F = \frac{\partial x}{\partial u}\frac{\partial x}{\partial v} + \frac{\partial y}{\partial u}\frac{\partial y}{\partial v} + \frac{\partial z}{\partial u}\frac{\partial z}{\partial v}$$

$$G = \left(\frac{\partial x}{\partial v}\right)^2 + \left(\frac{\partial y}{\partial v}\right)^2 + \left(\frac{\partial z}{\partial v}\right)^2$$

Show that the inequality (7-55) can also be expressed as $EG - F^2 > 0$.

13 A surface Σ is described by

$$\begin{cases} x = 2Au \cos v & 0 \le u \le L \\ y = 2Bu \sin v & 0 \le v \le \pi/2 \\ z = u^2(A \cos^2 v + B \sin^2 v) \end{cases}$$

Describe the trace of Σ and find its area.

14 Find the area of the portion of the upper hemisphere of the sphere with center (0, 0, 0) and radius R that obeys $x^2 + y^2 - Ry \le 0$.

15 Let Σ be a C'' surface defined on an open-connected set D in the UV plane. Suppose $d^2\Sigma = 0$ in D. Prove that Σ is a plane.

16 If two smooth surfaces Σ_1 and Σ_2 of class C' are parametrically equivalent, prove that the continuous functions that relate them are of class C'.

17 Let f be of class C'' on the plane, $f(0, 0) = 1$ and $f_1(0, 0) = f_2(0, 0) = 0$. The graph of f is a surface defined by

$$\Sigma \begin{cases} x = u \\ y = v \\ z = f(u, v) \end{cases}$$

Choose any θ, $0 \leq \theta < \pi$. Let $\gamma_\theta(t) = (t \cos \theta, t \sin \theta)$ be a line in the parameter space, and let $\Gamma_\theta(t) = \Sigma(\gamma_\theta(t))$ be the corresponding curve on Σ.

(*a*) Find the curvature k_θ of Γ_θ at the point $(0, 0, 1)$.

(*b*) What is the connection between k_θ and d^2f?

*(*c*) Show that the largest value of k_θ, as θ varies, is the largest eigenvalue of d^2f.

18 A two-dimensional surface in 4-space is a mapping from $\mathbf{R}^2$ into $\mathbf{R}^4$. If

$$\Sigma(u, v) = (x, y, z, w)$$

with domain D, then the area of Σ is defined by

$$A(\Sigma) = \iint_D \{|\Sigma_u|^2 |\Sigma_v|^2 - (\Sigma_u \cdot \Sigma_v)^2\}^{1/2}\, du\, dv$$

Find the area of the surface whose equation is $x = 2uv$, $y = u^2 - v^2$, $z = u + v$, $w = u - v$, for $u^2 + v^2 \leq 1$.

***19** A three-dimensional surface in 4-space is a transformation χ from a domain D in $\mathbf{R}^3$ into $\mathbf{R}^4$. The three-dimensional volume (analogous to surface area) of χ is defined to be $V(\chi) = \iiint_D \sqrt{K}$, where K is the sum of the squares of determinants of the 3-by-3 submatrices of $d\chi$. Thus, if χ is given by $\chi(\theta, \phi, \psi) = (x, y, z, w)$, then

$$K = \left|\frac{\partial(x, y, z)}{\partial(\theta, \phi, \psi)}\right|^2 + \left|\frac{\partial(y, z, w)}{\partial(\theta, \phi, \psi)}\right|^2 + \left|\frac{\partial(z, w, x)}{\partial(\theta, \phi, \psi)}\right|^2 + \left|\frac{\partial(w, x, y)}{\partial(\theta, \phi, \psi)}\right|^2$$

The boundary of the four-dimensional ball of radius R is a three-dimensional manifold called the 3-sphere, given either by the equation $x^2 + y^2 + z^2 + w^2 = R^2$ or parametrically by $x = R \cos \psi \sin \phi \cos \theta$, $y = R \cos \psi \sin \phi \sin \theta$, $z = R \cos \psi \cos \phi$, $w = R \sin \psi$, for $-(\pi/2) \leq \psi \leq \pi/2$, $0 \leq \phi \leq \pi$, $0 \leq \theta \leq 2\pi$. Show that the "area" of the 3-sphere is $2\pi^2 R^3$.

20 Check Theorem 11 for a surface Σ given by the equation $F(x, y, z) = 0$, by showing directly that if p_0 is a point on its graph [so that $F(p_0) = 0$], then the vector $dF|_{p_0} = (F_1(p_0), F_2(p_0), F_3(p_0))$ is orthogonal to the tangent at p_0 of any smooth curve lying on Σ and passing through p_0.

8.6 INTEGRALS OVER CURVES AND SURFACES

In Chap. 4, we discussed integration in some detail, based on a simple approach in which the fundamental idea is the integral of a function defined on an interval, a rectangle, or an n-dimensional cell. In this section, we deal with a direct generalization, the integral of a function along a curve or over a surface, done in such a way that the integral of the function $f \equiv 1$ yields the length of the curve or the area of the surface, respectively. This is also related to the discussion of set functions in Sec. 8.2, since many of the usual applications of this deal with the calculation of physical quantities such as mass, gravitational attraction, or moment of inertia, in which the function to be integrated plays the role of a density.

Let γ be a smooth curve, given as a C' mapping from $[a, b]$ into 3-space, and let f be a real-valued function that is defined and continuous on the trace of γ. Since the trace is the image of $[a, b]$ under a continuous mapping, it is compact, and we may therefore assume that f is in fact continuous everywhere in $\mathbf{R}^3$. We then define the integral of f along γ by

$$\int_\gamma f = \int_a^b f(\gamma(t))|\gamma'(t)|\,dt \tag{8-55}$$

Note that the integration is done back in the parameter space, although the values of f are determined on the points p that lie on γ. Note that if $f(p) = 1$ for all p, then (8-55) becomes $\int_\gamma 1 = L(\gamma)$, the length of γ. If γ and γ^* are equivalent smooth curves, then

$$\int_\gamma f = \int_{\gamma^*} f$$

(see Exercise 5). If we choose γ^* as the curve which is equivalent to γ and which has arc length as parameter, then $|\gamma^{*\prime}(s)| = 1$ for all $s \in [0, l]$, where l is the length of γ; by (8-55) we then have

$$\int_\gamma f = \int_0^l f(\gamma^*(s))\,ds \tag{8-56}$$

Since $\gamma^*(s)$ is also a point on the trace of γ, this is frequently written

$$\int_\gamma f = \int_\gamma f(x, y, z)\,ds \tag{8-57}$$

and for this reason, an integral along a curve γ is sometimes referred to as the result of "integrating with respect to arc length."

With arc length as the parameter, the integral of a function along γ can be given a different interpretation, much closer in format to that of the ordinary one-dimensional Riemann integral. If the length of γ is l, we choose points s_j with $0 = s_0 < s_1 < s_2 < \cdots < s_m = l$, and think of these as partitioning the curve γ into short sections $\gamma_1, \gamma_2, \ldots, \gamma_m$, where γ_j is the mapping obtained by restricting γ to the subinterval $[s_{j-1}, s_j]$. Choose a point p_j on γ_j, and form the sum $\sum f(p_j)L(\gamma_j)$. It is then easily seen that this will converge to $\int_\gamma f$ as the norm of the partition of $[0, l]$ approaches 0, in the same way as was the case for the usual integral of a continuous function on a real interval. For, in fact,

$$\sum_1^m f(p_j)L(\gamma_j) = \sum_1^m f(p_j) \int_{s_{j-1}}^{s_j} |\gamma'(s)|\, ds$$

$$= \sum_1^m f(p_j)(s_j - s_{j-1})$$

which converges to (8-57).

Suppose now that we interpret the function f as giving the linear density (mass per centimeter) of a wire, which we can suppose varies from point to point. Then, $f(p_j)L(\gamma_j)$ is an estimate for the mass of the section γ_j, and (8-57) becomes the definition of mass of the entire segment of wire γ.

As an illustration, let us find the center of gravity of a uniform wire which is bent into the shape of a semicircle. To describe this shape, we shall use the curve γ whose equation is $x = \cos t$, $y = \sin t$, $0 \le t \le \pi$. By definition, the center of gravity will be the point $(\bar{x}, \bar{y})$ where

$$M\bar{x} = \int_\gamma x\rho\, ds \qquad M\bar{y} = \int_\gamma y\rho\, ds$$

and
$$M = \int_\gamma \rho\, ds$$

The function ρ specifies the density of the wire (mass per unit length), M is the total mass of the wire, and the first two integrals are the moments of the wire about the lines $x = 0$ and $y = 0$, respectively. Since the wire is uniform, ρ is constant, and $M = \rho L(\gamma) = \rho\pi$. From the equation for γ,

$$|\gamma'(t)| = |(-\sin t)^2 + (\cos t)^2|^{1/2} = 1$$

and γ has arc length for its parameter. Substituting for x and y, and using (8-56),

$$\int_\gamma \rho x\, ds = \rho \int_0^\pi \cos t\, dt = 0 = M\bar{x}$$

$$\int_\gamma \rho y\, ds = \rho \int_0^\pi \sin t\, dt = 2\rho = M\bar{y}$$

so that the center of gravity is $(0, 2/\pi)$. Other examples involving integrals along curves will be found in the Exercises.

Turning to integrals over surfaces, let Σ be a smooth surface with domain D, and let f be a continuous function defined on the trace of Σ. A subdivision of D into sets D_{ij} produces a subdivision of Σ into small surface elements Σ_{ij}. Let p_{ij} be a point lying on Σ_{ij} and form the sum $\Sigma f(p_{ij})A(\Sigma_{ij})$. If these have a limit as the norm of the subdivision of D tends to 0, then this limit is called the value of the integral of f over the surface Σ, and is written as $\iint_\Sigma f$. By an argument which is analogous to that used for integrals along curves, we arrive at the equivalent expression

$$\iint_\Sigma f = \iint_D f(\Sigma(u, v))|\mathbf{n}(u, v)|\, du\, dv \tag{8-58}$$

where
$$|\mathbf{n}(u, v)| = \left\{\frac{\partial(x, y)^2}{\partial(u, v)} + \frac{\partial(y, z)^2}{\partial(u, v)} + \frac{\partial(z, x)^2}{\partial(u, v)}\right\}^{1/2}$$

Thus, evaluation of the integral of a function over a surface is reduced to the evaluation of an ordinary double integral. If f is constantly 1, then $\iint_\Sigma f = A(\Sigma)$, the area of Σ. For this reason, the general integral $\iint_\Sigma f$ is often written $\iint_\Sigma f(x, y, z)\, dA$, and one speaks of integrating f with respect to the element of surface area, over the surface Σ.

To illustrate this, let us find the center of gravity of a thin uniform sheet of metal which is in the shape of the paraboloid $z = x^2 + y^2$, with $x^2 + y^2 \le 1$. The center of gravity is the point $(0, 0, \bar{z})$, where

$$M\bar{z} = \iint_\Sigma \rho z\, dA$$

and $M = \iint_\Sigma \rho\, dA$. Since the sheet of metal is uniform, ρ is constant, and $M = \rho A(\Sigma)$. To compute this, we need the value of $|\mathbf{n}|$ on Σ. Using x and y as the parameters for Σ rather than introducing u and v, and Exercise 3, Sec. 8.5, we have $|\mathbf{n}(x, y)| = [1 + 4x^2 + 4y^2]^{1/2}$, so that

$$A(\Sigma) = \iint_D \sqrt{1 + 4(x^2 + y^2)}\, dx\, dy$$

where D is the unit disk, $x^2 + y^2 \le 1$. The form of the integrand and of D suggest that we transform to polar coordinates. Accordingly, the integrand becomes $(1 + 4r^2)^{1/2}$, $dx\, dy$ is replaced by $r\, dr\, d\theta$, and D is replaced by the rectangle $0 \le \theta \le 2\pi$, $0 \le r \le 1$. Thus,

$$A(\Sigma) = \int_0^{2\pi} d\theta \int_0^1 \sqrt{1 + 4r^2}\, r\, dr$$

$$= \frac{\pi}{6}(5\sqrt{5} - 1)$$

In the same fashion, we have

$$\iint_{\Sigma} z\, dA = \iint_{D} (x^2 + y^2)\sqrt{1 + 4(x^2 + y^2)}\, dx\, dy$$

$$= \int_0^{2\pi} d\theta \int_0^1 r^2\sqrt{1 + 4r^2}\, r\, dr$$

$$= \frac{\pi}{12}\left(5\sqrt{5} + \tfrac{1}{5}\right)$$

Hence
$$\bar{z} = \left(\frac{1}{2}\right)\frac{5\sqrt{5} + \frac{1}{5}}{5\sqrt{5} - 1} = .56$$

Thus, in all cases, the integral of a function f "along the curve γ," or "over the surface Σ" reduces to ordinary integration of the related function $f(\gamma(t))$ or $f(\Sigma(u, v))$ over the appropriate parameter domain. While we have illustrated only the case in which f is a real-valued function, the same thing is done if f is complex-valued, or vector-valued, or even matrix-valued, since in each case, one merely integrates component by component. Thus, if $F(p) = (f(p), g(p), h(p))$, then the integral of F along γ is

$$\int_{\gamma} F\, ds = \left(\int_{\gamma} f\, ds, \int_{\gamma} g\, ds, \int_{\gamma} h\, ds\right)$$

As we shall see in the next chapter, many physical problems lead to the consideration of a special class of integrals of functions along curves and over surfaces.

EXERCISES

1 Find the center of gravity of a homogeneous wire which has the shape of the curve $y = (e^x + e^{-x})/2$, $-1 \le x \le 1$.

2 The moment of inertia of a particle of mass m about an axis of rotation whose distance is l is defined to be ml^2. Show that a reasonable definition for the moment of inertia of a wire in the shape of the plane curve γ about the Y axis is

$$I = \int_{\gamma} \rho x^2\, ds$$

3 A wire has the shape of the curve $y = x^2$, $-1 \le x \le 1$. The density of the wire at (x, y) is $k\sqrt{y}$. What is the moment of inertia of the wire about the Y axis?

4 Formulate a definition for the moment of inertia of a wire in space about the Z axis.

5 Let γ_1 and γ_2 be smoothly equivalent curves, and let f be continuous on their trace. Show by (8-55) that $\int_{\gamma_1} f = \int_{\gamma_2} f$.

6 The force of attraction between two particles acts along the line joining them, and its magnitude is given by $km_1m_2r^{-2}$, where m_1 and m_2 are their masses, and r is the distance between them. Find the attraction on a unit mass located at the origin which is due to a homogeneous straight wire of infinite length, whose distance from the origin is l.

***7** Let Σ_1 and Σ_2 be smoothly equivalent surfaces, and let f be continuous on their trace. Show by (8-58) that

$$\iint_{\Sigma_1} f = \iint_{\Sigma_2} f$$

8 Find the moment of inertia of a homogeneous spherical shell about a diameter.

9 A sheet of metal has the shape of the surface $z = x^2 + y^2$, $0 \le x^2 + y^2 \le 2$. The density at (x, y, z) is kz. Find the moment of inertia about the Z axis.

10 What is the force of attraction upon a unit mass located at $(0, 0, l)$ which is due to a homogeneous circular disk of radius R, center $(0, 0, 0)$, and lying in the XY plane? What happens if R is allowed to become infinite?

***11** Show that the force of attraction within a spherical shell of constant density is everywhere 0.

CHAPTER

NINE

DIFFERENTIAL GEOMETRY AND VECTOR CALCULUS

9.1 PREVIEW

The topics in this chapter are a strange mixture of applications and sophisticated theory. The study of the vector calculus (Stokes' theorem, Green's theorem, Gauss' theorem) is sometimes justified by its role in the solution of Laplace's equation, Poisson's equation, the inhomogeneous wave equation, and the study of electromagnetic radiation and Maxwell's equations, as described in Sec. 9.6. On the other hand, the theory of differential forms, sketched in Secs. 9.2, 9.4, and 9.5, gives a view of the surprising interplay between analysis and geometry that has marked many of the most significant mathematical advances of the current century.

In Sec. 9.2, we introduce 1-forms and 2-forms as functionals that assign numerical values to curves and surfaces, and present some of the accompanying algebraic formalism. In Sec. 9.3, we relate this to the more traditional algebra of vectors in 3-space, showing how vector-valued functions give rise to the study of line and surface integrals and differential forms. We state the generalized Stokes' theorem in Sec. 9.4, and then restate it (and prove it) in the form appropriate for the plane and 3-space. We note that Theorem 8 and the related Lemma 1 are an important part of this treatment. Finally, in Sec. 9.5 we discuss exact forms and locally exact forms (or closed forms) and the geometric conditions under which they are equivalent. This leads us back to simple connectedness and its higher-dimensional analogs. Theorems 3 and 15 provide the framework for this, along with Lemma 1; unfortunately, we have had to leave the general theory at this incomplete stage, referring to more advanced texts on differential topology for the rest of the story.

9.2 DIFFERENTIAL FORMS

The integral of a function along a curve, or over a surface, is an example of a function whose domain of definition is not a set of points. For example, $\int_\gamma f$ is a number that is determined by the pair (γ, f) consisting of the smooth curve γ and the continuous function f, and defined by (8-55). Similarly, the value of $\iint_\Sigma f$ is determined by the pair (Σ, f), and can therefore be regarded as a function of the pair; $\phi(\Sigma, f) = \iint_\Sigma f$. If Σ is held fixed, then the value is determined by the choice of f, and one might write $\sigma(f) = \iint_\Sigma f$ to indicate that we are dealing with a numerical-valued function σ whose domain is the class of all continuous functions f defined on the trace of Σ. Again, if f is defined in an open region Ω of space, and we regard Σ as the variable, we write $F(\Sigma) = \iint_\Sigma f$, thus obtaining a function whose domain is the class of smooth surfaces Σ having their trace in the region Ω.

The name "functional" is sometimes used for functions such as F and σ to distinguish them from the more usual type of function. We shall say that F is a **curve functional** if its domain is a class of smooth curves with trace in a specific open set; F will then assign the numerical value $F(\gamma)$ to any such curve γ. Likewise, F is a **surface functional** on the open set Ω if F assigns the numerical value $F(\Sigma)$ to any smooth surface Σ whose trace lies in Ω. For emphasis, we will sometimes use "point function" for an ordinary numerical-valued function whose domain is the set of points Ω itself. Finally, for consistency, we introduce the term **region functional** for functions F whose domain is the class of compact sets $K \subset \Omega$. Note that these are precisely the set functions discussed in Sec. 8.2, of which the following is a typical example.

$$F(K) = \iiint_K g$$

where g is a point function defined in Ω.

We can also generalize this to n space. We would have functionals of each dimension $k = 0, 1, 2, \ldots, n$. If $k \neq 0$, $k \neq n$, then a k-dimensional functional F on an open set Ω would assign a numerical value $F(\Sigma)$ to any smooth k-dimensional surface Σ whose trace lies in Ω. (Recall that Σ itself is a C' mapping from a rectangle in $\mathbf{R}^k$ into n space.) When $k = 0$, we identify

F with an ordinary point function defined on Ω, and when $k = n$, we identify F with the general set function defined on compact subsets of Ω. (In order to preserve consistency, one can introduce zero-dimensional surfaces in n space as maps which send the origin 0 into a point p in n space, and n-dimensional surfaces as maps which send a region in n space into itself, point by point.)

In the theory of integration of functions of one variable, one writes $\int_a^b f$ rather than $\int_{[a, b]} f$ in order to effect the orientation of the integral. We speak of integrating "from a to b" rather than integrating "over the interval $[a, b]$." Moreover, we write $\int_b^a f = -\int_a^b f$. Similarly, we can introduce orientation into the integral along a curve. Let us use $-\gamma$ for the curve which is obtained by reversing the direction of γ, so that if $C = \int_\gamma f$, then $-C = \int_{-\gamma} f$.

Likewise, we can introduce orientation into double integrals, or more generally, into integration over surfaces. If D is a region in the (u, v) plane, then D can be assigned one of two orientations, corresponding intuitively to the two "sides" of D. Stated another way, a normal vector can point forward, toward the viewer, or backward, away from the viewer. This, in turn, leads to an accepted sense of "positive" rotation. It is conventional to assign the term "positive" to a counterclockwise rotation, and "negative" to a clockwise rotation. Once a direction of positive rotation has been assigned to D, it, in turn, induces an orientation in the boundary curves of D, as indicated in Fig. 9-1; in the standard description, "trace each boundary curve in a direction which keeps the left hand in D." (Note that turning D over has the effect of reversing the orientation.) When D has been oriented, $-D$ will indicate the region endowed with the opposite orientation. When D is positively oriented, $\iint_D f$ has its usual meaning. When the orientation of D is reversed, it is natural to require that the integral change sign; thus, $\iint_{-D} f = -\iint_D f$.

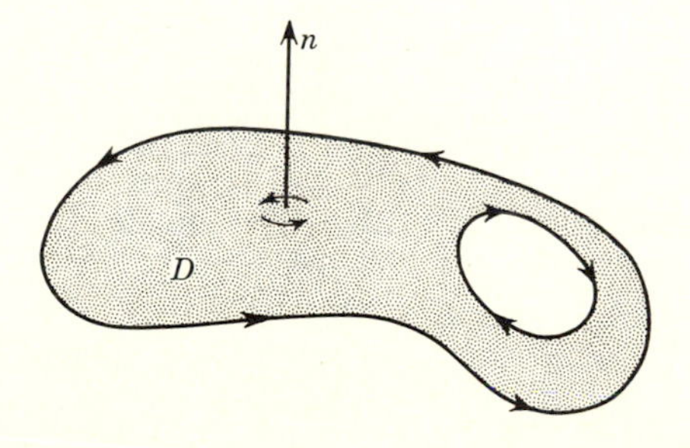

Figure 9-1 Orientation.

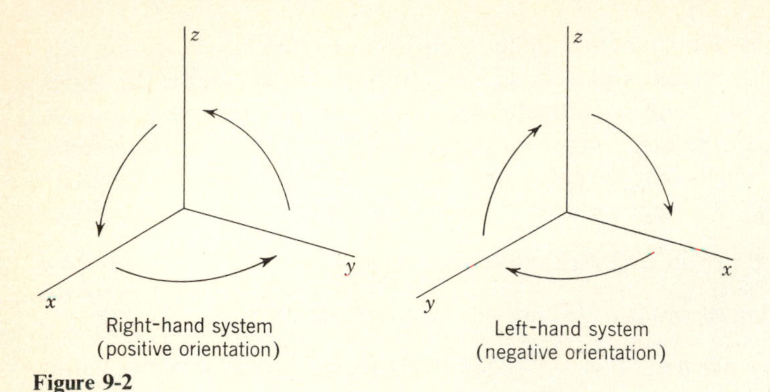

Figure 9-2

In 3-space, a region Ω may be assigned one of two orientations. The most familiar instance of this is in the choice of labels for the axes of a coordinate trihedral. There are only two essentially different systems. They are called "right-handed" or "left-handed"; as indicated in Fig. 9-2, they may be distinguished by viewing the origin from the first octant. If the trihedral is rotated so that the X axis moves to the position occupied by the Y, the Y to the Z, and the Z to the X, then the direction of rotation will be counterclockwise for a right-hand system, and clockwise for a left-hand system. We choose, by convention, to call the right-hand system **positive.** Any other labeling of the axes may be brought into one of the two that are shown, by a rotation. This has the effect of permuting the labels cyclically; thus the positive systems are XYZ, YZX, and ZXY, and the negative systems are YXZ, XZY, and ZYX. Just as the choice of orientation for a plane region whose boundary is a simple closed curve γ induces an orientation in γ, and vice versa, so orientation of a three-dimensional region whose boundary is a simple closed surface S induces an orientation in S, and conversely. If Ω is a solid ball and S is the sphere which forms its boundary, then, viewing S from outside, positive orientation of Ω gives counterclockwise orientation of S, and negative orientation of Ω gives clockwise orientation of S (see Fig. 9-3). Moreover, if the direction of the normals to S is chosen to correspond to the local orientation of S, then it will be seen that positive orientation of Ω goes with the fact that the normals to S always point out of the region, while negative orientation of Ω makes the normals point inward.

When a region Ω has positive orientation, we define the value of the oriented integral of a function f over Ω to be the same as before; when Ω has negative orientation, we define the value to be $-\iiint_\Omega f$. If $-\Omega$ indicates Ω with the opposite orientation, then we have

$$\iiint_{-\Omega} f = -\iiint_{\Omega} f$$

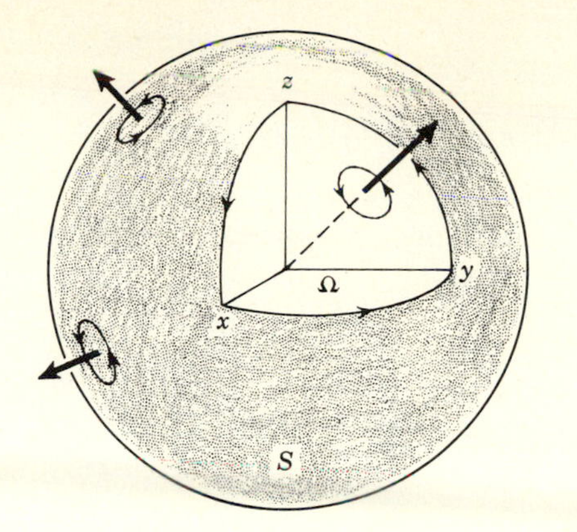

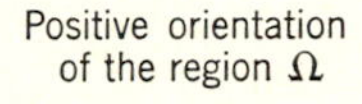
Positive orientation of the region Ω

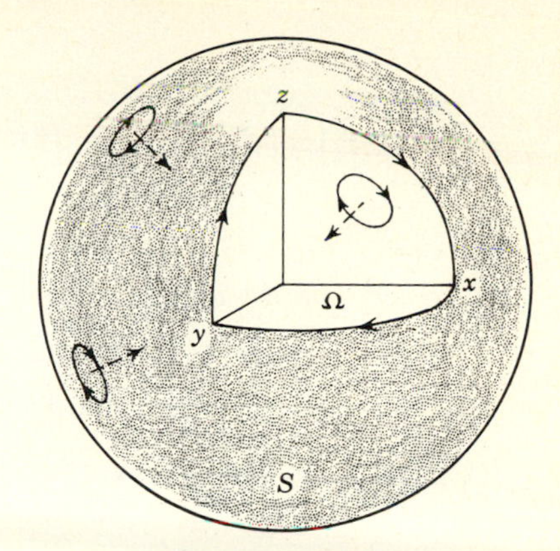

Negative orientation of the region Ω

Figure 9-3

In the integral of a function of one variable, the notation $\int_a^b f$ conveys the orientation of the interval of integration by means of the position of the limits, a and b. In discussing oriented multiple integrals, this device is no longer possible. However, another device is available. In the expression $\iint_D f(x, y)\, dx\, dy$, the symbol "$dx\, dy$" has been used so far only to convey the names of the variables of integration. Let us use it now to convey the orientation of D as well, writing "$dx\, dy$" when D has the same orientation as that of the XY plane, and "$dy\, dx$" when it has the opposite orientation. Accordingly, the equation governing reversal of orientation in a double integral becomes

$$\iint_D f(x, y)\, dy\, dx = -\iint_D f(x, y)\, dx\, dy \tag{9-1}$$

This relation is embodied in the symbolic equation

$$dy\, dx = -dx\, dy \tag{9-2}$$

A similar device may be used in triple integrals. The X, Y, and Z axes are positively oriented when they occur in one of the orders XYZ, YZX, ZXY, and negatively oriented in the orders XZY, ZYX, YXZ. Thus, we may convey the orientation of a region over which a triple integration is to be performed by writing "$dx\, dy\, dz$" (or "$dy\, dz\, dx$" or "$dz\, dx\, dy$") when the region has the same orientation as XYZ space, and "$dx\, dz\, dy$" (or "$dz\, dx\, dy$" or "$dy\, dx\, dz$") when it has the opposite orientation. For example, we would have

$$\iiint_\Omega f(x, y, z)\, dy\, dx\, dz = -\iiint_\Omega f(x, y, z)\, dx\, dy\, dz$$

We observe that this relation is also consistent with the symbolic equation (9-2). Since $dy\, dx = -dx\, dy$, $dy\, dx\, dz = -dx\, dy\, dz$. Again,

$$\begin{aligned} dy\, dz\, dx &= dy(dz\, dx) = dy(-dx\, dz) = -dy\, dx\, dz \\ &= -(-dx\, dy)\, dz = dx\, dy\, dz \end{aligned}$$

It is not possible to put all curve-functionals into the form

$$F(\gamma) = \int_\gamma f$$

where the function f is independent of γ. As we shall see, an example is furnished by the class of functionals that are known as line integrals. These form a special division of the more general category known as differential forms. In n space, there are $n + 1$ classes of differential forms, called in turn **0-forms, 1-forms, ..., n forms.** In 3-space we shall deal with 0-forms, 1-forms, 2-forms, and 3-forms, while in the plane, we shall meet only 0-forms, 1-forms, and 2-forms. For simplicity, we begin with the theory of differential forms in the plane ($n = 2$), starting with the class of 1-forms. These are also called **line integrals.**

Definition 1 *The general continuous* 1-*form in the* XY *plane is a curve-functional* ω *denoted by*

$$\omega = A(x, y)\, dx + B(x, y)\, dy$$

where A *and* B *are continuous functions defined in a region* Ω. *If* γ *is a smooth curve, with equation* $x = \phi(t)$, $y = \psi(t)$, $a \le t \le b$, *whose trace lies in* Ω, *then the value which* ω *assigns to* γ *is defined by the formula*

$$\omega(\gamma) = \int_a^b [A(\gamma(t))\phi'(t) + B(\gamma(t))\psi'(t)]\, dt \tag{9-3}$$

The rule for computing $\omega(\gamma)$ can be stated thus: Given the differential form ω and a curve γ, we "evaluate ω on γ" by substituting into ω the expressions for x, y, dx, and dy which are obtained from the equations of γ, following the familiar custom of replacing dx by

$$\frac{dx}{dt}\, dt = \phi'(t)\, dt$$

and dy by $\psi'(t)\, dt$; this results in an expression of the form $g(t)\, dt$, which is then used as an integrand to compute $\int_a^b g(t)\, dt$, where $[a, b]$ is the domain of γ.

To put this in a more evident way, the value of $\omega(\gamma)$ is usually written

$$\int_\gamma \omega = \int_\gamma [A(x, y)\,dx + B(x, y)\,dy]$$

and one speaks of *integrating the differential form* ω *along the curve* γ. If we choose $A = 1$, $B = 0$, then $\omega = dx$, so that dx (and dy) are special 1-forms. For example, dx assigns to the smooth curve γ the value

$$\int_\gamma dx = \int_a^b \phi'(t)\,dt = \phi(b) - \phi(a)$$

which is the horizontal distance between the endpoints of γ. Since every 1-form can be expressed as a linear combination of dx and dy with function coefficients, dx and dy are called the **basic 1-forms** in the XY plane.

Let $\omega = xy\,dx - y^2\,dy$ and let γ be given by $x = 3t^2$, $y = t^3$, for $0 \le t \le 1$. On γ, we have

$$\begin{aligned}\omega &= (3t^2)(t^3)(6t\,dt) - (t^3)^2(3t^2\,dt)\\ &= (18t^6 - 3t^8)\,dt\end{aligned}$$

Thus
$$\int_\gamma \omega = \int_0^1 (18t^6 - 3t^8)\,dt = \tfrac{18}{7} - \tfrac{1}{3} = \tfrac{47}{21}$$

Let us compute the integral of the same 1-form along the straight line from (0, 0) to (2, 4). The equation for this curve is $x = 2t$, $y = 4t$, $0 \le t \le 1$. On γ, $\omega = (2t)(4t)(2dt) - (4t)^2(4dt) = -48t^2\,dt$, so that

$$\int_\gamma \omega = \int_0^1 (-48t^2)\,dt = -16$$

As we shall show in Theorem 2, $\int_{\gamma_1} \omega = \int_{\gamma_2} \omega$ whenever γ_1 and γ_2 are smoothly equivalent curves. In this last example, for instance, we may also use the parameterization $y = 2x$, $x = x$, $0 \le x \le 2$. Making this substitution instead, we have on γ

$$\begin{aligned}\omega &= x(2x)\,dx - (2x)^2(2dx)\\ &= -6x^2\,dx\end{aligned}$$

and
$$\int_\gamma (xy\,dx - y^2\,dy) = \int_0^2 (-6x^2)\,dx = -16$$

in agreement with the previous calculation.

Certain general properties of 1-forms (line integrals) in the plane can be seen at once from (9-3). Reversal of the orientation of a curve changes the sign of the integral, since the direction of integration in the integral containing dt will be reversed. We may state this in the form

$$\int_{-\gamma} \omega = -\int_\gamma \omega$$

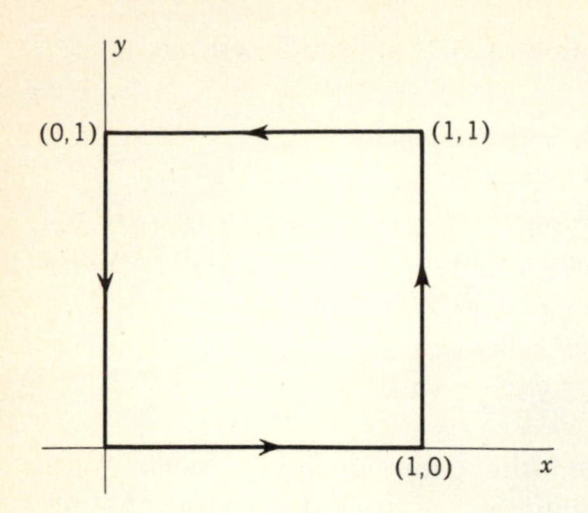

Figure 9-4

If the curve γ is the union of a finite set of curves

$$\gamma = \gamma_1 + \gamma_2 + \cdots + \gamma_n$$

then
$$\int_\gamma \omega = \int_{\gamma_1} \omega + \int_{\gamma_2} \omega + \cdots + \int_{\gamma_n} \omega$$

As an example, let us compute the integral of $\omega = xy\,dx + y^2\,dy$ along the closed polygonal path which forms the edge of the unit square, $0 \le x \le 1$, $0 \le y \le 1$ (see Fig. 9-4). We write

$$\gamma = \gamma_1 + \gamma_2 + \gamma_3 + \gamma_4$$

where these are the four sides of the square. We evaluate each of the line integrals $\int_{\gamma_j} \omega$ separately. On γ_1, $y = 0$ and $x = x$, with x going from 0 to 1; thus, on γ_1, $\omega = 0\,dx + 0 = 0$ and $\int_{\gamma_1} \omega = 0$. On γ_2, $x = 1$ and $y = y$, with y going from 0 to 1; thus,

$$\omega = (1)(y)0 + y^2\,dy$$

and $\int_{\gamma_2} \omega = \int_0^1 y^2\,dy = \frac{1}{3}$. On γ_3, $y = 1$ and $x = x$, with x going from 1 to 0 because of the orientation of γ_3; thus,

$$\omega = (x)(1)\,dx + (1)0 = x\,dx$$

and $\int_{\gamma_3} \omega = \int_1^0 x\,dx = -\frac{1}{2}$. Finally, on γ_4, $x = 0$ and $y = y$, with y going from 1 to 0, so that $\omega = 0 + y^2\,dy$ and

$$\int_{\gamma_4} \omega = \int_1^0 y^2\,dy = -\frac{1}{3}$$

Adding these, we have

$$\int_\gamma \omega = 0 + \tfrac{1}{3} - \tfrac{1}{2} - \tfrac{1}{3} = -\tfrac{1}{2}$$

Integration of a 1-form is especially simple if it involves only one of the variables. For example, let $\omega = A(x)\,dx$ and let γ be a smooth curve whose starting point is (x_0, y_0) and whose last point is (x_1, y_1), given by $x = \phi(t)$, $y = \psi(t)$, $0 \le t \le 1$. On γ we have

$$\omega = A(\phi(t))\phi'(t)\,dt$$

so that

$$\int_\gamma \omega = \int_0^1 A(\phi(t))\phi'(t)\,dt$$

Making the substitution $x = \phi(t)$, this becomes merely

$$\int_\gamma \omega = \int_{x_0}^{x_1} A(x)\,dx$$

(In particular, we note that the value of $\int_\gamma \omega$ in this case depends only upon the location of the endpoints of γ and not upon the rest of the trace of the curve.) Comparable statements can be made for 1-forms $\omega = B(y)\,dy$. (See Exercises 6 and 7.)

The general 1-form (line integral) in 3-space is

$$\omega = A(x, y, z)\,dx + B(x, y, z)\,dy + C(x, y, z)\,dz$$

and the method of evaluating $\omega(\gamma) = \int_\gamma \omega$ for a curve γ is the same. If γ is given by $x = \phi(t)$, $y = \psi(t)$, $z = \theta(t)$, for $0 \le t \le 1$, then we *define* the value of the line integral by

$$\int_\gamma \omega = \int_\gamma A\,dx + B\,dy + C\,dz$$

$$(9\text{-}4) \qquad = \int_0^1 A(\phi(t), \psi(t), \theta(t))\frac{dx}{dt}\,dt + \int_0^1 B(\phi(t), \psi(t), \theta(t))\frac{dy}{dt}\,dt + \int_0^1 C(\phi(t), \psi(t), \theta(t))\frac{dz}{dt}\,dt$$

As an illustration, let $\omega = xy\,dx - y\,dy + 3zy\,dz$ and γ be: $x = t^2$, $y = t$, $z = -t^3$. On γ, we have

$$\omega = (t^2)(t)(2t\,dt) - (t)(dt) + 3(-t^3)(t)(-3t^2\,dt) = (9t^6 + 2t^4 - t)\,dt$$

If the domain of γ is $[0, 1]$, then

$$\int_\gamma \omega = \int_0^1 (9t^6 + 2t^4 - t)\,dt = \tfrac{9}{7} + \tfrac{2}{5} - \tfrac{1}{2}$$

Let us turn now to the theory of 2-forms. These will be a special class of surface-functionals. In the case of 2-forms in the plane, the theory is especially simple. As defined in Chap. 8, a surface in n space is a mapping from a region D in the parameter plane into n space. When $n = 2$, the notion of surface degenerates into the ordinary notion of a transformation T of the plane into itself, and a smooth surface is such a transformation which is of class C' and whose Jacobian is never 0. A simple smooth surface (surface element) is thus a 1-to-1 continuous transformation with nonzero Jacobian, which maps a region D of the (u, v) plane into a region in the (x, y) plane. Since the Jacobian is continuous, it must be always positive in D, or always negative in D; in the former case, T is orientation preserving, and in the latter, orientation reversing. We may thus identify such a surface with its trace in the (x, y) plane, oriented according to the sign of the Jacobian of T. The notion of "surface" becomes simply that of an oriented region in the plane. We adopt the following definition for 2-forms in the plane.

Definition 2 *The general continuous 2-form in the XY plane is a region-functional denoted by*

(9-5) $$\omega = A(x, y)\,dx\,dy$$

where the function A is continuous. If Ω is a region (having area) in the domain of definition of A, then the value which ω assigns to Ω is defined by

$$\omega(\Omega) = \iint_{\Omega} A(x, y)\,dx\,dy$$

In line with our previous discussion, it is also natural to consider expressions of the form $A(x, y)\,dy\,dx$; this is also a 2-form, but by means of the convention that $dy\,dx = -dx\,dy$, we may replace it by $-A(x, y)\,dx\,dy$, which is again of the form (9-5). The general 2-form in the plane becomes merely the oriented double integral [see (9-1)].

The situation in 3-space is more complicated. In addition to the basic 2-form $dx\,dy$, we also have $dy\,dz$ and $dz\,dx$; moreover, their coefficients may be functions of three variables. Thus, the general 2-form in space becomes

$$\omega = A(x, y, z)\,dy\,dz + B(x, y, z)\,dz\,dx + C(x, y, z)\,dx\,dy$$

This is a surface-functional, and the value which it assigns to a surface Σ is denoted by $\iint_{\Sigma} \omega$. However, we cannot explain the meaning of this symbol further, nor the method for evaluating it, without some additional machinery.

We complete the roster of differential forms in the plane by defining a 0-form to be merely any continuous function (= point function). In space, we must also define 3-forms.

Definition 3 *The general continuous 3-form in XYZ space is region-functional denoted by*

$$\omega = A(x, y, z)\, dx\, dy\, dz$$

where the function A is continuous. If Ω *is a subset (having volume) of the domain of definition of A, then the value which* ω *assigns to* Ω *is defined by*

$$\omega(\Omega) = \iiint_{\Omega} A(x, y, z)\, dx\, dy\, dz$$

Thus, the 3-forms in space are oriented triple integrals; a 3-form in which the order is not $dx\, dy\, dz$ can be reduced to this form by permuting pairs. The 3-form $C(x, y, z)\, dy\, dz\, dx$ is the same as $-C(x, y, z)\, dy\, dx\, dz$ and as $C(x, y, z)\, dx\, dy\, dz$.

We next introduce an algebraic structure into the system of differential forms by defining **addition** and **multiplication** of forms. Any two forms of the same class are added by combining coefficients of like terms. For example,

$$(x^2\, dx + xy^2\, dy) + (dx - 3y\, dy) = (x^2 + 1)\, dx + (xy^2 - 3y)\, dy$$

$$(2x + y)\, dx\, dy + (xy - y)\, dx\, dy = (2x + xy)\, dx\, dy$$

$$(3dx + xyz\, dy - xz\, dz) + (y\, dx - dz) = (3 + y)\, dx + xyz\, dy - (xz + 1)\, dz$$

When two terms contain the same differentials, but in different orders, they must be brought into agreement before adding their coefficients, using the convention:

$$\begin{aligned} dx\, dy &= -dy\, dx \\ dy\, dz &= -dz\, dy \\ dz\, dx &= -dx\, dz \end{aligned} \tag{9-6}$$

For example,

$$\begin{aligned} &(x\, dy\, dz + y^2\, dx\, dy + z\, dx\, dz) + (3dy\, dz - z\, dy\, dx + x^2\, dz\, dx) \\ &\quad = (x\, dy\, dz + 3dy\, dz) + (y^2\, dx\, dy - z\, dy\, dx) + (z\, dx\, dz + x^2\, dz\, dx) \\ &\quad = (x + 3)\, dy\, dz + (y^2 + z)\, dx\, dy + (z - x^2)\, dx\, dz \end{aligned}$$

Multiplication of differential forms is governed by the rules set forth in (9-6) and the following:

$$dx\, dx = dy\, dy = dz\, dz = 0 \tag{9-7}$$

Except for these conventions, and the necessity of preserving the order of factors, multiplication can be performed as in elementary algebra. Stated in different terms, the system of differential forms in 3-space is a linear associative algebra whose basis elements are 1, dx, dy, dz, $dxdy$, $dydz$, $dzdx$, and $dxdydz$; whose coefficient space is the space of continuous functions of three variables; and whose multiplication table is specified by the relations (9-6) and (9-7).

For example,

$$\begin{aligned}(2dx + 3xy\,dy)(dx) &= 2(dx\,dx) + (3xy)(dy\,dx)\\ &= 0 + (3xy)(-dx\,dy)\\ &= -3xy\,dx\,dy\end{aligned}$$

$$\begin{aligned}&(2dx + 3xy\,dy)(x\,dx - y\,dy)\\ &\quad= (2x)(dx\,dx) + (3x^2y)(dy\,dx) + (-2y)(dx\,dy) + (-3xy^2)(dy\,dy)\\ &\quad= 0 + (3x^2y)(-dx\,dy) + (-2y)(dx\,dy) + 0\\ &\quad= -(3x^2y + 2y)\,dx\,dy\end{aligned}$$

Taking a somewhat more complicated example involving three variables,

$$\begin{aligned}&(x\,dx - z\,dy + y^2\,dz)(x^2\,dy\,dz + 2dz\,dx - y\,dx\,dy)\\ &\quad= x^3\,dx\,dy\,dz + 2x\,dx\,dz\,dx - xy\,dx\,dx\,dy\\ &\qquad - x^2z\,dy\,dy\,dz - 2z\,dy\,dz\,dx + yz\,dy\,dx\,dy\\ &\qquad + y^2x^2\,dz\,dy\,dz + 2y^2\,dz\,dz\,dx - y^3\,dz\,dx\,dy\\ &\quad= x^3\,dx\,dy\,dz - 2z\,dy\,dz\,dx - y^3\,dz\,dx\,dy\\ &\quad= (x^3 - 2z - y^3)\,dx\,dy\,dz\end{aligned}$$

Such computations may be shortened by observing that any term which contains a repetition of one of the basic differential forms dx, dy, dz will be 0. Accordingly, the product of two 2-forms in space, or in the plane, and the product of a 1-form and a 2-form in the plane are automatically 0. In general, the product of a k-form and an m-form is a $(k + m)$-form; if $k + m$ is larger than n, the number of variables, then there will be repetitions, and such a product will be 0. Since a 0-form is merely a function, multiplication by a 0-form does not affect the degree of a form. For example,

$$(3x^2y)(x\,dx - xy\,dy) = 3x^3y\,dx - 3x^3y^2\,dy$$

Finally we define a notion of **differentiation** for forms. In general, if ω is a k form, its derivative $d\omega$ will be a $(k + 1)$-form. We give the definition of $d\omega$ for forms in 3-space.

Definition 4

i. *If A is a 0-form (function) of class C', then dA is the 1-form*

$$dA = \frac{\partial A}{\partial x}\,dx + \frac{\partial A}{\partial y}\,dy + \frac{\partial A}{\partial z}\,dz$$

ii. *If ω is a 1-form $A\,dx + B\,dy + C\,dz$ whose coefficients are functions of class C', then $d\omega$ is the 2-form*

$$d\omega = (dA)\,dx + (dB)\,dy + (dC)\,dz$$

iii. *If ω is a 2-form $A\,dy\,dz + B\,dz\,dx + C\,dx\,dy$ whose coefficients are of class C', then $d\omega$ is the 3-form*

$$d\omega = (dA)\,dy\,dz + (dB)\,dz\,dx + (dC)\,dx\,dy$$

In (ii) and (iii), $d\omega$ is to be computed by evaluating dA, dB, and dC, and then computing the indicated products. Since dA is a 1-form, $(dA)\,dx$ is indeed a 2-form and $(dA)\,dy\,dz$ a 3-form. The derivative of a 3-form may be computed in the same fashion, but since it will be a 4-form in three variables, it will automatically be 0. Differentiation of forms in two variables is done in the same fashion. Some examples will illustrate the technique.

Form	*Derivative*
$A = x^2y$	$dA = 2xy\,dx + x^2\,dy$
$A = xy + yz$	$dA = y\,dx + (x+z)\,dy + y\,dz$
$\omega = x^2y\,dx$	$d\omega = d(x^2y)\,dx = (2xy\,dx + x^2\,dy)\,dx = -x^2\,dx\,dy$
$\omega = (xy + yz)\,dx$	$d\omega = d(xy+yz)\,dx = (y\,dx + (x+z)\,dy + y\,dz)\,dx = (x+z)\,dy\,dx + y\,dz\,dx$
$\omega = x^2y\,dy\,dz - xz\,dx\,dy$	$d\omega = d(x^2y)\,dy\,dz - d(xz)\,dx\,dy = (2xy\,dx + x^2\,dy)\,dy\,dz - (z\,dx + x\,dz)\,dx\,dy = 2xy\,dx\,dy\,dz - x\,dz\,dx\,dy = (2xy - x)\,dx\,dy\,dz$

This extensive barrage of definitions is partly justified and motivated by the following result.

Theorem 1 *If $u = f(x, y)$ and $v = g(x, y)$, then*

$$du\,dv = \frac{\partial(u, v)}{\partial(x, y)}\,dx\,dy \tag{9-8}$$

We have $du = f_1\,dx + f_2\,dy$ and $dv = g_1\,dx + g_2\,dy$, so that

$$\begin{aligned} du\,dv &= (f_1\,dx + f_2\,dy)(g_1\,dx + g_2\,dy) \\ &= f_1g_1\,dx\,dx + f_1g_2\,dx\,dy + f_2g_1\,dy\,dx + f_2g_2\,dy\,dy \\ &= f_1g_2\,dx\,dy + f_2g_1\,dy\,dx = (f_1g_2 - f_2g_1)\,dx\,dy \\ &= \begin{vmatrix} f_1 & f_2 \\ g_1 & g_2 \end{vmatrix}\,dx\,dy = \frac{\partial(u, v)}{\partial(x, y)}\,dx\,dy \quad \blacksquare \end{aligned}$$

Recalling that the theorem on transformation of multiple integrals requires one to replace $du\,dv$ by $\partial(u, v)/\partial(x, y)\,dx\,dy$ when making the substitution $u = f(x, y)$, $v = g(x, y)$, this theorem makes it possible to carry out such substitutions by the same routine procedure that one uses in singlefold integrals. For example, in Sec. 8.3 we considered the integral

$$\iint_{D^*} \exp\left((x - y)/(x + y)\right) dx\,dy,$$

where D^* is the triangle bounded by $x = 0$, $y = 0$, $x + y = 1$. We set $u = x - y$, $v = x + y$, which maps D^* onto another triangle D with vertices $(0, 0)$, $(1, 1)$, $(-1, 1)$. We have $du = dx - dy$, $dv = dx + dy$, so that

$$du\,dv = (dx - dy)(dx + dy) = dx\,dy - dy\,dx = 2dx\,dy$$

Accordingly, we have

$$\iint_{D^*} \exp\left(\frac{x - y}{x + y}\right) dx\,dy = \iint_{D} e^{u/v}\tfrac{1}{2}\,du\,dv$$

from which one may proceed as before to complete the problem.

It should be observed that (9-8) does not have the absolute value of the Jacobian. The reason is that we can now make good use of the orientation of double integrals (2-forms) in the plane; if $\partial(u, v)/\partial(x, y)$ is negative, then the corresponding mapping sending (x, y) into (u, v) will be orientation reversing, and the formalism of (9-8) will assign the correct sign to the result of the change of variable in the integral.

We are at last ready to explain the meaning to be attached to a 2-form in 3-space, and more generally, a k form in n space.

Definition 5 *The general continuous 2-form in space is a surface-functional denoted by*

$$\omega = A(x, y, z)\,dy\,dz + B(x, y, z)\,dz\,dx + C(x, y, z)\,dx\,dy$$

where A, B, and C are continuous functions defined in a region Ω. Let Σ be a smooth surface with domain D defined by

$$\begin{cases} x = \phi(u, v) \\ y = \psi(u, v) \qquad (u, v) \in D \\ z = \theta(u, v) \end{cases}$$

whose trace lies in Ω. Then, the value which ω assigns to Σ is defined by

$$\iint_{\Sigma} \omega = \iint_{D} \{A(\Sigma(u, v))\,d\psi\,d\theta + B(\Sigma(u, v))\,d\theta\,d\phi + C(\Sigma(u, v))\,d\phi\,d\psi\}$$

We note that again, $\iint_{\Sigma} \omega$ is computed by a straightforward process of substitution; ω is evaluated on the surface by using the equation for Σ and the operations of differentiation and multiplication for differential forms. For example, let $\omega = xy\,dy\,dz + x\,dz\,dx + 3zx\,dx\,dy$, and let Σ be the surface

$$\begin{cases} x = u + v & 0 \le u \le 1 \\ y = u - v & 0 \le v \le 1 \\ z = uv \end{cases}$$

On Σ, we have

$$dy\,dz = (du - dv)(v\,du + u\,dv) = (u + v)\,du\,dv$$
$$dz\,dx = (v\,du + u\,dv)(du + dv) = (v - u)\,du\,dv$$
$$dx\,dy = (du + dv)(du - dv) = -2du\,dv$$

and

$$\begin{aligned} \omega &= (u + v)(u - v)(u + v)\,du\,dv + (u + v)(v - u)\,du\,dv + 3uv(u + v)(-2)\,du\,dv \\ &= (u^3 - 5u^2v - 7uv^2 - v^3 + v^2 - u^2)\,du\,dv \end{aligned}$$

so that

$$\begin{aligned} \iint_\Sigma \omega &= \int_0^1 du \int_0^1 (u^3 - 5u^2v - 7uv^2 - v^3 + v^2 - u^2)\,dv \\ &= \int_0^1 (u^3 - \tfrac{7}{2}u^2 - \tfrac{7}{3}u + \tfrac{1}{12})\,du \\ &= -2 \end{aligned}$$

Again, let ω be the same 2-form, and let Σ be the surface given by $z = x^2 + y^2$, with $0 \le x \le 1$, $0 \le y \le 1$. On Σ, $dz = 2x\,dx + 2y\,dy$, so that

$$dy\,dz = dy(2x\,dx + 2y\,dy) = -2x\,dx\,dy$$
$$dz\,dx = (2x\,dx + 2y\,dy)\,dx = -2y\,dx\,dy$$

and on Σ we have

$$\begin{aligned} \omega &= xy\,dy\,dz + x\,dz\,dx + 3zx\,dx\,dy \\ &= (xy)(-2x\,dx\,dy) + x(-2y\,dx\,dy) + 3(x^2 + y^2)x\,dx\,dy \\ &= (3x^3 + 3xy^2 - 2x^2y - 2xy)\,dx\,dy \end{aligned}$$

Thus, the integral of this 2-form over Σ is

$$\begin{aligned} \iint_\Sigma \omega &= \int_0^1 dx \int_0^1 (3x^3 + 3xy^2 - 2x^2y - 2xy)\,dy \\ &= \int_0^1 (3x^3 - x^2)\,dx = \tfrac{5}{12} \end{aligned}$$

When two surfaces are parametrically equivalent, then a 2-form will assign the same value to both. This, together with the analogous statement for line integrals, is the content of the next theorem.

Theorem 2

i. *If γ_1 and γ_2 are smoothly equivalent curves and ω is a continuous 1-form defined on the trace of the curves, then* $\int_{\gamma_1} \omega = \int_{\gamma_2} \omega$.

ii. *If Σ_1 and Σ_2 are smoothly equivalent surfaces, and ω is a continuous 2-form defined on their trace, then* $\iint_{\Sigma_1} \omega = \iint_{\Sigma_2} \omega$.

For (i), let us consider the 1-form $\omega = A(x, y, z)\,dx$. We may assume that $\gamma_2(t) = \gamma_1(f(t))$, where f is of class C', and maps the interval $[\alpha, \beta]$, which is the domain of γ_2, onto the interval $[a, b]$, which is the domain of γ_1. If we have $x = \phi(t)$ on γ_1, then $x = \phi(f(t))$ on γ_2. Computing the integral of ω along γ_2, we have

$$\int_{\gamma_2} \omega = \int_\alpha^\beta A(\gamma_2(t)) \frac{dx}{dt}\,dt$$

$$= \int_\alpha^\beta A(\gamma_1(f(t)))\phi'(f(t))f'(t)\,dt$$

In this ordinary definite integral, make the substitution $s = f(t)$, and obtain

$$\int_{\gamma_2} \omega = \int_a^b A(\gamma_1(s))\phi'(s)\,ds = \int_{\gamma_1} \omega$$

Similarly, one can show the same relation for the 1-forms $B\,dy$ and $C\,dz$, and by addition, obtain $\int_{\gamma_2} \omega = \int_{\gamma_1} \omega$ for a general 1-form ω.

For (ii), we consider first a 2-form $\omega = A(x, y, z)\,dx\,dy$. Let Σ be a surface with domain D_1 on which we have $x = \phi(r, s)$, $y = \psi(r, s)$, $z = \theta(r, s)$, and let Σ_2 be a surface with domain D_2 which is smoothly equivalent to Σ_1; we may therefore assume that we have

$$\Sigma_2(u, v) = \Sigma_1(T(u, v))$$

where $T(u, v) = (r, s)$ describes a C' transformation T mapping D_2 onto D_1, 1-to-1, orientation preserving. Using the result of Theorem 1, we evaluate ω on Σ_2:

$$\omega = A(\Sigma_2(u, v))\,dx\,dy = A(\Sigma_2(u, v)) \frac{\partial(x, y)}{\partial(u, v)}\,du\,dv$$

$$= A(\Sigma_1(T(u, v))) \frac{\partial(x, y)}{\partial(u, v)}\,du\,dv$$

Thus, $$\iint_{\Sigma_2} \omega = \iint_{D_2} A(\Sigma_1(T(u, v))) \frac{\partial(x, y)}{\partial(u, v)}\, du\, dv$$

In this, let us make the transformation of coordinates $(r, s) = T(u, v)$. Again applying Theorem 1, we have

$$\iint_{\Sigma_2} \omega = \iint_{D_1} A(\Sigma_1(r, s)) \frac{\partial(x, y)}{\partial(u, v)} \frac{\partial(u, v)}{\partial(r, s)}\, dr\, ds$$

However, by the chain rule for differentiation,

$$\frac{\partial(x, y)}{\partial(u, v)} \frac{\partial(u, v)}{\partial(r, s)} = \frac{\partial(x, y)}{\partial(r, s)} = \begin{vmatrix} \phi_1 & \phi_2 \\ \psi_1 & \psi_2 \end{vmatrix}$$

so that $$\iint_{\Sigma_2} \omega = \iint_{D_1} A(\Sigma_1(r, s)) \frac{\partial(x, y)}{\partial(r, s)}\, dr\, ds$$

The expression $\partial(x, y)/\partial(r, s)\, dr\, ds$ is, again by Theorem 1, equivalent to $dx\, dy$ on Σ_1, so that we finally obtain

$$\iint_{\Sigma_2} \omega = \iint_{\Sigma_1} A(x, y, z)\, dx\, dy = \iint_{\Sigma_1} \omega$$

Again, the theorem may be completed by considering the 2-forms $B\, dy\, dz$ and $C\, dz\, dx$, adding the results to arrive at the general case. ▮

So far, we have given no motivation for the consideration of integrals of 1-forms and of 2-forms. This will be done in Sec. 9.3, against a background of classical vector analysis. We shall show that the value which a 1-form ω assigns to the curve γ can also be obtained by integrating a certain function f, constructed from both ω and γ, along the curve γ. Similarly, the value which a 2-form ω assigns to the surface Σ can be obtained by integrating a function F over the surface Σ, where again F depends both upon ω and Σ. These integrations are carried out as in Sec. 8.6. The functions f and F which arise are defined only on the curve and on the surface; however, they arise from "vector-valued" functions in a manner which is physically significant, and which leads to important physical interpretations of the integrals involved.

EXERCISES

1 Evaluate $\int_\gamma (x\, dx + xy\, dy)$ where

(*a*) γ is the line $x = t,\ y = t,\ 0 \le t \le 1$.
(*b*) γ is the portion of the parabola $y = x^2$ from $(0, 0)$ to $(1, 1)$.
(*c*) γ is the portion of the parabola $x = y^2$ from $(0, 0)$ to $(1, 1)$.
(*d*) γ is the polygon whose successive vertices are $(0, 0)$, $(1, 0)$, $(0, 1)$, $(1, 1)$.

2 Evaluate $\int_\gamma (y\,dx - x\,dy)$ where

(*a*) γ is the closed curve $x = t^2 - 1$, $y = t^3 - t$, $-1 \le t \le 1$.
(*b*) γ is the straight line from (0, 0) to (2, 4).
(*c*) γ is the portion of the parabola $y = x^2$ from (0, 0) to (2, 4).
(*d*) γ is the polygon whose successive vertices are (0, 0), (−2, 0), (−2, 4), (2, 4).

3 Evaluate $\int_\gamma (y\,dx + x\,dy)$ for the curves given in Exercise 2.

4 Evaluate $\int_\gamma (z\,dx + x^2\,dy + y\,dz)$ where

(*a*) γ is the straight line from (0, 0, 0) to (1, 1, 1).
(*b*) γ is the portion of the *twisted cubic* $x = t$, $y = t^2$, $z = t^3$ from (0, 0, 0) to (1, 1, 1).
(*c*) γ is the portion of the *helix* $x = \cos t$, $y = \sin t$, $z = t$, for $0 \le t \le 2\pi$.
(*d*) γ is the closed polygon whose successive vertices are (0, 0, 0), (2, 0, 0), (2, 3, 0), (0, 0, 1), (0, 0, 0).

5 Evaluate $\int_\gamma (2x\,dx + z\,dy + y\,dz)$ for the curves given in Exercise 4.

6 If $\omega = B(y)\,dy$, show that $\int_\gamma \omega = \int_{y_0}^{y_1} B(y)\,dy$ for any smooth curve γ which starts at (x_0, y_0) and ends at (x_1, y_1).

7 If $\omega = A(x)\,dx + B(y)\,dy$, show that $\int_\gamma \omega = 0$ for any closed curve γ.

8 Verify the following:

(*a*) $(3x\,dx + 4y\,dy)(3x^2\,dx - dy) = -(3x + 12x^2y)\,dx\,dy$.
(*b*) $(3x^2\,dx - dy)(3x\,dx + 4y\,dy) = (3x + 12x^2y)\,dx\,dy$.
(*c*) $(x\,dy - yz\,dz)(y\,dx + xy\,dy - z\,dz) = (xy^2z - xz)\,dy\,dz - y^2z\,dz\,dx - xy\,dx\,dy$.
(*d*) $(x^2\,dy\,dz + yz\,dx\,dy)(3dx - dz) = (3x^2 - yz)\,dx\,dy\,dz$.
(*e*) $(dx\,dy - dy\,dz)(dx + dy + dz) = 0$.
(*f*) $(dx - x\,dy + yz\,dz)(x\,dx - x^2\,dy + xyz\,dz) = 0$.

9 Show that

$$(A\,dx + B\,dy + C\,dz)(a\,dx + b\,dy + c\,dz)$$
$$= \begin{vmatrix} B & C \\ b & c \end{vmatrix} dy\,dz + \begin{vmatrix} C & A \\ c & a \end{vmatrix} dz\,dx + \begin{vmatrix} A & B \\ a & b \end{vmatrix} dx\,dy$$

10 Show that

$$(A\,dx + B\,dy + C\,dz)(a\,dy\,dz + b\,dz\,dx + c\,dx\,dy) = (aA + bB + cC)\,dx\,dy\,dz$$

11 Evaluate df if (*a*) $f(x, y, z) = x^2yz$; (*b*) $f(x, y) = \log(x^2 + y^2)$.

12 Let $x = \phi(u, v, w)$, $y = \psi(u, v, w)$, $z = \theta(u, v, w)$. Show that

$$dx\,dy\,dz = \frac{\partial(x, y, z)}{\partial(u, v, w)}\,du\,dv\,dw$$

13 Evaluate $d\omega$ where

(*a*) $\omega = x^2y\,dx - yz\,dz$
(*b*) $\omega = 3x\,dx + 4xy\,dy$
(*c*) $\omega = 2xy\,dx + x^2\,dy$
(*d*) $\omega = e^{xy}\,dx - x^2y\,dy$
(*e*) $\omega = x^2y\,dy\,dz - xz\,dx\,dy$
(*f*) $\omega = x^2z\,dy\,dz + y^2z\,dz\,dx - xy^2\,dx\,dy$
(*g*) $\omega = xz\,dy\,dx + xy\,dz\,dx + 2yz\,dy\,dz$

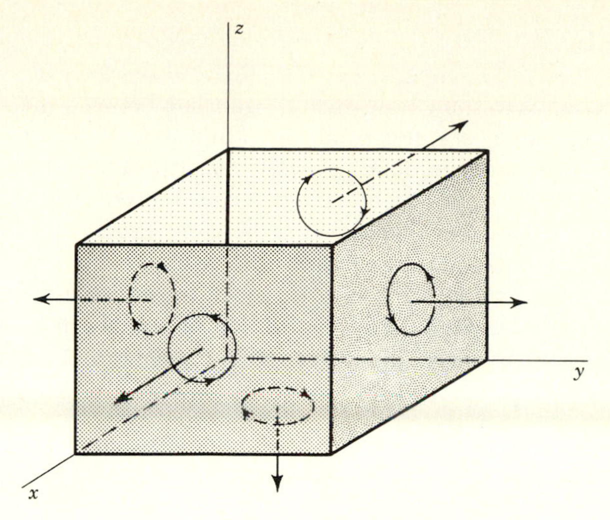

Figure 9-5

14 If $\omega = A(x, y, z)\,dx + B(x, y, z)\,dy + C(x, y, z)\,dz$, show that

$$d\omega = (C_2 - B_3)\,dy\,dz + (A_3 - C_1)\,dz\,dx + (B_1 - A_2)\,dx\,dy$$

15 If $\omega = A(x, y, z)\,dy\,dz + B(x, y, z)\,dz\,dx + C(x, y, z)\,dx\,dy$, show that

$$d\omega = (A_1 + B_2 + C_3)\,dx\,dy\,dz$$

16 Evaluate $\iint_\Sigma (x\,dy\,dz + y\,dx\,dy)$, where

(*a*) Σ is the surface described by $x = u + v$, $y = u^2 - v^2$, $z = uv$, $0 \le u \le 1$, $0 \le v \le 1$.

(*b*) Σ is the portion of the cylinder $x^2 + y^2 = 1$ with $0 \le z \le 1$, oriented so that the normal is outward (away from the Z axis).

(*c*) Σ is the boxlike surface which is the union of five squares, each side of length 1 (see Fig. 9-5).

17 Verify the following calculations with forms in four variables:

(*a*) $(dx + dy - dz + dw)(dx\,dy + 2\,dz\,dw)$

$$= 2\,dx\,dz\,dw + 2\,dy\,dz\,dw - dx\,dy\,dz + dx\,dy\,dw.$$

(*b*) $(dx\,dy + dy\,dz - dz\,dw)(x\,dx\,dy + y\,dz\,dw) = (y - x)\,dx\,dy\,dz\,dw.$

(*c*) $d(x^2y\,dy\,dw + yw^2\,dx\,dz + xyzw\,dx\,dy) = (xyw - w^2)\,dx\,dy\,dz$

$$+ (2xy + xyz)\,dx\,dy\,dw + 2yw\,dx\,dz\,dw.$$

(*d*) $dx\,dy\,dz\,dw = \dfrac{\partial(x, y, z, w)}{\partial(r, s, t, u)}\,dr\,ds\,dt\,du$

when $x = \phi(r, s, t, u)$, $y = \psi(r, s, t, u)$, $z = \theta(r, s, t, u)$, $w = \chi(r, s, t, u)$.

18 Evaluate $\iint_\Sigma (xy\,dy\,dz + yz\,dx\,dw)$, where Σ is the two-dimensional surface in 4-space described by $x = r^2 + s^2$, $y = r - s$, $z = rs$, $w = r + s$, and (r, s) obeys $0 \le r \le 1$, $0 \le s \le 1$.

9.3 VECTOR ANALYSIS

Since the purpose of this section is to show how one can translate between the language and notation of the system of differential forms and that of vector analysis, we shall not develop the latter in detail. Since the early years of this century, there have appeared many texts which explain vector analysis in the form used in engineering and physics; our discussion is therefore abbreviated.

Classical vector analysis is three-dimensional, and the reasons behind its notation and form are historical, having roots in a desire to introduce more algebraic structure into geometry. As we have seen, addition of points can be defined for points of the plane, or of space, or, in general, for points in $\mathbf{R}^n$. If $p = (x_1, x_2, \ldots, x_n)$ and $q = (y_1, y_2, \ldots, y_n)$, then

$$p + q = (x_1 + y_1, x_2 + y_2, \ldots, x_n + y_n)$$

We also defined the product of a point p by a real number λ as the point given by

$$\lambda p = (\lambda x_1, \lambda x_2, \ldots, \lambda x_n)$$

It is natural to ask if there is a useful definition of multiplication for points such that the product of two points in $\mathbf{R}^n$ is again a point in $\mathbf{R}^n$ and such that the customary rules which govern the algebra of real numbers still hold for points. The **inner product** (also called scalar or dot product) of two points, which was defined in Sec. 1.2 by

$$p \cdot q = x_1 y_1 + x_2 y_2 + \cdots + x_n y_n$$

is not suitable, since the inner product of two points is a number (scalar) and not a point.

It is well known that suitable multiplications exist when $n = 2$. For example, one may define the product of two points in the plane by the formula

$$(x_1, x_2)(y_1, y_2) = (x_1 y_1 - x_2 y_2, x_1 y_2 + x_2 y_1) \tag{9-9}$$

Upon checking the various algebraic rules, it is found that under this definition for multiplication, and the previous definition of addition, $\mathbf{R}^2$ becomes what is called a field (see Appendix 2). The motivation for (9-9) may be seen by making the correspondence $(a, b) \leftrightarrow a + bi$ between the plane and the field of all complex numbers. If $p = (x_1, x_2)$ corresponds to $z = x_1 + ix_2$ and $q = (y_1, y_2)$ corresponds to $w = y_1 + iy_2$, then we see that

$$zw = (x_1 + ix_2)(y_1 + iy_2) = (x_1 y_1 - x_2 y_2) + i(x_1 y_2 + x_2 y_1)$$

which corresponds to the point which is given as the product of p and q.

With this example in mind, one may attempt to find a similar definition for multiplication of points in 3-space. By algebraic methods, it can be shown that no such formula exists (still requiring that the ordinary algebraic rules remain valid). However, going to the next higher dimension, Hamilton (1843)

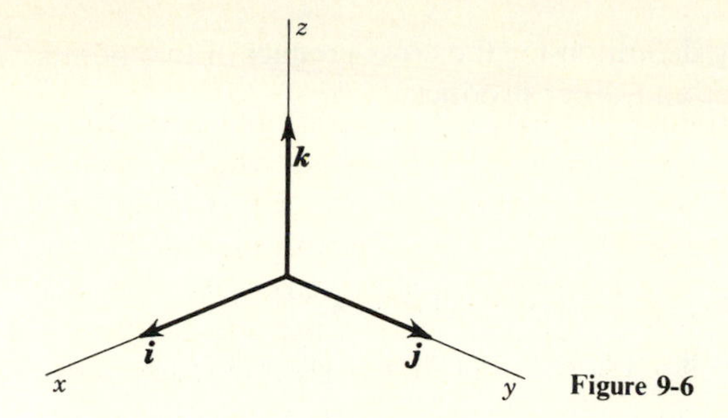

Figure 9-6

discovered that a definition for multiplication of points in $\mathbf{R}^4$ could be given which yields a system obeying all the algebraic rules which apply to real numbers (i.e., the field axioms) except one; multiplication is no longer commutative, so that (pq) and (qp) may be different points. This system is called the algebra of **quaternions.** It was soon seen that it could be used to great advantage in the theory of mechanics. By restricting points to a particular 3-space embedded in $\mathbf{R}^4$, Gibbs and others developed a modification of the algebra of quaternions which was called **vector analysis,** and which gained widespread acceptance and importance, particularly in physics and electromagnetic theory. (A discussion of the mathematical background of this is found in the book edited by Albert, and the historical details in the book by Crowe, both given in the Suggested Reading List.)

Let $\mathbf{i} = (1, 0, 0)$, $\mathbf{j} = (0, 1, 0)$, $\mathbf{k} = (0, 0, 1)$. Any point in 3-space can be expressed in terms of these three. If $p = (a, b, c)$, then

$$p = a\mathbf{i} + b\mathbf{j} + c\mathbf{k}$$

We define a multiplication operation $\times$ for points in $\mathbf{R}^3$ by first defining it on the basis elements $\mathbf{i}, \mathbf{j}, \mathbf{k}$:

$$\mathbf{i} \times \mathbf{i} = \mathbf{j} \times \mathbf{j} = \mathbf{k} \times \mathbf{k} = 0 \tag{9-10}$$

$$\begin{aligned} \mathbf{i} \times \mathbf{j} &= -\mathbf{j} \times \mathbf{i} = \mathbf{k} \\ \mathbf{j} \times \mathbf{k} &= -\mathbf{k} \times \mathbf{j} = \mathbf{i} \\ \mathbf{k} \times \mathbf{i} &= -\mathbf{i} \times \mathbf{k} = \mathbf{j} \end{aligned} \tag{9-11}$$

(Note that the cyclic order $\mathbf{i} : \mathbf{j} : \mathbf{k} : \mathbf{i} : \mathbf{j}$, in which the product of any neighboring pair in order is the next, is consistent with the positive orientation of the axes as shown in Fig. 9-6.)

We may use these equations, together with the distributive law, to obtain a product for any two points. We have

$$\begin{aligned} (x_1\mathbf{i} + x_2\mathbf{j} + x_3\mathbf{k}) \times (y_1\mathbf{i} + y_2\mathbf{j} + y_3\mathbf{k}) &= (x_1y_1)(\mathbf{i} \times \mathbf{i}) + (x_1y_2)(\mathbf{i} \times \mathbf{j}) \\ &\quad + (x_1y_3)(\mathbf{i} \times \mathbf{k}) + (x_2y_1)(\mathbf{j} \times \mathbf{i}) + (x_2y_2)(\mathbf{j} \times \mathbf{j}) + (x_2y_3)(\mathbf{j} \times \mathbf{k}) \\ &\quad + (x_3y_1)(\mathbf{k} \times \mathbf{i}) + (x_3y_2)(\mathbf{k} \times \mathbf{j}) + (x_3y_3)(\mathbf{k} \times \mathbf{k}) \\ &= (x_1y_2 - x_2y_1)\mathbf{k} + (x_3y_1 - x_1y_3)\mathbf{j} + (x_2y_3 - x_3y_2)\mathbf{i} \end{aligned}$$

We therefore adopt the following definition for the **cross product** of two points. (Other names are vector product and outer product.)

Definition 6 *If*

$$p = (x_1, x_2, x_3) = x_1\mathbf{i} + x_2\mathbf{j} + x_3\mathbf{k}$$

and
$$q = (y_1, y_2, y_3) = y_1\mathbf{i} + y_2\mathbf{j} + y_3\mathbf{k}$$

then
$$p \times q = (x_2 y_3 - x_3 y_2)\mathbf{i} + (x_3 y_1 - x_1 y_3)\mathbf{j} + (x_1 y_2 - x_2 y_1)\mathbf{k}$$

It is seen that the coordinates of $p \times q$ have simple expressions as determinants, so that we may also write

$$p \times q = \begin{vmatrix} x_2 & x_3 \\ y_2 & y_3 \end{vmatrix}\mathbf{i} + \begin{vmatrix} x_3 & x_1 \\ y_3 & y_1 \end{vmatrix}\mathbf{j} + \begin{vmatrix} x_1 & x_2 \\ y_1 & y_2 \end{vmatrix}\mathbf{k} \tag{9-12}$$

or in an even more abbreviated form,

$$p \times q = \begin{vmatrix} \mathbf{i} & \mathbf{j} & \mathbf{k} \\ x_1 & x_2 & x_3 \\ y_1 & y_2 & y_3 \end{vmatrix}$$

(where we expand by the top row).

As indicated, such a multiplication operation defined on $\mathbf{R}^3$ cannot obey all the rules of ordinary algebra. In particular, $\times$ is not an associative product; it is not in general true that $p \times (q \times r) = (p \times q) \times r$. For example, $(\mathbf{i} \times \mathbf{i}) \times \mathbf{k} = \mathbf{0} \times \mathbf{k} = \mathbf{0}$, but

$$\mathbf{i} \times (\mathbf{i} \times \mathbf{k}) = \mathbf{i} \times (-\mathbf{j}) = -\mathbf{k}$$

(We remark in passing that the algebra of quaternions is better behaved, for multiplication there is associative. Any point p in 4-space is represented in the form $p = x_1\mathbf{i} + x_2\mathbf{j} + x_3\mathbf{k} + x_4\mathbf{l}$, where $\mathbf{i}$, $\mathbf{j}$, $\mathbf{k}$, and $\mathbf{l}$ now denote the four basic unit vectors determining the axes. The rule for multiplying quaternions is described by giving (9-11), replacing (9-10) by $\mathbf{i}^2 = \mathbf{j}^2 = \mathbf{k}^2 = -\mathbf{l}$, and saying that $\mathbf{l}$ acts like a multiplicative unit, so that $\mathbf{l}p = p\mathbf{l} = p$ for every p. The connection between vector analysis and quaternions is the relation

$$pq = p \times q - (p \cdot q)\mathbf{l} \tag{9-13}$$

which holds for any choice of $p = x_1\mathbf{i} + x_2\mathbf{j} + x_3\mathbf{k}$, $q = y_1\mathbf{i} + y_2\mathbf{j} + y_3\mathbf{k}$ as "spacelike" quaternions.)

One reason for the marked usefulness of vector analysis is that all the operations have intrinsic geometrical interpretations. Thus, the sum of vectors can be found by taking a set of directed line segments, representing the vectors, and placing them in juxtaposition, head to tail, and constructing the segment which connects the initial point of the first and the terminal point of the last. The scalar product $\mathbf{a} \cdot \mathbf{b}$ of two vectors is the number $|\mathbf{a}|\,|\mathbf{b}| \cos\theta$, where θ is the angle between the vectors; thus, if $\mathbf{b}$ is a unit vector, $\mathbf{a} \cdot \mathbf{b}$ is the projection of **a** in the direction specified by **b**. Finally, the cross product of two vectors **a** and **b** is the vector $\mathbf{c} = \mathbf{a} \times \mathbf{b}$ which is orthogonal to both **a** and **b**, whose length is $|\mathbf{a}|\,|\mathbf{b}| \sin \theta$, and such that the trihedral **a**, **b**, **c** is right-handed. To prove this statement, we first observe that two vectors **u** and **v** are orthogonal if and only if $\mathbf{u} \cdot \mathbf{v} = 0$. Let **a** and **b** be represented by the points (a_1, a_2, a_3) and (b_1, b_2, b_3). Then, their cross product **c** will be represented, according to (9-12), by the point

$$\left(\begin{vmatrix} a_2 & a_3 \\ b_2 & b_3 \end{vmatrix}, \quad \begin{vmatrix} a_3 & a_1 \\ b_3 & b_1 \end{vmatrix}, \quad \begin{vmatrix} a_1 & a_2 \\ b_1 & b_2 \end{vmatrix} \right)$$

so that
$$\mathbf{c} \cdot \mathbf{a} = \begin{vmatrix} a_2 & a_3 \\ b_2 & b_3 \end{vmatrix} a_1 + \begin{vmatrix} a_3 & a_1 \\ b_3 & b_1 \end{vmatrix} a_2 + \begin{vmatrix} a_1 & a_2 \\ b_1 & b_2 \end{vmatrix} a_3$$
$$= \begin{vmatrix} a_1 & a_2 & a_3 \\ a_1 & a_2 & a_3 \\ b_1 & b_2 & b_3 \end{vmatrix} = 0$$

Similarly, $\mathbf{c} \cdot \mathbf{b} = 0$, so that $\mathbf{a} \times \mathbf{b}$ is orthogonal to **a** and to **b**. To find the length of **c**, we recall that

$$|\mathbf{c}|^2 = \begin{vmatrix} a_2 & a_3 \\ b_2 & b_2 \end{vmatrix}^2 + \begin{vmatrix} a_3 & a_1 \\ b_3 & b_1 \end{vmatrix}^2 + \begin{vmatrix} a_1 & a_2 \\ b_1 & b_2 \end{vmatrix}^2$$
$$= (a_1^2 + a_2^2 + a_3^2)(b_1^2 + b_2^2 + b_3^2) - (a_1 b_1 + a_2 b_2 + a_3 b_3)^2$$
$$= |\mathbf{a}|^2\,|\mathbf{b}|^2 - (\mathbf{a} \cdot \mathbf{b})^2$$

Using the formula for $\mathbf{a} \cdot \mathbf{b}$, we obtain

$$|\mathbf{c}|^2 = |\mathbf{a}|^2 |\mathbf{b}|^2 - (|\mathbf{a}|\,|\mathbf{b}| \cos \theta)^2$$
$$= |\mathbf{a}|^2 |\mathbf{b}|^2 (1 - \cos^2 \theta)$$
$$= |\mathbf{a}|^2 |\mathbf{b}|^2 \sin^2 \theta$$

so that $|\mathbf{c}| = |\mathbf{a} \times \mathbf{b}| = |\mathbf{a}|\,|\mathbf{b}| \sin \theta$. (We also note that this number is the area of the parallelogram having **a** and **b** for consecutive sides, as shown in Fig. 9-7.) When **a** and **b** are chosen from among the basis vectors **i**, **j**, **k**, it is clear from

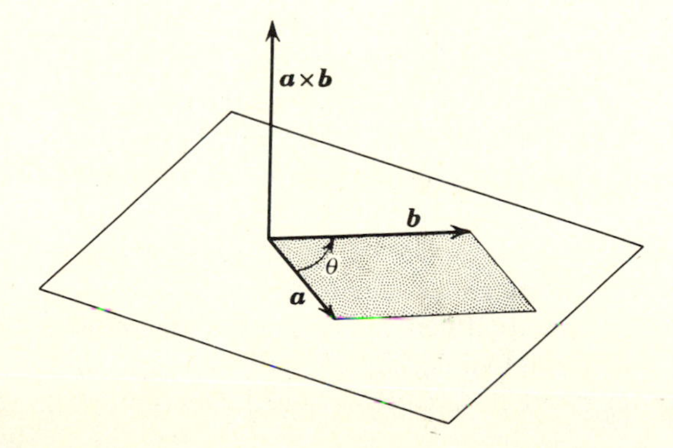

Figure 9-7

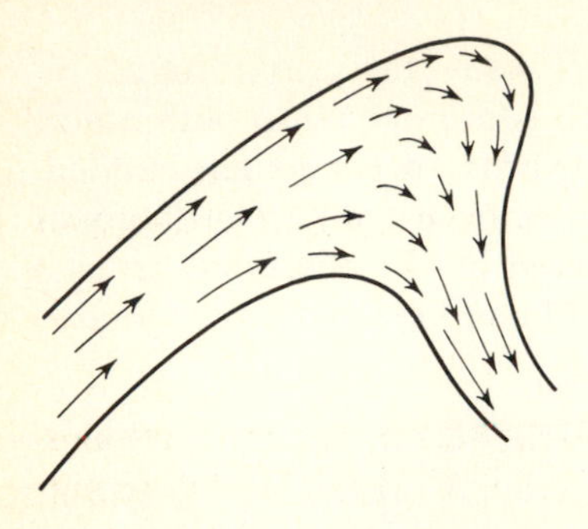

Figure 9-8 Velocity vector field.

the table (9-11) that **a**, **b**, and **a** × **b** form a positively oriented trihedral; for the general case, see Exercise 3.

The fact that the definitions of sum, inner product, and cross product, given originally in terms of the coordinates of the points, can also be given solely in geometrical terms means that these operations have an invariant quality which lends itself to the statement of physical laws. This invariance is inherent also in the coordinate definitions. The inner product of two points, p and q, is an algebraic function of their coordinates whose value is unchanged if the underlying 3-space is subjected to any orthogonal linear transformation (rotation) resulting in new coordinates for p and q.

Vectors are used in a variety of ways to represent physical quantities. With its initial point at the origin, a vector merely describes a position in 3-space. With its initial point at a point on a curve, the vector may represent the tangent to the curve, or the center of the circle of curvature there. A vector-valued function defined in a region D of space can be regarded as associating with each point $p \in D$ a specific vector $\mathbf{V}(p)$, whose initial point is placed at p. The resulting "vector field" or "direction field" in D might be used to model a variety of different physical situations. As an illustration, suppose that a flow of liquid is taking place throughout a region D. At each point of D, the liquid has a particular velocity which may be described by a directed line segment giving the direction and the speed of the motion. The vector-valued function so constructed is called the velocity field of the liquid (see Fig. 9-8). More generally, the directed line segment associated with the point p in the domain of $\mathbf{V}$ may have its initial point not at p, but at some other point determined by p. As an example of this, let $\mathbf{g}$ be a vector-valued function defined on an interval $a \le t \le b$, and whose endpoint describes a curve. Its derivative, $\mathbf{g}'(t)$, is a vector which we may represent by the directed line segment whose initial point is the point $\mathbf{g}(t)$ lying on the curve; this gives the customary picture of a curve and its tangent vectors (Fig. 9-9).

All the results dealing with differentiation of functions and transformations, presented in Chaps. 3 and 7, can be reinterpreted in terms of vectors when limited to 3-space, but the more general treatment is needed for functions of more than three variables. If F is a (scalar)-valued function of three variables, then its differential dF can be regarded as a vector-valued function; it is

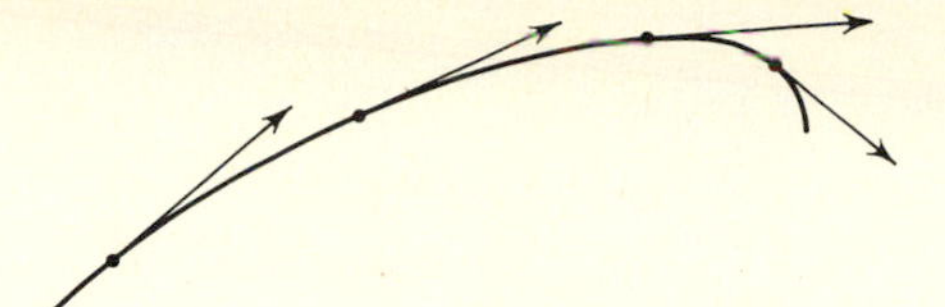

Figure 9-9 Graph of a curve and its associated tangent vectors.

then called the gradient of F and is denoted by ∇F. A vector-valued function T can be interpreted as a transformation of 3-space into itself. The differential dT of T is a matrix-valued function. Thus, the "derivative" of a scalar function is a vector function, and the "derivative" of a vector-valued function or vector field is a matrix-valued function. We put some of this in traditional vector form.

Definition 7 *If f is of class C', then its gradient is the vector-valued function*

$$\begin{aligned}\text{grad}\,(f) &= f_1\mathbf{i} + f_2\mathbf{j} + f_3\mathbf{k}\\ &= \frac{\partial f}{\partial x}\mathbf{i} + \frac{\partial f}{\partial y}\mathbf{j} + \frac{\partial f}{\partial z}\mathbf{k}\\ &= \nabla f\end{aligned}$$

In explanation of the last symbol, ∇ is to be thought of as the vector differential operator

$$\nabla = \mathbf{i}\frac{\partial}{\partial x} + \mathbf{j}\frac{\partial}{\partial y} + \mathbf{k}\frac{\partial}{\partial z} \tag{9-14}$$

The gradient of f at a point p may be represented by a directed line segment with initial point at p. If we take any unit vector

$$\mathbf{b} = b_1\mathbf{i} + b_2\mathbf{j} + b_3\mathbf{k}$$

then

$$\nabla f \cdot \mathbf{b} = f_1 b_1 + f_2 b_2 + f_3 b_3 \tag{9-15}$$

The differential of f (see Sec. 7.4) was $[f_1, f_2, f_3]$, so that the expression in (9-15) is the same as the value of the directional derivative of f at p in the direction (b_1, b_2, b_3). If $\mathbf{v}$ is any vector, and $\mathbf{b}$ is a unit vector, then $\mathbf{v}\cdot\mathbf{b} = |\mathbf{v}|\cos\theta$ is the component of $\mathbf{v}$ in the direction $\mathbf{b}$ (see Fig. 9-10). Its greatest value is $|\mathbf{v}|$ and is obtained when $\mathbf{b}$ is parallel to $\mathbf{v}$.

Applying this remark to the present case, we see that the gradient of f at p is a vector whose component in a general direction $\mathbf{b}$ is the derivative of f at p in the direction $\mathbf{b}$; accordingly, the gradient itself points in the direction of greatest increase of f, its magnitude is the value of this directional derivative, and it is normal to the level surface of f at p (Exercise 11).

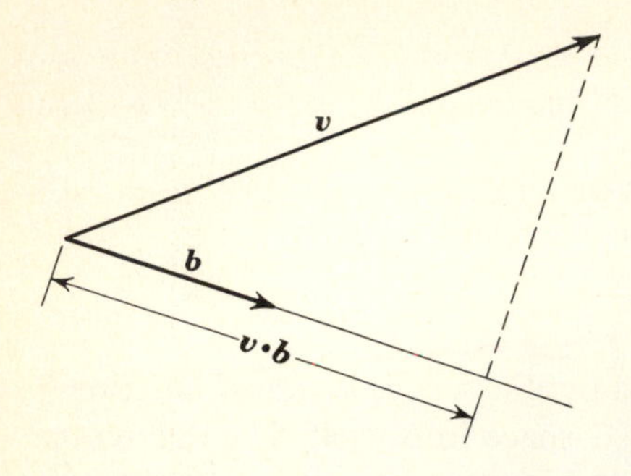

Figure 9-10

With any vector field **V** of class C', one may associate two other functions. The first is called the **divergence** of **V** and is an ordinary scalar function.

Definition 8 *If* $\mathbf{V} = A\mathbf{i} + B\mathbf{j} + C\mathbf{k}$, *where* A, B, *and* C *are (scalar) functions of class* C' *defined in a region* Ω, *then the divergence of* **V** *is*

$$\begin{aligned} \operatorname{div}(\mathbf{V}) &= A_1 + B_2 + C_3 \\ &= \frac{\partial A}{\partial x} + \frac{\partial B}{\partial y} + \frac{\partial C}{\partial z} \qquad (9\text{-}16) \\ &= \nabla \cdot \mathbf{V} \end{aligned}$$

We shall later recast this in a form which makes evident the geometrical invariance of the divergence (Sec. 9.5). At the moment, we give only the following algebraic justification. We may view **V** as defining a transformation from (x, y, z) space into (r, s, t) space by means of the equations

$$\begin{cases} r = A(x, y, z) \\ s = B(x, y, z) \\ t = C(x, y, z) \end{cases}$$

The differential of this transformation is represented by the matrix

$$\begin{bmatrix} \dfrac{\partial A}{\partial x} & \dfrac{\partial A}{\partial y} & \dfrac{\partial A}{\partial z} \\ \dfrac{\partial B}{\partial x} & \dfrac{\partial B}{\partial y} & \dfrac{\partial B}{\partial z} \\ \dfrac{\partial C}{\partial x} & \dfrac{\partial C}{\partial y} & \dfrac{\partial C}{\partial z} \end{bmatrix}$$

and the divergence of V is the **trace** of this matrix, that is, the sum of the diagonal entries. This is known to be one of the invariants of a matrix, under the orthogonal group of rotations of 3-space (see Exercise 12).

The second function associated with a vector field **V** is a vector-valued function and is called the **curl** of **V**.

Definition 9 *If* $\mathbf{V} = A\mathbf{i} + B\mathbf{j} + C\mathbf{k}$ *where* A, B, *and* C *are (scalar) functions of class* C' *defined in a region* Ω, *then the curl of* $\mathbf{V}$ *is the vector-valued function:*

$$\begin{aligned}\operatorname{curl}(\mathbf{V}) &= (C_2 - B_3)\mathbf{i} + (A_3 - C_1)\mathbf{j} + (B_1 - A_2)\mathbf{k} \\ &= \left(\frac{\partial C}{\partial y} - \frac{\partial B}{\partial z}\right)\mathbf{i} + \left(\frac{\partial A}{\partial z} - \frac{\partial C}{\partial x}\right)\mathbf{j} + \left(\frac{\partial B}{\partial x} - \frac{\partial A}{\partial y}\right)\mathbf{k} \\ &= \nabla \times \mathbf{V}\end{aligned} \tag{9-17}$$

The last expression is to be regarded as a convenient formula for the curl of $\mathbf{V}$ and may be written as

$$\nabla \times \mathbf{V} = \begin{vmatrix} \mathbf{i} & \mathbf{j} & \mathbf{k} \\ \dfrac{\partial}{\partial x} & \dfrac{\partial}{\partial y} & \dfrac{\partial}{\partial z} \\ A & B & C \end{vmatrix}$$

The invariant nature of this will also be shown later on.

Many of the most important postulates of physical theory find their simplest statements in vector equations. To cite only one example, one form of the Maxwell equations is:

$$\begin{aligned}&\operatorname{div}(\mathbf{E}) = \rho \qquad \operatorname{curl}(\mathbf{E}) = -\frac{\partial \mathbf{H}}{\partial t} \\ &\operatorname{div}(\mathbf{H}) = 0 \qquad \operatorname{curl}(\mathbf{H}) = 4\pi\left(\mathbf{J} + \frac{\partial \mathbf{E}}{\partial t}\right)\end{aligned} \tag{9-18}$$

One of the chief tools in the use of vector analysis in such applications is a knowledge of certain standard identities. In the list below, we give a number of these. All can be verified directly by substitution and computation.

If $\mathbf{a} = a_1\mathbf{i} + a_2\mathbf{j} + a_3\mathbf{k}$, $\mathbf{b} = b_1\mathbf{i} + b_2\mathbf{j} + b_3\mathbf{k}$, and $\mathbf{c} = c_1\mathbf{i} + c_2\mathbf{j} + c_3\mathbf{k}$, then

$$(\mathbf{a} \times \mathbf{b}) \cdot \mathbf{c} = (\mathbf{b} \times \mathbf{c}) \cdot \mathbf{a} = (\mathbf{c} \times \mathbf{a}) \cdot \mathbf{b} = \begin{vmatrix} a_1 & a_2 & a_3 \\ b_1 & b_2 & b_3 \\ c_1 & c_2 & c_3 \end{vmatrix} \tag{9-19}$$

$$\mathbf{a} \times (\mathbf{b} \times \mathbf{c}) = (\mathbf{a} \cdot \mathbf{c})\mathbf{b} - (\mathbf{a} \cdot \mathbf{b})\mathbf{c} \tag{9-20}$$

If f is a scalar function of class C'', then

$$\operatorname{curl}(\operatorname{grad}(f)) = \nabla \times \nabla f = \mathbf{0} \tag{9-21}$$

$$\operatorname{div}(\operatorname{grad}(f)) = \nabla \cdot \nabla f = \nabla^2 f = \frac{\partial^2 f}{\partial x^2} + \frac{\partial^2 f}{\partial y^2} + \frac{\partial^2 f}{\partial z^2} \tag{9-22}$$

If $\mathbf{V} = A\mathbf{i} + B\mathbf{j} + C\mathbf{k}$ is a vector field with components of class C'', then

$$\text{(9-23)} \qquad \operatorname{div}(\operatorname{curl}(\mathbf{V})) = \nabla \cdot (\nabla \times \mathbf{V}) = \mathbf{0}$$

$$\begin{aligned}\text{(9-24)} \qquad \operatorname{curl}(\operatorname{curl}(\mathbf{V})) &= \nabla \times (\nabla \times \mathbf{V}) \\ &= \nabla(\nabla \cdot \mathbf{V}) - [\nabla^2 A\mathbf{i} + \nabla^2 B\mathbf{j} + \nabla^2 C\mathbf{k}] \\ &= \operatorname{grad}(\operatorname{div}(\mathbf{V})) - \nabla^2\mathbf{V}\end{aligned}$$

If $\mathbf{F}$ and $\mathbf{G}$ are both vector-valued functions of class C', then

$$\text{(9-25)} \qquad \operatorname{div}(\mathbf{F} \times \mathbf{G}) = \nabla \cdot (\mathbf{F} \times \mathbf{G}) = \mathbf{G} \cdot (\nabla \times \mathbf{F}) - \mathbf{F} \cdot (\nabla \times \mathbf{G})$$

Let us now compare the system of vector analysis, as sketched above, with the system of differential forms in three variables. We notice first that there is a certain formal similarity between the multiplication table for the vector units $\mathbf{i}, \mathbf{j}, \mathbf{k}$ given in (9-10) and (9-11) and the corresponding table for the basic differential forms dx, dy, and dz given in (9-6) and (9-7). In the latter, however, we do not have the identification $dx\,dy = dz$ which corresponds to the relation $\mathbf{i} \times \mathbf{j} = \mathbf{k}$. This suggests that we correspond elements in pairs:

$$\begin{matrix} dx \\ dy\,dz \end{matrix} \leftrightarrow \mathbf{i} \qquad \begin{matrix} dy \\ dz\,dx \end{matrix} \leftrightarrow \mathbf{j} \qquad \begin{matrix} dz \\ dx\,dy \end{matrix} \leftrightarrow \mathbf{k}$$

To complete these, and take into account 0-forms and 3-forms, we adjoin one more correspondence:

$$\begin{matrix} 1 \\ dx\,dy\,dz \end{matrix} \leftrightarrow 1$$

With these, we can set up a 2-to-1 correspondence between differential forms and vector- and scalar-valued functions. To any 1-form or 2-form will correspond a vector function, and to any 0-form or 3-form will correspond a scalar function. The method of correspondence is indicated below:

$$\left.\begin{matrix} A\,dx + B\,dy + C\,dz \\ A\,dy\,dz + B\,dz\,dx + C\,dx\,dy \end{matrix}\right\} \leftrightarrow A\mathbf{i} + B\mathbf{j} + C\mathbf{k}$$

$$\left.\begin{matrix} f(x, y, z) \\ f(x, y, z)\,dx\,dy\,dz \end{matrix}\right\} \leftrightarrow f(x, y, z)$$

In the opposite direction, we see that a vector-valued function corresponds to both a 1-form and a 2-form, and a scalar function to a 0-form and a 3-form. To see the effect of this relationship, let

$$\mathbf{V} = A\mathbf{i} + B\mathbf{j} + C\mathbf{k} \qquad \text{and} \qquad \mathbf{W} = a\mathbf{i} + b\mathbf{j} + c\mathbf{k}$$

Corresponding to $\mathbf{V}$, we choose the 1-form

$$v = A\,dx + B\,dy + C\,dz$$

and corresponding to **W**, both the 1-form and the 2-form

$$(9\text{-}26)\qquad \begin{aligned} \omega &= a\,dx + b\,dy + c\,dz \\ \omega^* &= a\,dy\,dz + b\,dz\,dx + c\,dx\,dy \end{aligned}$$

As shown in Exercises 9 and 10, Sec. 9.2,

$$\nu\omega = \begin{vmatrix} B & C \\ b & c \end{vmatrix} dy\,dz + \begin{vmatrix} C & A \\ c & a \end{vmatrix} dz\,dx + \begin{vmatrix} A & B \\ a & b \end{vmatrix} dx\,dy$$

$$\nu\omega^* = (aA + bB + cC)\,dx\,dy\,dz$$

Reversing the direction of correspondence, and comparing these with (9-12), we see that $\nu\omega$ corresponds to $\mathbf{V} \times \mathbf{W}$ and $\nu\omega^*$ to $\mathbf{V} \cdot \mathbf{W}$. So a *single notion* of multiplication among differential forms corresponds both to the inner product and the cross product among vectors.

What vector operations correspond to differentiation of forms? Let us start with a scalar function f, go to the corresponding 0-form f, and apply d. We obtain the 1-form

$$df = f_1\,dx + f_2\,dy + f_3\,dz$$

which in turn corresponds to the vector function $f_1\mathbf{i} + f_2\mathbf{j} + f_3\mathbf{k}$, the gradient of f. Again, let us start from a vector-valued function

$$\mathbf{V} = A\mathbf{i} + B\mathbf{j} + C\mathbf{k}$$

go to the corresponding 1-form $\omega = \mathrm{A}\,dx + B\,dy + C\,dz$, and again apply d. We obtain a 2-form which was shown in Exercise 14, Sec. 9.2, to be

$$d\omega = (C_2 - B_3)\,dy\,dz + (A_3 - C_1)\,dz\,dx + (B_1 - A_2)\,dx\,dy$$

Upon comparing this with (9-17), we see that this corresponds to the vector function curl (**V**). Finally, if we correspond to **V** the 2-form $\omega^* = A\,dy\,dz + B\,dz\,dx + C\,dx\,dy$, and apply d, we obtain the 3-form $(A_1 + B_2 + C_3)\,dx\,dy\,dz$, which corresponds in turn to the scalar function div (**V**). Briefly, then, the single operation of differentiation in the system of differential forms corresponds in turn to the operations of taking the gradient of a scalar and taking the curl and the divergence of a vector. This is indicated schematically in Fig. 9-11.

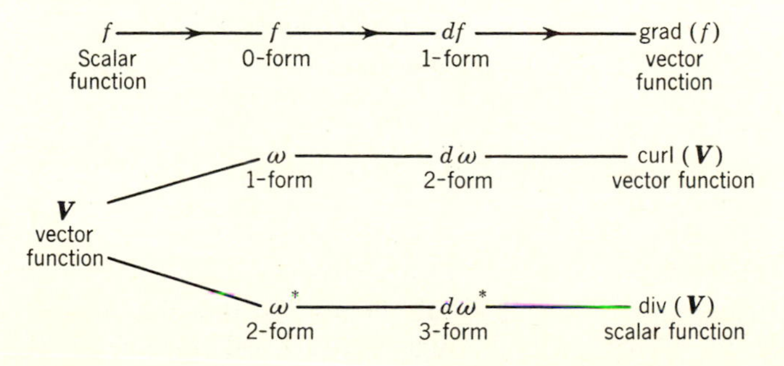

Figure 9-11

Some of the identities in the list given earlier correspond to simple statements about forms.

Theorem 3 *If ω is any differential form of class C'', then $dd\omega = 0$.*

This holds in general when ω is a k form in n variables. We shall prove it when ω is a 1-form in three variables. Let $\omega = A(x, y, z)\,dx$. Then

$$d\omega = d(A)\,dx = \frac{\partial A}{\partial x}\,dx\,dx + \frac{\partial A}{\partial y}\,dy\,dx + \frac{\partial A}{\partial z}\,dz\,dx$$

$$= \frac{\partial A}{\partial y}\,dy\,dx + \frac{\partial A}{\partial z}\,dz\,dx$$

and

$$dd\omega = d\left(\frac{\partial A}{\partial y}\right)dy\,dx + d\left(\frac{\partial A}{\partial z}\right)dz\,dx$$

$$= \frac{\partial^2 A}{\partial z\,\partial y}\,dz\,dy\,dx + \frac{\partial^2 A}{\partial y\,\partial z}\,dy\,dz\,dx = 0$$

using the equality of the mixed derivatives and the fact that

$$dy\,dz\,dx = -dz\,dy\,dx$$

A similar argument holds for $B\,dy$ and $C\,dz$. ▌ (See also Exercise 18.)

Using the relations shown in Fig. 9-11, we see that the statement $ddf = 0$, holding for a 0-form f, corresponds to the vector identity curl (grad f) = 0 [see (9-21)] and the statement $dd\omega = 0$, holding for a 1-form to the vector identity div (curl $\mathbf{V}$) = 0 [see (9-23)].

Our final connections between vector analysis and differential forms will be made by relating the integral of a form to integrals of certain scalar functions which are obtained by vector operations.

Theorem 4 *Let $\mathbf{F} = A\mathbf{i} + B\mathbf{j} + C\mathbf{k}$ define a continuous vector field in a region Ω of space, and let $\omega = A\,dx + B\,dy + C\,dz$ be the corresponding 1-form. Let γ be a smooth curve lying in Ω. Let F_T be the scalar function defined on the trace of γ whose value at a point p is the component of $\mathbf{F}$ in the direction of the tangent to γ at p. Then,*

$$\int_\gamma F_T\,ds = \int_\gamma \omega = \int_\gamma (A\,dx + B\,dy + C\,dz)$$

The first integral is the integral of a numerical-valued function along the curve γ, as discussed in Sec. 8.6. It should be noticed that the function F_T is defined only for points on γ, although $\mathbf{F}$ is defined throughout Ω. We may assume that γ is parametrized by arc length, so that the tangent vector to γ at $p = \gamma(s)$ is $\mathbf{v} = \gamma'(s)$ and $|\mathbf{v}| = 1$. The component of $\mathbf{F}$ along $\mathbf{v}$ is then

$$\begin{aligned} F_T &= \mathbf{F} \cdot \mathbf{v} \\ &= (A\mathbf{i} + B\mathbf{j} + C\mathbf{k}) \cdot \left(\frac{dx}{ds}\,\mathbf{i} + \frac{dy}{ds}\,\mathbf{j} + \frac{dz}{ds}\,\mathbf{k}\right) \\ &= A\,\frac{dx}{ds} + B\,\frac{dy}{ds} + C\,\frac{dz}{ds} \end{aligned}$$

and

$$\int_\gamma F_T\,ds = \int_0^l \left(A\,\frac{dx}{ds} + B\,\frac{dy}{ds} + C\,\frac{dz}{ds}\right) ds$$

Comparing this with the definition of the integral of a 1-form (9-4), we see that

$$\int_\gamma F_T\,ds = \int_\gamma \omega \quad \blacksquare$$

The situation for integrals of 2-forms is similar.

Theorem 5 *Let* $\mathbf{F} = A\mathbf{i} + B\mathbf{j} + C\mathbf{k}$ *define a continuous vector field in a region* Ω, *and let* $\omega = A\,dy\,dz + B\,dz\,dx + C\,dx\,dy$ *be the corresponding 2-form. Let* Σ *be a smooth surface lying in* Ω. *Let* F_N *be the scalar function defined on the trace of* Σ *whose value at a point p is the component of* $\mathbf{F}$ *in the direction of the normal to* Σ *at p. Then*

$$\iint_\Sigma F_N\,dA = \iint_\Sigma \omega = \iint_\Sigma (A\,dy\,dz + B\,dz\,dx + C\,dx\,dy)$$

Let Σ have domain D in the (u, v) plane. By (8-47), the normal to Σ at the point $p = \Sigma(u, v)$ is

$$\mathbf{n}(u, v) = \left(\frac{\partial(y, z)}{\partial(u, v)}, \frac{\partial(z, x)}{\partial(u, v)}, \frac{\partial(x, y)}{\partial(u, v)}\right)$$

so that the normal component of $\mathbf{F}$ at p is

$$F_N = \frac{\mathbf{F} \cdot \mathbf{n}}{|\mathbf{n}|} = \frac{A\,\dfrac{\partial(y, z)}{\partial(u, v)} + B\,\dfrac{\partial(z, x)}{\partial(u, v)} + C\,\dfrac{\partial(x, y)}{\partial(u, v)}}{|\mathbf{n}(u, v)|}$$

Thus, according to the definition given in (8-58),

$$\begin{aligned} \iint_\Sigma F_N\,dA &= \iint_D F_N(\Sigma(u, v))\,|\mathbf{n}(u, v)|\,du\,dv \\ &= \iint_D \left(A\,\frac{\partial(y, z)}{\partial(u, v)} + B\,\frac{\partial(z, x)}{\partial(u, v)} + C\,\frac{\partial(x, y)}{\partial(u, v)}\right) du\,dv \end{aligned}$$

Using Theorem 1, Sec. 9.2, this may be written as

$$\iint_\Sigma F_N\, dA = \iint_\Sigma A\, dy\, dz + B\, dz\, dx + C\, dx\, dy = \iint_\Sigma \omega \quad \blacksquare$$

To show how such integrals arise, let **V** be the vector-valued function which describes the velocity field of a liquid which is flowing throughout a region Ω, and let $\mathbf{F} = \rho\mathbf{V}$, where ρ is the scalar function which gives the density distribution of the liquid in Ω. Let Σ be a smooth orientable surface (for example, a portion of the surface of a sphere) lying in Ω. At any point on Σ, F_N measures the rate of flow of mass across Σ in the direction of the normal. The integral $\iint_\Sigma F_N\, dA$ is then the total mass of fluid which passes through Σ, per unit time. Again, let γ be a smooth curve lying in Ω. At a point lying on γ, V_T is the component of the velocity of the fluid taken in the direction of the tangent to γ. The integral $\int_\gamma V_T\, ds$ is then a measure of the extent to which the motion of the fluid is a flow *along* the curve γ. If γ is a closed curve, then $\int_\gamma V_T$ is called the circulation around γ.

Integrals of the same sort also arise in mechanics, especially in connection with the notion of work, in thermodynamics, and in electromagnetic theory.

(See J. C. Slater and N. H. Frank, "Introduction to Theoretical Physics," 1974; J. A. Stratton, "Electromagnetic Theory," 1941; and P. M. Morse and H. Feshbach, "Methods of Theoretical Physics," 1953, all published by McGraw-Hill Book Company, New York.)

EXERCISES

1 Let $\mathbf{a} = 2\mathbf{i} - 3\mathbf{j} + \mathbf{k}$, $\mathbf{b} = \mathbf{i} - \mathbf{j} + 3\mathbf{k}$, $\mathbf{c} = \mathbf{i} - 2\mathbf{j}$. Compute the vectors $(\mathbf{a} \times \mathbf{b}) \cdot \mathbf{c}$, $\mathbf{a} \times (\mathbf{b} \times \mathbf{c})$, $(\mathbf{a} \times \mathbf{b}) \times \mathbf{c}$, $\mathbf{a} \times (\mathbf{a} \times \mathbf{b})$, $(\mathbf{a} + \mathbf{b}) \times (\mathbf{b} + \mathbf{c})$, $(\mathbf{a} \cdot \mathbf{b})\mathbf{c} - (\mathbf{a} \cdot \mathbf{c})\mathbf{b}$.

2 If **a**, **b**, and **c** are position vectors, show that the vector $\mathbf{n} = \mathbf{a} \times \mathbf{b} + \mathbf{b} \times \mathbf{c} + \mathbf{c} \times \mathbf{a}$ is a normal to the plane through the points **a**, **b**, **c**.

3 Three points $p_j = (x_j, y_j, z_j)$ which do not lie in a plane through the origin determine a trihedral with sides $\overrightarrow{\mathbf{0}p_1}, \overrightarrow{\mathbf{0}p_2}, \overrightarrow{\mathbf{0}p_3}$ which has positive orientation if and only if

$$\begin{vmatrix} x_1 & y_1 & z_1 \\ x_2 & y_2 & z_2 \\ x_3 & y_3 & z_3 \end{vmatrix} > 0$$

Using this, show that the vectors **a**, **b**, and $\mathbf{a} \times \mathbf{b}$ form a trihedral having positive orientation, unless **a** and **b** are parallel.

4 Given vectors **a** and **b**, and a real number k, when is there a vector **v** such that $\mathbf{a} \times \mathbf{v} = \mathbf{b}$ and $\mathbf{a} \cdot \mathbf{v} = k$?

5 Define a sequence of vectors $\{p_n\}$ by $p_1 = \mathbf{a}$, $p_2 = \mathbf{i}$, $p_3 = \mathbf{a} \times \mathbf{i}$, $p_4 = \mathbf{a} \times p_3$, and in general, $p_{n+1} = \mathbf{a} \times p_n$. What is the ultimate behavior of the sequence?

6 If **f** and **g** are vector-valued functions of a single variable, show that

$$\frac{d}{dt}(\mathbf{f} \cdot \mathbf{g}) = \left(\frac{d}{dt}\mathbf{f}\right) \cdot \mathbf{g} + \mathbf{f} \cdot \left(\frac{d}{dt}\mathbf{g}\right)$$

and

$$\frac{d}{dt}(\mathbf{f} \times \mathbf{g}) = \left(\frac{d}{dt}\mathbf{f}\right) \times \mathbf{g} + \mathbf{f} \times \left(\frac{d}{dt}\mathbf{g}\right)$$

7 If **f** is a vector-valued function of one variable, and $|\mathbf{f}(t)| = 1$ for all t, show that $\mathbf{f}(t)$ and $\mathbf{f}'(t)$ are always orthogonal. Does this have a simple geometric interpretation?

8 Show that the curvature of a curve γ at the point $\gamma(c)$ is

$$k = \frac{|\gamma'(c) \times \gamma''(c)|}{|\gamma'(c)|^3}$$

9 Show that the normal to the surface Σ can be defined by [see (8-47)]

$$\mathbf{n} = \Sigma_u \times \Sigma_v$$

10 Let $p = (x, y, z)$ and $r = |p|$. Find the gradient of the functions r^2, r, $1/r$, r^m, and $\log r$.

11 Show that the gradient vectors for f are orthogonal to the level surfaces

$$f(x, y, z) = c$$

***12** Let A be the square matrix $[a_{ij}]$ and let B be a nonsingular matrix. Set $A^* = B^{-1}AB$. Show that the trace of A (the sum of the diagonal entries) is the same as that of A^*.

13 Verify identities (9-19) and (9-20).

14 Verify identities (9-21) and (9-23).

15 (*a*) Verify identity (9-22).

(*b*) Show that if f and g are scalar functions of class C'', then

$$\nabla^2(fg) = f\nabla^2 g + g\nabla^2 f + 2(\text{grad } f \cdot \text{grad } g)$$

16 Verify (9-24).

17 Verify (9-25).

18 Prove Theorem 3 when ω is a 0-form, and when ω is a 2-form in four variables.

19 Show that div (grad $f \times$ grad g) = 0.

20 Let $\mathbf{a} = (a_1, a_2, a_3)$, $\mathbf{b} = (b_1, b_2, b_3)$, and $\mathbf{c} = (c_1, c_2, c_3)$ be three vector functions defined in Ω. Put $\alpha = \sum a_i\, dx_i$, $\beta = \sum b_i\, dx_i$, $\gamma = \sum c_i\, dx_i$, and show that $\alpha\beta\gamma = \{(\mathbf{a} \times \mathbf{b}) \cdot \mathbf{c}\}\, dx_1\, dx_2\, dx_3$ [see (9-19)].

9.4 THE THEOREMS OF GREEN, GAUSS, AND STOKES

The important theorems which form the subject of this section deal with relations among line integrals, surface integrals, and volume integrals. In the language of general differential forms, they connect an integral of a differential form ω with an integral of its derivative $d\omega$. The theorems named in the title of the section are the special cases in 3-space of what is called the **generalized Stokes' theorem,** which takes the following form.

Theorem 6 *Let ω be a k form defined in an open region Ω of n space, and let M be a suitably well-behaved $k+1$–dimensional surface (manifold) in Ω. Then,*

$$\int \cdots \iint_{\partial M} \omega = \int \cdots \iint_{M} d\omega \tag{9-27}$$

where the integral on the left is k-fold, and that on the right is $k+1$–fold.

If $n = 2$, and k is chosen as 1, Theorem 6 becomes:

Green's Theorem *Let D be a suitably well-behaved region in the plane whose boundary is a curve ∂D. Let ω be a 1-form of class C' defined in D. Then,*

$$\int_{\partial D} \omega = \iint_{D} d\omega \tag{9-28}$$

If $n = 3$ and k is chosen first as 1 and then as 2, Theorem 6 becomes in turn:

Stokes' Theorem *Let Σ be a suitably well-behaved orientable surface whose boundary is a curve $\partial\Sigma$. Let ω be a 1-form of class C' defined on Σ. Then*

$$\int_{\partial \Sigma} \omega = \iint_{\Sigma} d\omega$$

Divergence Theorem (Gauss) *Let R be a suitably well-behaved region in space whose boundary ∂R is a surface. Let ω be a 2-form of class C' defined on R. Then*

$$\iint_{\partial R} \omega = \iiint_{R} d\omega$$

The qualification "suitably well behaved" which occurs in each statement is inserted to indicate that it is convenient to impose some restrictions upon the regions, surfaces, and curves on which the integration is carried out. The exact nature of these restrictions is chiefly dependent upon the tools which are employed to prove the theorem. In most of the simpler applications of these theorems, one encounters only the nicest of curves and surfaces; for these, the proofs which we shall give are sufficient.

Let us begin by proving a special case of Green's theorem.

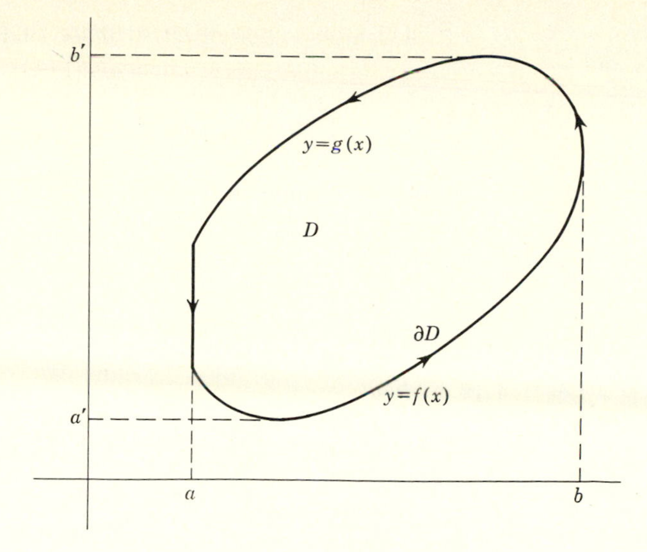

Figure 9-12

Theorem 7 *Let D be a closed convex region in the plane, and let* $\omega = A(x, y)\,dx + B(x, y)\,dy$ *with A and B of class C′ in D. Then,*

$$\int_{\partial D} A\,dx + B\,dy = \iint_D d\omega = \iint_D \left(\frac{\partial B}{\partial x} - \frac{\partial A}{\partial y}\right) dx\,dy$$

The assumption that D is convex allows us to describe D in two ways (see Fig. 9-12). If the projection of D onto the horizontal axis is the closed interval $[a, b]$, then D is the set of all points (x, y) such that

$$a \le x \le b \qquad (9\text{-}29)$$
$$f(x) \le y \le g(x)$$

where g and f are the continuous functions whose graphs form the top and bottom pieces of the boundary of D. Likewise, if $[a', b']$ is the projection of D upon the vertical axis, D is the set of points (x, y) such that

$$a' \le y \le b' \qquad (9\text{-}30)$$
$$F(y) \le x \le G(y)$$

Since $\int_{\partial D} \omega = \int_{\partial D} A\,dx + \int_{\partial D} B\,dy$, we may treat each part separately. Suppose that $\omega = A(x, y)\,dx$. On γ_1, the lower part of ∂D, $y = f(x)$, $a \le x \le b$, and $\omega = A(x, f(x))\,dx$.

Thus
$$\int_{\gamma_1} \omega = \int_a^b A(x, f(x))\,dx$$

On γ_2, the upper part of ∂D, $y = g(x)$ and x goes from b to a. Thus, $\omega = A(x, g(x))\,dx$ and

$$\int_{\gamma_2} \omega = \int_b^a A(x, g(x))\,dx = -\int_a^b A(x, g(x))\,dx$$

On the vertical parts of ∂D, if any, $\omega = 0$. Adding, we find

$$\int_{\partial D} \omega = \int_{\partial D} A\,dx = \int_a^b [A(x, f(x)) - A(x, g(x))]\,dx$$

On the other hand, $d\omega = d(A(x, y)\,dx) = A_2(x, y)\,dy\,dx$, so that

$$\iint_D d\omega = \iint_D A_2(x, y)\,dy\,dx$$

$$= -\iint_D A_2(x, y)\,dx\,dy$$

$$= -\int_a^b \mathrm{dx} \int_{f(x)}^{g(x)} A_2(x, y)\,dy$$

$$= -\int_a^b [A(x, g(x)) - A(x, f(x))]\,dx$$

Comparing the two results, we have $\int_{\partial D} \omega = \iint_D d\omega$. A similar computation, using (9-30), shows that the same relation holds if $\omega = B(x, y)\,dy$, and adding these, we obtain the formula for a general 1-form. ▮

In generalizing this theorem, we first remark that the proof used only the fact that D could be described both in the form (9-29) and in the form (9-30). Such regions need not be convex, as is shown by the region given by: $x \geq 0$, $y \leq x^2$, $y \geq 2x^2 - 1$. For such regions, we shall use the term "standard region." Suppose now that D is itself not standard, but is the union of a finite number of standard regions such as shown in Fig. 9-13. Green's theorem holds for each of the regions D_j so that

$$\int_{\partial D_j} \omega = \iint_{D_j} d\omega$$

and adding, we have

$$\int_{\partial D_1} \omega + \int_{\partial D_2} \omega + \cdots + \int_{\partial D_n} \omega = \iint_{D_1} d\omega + \iint_{D_2} d\omega + \cdots + \iint_{D_n} d\omega$$

$$= \iint_D d\omega$$

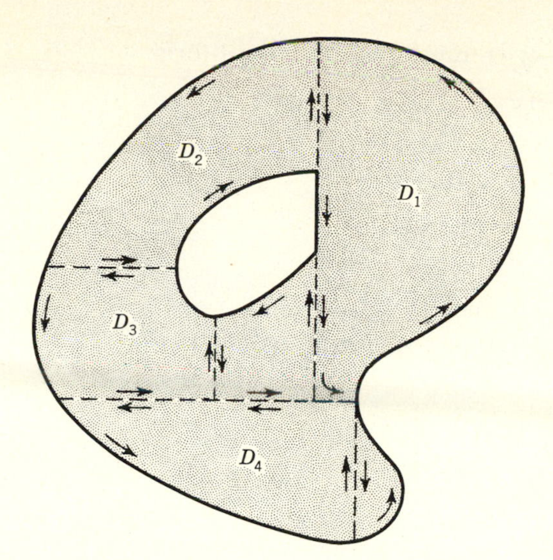

Figure 9-13

However, in adding the line integrals on the left, only the terms which arise from parts of the boundary of D will remain. A curve γ which forms a portion of ∂D_j but not of ∂D will also appear as part of the boundary of one of the other standard regions; moreover, it will appear with the opposite orientation, so that the sum of the corresponding line integrals will be $\int_\gamma \omega + \int_{-\gamma} \omega = 0$. This type of argument shows that Green's theorem is valid for regions D which can be expressed as the union of a finite number of standard regions. Finally, one may extend the theorem still further by considering regions D which are the limits, in a suitable sense, of such regions.

One may generalize Theorem 7 in another way. Let T be a continuous transformation from the (u, v) plane into the (x, y) plane which maps a set D onto a set D^*. Suppose that Green's theorem is valid for the set D. Does it then hold for the set D^*? To obtain an answer for this question, we must discuss the behavior of differential forms under a general transformation. Let us assume that T is 1-to-1 and of class C'' in D, and is given by

$$\begin{cases} x = \phi(u, v) \\ y = \psi(u, v) \end{cases}$$

We use these equations to transform any differential form in the XY plane into a differential form in the UV plane. To effect the transformation, we replace x and y wherever they appear by ϕ and ψ, respectively. For example, a 0-form $f(x, y)$ is transformed into the 0-form

$$f^*(u, v) = f(\phi(u, v), \psi(u, v))$$

a 1-form $\omega = A(x, y)\,dx + B(x, y)\,dy$ is transformed into the 1-form

$$\begin{aligned}\omega^* &= A^*(u, v)\,d\phi + B^*(u, v)\,d\psi \\ &= A^*(u, v)[\phi_1(u, v)\,du + \phi_2(u, v)\,dv] \\ &\quad + B^*(u, v)[\psi_1(u, v)\,du + \psi_2(u, v)\,dv] \\ &= [A^*(u, v)\phi_1(u, v) + B^*(u, v)\psi_1(u, v)]\,du \\ &\quad + [A^*(u, v)\phi_2(u, v) + B^*(u, v)\psi_2(u, v)]\,dv\end{aligned}$$

where A^* and B^* are the 0-forms obtained by transforming A and B.

As an illustration, let T be given by $x = u^2 + v$, $y = v$, and let

$$\omega = xy\,dx$$

Then,

$$\begin{aligned}\omega^* &= (u^2 + v)(v)\,d(u^2 + v) \\ &= (u^2v + v^2)(2u\,du + dv) \\ &= 2(u^3v + uv^2)\,du + (u^2v + v^2)\,dv\end{aligned}$$

Likewise, a 2-form such as $\sigma = xy^2\,dx\,dy$ in the (x, y) plane is transformed by substitution into a 2-form in the (u, v) plane given by

$$\begin{aligned}\sigma^* &= (u^2 + v)(v)^2(2u\,du + dv)(dv) \\ &= (2u^3v^2 + 2uv^3)\,du\,dv\end{aligned}$$

In this illustration, we have a relationship between an arbitrary transformation T mapping the (u, v) plane into the (x, y) plane, and another transformation, which we shall call T^*, that acts on differential forms in the (x, y) plane and changes them into differential forms in the (u, v) plane. This is sometimes indicated as in Fig. 9-14; observe that T^* acts in a direction opposite to T, and that T^* is not a point mapping, but is applied to k forms. T^* can be called the substitution mapping on forms induced by the point mapping T, with $T^*(\omega) = \omega^*$.

A key property of T^* is given in the next result, which shows that T^* and the differentiation operator d commute. We can illustrate it again with the transformation used above and the 1-form $\omega = xy\,dx$. Recall that we had found

$$T^*(\omega) = \omega^* = 2(u^3v + uv^2)\,du + (u^2v + v^2)\,dv$$

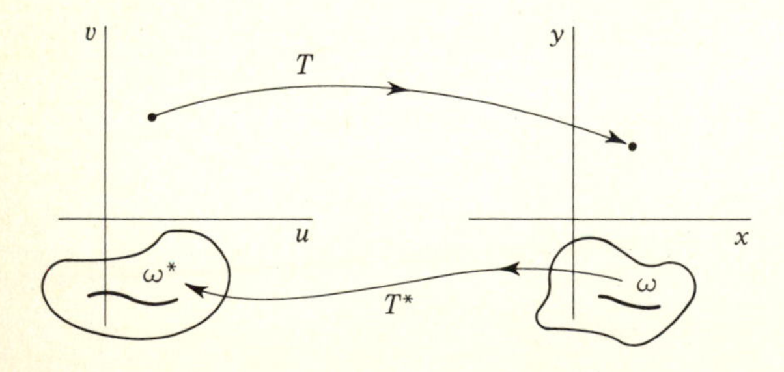

Figure 9-14

Suppose we compare $(d\omega)^*$ and $d(\omega^*)$.

$$\begin{aligned} d(\omega^*) &= 2(3u^2v\,du + u^3\,dv + v^2\,du + 2uv\,dv)\,du + (2uv\,du + u^2\,dv + 2v\,dv)\,dv \\ &= 2(u^3 + 2uv)\,dv\,du + (2uv)\,du\,dv \\ &= (2u^3 + 4uv - 2uv)\,dv\,du = (2u^3 + 2uv)\,dv\,du \end{aligned}$$

On the other hand, we have

$$\begin{aligned} d\omega &= d(xy\,dx) = d(xy)\,dx \\ &= (x\,dy + y\,dx)(dx) = x\,dy\,dx \end{aligned}$$

and applying T^*, we have

$$\begin{aligned} T^*(d\omega) &= (d\omega)^* \\ &= (u^2 + v)(dv)(2u\,du + dv) \\ &= 2u(u^2 + v)\,dv\,du \end{aligned}$$

which agrees with $d(\omega^*)$; $(d\omega)^* = d(\omega^*)$.

The next theorem proves that this is true in general.

Theorem 8 *If ω is a differential form of class C', then*

$$T^*(d\omega) = (d\omega)^* = d(\omega^*) = dT^*(\omega) \tag{9-31}$$

Let T be a transformation of class C'' from (u, v) space into (x, y, z) space, described by

$$\begin{cases} x = \phi(u, v) \\ y = \psi(u, v) \\ z = \theta(u, v) \end{cases}$$

In the same fashion as before, we may use T to transform differential forms in the variables x, y, z into forms in u, v, and we shall use ω^* to denote the form obtained by transforming ω. To begin with, suppose that ω is a 0-form $f(x, y, z)$. Its transform ω^* will then be the 0-form

$$\omega^* = f^*(u, v) = f(\phi(u, v), \psi(u, v), \theta(u, v))$$

Differentiating this, we obtain the 1-form

$$d(\omega^*) = (f_1\phi_1 + f_2\psi_1 + f_3\theta_1)\,du + (f_1\phi_2 + f_2\psi_2 + f_3\theta_2)\,dv$$

If we differentiate ω *before* transforming, we obtain the 1-form

$$d\omega = f_1\,dx + f_2\,dy + f_3\,dz$$

and the transform of this is the 1-form

$$\begin{aligned} (d\omega)^* &= f_1(\phi_1\,du + \phi_2\,dv) + f_2(\psi_1\,du + \psi_2\,dv) + f_3(\theta_1\,du + \theta_2\,dv) \\ &= (f_1\phi_1 + f_2\psi_1 + f_3\theta_1)\,du + (f_1\phi_2 + f_2\psi_2 + f_3\theta_2)\,dv \end{aligned}$$

This agrees with the expression for $d(\omega^*)$, and we have shown that the relation $d(\omega^*) = (d\omega)^*$ holds for 0-forms.

Let us suppose now that ω is the 1-form $A(x, y, z)\,dx$. Its transform will be the 1-form

$$\begin{aligned}\omega^* &= A[\phi_1\,du + \phi_2\,dv]\\ &= A\phi_1\,du + A\phi_2\,dv\end{aligned}$$

Differentiating this, we obtain the 2-form

$$\begin{aligned}d(\omega^*) &= \left[\frac{\partial}{\partial u}(A\phi_1)\,du + \frac{\partial}{\partial v}(A\phi_1)\,dv\right]du\\ &\quad+ \left[\frac{\partial}{\partial u}(A\phi_2)\,du + \frac{\partial}{\partial v}(A\phi_2)\,dv\right]dv\\ &= \left[\frac{\partial}{\partial u}(A\phi_2) - \frac{\partial}{\partial v}(A\phi_1)\right]du\,dv\\ &= \left[A\phi_{12} + \phi_2\frac{\partial A}{\partial u} - A\phi_{21} - \phi_1\frac{\partial A}{\partial v}\right]du\,dv\\ &= \left[\phi_2\frac{\partial A}{\partial u} - \phi_1\frac{\partial A}{\partial v}\right]du\,dv\end{aligned}$$

making use of the fact that T is of class C'' and thus $\phi_{12} = \phi_{21}$. We compute $\partial A/\partial u$ and $\partial A/\partial v$ by the chain rule.

$$\begin{aligned}\frac{\partial A}{\partial u} &= A_1\phi_1 + A_2\psi_1 + A_3\theta_1\\ \frac{\partial A}{\partial v} &= A_1\phi_2 + A_2\psi_2 + A_3\theta_2\end{aligned}$$

so that

$$\begin{aligned}\phi_2\frac{\partial A}{\partial u} - \phi_1\frac{\partial A}{\partial v} &= \phi_2[A_1\phi_1 + A_2\psi_1 + A_3\theta_1] - \phi_1[A_1\phi_2 + A_2\psi_2 + A_3\theta_2]\\ &= A_2(\psi_1\phi_2 - \psi_2\phi_1) + A_3(\theta_1\phi_2 - \theta_2\phi_1)\end{aligned}$$

Thus $$d(\omega^*) = \left\{A_2\begin{vmatrix}\psi_1 & \psi_2\\ \phi_1 & \phi_2\end{vmatrix} + A_3\begin{vmatrix}\theta_1 & \theta_2\\ \phi_1 & \phi_2\end{vmatrix}\right\}du\,dv$$

If, on the other hand, we differentiate ω before transforming, we have

$$\begin{aligned}d\omega &= (A_1\,dx + A_2\,dy + A_3\,dz)\,dx\\ &= A_2\,dy\,dx + A_3\,dz\,dx\end{aligned}$$

If we transform this, and use Theorem 1, we have

$$dy\,dx = \frac{\partial(y, x)}{\partial(u, v)}\,du\,dv = \begin{vmatrix} \psi_1 & \psi_2 \\ \phi_1 & \phi_2 \end{vmatrix} du\,dv$$

$$dz\,dx = \frac{\partial(z, x)}{\partial(u, v)}\,du\,dv = \begin{vmatrix} \theta_1 & \theta_2 \\ \phi_1 & \phi_2 \end{vmatrix} du\,dv$$

so that

$$(d\omega)^* = \left\{ A_2 \begin{vmatrix} \psi_1 & \psi_2 \\ \phi_1 & \phi_2 \end{vmatrix} + A_3 \begin{vmatrix} \theta_1 & \theta_2 \\ \phi_1 & \phi_2 \end{vmatrix} \right\} du\,dv$$

Since this agrees with the expression for $d(\omega^*)$, we have shown that $d(\omega^*) = (d\omega)^*$ holds when $\omega = A(x, y, z)\,dx$. A similar computation shows that the relation is valid also for the 1-forms $B\,dy$ and $C\,dz$, and we have thus shown that it holds for any 1-form.

A comparable direct verification may also be made when ω is a 2-form such as $A(x, y, z)\,dy\,dz$. Rather than carry this out, we shall use an inductive method which proves the theorem for differential forms in n variables. We make use of two special formulas. Forms are transformed by T by replacing $x, y, z, \ldots$ wherever they occur by $\phi, \psi, \theta, \ldots$, and then simplifying the result by means of the algebra of forms. Thus, if α and β are any two forms, then

$$T^*(\alpha\beta) = (\alpha\beta)^* = \alpha^*\beta^* = T^*(\alpha)T^*(\beta) \tag{9-32}$$

The second formula expresses the manner in which the differentiation operator acts on products of forms.

Lemma 1 *If α is a k form and β any differential form, then*

$$d(\alpha\beta) = (d\alpha)\beta + (-1)^k\alpha(d\beta) \tag{9-33}$$

We prove this by induction on the dimension of α. Suppose that α is a 0-form, and thus $\alpha = A(x, y, \ldots)$, an ordinary function, and suppose that $\beta = B\sigma$, where B is a function and σ is a product of pure differentials (e.g., $\sigma = dx\,dz\,dw$). Then, $\alpha\beta = AB\sigma$, and

$$\begin{aligned} d(\alpha\beta) = d(AB)\sigma &= \{(A_x B + AB_x)\,dx + (A_y B + AB_y)\,dy + \cdots\}\sigma \\ &= (A_x B\,dx + A_y B\,dy + \cdots)\sigma + (AB_x\,dx + AB_y\,dy + \cdots)\sigma \\ &= (A_x\,dx + A_y\,dy + \cdots)B\sigma + A(B_x\,dx + B_y\,dy + \cdots)\sigma \\ &= (dA)(B\sigma) + A(dB)\sigma = (d\alpha)\beta + \alpha(d\beta) \end{aligned}$$

verifying (9-33) for the case when α is a 0-form. We now verify (9-33) if α is the 1-form dw (or any other single basic differential). As before, take

$\beta = B\sigma$, where σ is a product of basic differentials. Then, $\alpha\beta = (dw)B(\sigma) = B(dw)(\sigma)$, and

$$\begin{aligned} d(\alpha\beta) &= (dB)(dw)(\sigma) = (B_x\,dx + B_y\,dy + \cdots)(dw)(\sigma) \\ &= -(dw)(B_x\,dx + B_y\,dy + \cdots)(\sigma) \\ &= -(dw)(dB)(\sigma) = -\alpha(d\beta) \end{aligned}$$

But, this verifies (9-33) for this case, since $k = 1$ and $d\alpha = d(dw) = 0$. More generally, if $\alpha = A\,dw$, then $\alpha\beta = AB(dw)(\sigma)$,

$$\begin{aligned} d(\alpha\beta) &= d(AB)(dw)(\sigma) = [(dA)B + A(dB)](dw)(\sigma) \\ &= (dA)B(dw)(\sigma) + A(dB)(dw)(\sigma) \\ &= (dA)(dw)B(\sigma) - A(dw)(dB)(\sigma) \end{aligned}$$

Thus,

$$d(\alpha\beta) = (d\alpha)\beta - \alpha(d\beta) \tag{9-34}$$

as required by (9-33).

Knowing that (9-33) holds for $k = 0, 1$, we use induction to complete the proof. Assume now that (9-33) has been verified when α is any differential form of dimension less than k, and let α be a k form. We may suppose that $\alpha = \gamma\,dw$, where γ is a form of dimension $k - 1$, and where dw is one of the basic differentials present in α. Then, by the inductive assumption,

$$\begin{aligned} d\alpha = d(\gamma\,dw) &= (d\gamma)\,dw + (-1)^{k-1}\gamma\,d(dw) \\ &= (d\gamma)(dw) + 0 = (d\gamma)(dw) \end{aligned}$$

Also, for any form β, we have, from (9-34),

$$\begin{aligned} d((dw)\beta) &= d(dw)\beta - (dw)(d\beta) \\ &= 0 - (dw)(d\beta) = -(dw)(d\beta) \end{aligned} \tag{9-35}$$

Finally, $\alpha\beta = (\gamma\,dw)\beta = \gamma((dw)\beta)$, so that, again by the inductive assumption and (9-35),

$$\begin{aligned} d(\alpha\beta) &= (d\gamma)(dw\beta) + (-1)^{k-1}\gamma\,d((dw)\beta) \\ &= (d\gamma)(dw)\beta + (-1)^{k-1}(-1)\gamma(dw)(d\beta) \\ &= (d\alpha)\beta + (-1)^k\alpha(d\beta) \end{aligned}$$

which is (9-33). ▮

With this lemma established, we complete the proof of Theorem 8. Suppose we know that the relation (9-31) holds if ω is either of the differential forms α and β; we will show that it also holds if $\omega = \alpha\beta$. By (9-32), $\omega^* = \alpha^*\beta^*$. Since (9-31) holds for α and β, $d(\alpha^*) = (d\alpha)^*$ and $d(\beta^*) = (d\beta)^*$.

Using these, together with the lemma, we have

$$\begin{aligned} d(\omega^*) = d(\alpha^*\beta^*) &= d(\alpha^*)\beta^* + (-1)^k\alpha^*\, d(\beta^*) \\ &= (d\alpha)^*\beta^* + (-1)^k\alpha^*(d\beta)^* \\ &= [(d\alpha)\beta]^* + [(-1)^k\alpha(d\beta)]^* \\ &= [(d\alpha)\beta + (-1)^k\alpha(d\beta)]^* \\ &= [d(\alpha\beta)]^* \end{aligned}$$

This shows that if (9-31) holds for any two differential forms ω, it holds for their product. Since it holds for 0-forms and for the basic 1-forms dx, dy, dz, etc., and since any differential form is built from these by multiplication and addition, (9-31) holds for any choice of ω. ▮

Having shown that differentiation of differential forms commutes with the substitution operator T^*, we at once obtain a useful extension of Green's theorem, which will be a guide for other extensions of various forms of the generalized Stokes' theorem.

Theorem 9 *Let T be a transformation which is 1-to-1 and of class C'' in a closed bounded region D, mapping D onto D^*. Then, if Green's theorem holds for D, it holds for D^*.*

We take D as a set in the (u, v) plane whose boundary is a curve γ. T carries D into the set D^* in the (x, y) plane, and γ into the boundary γ^* of D^*. Let ω be any 1-form in the (x, y) plane of class C', and let T transform ω into the 1-form ω^*. T will also transform the 2-form $d\omega$ into a 2-form $(d\omega)^*$ in u and v. The results in Sec. 8.3 dealing with transformation of integrals give at once the formulas

$$(9\text{-}36)\qquad \int_\gamma \omega^* = \int_{\gamma^*} \omega \qquad \iint_D (d\omega)^* = \iint_{D^*} d\omega$$

By assumption, Green's theorem holds in D; thus, applying it to the 1-form ω^*,

$$(9\text{-}37)\qquad \int_\gamma \omega^* = \iint_D d(\omega^*)$$

By the fundamental relation (9-31), $d(\omega^*) = (d\omega)^*$. Combining this with (9-36) and (9-37),

$$\int_{\gamma^*} \omega = \int_\gamma \omega^* = \iint_D (d\omega)^* = \iint_{D^*} d\omega$$

and Green's theorem holds for D^*. ▮

This same technique gives another proof of the formula for change of variable in double integrals in a more general form.

Theorem 10 *Let T be a transformation of class C'' defined by $x = \phi(u, v)$, $y = \psi(u, v)$, mapping a compact set D onto D^*. We assume that D and D^* are finite unions of standard regions and that T is 1-to-1 on the boundary of D and maps it onto the boundary of D^*. Let f be continuous in D^*. Then,*

$$\iint_{D^*} f(x, y)\, dx\, dy = \iint_{D} f(\phi(u, v), \psi(u, v)) \frac{\partial(x, y)}{\partial(u, v)}\, du\, dv$$

As we have seen in Sec. 8-3, the difficult step is to show that this holds when $f(p)$ is constantly 1:

$$\iint_{D^*} dx\, dy = A(D^*) = \iint_{D} \frac{\partial(x, y)}{\partial(u, v)}\, du\, dv$$

Consider the special 1-form $\omega = x\, dy$. This is chosen because $d\omega = dx\, dy$, so that by Green's theorem,

$$A(D^*) = \iint_{D^*} d\omega = \int_{\partial D^*} \omega$$

Writing this as an ordinary singlefold definite integral, and applying the formula for change of variable in this case (Sec. 4.3, Theorem 8), we obtain

$$A(D^*) = \int_{\partial D} \omega^*$$

Applying Green's theorem to D, and the fundamental invariance relation (9-31), we have

$$A(D^*) = \iint_{D} d(\omega^*) = \iint_{D} (d\omega)^*$$

Since $d\omega = dx\, dy$,

$$(d\omega)^* = d\phi\, d\psi = \begin{vmatrix} \phi_1 & \phi_2 \\ \psi_1 & \psi_2 \end{vmatrix} du\, dv = \frac{\partial(x, y)}{\partial(u, v)}\, du\, dv. \quad \blacksquare$$

We note that this form of the change-of-variable theorem is in certain respects more general than that obtained in Sec. 8.3. Here, the Jacobian of T is not required to have constant sign, provided that the integral for $A(D^*)$ is understood to be an oriented integral. To offset this, the transformation T was required to be of class C''.

Next, we proceed to give a proof of Stokes' theorem by reducing it to an application of Green's theorem.

Theorem 11 *Let Σ be a smooth surface of class C'' whose domain D is a standard region, or a finite union of standard regions, in the UV plane. Let*

$$\omega = A\,dx + B\,dy + C\,dz$$

where A, B, and C are of class C' on Σ. Then,

$$\int_{\partial\Sigma} A\,dx + B\,dy + C\,dz = \iint_{\Sigma} \left(\frac{\partial C}{\partial y} - \frac{\partial B}{\partial z}\right) dy\,dz + \left(\frac{\partial A}{\partial z} - \frac{\partial A}{\partial x}\right) dz\,dx$$

$$+ \left(\frac{\partial B}{\partial x} - \frac{\partial A}{\partial y}\right) dx\,dy$$

The line integral on the left is $\int_{\partial\Sigma} \omega$, while the surface integral on the right is $\iint_{\Sigma} d\omega$. It should be noticed that this becomes Green's theorem if the surface Σ is taken as the region D in the XY plane; $C = 0$ and $dz = 0$, so that the differential form in the right-hand integral becomes $(\partial B/\partial x - \partial A/\partial y)\,dx\,dy$. The proof of the theorem is almost immediate. By assumption, Σ is described by a set of equations $x = \phi(u, v)$, $y = \psi(u, v)$, $z = \theta(u, v)$ for $(u, v) \in D$ which define a transformation of class C''. Moreover, D is a region to which Green's theorem applies. If we use the transformation Σ to transform ω, we obtain a 1-form ω^* in u and v. Applying the method used in Theorem 9, Green's theorem in D, and the fundamental relation (9-31), we have

$$\int_{\partial\Sigma} \omega = \int_{\partial D} \omega^* = \iint_{D} d(\omega^*) = \iint_{D} (d\omega)^* = \iint_{\Sigma} d\omega$$

which, as we have seen, is the conclusion of Stokes' theorem. ∎

By allowing D to be a union of finite number of standard regions, we admit surfaces Σ which may have a finite number of "holes" (see Fig. 9-15). In computing $\int_{\partial\Sigma} \omega$, we must of course integrate around all the separate curves which compose the boundary of Σ, each in its proper orientation. The theorem may also be extended to a surface (differentiable manifold) which is obtained by piecing together simple surface elements; however, it is necessary that it be orientable. The Möbius strip is a nonorientable manifold M which can be

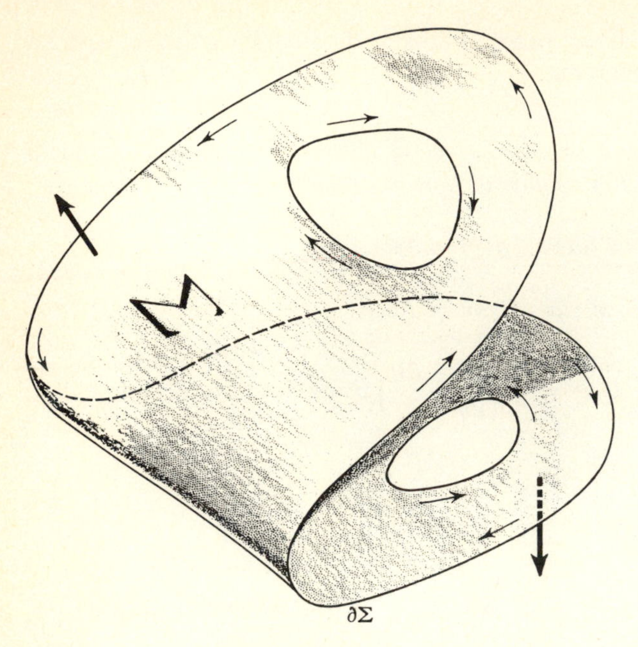

Figure 9-15

represented as the union of two simple surface elements Σ_1 and Σ_2. To each of these, Stokes' theorem may be applied, and for a suitable 1-form ω

$$\int_{\partial\Sigma_1} \omega = \iint_{\Sigma_1} d\omega$$

$$\int_{\partial\Sigma_2} \omega = \iint_{\Sigma_2} d\omega$$

Adding these, however, we do not get the integral of ω around the simple closed curve Γ which forms the edge of the Möbius strip (see Fig. 9-16). Since M is nonorientable, no consistent orientations of Σ_1 and Σ_2 can be found, and one of the "inside" edges of Σ_1 will be traced twice in the *same* direction and the integral of ω along it will not in general drop out. (Note also that the edge of the strip is not traced in a consistent fashion.)

Turning finally to the divergence theorem, we prove it first for the case of a cube.

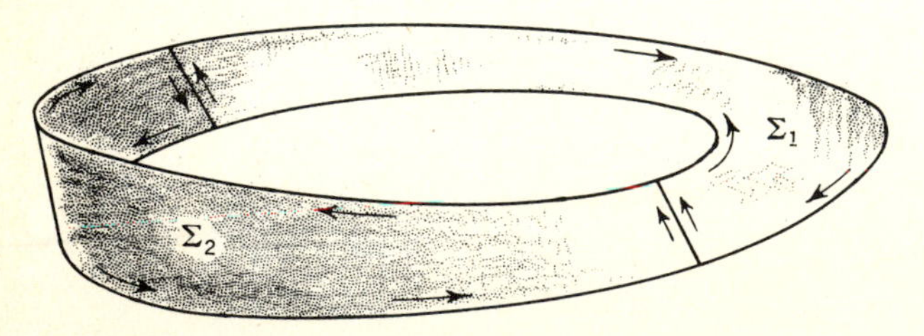

Figure 9-16 Möbius strip.

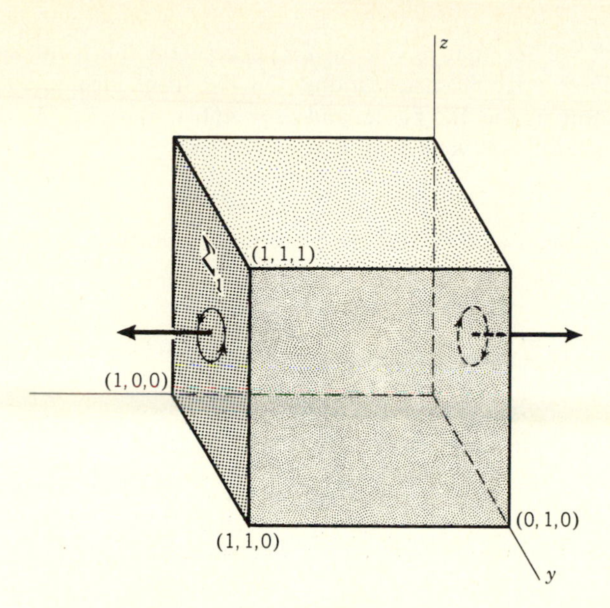

Figure 9-17

Theorem 12 *Let R be a cube in* (x, y, z) *space with faces parallel to the coordinate planes. Let* ω *be a 2-form*

$$\omega = A\,dy\,dz + B\,dz\,dx + C\,dx\,dy$$

with A, B, and C of class C′ in R. Then,

$$\iint_{\partial R} A\,dy\,dz + B\,dz\,dx + C\,dx\,dy = \iiint_R \left[\frac{\partial A}{\partial x} + \frac{\partial B}{\partial y} + \frac{\partial C}{\partial z}\right] dx\,dy\,dz \tag{9-38}$$

The surface integral on the left is $\iint_{\partial R} \omega$ and the volume integral on the right is $\iiint_R d\omega$. Let us suppose that R is the unit cube having as opposite vertices the origin and (1, 1, 1) (see Fig. 9-17). We may consider the separate terms in ω individually; suppose that

$$\omega = A(x, y, z)\,dy\,dz$$

Since $d\omega = A_1(x, y, z)\,dx\,dy\,dz$,

$$\begin{aligned}\iiint_R d\omega &= \iiint_R A_1(x, y, z)\,dx\,dy\,dz \\ &= \int_0^1 dy \int_0^1 dz \int_0^1 A_1(x, y, z)\,dx \\ &= \int_0^1 dy \int_0^1 dz[A(1, y, z) - A(0, y, z)]\end{aligned}$$

On the other hand, we see that $\omega = 0$ on ∂R, except on the front and back faces. These have opposite orientations. On the front face, Σ_1, the orientation is the same as the YZ plane, and $\omega = A(1, y, z)\,dy\,dz$. On the back face, the orientation is reversed, so that

$$\omega = A(0, y, z)\,dz\,dy = -A(0, y, z)\,dy\,dz$$

In each case, the parameter (y, z) covers the positively oriented unit square S in the YZ plane. Thus,

$$\begin{aligned}\iint_{\partial R} \omega &= \iint_{\Sigma_1} \omega + \iint_{\Sigma_2} \omega \\ &= \iint_S A(1, y, z)\,dy\,dz - \iint_S A(0, y, z)\,dy\,dz \\ &= \int_0^1 dy \int_0^1 dz[A(1, y, z) - A(0, y, z)] \quad \blacksquare\end{aligned}$$

The line of argument used in Theorem 9 may now be used again to establish the divergence theorem for a more general class of regions R (see also Exercise 5).

Theorem 13 *The divergence theorem* (9-38) *holds for any region R which is the image of a closed cube under a* 1-*to*-1 *transformation of class C″.*

Again, we can combine regions together into more complicated regions (e.g., having cavities) and apply the divergence theorem to each. The surface integrals over the interface surfaces will cancel each other, since the common boundary will have opposite orientations, and only the surface integral over the boundary of the region itself will be left (see Fig. 9-18).

If we make use of the results in Sec. 9.3, especially Theorems 4 and 5, then we can recast these integral theorems in vector form. Let **F** be a vector field of class C' throughout a region in space.

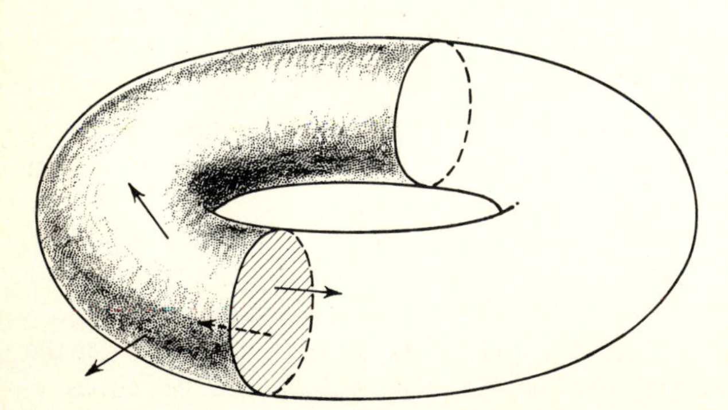

Figure 9-18

Divergence Theorem *Let R be a suitably well-behaved region in space whose boundary is a surface ∂R. Then*

$$\iiint_R \operatorname{div}(\mathbf{F}) = \iint_{\partial R} F_N$$

Stated verbally: The integral of div (**F**) *throughout R is equal to the integral of the normal component of* **F** *over the boundary of R.*

Stokes' Theorem *Let Σ be a suitably well-behaved orientable surface whose boundary is a curve ∂Σ. Then,*

$$\iint_\Sigma \operatorname{curl}(\mathbf{F})_N = \int_{\partial\Sigma} F_T$$

Stated verbally: The integral of the normal component of the curl of **F** *over a surface is equal to the integral of the tangential component of* **F** *around the boundary of the surface.*

These may be used to obtain physical interpretations for div (**F**) and curl (**F**). As in Sec. 9.3, let **V** be the velocity field of a fluid in motion, and let ρ be the density function; both may depend upon time. Let $\mathbf{F} = \rho\mathbf{V}$; this is the vector function which specifies the mass flow distribution. Let R be a closed bounded region of space to which the divergence theorem applies, e.g., a cube or a sphere. At any point p on the boundary of R, F_N is the normal component of **F**, and therefore measures the rate of flow of mass out of R at p. The surface integral $\iint_{\partial R} F_N$ is then the total mass per unit time which leaves R through ∂R. Let us suppose that there is no creation or destruction of mass within R, i.e., no "sources" or "sinks." Then, $\iint_{\partial R} F_N$ must also be exactly the rate of decrease of the total mass within R:

$$\iint_{\partial R} F_N = -\frac{d}{dt}\iiint_R \rho$$

If we apply the divergence theorem to the left side, and assume that $\partial\rho/\partial t$ is continuous so that we can move the time differentiation inside, we obtain

$$\iiint_R \operatorname{div}(\mathbf{F}) = -\iiint_\Sigma \frac{\partial\rho}{\partial t}$$

Since this relation must hold for all choices of R, we may conclude that the integrands are everywhere equal. Thus, we arrive at what is called the "equation

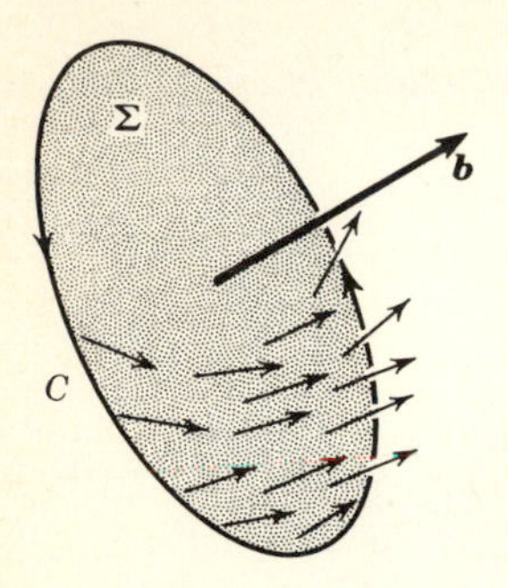

Figure 9-19

of continuity":

$$\operatorname{div}(\mathbf{F}) = \operatorname{div}(\rho\mathbf{V}) = -\frac{\partial \rho}{\partial t}$$

If the fluid is incompressible, then div $(\mathbf{V}) = 0$.

To obtain an interpretation for curl $(\mathbf{V})$, let $\mathbf{b}$ be a unit vector with initial point at a point p_0, and choose Σ as a circular disk of radius r, center p_0, having $\mathbf{b}$ as its normal (see Fig. 9-19). If p is a point of the circle C which forms the boundary of Σ, then V_T at p is the component of the velocity field along C, and $\int_C V_T = \int_{\partial\Sigma} V_T$ is a number which measures the extent to which the motion of the fluid is a rotation around C. By Stokes' theorem,

$$\int_C V_T = \iint_\Sigma \operatorname{curl}(\mathbf{V})_N$$

By the mean value theorem, the right side may be replaced by $\pi r^2 h$, where h is the value of curl $(\mathbf{V})_N$ at some point p of Σ. Divide both sides by πr^2, and let $r \to 0$. The point p must approach p_0, so that h approaches curl $(\mathbf{V})_N$ computed at p_0. Thus, we obtain

$$\operatorname{curl}(\mathbf{V})_N = \lim_{r \to 0} \frac{1}{\pi r^2} \int_{\partial\Sigma} V_T$$

The right side may be interpreted as a number which measures the rotation of the fluid at the point p_0 in the plane normal to $\mathbf{b}$, per unit area. Since the normal to Σ remains constantly $\mathbf{b}$,

$$\operatorname{curl}(\mathbf{V})_N = \operatorname{curl}(\mathbf{V}) \cdot \mathbf{b}$$

Thus, curl $(\mathbf{V}) \cdot \mathbf{b}$ measures the rotation of the fluid about the direction b. It will be greatest when $\mathbf{b}$ is chosen parallel to curl $(\mathbf{V})$. Thus, curl $(\mathbf{V})$ is a vector field which can be interpreted as specifying the axis of rotation and the magnitude (angular velocity) of that rotation, at each point in space. For example, if $\mathbf{V}$ is the velocity field for a rigid body rotating at constant angular velocity ω about a fixed axis **b**, then curl $(\mathbf{V}) = 2\omega\mathbf{b}$. The motion of a fluid is said to be "**irrotational**" if curl $(\mathbf{V}) \equiv 0$.

We may also use the integral theorems to arrive at expressions for div $(\mathbf{F})$ and curl $(\mathbf{F})$ having a form which is free of the coordinate appearance of the original definitions for these quantities. By the same line of argument that has been used above, one may arrive at the formulas

$$\operatorname{div}(\mathbf{F})\Big|_{p_0} = \lim_{r \downarrow 0} \frac{1}{\mathrm{v}(R)} \iint_{\partial R} F_N$$

$$\operatorname{curl}(\mathbf{F})\Big|_{p_0} \cdot \mathbf{b} = \lim_{\Sigma \downarrow p_0} \frac{1}{A(\Sigma)} \int_{\partial\Sigma} F_T$$

where $\mathrm{v}(R)$ is the volume of R and $A(\Sigma)$ the area of Σ, and R and Σ are thought of as closing down on the point p_0 in such a fashion that the normals to Σ are always parallel to **b**, and Stokes' theorem and the divergence theorem always apply.

Line integrals also arise in another natural way. Let **F** be a vector-valued function which describes a force field throughout a region of space. The work done by the force in moving a particle along a curve γ is defined to be the value of the line integral

$$W = \int_\gamma F_T$$

In particular, if γ is a closed path which forms the boundary of a smooth orientable surface Σ, then we may apply Stokes' theorem, and obtain

$$W = \iint_\Sigma \operatorname{curl}(\mathbf{F})_N$$

If **F** should be such that curl $(\mathbf{F}) \equiv 0$, then $W = 0$. In this case, the work done by **F** around any such closed path is 0. Such force fields are given the name **conservative;** an example is the newtonian gravitational field of a particle.

EXERCISES

1 Verify Green's theorem for $\omega = x\,dx + xy\,dy$ with D as the unit square with opposite vertices at (0, 0), (1, 1).

2 Apply Green's theorem to evaluate the integral of $(x - y^3)\,dx + x^3\,dy$ around the circle $x^2 + y^2 = 1$.

3 Verify Stokes' theorem with $\omega = x\,dz$ and with Σ as the surface described by $x = uv$, $y = u + v$, $z = u^2 + v^2$ for (u, v) in the triangle with vertices (0, 0), (1, 0), (1, 1).

4 Carry out the details needed to show that the special case of Green's theorem stated in Theorem 1 holds for $\omega = B(x, y)\,dy$.

5 Prove the divergence theorem directly when R is the solid sphere

$$x^2 + y^2 + z^2 \le 1$$

6 (*a*) Show that the area of a region D to which Green's theorem applies may be given by

$$A(D) = \tfrac{1}{2} \int_{\partial D} (x\,dy - y\,dx)$$

(*b*) Apply this to find the area bounded by the ellipse $x = a\cos\theta$, $y = b\sin\theta$, $0 \le \theta \le 2\pi$.

7 Use Exercise 6 to find the area inside the loop of the folium of Descartes, described by $x = 3at/(1 + t^3)$, $y = 3at^2/(1 + t^3)$.

8 Let D be the region inside the square $|x| + |y| = 4$ and outside the circle $x^2 + y^2 = 1$. Using the relation in Exercise 6, find the area of D.

9 Find a 1-form ω for which $d\omega = (x^2 + y^2)\,dx\,dy$, and use this to evaluate

$$\iint_D (x^2 + y^2)\,dx\,dy$$

when D is the region described in Exercise 8.

10 Show that the volume of a suitably well-behaved region R in space is given by the formula

$$\mathrm{v}(R) = \tfrac{1}{3} \iint_{\partial R} x\,dy\,dz + y\,dz\,dx + z\,dx\,dy$$

11 (a) Show that the moment of inertia of a solid R about the z axis can be expressed in the form

$$I = (\tfrac{1}{6}) \iint_{\partial R} (x^3 + 3xy^2)\,dy\,dz + (3x^2y + y^3)\,dz\,dx$$

(*b*) Use this to find the moment of inertia of a sphere about a diameter.

12 Verify the invariance relation $(d\omega)^* = d(\omega^*)$ when $\omega = x\,dy\,dz$ and T is the transformation $x = u + v - w$, $y = u^2 - v^2$, $z = v + w^2$.

13 Assuming Green's theorem for rectangles, prove it for a region of the type described by (9-29) with $\omega = A(x, y)\,dx$ by means of the transformation $x = u$, $y = f(u)v + g(u)(1 - v)$, $a \le u \le b$, $0 \le v \le 1$.

14 Verify the differentiation formula (9-33) when α and β are 0-forms in x, y, z.

15 Verify (9-33) for $\alpha = A(x, y, z)\,dz$ and $\beta = B(x, y, z)\,dy$.

16 Let $\mathbf{V}$ be the velocity field of the particles of a rigid body which is rotating about a fixed axis in the direction of the unit vector $\mathbf{b}$, at an angular velocity of ω. Show that div $(\mathbf{V}) = 0$ and curl $(\mathbf{V}) = 2\omega\mathbf{b}$.

17 Using $\mathbf{n}$ for a general normal vector on a surface and $\mathbf{T}$ for a general tangent vector to a curve, Stokes' theorem and the divergence theorem may be expressed by

$$\iint_\Sigma (\nabla \times \mathbf{F}) \cdot \mathbf{n} = \int_{\partial\Sigma} \mathbf{F} \cdot \mathbf{T} \qquad \iiint_R \nabla \cdot \mathbf{F} = \iint_{\partial R} \mathbf{F} \cdot \mathbf{n}$$

When D is a region in the plane, and $\boldsymbol{\eta}$ and $\mathbf{T}$ are used for the (outward) normal vector and the tangent vector for the curve ∂D, and when $\mathbf{F} = A\mathbf{i} + B\mathbf{j}$ is a vector field in the plane, show that Green's theorem may be put into either of the forms

$$\iint_D \nabla \cdot \mathbf{F} = \int_{\partial D} \mathbf{F} \cdot \boldsymbol{\eta}$$

$$\iint_D (\nabla \times \mathbf{F}) \cdot \mathbf{k} = \int_{\partial D} \mathbf{F} \cdot \mathbf{T}$$

18 Assuming the truth of the generalized Stokes' theorem, as given in (9-27), show that it results in the following special formulas dealing with differential forms in 4-space:

(*a*) If M is a region in 4-space, and ∂M is its three-dimensional boundary, then

$$\iiint_{\partial M} \{A\, dy\, dz\, dw + B\, dx\, dz\, dw + C\, dx\, dy\, dw + D\, dx\, dy\, dz\}$$

$$= \iiiint_M \left\{\frac{\partial A}{\partial x} + \frac{\partial B}{\partial y} + \frac{\partial C}{\partial z} + \frac{\partial D}{\partial w}\right\} dx\, dy\, dz\, dw$$

(*b*) If Σ is a two-dimensional surface in 4-space bounded by a curve $\partial\Sigma$, then

$$\int_{\partial\Sigma} \{A\, dx + B\, dy + C\, dz + D\, dw\}$$

$$= \iint_\Sigma \left\{\left(\frac{\partial B}{\partial x} - \frac{\partial A}{\partial y}\right) dx\, dy + \left(\frac{\partial C}{\partial x} - \frac{\partial A}{\partial z}\right) dx\, dz + \left(\frac{\partial D}{\partial x} - \frac{\partial A}{\partial w}\right) dx\, dw\right.$$

$$\left. + \left(\frac{\partial C}{\partial y} - \frac{\partial B}{\partial z}\right) dy\, dz + \left(\frac{\partial D}{\partial y} - \frac{\partial B}{\partial w}\right) dy\, dw + \left(\frac{\partial D}{\partial z} - \frac{\partial C}{\partial w}\right) dz\, dw\right\}$$

19 Use (9-33) to give an inductive proof that $dd\omega = 0$ for any k form in n variables.

9.5 EXACT FORMS AND CLOSED FORMS

In this section, we discuss the special properties of certain differential forms which are important for both their physical and mathematical significance. We start with 1-forms.

Definition 10 *We say that a* 1*-form* ω *is* **exact** *in an open set* Ω *if there is a function* f*, at least of class* C' *there, such that* $df = \omega$ *everywhere in* Ω.

For example, $\omega = 3x^2y^2\, dx + 2x^3y\, dy$ is exact in the entire plane, since it coincides there with $d(x^3y^2)$. On the other hand, the following simple result enables us to decide at once that $\omega = y^2\, dx + x^2\, dy$ is not an exact form.

Theorem 14 *If* ω *is an exact* 1*-form in* Ω *whose coefficients are at least of class* C' *there, then a necessary (but not sufficient) condition is that* $d\omega = 0$ *in* Ω.

This follows at once from Theorem 3. For if f is defined in Ω and $df = \omega$, then $d\omega = ddf = 0$. ∎ (The fact that this criterion is not sufficient will be shown later.)

The condition $d\omega = 0$ is often written in an expanded form that depends on the nature of ω. Thus, if we are dealing with 1-forms in the plane, and $\omega = A\,dx + B\,dy$, then ω cannot be exact in a set Ω unless

$$\frac{\partial A}{\partial y} = \frac{\partial B}{\partial x} \quad \text{in } \Omega \tag{9-39}$$

Likewise, in 3-space, if $\omega = A\,dx + B\,dy + C\,dz$, then ω cannot be exact in Ω unless

$$\frac{\partial C}{\partial y} = \frac{\partial B}{\partial z} \qquad \frac{\partial A}{\partial z} = \frac{\partial C}{\partial x} \qquad \frac{\partial B}{\partial x} = \frac{\partial A}{\partial y} \tag{9-40}$$

[These follow at once merely by calculating $d\omega$ in each case.] Since these conditions arise frequently and are important to the theory, a special name is applied.

Definition 11 *A 1-form ω of class C' is said to be* **closed** *in a region Ω if $d\omega = 0$ everywhere in Ω.*

The meaning of this [and a reason for the choice of the name] will appear later. Theorem 14 can thus be restated: *Any 1-form that is exact in Ω must be closed in Ω.*

There is also a connection between exactness and integration.

Theorem 15 *Let ω be exact in Ω, with $\omega = df$, and let γ be any smooth curve in Ω, starting at p_1 and ending at p_2.*

Then

$$\int_\gamma \omega = \int_\gamma df = f\Big|_{p_1}^{p_2} = f(p_2) - f(p_1)$$

Suppose we are dealing with 1-forms in 3-space, and suppose that γ is given by $x = \phi(t)$, $y = \psi(t)$, $z = \theta(t)$, for $0 \le t \le 1$. Since $\omega = df$,

$$\omega = f_1(x, y, z)\,dx + f_2(x, y, z)\,dy + f_3(x, y, z)\,dz$$

so that on γ,

$$\omega = \{f_1(\gamma(t))\phi'(t) + f_2(\gamma(t))\psi'(t) + f_3(\gamma(t))\theta'(t)\}\,dt$$

But, this can also be written as

$$\omega = \frac{d}{dt}\{f(\gamma(t))\}\,dt$$

so that
$$\int_\gamma \omega = \int_0^1 \frac{d}{dt}\{f(\gamma(t))\}\,dt$$
$$= f(\gamma(1)) - f(\gamma(0))$$
$$= f(p_2) - f(p_1) \quad \blacksquare$$

Corollary *If ω is an exact 1-form in a region Ω, then $\int_\gamma \omega = 0$ for every closed γ lying in Ω.*

For, if γ is closed, then $p_2 = p_1$, and $f(p_1) = f(p_2)$. $\blacksquare$

A special name is given to the type of behavior described in the corollary.

Definition 12 *A line integral $\int_\gamma \omega$ is said to be independent of path in a region if its value is the same along any other curve lying in Ω which joins the same points (in the same order). Equivalently, $\int_\gamma \omega = 0$ for every closed curve γ lying in Ω.*

The equivalence arises from the fact that if γ_1 and γ_2 are two curves lying in Ω which have common endpoints, then the union of γ_1 and $-\gamma_2$ is a closed curve lying in Ω. Using this terminology, the corollary above states that any exact 1-form yields line integrals that are independent of path. We may use this to test a 1-form for exactness. For example, $\omega = y\,dx$ is not exact in the first quadrant, since its integral along the straight line from (0, 0) to (1, 1) is $\int_0^1 x\,dx = \frac{1}{2}$, while its integral along the parabola $y = x^2$ which also joins (0, 0) to (1, 1) is $\int_0^1 x^2\,dx = \frac{1}{3}$. This would be a difficult test to apply in some cases, since it depends upon a comparison between the numerical value of line integrals.

While the physical interpretation of both "exactness" and "independence of path" will be discussed in more detail later, we remark at this point that one illustration arises in the study of fields of force which have the property that the work done in moving from p_1 to p_2 depends only upon the positions of the two points, and not upon the path chosen.

The mathematical significance of these concepts focuses on the connection between being an exact form and being a closed form. We note first that exactness is a global property. In order to test if ω is exact in Ω, we must examine the behavior of ω everywhere in Ω at the same time, for there must exist a single function f, defined everywhere in Ω, such that $d\omega = f$; alternatively, we must integrate ω over every closed curve γ in Ω, both large ones and small ones.

We can introduce a related notion as follows: Let us say that ω is **locally exact** in Ω if ω is exact in some neighborhood of any point in Ω. Near any point $p \in \Omega$, ω will have the form $\omega = df$, but the function f that is used near one point p may differ from that used near another. Again, if ω is locally exact in Ω, then $\int_\gamma \omega = 0$ for every sufficiently small closed path γ lying in Ω.

In contrast, the notion of "closure" for 1-forms is already a local concept, since differentiation of functions or forms requires knowledge and calculations that can be restricted to a neighborhood of each point under study.

These remarks set the stage for the proofs of the following assertions. Exactness and independence of path are equivalent for 1-forms; closure and local exactness are equivalent for 1-forms; whether or not every locally exact 1-form in Ω is globally exact in Ω depends upon geometrical properties of the region Ω—in particular upon whether it is or is not simply connected.

We begin by settling a simple special case.

Theorem 16 *If ω is a 1-form which is of class C' and obeys the condition $d\omega = 0$ in a spherical region Ω, then ω is exact in Ω.*

Let us assume that $\omega = A\,dx + B\,dy + C\,dz$, and that Eqs. (9-40) hold throughout the sphere Ω in space whose center is at the origin $(0, 0, 0)$. We construct a function f by integrating ω along a polygon joining $(0, 0, 0)$ to (x, y, z) within Ω (see Fig. 9-20). Let

$$f(x, y, z) = \int_0^x A(t, 0, 0)\,dt + \int_0^y B(x, t, 0)\,dt + \int_0^z C(x, y, t)\,dt$$

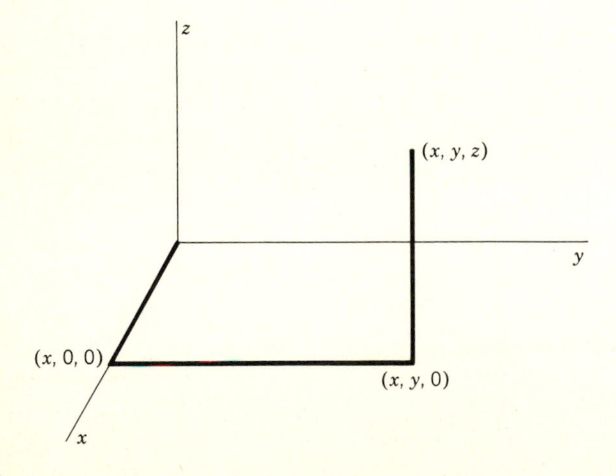

Figure 9-20

We wish to show that $\omega = df$. Computing the partial derivatives of f, and using Eqs. (9-10), we have

$$\begin{aligned}
f_3(x, y, z) &= C(x, y, z) \\
f_2(x, y, z) &= B(x, y, 0) + \int_0^z C_2(x, y, t)\, dt \\
&= B(x, y, 0) + \int_0^z B_3(x, y, t)\, dt \\
&= B(x, y, 0) + [B(x, y, z) - B(x, y, 0)] \\
&= B(x, y, z) \\
f_1(x, y, z) &= A(x, 0, 0) + \int_0^y B_1(x, t, 0)\, dt + \int_0^z C_1(x, y, t)\, dt \\
&= A(x, 0, 0) + \int_0^y A_2(x, t, 0)\, dt + \int_0^z A_3(x, y, t)\, dt \\
&= A(x, 0, 0) + [A(x, y, 0) - A(x, 0, 0)] + [A(x, y, z) - A(x, y, 0)] \\
&= A(x, y, z) \quad \blacksquare
\end{aligned}$$

For a general region, the best that this theorem gives is the following corollary.

Corollary *If $d\omega = 0$ in an open region Ω, then ω is locally exact in Ω, that is, about any point p there is a neighborhood in which ω has the form $\omega = df$.*

As the point p moves around in Ω, the function f may change, and it may not be possible to find a single function such that $\omega = df$ throughout all of Ω. For example, consider the 1-form

$$\omega = \frac{x}{x^2 + y^2}\, dy - \frac{y}{x^2 + y^2}\, dx \tag{9-41}$$

in the open ring $D = \{\text{all } (x, y) \text{ with } 1 \le x^2 + y^2 \le 4\}$. Direct computation shows that $d\omega = 0$ in D, so that ω is locally exact in D. However, ω is not exact in D, for if we compute the integral of ω around a circular path, $x = r\cos\theta$, $y = r\sin\theta$, lying in D, we obtain

$$\begin{aligned}
\int_\gamma \omega &= \int_0^{2\pi} \left\{ \frac{(r\cos\theta)}{r^2}(r\cos\theta) - \frac{(r\sin\theta)}{r^2}(-r\sin\theta) \right\} d\theta \\
&= \int_0^{2\pi} (\cos^2\theta + \sin^2\theta)\, d\theta = 2\pi \ne 0
\end{aligned}$$

If ω were exact in D, then by the corollary to Theorem 15, $\int_\gamma \omega$ would have to be 0. As we shall see later, the clue to this behavior lies in the nature of the set D.

Theorem 16 shows that the condition $d\omega = 0$ in a convex region Ω implies exactness, and therefore independence of path in Ω. A similar technique will prove the converse, for a general region Ω.

Theorem 17 *If ω is independent of path in an open connected set Ω, then ω is exact in Ω, and therefore obeys $d\omega = 0$.*

Choose any point $p_0 \in \Omega$. By assumption, Ω is connected so that any point $p \in \Omega$ can be joined to p_0 by a smooth curve γ lying in Ω. Define a function f in Ω by setting $f(p) = \int_\gamma \omega$. Since ω yields line integrals that are independent of path, it does not matter what curve γ we choose, so long as it lies in Ω and goes from p_0 to p. We again wish to show that $\omega = df$, and for this, we need the partial derivatives of f. If p_1 and $p_1 + \Delta p$ are points of Ω, and β is any curve in Ω from p_1 to $p_1 + \Delta p$, then choosing a curve γ from p_0 to p_1, we have

$$\begin{aligned} f(p_1 + \Delta p) &= \int_\gamma \omega + \int_\beta \omega \\ &= f(p_1) + \int_\beta \omega \end{aligned}$$

To compute $f_1(p_1)$, the partial derivative of f in the direction of the X axis, we take $\Delta p = (h, 0, 0)$ and compute

$$\lim_{h \to 0} \frac{f(p_1 + \Delta p) - f(p_1)}{h} = \lim_{h \to 0} \frac{1}{h} \int_\beta \omega$$

For β, we choose the straight line from p_1 to $p_1 + \Delta p$ whose equation is $x = x_1 + ht$, $y = y_1$, $z = z_1$; $0 \le t \le 1$. If $\omega = A\,dx + B\,dy + C\,dz$, then on β, $\omega = A(x_1 + ht, y_1, z_1)h\,dt$ and

$$\frac{f(p_1 + \Delta p) - f(p_1)}{h} = \int_0^1 A(x_1 + ht, y_1, z_1)\,dt$$

Letting h approach 0, and using the fact that A is continuous,

$$\begin{aligned} f_1(p_1) &= \lim_{h \to 0} \int_0^1 A(x_1 + ht, y_1, z_1)\,dt \\ &= \int_0^1 A(x_1, y_1, z_1)\,dt = A(x_1, y_1, z_1) \end{aligned}$$

In a similar fashion, we find $f_2(p_1) = B(p_1)$ and $f_3(p_1) = C(p_1)$, so that $\omega = df$ in Ω. ▮

We now invoke the generalized Stokes' theorem, in the form of either Green's theorem (for 2-space) or the original theorem of Stokes (for 3-space), to observe the following simple property of closed 1-forms.

Theorem 18 *Let ω be a 1-form defined in Ω, with $d\omega = 0$ there. Let γ be a closed curve in Ω that is the boundary $\gamma = \partial\Sigma$ of a smooth orientable surface Σ lying in Ω. Then,*

$$\int_\gamma \omega = 0 \tag{9-42}$$

The proof is immediate. We have

$$\int_\gamma \omega = \int_{\partial\Sigma} \omega = \iint_\Sigma d\omega = \iint_\Sigma 0 = 0 \quad ▮$$

In order to have this hold for every closed curve γ in Ω, which would then imply that ω is *exact* in Ω, we must have it true that every closed curve in Ω is a boundary for a surface lying in Ω. This is a special geometrical requirement imposed on the set Ω, and is given a special name.

Definition 13 *An open connected set Ω in n space is said to be* **simply connected** *if every closed curve γ in Ω is the boundary of an orientable surface lying in Ω.*

As is suggested in Fig. 8-19, n space itself is simply connected, as is any convex subset. However, the planar region shown in Fig. 9-21 is not, nor is the

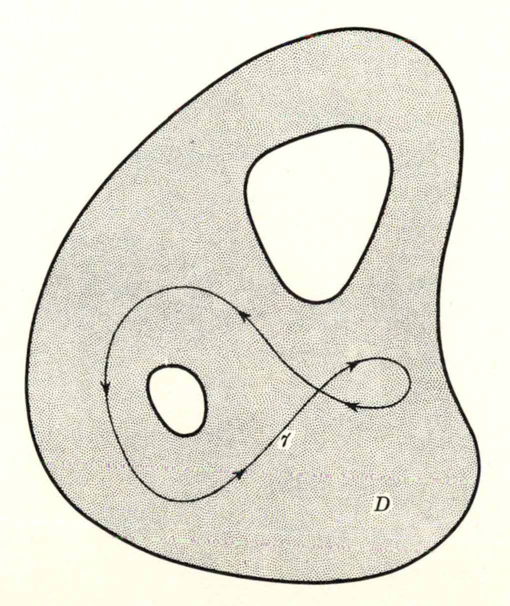

Figure 9-21 Multiply connected region in the plane.

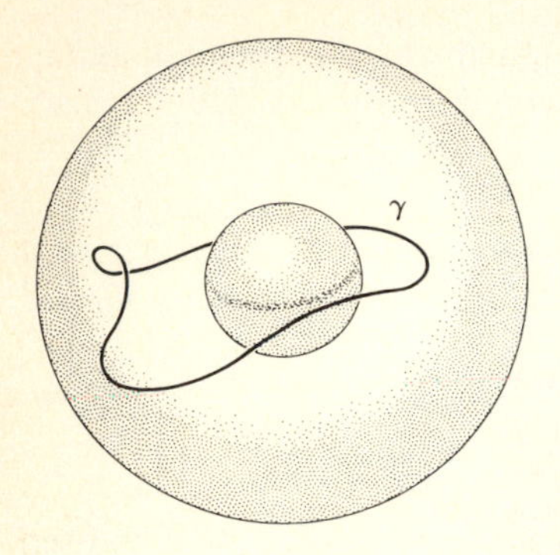

Figure 9-22 The region between two spheres is simply connected.

region in space shown in Fig. 8-27, consisting of the open set inside a torus. However, the region in Fig. 9-22, consisting of the points between two spheres, is simply connected.

This definition, taken in conjunction with Theorem 18, allows us to obtain the following at once.

Theorem 19 *In a simply connected region, every locally exact* 1-*form is exact. In particular, if* $d\omega = 0$ *in* Ω *and* Ω *is simply connected, then there is a function* f *defined in* Ω *such that* $\omega = df$ *everywhere in* Ω.

When a 1-form ω is known to be exact, and thus of the form df, the function f may be found by integrating ω along any convenient path from some point p_0 to $p = (x, y, z)$; it will be unique, up to an additive constant. For example, consider the 1-form $\omega = 2xy^3\,dx + 3x^2y^2\,dy$. Checking for exactness, we find that $d\omega = 6xy^2\,dy\,dx + 6xy^2\,dx\,dy = 0$. To find f, we integrate from (0, 0) to (x, y). If we choose the solid broken line shown in Fig. 9-23, we obtain

$$f(x, y) = \int_0^x 0\,dx + \int_0^y 3x^2y^2\,dy = x^2y^3$$

We may also use the dotted broken line, and obtain

$$f(x, y) = \int_0^y 0\,dy + \int_0^x 2xy^3\,dx = x^2y^3$$

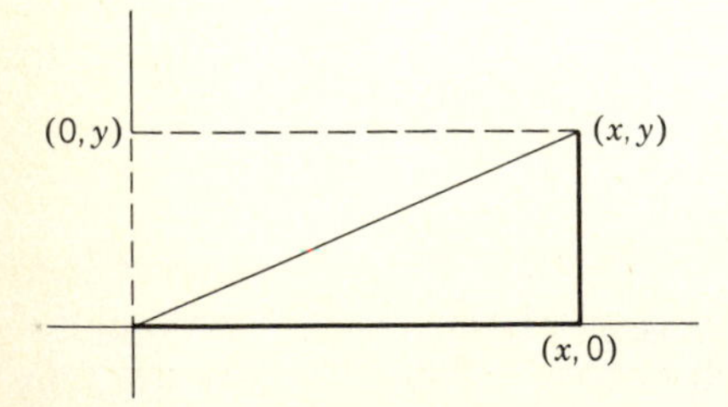

Figure 9-23

Finally, if we use the straight line joining (0, 0) and (x, y), we have

$$f(x, y) = \int_0^1 \{2(xt)(yt)^3\, d(xt) + 3(xt)^2(yt)^2\, d(yt)\}$$

$$= \int_0^1 \{(2xy^3)(x)t^4\, dt + (3x^2y^2)(y)t^4\, dt\}$$

$$= x^2y^3 \int_0^1 5t^4\, dt = x^2y^3$$

When the coefficients in ω have special properties, other methods may be used to find f (see Exercise 5).

The study of differential equations also leads to the consideration of exact differential forms. One says that a curve γ in the plane is a solution of the first-order "differential equation"

$$A(x, y)\, dx + B(x, y)\, dy = 0 \tag{9-43}$$

if the 1-form $\omega = A\, dx + B\, dy$ is 0 on γ. If ω is exact, and $\omega = df$, then the level curves of f are solutions for (9-43). An **integrating factor** for Eq. (9-43) is a function g such that $g\omega$ is an exact 1-form. Computing $d(g\omega)$, and requiring that this vanish, one is led to a partial differential equation for g. General existence theorems in the theory of differential equations show that, provided A and B are sufficiently well behaved, Eq. (9-43) always admits integrating factors. The situation is somewhat different in the case of forms in three or more variables. A curve γ in 3-space is said to be a solution of the "total differential equation" (or **Pfaffian** equation)

$$A(x, y, z)\, dx + B(x, y, z)\, dy + C(x, y, z)\, dz = 0 \tag{9-44}$$

if the 1-form $\omega = A\, dx + B\, dy + C\, dz$ is 0 on γ. When ω is exact, and has the form $\omega = df$, then any curve lying on the level surfaces of f will be a solution of (9-44). Since this is a very special type of possible behavior for the solution curves of such an equation, it is plausible that only certain Pfaffian equations admit integrating factors. The analysis supports this. If g is an integrating factor for ω, then we may assume that $d(g\omega) = 0$. Using the formula (9-33) for differentiating a product, we obtain, since g is a 0 form,

$$(dg)\omega + g(d\omega) = 0$$

It is easily seen that $\omega\omega = 0$; thus, if we multiply this equation on the right by ω, we obtain as a necessary condition

$$g(d\omega)\omega = 0$$

Dividing by the function g, we obtain the condition $(d\omega)\omega = 0$. In terms of the coefficients of ω, this shows that a necessary condition for (9-44) to have an integrating factor is that

$$\left(\frac{\partial C}{\partial y} - \frac{\partial B}{\partial z}\right)A + \left(\frac{\partial A}{\partial z} - \frac{\partial C}{\partial x}\right)B + \left(\frac{\partial B}{\partial x} - \frac{\partial A}{\partial y}\right)C = 0$$

It may also be shown that if this condition holds, and one imposes reasonable restrictions on A, B, and C, then an integrating factor for (9-44) exists.

The notions of exactness and of independence of path may also be given in vector form. Using the "dictionary" (Fig. 9-11) in Sec. 9.3, we may recast our results in the following form.

Theorem 20 *Let* **F** *be a vector field which is of class* C'' *and obeys the condition* curl (**F**) = 0 *throughout a simply connected region of space* Ω. *Then, there is a scalar function* f, *unique up to additive constants, such that in* Ω

$$\mathbf{F} = \text{grad}\,(f)$$

Moreover, if γ *is any smooth curve lying in* Ω *and going from* p_0 *to* p_1, *then*

$$\int_\gamma F_T = f(p_1) - f(p_0)$$

The function f is called the **potential** for the vector field **F**. When the condition curl (**F**) = 0 is translated by saying that **F** specifies an "irrotational" velocity field, then f is called the velocity potential. If **F** specifies a force field, then **F** is said to be **conservative** when curl (**F**) = 0; in this case, it is customary to set $U = -f$ and call this the **potential energy.** If γ is a curve from p_0 to p_1, then

$$\int_\gamma F_T = U(p_0) - U(p_1)$$

This is interpreted as meaning that the work done by the field in moving a particle from p_0 to p_1 is independent of the path chosen, and is equal to the loss in potential energy.

Before leaving 1-forms, we digress to point out that there exists a form of duality between the analysis and the geometrical notions we have been discussing. Consider a curve γ that starts at p and ends at q; it is reasonable to define its boundary to be the set consisting of p and q, taken in some order or assigned numerical weights to indicate which is the starting point and which the endpoint. It is then convenient to say that a *closed* curve (for which $p = q$) has no boundary, and thus to write $\partial\gamma = 0$. Likewise, one may say that a curve γ is bounding in Ω if there exists a surface Σ in Ω with $\gamma = \partial\Sigma$. Compare the following statements:

curves	**1-forms**
γ is closed if and only if $\partial\gamma = 0$.	ω is closed if and only if $d\omega = 0$.
γ is bounding in Ω if and only if $\gamma = \partial\Sigma$ for some Σ in Ω.	ω is exact in Ω if and only if $\omega = df$ for some f in Ω.

In the light of Theorem 19 and Definition 13, and this table, the following statement should not seem surprising: An open region Ω has the property that every closed 1-form is exact if and only if every closed curve is bounding—i.e., if and only if Ω is simply connected.

This connection between analytic notions dealing with the local and global behavior of 1-forms and geometric notions dealing with the behavior of curves is only one piece of a much larger picture, extending to general k forms, n space, and even general n-dimensional manifolds, for which reference should be made to recent books on modern differential topology. We illustrate this by discussing 2-forms briefly.

Definition 14 *A 2-form σ is exact in Ω if there is a 1-form ω of class C' in Ω such that $\sigma = d\omega$.*

There is more freedom in the choice of ω in this case. If σ is exact, and ω is a 1-form with $\sigma = d\omega$, then any exact 1-form may be added to ω. For example, σ is also given as $d(\omega + df)$, where f is a function of class C''.

What corresponds to the notion of independence of path for 2-forms? An answer is supplied by Stokes' theorem.

Theorem 21 *If σ is exact in Ω and Σ_1 and Σ_2 are two smooth orientable surfaces lying in Ω and having the same curve γ as boundary, then*

$$\iint_{\Sigma_1} \sigma = \iint_{\Sigma_1} \sigma$$

For, if $\sigma = d\omega$, then, by Stokes' theorem, $\iint_{\Sigma} \sigma = \iint_{\Sigma} d\omega = \int_{\partial\Sigma} \omega$ for both choices of Σ. ▌

Corollary *If σ is exact in Ω, then $\iint_{\Sigma} \sigma = 0$ for any smooth orientable closed surface Σ in Ω.*

We can also test a 2-form for exactness by examining its coefficients.

Theorem 22 *If $\sigma = A\,dy\,dz + B\,dz\,dx + C\,dx\,dy$ is exact and of class C'' in Ω, then $d\sigma = 0$ there; that is,*

$$\frac{\partial A}{\partial x} + \frac{\partial B}{\partial y} + \frac{\partial C}{\partial z} = 0$$

This also comes at once from Theorem 3; if $\sigma = d\omega$, then

$$d\sigma = dd\omega = 0 \quad \blacksquare$$

We are again able to prove a local converse to this.

Theorem 23 *If σ is a 2-form which is of class C' and obeys the condition $d\sigma = 0$ in a convex region Ω, then σ is exact in Ω.*

If $\sigma = A\,dy\,dz + B\,dz\,dx + C\,dx\,dy$, then we shall assume that $A_1 + B_2 + C_2$ is identically 0 in a convex region Ω containing the origin (0, 0, 0). We shall find a 1-form ω such that $d\omega = \sigma$. Because of the latitude that exists in the choice of ω, it will turn out to be possible to find a solution of the form $\omega = a(x, y, z)\,dx + b(x, y, z)\,dy$. The requirement $d\omega = \sigma$ imposes the following three conditions on the coefficients of ω:

$$\frac{\partial a}{\partial z} = B \qquad \frac{\partial b}{\partial z} = -A \qquad \frac{\partial b}{\partial x} - \frac{\partial a}{\partial y} = C \tag{9-45}$$

Integrating the first two with respect to z, and supplying an arbitrary function of (x, y) in the integral of the first, we are led to try the following:

$$a(x, y, z) = \int_0^z B(x, y, t)\,dt - \int_0^y C(x, s, 0)\,ds$$

$$b(x, y, z) = -\int_0^z A(x, y, t)\,dt$$

These clearly satisfy the first two equations in (9-45). To see that the third holds as well, we differentiate with respect to y and x, respectively, obtaining

$$\frac{\partial a}{\partial y} = \int_0^z B_2(x, y, t)\,dt - C(x, y, 0)$$

$$\frac{\partial b}{\partial x} = -\int_0^z A_1(x, y, t)\,dt$$

Accordingly,

$$\frac{\partial b}{\partial x} - \frac{\partial a}{\partial y} = -\int_0^z [A_1(x, y, t) + B_2(x, y, t)]\,dt + C(x, y, 0)$$

But, $A_1 + B_2 + C_3 = 0$, so that the right side becomes

$$\begin{aligned}\int_0^z C_3(x, y, t)\,dt + C(x, y, 0) &= [C(x, y, z) - C(x, y, 0)] + C(x, y, 0)\\ &= C(x, y, z)\end{aligned}$$

and all the equations in (9-45) are satisfied. $\blacksquare$ (We shall also give another treatment of this in Exercise 2 Sec. 9.6.)

Corollary *If a 2-form σ satisfies the condition $d\sigma = 0$ throughout an open region Ω, then σ is locally exact in Ω.*

This means again that in some neighborhood of any point of Ω, there is defined a 1-form ω such that σ is the derivative of ω in this neighborhood. The various 1-forms ω may not piece together to give a 1-form whose derivative is σ everywhere in Ω. For example, the 2-form

$$\sigma = \frac{x\,dy\,dz + y\,dz\,dx + z\,dx\,dy}{(x^2 + y^2 + z^2)^{3/2}} \tag{9-46}$$

is of class C' in all of space except at the origin, and direct calculation shows that $d\sigma = 0$ in this whole set. Thus, σ is locally exact everywhere except at the origin. However, σ is not exact in this set, for if one computes the surface integral $\iint_\Sigma \sigma$, where Σ is the unit sphere: $x = \sin\phi\cos\theta$, $y = \sin\phi\sin\theta$, $z = \cos\phi$, $0 \le \phi \le \pi$, $0 \le \theta \le 2\pi$, the value does not turn out to be zero! It will be noticed that the region Ω involved in this example is simply connected. It is natural to ask for the corresponding property, which, if possessed by Ω, will ensure that any 2-form that is locally exact in Ω must also be exact. By analogy, we would expect the condition to be the geometric dual, namely, that every closed surface in Ω is bounding. The region shown in Fig. 9-22 fails to have this property, but the interior of the torus (inner tube) in Fig. 8-27 does; note that this geometric condition is not the same as being simply connected!

Again, appealing to the Stokes' theorem for 2-forms in 3-space (Gauss' theorem), we can prove part of the conjectured result.

Theorem 24 *If every smooth surface in Ω is the boundary of a region D lying in Ω, and σ is a closed 2-form, then $\iint_\Sigma \sigma = 0$ for every closed surface Σ in Ω.*

For, we have

$$\iint_\Sigma \sigma = \iint_{\partial D} \sigma = \iiint_D d\sigma = \iiint_D 0 = 0 \quad \blacksquare$$

In Exercise 8, we prove that if the region Ω is "star-shaped," which is slightly more general than "convex," then one can go further and conclude that σ is necessarily exact in Ω. (This is far from the best possible result here, but an adequate discussion would take us in other directions.)

These results on exact 2-forms may also be cast into vector form. Let $\mathbf{F}$ be a vector field described by $\mathbf{F} = A\mathbf{i} + B\mathbf{j} + C\mathbf{k}$. Referring again to Sec. 9.3, we see that its corresponding 2-form is exact if and only if we can write $\mathbf{F} = \text{curl}\,(\mathbf{V})$, where $\mathbf{V}$ is another vector field. Since div (curl ($\mathbf{V}$)) = 0, a necessary condition on $\mathbf{F}$ is that div ($\mathbf{F}$) = 0. Such a field is said to be **solenoidal,** or divergence-free. The velocity field of an incompressible fluid is solenoidal.

Theorem 25 *If* **F** *is a solenoidal vector field of class* C', *then in any convex set,* **F** *is of the form* $\mathbf{F} = \text{curl}\,(\mathbf{V})$.

EXERCISES

1 For each of the following 1-forms ω, find if possible a function f such that $\omega = df$.

(*a*) $\omega = (3x^2y + 2xy)\,dx + (x^3 + x^2 + 2y)\,dy$

(*b*) $\omega = (xy\cos xy + \sin xy)\,dx + (x^2\cos xy + y^2)\,dy$

(*c*) $\omega = (2xyz^3 + z)\,dx + x^2z^3\,dy + (3x^2yz^2 + x)\,dz$

(*d*) $\omega = x^2\,dy + 3xz\,dz$

2 (*a*) By differentiating $tg(xt, yt, zt)$ with respect to t, show that

$$g(x, y, z) = \int_0^1 g(xt, yt, zt)\,dt + \int_0^1 [xg_1(xt, yt, zt) + yg_2(xt, yt, zt) + zg_3(xt, yt, zt)]t\,dt$$

(*b*) A region Ω in space is said to be star-shaped with respect to the origin if the line segment $\overline{0P}$ lies in Ω whenever the endpoint P lies in Ω. Let Ω be star-shaped, and let $\omega = A\,dx + B\,dy + C\,dz$ obey $d\omega = 0$ in Ω. Obtain an alternative proof of Theorem 16 by showing that $\omega = df$, where f is defined in Ω by the explicit formula

$$f(x, y, z) = \int_0^1 [xA(xt, yt, zt) + yB(xt, yt, zt) + zC(xt, yt, zt)]\,dt$$

(*c*) Can this integral be regarded as a line integral of ω along a curve joining $(0, 0, 0)$ to (x, y, z)?

3 Verify that the differential form given in Eq. 9-41 obeys $d\omega = 0$ throughout the ring D.

4 Consider the differential form

$$\omega = \frac{x\,dx + y\,dy}{x^2 + y^2}$$

Show that $d\omega = 0$ in the ring D. Is ω exact in D?

5 Recall that a function f is said to be homogeneous of degree k if

$$f(xt, yt, zt) = t^k f(x, y, z)$$

for all $t \geq 0$ and all (x, y, z) in a sphere about the origin. Let

$$\omega = A\,dx + B\,dy + C\,dz$$

be an exact 1-form whose coefficients are all homogeneous of degree k, $k \geq 0$. Show that $\omega = df$, where

$$f(x, y, z) = \frac{xA(x, y, z) + yB(x, y, z) + zC(x, y, z)}{k + 1}$$

6 If such exist, find integrating factors for the following differential forms:

(*a*) $(x^2 + 2y)\,dx - x\,dy$

(*b*) $3yz^2\,dx + xz^2\,dy + 2xyz\,dz$

(*c*) $xy\,dx + xy\,dy + yz\,dz$

7 (*a*) Show that the 2-form $\sigma = A\,dy\,dz + B\,dz\,dx + C\,dx\,dy$ can also be expressed in the form $\sigma = \alpha\,dx + \beta\,dy$, where $\alpha = B\,dz - C\,dy$, $\beta = -A\,dz$.

(*b*) In the proof of Theorem 23 a 1-form $\omega = a\,dx + b\,dy$ was found such that $d\omega = \sigma$. Show that the coefficients of ω are given by the equations:

$$a(x, y, z) = \int_\gamma \alpha \qquad b(x, y, z) = \int_\gamma \beta$$

where γ is the polygonal path from the origin to (x, y, z) that is shown in Fig. 9-20.

8 (*a*) By differentiating $t^2 g(xt, yt, zt)$ with respect to t, show that $g(x, y, z)$ may be given by

$$2\int_0^1 g(xt, yt, zt)t\,dt + \int_0^1 [xg_1(xt, yt, zt) + yg_2(xt, yt, zt) + zg_3(xt, yt, zt)]t^2\,dt$$

(*b*) If $\sigma = A\,dy\,dz + B\,dz\,dx + C\,dx\,dy$, show that

$$2\sigma = (B\,dz - C\,dy)\,dx + (C\,dx - A\,dz)\,dy + (A\,dy - B\,dx)\,dz$$

(*c*) Let Ω be star-shaped with respect to the origin, and let $d\sigma = 0$ in Ω. Obtain an alternative proof of Theorem 23 by showing that $\sigma = d\omega$, where $\omega = a\,dx + b\,dy + c\,dz$, and where

$$a(x, y, z) = \int_0^1 [zB(xt, yt, zt) - yC(xt, yt, zt)]t\,dt$$

$$b(x, y, z) = \int_0^1 [xC(xt, yt, zt) - zA(xt, yt, zt)]t\,dt$$

$$c(x, y, z) = \int_0^1 [yA(xt, yt, zy) - xB(xt, yt, zt)]t\,dt$$

9 Let $\sigma = A\,dy\,dz + B\,dz\,dx + C\,dx\,dy$, where the functions A, B, and C are homogeneous of degree k in a neighborhood of the origin. If σ is exact, show that $\sigma = d\omega$, where

$$\omega = \frac{(zB - yC)\,dx + (xC - zA)\,dy + (yA - xB)\,dz}{k + 2}$$

10 Show that the following 2-forms are exact by exhibiting each in the form $\sigma = d\omega$:

(*a*) $(3y^2z - 3xz^2)\,dy\,dz + x^2y\,dz\,dx + (z^3 - x^2z)\,dx\,dy$

(*b*) $(2xz + z)\,dz\,dx + y\,dx\,dy$

11 Formulate a necessary condition that a 1-form in n variables be exact.

12 Verify the assertions made in connection with the 2-form given in Eq. (9-46).

13 Formulate a definition for "integrating factor" for 2-forms. Obtain a differential equation which must be satisfied by any integrating factor for the 2-form

$$\sigma = A\,dy\,dz + B\,dz\,dx + C\,dx\,dy$$

Using this, and Euler's differential equation for homogeneous functions (Exercise 11, Sec. 3.4), obtain the most general integrating factor for

$$x\,dy\,dz + y\,dz\,dx + z\,dx\,dy$$

14 Using Theorem 8, Sec. 9.4, show that a transformation of class C'' carries exact forms into exact forms.

15 Let Ω be a region in space which can be mapped onto a star-shaped set by a 1-to-1 transformation of class C''. Show that any 2-form σ which satisfies the equation $d\sigma = 0$ in Ω is exact in Ω.

9.6 APPLICATIONS

Many of the classical applications of differential forms occur in the study of the vector fields that arise in physics. We illustrate some of the results of the previous sections in this context, translating them into vector form. We start with techniques that are derived from the treatment of exact 1-forms and exact 2-forms in Sec. 9.5, which permit us to work with vector functions that are irrotational or are solenoidal (divergence-free), expressing them in special forms at least locally.

Our first illustration is a standard one, and is based on Maxwell's equations for electromagnetic radiation, given earlier in (9-18). If we drop the factor 4π to simplify the discussion, let us suppose that **E** and **H** are vector functions of (x, y, z, t), representing the electric and magnetic fields, which obey the four equations:

$$\nabla \cdot \mathbf{E} = \rho \tag{9-47}$$

$$\nabla \cdot \mathbf{H} = 0 \tag{9-48}$$

$$\nabla \times \mathbf{E} + \frac{\partial \mathbf{H}}{\partial t} = 0 \tag{9-49}$$

$$\nabla \times \mathbf{H} - \frac{\partial \mathbf{E}}{\partial t} = \mathbf{J} \tag{9-50}$$

The scalar function ρ describes the distribution of charge, and the vector function **J** describes the distribution of current. If we regard ρ and **J** as known, we may seek a solution of these equations for **E** and **H**. We shall show that this may be reduced to the solution of a standard partial differential equation, the inhomogeneous wave equation.

By (9-48) and Theorem 25, **H** can be represented locally in the form

$$\mathbf{H} = \nabla \times \mathbf{A} \tag{9-51}$$

where **A**, the vector potential, may be altered by adding any vector function of the form ∇g, where g is a suitable scalar function. Putting this into (9-49), we have

$$0 = \nabla \times \mathbf{E} + \frac{\partial}{\partial t}(\nabla \times \mathbf{A})$$

$$= \nabla \times \left(\mathbf{E} + \frac{\partial \mathbf{A}}{\partial t}\right)$$

By Theorem 20, a scalar function ϕ exists such that, locally at least,

$$\mathbf{E} + \frac{\partial \mathbf{A}}{\partial t} = \nabla(-\phi) = -\nabla \phi$$

(The negative sign is chosen only to improve the final appearance of our equations.) We may thus assume that **E** and **H** have the representations given in (9-51) and in

(9-52) $$\mathbf{E} = -\nabla\phi - \frac{\partial \mathbf{A}}{\partial t}$$

Put these into (9-47) and (9-50), obtaining

(9-53) $$\begin{aligned}\rho &= \nabla \cdot \left(-\nabla\phi - \frac{\partial \mathbf{A}}{\partial t}\right) \\ &= -\nabla^2\phi - \frac{\partial}{\partial t}(\nabla \cdot \mathbf{A})\end{aligned}$$

and $$\begin{aligned}\mathbf{J} &= \nabla \times (\nabla \times \mathbf{A}) - \frac{\partial}{\partial t}\left(-\nabla\phi - \frac{\partial \mathbf{A}}{\partial t}\right) \\ &= \nabla \times (\nabla \times \mathbf{A}) + \nabla\frac{\partial \phi}{\partial t} + \frac{\partial^2 \mathbf{A}}{\partial t^2}\end{aligned}$$

Making use of the vector identity (9-24), this can be written as

(9-54) $$\mathbf{J} = \nabla(\nabla \cdot \mathbf{A}) - \nabla^2\mathbf{A} + \nabla\frac{\partial \phi}{\partial t} + \frac{\partial^2 \mathbf{A}}{\partial t^2}$$

Turning (9-53) and (9-54) around, we have two equations to be satisfied by **A** and ϕ:

$$\nabla^2\phi = -\rho - \frac{\partial}{\partial t}(\nabla \cdot \mathbf{A})$$

$$\nabla^2\mathbf{A} - \frac{\partial^2 \mathbf{A}}{\partial t^2} = -\mathbf{J} + \nabla\left(\nabla \cdot \mathbf{A} + \frac{\partial \phi}{\partial t}\right)$$

In the second, $\nabla^2\mathbf{A}$ is to be interpreted by writing $\mathbf{A} = a\mathbf{i} + b\mathbf{j} + c\mathbf{k}$ and applying ∇^2 to each component, obtaining

$$\nabla^2\mathbf{A} = \nabla^2(a)\mathbf{i} + \nabla^2(b)\mathbf{j} + \nabla^2(c)\mathbf{k}$$

The second equation would assume a much simpler form if **A** and ϕ could be chosen so that

(9-55) $$\nabla \cdot \mathbf{A} + \frac{\partial \phi}{\partial t} = 0$$

In this case, our two equations would become

$$\nabla^2\phi - \frac{\partial^2 \phi}{\partial t^2} = -\rho$$

and $$\nabla^2\mathbf{A} - \frac{\partial^2 \mathbf{A}}{\partial t^2} = -\mathbf{J}$$

In the choice of $\mathbf{A}$, we have a considerable amount of freedom. Suppose that $\mathbf{A}_0$ and ϕ_0 are particular choices which satisfy (9-51) and (9-52); if we add a vector of the form ∇g to $\mathbf{A}_0$, (9-51) will still hold, since

$$\begin{aligned}\nabla \times \mathbf{A} &= \nabla \times (\mathbf{A}_0 + \nabla g)\\ &= \nabla \times \mathbf{A}_0 + \nabla \times (\nabla g)\\ &= \nabla \times \mathbf{A}_0 = \mathbf{H}\end{aligned}$$

where we have used (9-21). If we modify $\mathbf{A}_0$ in this way, we must also modify ϕ in order to preserve (9-52). Putting $\phi = \phi_0 - \partial g/\partial t$, we have

$$\begin{aligned}-\nabla\phi - \frac{\partial \mathbf{A}}{\partial t} &= -\nabla\left(\phi_0 - \frac{\partial g}{\partial t}\right) - \frac{\partial}{\partial t}(\mathbf{A}_0 + \nabla g)\\ &= -\nabla\phi_0 + \nabla\left(\frac{\partial g}{\partial t}\right) - \frac{\partial}{\partial t}\mathbf{A}_0 - \frac{\partial}{\partial t}(\nabla g)\\ &= -\nabla\phi_0 - \frac{\partial}{\partial t}\mathbf{A}_0\\ &= \mathbf{E}\end{aligned}$$

We next choose g, which is thus far unrestricted, so that (9-55) holds. This requires that we have

$$\nabla \cdot (\mathbf{A}_0 + \nabla g) + \frac{\partial}{\partial t}\left(\phi_0 - \frac{\partial g}{\partial t}\right) = 0$$

which implies that g must be chosen to satisfy the equation

$$\nabla^2 g - \frac{\partial^2 g}{\partial t^2} = -\left(\nabla \cdot \mathbf{A}_0 + \frac{\partial \phi_0}{\partial t}\right)$$

With this line of argument, the solution of Maxwell's equations may be reduced to the consideration of the special partial differential equation

$$\nabla^2\psi - \frac{\partial^2\psi}{\partial t^2} = -h \tag{9-56}$$

In coordinate form, this is

$$\frac{\partial^2\psi}{\partial x^2} + \frac{\partial^2\psi}{\partial y^2} + \frac{\partial^2\psi}{\partial z^2} - \frac{\partial^2\psi}{\partial t^2} + h = 0 \tag{9-57}$$

and is called the **inhomogeneous wave equation.**

If ψ is independent of time, then equation (9-57) reduces to an equation $\nabla^2\psi = -h$, or

$$\frac{\partial^2\psi}{\partial x^2} + \frac{\partial^2\psi}{\partial y^2} + \frac{\partial^2\psi}{\partial z^2} + h = 0 \tag{9-58}$$

which is called **Poisson's equation.**

When h is identically 0, this in turn becomes **Laplace's equation**

$$\nabla^2 \psi = 0 \tag{9-59}$$

whose solutions in an open region Ω are called **harmonic functions.**

Vector methods are used in the solution of each of these equations. Before we can use them, we need certain relations which may be derived from the divergence theorem, and which are called **Green's identities.** To obtain these, we start from the vector form of the divergence theorem

$$\iiint_{\Omega} \nabla \cdot \mathbf{F} = \iint_{\partial\Omega} \mathbf{F} \cdot \mathbf{n} = \iint_{\partial\Omega} F_N \tag{9-60}$$

where the vector $\mathbf{n}$ in the second integral is the general unit normal on $\partial\Omega$. In our applications, Ω will be the region lying between two spheres (Fig. 9-22). If we choose the vector function $\mathbf{F}$ to have a particular form, then we can obtain a number of special cases of this result which are quite useful. To illustrate this, let us first take $\mathbf{F}$ as a gradient field, $\mathbf{F} = \nabla g$. Since $\nabla \cdot (\nabla g) = \nabla^2 g$, and $\nabla g \cdot \mathbf{n}$ is the directional derivative of g in the direction $\mathbf{n}$, (9-60) becomes

$$\iiint_{\Omega} \nabla^2 g = \iint_{\partial\Omega} D_{\mathbf{n}} g = \iint_{\partial\Omega} \frac{\partial g}{\partial \mathbf{n}} \tag{9-61}$$

where $\partial g/\partial \mathbf{n}$ is understood to mean the scalar function whose value at a point on the boundary of Ω is the directional derivative of g in the direction normal to the surface. (Briefly, $\partial g/\partial \mathbf{n}$ is the normal derivative of g.)

Again, if we choose $\mathbf{F} = f\nabla g$, we use the vector identity

$$\nabla \cdot (f\mathbf{V}) = \nabla f \cdot \mathbf{V} + f(\nabla \cdot \mathbf{V})$$

with $\mathbf{V} = \nabla g$, then

$$\nabla \cdot \mathbf{F} = \nabla f \cdot \nabla g + f\nabla^2 g$$

and (9-60) becomes Green's **first identity:**

$$\iiint_{\Omega} \nabla f \cdot \nabla g + \iiint_{\Omega} f\nabla^2 g = \iint_{\partial\Omega} f\frac{\partial g}{\partial \mathbf{n}} \tag{9-62}$$

We illustrate the use of these identities by proving certain properties of the class of **harmonic functions.** Recall that a function g is said to be harmonic in a region Ω if it satisfies Laplace's equation, (9-59), in Ω. In Theorem 20, Sec. 3.6, we proved by a different method that these functions were conditioned in a region by their values on the boundary; if g is harmonic in a bounded closed region Ω, and $g(p) \le M$ for all p on the boundary of Ω, then $g(p) \le M$ for all points p in Ω itself. As a deduction from this, in Exercise 18, Sec. 3.6, we found that two functions which are harmonic in Ω and which coincide on the boundary of Ω must coincide throughout Ω. The second of these results comes at once from (9-62). If we take $f = g$ and assume that g is harmonic in Ω, and that $g = 0$ on the boundary of Ω, then we have

$$\iiint_{\Omega} \nabla g \cdot \nabla g = \iiint_{\Omega} |\nabla g|^2 = 0$$

Since $|\nabla g| \geq 0$, we must have $\nabla g = 0$ (assuming that g is of class C''), and g is constant in Ω. Since the boundary value of g is 0, $g = 0$ in Ω. If g and g^* were two functions, both harmonic in Ω, with $g(p) = g^*(p)$ for all points p on the boundary of Ω, then $g - g^*$ would be harmonic in Ω, and would have boundary value 0; by the previous argument, $g - g^*$ would be identically 0 throughout Ω, so that $g = g^*$. By an analogous argument, it may be shown that if g and g^* are harmonic in Ω, and if their normal derivatives, $\partial g/\partial \mathbf{n}$ and $\partial g^*/\partial \mathbf{n}$, are equal on the boundary of Ω, then g and g^* differ at most by a constant in Ω (see Exercise 1). These results show that for a suitably well-behaved region Ω, the values of a harmonic function in Ω are determined completely by its values on $\partial\Omega$, and up to an additive constant, by the values of its normal derivative on $\partial\Omega$.

One may turn this around and ask the following question: Given a function f, defined on $\partial\Omega$, is there a function g which is harmonic in Ω and such that $g = f$ on $\partial\Omega$? This is called the **Dirichlet problem** for the region Ω. If Ω is suitably well behaved and f is continuous, then it has a solution, which, by the previous discussion, is unique. Correspondingly, given a function f defined on $\partial\Omega$, one may ask for a function g, harmonic in Ω, such that $\partial g/\partial \mathbf{n} = f$ on $\partial\Omega$. This is called **Neumann's problem** for Ω. There may exist no solutions, even though Ω is very well behaved and f is continuous; to see this, we resort again to Green's identities. If we apply (9-61) when g is harmonic in Ω, then we obtain

$$\iiint_{\Omega} \nabla^2 g = 0 = \iint_{\partial\Omega} \frac{\partial g}{\partial \mathbf{n}}$$

Thus, a necessary condition on the boundary-value function f in Neumann's problem is that its integral over the boundary of Ω be 0. Again, the previous discussion shows that if Neumann's problem has a solution g, and if Ω is sufficiently well behaved, then g is unique to within an additive constant.

Let us turn now to Poisson's equation:

$$\nabla^2 \psi = -h$$

where h is assumed to be of class C'', and, for convenience, to vanish outside a bounded region. We shall seek a solution of this equation which obeys two additional conditions:

$$\lim_{|p| \to \infty} |\psi(p)| = 0 \tag{9-63}$$

$$\lim_{|p| \to \infty} |\text{grad } \psi(p)|\,|p| = 0 \tag{9-64}$$

The first may be translated as saying that ψ vanishes at infinity, and the second as saying that the partial derivatives of ψ vanish more rapidly than $1/|p| = (x^2 + y^2 + z^2)^{-1/2}$. If we can obtain such a solution of Poisson's equation, then any harmonic function can be added to it; conversely, if ψ and ψ^* are two solutions of Poisson's equation, then

$$\nabla^2(\psi - \psi^*) = \nabla^2\psi - \nabla^2\psi^* = h - h = 0$$

so that $\psi - \psi^*$ is harmonic.

We begin by deriving **Green's second identity.** Write (9-62) with f and g interchanged:

$$\iiint_\Omega \nabla g \cdot \nabla f + \iiint_\Omega g\nabla^2 f = \iint_{\partial\Omega} g \frac{\partial f}{\partial \mathbf{n}}$$

Subtract this from (9-62); the first integral drops out, leaving

$$\iiint_\Omega \{f\nabla^2 g - g\nabla^2 f\} = \iint_{\partial\Omega} \left\{f \frac{\partial g}{\partial \mathbf{n}} - g \frac{\partial f}{\partial \mathbf{n}}\right\} \tag{9-65}$$

Theorem 26 *If ψ is a solution of the equation $\nabla^2\psi + h = 0$ which obeys the boundary conditions* (9-63) *and* (9-64), *then ψ is given at the point $p_0 = (x_0, y_0, z_0)$ by*

$$4\pi\psi(p_0) = \iiint \frac{h(p)}{|p - p_0|}\, dp = \iiint \frac{h(x, y, z)\, dx\, dy\, dz}{[(x - x_0)^2 + (y - y_0)^2 + (z - z_0)^2]^{1/2}}$$

Let $\phi(p) = 1/|p - p_0|$. By direct computation, it is seen that $\nabla^2\phi = 0$ everywhere, except at the singularity p_0. Also,

$$\text{grad}\, \phi(p) = (p - p_0)/|p - p_0|^3$$

so that $|\text{grad}\, \phi(p)| = 1/|p - p_0|^2$. The function ϕ thus obeys the conditions (9-63) and (9-64) "at infinity." Let Ω be the region between two spheres with center p_0; we suppose that the larger has radius R, and the smaller has radius r. We apply Green's identity (9-65) with this choice of Ω, and with $f = \phi$ and $g = \psi$. Since ψ is assumed to be a solution of Poisson's equation, $\nabla^2 g = -h$. Since ϕ is harmonic in Ω, $\nabla^2 f = 0$. We therefore have

$$-\iiint_\Omega \phi h = \iint_{\partial\Omega} \phi \frac{\partial\psi}{\partial \mathbf{n}} - \iint_{\partial\Omega} \psi \frac{\partial\phi}{\partial \mathbf{n}} \tag{9-66}$$

The boundary of Ω consists of two spherical surfaces, Σ_R and Σ_r. On the outside one, Σ_R, the normal $\mathbf{n}$ is directed away from p_0. If we employ spherical coordinates (ρ, θ, ϕ), with p_0 as origin, then "derivative in the direction of the normal" means $\partial/\partial\rho$ so that on Σ_R,

$$\phi(p) = \rho^{-1} = R^{-1}$$

$$\frac{\partial \phi}{\partial \mathbf{n}} = \frac{\partial \phi}{\partial \rho} = \frac{\partial}{\partial \rho}(\rho^{-1}) = -\rho^{-2} = -R^{-2}$$

Similarly, on Σ_r, the normal is directed *toward* p_0, and $\partial/\partial\mathbf{n} = -\partial/\partial\rho$; thus $\phi(p) = \rho^{-1} = r^{-1}$ and

$$\frac{\partial \phi}{\partial \mathbf{n}} = -\frac{\partial \phi}{\partial \rho} = -\frac{\partial}{\partial \rho}(\rho^{-1}) = \rho^{-2} = r^{-2}$$

Using these, (9-66) becomes

$$0 = \iiint_{\Omega} h\phi + \frac{1}{R}\iint_{\Sigma_R} \frac{\partial \psi}{\partial \mathbf{n}} + \frac{1}{r}\iint_{\Sigma_r} \frac{\partial \psi}{\partial \mathbf{n}} + \frac{1}{R^2}\iint_{\Sigma_R} \psi - \frac{1}{r^2}\iint_{\Sigma_r} \psi \tag{9-67}$$

We examine the behavior of each of these integrals as r approaches 0 and R increases. Estimating the first three surface integrals, we have

$$\left| \frac{1}{R}\iint_{\Sigma_R} \frac{\partial \psi}{\partial \mathbf{n}} \right| \le \frac{1}{R} \text{ (area of } \Sigma_R\text{)(maximum of } |\text{grad } \psi| \text{ on } \Sigma_R\text{)}$$

$$\le 4\pi R \text{ (max } |\text{grad } \psi| \text{ on } \Sigma_R\text{)}$$

$$\left| \frac{1}{r}\iint_{\Sigma_r} \frac{\partial \psi}{\partial \mathbf{n}} \right| \le \frac{1}{r} \text{ (area of } \Sigma_r\text{)(maximum of } |\text{grad } \psi| \text{ on } \Sigma_r\text{)}$$

$$\le 4\pi r \text{ (maximum of } |\text{grad } \psi| \text{ on } \Sigma_r\text{)}$$

$$\left| \frac{1}{R^2}\iint_{\Sigma_R} \psi \right| \le \frac{1}{R^2} \text{ (area of } \Sigma_R\text{)(maximum of } |\psi| \text{ on } \Sigma_R\text{)}$$

$$\le 4\pi \text{ (maximum of } |\psi| \text{ on } \Sigma_R\text{)}$$

Using (9-63) and (9-64), we see that all three approach 0, as $r \to 0$ and $R \to \infty$. Applying the mean value theorem to the remaining surface integral,

$$-\frac{1}{r^2}\iint_{\Sigma_r} \psi = -\frac{4\pi r^2}{r^2}\psi(p^*)$$

$$= -4\pi\psi(p^*)$$

where p^* is some point lying on Σ_r. As $r \to 0$, $p^* \to p_0$, so that

$$\lim_{r \to 0} \frac{1}{r^2}\iint_{\Sigma_r} \psi = 4\pi\psi(p_0)$$

Accordingly, if we return to (9-67), we have

$$4\pi\psi(p_0) = \lim_{\substack{r \to 0 \\ R \to \infty}} \iiint_\Omega h\phi = \iiint \frac{h(p)}{|p - p_0|}\, dp$$

Having thus obtained the necessary form of a solution of the Poisson equation, one may proceed to verify directly that this is a solution.

The general wave equation (9-57) may be treated in the same way. The solution which is obtained has the form

$$\psi(x_0, y_0, z_0, t) = \iiint \frac{h(x, y, z, t - \rho)\, dx\, dy\, dz}{\rho}$$

where $\rho = |p - p_0|$, assuming that ψ and h are independent of time for $t \le 0$. The form of the integrand has a simple interpretation; at time t, the value of ψ is not determined by the *simultaneous* values of h throughout space, but rather by those values which could be "communicated" to p_0 by a signal which travels at speed 1 (i.e., the speed of light), and which therefore left the point p at time $t - |p - p_0| = t - p$.

For details of this, and other applications of Green's identities, the reader is referred to the treatise by P. M. Morse and H. Feshbach, "Methods of Theoretical Physics," McGraw-Hill Book Company, New York, 1953.

EXERCISES

1 (*a*) If g is harmonic in Ω and the normal derivative of g on the boundary is 0, show that $\nabla g = 0$ in Ω.

(*b*) Let g and g^* be harmonic in Ω, and let $\partial g/\partial \mathbf{n} = \partial g^*/\partial \mathbf{n}$ on $\partial\Omega$. Show that $g^* = g + K$, where K is constant.

2 Let $\nabla \cdot \mathbf{F} = 0$ in a convex region Ω. Show that $\mathbf{F}$ can be expressed there in the form $\mathbf{F} = \nabla \times \mathbf{V}$, where $\nabla \cdot \mathbf{V} = 0$ and $\nabla^2 \mathbf{V} = -\nabla \times \mathbf{F}$. (This reduces the problem of finding a 1-form ω with $d\omega = \sigma$ for a given exact 2-form σ, to the solution of Poisson's equation.)

3 Prove the following vector analog for Green's identities.

(*a*) $$\iint_{\partial\Omega} (\mathbf{F} \times [\nabla \times \mathbf{G}])_N = \iiint_\Omega (\nabla \times \mathbf{F}) \cdot (\nabla \times \mathbf{G}) - \iiint_\Omega \mathbf{F} \cdot (\nabla \times [\nabla \times \mathbf{G}])$$

(*b*) $$\iiint_\Omega \{\mathbf{G} \cdot (\nabla \times [\nabla \times \mathbf{F}]) - \mathbf{F} \cdot (\nabla \times [\nabla \times \mathbf{G}])\} = \iint_{\partial\Omega} \{(\mathbf{F} \times [\nabla \times \mathbf{G}])_N - (\mathbf{G} \times [\nabla \times \mathbf{F}])_N\}$$

CHAPTER

TEN

NUMERICAL METHODS

10.1 PREVIEW

While many mathematical questions are regarded as solved when a solution has been proved to exist, there are times when the goal is to find a concrete numerical answer. We have illustrated this already in Sec. 5.5, where we discussed finding the sum of a convergent series, and in many of the exercises in Secs. 3.5 and 4.3, dealing with applications of the mean value theorem and with numerical integration.

In this chapter, we concentrate on the numerical aspect and point out a small number of algorithms and techniques, choosing those where results from earlier chapters of a theoretical nature are used to explain or justify the algorithm. The first three sections deal mainly with the following two problems:

i. Given a set D, a function T defined on D and taking values in S, and a point $q_0 \in S$, how can we find a point $\bar{p} \in D$ such that $T(\bar{p}) = q_0$? In particular, given an interval I and a real-valued function f, how do we solve $f(x) = 0$?
ii. Given a set D and a real-valued function f defined on D, how can we find the minimum (or maximum) value of $f(p)$ for $p \in D$?

For each of these, the approach is often to obtain a sequence $\{p_n\}$ of points in D which may be convergent to the desired solution point $\bar{p}$. Thus, in

Sec. 10.2 we discuss Newton's method for functions of one variable and for transformations; the latter depends upon some of the results in Chap. 7 dealing with differentiation of transformations. In Sec. 10.3, we discuss another very useful iterative procedure that depends on the properties of a contraction mapping and the notion of an attractive fixed point for a transformation. Section 10.4 concentrates on problem (ii), discussing iteration methods for solving extremal problems, and ends with a brief treatment of Lagrange multipliers.

The last section of the chapter touches upon several topics in numerical integration and differentiation, again picking some that use mathematical results obtained in earlier chapters. Thus, we use the mean value theorem to obtain formulas for approximate differentiation, and we obtain useful estimates for certain simple numerical integration methods. We have also included a description of the ingenious iterative procedure used by Gauss to evaluate a class of elliptic integrals.

This brief chapter is not intended to be a survey of the entire field of numerical analysis, but rather a sampling only, omitting all mention of some of the most important topics. For example, there is no mention of the solution of ordinary or partial differential equations.

We have also omitted any discussion of the computer aspects of these topics, although these will in fact be an important part of any solution procedure. For example, if we are attempting to solve the equation $f(x) = 0$, it should be observed that there are in fact two possible questions to be answered. Do we want to find a point x where $f(x)$ is computationally indistinguishable from 0, or do we want a point x that is computationally indistinguishable from the true solution $\bar{x}$ of the equation? Here, we are recognizing that it is usually impossible to tell if a computed number is the same as a given number, or even to be sure that it is positive or negative. Similar remarks apply to problem (ii), for we may be interested only in finding a point where $f(p)$ is nearly minimum, or we may want p to be near the actual minimum point $\bar{p}$, even if $f(p)$ itself is large.

Such questions are important when it comes to actual computational matters, and we leave them to the growing collection of treatises on the subject.

10.2 LOCATING ZEROS

The simplest problem of this type is to locate a solution of the equation $f(x) = 0$ on an interval $I = [a, b]$ where f is continuous, and where we have $f(a)f(b) < 0$. By the intermediate value theorem, we know that there must exist at least one point $\bar{x} \in I$ such that $f(\bar{x}) = 0$. A harder problem would be to locate all the zeros of f in I, since part of this problem is knowing that we have found them all. The curve shown in Fig. 10-1 will also suggest some of the pitfalls that enter into such problems. Inadequate sampling of the function

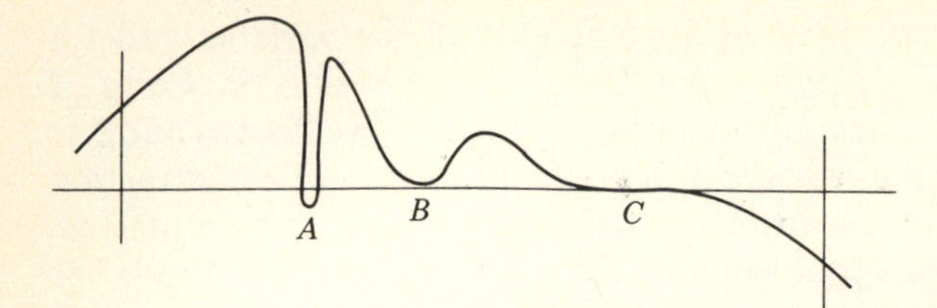

Figure 10-1

values may lead us to miss a zero entirely, as at A; round-off errors may lead us to confuse a positive minimum point B with a zero; a nearly horizontal portion near a zero point C may mislead us as to the true location of the zero. All of these are more the concern of a study of practical numerical analysis than of one directed toward the theoretical background, and for this reason we are content with this brief mention.

There always seem to be trade-offs in the choice of a numerical procedure. Simple procedures often tend to be reliable, while fancier ones may be faster but likely to go wrong with special classes of problems. The simplest procedure for solving the problem posed above is the **bisection process.** We bisect the interval I and calculate the value of f at the midpoint, noting the sign of this value; we now have a new interval of half the length in which f must have a zero, or we will have observed that the midpoint itself is a solution. Continuing this, we obtain a nested sequence of intervals which close down on the desired solution $\bar{x}$. In twenty steps we have arrived at an interval of length $2^{-20}\text{L}(I)$, which is a reduction by about a million. This method is effective, and we can predict in advance how many steps are needed to achieve a specified accuracy in the location of $\bar{x}$. Note that computer noise might give us the wrong sign for some value of f at a bisection point, but since this value would of necessity be very small, the ultimate point $\bar{x}$ provided by the algorithm would not be too far off.

Another simple method is linear interpolation. Rather than choosing the midpoint each time, one might choose that value of x where simple linear interpolation between the current values of f of opposite sign yields 0 as the predicted value of f. Another variant of this is called the **secant method,** and provides a sequence $\{x_n\}$ determined by the recursion

$$
\begin{aligned}
x_0 &= a \\
x_1 &= b \\
x_{n+1} &= \frac{x_{n-1}f(x_n) - x_n f(x_{n-1})}{f(x_n) - f(x_{n-1})}
\end{aligned}
\tag{10-1}
$$

The reason for the name and the theory behind this algorithm are both seen in Fig. 10-2, and it is clear that in many cases, the sequence $\{x_n\}$ will converge to one of the zeros of f. [However, note that it can be catastrophic to have $f(x_{n-1})$ and $f(x_n)$ nearly equal while x_{n-1} and x_n are far apart; see Exercise 5.]

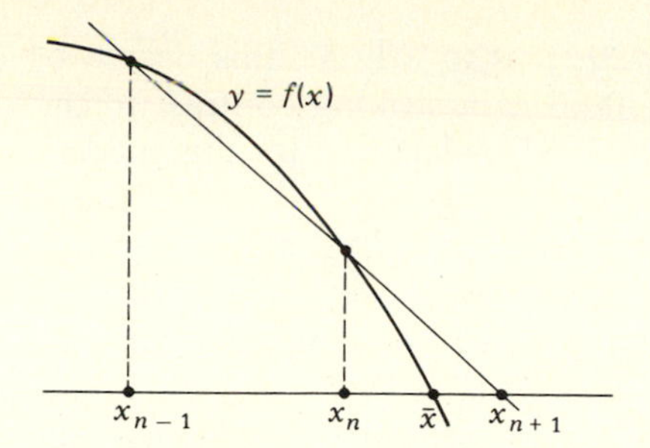

Figure 10-2

Elementary calculus texts often give **Newton's method,** which can be regarded as a limiting form of the secant method. This algorithm also produces a sequence $\{x_n\}$ which (hopefully) converges to $\bar{x}$ and is defined by the recursion

$$x_0 = a$$

(10-2)
$$x_{n+1} = x_n - \frac{f(x_n)}{f'(x_n)}$$

The rationale for this is seen in Fig. 10-3.

It is interesting to compare the effectiveness of these different methods on the same example. Let us solve $x^3 - 5x + 3 = 0$ on the interval [1, 2]. By the secant method we obtain

n	0	1	2	3	4	5	6
x_n	1	2	1.5	1.76	1.8736	1.8312	1.8341
$f(x_n)$	−1	1	−1.125	−0.328	0.2090	−0.0154	−0.000 64

and by Newton's method we obtain

n	0	1	2	3	4
x_n	2	1.8571	1.834 79	1.834 243 5	1.834 243 2
$f(x_n)$	1	0.1195	0.002 77	0.000 001 6	0.000 00 ···

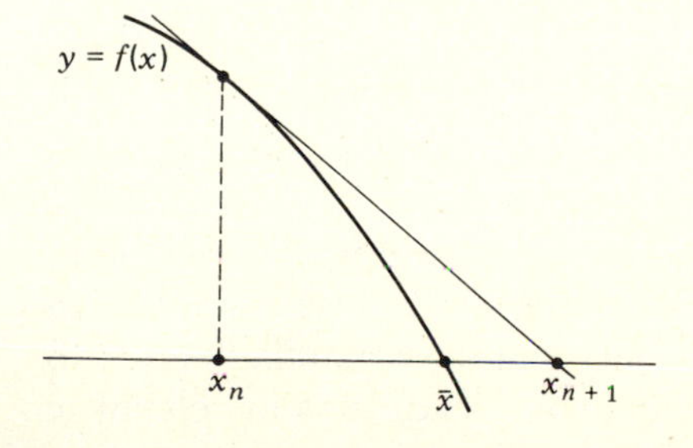

Figure 10-3

The bisection method requires more than 20 steps to achieve this degree of accuracy. The effectiveness of Newton's method is in striking contrast to the other two, and it is natural to seek a mathematical explanation for this success.

Theorem 1 *Let f be of class C'' on an interval I, and let $\bar{x} \in I$ be a simple zero of f, meaning that $f'(\bar{x}) \neq 0$. Then, there is an interval $I^* \subset I$ containing $\bar{x}$ and a constant M such that if $x_0 \in I^*$ and $\{x_n\}$ is defined by the recursion* (10-2), *then $\{x_n\}$ converges to $\bar{x}$ and*

(10-3) $$|x_{n+1} - \bar{x}| \leq M|x_n - \bar{x}|^2 \qquad \text{for } n = 1, 2, 3, \ldots$$

A converging sequence which obeys a condition such as (10-3) is often called "superconvergent," meaning that the errors decrease very rapidly, behaving something like the sequence .1, .01, .0001, .00000001, 10^{-16}, 10^{-32}, etc. To prove Theorem 1, let I_0 be an interval about $\bar{x}$ in which $|f'(x)| \geq C > 0$. Then, let x_n belong to I_0 and expand f about x_n by Taylor's theorem, obtaining

$$f(x) = f(x_n) + f'(x_n)(x - x_n) + \frac{f''(\tau)(x - x_n)^2}{2}$$

where τ lies between x and x_n. In this, set $x = \bar{x}$ where $f(\bar{x}) = 0$, and divide by $f'(x_n)$ to obtain

$$\frac{f''(\tau)}{2f'(x_n)}(\bar{x} - x_n)^2 = x_n - \bar{x} - \frac{f(x_n)}{f'(x_n)}$$

The right side of this expression is $x_{n+1} - \bar{x}$, so that we have (10-3) with $M = B/(2C)$ where B is the maximum of $|f''(x)|$ on I_0. Clearly, if x_0 is sufficiently close to $\bar{x}$, (10-3) will imply that x_1, x_2, and all later x_n lie in I_0. ∎

It is also natural to wonder if Newton's method can be applied when $\bar{x}$ is not a simple zero of f. Experimentation suggests that the method again produces a sequence converging to a zero $\bar{x}$ of f if the starting point is sufficiently close to $\bar{x}$, but that the rate of convergence is far slower. A mathematical justification of this is given in Exercise 6. (Of course, in order to be able to calculate x_{n+1} by (10-2), it is necessary to be sure that $f'(x_n) \neq 0$; a convenient way to achieve this is to require that $f''(x) \neq 0$ on a neighborhood of $\bar{x}$.)

Suppose we now turn to problems involving functions of several variables. The task of solving $f(x, y) = 0$ for points (x, y) in a region D is really that of constructing the 0-height level curve Γ for f in D. An obvious approach is to choose a set of discrete values of y, say y_i for $i = 1, 2, \ldots, N$, and for each to solve the equation $f(x, y_i) = 0$ for x. The result will be a collection of points on each of the lines $y = y_i$, each lying on the desired level curve (which may have more than one connected component), and the remaining task is to join these in an appropriate way to obtain an approximation to the complete level curve Γ. Accuracy can be increased by choosing the y_i close together, but at the expense of increasing the effort considerably. Any adequate discussion of the methods used to obtain smooth approximations to Γ from the discrete point data would take us far afield.

We therefore turn instead to the problem of solving a pair of simultaneous equations such as

$$\begin{aligned} f(x, y) &= 0 \\ g(x, y) &= 0 \end{aligned} \tag{10-4}$$

We can restate this in terms of the transformation T defined by $T(x, y) = (f(x, y), g(x, y))$ in the simpler form $T(p) = \mathbf{0}$. In either case, we see that a solution $(\bar{x}, \bar{y}) = \bar{p}$ will be a point that lies on the intersection of the 0-level curves for f and g. If it is possible to obtain sufficient information about these curves by the methods described in the preceding paragraph, it may be possible to obtain a nested sequence of rectangles $\{D_n\}$ known to contain $\bar{p}$. In simple cases, such a discrete search method can succeed. However, analytic methods based on the study of the mapping T are usually much more efficient. For example, there is a several-variable form of Newton's method, defined as follows: Choose a trial solution point p_0, which should be not too far from the actual solution point $\bar{p}$, and then use the following recursion to obtain a sequence $\{p_n\}$:

$$p_{n+1} = p_n - \left(dT\Big|_{p_n}\right)^{-1} T(p_n) \tag{10-5}$$

[The analogy between this formula and that for the usual one-variable Newton's method is more evident if the notation $T'(p_n)$ is used instead of the differential of T; compare with (10-2).]

With appropriate requirements on T and on the point $\bar{p}$, the sequence $\{p_n\}$ will again be superconvergent to $\bar{p}$ if the starting point p_0 lies in a sufficiently small neighborhood of $\bar{p}$.

Theorem 2 *Assume that T is of class C'' and that the differential of T is nonsingular at $\bar{p}$. Then, there is a neighborhood $\mathcal{N}$ about $\bar{p}$ and a constant M such that if p_0 lies in $\mathcal{N}$ and p_n is defined by* (10-5), *then*

$$|p_{n+1} - \bar{p}| \le M|p_n - \bar{p}|^2 \tag{10-6}$$

The proof follows the same pattern as that of Theorem 1. Let dT be the differential of T at p_n. Then, for any p sufficiently close to p_n,

$$T(p) = T(p_n) + (dT)(p - p_n) + R \tag{10-7}$$

where the remainder term R obeys $|R| \le B|p - p_n|^2$, for some suitable constant B. In (10-7), choose $p = \bar{p}$ and multiply through by the inverse of dT to obtain

$$\begin{aligned}(dT)^{-1}(R) &= -(dT)^{-1}(T(p_n)) - (\bar{p} - p_n)\\ &= p_n - (dT)^{-1}T(p_n) - \bar{p}\\ &= p_{n+1} - \bar{p}\end{aligned} \tag{10-8}$$

By the boundedness theorem for linear transformations, and the fact that the entries in $(dT)^{-1}$ are uniformly bounded on a neighborhood of $\bar{p}$, it follows that there is a constant A such that $|(dT)^{-1}(R)| \le A|R| \le AB|\bar{p} - p_n|^2$, so that from (10-8) we obtain (10-6). ∎

It is clear that the proof of Theorem 2 was not restricted to a pair of equations in two variables, but instead applied to a nonlinear system of n equations in n variables.

As a numerical procedure, Newton's method is not an easy one to apply, especially if the number of equations is even moderately large. To illustrate its use, consider the pair of equations

$$\begin{cases} x^2y - x - 2 = 0 \\ xy^2 - y - 6 = 0 \end{cases} \tag{10-9}$$

It is easy to see that this system has exactly one real solution, $x = 1$, $y = 3$. To apply Newton's method, we find

$$(dT)^{-1} = \frac{1}{3x^2y^2 - 4xy + 1}\begin{bmatrix} 2xy - 1 & -x^2 \\ -y^2 & 2xy - 1 \end{bmatrix}$$

and then define the sequence $\{p_n\}$ by the recursion

$$p_0 = (x_0, y_0) \qquad p_{n+1} = G(x_n, y_n) \tag{10-10}$$

where $G(x, y) = (u, v)$ and

$$\begin{aligned} u &= x - \frac{x(xy-1)^2 + 6x^2 - 4xy + 2}{3x^2y^2 - 4xy + 1} \\ v &= y - \frac{y(xy-1)^2 + 2y^2 - 12xy + 6}{3x^2y^2 - 4xy + 1} \end{aligned} \tag{10-11}$$

In practice, the convergence of $\{p_n\}$ to the desired solution $\bar{p}$ depends on a "good" choice for the start p_0. As Theorem 2 indicates, any point sufficiently near $\bar{p}$ is good, but others may be also. In our example, the choice $p_0 = (7, 7)$ yields the sequence

$$\begin{aligned} p_1 &= (4.684,\ 4.768) \\ p_2 &= (3.139,\ 3.380) \\ &\cdots\cdots\cdots\cdots\cdots \\ p_5 &= (1.0488,\ 2.8288) \\ p_6 &= (.99797,\ 2.99965) \\ p_7 &= (1.000004,\ 2.99994) \approx (1, 3) \end{aligned}$$

On the other hand, with $p_0 = (3, 1)$ the sequence obtained is

$$\begin{aligned} p_1 &= (-.5,\ 2.5) \\ p_2 &= (-1.058,\ -1.819) \\ p_3 &= (-1.085,\ -.907) \\ p_4 &= (-2.197,\ -2.029) \end{aligned}$$

and the sequence does not converge to $\bar{p}$. (Of course, one cannot choose a starting point p_0 such as (1, 1) where the Jacobian vanishes or is too small.)

We will reexamine Newton's method from a different point of view in the next section.

EXERCISES

1 Verify formula (10-1) for the secant method.

2 Verify formula (10-2) for Newton's method.

3 (*a*) Determine the number of real roots of the equation

$$\sin x + \tfrac{1}{6}x = 1$$

(*b*) Find each root, accurate to .0001.

4 Discuss the behavior of Newton's method for the function shown in Fig. 10-1, assuming that the starting point x_0 is chosen near to but not at the points A, B, and C.

5 Explain why the secant method encounters difficulty if $|f(x_n) - f(x_{n-1})|$ is small in comparison with $|x_n - x_{n-1}|$. (It may help to refer to Fig. 10-2.)

6 (*a*) Show that if $f'(\bar{x}) = 0$ but $f''(\bar{x}) \neq 0$, then the behavior of the error in Newton's method is governed by the growth estimate $|x_{n+1} - \bar{x}| \le B|x_n - \bar{x}|$, where B is approximately $\frac{1}{2}$.

(*b*) What effect does this have on the rate of convergence of $\{x_n\}$?

(*c*) Apply Newton's method to solve $x^4 - 6x^2 + 9 = 0$, and observe the rate of convergence of the sequence $\{x_n\}$.

(*d*) What happens if $f'(\bar{x}) = f''(\bar{x}) = 0$ but $f'''(\bar{x}) \neq 0$?

7 Sketch the set of points (A, B) for which the equation $x^5 - 5x^3 + Ax + B = 0$ has (*a*) exactly one real root; (*b*) exactly three real roots; (*c*) exactly five real roots.

8 Find the number of real solutions of the system of equations

$$\begin{cases} \sin(y) - x = 0 \\ \cos(x) - y = 0 \end{cases}$$

and apply Newton's method (10-5) to find one solution.

10.3 FIXED-POINT METHODS

Let D be a set in n space, and let T be a transformation defined on D and mapping D into itself, so that $T(p) \in D$ for each $p \in D$. A point $\bar{p}$ is called a **fixed point** for T if $T(\bar{p}) = \bar{p}$, and the problem to be discussed in this section is that of finding such points $\bar{p}$ for a given T and D.

There is a very close relationship between the topics in the preceding section and those in the present section. Clearly, if we define another transformation S by $S(p) = p - T(p)$, then the fixed points of T are exactly the zero points for S, and vice versa. Thus, any process which can be used to find zeros can be converted into a process for finding fixed points. However, the simple change in viewpoint turns out to be quite helpful in that it suggests other procedures and also helps to clarify the theory behind some of the procedures for locating zeros.

We start with a very useful general result for which we need a special definition.

Definition 1 *A transformation T mapping a set D into itself is called a* **contraction** *if there is a number $\lambda < 1$ such that*

$$|T(p) - T(q)| \le \lambda |p - q| \tag{10-12}$$

for all points p and q in D.

Note that a contraction mapping is necessarily continuous, and that $T(D)$ is usually smaller than D. The following result is usually called the **contraction theorem.**

Theorem 3 *Let T be a contraction mapping on a closed set D, $D \subset \mathbf{R}^n$. Then, T has exactly one fixed point $\bar{p}$ in D, and $\bar{p} = \lim_{n\to\infty} p_n$, where $\{p_n\}$ is the sequence defined recursively by*

$$p_{n+1} = T(p_n) \qquad n \ge 0 \tag{10-13}$$

where the initial point p_0 is arbitrarily chosen in D.

The first step is to prove $\{p_n\}$ a Cauchy sequence. From (10-13), we see that

$$p_{n+1} - p_n = T(p_n) - T(p_{n-1})$$

so that since T is a contraction, and (10-12) holds,

$$|p_{n+1} - p_n| \le \lambda |p_n - p_{n-1}|$$

for all $n = 1, 2, \ldots$. This yields $|p_2 - p_1| \le \lambda|p_1 - p_0|$ and $|p_3 - p_2| \le \lambda|p_2 - p_1| \le \lambda^2|p_1 - p_0|$, and in general

$$|p_{n+1} - p_n| \le \lambda^n|p_1 - p_0| \tag{10-14}$$

Take any indices N and $N + k$. Then

$$p_{N+k} - p_N = p_{N+k} - p_{N+k-1} + p_{N+k-1} - p_{N+k-2} + \cdots + p_{N+1} - p_N$$

so that, by (10-14),

$$\begin{aligned} |p_{N+k} - p_N| &\le |p_{N+k} - p_{N+k-1}| + |p_{N+k-1} - p_{N+k-2}| + \cdots + |p_{N+1} - p_N| \\ &\le \{\lambda^{N+k-1} + \lambda^{N+k-2} + \cdots + \lambda^N\}|p_1 - p_0| \\ &\le \lambda^N(1 + \lambda + \lambda^2 + \cdots + \lambda^{k-1})|p_1 - p_0| \end{aligned}$$

Since $\lambda < 1$, $\sum_0^\infty \lambda^n$ converges to $1/(1 - \lambda)$. Thus, we have

$$|p_{N+k} - p_N| \le \lambda^N \frac{|p_1 - p_0|}{1 - \lambda}$$

holding for all N and k. Since $\lim_{N\to\infty} \lambda^N = 0$, we see that $\{p_n\}$ is a Cauchy sequence. Since n space is complete, and D closed, $\{p_n\}$ must converge to some point $\bar{p} \in D$. Since T is continuous, $\lim T(p_n) = T(\bar{p})$. But, by (10-13), $T(p_n) = p_{n+1}$, which we know converges to $\bar{p}$. Accordingly, $T(\bar{p}) = \bar{p}$, and $\bar{p}$ is a fixed point for T in D. Suppose now that T were to have another fixed point, q. Then, because T is a contraction, $|T(\bar{p}) - T(q)| \le \lambda|\bar{p} - q|$, which becomes $|\bar{p} - q| \le \lambda|\bar{p} - q|$. But, since $\lambda < 1$, this can only occur if $q = \bar{p}$, showing that $\bar{p}$ is the only fixed point for T in D. ▌

The hypothesis that a transformation T is a contraction is therefore a very powerful one. The definition can be restated in a useful way: T is a contraction on a region D if T obeys a uniform Lipschitz condition on D, with a constant that is strictly less than 1. In earlier chapters, we have found that various versions of the mean value theorem provide ways to find a Lipschitz constant for sufficiently smooth functions or mappings (see Exercise 13 of Sec. 3.2, Exercise 18 of Sec. 3.5, and the corollary to Theorem 12 in Sec. 7.4). As a result, we immediately obtain the following two simple results.

Theorem 4 *Let f be a real-valued function of class C' on an interval I, and suppose that $f(I) \subset I$ and that $|f'(x)| \le \lambda < 1$ for all $x \in I$. Then, f is a contraction mapping on I.*

Theorem 5 *Let T be a transformation of class C' on a convex set D in n space, and suppose that $T(D) \subset D$. Suppose further that T is described in coordinate form as*

$$\begin{cases} y_1 = {}^1f(x_1, x_2, \ldots, x_n) \\ y_2 = {}^2f(x_1, x_2, \ldots, x_n) \\ \cdots\cdots\cdots\cdots\cdots\cdots\cdots\cdots \\ y_n = {}^nf(x_1, x_2, \ldots, x_n) \end{cases}$$

and that the partial derivatives of the functions ${}^1f$, ${}^2f$, ..., ${}^nf$ are bounded on D:

$$|{}^jf_i(p)| \le c_{ij} \qquad \text{all } p \in D$$

Then, if

$$\sum_{i,j=1}^{n} (c_{ij})^2 < 1 \tag{10-15}$$

T is a contraction on D.

The first of these theorems follows from the ordinary one-variable mean value theorem, and the second from the corresponding n-variable mean value theorem (Theorem 12, Sec. 7.4). In practice, the following two results are more often used.

Theorem 6 *Let f be of class C' on an open interval I containing a point $\bar{x}$ which obeys $f(\bar{x}) = \bar{x}$, and suppose that $|f'(\bar{x})| < 1$. Then, there is a subinterval I_0 about $\bar{x}$ on which f is a contraction.*

Theorem 7 *Let T be a transformation of class C' defined on an open set $D \subset \mathbf{R}^n$, mapping D into n space, and suppose that D contains a fixed point $\bar{p}$ for T. Suppose also that $dT|_{\bar{p}} = A = [a_{ij}]$, where $\sum_{i,j=1}^{n} (a_{ij})^2 < 1$. Then, there is a neighborhood $D_0 \subset D$ about $\bar{p}$ on which T is a contraction.*

Both of these arise from the observation that expressions involving the derivatives of the given functions will be continuous, and that the properties which hold at the point $\bar{x}$ or $\bar{p}$ must also hold when all the variables lie in an appropriately small neighborhood of the point. This guarantees a uniform Lipschitz condition on such a neighborhood, with a Lipschitz constant λ smaller than 1. The final step is to observe that this in turn implies that a smaller neighborhood of $\bar{x}$ (or of $\bar{p}$) is mapped into itself. ▮

We give several illustrations of these. Suppose we wish to solve the equation

$$\cos x + 3xe^{-x} = 0 \tag{10-16}$$

A sketch of $y = \cos x$ and of $y = -3xe^{-x}$ shows at once that there is one negative root and an infinite number of positive roots, with all the larger ones given approximately by $x = (2m + 1)\pi/2$. Equation (10-16) can be converted into a fixed-point problem in many ways; one obvious way is to restate it in the form

$$\begin{aligned} x &= -\tfrac{1}{3}e^x \cos x \\ &= f(x) \end{aligned} \tag{10-17}$$

If x_0 is chosen as $x_0 = 0$, and the sequence $\{x_n\}$ defined by $x_{n+1} = f(x_n)$, we obtain $x_1 = -\frac{1}{3}$, $x_2 = -.2257$, etc., which converges to one of the desired roots

$$\bar{x} = -.251\,162\,83 \cdots \tag{10-18}$$

(given with reasonable accuracy by x_{20}). If a different choice of the starting point x_0 is made, in almost all cases the same fixed point will be reached, in spite of the fact that there are infinitely many other fixed points—as shown by the graphical analysis of (10-16). For example, the first few are given by

2.322 613 ⋯, 4.569 887 ⋯, 7.863 056 ⋯, 10.995 020 ⋯, etc., matching approximately the claimed formula $(2m + 1)\pi/2$ for large m.

The explanation for this behavior is seen in Theorem 6. It is readily seen that $|f'(\bar{x})|$ is smaller that 1 for the fixed point given in (10-18), while this fails to be true for any of the other fixed points. Thus, $-.2511 \cdots$ is an *attractive* fixed point, and is surrounded by a neighborhood I_0 on which f is a contraction and in which every starting point x_0 leads to $\bar{x}$ when f is iterated. On the other hand, each of the other fixed points of f is a *repulsive* fixed point having the property that unless the initial choice x_0 coincides *exactly* with the fixed point, the sequence of iterates generated by it moves away from it, eventually enters the attractive region I_0, and thus once more converges to $-.2511 \cdots$. Indeed, it is immediately evident from (10-17) that a negative fixed point for f must be attractive, since

$$|f'(x)| = \tfrac{1}{3}e^x|\cos x - \sin x| \le \tfrac{2}{3}e^x \le \tfrac{2}{3}$$

We remark that in general, the behavior of sequences generated in this fashion by a given function f can be extremely erratic, producing both divergence to infinity and strange periodicities. (In this connection, we recommend an article by Li and Yorke in the *American Mathematical Monthly*, vol. 82, p. 985, 1975.)

There is a simple way to overcome this behavior by introducing a different function f. We start again from the given equation (10-16), and consider instead the function

$$g(x) = x - A(\cos x + 3xe^{-x}) \tag{10-19}$$

It is evident that the roots of (10-16) will again be fixed points of g. If we choose A appropriately, so that $|g'(x)|$ is small, iteration of g will yield the positive roots of (10-16). In particular, if $A = 1$, then $4.5698 \cdots$, and $10.995\,02 \cdots$ are attractive fixed points, and if $A = -1$, so are $2.3226 \cdots$ and $7.863\,05 \cdots$, and both can be calculated by iterating g from any sufficiently nearby starting point.

Our second example is the transformation

$$T\colon \begin{cases} u = x - .4x^2 - .2y + .6 \\ v = y + .3x - .5y^2 + .2 \end{cases} \tag{10-20}$$

and we seek a fixed point $\bar{p}$ for T. It is easily seen that there are two such points, (1, 1) and $(1.435\,82 \cdots, -1.123\,16 \cdots)$. The first is attractive, and if p_0 is chosen as (0, 0), the resulting sequence $\{p_n\}$, defined by $p_{n+1} = T(p_n)$, is convergent to (1, 1): $p_1 = (.6, .2)$, $p_2 = (1.016, .56)$, ..., $p_8 = (1.000015, .999876)$, etc. The second fixed point is not attractive. In confirmation of these statements, note that

$$dT\Big|_{(1,1)} = \begin{bmatrix} 1 - .8x & -.2 \\ .3 & 1 - y \end{bmatrix}\Bigg|_{(1,1)} = \begin{bmatrix} .2 & -.2 \\ .3 & 0 \end{bmatrix}$$

and that $(.2)^2 + (-.2)^2 + (.3)^2 < 1$. A similar computation at the second fixed point yields a number larger than 1. (We remark that the estimate $\sum_{i,j=1}^{n} (c_{ij})^2$ is not the best possible bound for the constant associated with the matrix $[c_{ij}]$ and the boundedness theorem, Theorem 8, Sec. 7.3; this means that the condition $\sum_{i,j=1}^{n} (c_{ij})^2 < 1$ is sufficient but not necessary, in order for a fixed point to be attractive.)

The contraction theorem also throws light on the general Newton's method for solving equations. Suppose that we wish to solve the equation $f(x) = 0$ on an interval I. Choose a constant A, and define a function g by

$$g(x) = x - Af(x) \tag{10-21}$$

Then, any root $\bar{x}$ of $f(x) = 0$ will be a fixed point for g. If we can choose A so that $|g'(\bar{x})| < 1$, then $\bar{x}$ will be an attractive fixed point and can be found by iterating g; moreover, smaller values for $|g'(\bar{x})|$ lead to faster convergence. Noting that $g'(x) = 1 - Af'(\bar{x})$, we see that an optimal choice for A is $1/f'(\bar{x})$. This analysis leads to the following procedure. To find $\bar{x}$, already located within an interval I, estimate the number $f'(\bar{x})$, choose A as $1/f'(\bar{x})$ (using the estimate), and then iterate the function g given in (10-21).

For example, if $f(x) = x^3 - 5$, and we know that there is a root $\bar{x}$ in the interval $[1, 2]$, then $f'(\bar{x})$ lies between 3 and 12. We estimate the correct value to be 7, and therefore set $g(x) = x - (x^3 - 5)/7$. If x_0 is chosen as any point in $[1, 2]$, then iteration will yield a sequence converging to $5^{1/3}$. (For example, $x_{10} = 1.709\,976\,3$ if $x_0 = 1$.)

The connection between this and Newton's method is now evident. The recursion just described has the form

$$x_{n+1} = x_n - \frac{f(x_n)}{f'(\bar{x})} \tag{10-22}$$

(where in practice an estimate for $f'(\bar{x})$ is used), while the recursion in Newton's method is

$$x_{n+1} = x_n - \frac{f(x_n)}{f'(x_n)} \tag{10-23}$$

automatically providing better and better estimates for $f'(\bar{x})$ as the sequence $\{x_n\}$ converges to $\bar{x}$. In both cases, it is now clear why multiple roots, for which $f'(\bar{x}) = 0$, lead to different behaviors. It is also evident that (10-22) can be a preferable method, especially if $f'(x)$ is difficult to evaluate, since in (10-22) this need be done only once.

A similar treatment can be given for systems of nonlinear equations in several variables. If we wish to solve

$$\begin{cases} f(x, y) = 0 \\ g(x, y) = 0 \end{cases}$$

knowing that there is a solution $\bar{p} = (\bar{x}, \bar{y})$ in a set D, consider the transformation S defined by

$$S(p) = p - AT(p) \tag{10-24}$$

where $T(p) = (f(p), g(p))$ and where A is an arbitrary nonsingular linear transformation. It is again clear that $\bar{p}$ is a fixed point for S. Calculating the derivative of S, we have

$$dS = I - A\,dT$$

If this is to be a matrix which at $\bar{p}$ has entries c_{ij} with $\sum_{i,j=1}^{n} (c_{ij})^2 < 1$, in order that $\bar{p}$ be an attractive fixed point for the transformation S, one choice for the constant matrix A is $(dT|_{\bar{p}})^{-1}$, since we will then have $dS = 0$. This leads to the following algorithm for finding $\bar{p}$, which is seen to be the direct analog of the one described above in formula (10-22). Knowing that $\bar{p}$ lies in D, we estimate the matrix $dT|_{\bar{p}}$, then choose A as its inverse. We then define S by (10-24), choose a starting point p_0, and define a sequence p_n by

$$p_{n+1} = p_n - AT(p_n) \tag{10-25}$$

If this is compared with Newton's method, as given in (10-5), we see that A is replaced by $(dT|_{p_n})^{-1}$, which is merely a moving estimate for $(dT|_{\bar{p}})^{-1}$. The success of (10-25) depends on how good the estimate is that leads to the matrix A, and how well p_0 is chosen. Its advantage over Newton's method lies in the fact that A is calculated only once and used in this form at each step of the iteration, while the corresponding matrix must be recalculated at each step in Newton's method.

We illustrate this with the same problem used as an illustration in the last section; we seek a solution for the system

$$\begin{cases} x^2y - x - 2 = 0 \\ xy^2 - y - 6 = 0 \end{cases}$$

Following the procedure outlined above, we must select a matrix $A = [a_{ij}]$ and then define a mapping S by $S(x, y) = (u, v)$, where

$$\begin{bmatrix} u \\ v \end{bmatrix} = \begin{bmatrix} x \\ y \end{bmatrix} - \begin{bmatrix} a_{11} & a_{12} \\ a_{21} & a_{22} \end{bmatrix} \begin{bmatrix} x^2y - x - 6 \\ xy^2 - y - 2 \end{bmatrix}$$

Suppose that we believe that the desired solution is in a disk centered on (2, 2). [Recall that the solution is actually at (1, 3).] Using this point as an estimate of $\bar{p}$ and computing $(dT|_{(2,2)})^{-1}$, we are led to choose

$$A = \begin{bmatrix} .21 & -.12 \\ -.12 & .21 \end{bmatrix}$$

If we iterate S, with (2, 2) as p_0, we obtain a sequence of points that converges rapidly to the correct point $\bar{p} = (1, 3)$; in fact, $p_5 = (.9220, 3.1911)$, $p_{10} = (1.0040, 2.9608)$, and $p_{30} = (.99993, 3.00019)$.

Specialized methods have been developed for selecting good starting points p_0 and for choosing the auxiliary matrix A, including some which modify A in the light of current values of p_n. Along with many other aspects of numerical analysis, this is left for texts devoted to this topic.

We close this topic with one final observation. The proof of the contraction theorem depended essentially only upon the sequence $\{p_n\}$ being convergent, and this in turn depended on the fact that in $\mathbf{R}^n$, any Cauchy sequence converges. Accordingly, a similar theorem holds for any mapping of a complete metric space into itself which is a contraction, in the sense of Definition 1. In Chap. 7, we studied a number of complete metric spaces whose points were functions. In these spaces, the contraction theorem can therefore be used to prove the existence of (and indeed to find) the solution of differential or integral equations, as well as other functional equations. This technique is of wide usefulness. In addition, there are also other theorems in which it is not required that the transformation under study be a contraction mapping, but in which the topological nature of the set D enables one to conclude that T must have a fixed point in D.

EXERCISES

1 Show that the iteration theorem (Theorem 4) fails to hold if f merely satisfies the condition $|f'(x)| < 1$ for all x, by considering the function $f(x) = x + 1/(1 + x)$ on the interval $0 \le x < \infty$.

2 Let $0 \le A \le 1$ and define a sequence $\{x_n\}$ by

$$x_0 = 0 \qquad x_{n+1} = x_n + \tfrac{1}{2}(A - (x_n)^2)$$

(*a*) Prove $\lim x_n = \sqrt{A}$.
(*b*) Estimate the rate of convergence.

3 Apply Exercise 2 to show that $P_0(x) = 0$, $P_{n+1}(x) = P_n(x) + \frac{1}{2}(x^2 - P_n(x))$ defines a sequence of polynomials which converge uniformly to $|x|$ on $-1 \le x \le 1$.

4 Let $A > 0$ and consider the following two sequences:

$$x_0 = A \qquad x_{n+1} = \tfrac{1}{2}[x_n + A/(x_n)^2]$$

$$y_0 = A \qquad y_{n+1} = \tfrac{1}{3}[2y_n + A/(y_n)^2]$$

(*a*) Prove that $\lim x_n = \lim y_n = \sqrt[3]{A}$.
(*b*) Which of these is a better algorithm for calculating $\sqrt[3]{A}$? Explain why.

5 Solve the system

$$\begin{cases} \sin(y) = x \\ \cos(x) = y \end{cases}$$

by fixed point methods.

6 Define a sequence of functions $\{f_n\}$ by $f_0(x) =$ any function of class C^∞ on $-\infty < x < \infty$, $f_{n+1}(x) = (d/dx) f_n(x)$, and suppose that $f_n(x)$ converges uniformly to $F(x)$ on $0 \le x \le 1$. What can you find out about $F(x)$?

10.4 EXTREMAL PROBLEMS

The general problem to be examined in this section is that of finding a point $\bar{p}$ which yields the minimum value of $f(p)$ for p in a given set S. This has already been examined from several viewpoints in earlier sections, especially Secs. 2.4 and 3.6. In the present section, we will look briefly at certain other aspects of this problem that are connected with the practical problem of finding $\bar{p}$, rather than the theoretical problem of proving that $\bar{p}$ exists.

For smooth functions f and regions S, it is evident that our problem can be reduced to another problem like those studied in Secs. 10.2 and 10.3. For, if the minimum of f occurs at an interior point of S, then $\bar{p}$ must be one of the critical points of f in S, and these are the points p at which $\mathbf{D}f = 0$, while if the minimum is on the boundary of S, then we have reduced the original problem to another similar problem with a different set S, namely bdy (S). If the boundary of S consists of one or more smooth curves, then substitution will yield a new function to be minimized on a domain of lower dimension, and the same process applies here.

However, in many cases this procedure is difficult to carry out, even if one is dealing with a one-dimensional problem (see Exercise 1). In such cases, other methods often turn out to be more effective, even some that are mathematically much more naive. The study of numerical methods for dealing with such problems has given rise to a new subject, optimization theory. In Sec. 3.6, we described one of these methods which made use of the gradient of f, $\mathbf{D}f$, identified as a vector at a point p in the domain of f which points in the direction of greatest increase in f. As explained there, one uses this to generate a sequence of points $\{p_n\}$ chosen so that along this sequence, the values of f constantly decrease—hopefully toward the minimum of f. One such formula was given there as

$$p_{n+1} = p_n - h\,\mathbf{D}f(p_n) \tag{10-26}$$

where h is a preselected constant which helps to determine the step size; each succeeding point is in the direction of decreasing values of f from the previous point.

It is also possible to take into account the behavior of the second derivative of f in choosing the next point after p_n. We illustrate this with a function of one variable. Suppose that $\bar{x}$ is a local minimum point for $f(x)$, and that x_n is a trial point near $\bar{x}$. In a neighborhood of x_n we can write

$$f(x_n + \lambda) = f(x_n) + \lambda f'(x_n) + \lambda^2 f''(x_n)/2 + \cdots$$

If we drop all the later terms in the Taylor series after the first three, we have a quadratic approximation to f near $x = x_n$, and we can ask for the choice of λ which will minimize this. The solution is seen to be $\lambda = -f'(x_n)/f''(x_n)$, and we are led to the formula

$$x_{n+1} = x_n - \frac{f'(x_n)}{f''(x_n)} \tag{10-27}$$

which generates a sequence $\{x_n\}$ that, with a good choice for the starting point x_0, may converge to the minimum point $\bar{x}$. (In Exercise 5, we point out another approach which also leads to the same formula, thus providing a second justification.)

There are also versions of this method which apply to extremal problems for functions of several variables. The role of $f'(x)$ is replaced by $\mathbf{D}f$, and that of $f''(x)$ by the matrix-valued second derivative of f, d^2f, discussed in Sec. 8.5 and defined in formula (8-48). Further discussion of this will be found in most of the recent texts on optimization, such as Kowalik and Osborne, "Methods for Unconstrained Optimization," American Elsevier Publishing Company, Inc., New York, 1968; and R. P. Brent, "Algorithms for Minimization without Derivatives," Prentice-Hall, Inc., Englewood Cliffs, N.J., 1973.

The set S on which a function f is to be minimized is often characterized by auxiliary equations; thus, S might be the set of all p in a given region Ω which also obey the restrictions $g(p) = h(p) = 0$. In such cases, the problem can be called a **constrained** minimum problem on Ω. The remainder of this section is devoted to two methods for solving such extremal problems.

The first method is one which replaces a constrained problem by a different *unconstrained* problem on the same set Ω. Suppose we wish to minimize $f(p)$ among those points $p \in \Omega$ such that $g(p) = 0$. The key idea is to consider instead the problem of minimizing on Ω the function

$$F(p) = f(p) + A|g(p)|^2 \tag{10-28}$$

where A is a suitable large constant, with the conjecture that a point p_A that is optimal for F will be one for which $g(p_A)$ is small, and for which $f(p_A)$ is therefore nearly equal to $f(\bar{p})$, the true minimum of $f(p)$ among those points p where $g(p) = 0$. Then, as A increases, the hope is that p_A will converge to $\bar{p}$.

For example, suppose we want to minimize $f(x, y) = 2x + y$ among those (x, y) with $x \geq 0$, $y \geq 0$ such that $xy - 18 = 0$. It is easily seen that $\bar{p} = (3, 6)$, and that the minimum of $f(p)$ is 12. We set

$$F(x, y) = 2x + y + A(xy - 18)^2$$

and look for the minimum of $F(p)$ for $x \geq 0$, $y \geq 0$. It is seen that this occurs approximately at the point $p_A = (3 - 1/(72A),\ 6 - 1/(36A))$, for large A, and if $A \to \infty$, $p_A \to (3, 6) = \bar{p}$.

The most direct approach to the solution of a constrained extremal problem is to make appropriate substitutions which convert it into an unconstrained problem. If we are to minimize $f(x, y, z)$, subject to the condition $g(x, y, z) = 0$, where p is confined to a set Ω and we can solve $g(x, y, z) = 0$ for $z = \phi(x, y)$, then we can instead consider the function $F(x, y) = f(x, y, \phi(x, y))$ and look for the (unconstrained) minimum of the function F. We would first look for the critical points of F, which would be found by solving $F_1 = F_2 = 0$ in Ω.

For example, let us find $(x, y, z) \in \mathbf{R}^3$ obeying

$$g(x, y, z) = 2x + 3y + z - 12 = 0 \tag{10-29}$$

for which $4x^2 + y^2 + z^2 = f(x, y, z)$ is minimum. Solving (10-29) for z, we arrive at the unconstrained extremal problem of minimizing

$$F(x, y) = 4x^2 + y^2 + (12 - 2x - 3y)^2$$

whose critical points are the solutions of

$$\begin{aligned} 8x - 4(12 - 2x - 3y) &= 0 \\ 2y - 6(12 - 2x - 3y) &= 0 \end{aligned} \tag{10-30}$$

which have only one solution, $x = \frac{6}{11}$, $y = \frac{36}{11}$, which clearly must be the desired minimum. Thus, the solution of the original constrained problem is the point $\bar{p} = (\frac{6}{11}, \frac{36}{11}, \frac{12}{11})$.

There are cases where it may be undesirable to carry out this direct approach explicitly, in part because it is not immediately obvious which variable should be the one to eliminate. One would therefore like to have an alternative way to explore such problems which treats all the variables alike. This can be done in terms of the critical points for general transformations; recall that p is a critical point for T if the rank of dT at p is smaller than the integer that is the lesser of the height and width of the matrix for dT.

Theorem 8 *The points $p = (x, y, z)$ which lie in the set S described by $g(x, y, z) = 0$ and at which $f(x, y, z)$ is locally a maximum or a minimum are among the critical points for the transformation*

$$T: \begin{cases} u = f(x, y, z) \\ v = g(x, y, z) \end{cases}$$

Since

$$dT = \begin{bmatrix} f_1 & f_2 & f_3 \\ g_1 & g_2 & g_3 \end{bmatrix}$$

an equivalent assertion is to say that p is among the simultaneous solutions of the equations

$$0 = \begin{vmatrix} f_1 & f_2 \\ g_1 & g_2 \end{vmatrix} \qquad 0 = \begin{vmatrix} f_2 & f_3 \\ g_2 & g_3 \end{vmatrix} \qquad 0 = \begin{vmatrix} f_1 & f_3 \\ g_1 & g_3 \end{vmatrix} \tag{10-31}$$

as well as $g(x, y, z) = 0$. (It should be noticed that any two of the determinant equations imply the third; the assertion that dT has rank less than 2 means that (f_1, f_2, f_3) and (g_1, g_2, g_3) are multiples of one another.) For example, let us apply this to the last illustrative example. Since $f(x, y, z) = 4x^2 + y^2 + z^2$ and $g(x, y, z) = 2x + 3y + z - 12$

we have

$$dT = \begin{bmatrix} 8x & 2y & 2z \\ 2 & 3 & 1 \end{bmatrix}$$

If the top row is a multiple of the bottom row, then $y = 6x$ and $z = 2x$. Substituting these into $g(x, y, z) = 0$, we again obtain the point $(\frac{6}{11}, \frac{36}{11}, \frac{12}{11})$.

We sketch two proofs of Theorem 8. For the first, we assume that $g_3 \neq 0$ and solve $g(x, y, z) = 0$ for z, obtaining $z = \phi(x, y)$. We then seek extremal values for $F(x, y) = f(x, y, \phi(x, y))$. The critical points for F are the solutions of

$$0 = F_1(x, y) = f_1 + f_3 \phi_1$$
$$0 = F_2(x, y) = f_2 + f_3 \phi_2$$

Since $g(x, y, \phi(x, y)) = 0$ for all x and y, we also have

$$0 = g_1 + g_3 \phi_1$$
$$0 = g_2 + g_3 \phi_2$$

Since g_3 does not vanish, we eliminate ϕ_1 and ϕ_2 and have

$$\begin{vmatrix} f_1 & f_3 \\ g_1 & g_3 \end{vmatrix} = 0 \qquad \text{and} \qquad \begin{vmatrix} f_2 & f_3 \\ g_2 & g_3 \end{vmatrix} = 0$$

so that (f_1, f_2, f_3) and (g_1, g_2, g_3) are proportional, and dT has rank 1 or 0. ▌

For the second proof, let us assume that p^* is a point on a surface element S defined by $g(x, y, z) = 0$ at which $f(p)$ takes its maximum value. The transformation T maps a portion of 3-space containing S into the (u, v) plane, Since $v = g(x, y, z)$, the points on S map into an interval of the u axis. Since $u = f(x, y, z)$, the point p^* at which f is greatest must map into the right-hand endpoint B of this interval (see Fig. 10-4). If the rank of dT at p^* were 2, then by Theorem 21, Sec. 7.7, T would carry a full neighborhood of p^* into a neighborhood of B. There would then be points p on S whose images are on the u axis and are to the right of B. Since this is impossible, the rank of dT at p^* must be less than 2. ▌

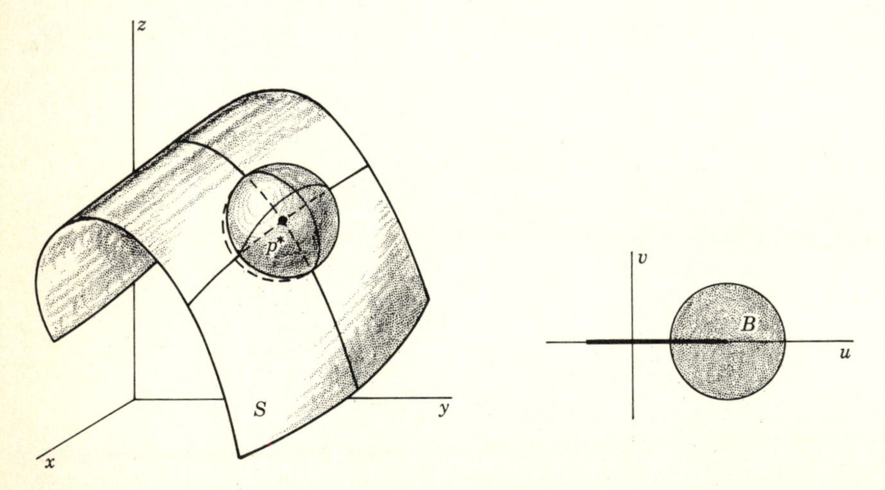

Figure 10-4 $T(x, y, z) = (u, v)$

The method of this theorem can be given in another form, due to Lagrange. The key to this is the observation that if one forms the special function

$$H(x, y, z, \lambda) = f(x, y, z) + \lambda g(x, y, z) \tag{10-32}$$

then its critical points are the solutions of the equations

$$\begin{aligned} 0 &= H_1 = f_1 + \lambda g_1 \\ 0 &= H_2 = f_2 + \lambda g_2 \\ 0 &= H_3 = f_3 + \lambda g_3 \\ 0 &= H_4 = \qquad g \end{aligned}$$

The first three equations merely assert that (f_1, f_2, f_3) is a multiple of (g_1, g_2, g_3), so that we again obtain Eqs. (10-31). Thus, the extreme points for f, constrained to satisfy $g(p) = 0$, are among the critical points of the function H.

Note that this resembles the procedure in (10-28) by introducing the constraint function g. However, there we were trying to minimize F on the original set Ω, while in (10-32) we are looking for critical points for H on a space of one higher dimension.

These results may be extended also to the case in which several constraints are imposed. If we wish to maximize $f(x, y, z)$ for points

$$p = (x, y, z)$$

which are required to obey $g(p) = h(p) = 0$, then the method of Lagrange multipliers would be to consider the special function

$$H(x, y, z, \lambda, \mu) = f(x, y, z) + \lambda g(x, y, z) + \mu h(x, y, z)$$

The critical points of H are the solutions of the equations

$$\begin{aligned} 0 &= f_1 + \lambda g_1 + \mu h_1 \\ 0 &= f_2 + \lambda g_2 + \mu h_2 \\ 0 &= f_3 + \lambda g_3 + \mu h_3 \\ 0 &= \qquad g \\ 0 &= \qquad\qquad h \end{aligned}$$

Alternatively, we may find the critical points for the transformation

$$T: \begin{cases} u = f(x, y, z) \\ v = g(x, y, z) \\ w = h(x, y, z) \end{cases}$$

which also satisfy the equations $g(p) = h(p) = 0$.

As an illustration, let us maximize $f(x, y, z) = z$ among the points satisfying $2x + 4y = 5$ and $x^2 + z^2 = 2y$. (In intuitive geometric language, we are asking for the highest point on the curve of intersection of a certain plane and a paraboloid of revolution.) Setting up T and dT, we have

$$dT = \begin{bmatrix} 0 & 0 & 1 \\ 2 & 4 & 0 \\ 2x & -2 & 2z \end{bmatrix}$$

This has rank less than 3 if its determinant vanishes. Expanding it, we obtain $x = -\frac{1}{2}$. If we use the constraints to find y and z, the desired point is $(-\frac{1}{2}, \frac{3}{2}, \frac{11}{2})$.

Since the method of Lagrange multipliers is intended to provide an alternative way to determine the critical points of the function whose minimum (or maximum) is sought, after certain of the variables have been eliminated, it is also necessary to check points on the appropriate boundaries, in case the desired extremal point is not interior to the natural domain of this function. In practice, it will often be found that the numerical and algebraic work involved is much the same, whether one uses Lagrange multipliers or whether one proceeds by substitution (see Exercise 4).

EXERCISES

1 A wheel of unit radius is revolving uniformly. Attached to the perimeter is a rod of length 3, and the other end P of the rod slides along a horizontal track (see Fig. 10-5).

(*a*) Outline a numerical procedure for finding where the maximum speed and the maximum acceleration of the point P occur.

(*b*) Compute these locations.

2 Use the method suggested in Formula (10-28) to find the minimum value of $f(x, y) = x^2 + y^2$ subject to the constraint $3x + 5y = 8$.

3 Find the minimum value of $F(x, y) = 3(x + 2y - \frac{3}{2})^2 + 4x^3 + 12y^2$ on the rectangle $|x| \le 1$, $0 \le y \le 1$.

4 Find the maximum and minimum values of $f(x, y, z) = xy + z$ subject to the constraints $x \ge 0$, $y \ge 0$, $xz + y = 4$, and $yz + x = 5$.

5 (*a*) Apply Newton's method (as in Sec. 10.2) to locate the zeros of $f'(x)$.

(*b*) Compare the resulting algorithm with the process given in (10-27) for finding an extreme value for $f(x)$.

6 What is the maximum value of $x - 2y + 2z$ among the points (x, y, z) with $x^2 + y^2 + z^2 = 9$?

7 Find the minimum of $xy + yz$ for points (x, y, z) which obey the relations $x^2 + y^2 = 2$, $yz = 2$.

8 What is the volume of the largest rectangular box with sides parallel to the coordinate planes which can be inscribed in the ellipsoid $(x/a)^2 + (y/b)^2 + (z/c)^2 = 1$?

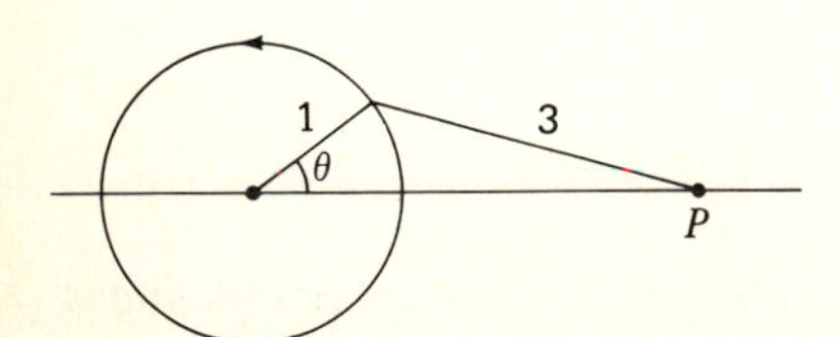

Figure 10-5

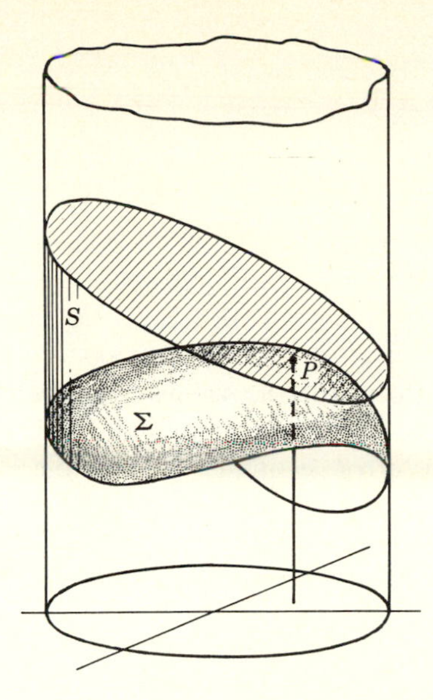

Figure 10-6

9 Let $z = f(x, y)$ be the equation of a convex surface Σ lying above the unit disk $x^2 + y^2 \leq 1$, and let P be any point on Σ. Let S be the region which is bounded below by Σ, on the sides by the cylinder $x^2 + y^2 = 1$, and on top by the tangent plane to Σ at P (see Fig. 10-6). For what position of P will S have minimum volume?

10.5 MISCELLANEOUS APPROXIMATION METHODS

Since it is not possible to cover even superficially the whole field of numerical methods, we have chosen in this last section to discuss certain specific topics dealing with integration and differentiation, more as an indication of the use of other techniques in analysis than for their own merit.

We start with the problem of finding a numerical estimate for an integral. Other aspects of this have been treated in Chap. 4, specifically at the end of Sec. 4.3, where we described the use of the trapezoidal rule and Simpson's rule for integrals in one variable. We did not, however, discuss the estimation of the error involved.

Suppose that $f(x)$ is defined and continuous for all x, $0 \leq x \leq 1$, choose an integer n, and set $h = 1/n$. We adopt the special notation f_k for the value $f(kh)$. Then, the trapezoidal estimate for $\int_0^1 f = V$ is

$$T = \frac{h}{2}(f_0 + 2f_1 + 2f_2 + \cdots + 2f_{n-1} + f_n) \tag{10-33}$$

It is easily seen that this is in fact the integral on [0, 1] of the piecewise linear

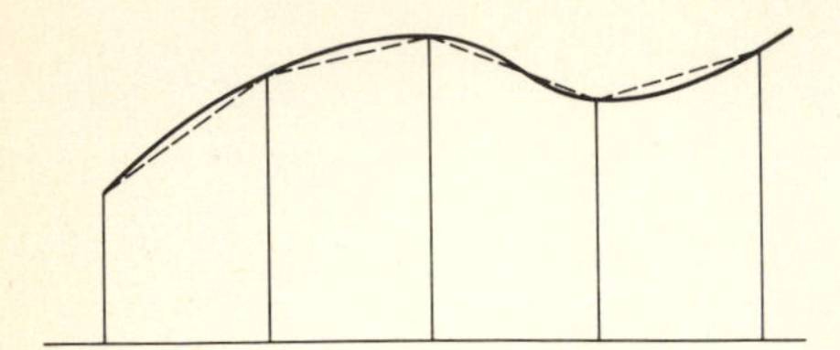

Figure 10-7

function F which matches f at the points $x_k = k/n$ for $k = 0, 1, 2, \ldots, n$, as suggested in Fig. 10-7.

With no information about the function f, except for the values f_k at the data points x_k, nothing can be said about the size of the error, $|T - V|$. However, if a bound is known for $f'(x)$, an error estimate is possible.

Theorem 9 *If* $|f'(x)| \le B$ *for* $x \in [0, 1]$, *then*

$$\left| \int_0^1 f - T \right| \le \frac{1}{4} Bh = \frac{B}{4n}$$

At each point (x_k, f_k) on the graph of f, construct lines with slope B and $-B$. Then, the mean value theorem guarantees that on each interval $[x_k, x_{k+1}]$, the graph of f lies in the parallelogram formed by these lines (see Fig. 10-8). The area of the parallelogram is at most $Bh^2/2$, obtained when $f_k = f_{k+1}$, and since the piecewise linear approximation to f is the diagonal of this parallelogram, the error made in each interval cannot exceed $Bh^2/4$, for a total of $n(Bh^2/4) = Bh/4$. ▮

Another simple procedure for numerical integration is the **midpoint rule.** For this we need the values $f_{k+1/2}$ for $k = 0, 1, \ldots, n-1$, and the estimate for V is simply

$$M = h(f_{.5} + f_{1.5} + \cdots + f_{n-.5}) \tag{10-34}$$

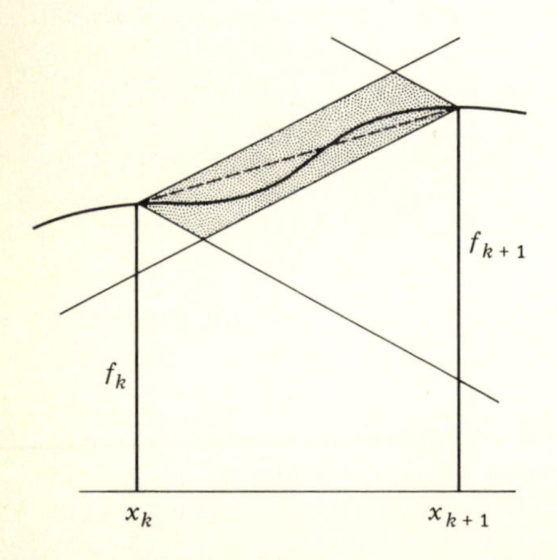

Figure 10-8

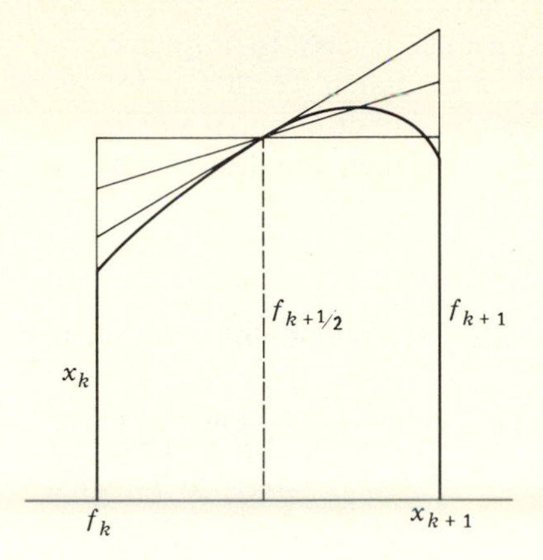

Figure 10-9

As suggested in Fig. 10-9, this is equivalent to approximating f on each interval by any line through the point $(kh + \frac{1}{2}, f_{k+1/2})$. Of these lines, the tangent is the most appropriate, and provides the following useful observation: if f'' does not change sign on $[0, 1]$, then the exact value V lies between M and T.

Finally, Simpson's rule can be described very simply in terms of these, by writing

$$S = \tfrac{1}{3}T + \tfrac{2}{3}M$$

As shown in Exercise 26 of Sec. 4.3, the Simpson value S coincides with the exact value V if f is a polynomial of degree at most 3.

It is natural to ask for analogous methods for multiple integrals. We deal only with that corresponding to the trapezoidal method. Suppose that $f(x, y)$ is defined and continuous for (x, y) in the unit square D, $0 \leq x \leq 1$, $0 \leq y \leq 1$, and that we choose n and subdivide the square into n^2 subsquares of side $1/n$. Suppose that we know the values $f_{ij} = f(x_i, y_j)$ at the $(n + 1)^2$ vertices of these squares, and wish to estimate $\iint_D f$. We cannot expect to be able to choose on each subsquare a linear function $F(x, y)$ of the form $Ax + By + C$ that matches f on the four vertices, since there are only three coefficients of F to be determined. However, such a match can be obtained on each of the triangles formed by the diagonals of the subsquare. If we make such a choice and replace the integral of f on the subsquare by the integral of F, the resulting number depends on which diagonal we choose.

It is therefore simpler to abandon linear interpolation, and instead match f by a hyperboloid function of the form

$$F(x, y) = A_0 + A_1 x + A_2 y + A_3 xy \tag{10-35}$$

The answer obtained by integrating this is the average of those obtained from the two competing linear matches.

The resulting approximate integration formula is easy to implement in terms of the data values f_{ij}. If $P_{ij} = (x_i, y_j)$, then we assign weights as follows: If P_{ij} is interior to D, let $w_{ij} = 1$; if P_{ij} lies on an edge of D but is not one of the corners, set $w_{ij} = \frac{1}{2}$; if P_{ij} is a corner of D, $w_{ij} = \frac{1}{4}$; then, the desired estimate for the exact value V of the integral is

$$T = h^2 \sum w_{ij} f(P_{ij})$$

As an illustration, if $f(x, y) = 1/(x + y + 10)$, this method with $n = 2$ yields

$$\begin{aligned} T &= \left(\frac{1}{2}\right)^2 \left\{\frac{1}{4}\left(\frac{1}{10} + \frac{1}{11} + \frac{1}{11} + \frac{1}{12}\right) + \frac{1}{2}\left(\frac{1}{10.5} + \frac{1}{10.5} + \frac{1}{11.5} + \frac{1}{11.5}\right)\right. \\ &\quad \left. + (1)\frac{1}{10 + .5 + .5}\right\} \\ &= .091\,097\,9 \end{aligned}$$

while the exact value can be shown to be $.091\,034\,7 \cdots$.

Turning to triple integrals, the same approach uses the approximating function

$$F(x, y, z) = A_0 + A_1 x + A_2 y + A_3 z + A_4 xy + A_5 yz + A_6 xz + A_7 xyz$$

and in implementing this, the data points can be weighted in a similar fashion, interior points having weight 1, interior points on a face having weight $\frac{1}{2}$, interior points on an edge having weight $\frac{1}{4}$, and each of the four vertices of the cube having weight $\frac{1}{8}$.

At times, much simpler methods can be useful. We have already observed in some of the exercises for Sec. 4.3 that the Schwarz inequality can provide useful estimates for one-dimensional integrals. The same is true of multiple integrals. For example, consider the integral

$$I = \iint_D \sqrt{x^2 - y^2}\, dx\, dy$$

where D is the triangle with vertices at (0, 0), (1, 1), (1, 0). Schwarz' inequality gives us

$$\begin{aligned} I^2 &= \left\{\iint_D \sqrt{x - y}\sqrt{x + y}\, dx\, dy\right\}^2 \\ &\le \iint_D (\sqrt{x - y})^2\, dx\, dy \iint_D (\sqrt{x + y})^2\, dx\, dy \\ &\le \iint_D (x - y)\, dx\, dy \iint_D (x + y)\, dx\, dy \\ &= \int_0^1 \tfrac{1}{2}x^2\, dx \int_0^1 \tfrac{3}{2}x^2\, dx = \tfrac{3}{36} \end{aligned}$$

Thus, $I \le \sqrt{3}/6 = .288\,675$. (Since the exact value turns out to be $\pi/(12) = .261\,799\cdots$, we see that this approximation is not too far off.)

We close this brief discussion of integration methods by illustrating one discovered by Gauss. The period of a simple pendulum, released from rest with an amplitude of θ_0 is

(10-36) $$T = 4\sqrt{\frac{L}{2g}}\int_0^{\theta_0} \frac{d\varphi}{\sqrt{\cos\varphi - \cos\theta_0}}$$

If θ_0 is 90°, we must evaluate the integral

(10-37) $$C = \int_0^{\pi/2} \frac{d\varphi}{\sqrt{\cos\varphi}}$$

If we substitute $\sin^2\theta$ for $\cos\varphi$, this becomes

(10-38) $$\frac{C}{2} = \int_0^{\pi/2} \frac{d\theta}{\sqrt{\cos^2\theta + 2\sin^2\theta}}$$

Gauss considered the general integral

(10-39) $$G(a, b) = \int_0^{\pi/2} \frac{d\theta}{\sqrt{a^2\cos^2\theta + b^2\sin^2\theta}}$$

This can be converted easily into an elliptic integral of the first kind or expressed in terms of the hypergeometric function, ${}_2F_1(\frac{1}{2}, \frac{1}{2}; 1; x)$. By using what is called a "quadratic transformation" for the functions ${}_2F_1$, Gauss was able to show that G satisfied a rather unusual identity:

(10-40) $$G(a, b) = G\left(\frac{a+b}{2}, \sqrt{ab}\right)$$

This in turn led him to the following numerical method for calculating the values of G. Define a pair of sequences $\{a_n\}$, $\{b_n\}$ by the recursion

(10-41) $$\begin{aligned} a_0 &= a & b_0 &= b \\ a_{n+1} &= \tfrac{1}{2}(a_n + b_n) & b_{n+1} &= \sqrt{a_n b_n} \end{aligned}$$

Then, both sequences will converge to the same limit (Exercise 9), called by Gauss the **arithmetic-geometric mean** of a and b. Denote this common limit by L. Then, from (10-40), we see that $G(a, b) = G(a_n, b_n)$ for all n, and since G is continuous, $G(a, b) = \lim G(a_n, b_n) = G(L, L)$. However, this can be calculated from (10-39), resulting in the formula given by Gauss:

(10-42) $$G(a, b) = \frac{\pi}{2L}$$

where L is the arithmetic-geometric mean of the pair a, b. The number L can be found very easily by calculating terms in the sequences $\{a_n\}$ and $\{b_n\}$, since the convergence is very rapid. To evaluate (10-37), we note from (10-38) that

$C/2 = G(1, \sqrt{2})$, and then obtain

a_n	1	1.2071	1.198 157	1.198 140
b_n	$\sqrt{2}$	1.1892	1.198 124	1.198 140

giving $C = 2.622\,057$.

We remark that the methods developed in Sec. 6.5 can also be applied to (10-37), and yield $C = \frac{1}{2}\sqrt{\pi}\,\Gamma(\frac{1}{4})/\Gamma(\frac{3}{4})$.

Numerical differentiation is far less used than numerical integration, although closely related formulas lie at the heart of the numerical methods for the solution of differential equations. Suppose we have available the values of f at a discrete set of points near $x = a$; can we use these to estimate the values of $f'(a)$ and $f''(a)$?

Sketches and analogies suggest the following:

$$f'(a) \approx \frac{f(a+h) - f(a-h)}{2h} \tag{10-43}$$

$$f''(a) \approx \frac{f(a+h) - 2f(a) + f(a-h)}{h^2} \tag{10-44}$$

We wish to know how good these estimates are. The following result supplies an answer, assuming that we have some a priori information about $f^{(3)}$ near $x = a$.

Theorem 10 *Let* $|f^{(3)}(x)| \le M$ *on* $[a - h, a + h]$. *Then, the error in formula* (10-43) *is less than* $h^2M/6$, *and the error in* (10-44) *is less than* $hM/3$.

We start from Taylor's formula, with remainder,

$$f(a+x) = f(a) + f'(a)x + f''(a)x^2/2 + f^{(3)}(t)x^3/6$$

Using this with $x = h$ and $x = -h$, we arrive at

$$f(a+h) - f(a-h) = 2hf'(a) + \{f^{(3)}(t_1) + f^{(3)}(t_2)\}h^3/6$$

and then, dividing by $2h$,

$$\left| \frac{f(a+h) - f(a-h)}{2h} - f'(a) \right| \le \frac{h^3}{12h} |f^{(3)}(t_1) + f^{(3)}(t_2)|$$

$$\le \frac{h^2}{6} M$$

In the same way,

$$f(a+h) - 2f(a) + f(a-h) = h^2 f''(a) + \{f^{(3)}(t_1) - f^{(3)}(t_2)\}h^3/6$$

so that

$$\left| \frac{f(a+h) - 2f(a) + f(a-h)}{h^2} - f''(a) \right| \le \frac{h}{6} |f^{(3)}(t_1) - f^{(3)}(t_2)|$$

$$\le \frac{h}{3} M \quad \blacksquare$$

We note that this can be replaced by $h^2M^*/3$, where M^* is an upper bound on $f^{(4)}$ (Exercise 11).

If more points are used, better estimates for $f'(a)$ and $f''(a)$ can be found; several illustrations are found in the exercises. Generalizations for functions of several variables are also easily obtained.

More detailed and more comprehensive information about numerical methods in analysis should be sought in the rapidly growing literature in this area.

EXERCISES

1 Fill in the missing details in the proof of Theorem 9.

2 Estimate the values of the following integrals within .05.

(a) $\int_0^1 e^{-x^2}\,dx$ (b) $\int_0^1 \frac{\sin x}{x}\,dx$ (c) $\int_1^2 \frac{dx}{1+\log x}$

3 Use Simpson's rule to estimate the value of $\int_0^1 dx/(1+x)$, using the values of f at the points 0, .25, .50, .75, 1.0.

4 Show that

$$\iint_D \sqrt{4x^2 - y^2}\,dx\,dy < \frac{\sqrt{15}}{6}$$

where D is the triangle with vertices at (0, 0), (1, 0), (1, 1).

5 Show that $\int_0^\pi e^{-x}\sqrt{\sin x}\,dx < 1$.

6 Suppose that $f(0) = 1, f(1) = 4, f(2) = 4, f(3) = 3$, and that $f''(x) \le 0$ for $0 \le x \le 3$. What is the best estimate you can give for $\int_0^3 f$?

7 Estimate the value of

$$\iiint_D \frac{dx\,dy\,dz}{5 + x + y + z}$$

where D is the unit cube with opposite vertices at (0, 0, 0) and (1, 1, 1), using a decomposition of D into 8 subcubes and the trapezoidal method.

8 Estimate $\int_0^{.2} \sin(1/x)\,dx$ to within .01.

9 Prove that the sequences $\{a_n\}$, $\{b_n\}$ in (10-41) are superconvergent to a common limit.

10 Use the method of Gauss (10-41) to verify the following computation:

$$\int_0^{\pi/2} \frac{d\theta}{\sqrt{5 - \sin^2 \theta}} = .742\,206\,24$$

11 Show that $hM/3$ can be replaced by $h^2M^*/3$ if it is known that $|f^{(4)}(x)| \le M^*$.

12 Given that $f(.5) = 2.0, f(.6) = 2.3, f(.8) = 2.7$, and $|f'''(x)| \le 4$, estimate the value of $f'(.6)$ and $f''(.6)$ with error terms.

13 Obtain an approximate formula for $f'(a)$ and $f''(a)$, making use of the five values $f(a + kh)$ for $k = 0, \pm 1, \pm 2$, and estimate its accuracy in terms of the maximum value of $|f^{(5)}(x)|$ or $|f^{(6)}(x)|$ on the interval $[a - 2h, a + 2h]$.

APPENDIX

ONE

LOGIC AND SET THEORY

There are many excellent books which deal with mathematical logic and its dual position as an aid to mathematics and a part of mathematics. Historically, the discovery of paradoxes in set theory led to a renewed study of logic in order to "lay a firm foundation" for classical mathematics, and this in turn led to an emphasis upon axiomatic approaches, constructivity, and deduction. More recently, of course, the growth of interest in recursive functions and in algebraic logic has made renewed contact with the main body of mathematics, and on an advanced level, there has been much cross fertilization. The purpose of this brief section is not to present a survey of the status of logic, but only to mention certain simple techniques and concepts which are useful for a person who is working with elementary analysis; the emphasis is upon informalism. For a more sophisticated point of view, we recommend the book "Naive Set Theory" by Halmos.

In the beginning stages of analysis, there seems little reason to insist upon the adoption of a formal syntax for presentation of statements and arguments. It seems easier to work with statements such as

(*) *each point p is the center of a neighborhood in which the values of f are less than C*

than with its formalized equivalent

(**) $$(p)_D(\exists\delta)_P(q)_D : |q - p| < \delta \rightarrow f(p) < C$$

Here, we use the convention that "(x)" means "for all x", that "$(x)_E$" means "for all $x \in E$", that "$(\exists y)$" means "there exists a y", that "$(\exists y)_E$" means that "there exists a $y \in E$", and that "$A \to B$" can be read "if A then B". The sets D and P are understood to be a set in the plane and the positive real numbers, respectively.

However, there is one type of problem in which the use of a formalized language is very helpful. Suppose that we wish to prove a proposition of the form $U \to V$. The device of "proof by contradiction" amounts to the statement that it is equivalent to prove **not** $V \to$ **not** U. (The fact that some persons have doubts about this will be discussed later.) The problem then is, knowing the statement "V", how can we formulate the statement "**not** V" in a simple and useful way?

The simple routine which can be used is based upon the following semantic rules:

$$\begin{aligned}
\textbf{not } (x)_E &= (\exists x)_E \textbf{ not}\\
\textbf{not } (\exists x)_E &= (x)_E \textbf{ not}\\
\textbf{not } (A \to B) &= A \,\&\, \textbf{not } B\\
\textbf{not } (A \,\&\, B) &= \textbf{not } A \text{ or } \textbf{not } B\\
\textbf{not } (A \text{ or } B) &= \textbf{not } A \,\&\, \textbf{not } B\\
\textbf{not } (\textbf{not } A) &= A
\end{aligned}$$

As a sample, let us apply these to find "**not** V", when V is the statement (*) above. Using (**), we first have

$$\textbf{not } (p)_D(\exists\delta)_P(q)_D : |p - q| < \delta \to f(q) < C$$

Then, we move "**not**" past each quantifier in turn. The steps appear thus:

$$\begin{aligned}
&(\exists p)_D \textbf{ not } (\exists\delta)_P(q)_D : |p - q| < \delta \to f(q) < C\\
&(\exists p)_D(\delta)_P \textbf{ not } (q)_D : |p - q| < \delta \to f(q) < C\\
&(\exists p)_D(\delta)_P(\exists q)_D \textbf{ not}: \ |p - q| < \delta \to f(q) < C\\
&(\exists p)_D(\delta)_P(\exists q)_D : |p - q| < \delta \,\&\, \textbf{not} f(q) < C
\end{aligned}$$

Finally,

$$(\exists p)_D(\delta)_P(\exists q)_D : |p - q| < \delta \,\&\, f(q) \geq C$$

which could be translated into the following less formal statement:

there is a point p such that every neighborhood about p
contains at least one point where f has a value as big as C

As mentioned above, some logicians would object to certain of the equivalences listed above. Without entering into this in depth, this point of view can be appreciated in part by examining the first; we can translate

$$\textbf{not } (x)_E A(x)$$

to mean

it is false that every $x \in E$ makes $A(x)$ true

The "equivalent" statement,

$$(\exists x)_E \text{ not } A(x)$$

would translate

there is an $x \in E$ for which $A(x)$ is false

But this statement might seem to convey that indeed you *know* an $x_0 \in E$ and *know* that $A(x_0)$ is false, rather than that you merely know it is impossible for there *not* to exist such an x. The point here is one of existence vs. constructibility. In most cases, it is clearly better to have some algorithm for finding such an x_0, rather than a theoretical proof that such an x_0 exists.

Another topic that is often regarded as part of logic is the use of mathematical induction as a mode of proof. Strictly speaking, we are here dealing with properties of the system of whole numbers and similar systems. If S is a set of whole numbers which contains 1 and is such that, whenever it contains x, it must also contain $x + 1$, then S contains *all* the whole numbers. A familiar use of this is the proof that $1 + 2 + \cdots + n = n(n + 1)/2$, for all $n = 1, 2, \ldots$. Calling this proposition P_n, we observe that P_1 is true; then, let $S = \{\text{all } n \text{ with } P_n \text{ true}\}$, and show that S is inductive,

$$x \in S \rightarrow x + 1 \in S$$

and thus contains all whole numbers n. However, this same proposition is also proved in the following way: Let

$$L = 1 + \quad 2 \quad + \quad 3 \quad + \cdots + n$$

Then

$$L = n + (n - 1) + (n - 2) + \cdots + 1$$

and adding, we have

$$2L = (n + 1) + (n + 1) + \cdots + (n + 1) = n(n + 1)$$

whence $L = n(n + 1)/2$. Is this argument less convincing than the appeal to mathematical induction? Is it logically valid? What is the role of "$\cdots$"?

Finally, let us examine briefly some of the aspects of set theory. Although its importance to the foundations of analysis is undeniable, there is much of the theory of sets that has little relevance to advanced calculus. In particular, it does not seem to be essential at this level to understand the formal axiomatic approach. As an example, it does not seem necessary to examine the notion of ordered pair or triple, or to show that $\{a, \{a, b\}\}$ is an adequate model for $\langle a, b \rangle$.

The notion of infinite cardinal number is useful, however, and should be part of a student's experience. Excellent treatments can be found in many places, for example, in the book "A Primer of Real Functions," by R. P. Boas, Jr., which is No. 13 in the *Carus Monographs*. Here, we shall mention only some aspects of countability and noncountability.

Two sets A and B are said to have the same number of members if there is a function $f: A \to B$ which maps A onto B, 1-to-1. Looked at differently, this is the same as saying that we have used the members of B to label the members of A. A set that has the same number of members as the set $I = \{1, 2, 3, \ldots\}$ of all whole numbers is said to be *countable*. Any infinite set that is not countable is said to be *noncountable*. I, of course, is countable, as is the set of all rational numbers, but the set of all real numbers is not countable. If we identify the latter with unending decimals, then it is easy to prove noncountability. If f is a function from I into the real numbers in the closed interval $[0, 1]$, then we can exhibit a number x_0 which is not in the range of f. All we need do is to take x_0 so that it differs at the nth decimal from the nth decimal of the number $f(n)$, with appropriate care to avoid decimals that terminate in a string of "9".

More generally, any set can be shown to have more subsets than it has members; in particular, the collection of all sets of integers is noncountable, while the collection of all sets of real numbers has a cardinal number larger than that of the set of real numbers itself.

One says that a set A has at least as many members as the set B if B can be paired, 1-to-1, with a subset of A. A fundamental result about infinite sets is the Schroeder-Bernstein theorem, asserting that, if A has at least as many members as B, and B has at least as many as A, then A and B have the same number of members. A simple proof of this may be constructed along the following lines. We are given two functions f and g,

$$f: A \to B$$
$$g: B \to A$$

each 1-to-1, but neither onto. We wish to produce a function $h: A \to B$ which is onto and 1-to-1. Suppose we could split each of the sets A and B, so that A is the union of a subset C and the set $A - C$ and $B = D \cup (B - D)$, in such a way that, as indicated in Fig. A1-1, $g(D) = A - C$ and $f(C) = B - D$. Then, the desired mapping h has been obtained,

$$h(p) = \begin{cases} f(p) & \text{when } p \in C \\ g^{-1}(p) & \text{when } p \in A - C \end{cases}$$

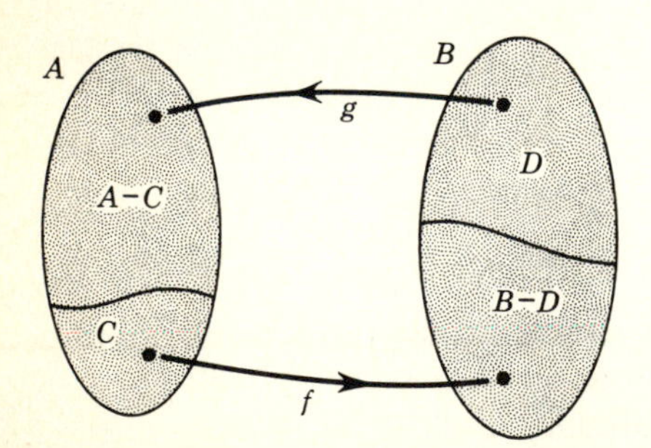

Figure A1-1

It is easy to show that we need only find the set C and can then determine D from it and that C must satisfy the identity

$$C = A - g(B - f(C))$$

Define a function T on the subsets of A by

$$T(S) = A - g(B - f(S))$$

For example, $T(\phi) = A - g(B)$, and $T(A) = A - g(B - f(A)) \subset A$. We are searching for a set $C \subset A$ such that $T(C) = C$. The following steps complete the proof that such a set exists and thus prove the Schroeder-Bernstein theorem:

i. Show that the mapping T has the property that if $S_1 \subset S_2 \subset A$, then $T(S_1) \subset T(S_2) \subset A$ (T is monotone).
ii. Let $\mathcal{S}$ be the class of subsets $S \subset A$ such that $T(S) \subset S$. (For example, $A \in \mathcal{S}$.) Show that, if $S \in \mathcal{S}$, then $T(S) \in \mathcal{S}$.
iii. Show that any monotone mapping has a fixed set C, with $T(C) = C$. (*Method:* Define C to be the intersection of all sets S in $\mathcal{S}$.)

APPENDIX

TWO

FOUNDATIONS OF THE REAL NUMBER SYSTEM

The real-number system underlies all analysis. In the next few pages, we shall set forth its characteristic properties. We start with a definition: *The real numbers constitute a complete simply ordered field.*

We expand this concise statement by explaining the meaning of the terms involved.

Definition 1 *A field is a set* **K** *of elements* $a, b, x, \ldots$, *together with two functions* $+$ *and* $\cdot$, *called "sum" and "product," which satisfy the following requirements:*

(F_1) **closure** *If* a *and* b *are in* **K**, *then their sum* {*product*} *is defined and is a unique element of* **K** *denoted by* $a + b$ $\{a \cdot b\}$.

(F_2) **commutative** *If* a *and* b *are in* **K**, *then*

$$a + b = b + a$$
$$a \cdot b = b \cdot a$$

(F_3) **associative** *If* a, b, *and* c *are in* **K**, *then*

$$a + (b + c) = (a + b) + c$$
$$a \cdot (b \cdot c) = (a \cdot b) \cdot c$$

(F_4) **distributive** *If* a, b, *and* c *are in* **K**, *then*

$$a \cdot (b + c) = (a \cdot b) + (a \cdot c)$$

(F_5) **existence of neutral elements** *There are two special elements of* **K**, *denoted by* 0 *and* 1, *such that for any* $x \in \mathbf{K}$,

$$x + 0 = x \qquad \text{and} \qquad x \cdot 1 = x$$

(F_6) **inverses** *For any* $a \in \mathbf{K}$ *the equation* $a + x = 0$ *has a solution, and for any* $a \in \mathbf{K}$ *except* 0, *the equation* $a \cdot x = 1$ *has a solution.*

In working with a field, it is customary to use certain abbreviations. For example, according to (F_6), there is an element x such that $a + x = 0$; moreover, this element is unique, for if $a + x' = 0$, then

$$\begin{aligned} x = x + 0 = x + (a + x') &= (x + a) + x' \\ &= (a + x) + x' = 0 + x' = x' + 0 = x' \end{aligned}$$

Thus, the element depends solely upon a, and we denote it by $-a$. The function "difference" is then defined by $a - b = a + (-b)$. Likewise, if $a \neq 0$, then the solution of $a \cdot x = 1$ is unique and may be denoted by a^{-1} or $1/a$. We then define "quotient" by $a/b = a \cdot (1/b)$. On the basis of these postulates on **K**, all the familiar algebraic rules follow. To cite a few, one may prove that:

$$(-a) \cdot (-b) = a \cdot b$$

$$a \cdot 0 = 0$$

$$(a/b)(c/d) = (a \cdot c)/(b \cdot d)$$

$$(a/b) + (c/d) = (a \cdot d + b \cdot c)/(b \cdot d)$$

$$a \cdot b = 0 \qquad \text{only if } a = 0 \text{ or } b = 0$$

The notion of an ordering relation may be introduced most easily by means of a set of **positive elements.**

Definition 2 *A field* **K** *with a simple order is one in which there is a subset* **P** (*called the set of positive elements*) *such that:*

(O_1) *If* a *and* b *are in* **P**, *then so are* $a + b$ *and* $a \cdot b$.
(O_2) *The zero element*, 0, *is not in* **P**.
(O_3) *If* x *is any element not in* **P**, *then* $x = 0$, *or* $-x \in \mathbf{P}$.

To convert these postulates into more familiar properties, one defines a relation $>$ on **K** by: $a > b$ *if and only if* $(a - b) \in \mathbf{P}$. The conditions (O_1), (O_2), (O_3) imply the usual properties of $>$. For example, $a > 0$ is equivalent to saying that a belongs to **P**. Requirement (O_1) then implies that if $a > b$ and $b > c$, then $a > c$, and if $d > 0$, then $a \cdot d > b \cdot d$. The equivalent form of

(O_3) is the statement that if a and b are any elements of $\mathbf{K}$, then either $a = b$, $a > b$, or $b > a$. An ordering relation $>$ which has this last property is said to be **simple** or linear; many important ordering relations are partial orderings in which this law fails, and two elements a and b may be incomparable. We note one important consequence of the existence of a simple order in $\mathbf{K}$, namely, that the sum $x_1^2 + x_2^2 + \cdots + x_n^2$ is always positive (> 0) unless $x_1 = x_2 = \cdots = x_n = 0$.

Finally, we come to the notion of **completeness** of an order relation. As we shall see, this has many equivalent formulations. One of these may be described in terms of what are called **Dedekind cuts.** A cut in $\mathbf{K}$ is a pair A, B, where A and B are nonempty subsets of $\mathbf{K}$ whose union is $\mathbf{K}$ and such that $a \le b$ whenever $a \in A$ and $b \in B$. Any element c of $\mathbf{K}$ can be used to generate a cut by taking A as the set of all $x \in \mathbf{K}$ with $x \le c$, and B as the set of all $x \in \mathbf{K}$ with $x \ge c$.

Definition 3 *A simple ordering on a field* $\mathbf{K}$ *is complete if every cut in* $\mathbf{K}$ *is generated by an element of* $\mathbf{K}$.

Some additional comments are in order. The real field $\mathbf{R}$ is not the only example of a field. One familiar example is the **complex field** $\mathbf{C}$.

Definition 4 *The complex field* $\mathbf{C}$ *is the class of all ordered pairs* (a, b) *with* a *and* b *real numbers, and with sum and product defined by*

$$(a, b) + (x, y) = (a + x, b + y)$$

$$(a, b) \cdot (x, y) = (ax - by, ay + bx)$$

One may verify that $\mathbf{C}$ satisfies the field axioms $F_1, \ldots, F_6$. For example, $0 = (0, 0)$, $1 = (1, 0)$, $-(x, y) = (-x, -y)$, and

$$(x, y)^{-1} = \left(\frac{x}{(x^2 + y^2)}, \frac{-y}{(x^2 + y^2)}\right)$$

If we write $(0, 1) = i$ and $(a, 0) = a$, then $(x, y) = x + iy$, where $i^2 = -1$. The usual algebraic operations may now be used. The complex field cannot be given a simple order; as we have seen, in any simply ordered field, it is necessary that $x^2 + y^2 \ne 0$ except when $x = y = 0$. This fails in $\mathbf{C}$, since $1^2 + i^2 = 0$.

Other examples of fields may be obtained from $\mathbf{R}$. A subset S of $\mathbf{R}$ is called a **subfield** if S is a field under the operations of $\mathbf{R}$; one need only verify that if a and b are any two elements of S, then $a + b$, $a \cdot b$, $-a$, and if $b \ne 0$, $1/b$ are all members of S. The smallest subfield of the real field is called the **rational field** R_0. The elements of the rational field are the rational numbers; each may be represented in the form a/b, where a and b belong to the special subsystem called the ring of **integers** Z. Every element in Z is expressible as m or $-m$ or 0, where m is in Z^+, the system of positive integers (whole numbers). As a subfield of **R**, the rational field R_0 is simply ordered. The class of positive elements of R_0 can be taken as $P_0 = P \cap R_0$. However, unlike the real field, R_0 is not a completely ordered field. The Dedekind cut (A, B) in R_0 defined by

$$A = \{\text{all } x \in R_0 \text{ with } x^3 < 2\}$$

$$B = \{\text{all } x \in R_0 \text{ with } x^3 > 2\}$$

is one which cannot be generated from an element in R_0. In R, the corresponding cut is generated by $\sqrt[3]{2}$, but this number is not present in R_0.

With this preparation, we state without proof the fundamental result dealing with the real field: *any two fields which are completely ordered are isomorphic*. This gives one the right to speak of *the* real field, and guarantees that the description we have chosen is adequate. To explain by illustration the meaning of "isomorphic," we give another construction of the complex field. Let C^* be the collection of all 2-by-2 matrices with real entries and having the form

$$\begin{bmatrix} a & b \\ -b & a \end{bmatrix}$$

For multiplication, use the customary product operations for matrices, and add matrices by adding corresponding entries. Then, one may again verify that C^* is a field. The correspondence

$$\begin{bmatrix} a & b \\ -b & a \end{bmatrix} \leftrightarrow a + bi = (a, b)$$

is an isomorphism between C^* and **C**. (This class of matrices also appears in another connection in Theorem 10, Sec. 8.4. Nor is this entirely a coincidence; conformal transformations are locally expressible as power series with complex coefficients.)

Having obtained a categorical description of the real numbers, is it sufficient to proceed by fiat; "let **R** be a completely ordered field," and then continue from there? This would leave open the possibility that there might not exist such a field, due perhaps to some undiscovered inconsistency in the postulates. To avoid this, we may attempt to construct an example of such a field, building up from a simpler system whose existence we are willing to accept. This process has been set forth in great detail in Landau, "Grundlagen der Analysis," Leipzig, 1930. We sketch such a construction in a sequence of steps.

Step 1. *Construction of* **R** *from* R_0. Two methods have been used here. The first method is conceptually simpler. R_0 is an ordered field which is not complete. Some of its Dedekind cuts can be generated by elements of R_0 and some cannot. Define **R** to be the collection of all cuts. Those that can be generated from elements of R_0 are identified with the corresponding element of R_0 and are rational numbers; those that cannot be so generated are the irrational real numbers. One then defines sum and product of cuts, and verifies all the postulates for **R**.

The second method is easier to carry out. Working only with rational numbers, we define convergence of sequences of rational numbers, and Cauchy sequences of rational numbers. It is no longer true in R_0 that every Cauchy sequence converges. Consider the collection $\mathcal{S}$ of all Cauchy sequences of rational numbers, and introduce a notion of equivalence by saying that $\{a_n\} \approx \{b_n\}$ if and only if $\lim_{n\to\infty} (a_n - b_n) = 0$. This splits $\mathcal{S}$ into equivalence classes. Each of these is now called a real number. The product and sum functions are now defined, and the postulates verified.

Step 2. *Construction of* R_0 *from* Z. Consider the collection $\mathcal{P}$ of all ordered pairs of integers (m, n) with $m \neq 0$. Thinking of (m, n) as representing the rational number n/m, a notion of equivalence is defined in $\mathcal{P}$ by $(m, n) \approx (m', n')$ *if and only if* $mn' = m'n$. This again divides $\mathcal{P}$ into equivalence classes. Each of these classes is called a rational number; with appropriate definitions for sum, $>$, and product, R_0 is seen to be an ordered field.

Step 3. *Construction of* Z *from* Z^+, *the positive whole numbers.* Again, one considers a collection of ordered pairs (m, n) where m and n are in Z^+. With (m, n) being thought of as corresponding to the integer $m - n$, one says that $(m, n) \approx (m', n')$ if and only if

$$m + n' = n + m'$$

The resulting equivalence classes are called the integers. The rest goes as before.

Step 4. *Construction of* Z^+ *from axiomatic set theory.* Within an adequate system of logic, one may define the notion of cardinal number for finite sets. From this, one may then obtain Z^+.

We conclude this discussion of the real-number system by giving some of the equivalent ways in which the completeness property, which is so characteristic of **R**, can be obtained. Most of these involve results which we have assumed, or which have been proved from others which we have assumed. In the chart in Fig. A2-1, we have indicated by arrows certain mutual implications which hold among these. The branched arrows on the right side of this diagram indicate that the combination of the Archimedean and Cauchy convergence property jointly implies all the rest (which are seen from this diagram to all be equivalent to one another). It may seem surprising that the Cauchy property alone is not enough. This is true because the statement that a given sequence has the Cauchy property is actually a very strong statement, and one cannot prove that a bounded monotonic sequence is Cauchy without using the archimedian property—which is itself equivalent to the assertion that the sequence $\{1/n\}$ converges to 0. In fact, there exist non-archimedean fields in which the only Cauchy sequences are ultimately constant, while there are bounded monotonic sequences that diverge.

This discussion should not end without a mention of ∞ and $-\infty$. These are to be regarded as "ideal" points adjoined to the real field **R**. The

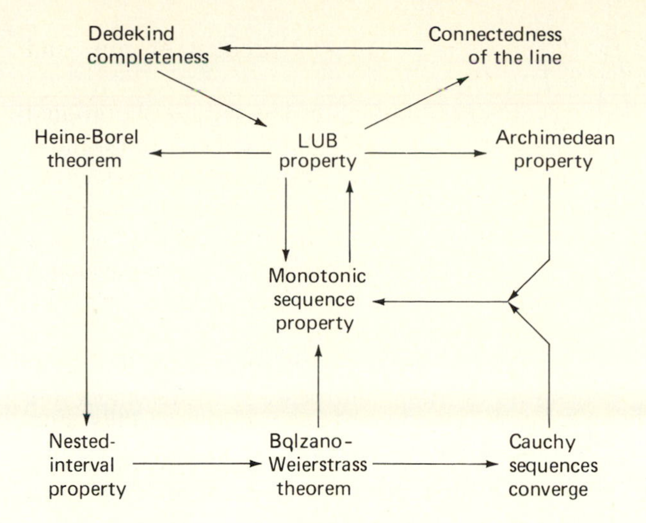

Figure A2-1

neighborhoods of ∞ are defined to be the intervals

$$\{\text{all } x \text{ with } x > b\}$$

and the neighborhoods of $-\infty$ are the intervals

$$\{\text{all } x \text{ with } x < b\}$$

Using these, we see that the sequence $\{x_n\}$ with $x_n = n$ is convergent to ∞ and that every sequence of real numbers has a limit point in the extended system. The new points ∞ and $-\infty$ do not participate fully in the algebra of $\mathbf{R}$; in particular, the new system is no longer a field. One may attach meaning to some of the operations, but not all; for example, we may define $\infty + c$ to be ∞ for all $c \in \mathbf{R}$, but no meaning within the new system is attached to $\infty + (-\infty)$.

The adjunction of ∞ and $-\infty$ to $\mathbf{R}$ is done with the aim of achieving compactness. This may also be done in other ways. If ϕ is a transformation which is 1-to-1, bicontinuous, and which maps $\mathbf{R}$ onto a set A in a compact set S, then the closure of A in S is a compactification of $\mathbf{R}$. The boundary points for A which do not lie in $\mathbf{R}$ form the ideal points which are to be adjoined to $\mathbf{R}$. As a simple illustration of this, let S be the circle of radius 1, center $C = (0, 1)$, and let ϕ be the transformation mapping the point $(x, 0)$ on the horizontal axis ($= \mathbf{R}$) into the point p on S which lies on the line through $(x, 0)$ and $Q = (0, 2)$. It is easily seen that the image of $\mathbf{R}$ is the set A consisting of all points on S except Q. The closure of A is S itself, and Q is the only boundary point of A which does not lie in A. Thus, $\mathbf{R}$ may be compactified by adjoining a single point "at infinity" whose neighborhoods are the sets $\{\text{all } x \text{ with } |x| > c\}$. With the same choice of S, we may take ϕ as the transformation mapping $(x, 0)$ into the point of S lying on the line through $(x, 0)$ and C; the closure of the image set A is a semicircle whose endpoints

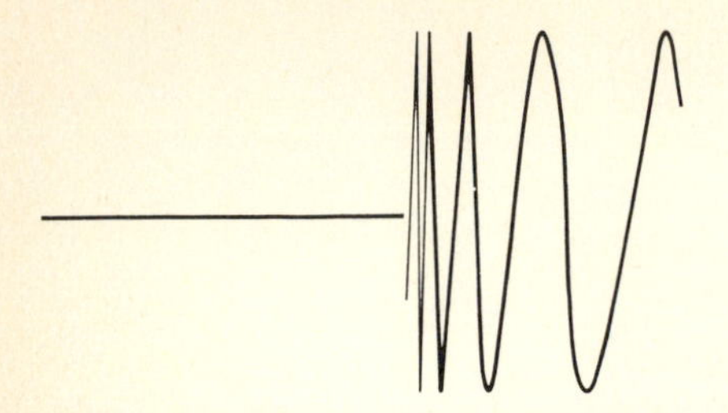

Figure A2-2

correspond to the ideal points ∞ and $-\infty$. Other choices of S and ϕ may yield an infinite number of ideal points "at infinity."

Connectedness is a difficult topological property to understand fully. For open sets in $\mathbf{R}^n$, it is equivalent to being pathwise connected, which is certainly a more easily visualized concept. However, there is a standard example of a connected set in the plane that is not pathwise connected, namely

$$E = \text{all } (x, y) \text{ with } y = \begin{cases} \sin(1/x) & \text{if } \ 0 < x \le 1 \\ 0 & \text{if } -1 \le x \le 0 \end{cases}$$

shown (in part) in Fig. A2-2.

A more dramatic example can be given. Consider a rectangle S with vertices A, B, C, D, and suppose that we have two connected subsets of S, α and β, such that α contains the pair A and C and β contains B and D. If each of these connected sets were curves, we would know that there would have to be a point that belonged to both α and β. However, Fig. A2-3 indicates how it is possible for α and β to be connected and still disjoint.

Another nonintuitive consequence of the careful analysis of the meaning of continuity was the demonstration by Peano that a continuous curve could pass through *every* point of a solid square. In Figs. A2-4 and A2-5 we illustrate several stages in the construction of a "space-filling" curve.

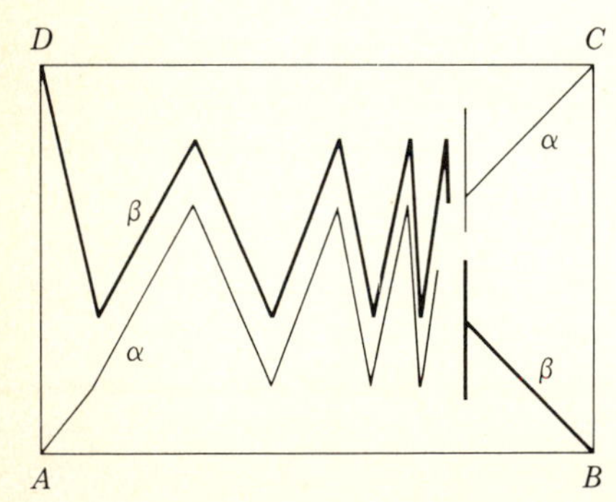

Figure A2-3

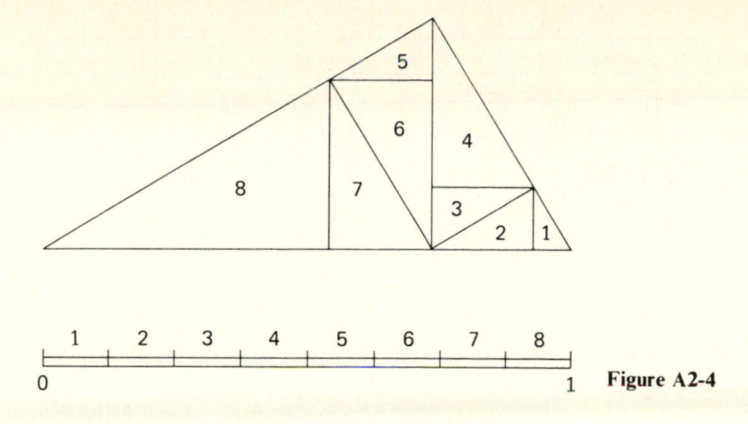

Figure A2-4

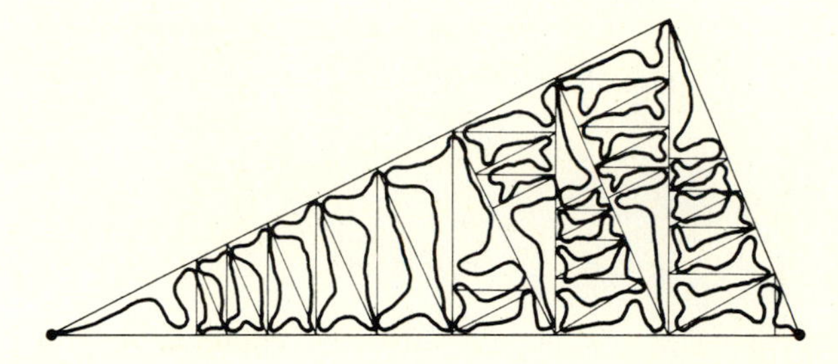

Figure A2-5

APPENDIX

THREE

LINEAR ALGEBRA

This is intended to be a brief synopsis or review of those aspects of linear algebra and matrices which are most needed for the topics in analysis that are studied in Chaps. 7, 8, and 9. Proofs and complete discussions can be found in any of the large number of current texts in linear algebra.

The definition of a linear (or vector) space has been given in Sec. 1.2; the associated field **K** of scalars is usually either the real numbers **R** or the complex numbers **C**. Standard examples of real linear spaces are $\mathbf{R}^n$ and $C[I]$. We use $\mathbf{K}^n$ for the space whose points are n-tuples of scalars $(c_1, c_2, \ldots, c_n)$ from **K**. If V is a linear space and $M \subset V$, and if λu and $u - v$ belong to M for every u and v in M and every scalar $\lambda \in \mathbf{K}$, then M is a linear subspace of V. If $U = \{u_1, u_2, \ldots\}$ is a collection of points in a linear space V, then the (linear) span of the set U is the set of all points of the form $\sum c_i u_i$, where $c_i \in \mathbf{K}$, and all but a finite number of the scalars c_i are 0. The span of U is always a linear subspace of V.

A key concept in linear algebra is independence. A finite set $u_1, u_2, \ldots, u_k$ is said to be linearly independent in V if the only way to write $\mathbf{0} = \sum c_i u_i$ is by choosing all the $c_i = 0$. An infinite set is linearly independent if every finite subset is independent. If a set is not independent, it is linearly dependent, and in this case, some point in the set can be written as a linear combination of other points in the set. A basis for a linear space M is an independent set that spans M. A space M is finite-dimensional if it can be spanned by a finite set;

it can then be shown that every spanning set contains a basis, and every basis for M has the same number of points in it. This common number is called the dimension of M.

Another key concept is that of linear transformation. If V and W are linear spaces with the same scalar field $\mathbf{K}$, a mapping L from V into W is called linear if $L(u + v) = L(u) + L(v)$ and $L(\lambda u) = \lambda L(u)$ for every u and v in V and λ in $\mathbf{K}$. With any L are associated two special linear spaces:

$$\begin{aligned}\mathcal{N}(L) = \ker(L) &= \text{null space of } L = L^{-1}(\mathbf{0}) \\ &= \{\text{all } x \in V \text{ such that } L(x) = \mathbf{0}\}\end{aligned}$$

$$\begin{aligned}\operatorname{Im}(L) &= \text{image or range of } L = L(V) \\ &= \{\text{all } L(x) \text{ for } x \in V\}\end{aligned}$$

The first is a subspace of V, and the second a subspace of W. L is a 1-to-1 mapping (injective) if and only if $\mathcal{N}(L) = \{\mathbf{0}\}$, and L is an onto mapping (surjective) if and only if $\operatorname{Im}(L) = W$. A linear transformation that is both 1-to-1 and onto is called an isomorphism; two spaces V and W are called isomorphic if there is an isomorphism L mapping V onto W. The inverse L^{-1} then exists and is an isomorphism mapping W onto V. Any space M of dimension n is isomorphic to the space $\mathbf{K}^n$. If $\{u_1, u_2, \ldots, u_n\}$ is a basis for M, then every point x in M can be written uniquely in the form $\sum c_i u_i$, $c_i \in \mathbf{K}$, and the required isomorphism L is defined by $L(x) = (c_1, c_2, \ldots, c_n)$. All linear spaces of the same dimension are isomorphic. Henceforth, all the linear spaces discussed will be finite-dimensional.

With any linear L mapping V into W, two integers are useful:

$$r = \text{rank } (L) = \text{dimension of } \operatorname{Im}(L)$$

$$k = \text{nullity of } L = \text{dimension of } \mathcal{N}(L)$$

If V has dimension n, it can be shown that $r + k = n$. If W also has dimension n, then the following useful criterion results: *L is 1-to-1 if and only if L is onto.* In particular, if L is a linear map of V into itself, and the only solution of $L(x) = \mathbf{0}$ is $\mathbf{0}$, then L is onto and is therefore an isomorphism of V onto V, and has an inverse L^{-1}. Such a transformation of V onto itself is also said to be nonsingular.

Suppose now that L is a linear transformation from V into W where $\dim(V) = n$ and $\dim(W) = m$. Choose a basis $\{v_1, v_2, \ldots, v_n\}$ for V and a basis $\{w_1, w_2, \ldots, w_m\}$ for W. Then, these define isomorphisms of V onto $\mathbf{K}^n$ and W onto $\mathbf{K}^m$, respectively, and these in turn induce a linear transformation A between these, as shown in Fig. A3-1, such that $L(v) = \psi^{-1} A \phi(v)$. Any linear transformation (such as A) between $\mathbf{K}^n$ and $\mathbf{K}^m$ is described by means of a matrix $[a_{ij}]$, according to the formula $A(x) = y$, where $x = (x_1, x_2, \ldots, x_n)$,

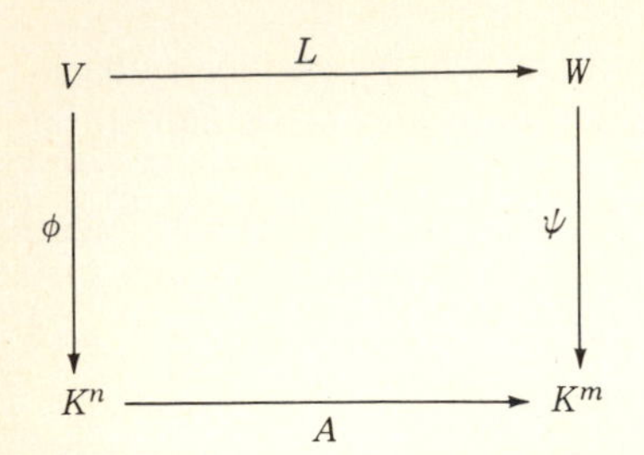

Figure A3-1

$y = (y_1, y_2, \ldots, y_m)$, and

$$\text{(A3-1)} \qquad y_i = \sum_{j=1}^{n} a_{ij} x_i \qquad i = 1, 2, \ldots, m$$

The matrix A is said to represent the transformation L and to be the representation induced by the particular basis chosen for V and for W. (A different choice of basis would produce a different numerical matrix A.)

All the properties of the transformation L are mirrored in terms of properties of the system of linear equations (A3-1) and the numerical matrix A, and do not depend on the choice of bases. Thus, the rank of L is the dimension of the subspace spanned by the columns of A, which is the same as the number of columns of A that are linearly independent; if the rank is m, then equations (A3-1) have a solution for any choice of y, since T is then onto, and Theorems 6 and 7 of Sec. 7.3 follow.

When $V = W$, so that $n = m$, then the matrix A is square and (A3-1) is a system of n linear equations in the n unknowns x_j. The transformation L is nonsingular if and only if A is of rank n, which means that its columns are linearly independent, and which is therefore equivalent to the assertion that (A3-1) has a unique solution for any choice of y_i; in particular, this must hold if and only if L is 1-to-1, which means that the null space of L is $\{\mathbf{0}\}$, and which is equivalent to the assertion that the homogeneous equations $0 = \sum_{j=1}^{n} a_{ij} x_j$ have only the obvious solution $x_1 = x_2 = \cdots = x_n$.

If S and T are linear transformations of V into itself, so is the composite transformation ST. If we choose a basis in V, and use this to obtain matrix representations for these, with A representing S and B representing T, then ST must have a matrix representation C. This is defined to be the product AB of the matrices A and B, and leads to the standard formula for matrix multiplication.

The least satisfactory aspect of linear algebra is still the theory of determinants—even though this is the most ancient portion of the theory, dating back to Leibniz if not to early China. One standard approach to determinants is to regard an n-by-n a matrix as an ordered array of vectors $(u_1, u_2, \ldots, u_n)$, and then its determinant det (A) as a function $F(u_1, u_2, \ldots, u_n)$ of these n vectors which obeys certain rules. F is required to be **multilinear** (i.e., linear in each position), so that, for example,

$$F(\lambda u_1, u_2, \ldots, u_n) = \lambda F(u_1, u_2, \ldots, u_n)$$

$$F(v + w, u_2, u_3, \ldots, u_n) = F(v, u_2, \ldots, u_n) + F(w, u_2, u_3, \ldots, u_n)$$

and **alternating,** so that

$$F(u_k, u_2, \ldots, u_{k-1}, u_1, \ldots, u_n) = -F(u_1, u_2, \ldots, u_n)$$

for every $k \neq 1$, and normalized by $F(e_1, e_2, \ldots, e_n) = 1$, where $e_1 = (1, 0, 0, \ldots, 0)$, $e_2 = (0, 1, 0, 0, \ldots, 0)$, etc. One is then able to show that there is exactly one such function, which not surprisingly turns out to coincide with the college algebra notion of determinant for $n = 2$ and $n = 3$.

The determinant of such an array A turns out to be a convenient criterion for characterizing the nonsingularity of the associated linear transformation, since $\det(A) = F(u_1, u_2, \ldots, u_n) = 0$ if and only if the set of vectors u_i are linearly dependent. There are many other useful and elegant properties of determinants, most of which will be found in any classical book on linear algebra. Thus, $\det(AB) = \det(A)\det(B)$, and $\det(A) = \det(A^t)$, where A^t is the transpose of A, obtained by the formula $A^t = [a_{ji}]$, thereby rotating the array about the main diagonal. If a square matrix A is triangular, meaning that all its entries above the main diagonal are 0, then $\det(A)$ turns out to be exactly the product of the diagonal entries.

Even though the determinant is calculated from a matrix representation of a transformation L, it does not depend upon which representation is chosen. For, if a particular choice of basis for V yields the matrix A, then a different choice of basis will merely give rise to an isomorphism of V, and replace A by the matrix $B = PAP^{-1}$, where P induces the required isomorphism. However, $\det(B) = \det(P)\det(A)\det(P^{-1}) = \det(A)\det(P)\det(P^{-1}) = \det(A)$. Thus, one is justified in speaking of the determinant of a transformation $L: V \to V$ rather than the determinant of matrices alone.

Another useful concept is that of eigenvalue. A scalar is said to be an eigenvalue for a transformation T if there is a nonzero vector v with $T(v) = \lambda v$. It is then clear that the eigenvalues will be those numbers $\lambda \in \mathbf{K}$ such that $T - \lambda I$ is a singular transformation. Any vector in the null space of $T - \lambda I$ is called an eigenvector of T associated with the eigenvalue λ, and their span the eigenspace, E_λ. It is invariant under the action of T, meaning that T carries E_λ into itself. The eigenvalues of T are then exactly the set of roots of the polynomial $p(\lambda) = \det(T - \lambda I)$. If A is a matrix representing T, then one has $p(\lambda) = \det(A - \lambda I)$, which permits one to find the eigenvalues of T easily if the dimension of V is not too large, or if the matrix A is simple enough. The eigenvalues and eigenspaces of T provide a means by which the nature and structure of the linear transformation T can be examined in detail.

The theory of linear algebra is rich, and cannot be summarized with any brevity. We conclude this sketch by mentioning two other topics. Given a linear space V, we can consider the scalar-valued linear function f on V. Denote the collection of all these by V^*; this is itself a linear space, usually called the (algebraic) dual of V. If V has dimension n, then V^* also has dimension n; if u_i is a basis for V, then a basis for V^* consists of the functions f_i where $f_i(u_j) = \delta_{ij}$, using the special symbol introduced in Sec. 6.6 in connection with orthogonality. Borrowing another notation from that section, we write $\langle f, v \rangle$ for $f(v)$ for any $f \in V^*$ and $v \in V$.

Suppose now that we have a second linear space W and its dual space W^*, and suppose that T is a linear transformation from V into W. Given any $g \in W^*$, we can construct a linear function f on V by writing $f(x) = g(T(x))$. We use this to define a transformation T^* of W^* into V^* by $T^*(g) = f$. T^* is the adjoint of T, and is linear. The relation between T and T^* can also be given more simply by the equation

$$\langle g, Tv \rangle = \langle T^*g, v \rangle$$

which holds for all $g \in W^*$ and $v \in V$. These relations are especially useful when $V = W$ and $V^* = W^*$. Every linear mapping of V into itself induces a dual mapping of V^* into itself. It is easily seen, for example, that T is 1-to-1 if and only if T^* is onto.

If $V = \mathbf{R}^n$, and if V^* is also identified with $\mathbf{R}^n$, then every linear functional f has the form of an inner product $f(v) = \langle u, v \rangle = u \cdot v$. Given any symmetric matrix A (for which $A = A^t$), one may introduce the bilinear function $B(u, v) = \langle u, Av \rangle$ and the associated quadratic form $Q(x) = \langle x, Ax \rangle = B(x, x)$. The study of the canonical forms of quadratic forms and their corresponding geometry is an important part of the theory of linear algebra.

In analysis, it is useful to consider the class of positive definite quadratic forms $Q(x)$ for which $Q(x) > 0$ for all $x \neq \mathbf{0}$ in $\mathbf{R}^n$. It is possible to give a criterion by which one can test a matrix A to see if it yields a positive definite quadratic form. Let A_k denote the matrix obtained from A by deleting the last $n - k$ rows and columns. Thus, if $A = [a_{ij}]$ with i, j running from 1 to n, then $A_k = [a_{ij}]$ where i and j now run only from 1 to k. Then $Q(x)$ is positive definite if and only if $\det(A_k) > 0$ for $k = 1, 2, \ldots, n$.

Proofs of these, and a much more complete discussion of all these topics, may be found in references [17], [19], [21], and [22] in the Reading List that follows these appendices.

APPENDIX
FOUR

APPLICATIONS OF MATHEMATICS

Regardless of their ultimate interests and careers, students of mathematics ought to understand something about the way in which mathematics is used in applications and the complicated interaction between mathematics and the sciences. For many mathematicians engaged in pure research, contact with other sciences may be infrequent; they are apt to see their subject as one that is largely self-sufficient, breeding its own subdisciplines and creating its own research problems, with only occasional stimulus from outside. Others who are more directly involved in neighboring disciplines may find it difficult to agree with this position, and indeed may lay great stress upon the role of the physical sciences as a source for mathematical ideas and techniques.

Both attitudes of course are wrong, because both are incomplete; at the same time, both are at least partially correct. If there is any basic error of the applied mathematician, it is in an attitude that segregates certain areas of mathematics as being those which "are applied" and dismissing the rest as belonging to the nonuseful arts. Many mathematicians share with us the belief that any branch of mathematics could become applied overnight in the right hands.

What is then the role of mathematics in the sciences? Why is it important, and what part does it play in the development of a subject? We think that mathematics offers the scientist a vast warehouse full of objects, each available as a model for various aspects of physical reality. The richness and diversity of this supply are central reasons for the importance of mathematics; another

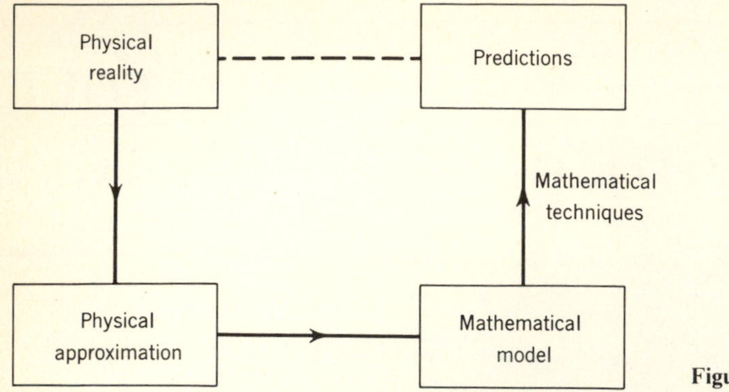

Figure A4-1

is that, along with the objects, mathematics offers a system for using the models, to help raise or answer questions about physical reality, and techniques for exploring the behavior of the models themselves.

Figure A4-1 may help in visualizing this relationship. We start from physical reality (whatever that may be) and proceed from this to what we have called a physical approximation. To think of a specific situation, suppose that the physical reality is a simple pendulum consisting of a spherical weight mounted at the end of a metal bar and swung from a pivot. In creating the physical approximation, we may make statements such as the following: "I think we should assume Newton's laws, disregard air resistance, assume that the pivot has zero friction and that the mass of the bar is negligible in comparison with the weight."

At this stage, the next step is to construct (or select) an appropriate mathematical model. The model could be something as simple as a quadratic equation, or it could be a complicated differential equation, or it could be a topological manifold with a group of measure-preserving mappings acting on it. At the next stage, the scientist or mathematician begins to explore the properties of the model, using mathematical techniques, and to answer specific questions that have been translated from questions about the physical reality or the physical approximation, so that they become meaningful questions about the mathematical model. The final step, of course, is the comparison of these answers with what is known about physical reality from observation or experiment and the evaluation of the success of the model and the correctness of the scientist's intuition.

It is very important not to confuse the model with reality itself. A model is most useful when it resembles reality closely, but there will always be aspects of reality that it does not reproduce, and it will always predict behaviors that do not in fact occur. The skill of the scientist is in knowing how far, and in which contexts, to use a particular mathematical model.

One simple illustration may help here. Physicists sometimes speak of light as a wave, and sometimes as a particle. Which is it? Obviously, neither, for

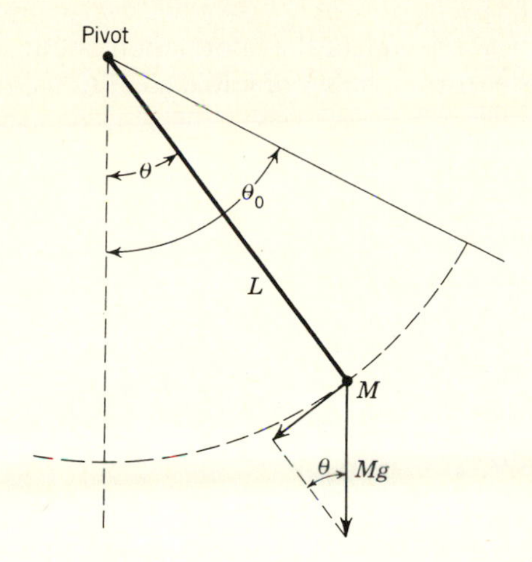

Figure A4-2

these are names for specific mathematical models, and neither one is the physical reality that is light itself. Both successfully predict ("explain") some of the observed phenomena of light; each is unsuccessful in predicting others.

The moral is clear: One should never expect to have only one correct mathematical model for any aspect of reality.

Let us continue with the specific example of the pendulum. We shall raise two questions about its motion and attempt to obtain answers from appropriate mathematical models. We move the weight out to form an initial angle θ_0 and then release it from rest. (i) What is the speed of the weight as it passes the lowest point of its swing? (ii) How long will it take to return (approximately?) to its original position? (See Fig. A4-2.) The various assumptions that underlie the physical approximation can be listed:

a. We assume that the rod is rigid, of constant length L, and of zero mass.
b. We assume that the weight can be treated as a particle of mass M.
c. We assume that there is no air resistance and no friction at the pivot and that the only external force present is a constant vertical gravitational attraction.
d. We assume Newton's laws.

From these, in the usual way, we arrive at a differential equation for the pendulum motion. We assume that the mathematical description of the motion will be a function ϕ such that $\theta = \phi(t)$ will specify its position at time t. Then, we arrive at

$$-Mg \sin \theta = \frac{d}{dt}(Mv) = \frac{d}{dt}\left(ML\frac{d\theta}{dt}\right)$$

or

$$\frac{d^2\theta}{dt^2} = -\frac{g}{L}\sin \theta$$

We must also translate into its mathematical equivalent the information about the mode of release of the pendulum. This becomes the set of initial conditions:

$$t = 0 \qquad \theta = \theta_0 \qquad \frac{d\theta}{dt} = 0$$

Finally, we translate the pair of questions into the following:

i. What is the value of $d\theta/dt$ when $\theta = 0$?
ii. What is the value of t when $\theta = \theta_0$?

We could now forget completely that we are concerned about a pendulum and pose the following "pure" mathematics problem:

Find a function ϕ defined for all $t \geq 0$ such that $\phi(0) = \theta_0$, $\phi'(0) = 0$, and

$$\phi''(t) + \frac{g}{L} \sin \phi(t) = 0 \qquad t \geq 0$$

Specifically, find $\phi'(t)$ at the first value of t for which $\phi(t) = 0$, and find the first value of $t > 0$ for which $\phi(t) = \theta_0$.

This can be treated in this form; for example, we have

$$2\phi'(t)\phi''(t) + \frac{2g}{L} \phi'(t) \sin \phi(t) = 0$$

and, integrating,

$$(\phi'(t))^2 - \frac{2g}{L} \cos \phi(t) = \text{const} = C$$

Since we are to have $\phi(0) = \theta_0$, $\phi'(0) = 0$, we have

$$0 - \frac{2g}{L} \cos \theta_0 = C$$

and finally arrive at

$$(\phi'(t))^2 = \frac{2g}{L} \{\cos \phi(t) - \cos \theta_0\}$$

from which we obtain an answer to the first question: putting $t = t_1$, where $\phi(t_1) = 0$, we find the speed at the bottom of the swing to be

$$V = \sqrt{\frac{2g}{L}} \sqrt{1 - \cos \theta_0} = 2\sqrt{\frac{g}{L}} \sin\left(\frac{\theta_0}{2}\right)$$

[Notice that this solution is predicated on several *mathematical* assumptions: (i) that indeed there is such a function ϕ, (ii) that there is such a value t_1 where $\phi(t_1) = 0$. Both of these can be proved by more detailed mathematical analysis of the differential equation; both are also implied by the physical reality, if we could be sure in advance that the model imitates reality.]

In the same way, an answer can be found for the second question, and we find the length of the period (the time taken from release to first return) to be

(A4-1) $$T = 4\sqrt{\frac{L}{2g}}\int_0^{\theta_0} \frac{d\theta}{\sqrt{\cos\theta - \cos\theta_0}}$$

More generally, some of the other predictions which arise from this model are as follows: The motion is periodic, and never ceases. The period depends in a simple way upon the length L and in a more complicated way upon the initial amplitude θ_0 of the swing.

If we were to modify the physical approximation by altering some of the assumptions (a) to (d), we would arrive at a new mathematical model. We could, for example, bring in resistance, or we could treat the rod as having mass, or we could assume the rod slightly flexible or slightly elastic so that it could change its length, etc. Each change would lead to a new and more complicated mathematical model. Our intuition tells us that these would be more likely to reproduce reality better; we would, for example, expect that the model will not exhibit complete periodicity, and that the amplitude will now decrease with time.

There is, however, another way in which a mathematical model such as this can come to be modified. When a model has been constructed, it may be of such a nature that present mathematical techniques do not suffice to explore its behavior satisfactorily. In this case, one might wish to replace the model with another that is similar but easier to work with; moreover, the choice may be made for purely mathematical reasons, divorced completely from physical motivations. This is the case with the pendulum problem we have just treated, and one often finds it stated that the following differential equation is the equation of motion of a pendulum:

$$\frac{d^2\theta}{dt^2} = -\frac{g}{L}\theta \qquad \begin{cases} t = 0 \\ \theta = \theta_0 \\ d\theta/dt = 0 \end{cases}$$

in which we have replaced $\sin\theta$ by θ.

Being a linear equation with constant coefficients, this is more easily solved, and we find

$$\theta = \phi(t) = \theta_0 \cos\left(\sqrt{\frac{g}{L}}\,t\right)$$

$$V = \sqrt{\frac{g}{L}}\,\theta_0$$

$$T = 2\pi\sqrt{L/g}$$

It should be noted that the predictions from *this* model are somewhat different. For example, the length of the period is now independent of the initial amplitude θ_0, and the values of V are not the same. (For $\theta_0 = \pi/2$, the difference is 10 percent.)

Would it be correct to say that the second model is bad and the first good, or that only the first deserves to be called the motion law for a pendulum? To agree with these would be to overlook the moral; both are models, and neither is *the* model; both have limited usefulness, and both represent different approximations to various aspects of reality.

The rapid growth in the potential and use of high-speed computers is having a striking effect upon this whole picture. Already, scientists are able to use them to explore the behavior of mathematical models of great complexity, and thus to simulate large-scale systems that were formerly quite unmanageable. At the same time, mathematicians are able to acquire some intuitive understanding of the qualitative nature of these systems, which is the first step toward their complete analysis.

While the main concern of this book has been analysis, in the mathematical sense, many topics and illustrative examples have been chosen because of their connection with applicable mathematics. This is evident in Chaps. 7, 8, and 9, and particularly so in Sec. 9.6 and Chap. 10.

Suppose that we wish to study the interaction of several charged particles. It is then convenient to think of one particle creating, by its presence, a "field" throughout space; the second particle is then regarded as interacting with this field, rather than with the particle itself, directly. From this viewpoint, it is only a step to the observation that the second particle also creates a field, and that we are really concerned with the interaction of two fields; the particles may in fact be regarded as unnecessary fictions. The result can be a theory of physics based upon fields alone. Some fields are "particle-like" and may have other attributes associated with them such as "charge," "mass," "momentum," etc. Other fields may lack these attributes, or have others called "spin," "charm," etc. The simplest fields will be numerical- or vector-valued functions defined on all of 4-space (= space-time), but one may also wish to work with fields whose values are matrices, differential operators, probability distributions, measures, points in an infinite-dimensional function space, etc.

The boundaries of the realm of applicable mathematics are indeed flexible, and change with each generation.

APPENDIX
FIVE
INTRODUCTION TO COMPLEX ANALYSIS

Although it has great importance as an active branch of pure mathematics, the theory of **holomorphic** (also called analytic) functions owes its existence and much of its prestige to the success which it has had in dealing with problems in the field of differential equations, hydrodynamics, and potential theory. It came into independent existence late in the nineteenth century, when the traditional "theory of functions" was separated into "real variable theory" and "complex variable theory," under the pressure of a more rigorous consideration of the concept of set, number, integral, and derivative. To a historian of mathematics, complex analysis is especially interesting because of its role in the birth of algebraic topology and because of its surprising connections with the theory of numbers and with algebra.

In a brief summary such as this, it is impossible to do more than indicate some of these connections and to describe a few of the main results and techniques. Most of the details are omitted but may be sought in the references cited at the end.

In the present book, the central subject of study has been the class of continuous real-valued functions defined on sets D in n space and, more generally, the class of continuous transformations T mapping a set D in n space into a set in m space. What would be the effect on the former if we were to allow functions f which could take complex values? Such a function might be

$$F(t) = 3t \sin t - 3i \cos t$$

or

$$G(s, t) = 3se^{-t} + i \frac{4st}{s^2 + t^2}$$

More generally, any complex-valued function F defined on a region D in $\mathbf{R}^n$ has the form

$$F(p) = f(p) + ig(p)$$

where f and g are real-valued functions defined on D. In this case, we speak of f and g as the real and imaginary parts of F, respectively; the complex conjugate of F is the function $\bar{F}$ defined by

$$\bar{F}(p) = f(p) - ig(p)$$

Note that $f = (F + \bar{F})/2$ and $g = (F - \bar{F})/2i$.

A complex-valued function F is continuous if and only if its real and imaginary parts are continuous. Similarly, we can speak of the differentiability of F and define partial derivatives of F in terms of the derivatives of f and g. For example, with $F(t)$ as given above,

$$F'(t) = 3 \sin t + 3t \cos t + 3i \sin t$$

and
$$\frac{\partial}{\partial s} G(s, t) = 3e^{-t} + i\,\frac{(s^2 + t^2)(4t) - (4st)(2s)}{(s^2 + t^2)^2}$$

From one point of view, the introduction of complex-valued functions brings in nothing basically new. If we identify the complex number $a + bi$ with the point (a, b), then a complex-valued function F defined on a region $D \subset \mathbf{R}^n$ becomes merely a mapping from D into $\mathbf{R}^2$. Indeed, if $F(p) = f(p) + ig(p)$, then we can write

$$F(p) = u + iv = (u, v)$$

and identify F with the transformation

$$\begin{cases} u = f(p) \\ v = g(p) \end{cases} \qquad p \in D$$

Likewise, the function $h(t) = t + it^2$, $0 \le t \le 1$, can be regarded simply as a curve (parabola) in the plane.

However, if we look at certain functions that are complex-valued and defined on subsets of the plane (or, more generally, defined on subsets of $\mathbf{R}^{2k}$ for some k), then something new emerges. The reason for this is that the complex numbers are more than $\mathbf{R}^2$ and in fact have the algebraic structure of a field. Writing $x + iy = z$, we can single out a special class of complex functions defined on $\mathbf{R}^2$. Let $c_0, c_1, c_2, \ldots, c_n$ be complex numbers, and define a complex-valued function F by $F(x, y) = w$, where

$$w = c_0 + c_1 z + c_2 z^2 + \cdots + c_n z^n = P(z)$$

Here, P is a polynomial with complex coefficients. If we separate into real and imaginary parts, we arrive at a formula for F of the form

$$\begin{cases} u = A(x, y) \\ v = B(x, y) \end{cases}$$

where A and B are real polynomials in x and y. For example, the equation $w = z^3 - 3iz + (1 + i)$ can be written as

$$T: \begin{cases} u = x^3 - 3xy^2 + 3y + 1 \\ v = 3x^2y - y^3 - 3x + 1 \end{cases} \tag{A5-1}$$

In a similar manner, writing $z_1 = x_1 + iy_1$, $z_2 = x_2 + iy_2$, etc., we can define certain special complex functions on the space $\mathbf{R}^{2m}$ of points $p = (x_1, y_1, x_2, y_2, \ldots, x_m, y_m)$ by

$$w = P(z_1, z_2, \ldots, z_m)$$

where P is a polynomial in m variables over the complex field. This, too, can be separated into real and imaginary parts, obtaining a transformation from $\mathbf{R}^m$ into $\mathbf{R}^2$ of the form

$$\begin{cases} u = A(x_1, y_1, x_2, y_2, \ldots, x_m, y_m) = A(p) \\ v = B(x_1, y_1, x_2, y_2, \ldots, x_m, y_m) = B(p) \end{cases}$$

Such functions are special examples of the class of functions to which the name "holomorphic" is given, and the study of complex analysis is very largely the study of properties of holomorphic functions; the name "analytic" is also used, even though this is also used in another, closely related sense.

Before defining the general class of holomorphic functions, we could ask if there is some special property which will enable us to identify one of these special holomorphic polynomial mappings, within the class of *all* polynomial mappings. We shall henceforth consider only functions of one complex variable, and thus deal with mappings from $\mathbf{R}^2$ (the Z plane or the XY plane) into $\mathbf{R}^2$ (the W plane or the UV plane). The answer to this question is given by the following theorem:

Theorem 1 *A mapping F, given by the equations*

$$\begin{cases} u = A(x, y) \\ v = B(x, y) \end{cases}$$

where A and B are real polynomials, has the form $w = P(z)$ for a complex polynomial P if and only if the differential of F has the form

$$\begin{bmatrix} a & -b \\ b & a \end{bmatrix}$$

at each point z.

Since the differential of F is given by

$$dF = \begin{bmatrix} \dfrac{\partial u}{\partial x} & \dfrac{\partial u}{\partial y} \\ \dfrac{\partial v}{\partial x} & \dfrac{\partial v}{\partial y} \end{bmatrix}$$

this is equivalent to asserting that the functions u and v obey the system of differential equations

$$\text{(A5-2)} \qquad \begin{cases} \dfrac{\partial u}{\partial x} = \dfrac{\partial v}{\partial y} \\[2ex] \dfrac{\partial u}{\partial y} = -\dfrac{\partial v}{\partial x} \end{cases}$$

(These equations are called the Cauchy-Riemann equations; they arose in another connection at the end of Sec. 8.4.)

To check this result, we observe that the differential of the mapping given in Eq. (A5-1) is

$$dT = \begin{bmatrix} 3x^2 - 3y^2 & -6xy + 3 \\ 6xy - 3 & 3x^2 - 3y^2 \end{bmatrix}$$

The proof of Theorem 1 may be based upon the following observation: The "remainder theorem" of elementary algebra implies that a polynomial P has the property that, for any z_0,

$$P(z) - P(z_0) = (z - z_0)Q(z) = (z - z_0)Q(z_0) + (z - z_0)^2R(z)$$

where Q and R are polynomials. On the other hand, the approximation property of the differential of F shows that

$$P(z) - P(z_0) = dF\Big|_{z_0}(z - z_0) + R(z, z_0)$$

Comparison of these shows that for any $\Delta z = (\Delta x, \Delta y)$ one must have

$$dF\Big|_{z_0}(\Delta z) = Q(z_0)\,\Delta z$$

If $Q(z_0) = a + bi$, then this yields

$$dF\Big|_{z_0} = \begin{bmatrix} a & -b \\ b & a \end{bmatrix}$$

as asserted in the theorem.

From this point, there are several ways to arrive at the class of general holomorphic functions. One is based on the behavior of power series with complex coefficients. Certainly, the most obvious generalization of a polynomial is the function defined by

$$f(z) = a_0 + a_1(z - c) + a_2(z - c)^2 + a_3(z - c)^3 + \cdots$$

If this series converges for some value of z other than c, then it is easy to see that it converges for all z in the neighborhood of c and that the real and imaginary parts of the resulting function obey the Cauchy-Riemann equations (A5-2). For example, $\exp(z) = \sum_0^\infty z^n/n!$ converges for all z and has real part

$u = e^x \cos y$, imaginary part $v = e^x \sin y$. It is then possible to adopt the following characterization: a complex-valued function F, defined on an open region D, is said to be holomorphic (or analytic) there if F can be represented locally in D by convergent power series. (It should be noted that this is a direct extension of the use of the term analytic as it was applied to real-valued functions of one or more variables; see Secs. 3.4, 3.5, and 6.3.)

Another approach is conceptually even simpler. It is easy to extend Theorem 1 to rational functions $R(z) = P(z)/Q(z)$, where P and Q are polynomials with complex coefficients, provided that we exclude from consideration points z, where $Q(z) = 0$. We can then obtain the most general holomorphic function by taking appropriate limits of these. Specifically, Runge showed that a function F is holomorphic on a region D if and only if it can be approximated uniformly on each compact subset of D by rational functions.

A third approach to the holomorphic functions is by means of differentiation. Returning to a polynomial $P(z)$, given by

$$P(z) = c_0 + c_1 z + c_2 z^2 + \cdots + c_n z^n$$

we can define its derivative

$$\begin{aligned} P'(z_0) &= \lim_{z \to z_0} \frac{P(z) - P(z_0)}{z - z_0} \\ &= c_1 + 2c_2 z_0 + 3c_3 z_0^2 + \cdots + nc_n z_0^{n-1} \end{aligned}$$

Not all complex-valued functions have derivatives; if we were to ask for the same limit for the function f defined by

$$f(z) = f(x, y) = x^2 + y^2 + 2xyi$$

it would fail to exist. Indeed, the following result is true:

Theorem 2 *If $F(z) = w = u + iv$, where u and v are real polynomials in x and y, and if*

$$\lim_{z \to z_0} \frac{F(z) - F(z_0)}{z - z_0}$$

exists at each point z_0, then there is a polynomial P with complex coefficients such that $w = P(z)$; moreover, u and v satisfy the Cauchy-Riemann equations.

It remained for Cauchy, Riemann, Weierstrass, and others to bring all these aspects together into a coherent structure of striking beauty. A key result is the demonstration of the equivalence of the three approaches: holomorphic functions are those which have derivatives, which can be represented locally by their Taylor series, and which can be approximated arbitrarily closely on compact sets by rational functions.

Most proofs of this rest upon the discovery that holomorphic functions can also be characterized by integral properties. Let

$$w = f(z) = u + iv$$

and write $dz = dx + i\,dy$. Then, the complex 1-form $f(z)\,dz$ can also be written as

$$\begin{aligned} f(z)\,dz &= (u + iv)(dx + i\,dy) \\ &= (u\,dx - v\,dy) + i(v\,dx + u\,dy) \\ &= \alpha + i\beta \end{aligned}$$

where α and β are real 1-forms. Following the techniques of Chap. 9, we can compute the derivative of each of these forms,

$$\begin{aligned} d\alpha &= d(u\,dx - v\,dy) \\ &= du\,dx - dv\,dy \\ &= (u_x\,dx + u_y\,dy)\,dx - (v_x\,dx + v_y\,dy)\,dy \\ &= u_y\,dy\,dx - v_x\,dx\,dy \\ &= -(u_y + v_x)\,dx\,dy = -\left(\frac{\partial u}{\partial y} + \frac{\partial v}{\partial x}\right) dx\,dy \end{aligned}$$

and, in the same fashion,

$$\begin{aligned} d\beta &= dv\,dx + du\,dy \\ &= v_y\,dy\,dx + u_x\,dx\,dy = \left(\frac{\partial u}{\partial x} - \frac{\partial v}{\partial y}\right) dx\,dy \end{aligned}$$

Suppose now that the function f is holomorphic, so that u and v satisfy the Cauchy-Riemann equations. Then, each of the forms $d\alpha$ and $d\beta$ is 0. Thus, the fact that f is holomorphic in a region D is reflected in the fact that the complex differential form $f(z)\,dz$ is closed. Appealing to Green's theorem, we conclude that

$$0 = \int_\Gamma f(z)\,dz = \int_\Gamma \alpha + i \int_\Gamma \beta$$

where Γ is any closed path in D, bounding a region in which f is holomorphic. This important result is known as **Cauchy's theorem.** From this, one may derive Cauchy's integral representation formula,

$$f(z) = \frac{1}{2\pi i} \int_\Gamma \frac{f(t)\,dt}{t - z}$$

which expresses the value of f at points inside a simple closed path Γ, where f is holomorphic, in terms of the values which f takes on Γ, as a complex line integral (also called a **contour integral**).

In contrast with the general class of functions mapping the plane into the plane, holomorphic functions have many striking properties. If F is holomorphic on an open region D and not constant, then F is open—i.e., carries open subsets of D into open sets. From this, one can immediately infer the **maximum modulus** theorem: The real-valued function $|F(z)|$ cannot have a local maximum anywhere in D. Alternatively, if E is a compact subset of D and $|F(z)| \leq M$ for all $z \in \partial E$, then $|F(z)| \leq M$ at all interior points of E.

Even though such a mapping F is open, it need not be 1-to-1 in D. However, if we remove from D all the points (at most countable) where f' takes the value 0, then f is locally 1-to-1 in the remaining set and is therefore locally a homeomorphism on this set. Furthermore, examine the set S of all points z where $F(z) = c$. This set might be empty; however, if F is not constant, it cannot be more than countable, and if it is not finite, then all limit points of S lie on the boundary of D. This in turn implies the surprising uniqueness theorem for holomorphic functions: If F and G are both holomorphic in an open region D, and if $F(z) = G(z)$ for all z in a neighborhood, or an arc, or even on a converging sequence of points in D, then F and G must agree everywhere in D.

People often judge the significance of a new branch of mathematics, not by its elegance and unity, but by its ability to solve problems that had been raised previously. In this direction, one of the early successes of complex analysis was the ease with which it led to the evaluation of certain real integrals. We shall illustrate this with one example, representative of a class of similar definite integrals that can be treated by contour integration.

Let us find the value of the integral

$$V = \int_{-\pi}^{\pi} \frac{d\theta}{(5 + 3\cos\theta)(5 + 4\sin\theta)}$$

Put $t = e^{i\theta}$, observing that as θ goes from $-\pi$ to π, t moves around the unit circle $|t| = 1$, starting and ending at the point -1; we have $\cos\theta = (e^{i\theta} + e^{-i\theta})/2 = (t + t^{-1})/2$ and $\sin\theta = (t - t^{-1})/2i$, while $dt = ie^{i\theta}\,d\theta = it\,d\theta$. Making these substitutions, we find that

$$V = \frac{1}{3}\int_{\mathscr{C}} \frac{t\,dt}{(t + \frac{1}{3})(t + 3)(t + \frac{1}{2}i)(t + 2i)}$$

where $\mathscr{C}$ is the circumference of the unit circle, traced counterclockwise. To this, the Cauchy integral formula can be applied twice by replacing $\mathscr{C}$ by two paths enclosing the points $-\frac{1}{3}$ and $-i/2$ separately; here, we have used the fact that the integrand is a closed form in regions which exclude the points where the denominator becomes 0. We therefore arrive at the evaluation

$$V = \frac{1}{3}2\pi i\left\{\frac{-\frac{1}{2}i}{(-\frac{1}{2}i + \frac{1}{3})(-\frac{1}{2}i + 3)(-\frac{1}{2}i + 2i)} + \frac{(-\frac{1}{3})}{(-\frac{1}{3} + 3)(-\frac{1}{3} + \frac{1}{2}i)(-\frac{1}{3} + 2i)}\right\}$$

$$= \frac{\pi}{6}\left(\frac{35}{37}\right)$$

In a very similar way, one may also use contour integration to prove the **fundamental theorem** of **algebra.** The following elegant argument is due to R. P. Boas. We wish to show that any polynomial of degree n must have at least one root. The general case can be reduced to polynomials with real coefficients; so we suppose that $P(z) = a_0 + a_1 z + \cdots + a_n z^n$, with $a_n \neq 0$, a_k real. The proof rests on the observation that if $P(z)$ is never 0, for any complex number z, then the same statement holds for the polynomial $Q(z)$, where $Q(z) = z^n P(z + z^{-1})$; note that $Q(0) = a_n \neq 0$. Since the coefficients of P are assumed real, we know that the value of the following definite integral

$$V = \int_{-\pi}^{\pi} \frac{d\theta}{P(2\cos\theta)}$$

is real and different from 0. Make the same substitution $t = e^{i\theta}$, and obtain

$$V = \int_{\mathscr{C}} \frac{dt}{it\, P(t + t^{-1})} = \frac{1}{i} \int_{\mathscr{C}} \frac{t^{n-1}\, dt}{Q(t)}$$

But, as observed above, the function $f(t) = t^{n-1}/Q(t)$ is holomorphic everywhere, since $Q(t)$ is never 0, and by the Cauchy theorem the value of this last integral must be 0. This contradiction proves the theorem.

The theory of functions of a complex variable has close connections with both hydrodynamics and the study of heat conduction. One point of contact is the class of harmonic functions of two real variables. If $f(z) = u(z) + i\, v(z)$ is holomorphic in a region D, then, by differentiating the Cauchy-Riemann equations (A5-2), one finds that the real functions u and v are both solutions of the Laplace equation

$$\nabla^2 H = \frac{\partial^2 H}{\partial x^2} + \frac{\partial^2 H}{\partial y^2} = 0$$

Any solution of this equation in a region D is said to be harmonic in D; so this shows that the real and imaginary parts of a holomorphic function are always harmonic. The converse is partially true: if D is a simple connected region and H is harmonic in D, then H is the real part of some function f that is holomorphic in D. (For more general regions D, this statement is not always true.)

One model for the stable temperature distribution throughout a plane region produced by a variable boundary temperature leads to the Laplace equation and thus to harmonic functions. The same equation arises in diffusion problems, and in the study of the steady-state flow of a two-dimensional incompressible irrotational fluid. Here, the velocity vector of the fluid at the point $z = (x, y)$ can be written as $a(x, y) + i\, b(x, y)$, and there must exist a harmonic function u with $u_x = a$ and $u_y = b$. The function u is called the velocity potential of the flow. If one constructs a holomorphic function f whose real part is u, then techniques of complex analysis can be applied to f

to predict the nature of the streamlines past a cross section immersed in the flow or to compute the theoretical drag or lift of a proposed cross section.

Further information about the theory and application of holomorphic functions must be sought elsewhere.

APPENDIX

SIX

FURTHER TOPICS IN REAL ANALYSIS

In writing a book to be used at the junior or senior level and intended for students with a wide range of prior mathematical experience, many compromises must be made. Elegance must at times be outweighed by the need for clarity, and inviting digressions withheld. It is frustrating for the teacher, and possibly harmful for the best students, to be forced to pass by so many instances where one could so easily step aside and make contact with the frontiers of mathematical research. The purpose of this appendix is to restore some of this by a reexamination of certain topics in the text. No effort is made to be complete; for this, one should refer to the books listed at the end.

The two threads that we wish to emphasize come under the headings of algebraic analysis and linear analysis. The former deals with the algebraic aspects of what has been covered, the latter with the vector spaces that are pandemic to analysis. We assume that readers are familiar with the concepts of ring, field, homomorphism, ideal, etc., as they might be presented in an undergraduate algebra course.

The first example we shall use is the class $C[D]$ of all continuous real-valued functions on a compact set D. As a mathematical object, what is its structure? One observes first that it is a linear space, with the usual notion of addition of functions. Moreover, with the norm $\|f\| = \sup_{p \in D} |f(p)|$, we can introduce a (metric) topology into $C[D]$. A neighborhood of a function $f_0 \in C[D]$ consists of all functions $f \in C[D]$ such that $\|f - f_0\| < \delta$, and convergence of a sequence of functions $\{f_n\}$ means uniform convergence on D. Theorem 1 of Sec. 6.2 asserts that the resulting normed linear space is complete; every Cauchy sequence converges to a "point" in the space. If D is a compact set in n space, then the class of polynomials with rational coefficients forms a countable dense subset of $C[D]$.

Whenever one has a topological linear space, a central question is to ask for the dual space of continuous linear functions. In our case, some are easily found. If g is any continuous function on D, define a function L on $C[D]$ by

$$L(f) = \int_D fg$$

Then, one may see that L is continuous; if $\{f_n\}$ converges to f, then $L(f_n)$ converges to $L(f)$. This follows at once from the fact that there must exist a number B such that $|L(f)| < B\|f\|$ for all $f \in C[D]$. These are not *all* the linear functions L; the remaining ones are also integrals, but of a more general variety. This is also closely related to Theorem 2 in Sec. 8.2, and to formula (6-61) in Sec. 6.6.

But the linear space $C[D]$ is also an algebra. If f and g belong to $C[D]$, so does fg. As an algebra, it has a more elaborate structure, which suggests further questions that can be asked. What are the ideals in $C[D]$? Are there any interesting subalgebras? What types of homomorphic images can $C[D]$ have? What are the automorphisms? And what types of algebras can be represented isomorphically (or homomorphically) in $C[D]$? A few examples might be of interest. If a point $p_0 \in D$ is selected, then one may consider the mapping h from $C[D]$ into the real numbers, defined by $h(f) = f(p_0)$. This is easily seen to be a homomorphism of $C[D]$ onto the real field. As such, its kernel $M = \{\text{all } f \in C[D] \text{ with } f(p_0) = 0\}$ must be a maximal ideal. Furthermore, all the maximal ideals are obtained in exactly this fashion. This important fact leads in turn to the observation that if two algebras $C[D_1]$ and $C[D_2]$ of this sort are algebraically isomorphic, then the compact sets D_1 and D_2 must be topologically the same, i.e., must be homeomorphic.

Again, $C[D]$ has an additional type of structure as well. If we introduce an order relation by saying $f \le g$ whenever $f(p) \le g(p)$ for all $p \in D$, then we have an important example of what is called a partially ordered vector space. [The term "partially" is used because it is no longer true that a function f must obey either $f \ge 0$ or $f \le 0$; for example, $f(x) = \sin x$.] The order properties of $C[D]$ are as interesting as its algebraic properties and are also closely related to the properties of the integral as a functional. For example, if $g \ge 0$, then the function L defined above has the property that $L(f) \ge 0$ whenever $f \ge 0$; for this reason, it is said to be a positive functional.

There are many other attractive roads branching off in other directions; some of these are suggested by the treatment of function spaces, and especially Hilbert space, given in Sec. 6.6.

One final topic which we think should be mentioned is that of generalized differentiation and generalized functions. In elementary calculus, much attention is paid to the problem of when a function can or cannot be differentiated. In particular, we say that $f(x) = |x|$ does not have a derivative at the point $x = 0$, even though it is continuous there. As a matter of fact, there is a perfectly good sense in which this function *has* a derivative, and indeed in which badly discontinuous functions *can* have derivatives.

To explain this, we start with the observation that the use of functions in physical applications is not always (indeed, is hardly ever) as point-to-number mappings. If we measure the value of some physical variable such as velocity or charge, we are forced to read instead the time average of something over a short interval. When one analyzes (à la Bridgeman) the operational approach to physics, one sees that functions are used in physics, not for their point-to-point numerical values, but for their effect on other functions. This brings us back to integration, where we write (with a different notation now)

$$\langle f, g \rangle = \int_{-\infty}^{\infty} fg$$

Suppose that we put no continuity restrictions on f at all but ask that g be a function of class C^∞, which in addition vanishes off some compact set G. It is not difficult to see that we can examine f by means of the values of $\langle f, g \rangle$ for all choices of g and that, for example, if $\langle f, g \rangle = \langle h, g \rangle$ for all g, and f and h are reasonably well behaved, then $f = h$.

Now, suppose that f is itself of class C'; we can ask how the function f' acts on sample functions g, in comparison with f. Integration by parts gives the answer immediately.

$$\begin{aligned}
\langle f', g \rangle &= \lim_{r \to \infty} \int_{-r}^{r} f'(t)g(t)\,dt \\
&= \lim_{r \to \infty} \left\{ f(t)g(t)\Big|_{-r}^{r} - \int_{-r}^{r} f(t)g'(t)\,dt \right\} \\
&= -\int_{-\infty}^{\infty} f(t)g'(t)\,dt = -\langle f, g' \rangle = \langle -f, g' \rangle
\end{aligned}$$

In words, the "value" of f' at g is the negative of the "value" of f at g'. We could therefore use this relation to find something about f', even if we did not know f' initially.

Suppose that we examine this for a more general function f. Even though f' does not exist in the usual sense, we can define a "generalized" derivative f' by saying: f' is the functional L which acts on functions g in the following manner:

$$\begin{aligned}
L(g) &= \langle -f, g' \rangle \\
&= -\int_{-\infty}^{\infty} fg'
\end{aligned}$$

Even though L is not representable in the form $\langle f', g \rangle$, we can still work with it as though it *were* a function, except that we cannot ask for it to have values at points; we study f' only by the way it affects functions g.

One example will show what this means. Let us take f to be the step function

$$f(t) = \begin{cases} 1 & \text{when } t \geq 0 \\ 0 & \text{when } t < 0 \end{cases}$$

Then, we wish to find out what f' is. As above, f' is characterized by the relation

$$\langle f', g \rangle = -\int_{-\infty}^{\infty} f(t)g'(t)\, dt$$

for all g. However, using the definition of f, we have

$$\langle f', g \rangle = -\int_{0}^{\infty} g'(t) = -g(t) \Big|_{0}^{\infty}$$
$$= g(0)$$

Thus, the derivative f' of the step function f is a generalized function which acts on functions g, so that $\langle f', g \rangle = g(0)$. If we set $f' = \delta$, then the integral notation (if applicable) would have to give

$$g(0) = \int_{-\infty}^{\infty} \delta(t)g(t)\, dt$$

for all smooth functions g. There is no actual function δ that has this property, although there are functions δ_n that almost achieve this. If $\delta_n(t)$ is positive, even, and has a very tall peak at $t = 0$, so that $\int \delta_n(t)\, dt = 1$ while δ_n is very small away from the origin, then the value of $\langle \delta_n, g \rangle$ is almost $g(0)$.

We have thus been led to the generalized function that is customarily called the **Dirac delta function;** it is in fact nothing more than a measure of mass 1 located at the origin. As above, δ is the limit, in a suitable sense, of honest functions δ_n, explaining the intuitive picture of δ as a function which is zero-valued except at $t = 0$ but which has integral 1.

This process for differentiating functions can also be applied to generalized functions, and one may find δ'; this turns out to be a generalized function q such that $\langle q, g \rangle = -g'(0)$ for all g. It is the limit of the sequence of derivatives δ_n' and can be regarded as a **dipole** at $t = 0$. It is a generalized function (or distribution) which is not even a measure. It is by introducing such extensions of the classical concept of function that the formal differentiation of Fourier series, discussed in Sec. 6.6, can be justified and used.

This example is a good illustration of the way in which research in mathematics comes about. Clearly, the original step function f does not have a derivative; however, it turns out to be profitable to ask what would happen if we could differentiate it! One recalls the similar question "what would happen if -1 did have a square root?" in the light of the productive theory outlined in the preceding appendix.

In Sec. 6.2, we stated Theorem 6, the bounded convergence theorem, without a proof, mentioning that this would require an excursion into some aspects of the theory of Lebesgue measure and integration. We end this appendix by sketching such a proof, omitting all the complicating details. We begin by assuming a related result.

Theorem 1 *Let $\{f_n\}$ be a sequence of functions that converge pointwise to a function f on an interval I, and are such that $0 \le f_1 \le f_2 \le f_3 \le \cdots$. (That is, $\{f_n\} \uparrow f$ on I.) Then*

$$\lim \int_I f_n = \int_I f \tag{A6-1}$$

[Note that if each f_n, as well as f, is continuous and I is compact, then by Exercise 11 in Sec. 6.2, $\{f_n\}$ in fact converges uniformly to f.]

We can now weaken this by removing the monotonicity, at the expense of a weaker conclusion.

Theorem 2 *Let $0 \le f_n$ for all n, and let $\{f_n\}$ converge pointwise on I to f. Then,*

$$\int_I f \le \liminf \int_I f_n \tag{A6-2}$$

To prove this, set $g_n(x) = \inf\{f_n(x), f_{n+1}(x), f_{n+2}(x), \ldots\}$ and observe that $0 \le g_n(x) \le f_n(x)$ for all n and x, and that the sequence $\{g_n\}$ is monotonic increasing with limit f. Integrating, we have

$$\int_I g_n \le \int_I f_n$$

and thus

$$\liminf \int_I g_n \le \liminf \int_I f_n$$

But, by Theorem 1, $\lim \int_I g_n$ exists and is $\int_I f$, which therefore gives us (A6-2).

We are now ready for the final result.

Theorem 3 *Let $\{f_n\}$ converge pointwise on I to f, and assume that $|f_n(x)| \le g(x)$ for all x and n, where $\int_I g$ is finite. Then,*

(A6-3) $$\lim \int_I f_n = \int_I f$$

For, we have $0 \leq g + f_n$ and $0 \leq g - f_n$, and these converge pointwise to $g + f$ and $g - f$, respectively. Accordingly, by Theorem 2, we have

$$\int_I g \pm \int_I f \leq \liminf \int_I (g \pm f_n)$$

$$\leq \int_I g + \liminf \int_I \pm f_n$$

But, $\int_I g$ is finite, and so it can be cancelled from both sides, and the result rewritten as

$$\limsup \int_I f_n \leq \int_I f \leq \liminf \int_I f_n$$

which clearly implies (A6-3).

(It should be obvious that many of the omitted details have to do with being sure that each of the functions appearing in this proof is integrable on the set I.)

SUGGESTED READING

This list of books and articles is intended for random browsing, or for background reference, or to cast further light on later developments. It is meant neither to be exclusive nor exhaustive.

Analysis

1. L. V. Ahlfors, "Complex Analysis," 2d ed., McGraw-Hill Book Company, New York, 1966.
2. R. P. Boas, Jr., "Primer of Real Functions," *Carus Monograph* 13, Math. Assoc. of America and John Wiley & Sons, Inc., New York, 1960.
3. L. Brand, "Vector Analysis," John Wiley & Sons, Inc., New York, 1957.
4. R. C. Buck (ed.), "Studies in Modern Analysis," vol. 1 in Math. Assoc. of America, "Studies in Mathematics," Prentice-Hall, Inc., Englewood Cliffs, N.J., 1962.
5. P. J. Davis, "Leonhard Euler's Integral: An Historical Profile of the Gamma Function," *Am. Math. Monthly*, vol. 66, pp. 849–869, 1959.
6. J. A. Dieudonne, "Foundations of Modern Analysis," Academic Press, Inc., New York, 1960.
7. W. F. Eberlein, "The Elementary Transcendental Functions," *Am. Math. Monthly*, vol. 74, pp. 1223–1225, 1967.
8. H. Flanders, "Differential Forms with Applications to the Physical Sciences," Academic Press, Inc., New York, 1963.
9. B. R. Gelbaum and J. M. H. Olmsted, "Counterexamples in Analysis," Holden-Day, Inc., Publisher, San Francisco, 1964.
10. P. Halmos, "Measure Theory," D. Van Nostrand Company, Inc., Princeton, N.J., 1950.
11. W. F. Osgood, "A Jordan Curve of Positive Area," *Trans. Am. Math. Soc.*, vol. 4, pp. 107–112, 1903.
12. A. W. Roberts and D. E. Varberg, "Convex Functions," Academic Press, Inc., New York, 1973.
13. W. Rudin, "Principles of Mathematical Analysis," 3d ed., McGraw-Hill Book Company, New York, 1976.

14. G. Springer, "Introduction to Riemann Surfaces," Addison-Wesley Publishing Company, Inc., Reading, Mass., 1957.
15. D. J. Struik, "Lectures on Classical Differential Geometry," Addison-Wesley Publishing Company, Inc., Cambridge, Mass., 1950.

Algebra

16. A. A. Albert (ed.), "Studies in Modern Algebra," vol. 2 in Math. Assoc. of America, "Studies in Mathematics," Prentice-Hall, Inc., Englewood Cliffs, N.J., 1963.
17. S. Barnard and J. M. Child, "Introduction to Higher Algebra" and "Advanced Algebra," The Macmillan Company, New York, 1936, 1939.
18. G. Birkhoff and S. MacLane, "Survey of Modern Algebra," rev. ed., The Macmillan Company, New York, 1965.
19. C. W. Curtis, "Linear Algebra, an Introductory Approach," Allyn and Bacon, Inc., Englewood Cliffs, N.J., 1963.
20. I. N. Herstein, "Topics in Algebra," Blaisdell Publishing Company, New York, 1963.
21. A. I. Mal'cev, "Foundations of Linear Algebra," W. H. Freeman and Company, San Francisco, 1963.
22. B. Noble and J. W. Daniel, "Applied Linear Algebra," 2d ed., Prentice-Hall, Inc., Englewood Cliffs, N.J., 1977.

Topology-Geometry

23. S. S. Chern (ed.), "Studies in Global Geometry and Analysis," vol. 4 in Math. Assoc. of America, "Studies in Mathematics," Prentice-Hall, Inc., Englewood Cliffs, N.J., 1967.
24. P. J. Hilton (ed.), "Studies in Modern Topology," vol. 5 in Math. Assoc. of America, "Studies in Mathematics," Prentice-Hall, Inc., Englewood Cliffs, N.J., 1968.
25. J. L. Kelley, "General Topology," D. Van Nostrand Company, Inc., Princeton, N.J., 1955.
26. H. P. Manning, "Geometry of Four Dimensions," The Macmillan Company, New York, 1914.
27. I. M. Singer and J. A. Thorpe, "Lecture Notes on Elementary Topology and Geometry," Scott, Foresman and Company, Glenview, Ill., 1967.
28. M. Spivak, "Calculus on Manifolds," W. A. Benjamin, Inc., New York, 1965.

Logic and Foundations

29. P. R. Halmos, "Naive Set Theory," D. Van Nostrand Company, Inc., Princeton, N.J., 1960.
30. J. Olmsted, "The Real Number System," Appleton-Century-Crofts, Inc., New York, 1962.
31. R. R. Stoll, "Introduction to Set Theory and Logic," W. H. Freeman and Company, San Francisco, 1963.
32. R. L. Wilder, "Introduction to the Foundations of Mathematics," John Wiley & Sons, Inc., New York, 1952.

Applications

33. R. C. Buck with E. F. Buck, "Introduction to Differential Equations," Houghton Mifflin Company, Boston, Mass., 1976.
34. S. D. Conte and C. W. deBoor, "Elementary Numerical Analysis—An Algorithmic Approach," 2d ed., McGraw-Hill Book Company, New York, 1972.
35. A. Friedman, "Generalized Functions and Partial Differential Equations," Prentice-Hall, Inc., Englewood Cliffs, N.J., 1963.
36. A. Hochstadt, "Differential Equations, a Modern Approach," Holt, Rinehart and Winston, Inc., New York, 1963.

37. C. C. Lin and L. A. Segel, "Mathematics Applied to Deterministic Problems in the Natural Sciences," The Macmillan Company, New York, 1974.
38. R. D. Luce and H. Raiffa, "Games and Decisions," John Wiley & Sons, Inc., New York, 1957.
39. B. Noble, "Applications of Undergraduate Mathematics in Engineering," The Macmillan Company, New York, 1967.
40. E. M. Purcell, "Electricity and Magnetism," Berkeley Physics Course, vol. 2, McGraw-Hill Book Company, New York, 1965.
41. I. S. Sokolnikoff and R. M. Redheffer, "Mathematics of Physics and Modern Engineering," 2d ed., McGraw-Hill Book Company, New York, 1966.
42. D. J. Wilde, "Optimum Seeking Methods," Prentice-Hall, Inc., Englewood Cliffs, N.J., 1964.

Miscellaneous

43. Edwin A. Abbott, "Flatland, a Romance of Many Dimensions," 6th ed., Dover Publications, Inc., New York, 1952.
44. C. Boyer, "History of Mathematics," John Wiley & Sons, Inc., New York, 1968.
45. R. Courant and H. Robbins, "What Is Mathematics?," Oxford University Press, New York, 1941.
46. Michael J. Crowe, "A History of Vector Analysis," University of Notre Dame Press, Notre Dame, Ind., 1967.
47. M. Gardner, "Can time go backward?," *Scientific American*, January 1967, pp. 98–108.
48. I. J. Good, "The Scientist Speculates," Capricorn Books, G. P. Putnam's Sons, New York, 1965.
49. Robert Heinlein, "Lifeline," *Astounding Science Fiction*, August 1939, vol. 23, pp. 83–100.
50. Robert Heinlein, "——— and He Built a Crooked House," *Astounding Science Fiction*, February 1941, pp. 68–83.
51. H. Petard, "Contributions to the Mathematical Theory of Big Game Hunting," *Am. Math. Monthly*, vol. 45, pp. 446–447, 1938.
52. H. Poincaré, "Science and Hypothesis," Dover Publications, Inc., New York, 1952.
53. T. Rado, "What Is the Area of a Surface?," *Am. Math. Monthly*, vol. 50, pp. 139–141, 1943.
54. S. Ulam, "A Collection of Mathematical Problems," Interscience Publishers, Inc., New York, 1960.
55. H. Weyl, "Symmetry," Princeton University Press, Princeton, N.J., 1952.
56. COSRIMS, "The Mathematical Sciences—A Collection of Essays," The M.I.T. Press, Cambridge, Mass., 1969.

HINTS AND ANSWERS

"For emergency use only"

Section 1.2

3 If you make a two-dimensional picture, time vs. one space axis, the resulting diagram resembles a Z. It might be the history of one particle, if a positron is "really" an electron moving "backwards through time" (see [47] in Reading List).

7 (*b*); $(-1, -1, -1)$.

8 Don't use coordinates; instead eliminate P or Q.

10 Any multiple of $(4, 5, -3)$.

12 This is easier if you do not use coordinates.

16 The three cases depend on how the vertices are labeled. In one case, the vector AB is equal and parallel to the vector CD only if $B - A = D - C$.

19 (*b*) If $x < 0$, $-x > 0$ and $(-x)^2 > 0$. But, $-x = (-1)x$ and $(-1)^2 = 1$.

21 Start with $(a + b)/(A + B) = a/(A + B) + b/(A + B)$.

23 Map n to $2n$ if $n > 0$, and to $-2n + 1$ if $n \leq 0$.

Section 1.3

2 It helps to read $|p - A|$ as "the distance from p to A."

3 (*a*) Write the relation as $y - x \leq x + 2y \leq x - y$.

5 Use $|p| = |(p - q) + q|$. **10** Yes.

11 One component of the equation is $y = 2t + 3$.

12 One point is $(\frac{1}{2}, \frac{1}{2}, \frac{1}{2}, \frac{1}{2})$.

13 Express R as a weighted average of each vertex and the opposite midpoint.

15 Two of the vertices are $(A + 4B + 2C)/7$ and $(2A + B + 4C)/7$.

17 Pictures will help!

Section 1.4

1 (*b*) A paraboloid opening downward. (*d*) Hard to sketch; domain omits the lines $y = x$ and $y = -x$, and graph turns abruptly upward or downward as you approach one of these lines, depending upon which side you are on. (*e*) Graph is a cliff, with a crevasse at its foot.

2 (*a*) 6
(*b*) 13
(*c*) 6
(*d*) $(x^2 + x)(y + 1)$
(*e*) $x(x + 1)^2$
(*f*) $x^2(y + 1)^2 + x(y + 1)$
(*g*) $x^4 + 2x^3 + 2x^2 + x$

3 The answers to (*a*) and (*b*) are different.

4 (*b*) $f(x)$ is not defined for any real x.

5 Graph $y = F(t)$ by putting $t = 1/(x - 1)$ and solving for y.

7 (*c*) One of the level "curves" contains an entire open disk.

9 Among other observations, one could say that at any time, the temperature becomes arbitrarily large as one moves away from the origin in any direction, and also that at any one spot, the temperature will become arbitrarily large as one waits.

12 There are infinitely many, but only four that are continuous everywhere.

14 (*b*) Look for those choices of k and m such that $4k + 14 = 8m + 2$.

Section 1.5

2 (*a*) Closed, bounded, connected, and boundary is itself. (*c*) Closed, unbounded, connected, and is its own boundary. (*e*) Open, unbounded, disconnected, and the boundary is a pair of intersecting lines.

3 (*a*) Open, unbounded, connected, boundary is sphere. (*c*) Open, unbounded, connected, boundary is surface.

4 (*a*) It may help to look at the graph of the polynomial $P(x) = x(x - 1)^2$.

5 (*b*) The entire plane.

6 Cluster points include those such as $(0, 1/n)$ for each n and others on the horizontal axis. Be sure to examine the origin, which is also a cluster point.

11 Yes.

12 Is it true in general that $\text{bdy}(A) \cap \text{bdy}(B) \subset \text{bdy}(A \cap B)$?

13 (*b*) Because its complement is $\varnothing$.

17 (*a*) No; (*b*) no, give examples.

19 It is helpful to ask which values of x are acceptable at each value of y.

Section 1.6

3 Show that $|p_{n+1} - q| \le C^n |p_1 - q|$.

4 Use coordinates.

5 For example, $P_{4n+1} = (1, 0) + n(2, -2)$.

6 Note that (*a*) and (*b*) are subsequences of each other.

8 It has just one limit point, but is divergent.

13 Use $c = (1 + a_n)^n = 1 + na_n + \cdots$.

15 Use mathematical induction.

16 (*e*) Let $A = \liminf (x_n/n^2)$. Show that $A \le \frac{1}{4}$ and that $x_{n+1} - x_n \ge n\sqrt{A - \varepsilon}$ for all large n, and infer that $A \ge \sqrt{A - \varepsilon}/2$ and then that $A \ge \frac{1}{4}$.

17 (*b*) Put $b_n = [2 \cdot 4 \cdot 6 \cdots (2n)]/[3 \cdot 5 \cdot 7 \cdots (2n+1)]$ and show that $(a_n)^2 < a_n b_n < 1/(2n+1)$.

19 Show that $a_{n+2} - a_{n+1} = (a_{n+1} - a_n)/3$, which enables one to calculate $\lim a_n$.

21 (*c*) Put $d_k = a_{k+1} - a_k$. Prove that $d_{n+1} = -d_n/(n+1)$, evaluate d_k, and then show that $\lim a_n = a_1 + \sum_1^\infty d_k$.

26 (*a*) $\limsup a_n = 1$. (*b*) $\liminf a_n = -2$. (*c*) $\limsup a_n = 3$, $\liminf a_n = -1$. (*d*) $\limsup a_n = \sqrt{3}/2$. What do you suppose would happen in this if $a_n = \sin(n/3)$?

27 Construct the sequences so that the large terms of $\{a_n\}$ are in phase with the small terms of $\{b_n\}$.

28 Another way to say that $L = \limsup_{n\to\infty} a_n$ is that, for any $\varepsilon > 0$, we have $a_n < L + \varepsilon$ for all large n, while $a_n > L - \varepsilon$ for infinitely many indices n. Similarly, if $l = \liminf_{n\to\infty} a_n$, then, for any $\varepsilon > 0$, $a_n > l - \varepsilon$ for all large indices n and $a_n < l + \varepsilon$ for an infinite number of indices n.

29 Given $\varepsilon > 0$, choose N so that $|a_n| < \varepsilon$ when $n \ge N$. Then, write

$$|\sigma_n| \le \frac{|a_1 + a_2 + \cdots + a_N|}{n} + \frac{(n-N)\varepsilon}{n}$$

and from this, get $\limsup |\sigma_n| \le \varepsilon$. Since this holds for any $\varepsilon > 0$, $\lim \sigma_n = 0$.

31 $(\frac{4}{3}, \frac{2}{3})$.

33 Obtain the formula $P_{n+2} - P_{n+1} = (P_{n+1} - P_n)/4$, and conclude that $\{P_n\}$ is convergent to $(\frac{2}{3}, \frac{1}{3})$.

36 (*b*) No. **37** Work with $b_k = \log a_k$.

Section 1.7

1 For any $\varepsilon > 0$, $L - \varepsilon < L$, so that there must exist an N with $L - \varepsilon < a_N$.

3 If $b = \sup(S)$ and $b \notin S$, then it is seen to be true that $b \in \text{bdy}(S) \subset \text{closure}(S)$.

4 If $p \in A$, $q \in B$, then $|p - q| \le \text{diam}(B)$.

5 If $c \in C$, then $\text{dist}(B, C) \le |p - c|$ for every $p \in A$.

6 (*c*) No. **7** Modify Exercise 6.

9 Work with the nested intervals $[a_n, b_n]$ and the nested intervals $[c_n, d_n]$.

11 Count the number of points from the given set that lie in $D_n = \{\text{all } p,\ |p| \le n\}$, and then note that the collection of all sets D_n is a countable collection.

Section 1.8

1 Cover the given set by the open sets D_n, where $D_n = \{\text{all } p,\ |p| < n\}$.

3 If F is closed and C compact and $F \subset C$, then any open covering of F leads to an open covering of C if $C - F$ is adjoined.

6 Let $(a_n, b_n) \in A \times B$ and choose n_k so that $\lim a_{n_k}$ exists, then choose a subsequence of b_{n_k} which converges.

9 (*a*) Yes. (*b*) No. (*c*) If $x_n \in X(S)$, choose y_n with $p_n = (x_n, y_n) \in S$. Let p be a limit point of $\{p_n\}$, and show that $X(\{p\})$ is a limit point of $\{x_n\}$.

10 (*a*) Each point in the open disc D is the center of a closed square that lies in the disc. (*b*) The closed disc D_N of radius $1 - 1/N$ can be covered by a finite number of closed squares contained in D, and $\bigcup_1^\infty D_N = D$, so one can succeed with a countable number of squares.

Section 2.2

2 Observe that
$$\frac{x}{y} - \frac{x_0}{y_0} = \frac{(x - x_0)y_0 + (y_0 - y)x_0}{yy_0}$$
and estimate this when $|y| > \rho$, $|y_0| > \rho$, $\rho > 0$.

4 Put $s = r \cos \theta$, $y = r \sin \theta$.

10 Use the fact that the complement of any closed set is open.

12 $f^{-1}(A \cap B) \supset f^{-1}(A) \cap f^{-1}(B)$

Section 2.3

1 You must prove that there are pairs of points, arbitrarily close together, on which the variation of F is large, for example, $(n, 0)$ and $(n + 1/n, 0)$.

5 (*a*) The key idea is that dist $(A, B) > 0$.

7 Choose a finite number of points $p_k \in D$ which are ε-dense in D and let $M = \max |f(p_k)|$, and use this to estimate $\sup_{p \in D} |f(p)|$.

10. (*a*) Draw a line through p and p_3 to meet the edge connecting p_1 and p_2. (*b*) Observe that $f(p) = \sum \alpha_i f(p)$ and estimate $|f(p) - F(p)|$.

Section 2.4

2 (*b*) The function $3 - x/y - y/x$ is not continuous everywhere in the plane.

3 (*b*) Construct a polynomial of the form $f(x, y) = [A(x, y)]^2 B(x, y)$ and examine the set where $A(x, y) = 0$.

4 Consider
$$f(x) = \begin{cases} 1/x & \text{for } x > 0 \\ 2 & \text{for } x = 0 \end{cases}$$

5 Look at $h(x) = f(x) - g(x)$.

6 Show that for sufficiently large c, $P(c)P(-c) < 0$.

7 Sketch the graphs of f and g, and then examine $h(x) = f(x) - g(x)$.

8 If f is constant on two overlapping intervals, it is the same constant on both.

9 Look at the inverse images of neighborhoods of the values 2 and 3.

14 Arrange the temperatures of the vertices in increasing order, and then examine their respective positions in the tetrahedron.

17 Pick any value y_0. Then, for any x, $\min_y F(x, y) \le F(x, y_0)$. Hence $A \le \max_x F(x, y_0)$.

18 In the complex field, a continuous path can go from 1 to -1 without going through 0.

Section 2.5

1 (*b*) No; $\lim_{x \to 0} g(f(x)) = 2$. **2** Put $x = t^6$.

3 $\frac{14}{15}$ (note that this cannot be found merely by substitution). **4** $\frac{9}{10}$.

8 Note that $\sqrt{1 + x^2} = |x|\sqrt{1 + 1/x^2}$.

14 Note that it is not permissible to assume that $\lim f(x)$ exists.

15 For any $\varepsilon > 0$, $f(x) < L + \varepsilon$ on a deleted neighborhood of b, but $f(x_n) > L - \varepsilon$ for a sequence approaching b.

18 Prove that f is "Cauchy" at p_0, meaning that $\lim |f(p) - f(q)| = 0$ as p and q simultaneously approach p_0.

Section 2.6

1 (*a*) Continuous everywhere. (*b*) Removable discontinuity at the origin. (*c*) Essential discontinuity at the origin. (*d*) Removable discontinuities on the line $y = x$, except at the origin, where it is continuous.

2 The function is continuous nowhere.

3 The function is continuous at each irrational point and discontinuous (essential) at each rational point. This depends upon the fact that any sequence of rationals that approach an irrational must have larger and larger denominators.

4 (*b*) Removable at $x = 1$; essential at 2 and -1.

5 (*b*) All removable.

6 (*a*) Removable at (0, 0). (*b*) Essential at (0, 0), since it depends upon which parabola is considered.

7 No. **8** (*a*) Yes. (*b*) No.

9 All discontinuities arise as jumps, due to unequal left- and right-hand limits. There cannot be more than a finite number of discontinuities with jump more than $1/n$, and thus at most a countable number *in toto*.

Section 2.7

2 (*a*) Graph f. (*b*) Recognize $f(x) - 1$.

7 Let $f(a) = f(b) = 1$ with $a < b$ and prove that $a = 0$, $b = 1$.

8 The result depends on whether or not $|A| < 1$.

Section 3.2

4 These are intuitively obvious; what is asked for is a proof based on the mean value theorem.

5 First study the zeros of F.

6 Study $P'(x)$ at and between its zeros.

7 Apply the mean value theorem to $f(x + 1) - f(x)$.

8 First method: Prove $f'(x)$ exists and evaluate it.
Second method: Estimate $|f(b) - f(a)| = \left|\sum_1^N (f(x_{j+1}) - f(x_j))\right|$ where $x_0 = a$, $x_N = b$.

10 What must be shown is that $(b - a)f(x) \geq (b - x)f(a) + (x - a)f(b)$. Apply the mean value theorem to $f(x) - f(a)$ and to $f(x) - f(b)$, and use the result to express $(b - a)f(x)$. Finally, use the mean value theorem on f' and make use of the hypothesis on f''.

12 (*a*) The discontinuity of f' at $x = 0$ is removable.

14 (*a*) This is intuitively clear from a picture, but a rigorous proof is called for. Show by the mean value theorem that if $f'(x_0) < 0$, then f must be negative somewhere to the right of x_0. (*b*) No.

15 (*a*) Never. (*b*) Not necessarily. (*c*) Not necessarily.

16 Show that $\int_0^2 (1 - P'(t))\, dt = 0$. **18** Look at a helix.

19 Set $f(\theta) = 3b/(2H + B)$ and express b, H, and B in terms of θ; then use the mean value theorem to estimate $f(\theta) - f(0)$ and then to estimate $|f(\theta) - \theta|$.

20 Use the mean value theorem on $f(x) = \arctan(x)$, and then use $\pi/2 - f(x) = \arctan(1/x)$.

22 (*b*) No. **23** (*a*) $\frac{1}{2}$. (*b*) 1. (*c*) -1.

24 It is not legal to cancel the factor $\cos x$, and with this, the hypothesis of Theorem 6 is not satisfied.

25 $\lim f'(x)/g'(x)$ fails to exist, but $\lim f(x)/g(x)$ does.

27 (*b*) No; try $f(x) = xe^x$.

28 $f(0) = 0$, and $f'(0)$, calculated from the original definition of derivative, is 0.

29 First case: If $f(a) < f(b)$, then prove that there must be a point u with $f(a) = f(u)$, and thus c with $f'(c) = 0$.
Second case: $f(a) \geq f(b)$.

Section 3.3

1 (*a*) $f_1(x, y) = 2x \log(x^2 + y^2) + 2x^3/(x^2 + y^2)$
$f_2(x, y) = 2x^2y/(x^2 + y^2)$, $f_{12}(x, y) = 4xy^3/(x^2 + y^2)^2$.

2 $f_2(x, y) = 3x^2y^2 - 2$, $f_2(2, 3) = 106$, $f_2(y, x) = 3x^2y^2 - 2$.

3 (*c*) $[-1, -1, -4]$.

4 (*b*) For no direction, except those of the axes; (*c*) no.

5 Using the mean value theorem, show that $|f(p) - f(p_0)| \leq M|p - p_0|$.

6 Assume $f = 0$ on the boundary of a bounded open set.

7 Use Exercise 6. **8** 0. **13** $\frac{25}{7}$.

14 "Nonsense. F must have been constant." Show why.

Section 3.4

1 (*b*) $$\frac{dw}{dt} = \frac{\partial w}{\partial t} + \frac{\partial w}{\partial u}\frac{\partial u}{\partial t} + \frac{\partial w}{\partial u}\frac{\partial u}{\partial x}\frac{dx}{dt} + \frac{\partial w}{\partial x}\frac{dx}{dt}.$$

(*c*) $$\frac{\partial w}{\partial x} = F_1 + F_2 f_1 + F_3 g_1; \quad \frac{\partial w}{\partial y} = F_2 f_2; \quad \frac{\partial w}{\partial z} = F_3 g_2.$$

2 $$\frac{d^2w}{dx^2} = \frac{\partial^2 w}{\partial x^2} + 2\frac{\partial^2 w}{\partial x\,\partial y}\frac{dy}{dx} + \frac{\partial^2 w}{\partial y^2}\left(\frac{dy}{dx}\right)^2 + \frac{\partial w}{\partial y}\frac{d^2y}{dx^2}.$$

3 At $(-1, 2, 1)$, $\partial y/\partial x = \frac{2}{3}$, $\partial y/\partial z = \frac{4}{3}$.

4 $$\frac{\partial u}{\partial x} = -\frac{(2x^3u + 2xy^2v + x^2y - 3y^2)}{x^4 + 4y^3} \quad \text{and also,} \quad \frac{\partial u}{\partial x} = \frac{(12y^5 - 12x^2y^4 - 9x^4y^2 + x^6y)}{(x^4 + 4y^3)^2}.$$

5 $$\frac{\partial z}{\partial x} = -\frac{F_1}{F_3}, \qquad \frac{\partial z}{\partial y} = -\frac{F_2}{F_3}.$$

6 $$\frac{\partial^2 z}{\partial x^2} = \frac{\{2F_1F_3F_{13} - F_1{}^2F_{33} - F_3{}^2F_{11}\}}{F_3{}^3}$$

$$\frac{\partial^2 z}{\partial x \partial y} = \frac{\{F_1F_2F_{23} + F_2F_3F_{13} - F_1F_2F_{33} - F_3{}^2F_{12}\}}{F_3{}^3}$$

8 $$\frac{dx}{dt} = \frac{\partial(F, G)}{\partial(y, t)} \bigg/ \frac{\partial(F, G)}{\partial(x, y)}, \qquad \frac{dy}{dt} = -\frac{\partial(F, G)}{\partial(x, t)} \bigg/ \frac{\partial(F, G)}{\partial(x, y)}.$$

11 Differentiate the equation with respect to t, and set $t = 1$.

19 Apply Taylor's theorem to express $f(p + \Delta p)$ in the form $f(p) + R(\Delta p)$ for any point p in S. Then use the fact that f_{11}, f_{12}, f_{22} are continuous (and therefore bounded) on S to estimate $R(\Delta p)$.

20 Let p and q belong to S, with $p = (\phi(a), \psi(a))$, $q = (\phi(b), \psi(b))$. Subdivide $[a, b]$ by points t_j, with $t_0 = a$, $t_n = b$. Let $P_j = (\phi(t_j), \psi(t_j))$. Then, $|f(p) - f(q)| \leq \sum |f(P_{j+1}) - f(P_j)| \leq M \sum |P_{j+1} - P_j|^2$. Show that the latter sum can be made arbitrarily small by proper choice of the t_j. (For a generalization of this, see a paper by Kakutani, *Proc. Am. Math. Soc.*, vol. 3, pp. 532–542, 1952.)

21 $\phi(1) = 1$, $\phi'(1) = a(1 + b + b^2) + b^3$.

Section 3.5

2 The error does not exceed $1/6! = 1/720 < .0014$.

3 By Taylor's theorem, we obtain $|\text{error}| < \frac{1}{5}$. The actual error is no more than .07.

5 6, 11.

6 Use Taylor's theorem at $x = 0$ and estimate $f(1)$ and $f(-1)$, and obtain $f^{(3)}(\tau_1) + f^{(3)}(\tau_2) = 6$.

7 Use Taylor's theorem at $t = x$; estimate $|f'(x)|$ and then estimate $\limsup |f'(x)|$ as $x \uparrow \infty$. Note that many functions obey $f(x) \to 0$ without $f'(x) \to 0$, as $x \uparrow \infty$.

9 We have $1/x < 1/2\{1/x^{1+\delta} + 1/x^{1-\delta}\}$ for any δ, $0 < \delta < 1$. Integrate, and obtain $\log x < (1/2\delta)(x^\delta - x^{-\delta})$. Then take $\delta = \frac{1}{2}$.

10 Yes. $e^x \geq (e^2/4)x^2$, and this is best possible.

11 $|R_n(x)| \leq B^n(x - a)^n/n! \to 0$.

12 Write $f(x) = f(a) + f'(a)(x - a) + f''(\tau)(x - a)^2/2$, set $x = 0$, $x = 1$, and subtract. One then obtains $|f'(a)| \leq (A/2)[a^2 + (1 - a)^2]$.

13 $|f(a)| = \left| \int_0^a f' \right| \leq M \int_0^a |f|$.

14 Show that for any x_0, there is an m and a neighborhood of x_0 on which $(d^m/dx)(f(x)) \equiv 0$.

Section 3.6

1 3 and $\frac{11}{6}$. **2** (*a*) Examine $P'(x)$.

3 The critical point is $(\frac{1}{2}, 0)$ and is a saddle point. The maximum and minimum occur on the boundary, and are 4 and $-\frac{16}{5}$.

4 Maximum = 4; minimum = $-\frac{1}{2}$.

5 (*a*) Saddle; (*b*) saddle; (*c*) $(0, 0)$ is a maximum, $(\pm\frac{1}{2}, \pm 1)$ are minima, and $(0, \pm 1)$ and $(\pm\frac{1}{2}, 0)$ are saddle points; (*d*) all the points on the line $y = (\frac{2}{3})x$ are minima; (*e*) $(0, 0)$ is a minimum; (*f*) $(0, 0)$ is a saddle point.

9 Look at the behavior of f when $y = mx$ and when $y = 3x^2/2$.

10 The centroid.

11 An open set where $df = 0$ must contain a critical point.

12 One point is $(\frac{1}{2}, \frac{1}{2}, 0)$. **13** The maximum is $106\frac{1}{4}$.

14 There is an endpoint maximum, with an interior local extreme.

17 The line L goes through the centroid of the given points. The slope is determined by $\alpha = E(x^2) - [E(x)]^2$, $\beta = E(y^2) - [E(y)]^2$, and $\gamma = E(xy) - E(x)E(y)$, where $E(u) = \sum u_i/n$ for any numbers $u_i, u_2, \ldots, u_n$.

18 $F = f - g$ is harmonic in D and $F \geq 0$ (or $F \leq 0$) on bdy (D).

21 Let $k = [A]$, the largest integer obeying $k \leq A$. Then, the maximum occurs for the choice $x_1 = x_2 = \cdots = x_k = 1$, $x_{k+1} = \beta$, $x_{k+2} = \cdots = x_n = 0$, where $k + \beta = A$. This type of problem arises in many linear programming situations.

Section 4.2

3 $M - f(p)$ is positive, so that $\iint_D (M - f) = M\text{A}(D) - \iint_D f \geq 0$

5 As in Exercise 3, show that $\iint_D fg$ lies between $M \iint_D g$ and $m \iint_D g$. Then use the intermediate value theorem to find the point $\bar{p}$ with

$$f(\bar{p}) = \left\{\iint_D fg\right\} \left\{\iint_D g\right\}^{-1}$$

6 There is a neighborhood where $f(p) \geq \delta$.

11 Yes, 0.

13 Draw an appropriate picture and interpret the two estimates in geometric form.

14 Use the uniform continuity of f.

15 Subdivide $[0, 1]$ and cover the curve by rectangles with a small total area, again using the uniform continuity of f and B as a Lipschitz constant for g.

Section 4.3

2 (*a*) $F(x) = \begin{cases} x^2 - x + 1 & x \geq 1 \\ x & x < 1 \end{cases}$ (*b*) $F(x) = (e^x - 1)\exp(e^x)$

3 (*c*) No. **5** Use the substitution $v = 1/u$. **6** Let $F' = f$, and evaluate each side.

7 There are infinitely many solutions, of which one is

$$f(x) = \begin{cases} 0 & 0 \leq x \leq 1 \\ x/2 - 1/2 & 1 < x \end{cases}$$

9 $\int_0^1 dy \int_1^2 f(x, y)\, dx + \int_1^2 dy \int_y^2 f(x, y)\, dx.$

11 The first blank should contain $\sqrt{y/2}\, f(y)$.

12 1/20.

13 $\int_0^2 dx \int_x^2 dz \int_1^{2-(1/2)x} f(x, y, z)\, dy$

$$\int_1^2 dy \int_0^{4-2y} dx \int_x^2 f(x, y, z)\, dz$$

$$\int_0^2 dz \int_0^z dx \int_1^{2-(1/2)x} f(x, y, z)\, dy$$

$$\int_1^2 dy \int_0^{4-2y} dz \int_0^z f(x, y, z)\, dx + \int_1^2 dy \int_{4-2y}^2 dz \int_0^{4-2y} f(x, y, z)\, dx$$

$$\int_0^2 dz \int_1^{2-(1/2)z} dy \int_0^z f(x, y, z)\, dx + \int_0^2 dz \int_{2-(1/2)z}^2 dy \int_0^{4-2y} f(x, y, z)\, dx.$$

14 44/15. **16** (*a*) $(e - 1)/2$; (*b*) $\frac{1}{2}e^4 - e^2$.

17 The reversed order is $\int_{-\sqrt[3]{6}}^0 dy \int_{-6}^{y^3} xy\, dx + \int_0^2 dy \int_{7y-6}^{y^3} xy\, dx.$

18 Use the uniform continuity of f.

22 Dividing the rectangle in half one finds the estimate .175 58, good to .015. A little more precision gives .174 47. The "exact" answer is

$$13 \log 13 + 10 \log 10 - 12 \log 12 - 11 \log 11$$

Writing this instead as $2 \log (\frac{13}{12}) + \log (\frac{13}{11}) - 10 \log [1 + (\frac{2}{130})]$, and using power series, one gets .174 465, good to the last digit.

23 Consider $\int_a^b (f + \lambda g)^2$, which cannot be negative.

25 (*a*) Less than $\sqrt{5/2}$; (*b*) less than $\sqrt{2\pi}$.

26 Prove the identity for $P(x) = 1, x, x^2$, and x^3 in turn.

28 (*a*) $(3e^{x^3} - 2e^{x^2})/x$; (*b*) $3 \cos (12x) - 2 \cos (8x)$.

31 Calculate $\int_0^1 dx \int_0^1 |y - x| \, dy$.

Section 4.5

1 Area $= \int_0^\infty e^{-x} \, dx = 1$.

2 (*a*) Diverge; (*b*) diverge; (*c*) diverge; (*d*) converge; (*e*) converge; (*f*) converge; (*g*) converge; (*h*) diverge.

3 (*a*) Diverge; (*b*) converge; (*c*) converge (not improper); (*d*) converge; (*e*) converge; (*f*) converge; (*g*) diverge; (*h*) converge.

4 Either $\alpha < 1$ and $\beta > 1$, or $\alpha > 1$ and $\beta < 1$.

5 All (α, β) with $\alpha > -1$, $\beta > -1$ and $\alpha + \beta < -1$.

6 Yes. Evaluate the first half of $\int_r^R dx \int_0^x x^{-3/2} e^{y-x} \, dy$, and examine the resulting integral as $r \downarrow 0$, $R \uparrow \infty$.

8 Divergent. **9** Convergent.

11 Use the intermediate value theorem. **12** (*b*) The order $\int dx \int dy$ is hard.

Section 5.2

1 (*a*) Diverges; (*b*) diverges; (*c*) converges.

3 (*a*) $a_n = 1/(n + 1)(\sqrt{n+1} + \sqrt{n}) \approx 1/n^{3/2}$, convergent.
(*b*) $a_n \approx 1/n^{3/4}$, divergent.

5 (*b*) Take $b_n = 1/(n - 1)$ for $n \geq 2$.
(*c*) Take $b_n = 1/(n - A - 1)$ for $n \geq A + 2$.

7 Bracket the series in blocks of successive length 1, 2, 4, 8, etc.

8 Take logs and use Exercise 29, Sec. 1.6.

9 (*b*) False; (*c*) true; (*d*) true, use Schwarz inequality; (*f*) sum from $n + 1$ to $2n$.

14 Write $a_k = (a_k/\sqrt{k})\sqrt{k}$ and use Schwarz inequality.

Section 5.3

1 Neither is alternating so use the Dirichlet test.

2 (*a*) $-1 \leq r < 1$ (*b*) $x = 0$ (*c*) $-3 \leq x \leq -1$

(*d*) $|x| \le \frac{1}{4}$
(*e*) $|x-1| \le \sqrt{3}$
(*f*) $s > 0$
(*g*) $-e^{-1} \le \beta < e^{-1}$
(*h*) $\gamma > \alpha + \beta$
(*i*) $-1 < x \le 1$
(*j*) $-\infty < x < \infty$
(*k*) $x > -1$ or $x \le -2$
(*l*) $x \le 0$ or $x > 2$
(*m*) All x

5 (*a*) 0; (*b*) e; (*c*) ∞ if $c < 1$, 1 if $c = 1$, 0 if $c > 1$.
9 Consider $\sum (p_{n+1} - p_n)$.

Section 5.4

2 $[2^{n+1} + (-1)^n]/3$.
7 $-2 < x < 1$.

Section 5.5

1 Approx. 10^{9559} terms.
2 (*b*) $(\log n)^2/2 + O(1)$, where $O(1)$ means a bounded term.
3 Allow a radius of rotation of about $2L$.
5 (*a*) $S = -.0826$, error less than .0005; (*b*) $S = .904\,412$, error less than .005.
9 $\sum_1^N k^3 = P(N)$, where P is a polynomial of degree 4.
10 Write the general term as $1/(n-a) - 1/(n-b)$.
12 Use Stirling's formula.
16 Examine some "partial products."

Section 6.2

3 Yes; no; yes.
5 $\lim_{x \downarrow 0} F(x)/x^2 = \sum 1/n^2$.
9 Examine the behavior of $\int_0^L f_n$ and $\int_L^\infty f_n$.
10 Split the interval of integration into $[-1, -c]$, $[-c, c]$, and $[c, 1]$, and show that the first and last lead to contributions that are very small if n is large, while the remaining integral is close to $g(0)$.
11 If q lies in all the sets C_n, then a contradiction arises. Thus the intersection of all the sets (and thus the intersection of finitely many) is empty.

Section 6.3

1 (*b*) $$\frac{1}{x} = \frac{1}{1 + (x-1)} = 1 - (x-1) + (x-1)^2 - (x-1)^3 + \cdots.$$

(*c*) $$\log(1 + x^2) = x^2 - \frac{(x^2)^2}{2} + \frac{(x^2)^3}{3} - \cdots.$$

2 (*a*) $x \dfrac{d}{dx}\left\{x \dfrac{d}{dx} \dfrac{1}{1-x}\right\} = \dfrac{x + x^2}{(1-x)^3}$; (*c*) $\dfrac{1}{2} \log\left(\dfrac{1+x}{1-x}\right)$

3 (*a*) No; (*b*) yes.
5 Apply the corollary to Theorem 11.
9 If $\omega = -\frac{1}{2} + i\frac{\sqrt{3}}{2}$, then $\omega^2 + \omega + 1 = 0$.
10 $1/F(x) = 1 - 2x + 3x^2 - \cdots - 96x^7 + \cdots$
11 $1/F(x)$ is a polynomial.

13 If $\phi = \angle QPC$, then $\tan \phi = 2 \sin (\theta/2)/[1 + 2 \cos (\theta/2)] = \theta/3 + \theta^3/72 + \cdots$, from which one may find that $\phi = \theta/3 + \theta^3/648 + \cdots$.

15 See Sec. 7.4 of Reference [33] in the Reading List.

Section 6.4

4 0. **5** ∞. **6** $(\pi/2)|x|$.

7 $\pi/2$ (try simple substitution).

8 Differentiate and then integrate by parts.

10 Find the smallest value for $g(t)$, noting that any $x \geq 1$ is allowed.

16 Evaluate each side. (Nevertheless, $|G(x, y)| \leq 1/x^2$ for $1 \leq y$, so the inside integrals converge uniformly.)

Section 6.5

1 Put $t = s/y$. **2** (*b*) Put $A(x) = \theta$.

4 (*a*) $\Gamma(\frac{4}{3})\Gamma(\frac{1}{2})/3\Gamma(\frac{11}{6})$. (*b*) $\sqrt{2}\,\Gamma(\frac{1}{2}) = \sqrt{2\pi}$.
(*c*) $-\Gamma(\frac{4}{3})\Gamma(\frac{2}{3})$. (*d*) $\frac{1}{2}\Gamma(\frac{1}{4})\Gamma(\frac{3}{4})$.

6 $\dfrac{1}{q}\Gamma((p + 1)/q)$.

7 $\Gamma(s + 1)/(r + 1)^{s+1}$.

8 (*a*) $Le^{-1/L^2} + \sqrt{\pi} \operatorname{erf} (1/L) - \sqrt{\pi}$. (*b*) $\sqrt{\pi}/4 \operatorname{erf} (1) - 1/(2e)$.
(*c*) $F(x)$ has the form $A/\sqrt{x}$.

Section 6.6

1 $13.86^\circ = \arccos (.97089)$.

3 Calculate $(\|f_n - f_m\|_{[-1, 1]})^2$ directly.

9 $e^x \sim (2/\pi) \sinh (\pi)\left\{\frac{1}{2} + \sum_1^\infty [(-1)^n/(1 + n^2)](\cos nx - n \sin nx)\right\}$.

10 (*h*) $p \cdot q = \sum a_n \bar{b}_n$.

11 (*a*) $1/\sqrt{2\pi}$; (*b*) use deMoivre's identity.

13 (*b*) Use the recursion relation for the polynomials T_n.

Section 7.2

1 (*a*) A translation sending the origin into $(3, -1)$. (*b*) A reflection about the line $y = x$. (*c*) A rotation by $\pi/4$, followed by a radial expansion which multiplies points by $\sqrt{2}$.

3 The line $u + v = 0$, in the plane $w = 0$.

4 $T(p) = T(-p)$. If $(x, y) = x + iy = z$, then $T(z) = z^2$.

10 (*c*) The XY plane is mapped onto the whole UV plane, except for certain points on one of the axes. The mapping is mostly 1-to-1.
(*d*) The image of the XY plane is only a curve in the UV plane.
(*e*) This is, of course, the familiar polar coordinate mapping; it should be analyzed in detail.

12 This will be treated in detail in Chap. 10.

13 (*d*) $T(x, y) = (x', y')$ where $x' = -3x/5 + 4y/5 + \frac{4}{5}$, $y' = 4x/5 + 3y/5 - \frac{2}{5}$.

Section 7.3

2 $[2, -1, 3]$. **4** $(0, -3), (-5, 6), (2, -3), (-1, 0)$.

6 (*a*) $\begin{bmatrix} 0 & 1 & 2 \\ 1 & 4 & 3 \\ 1 & 0 & 1 \end{bmatrix}$ **7** The ranks are (*c*) 2; (*d*) 3.

8 (*a*) All of XYZ space is mapped onto the plane whose equation is $u + v - w = 0$, in a many-to-one fashion.

9 (*a*) $\begin{bmatrix} -5 & -1 \\ 1 & 3 \end{bmatrix}$.

10 $(ST)(x, y) = (2x - 2y, -2x + 6y)$, and
$(TS)(x, y, z) = (4x - y - 2z, 12x - 2y + 4z, -2x + y + 6z)$.

11 (*c*) Yes.

Section 7.4

1 (*b*) $\begin{bmatrix} -18 & 25 & -12 \\ 12 & 0 & -4 \end{bmatrix}$ (*c*) $\begin{bmatrix} 1 & 6 \\ 3 & 3 \\ 2 & -6 \end{bmatrix}$

5 Take $\Delta p = (h, 0)$ and $(0, h)$ in turn, and compute

$$\lim_{h \to 0} \frac{T(p + \Delta p) - T(p)}{h}$$

12 Estimate the maxima of the absolute values of the derivatives, arriving at the matrix $\begin{bmatrix} 4 & 6 \\ 3 & 2 \end{bmatrix}$.

Section 7.5

2 (*a*) e^{2x}; (*b*) 2, if $x \neq 0$.

3 (*a*) Locally 1-to-1 but not 1-to-1 in the whole plane, even though J is never 0. (*b*) 1-to-1 in the right half plane and in the left half plane. (*c*) Never 1-to-1 locally, since it maps the (x, y) plane onto the parabola $v^2 = 4u$.

10 (*a*) Use $y = x^3$. **11** The image set is a curve.

12 Consider separately the case where $x^2 > y^2$ and where $x^2 \leq y^2$.

16 This depends on the multiplication property for determinants [see Appendix 3].

Section 7.6

1 Yes; no. At $(0, 0)$, $\partial F/\partial y = 1$, $\partial F/\partial x = 0$; moreover, for $|xy|$ small, the equation is approximately $x^2 + y = 0$.

2 No; yes.

4 No such representation exists. (Examine the graph of each.)

6 (*a*) Assume f of class C' near 1, and $f'(1) \neq 0$.

7 There is always the solution $y = 1$, identically. In some cases, this is not the only solution; when $f(t) = (t - 1)^2$, $y = (1 - x)/(1 + x)$ is also a solution.

10 It is sufficient if $F_1(0, 0) \neq -1$ and $F_2(0, 0) \neq 0$.

Section 7.7

2 Evaluate the Jacobians first.

4 (*b*) Solve for y in terms of u, v, and x and substitute this in the second equation.

Section 8.2

1 (*a*) No; (*b*) no; (*c*) yes; (*d*) no.

2 (*b*) Look at the seven disjoint sets that compose $S_1 \cup S_2 \cup S_3$, and write $F(S_1 \cup S_2 \cup S_3)$ in terms of $F(S_i)$, $F(S_i \cap S_j)$, and $F(S_1 \cap S_2 \cap S_3)$.

5 Use Theorem 8 of Sec. 2.4. **7** See also Exercise 10 below.

8 Not necessarily; try $f(x, y) = xy/(x^2 + y^2)$.

10 (*c*) If f is of class C'.

Section 8.3

3 (*a*) $\int_0^{1/2} dv \int_v^{1-v} 2(u^2 - v^2)\, du = \frac{1}{8}$.

(*b*) $\int_0^1 du \int_{1-2u}^1 (u^2 + uv)\, dv = \frac{2}{3}$.

5 Setting $u = xy$, $v = x^2 - y^2$, the integral transforms to $(\frac{1}{2}) \int_1^3 du \int_1^4 dv = 3$.

8 The areas of the faces are $\sqrt{6}$, $\sqrt{29}$, $\sqrt{11}$, and the volume is 5.

9 The upper half of the XY plane is mapped onto the portion of the (s, t) plane lying inside the parabola $(s - t)^2 - 8(s + t) + 16 = 0$. Lines in the (s, t) plane correspond to conics in the XY plane. The Jacobian is 0 only on the boundary $y = 0$.

Section 8.4

1 The double point corresponds to $t = 2$, $t = -2$, and the slopes of the tangents are 2 and -2.

2 $k = 4\pi(1 + \pi^2 + 4\pi^4)^{1/2}(1 + 4\pi^2)^{-3/2}$.

3 $L = \int_0^1 (9t^4 + 2t^2 - 4t + 2)^{1/2}\, dt$. **4** $k = (76)^{1/2}(14)^{-3/2}$.

7 The differential has the requisite form, except at the origin. There, angles are tripled.

8 $L = a \int_0^{2\pi} \sqrt{2 - 2 \cos t}\, dt = 4a \int_0^{\pi} |\cos \theta|\, d\theta = 8a$.

9 The curves are perpendicular.

13 Use the mean value theorem. **16** Consider $|f'| + f$.

17 One choice is $y = 2t/(t^2 - 1)$.

Section 8.5

1 The tangent plane may be given by $x = -\frac{1}{4} - v$, $y = \frac{1}{2} + u + v$, $z = 2 + 4v$, or by $4x + z - 1 = 0$.

4 The curve meets the surface in three points, $(0, 0, 0)$, $(1, 1, 2)$, and $(\frac{1}{3}, \frac{1}{9}, \frac{2}{27})$. The angles between the normal to the surface and the tangent to the curve are $\pi/2$, and $\arccos [6/(\sqrt{17}\sqrt{97})]$ $\arccos [-2/(\sqrt{41}\sqrt{17})]$.

7 The area of the ellipsoid is

$$\int_0^{\pi} \sin\phi\, d\phi \int_0^{2\pi} \{a^2b^2\cos^2\phi + c^2\sin^2\phi\,[a^2\sin^2\theta + b^2\cos^2\theta]\}^{1/2}\, d\theta$$

8 The area of the Möbius strip is the same as that of Σ_1, which is

$$\int_0^{2\pi} du \int_{-1}^{1} \left\{\left(\frac{v}{2}\right)^2 + \left[2 - v\sin\left(\frac{v}{2}\right)\right]^2\right\}^{1/2} dv$$

10 $(\frac{49}{3})\pi$.

13 The area is $(2AB/3)((1 + L^2)^{3/2} - 1)$.

14 $\pi R^2 - 2R^2$.

18 4.

20 If the curve is $p = \gamma(t)$ with $\gamma(0) = p_0$, then $F(\gamma(t)) = 0$ on a neighborhood of $t = 0$. Apply the chain rule, and interpret the result.

Section 8.6

1 $\bar{x} = 0$, $\bar{y} = (\frac{1}{4})(e + e^{-1}) + (e - e^{-1})^{-1}$.

3 $I = \dfrac{k}{4}\left[\left(\dfrac{5}{3}\right)\sqrt{5} + \left(\dfrac{1}{15}\right)\right]$.

6 $F = \dfrac{2\rho k}{l}$.

8 $I = 2\rho \displaystyle\iint_D \frac{(x^2 + y^2)R\,dx\,dy}{\sqrt{R^2 - x^2 - y^2}} = \frac{8\pi\rho R^4}{3}$, where D is the disk of radius R.

9 $I = \dfrac{\pi k}{16}\left(\dfrac{1093}{7} - \dfrac{242}{5} + \dfrac{13}{3}\right)$.

10 $F = 2\pi k\rho\left(1 - \dfrac{l}{\sqrt{l^2 + R^2}}\right)$, which approaches $2\pi k\rho$ as $R \uparrow$. The attraction of an infinite plate is independent of the distance from it.

11 Describe the shell by $x = \sin\phi\cos\theta$, $y = \sin\phi\sin\theta$, $z = \cos\phi$, $0 \le \phi \le \pi$, $0 \le \theta \le 2\pi$, and let $P = (0, 0, a)$ with $0 \le a < 1$. With ρ = density (mass per unit area), the component of the force at P in the vertical direction is

$$F = -\int_0^{2\pi} d\theta \int_0^{\pi} \frac{(\cos\phi - a)(\rho\sin\phi)\,d\phi}{(1 + a^2 - 2a\cos\phi)^{3/2}}$$

This may be integrated easily; for example, put $u^2 = 1 + a^2 - 2a\cos\phi$. One finds that $F = 0$.

Section 9.2

1 (*a*) $\frac{5}{6}$; (*b*) $\frac{9}{10}$; (*d*) $\frac{1}{2} - \frac{1}{3} + \frac{1}{2} = \frac{2}{3}$.

3 0, 8, 8, 8.

4 (*b*) $\frac{27}{20}$; (*c*) 2π; (*d*) $0 + 12 - \frac{7}{2} + 0 = \frac{17}{2}$.

5 2, 2, 0, 0.

11 (*a*) $2xyz\,dx + x^2z\,dy + x^2y\,dz$; (*b*) $(x^2 + y^2)^{-1}(2x\,dx + 2y\,dy)$.

13 (*a*) $x^2\,dy\,dx - z\,dy\,dz$; (*c*) 0; (*e*) $(2xy - x)\,dx\,dy\,dz$; (*g*) 0.

16 (*a*) $4\displaystyle\int_0^1 du \int_0^1 (uv^2 + v^3)\,dv = \frac{5}{3}$; (*b*) π; (*c*) $1 - \frac{1}{2} = \frac{1}{2}$.

18 $\displaystyle\int_0^1 ds \int_0^1 dr\,(r^4 + 2r^3s - 4r^2s^2 + 2rs^3 - s^4) = \frac{1}{18}$.

Section 9.3

1 $(\mathbf{a} \times \mathbf{b}) \cdot \mathbf{c} = 2$, $\mathbf{a} \times (\mathbf{b} \times \mathbf{c}) = 8\mathbf{j} + 24\mathbf{k}$, $(\mathbf{a} \times \mathbf{b}) \times \mathbf{c} = 2\mathbf{i} + \mathbf{j} + 21\mathbf{k}$,

$$(\mathbf{a} \cdot \mathbf{b})\mathbf{c} - (\mathbf{a} \cdot \mathbf{c})\mathbf{b} = -8\mathbf{j} - 24\mathbf{k}$$

3 With $\mathbf{c} = \mathbf{a} \times \mathbf{b}$, show that

$$\begin{vmatrix} a_1 & a_2 & a_3 \\ b_1 & b_2 & b_3 \\ c_1 & c_2 & c_3 \end{vmatrix} = c_1{}^2 + c_2{}^2 + c_3{}^2$$

4 If $\mathbf{a} \times \mathbf{v} = \mathbf{b}$, then $\mathbf{a} \times \mathbf{b} = (\mathbf{a} \cdot \mathbf{v})\mathbf{a} - (\mathbf{a} \cdot \mathbf{a})\mathbf{v} = k\mathbf{a} - |a|^2\mathbf{v}$. Thus, if such a vector exists, it must be $(k\mathbf{a} - (\mathbf{a} \times \mathbf{b}))/|\mathbf{a}|^2$ (provided $|\mathbf{a}| \neq 0$). However, a solution does not always exist; a necessary condition is that $\mathbf{a} \cdot \mathbf{b} = 0$. This may be seen to be sufficient.

5 The vectors p_n lie (for $n \geq 3$) in a plane normal to $\mathbf{a}$, and rotate about it, each being orthogonal to its predecessor. If $|\mathbf{a}| < 1$, $\lim p_n = 0$.

7 Since $\mathbf{f}(t) \cdot \mathbf{f}(t) = 1$, $\mathbf{f}'(t) \cdot \mathbf{f}(t) = 0$. The curve described by $p = \mathbf{f}(t)$ lies on the sphere $|p| = 1$; this therefore states that the tangent vector at p is orthogonal to the vector from 0 to p.

13 For the second, let $\mathbf{b} = b_1\mathbf{i} + b_2\mathbf{j} + b_3\mathbf{k}$, $\mathbf{c} = c_1\mathbf{i} + c_2\mathbf{j} + c_3\mathbf{k}$, and take $\mathbf{a} = \mathbf{i}$. By direct computation, it is seen that

$$\mathbf{i} \times (\mathbf{b} \times \mathbf{c}) = c_1(b_2\mathbf{j} + b_3\mathbf{k}) - b_1(c_2\mathbf{j} + c_3\mathbf{k}) = c_1\mathbf{b} - b_1\mathbf{c} = (\mathbf{i} \cdot \mathbf{c})\mathbf{b} - (\mathbf{i} \cdot \mathbf{b})\mathbf{c}$$

Similarly, one shows that $\mathbf{j} \times (\mathbf{b} \times \mathbf{c}) = (\mathbf{j} \cdot \mathbf{c})\mathbf{b} - (\mathbf{j} \cdot \mathbf{b})\mathbf{c}$ and $\mathbf{k} \times (\mathbf{b} \times \mathbf{c}) = (\mathbf{k} \cdot \mathbf{c})\mathbf{b} - (\mathbf{k} \cdot \mathbf{b})\mathbf{c}$. Putting these together, with coefficients a_1, a_2, a_3, one arrives at the general formula.

14 Assuming that f and $\mathbf{V}$ are of class C'', these follow by direct calculation. Formally, they may also be obtained from the relation $\nabla \times \nabla = 0$; thus,

$$\text{curl}\,(\text{grad}\, f) = \nabla \times \nabla f = (\nabla \times \nabla)f = 0$$

and div (curl $\mathbf{V}$) $= \nabla \cdot (\nabla \times \mathbf{V}) = (\nabla \times \nabla) \cdot \mathbf{V} = 0$. This can be made acceptable by discussing vector systems whose components are elements from an arbitrary noncommutative ring.

15 For the second, observe that $\partial^2(fg)/\partial x^2 = g\partial^2 f/\partial x^2 + 2(\partial f/\partial x)(\partial g/\partial x) + f\partial^2 g/\partial x^2$.

16 As an alternative to direct computation, one may use the relation

$$\mathbf{a} \times (\mathbf{b} \times \mathbf{c}) = \mathbf{b}(\mathbf{a} \cdot \mathbf{c}) - (\mathbf{a} \cdot \mathbf{b})\mathbf{c}$$

with $\mathbf{a} = \mathbf{b} = \nabla$, and $\mathbf{c} = \mathbf{V}$, obtaining $\nabla \times (\nabla \times \mathbf{V}) = \nabla(\nabla \cdot \mathbf{V}) - (\nabla \cdot \nabla)\mathbf{V}$. This requires the additional consideration indicated in Exercise 14.

17 An alternative to direct computation is the following. The analog of the rule for differentiation of a product is $\nabla \cdot (\mathbf{F} \times \mathbf{G}) = \nabla \cdot (\dot{\mathbf{F}} \times \mathbf{G}) + \nabla \cdot (\mathbf{F} \times \dot{\mathbf{G}})$, where the dot indicates the function to which the differentiation is applied. Using the relation $\mathbf{a} \cdot (\mathbf{b} \times \mathbf{c}) = \mathbf{c} \cdot (\mathbf{a} \times \mathbf{b})$, we have
have

$$\begin{aligned} \nabla \cdot (\mathbf{F} \times \mathbf{G}) &= \mathbf{G} \cdot (\nabla \times \mathbf{F}) - \nabla \cdot (\dot{\mathbf{G}} \times \mathbf{F}) \\ &= \mathbf{G} \cdot (\nabla \times \mathbf{F}) - \mathbf{F} \cdot (\nabla \times \mathbf{G}) \end{aligned}$$

18 If $\omega = f(x, y, z)$, then $d\omega = f_1\,dx + f_2\,dy + f_3\,dz$ and

$$dd\omega = (f_{12}\,dy + f_{13}\,dz)\,dx + (f_{21}\,dx + f_{23}\,dz)\,dy + (f_{31}\,dx + f_{32}\,dy)\,dz$$

Assuming that $f \in C''$, $dd\omega = 0$. Likewise, if $\omega = A(x, y, z, w)\,dx\,dy$, then $d\omega = A_3\,dz\,dx\,dy + A_4\,dw\,dx\,dy$ and $dd\omega = A_{34}\,dw\,dz\,dx\,dy + A_{43}\,dz\,dw\,dx\,dy$, which is again 0, if $A \in C''$.

19 Use Exercises 14 and 17.

Section 9.4

2 $d\omega = 3(x^2 + y^2)\,dx\,dy$, so that

$$\int_\gamma \omega = 3 \iint_D (x^2 + y^2)\,dx\,dy = 3\int_0^{2\pi} d\theta \int_0^1 r^2 r\,dr = 3\pi/2$$

5 If V is the solid ball, $x^2 + y^2 + z^2 \le 1$, and Σ is the sphere which is its boundary, we indicate the proof of the relation $\iint_\Sigma A\,dx\,dy = \iiint_V A_3$. Let D be the disk $x^2 + y^2 \le 1$. Then,

$$\iiint_V A_3 = \iint_D \{A(x, y, \sqrt{1 - x^2 - y^2}) - A(x, y, -\sqrt{1 - x^2 - y^2})\}\,dx\,dy$$

Using the parametrization $z = \sqrt{1 - x^2 - y^2}$ on the top half of Σ, and its negative on the bottom (with reversed orientation), we obtain for the surface integral

$$\iint_\Sigma A\,dx\,dy = \iint_D A(x, y, \sqrt{1 - x^2 - y^2})\,dx\,dy + \iint_D A(x, y, -\sqrt{1 - x^2 - y^2})\,dy\,dx$$

$$= \iint_D \{A(x, y, \sqrt{1 - x^2 - y^2}) - A(x, y, -\sqrt{1 - x^2 - y^2})\}\,dx\,dy$$

verifying the relation.

7 Area $= 9a^2 \int_0^\infty \frac{t^2\,dt}{(1 + t^3)^2} = 3a^2/2.$

9 With $\omega = xy^2\,dy - x^2y\,dx$, $d\omega = (x^2 + y^2)\,dx\,dy$. Let D_0 be the portion of D in the first quadrant bounded by the closed curve γ formed of the lines $x + y = 4$, $x = 0$, $y = 0$, and part of the circle $x^2 + y^2 = 1$. Then, $\iint_{D_0} d\omega = \int_\gamma \omega = 128/3 - \pi/8$. By symmetry, this is the same as the integral of $d\omega$ over the other three pieces of D, so that the result is $512/3 - \pi/2$.

10 $d(x\,dy\,dz + y\,dz\,dx + z\,dx\,dy) = 3\,dx\,dy\,dz$.

11 (*b*) Using $z = [R^2 - x^2 - y^2]^{1/2}$, we have $dz = -(x/z)\,dx - (y/z)\,dy$, so that

$$I = \frac{2}{6}\iint_D \frac{(x^2 + y^2)^2 + 4x^2y^2}{\sqrt{R^2 - x^2 - y^2}}\,dx\,dy$$

$$= \frac{1}{3}\int_0^R \frac{r^5\,dr}{\sqrt{R^2 - r^2}} \int_0^{2\pi} (1 + 4\cos^2\theta\sin^2\theta)\,d\theta = \frac{8}{15}\pi R^5$$

12 With $\omega = x\,dy\,dz$, $d\omega = dx\,dy\,dz$, and $(d\omega)^* = -(4vw + 4uw + 2u)\,du\,dv\,dw$. Also,

$$\omega^* = (2u^2 + 2uv - 2uw)\,du\,dv + (4u^2w + 4uvw - 4uw^2)\,du\,dw - (4uvw + 4v^2w - 4vw^2)\,dv\,dw$$

and it is seen that $d(\omega^*) = (d\omega)^*$.

17 For the first, apply the divergence theorem to the cylindrical region R obtained by erecting lines of height 1 on the set D. One then sees that

$$\iint_{\partial R} \mathbf{F}\cdot\mathbf{n} = \int_{\partial D} \mathbf{F}\cdot\boldsymbol{\eta}$$

and that

$$\iiint_R \nabla\cdot\mathbf{F} = \iint_D \nabla\cdot\mathbf{F}$$

Section 9.5

1 (*a*) $f(x, y) = x^3y + x^2y + y^2 + C$; (*c*) $f(x, y, z) = x^2yz^3 + xz$; (*d*) since $d\omega \neq 0$, no function f exists.

2 (*b*) $$f_1(x, y, z) = \int_0^1 A + \int_0^1 t(xA_1 + yB_1 + zC_1) = \int_0^1 A + \int_0^1 t(xA_1 + yA_2 + zA_3) = A(x, y, z)$$

using part (*a*) and the relations $B_1 = A_2$, $C_1 = A_3$ which come from $d\omega = 0$. (*c*) Along the straight line from (0, 0, 0) to (x, y, z).

4 Yes. $f(x, y) = (\frac{1}{2}) \log (x^2 + y^2)$.

5 Use Exercise 2. The homogeneity of A, B, and C enables one to factor out t^k, and carry out the integration.

6 (*a*) One possible factor is x^{-3}. (*b*) Since $\omega\, d\omega = 0$, integrating factors exist. One is x^2. (*c*) No integrating factor exists.

8 (*c*) If $\sigma = A\,dy\,dz + B\,dz\,dx + C\,dx\,dy$ and $\sigma = d\omega$ where $\omega = a\,dx + b\,dy + c\,dz$, then $C = \partial b/\partial x - \partial a/\partial y$. To verify that the given functions have this property, we differentiate $a(x, y, z)$ and $b(x, y, z)$, obtaining

$$\frac{\partial b}{\partial x} = \int_0^1 tC + \int_0^1 t^2(xC_1 - zA_1)$$

$$\frac{\partial a}{\partial y} = -\int_0^1 tC + \int_0^1 t^2(zB_2 - yC_2)$$

By assumption, $d\sigma = 0$, so that $A_1 + B_2 + C_3 = 0$. Using this, replace $z(A_1 + B_2)$ by $-zC_3$, obtaining

$$\frac{\partial b}{\partial x} - \frac{\partial a}{\partial y} = 2\int_0^1 tC + \int_0^1 t^2(xC_1 + yC_2 + zC_3) = C(x, y, z)$$

by part (*a*).

10 (*a*) Using Exercise 9, one obtains

$$\omega = (\tfrac{1}{5})(2x^2yz - z^3y)\,dx + (\tfrac{1}{5})(4xz^3 - 3y^2z^2 - x^3z)\,dy + (\tfrac{1}{5})(3y^3z - 3xyz^2 - x^3y)\,dz$$

To this, any exact 1-form df may be added. With a judicious choice, one obtains the simpler solution $\omega = x^2yz\,dx + xz^3\,dy + y^3z\,dz$. (*b*) Exercise 8 yields the solution

$$\omega = [(\tfrac{1}{2})xz^2 + (\tfrac{1}{3})(z^2 - y^2)]\,dx + (\tfrac{1}{3})xy\,dy - [(\tfrac{1}{2})x^2z + (\tfrac{1}{3})xz]\,dz$$

However, by inspection, we see that $\sigma = [(2xz + z)\,dz - y\,dy]\,dx = \beta\,dx$. Moreover, β itself is an exact 1-form. Thus, we obtain the simpler solution

$$\omega = (xz^2 + \tfrac{1}{2}z^2 - \tfrac{1}{2}y^2)\,dx$$

13 If ϕ is an integrating factor for the 2-form, then

$$A\phi_1 + B\phi_2 + C\phi_3 = -(A_1 + B_2 + C_3)\phi$$

When $\sigma = x\,dy\,dz + y\,dz\,dx + z\,dx\,dy$, this differential equation is

$$x\phi_1 + y\phi_2 + z\phi_3 = -3\phi$$

which is satisfied by any function ϕ which is homogeneous of degree -3. For example, we may take $\phi(x, y, z) = x^{-3}$ and have $\phi\sigma$ an exact 2-form, which is $d\omega$ for $\omega = (\frac{1}{2})(yx^{-2}\,dz - zx^{-2}\,dy)$. It is interesting to notice that in this example, the form $\phi\sigma$ has homogeneous coefficients, but that the methods of Exercise 9 (and also of Exercise 8) fail.

14 If ω is exact, then $\omega = d\beta$. Hence $\omega^* = (d\beta)^* = d(\beta^*)$, and ω^* is also exact.

Section 9.6

1 (*a*) In (9-62), take $f = g$. (*b*) Since $(\partial/\partial\mathbf{n})(g^* - g) = 0$ on $\partial\Omega$, $\nabla(g^* - g) = 0$ throughout Ω. If Ω is a connected set, we may conclude that $g^* - g$ is constant.

3 (*a*) Apply the divergence theorem to $\mathbf{F} \times [\nabla \times \mathbf{G}]$, and use the relation

$$\nabla \cdot (\mathbf{F} \times [\nabla \times \mathbf{G}]) = (\nabla \times \mathbf{G}) \cdot (\nabla \times \mathbf{F}) - \mathbf{F} \cdot (\nabla \times [\nabla \times \mathbf{G}])$$

(*b*) In (*a*), interchange **F** and **G**, and subtract the two formulas. These relations may be used to solve the vector analog of Poisson's equation,

$$\nabla \times [\nabla \times \mathbf{V}] = \mathbf{F}$$

Section 10.2

3 (*a*) It helps to look at $y = \sin x$ and $y = 1 - x/6$. (*b*) The first three are .988 567, 2.523 600, 6.242 721.

6 (*a*) Use Taylor's theorem about $x = \bar{x}$, where $f(\bar{x}) = f'(\bar{x}) = 0$. (*c*) If $x_1 = 2$, you should have $x_5 = 1.751\,18$, and $x_{10} = 1.732\,655$. (*d*) $\frac{1}{2}$ is replaced by $\frac{2}{3}$.

7 The diagram is symmetric about the A axis and consists of three regions where three roots are found, the left region being unbounded, and a bounded region where five roots are found. In the connected set that remains, only one root is found.

Section 10.3

1 Sketch $y = x$ and $y = f(x)$ and locate the points (x_n, x_{n+1}).

2 (*a*) Show that $a_b \le a_{n+1} < \sqrt{A}$. (*b*) Show that $\sqrt{A} - a_n \le 2/(n + 1)$.

3 Note that $|x| = \sqrt{x^2}$. (This result is the key to the Weierstrass approximation theorem. See Reference [4] in the Reading List.)

4 (*b*) Let $A = a^3$, and suppose $x_n = a - \varepsilon$. Then estimate x_{n+1} according to each algorithm.

5 Starting with $x = 1$, $y = 1$, $P_5 = (.7277, .7216)$, $P_{20} = (.694\,967, .768\,263)$, $P_{50} = (.694\,819\,689\,4, \ .768\ \ 169\,155\,9)$.

6 Consider $\int_0^x f_n$.

Section 10.4

1 (*a*) This is an instance when it may be easier to do a direct search, rather than to attempt to solve $d^2x/dt^2 = 0$ or $d^3x/dt^3 = 0$. (*b*) The maximum velocity occurs when $\theta = 1.277$ radians.

2 The minimum of $f + Ag$ occurs when $x = 24/(34 + 1/A)$, so that for large A, $x \to \frac{12}{17}$.

3 $(-\frac{3}{4}, \frac{9}{16})$ is one critical point.

4 The Lagrange equations have a solution for which $\beta = -1$, $x = 1$.

6 9. **7** Minimum 1; maximum 3. **8** $8abc/3\sqrt{3}$.

9 $P = (0, 0, f(0, 0))$. A nonanalytic treatment is possible. The volume of the region bounded by the tangent plane, the cylinder, and the XY plane is πh, where h is the height of the point on the tangent plane which lies directly above $(0, 0, 0)$. Thus, one need only minimize h.

Section 10.5

2 (*a*) .743; (*b*) .944; (*c*) .736. **3** $I \approx .693\,25$.

4 Write $\sqrt{4x^2 - y^2} = \sqrt{2x - y}\sqrt{2x + y}$ and use Schwarz' inequality.

6 $10 \le I \le 12$, and these bounds cannot be improved.

7 $.156\,63 \pm .0182$.

8 Put $t = 1/x$, and then write the interval of integration as $[5, 2\pi]$, $[2\pi, 3\pi]$, etc., and integrate by parts.

9 Show that $a_{n+1} - b_{n+1} = (\sqrt{a_n} - \sqrt{b_n})^2/2$.

12 $f'(.6) = 2.66 \pm .0133$.

LIST OF SYMBOLS

$\mathrm{A}(S)$, $\mathrm{V}(S)$	area, volume, 169
$\mathscr{A}$	class of sets, 376
C^k, C', C''	differentiable functions, 128
$\mathbf{C}$	complex field, 556
$\mathscr{C}$, $*\mathscr{C}$	class of continuous functions, 304, 307
$\mathbf{D}$, $\mathbf{D}_\beta$	derivative operators, 130
$\mathbf{K}$	field, 554
$\mathscr{L}$	lower bounds, 58
$\mathscr{M}$	metric space, 304
$\mathscr{N}$	neighborhood, 31
$\mathscr{O}$	open set, 35
Q	rational field, 6
$\mathbf{R}$, $\mathbf{R}^n$	real field, n space, 2
$\mathscr{S}$	class of sets, 64
$\mathscr{U}$	upper bounds, 58
Z	set of integers, 6
d, d^2	differential operators, 341, 425
γ	curve, 399
∂	boundary operator, 417
ℓ^2	Hilbert space, 305
ω, σ	differential form, 450
Σ	surface, 417
inf, sup	greatest lower bound, lub, 58
bdy	boundary, 30
f_j, f_x, f_y	partial derivatives, 127

$\subset$ — set inclusion, 4
$\in, \notin$ — set membership, 4
ε — epsilon, 39
$\cup \; \bigcup$ — set union, 5, 33
$\cap \; \bigcap$ — set intersection, 5, 33
$\varnothing$ — empty set, 5
$\mathbf{0}$ — origin of $\mathbf{R}^n$, 3
$\bar{S}$ — closure of S, 31
$[x]$ — greatest integer in x, 6
$[a_{ij}]$ — matrix, 335
$[a, b]$ — closed interval, 30
$|p|$ — length (norm) of p, 11
$\|f\|_E$ — norm of function, 265
$^*\|f\|$, $^*\lim$, $^*\sum$ — operations in $^*\mathscr{C}$, 307, 311
$\langle f, g \rangle$ — inner product in $^*\mathscr{C}$, 308
$d(p, q)$ — metric, 304
$d(A, B)$ — distance between sets, 60
$B(p_0, r)$ — ball of radius r at p_0, 18
$\cdot$ — inner (scalar) product, 8
$\times$ — vector product, 465
$\mathbf{n}$ — normal to surface, 422
∇f — gradient, 130
$\approx$ — approximate equality, 124
$\mathbf{e}_j$ — basis vectors, 126, 306
$\sim$ — Fourier mapping, 311
$\blacksquare$ — end of proof, 14

Index

Index